THE CARBOHYDRATES

Chemistry and Biochemistry

SECOND EDITION

VOLUME IIA

CONTRIBUTORS

F. A. Bettelheim

Jean Émile Courtois

I. Danishefsky

C. A. Dekker

L. Goodman

Stephen Hanessian

Y. Hashimoto

Theodore H. Haskell

W. Z. Hassid

K. Nisizawa

John H. Pazur

François Percheron

E. L. Richards

P. A. Seib

K. Ward, Jr.

Roy L. Whistler

THE CARBOHYDRATES

Chemistry and Biochemistry

SECOND EDITION

EDITED BY

Ward Pigman

Department of Biochemistry
New York Medical College
New York, New York

Derek Horton

Department of Chemistry
The Ohio State University
Columbus, Ohio

ASSISTANT EDITOR

Anthony Herp

Department of Biochemistry
New York Medical College
New York, New York

VOLUME IIA

1970

ACADEMIC PRESS   New York   San Francisco   London

A Subsidiary of Harcourt Brace Jovanovich, Publishers

ACADEMIC PRESS, INC.
111 Fifth Avenue, New York, New York 10003

United Kingdom Edition published by
ACADEMIC PRESS, INC. (LONDON) LTD.
24/28 Oval Road, London NW1

LIBRARY OF CONGRESS CATALOG CARD NUMBER: 68–26647

PRINTED IN THE UNITED STATES OF AMERICA

We dedicate this work to the persons most responsible for our professional development

HORACE S. ISBELL AND THE LATE MELVILLE L. WOLFROM

CONTENTS

LIST OF CONTRIBUTORS xv

PREFACE xvii

29. Nucleosides

C. A. Dekker and L. Goodman

   I. Introduction by C. A. Dekker 1
     A. Definition of Term 1
     B. Structures and Nomenclature 2
  II. Isolation and Characterization by C. A. Dekker 3
     A. Isolation and Fractionation 3
     B. Characterization 5
 III. Synthesis of Nucleosides by L. Goodman 9
     A. Direct Methods with Protected Sugar Derivatives . . . 9
     B. Direct Methods with Unprotected Sugars 15
     C. Synthesis by Construction of a Heterocyclic Base after Glycosyla-
        tion 16
     D. Developments in Conventional Synthetic Methods . . . 17
     E. *C*-Nucleosides 20
 IV. Chemical Reactions of the Sugar Portion of Nucleosides by L.
    Goodman 22
     A. Blocking Groups 22
     B. Cyclonucleosides and Their Reactions 28
     C. Neighboring-Group Reactions 35
     D. Oxidation 38
     E. Transglycosylation 39
     F. Substitution and Elimination Reactions 41
     G. Hydrolysis by C. A. Dekker 43
  V. Physical Properties of Nucleosides by L. Goodman . . . 52
     A. Proton Magnetic Resonance Spectra 52
     B. Optical Rotatory Dispersion Measurements 55
     C. Ultraviolet Spectra 57
     D. Miscellaneous Physical Data 58
     References 59

30.   Oligosaccharides

John H. Pazur

I. Introduction 69
II. Classification and Nomenclature 70
   A. Classification 70
   B. Nomenclature 71
III. Methods of Synthesis 73
   A. General Remarks 73
   B. Biosynthesis 74
   C. Hydrolysis of Polymers 74
   D. Chemical Synthesis 79
IV. Methods of Isolation 83
   A. Types of Impurities 83
   B. Precipitation and Extraction 84
   C. Chromatography 85
   D. Concentration and Crystallization 86
V. Determination of Structure 88
VI. General Properties 91
   A. Physical Properties 91
   B. Organoleptic Properties 94
   C. Susceptibility to Hydrolysis by Enzymes 95
   D. Susceptibility to Hydrolysis by Acids 96
   E. Chemical Reactions 99
VII. Important Oligosaccharides of Biological Origin 101
   A. Disaccharides 101
   B. Trisaccharides 121
   C. Other Oligosaccharides 127
   References 129

31.   Antibiotics Containing Sugars

Stephen Hanessian and Theodore H. Haskell

I. Introduction 139
II. Classification and Chemistry of Antibiotics Containing Sugars . . 140
   A. Macrolide Antibiotics 140
   B. Cyclitol Antibiotics 159
   C. Nucleoside Antibiotics 172
   D. Antibiotics of the Glycosylamine Type 180
   E. Antibiotics Containing Aromatic Groups 183
   F. Miscellaneous Antibiotics 190
III. Characterization of Antibiotics and Their Sugar Components by
   Physical Means 194
   A. Nuclear Magnetic Resonance Spectroscopy 194
   B. Mass Spectrometry 196

IV. Biological Concepts 198
   References 200

32. Complex Glycosides

Jean Émile Courtois and François Percheron

    I. Introduction 213
   II. Carbohydrate Constituents of Naturally Occurring Glycosides . . 214
  III. Glycosides of Aliphatic Alcohols and of Alditols 216
  IV. Cyanogenetic Glycosides 217
      A. Heterosides Yielding an Aromatic Aldehyde on Hydrolysis . 217
      B. Heterosides Yielding an Aliphatic Ketone on Hydrolysis . . 218
   V. Glycosides of Phenols 218
  VI. Coumarin Glycosides 220
 VII. Glycosides of Anthracene Derivatives 220
VIII. Phenanthrene Glycosides 221
      A. Cardiac Glycosides 221
      B. Steroid Glycosides 222
      C. Steroid Saponins 223
      D. Triterpenoid Saponins 224
      E. Glyco-alkaloids of *Solanum* 225
  IX. Glycosides of Natural Pigments 225
      A. Indole Glycosides 226
      B. Anthocyanidin Glycosides 226
      C. Hydroxyflavone Glycosides 227
      D. Carotenoid Glycosides 229
   X. 1-Thioglycosides 230
  XI. Miscellaneous Glycosides 232
 XII. *C*-Glycosyl Compounds 232
      A. Anthraquinone Derivatives 233
      B. *C*-Glycosylflavones and Related Compounds 234
XIII. Biogenesis and Metabolism 235
      A. Biogenesis 235
      B. Metabolism 236
     References 237

33. Glycoside Hydrolases and Glycosyl Transferases

K. Nisizawa and Y. Hashimoto

    I. Introduction 242
   II. Classification 242
  III. Mechanism of Glycosidase Action 243
      A. Cleavage of Glycosidic Bonds 243

B. Retention or Inversion of the Configuration of the Potential
Reducing Group during Enzyme Action   243
C. Hydrolysis and Transfer   244
D. Reversibility of Glycosidase Action.   244
E. Mechanism of Fission of Glycosidic Bonds   245
F. Role of Acceptor and Transglycosylation.   247
IV. Chemical Nature of Glycoside Hydrolases and Glycosyl Transferases 248
V. Individual Glycosidases and Transglycosylases   .   .   . 251
A. α-D-Glucosidases (α-D-Glucoside Hydrolases and α-D-Glucosyl
Transferases)   251
B. β-D-Glucosidases (β-D-Glucoside Hydrolases and β-D-Glucosyl
Transferases)   258
C. α-D-Galactosidases (α-D-Galactoside Hydrolases and α-D-Galacto-
syl Transferases) 266
D. β-D-Galactosidases (β-D-Galactoside Hydrolases and β-D-Gal-
actosyl Transferases)   268
E. α,β-D-Mannosidases (α,β-D-Mannoside Hydrolases) .   .   . 274
F. β-D-Fructofuranosidases (β-D-Fructofuranoside Hydrolases and
β-D-Fructofuranosyl Transferases)   275
G. 2-Acetamido-2-deoxy-α,β-D-hexosidases and Related Enzymes   . 280
H. β-D-Glucosiduronases   284
I. β-D-Xylosidases (β-D-Xyloside Hydrolases)   286
J. Sialidases   287
K. α,β-D and L-Fucosidases and α,β-L-Rhamnosidases (α,β-D- and
L-Fucoside Hydrolases and α,β-L-Rhamnoside Hydrolases).   . 288
L. Cyclodextrin Transglucosylases   288
M. D-Enzyme 289
N. Amylomaltase   290
References   290

34.   Biosynthesis of Sugars and Polysaccharides

W. Z. Hassid

I. Introduction   302
II. Monosaccharides, Phosphorylated Sugars, and Nucleoside Ortho- and
Pyrophosphate Derivatives 304
III. Mechanisms Involved in Enzymic Formation of Complex Saccharides 312
IV. Reactions Catalyzed by Disaccharide Phosphorylases (D-Glucosyl-
transferases)   314
A. Sucrose and Sucrose Analogs 314
B. Maltose   315
C. Cellobiose 316
D. Laminarabiose   316
V. Synthesis of Disaccharides by Transglucosylation   .   .   . 316
VI. Synthesis of Oligosaccharides from Sugar Nucleotides   .   . 319
A. Sucrose   319
B. Raffinose   320

C. Stachyose 321
D. Lactose 322
E. Trehalose 324
VII. Synthesis of Glycosides from Sugar Nucleotides 325
VIII. Synthesis of Polysaccharides by Phosphorolysis and Transglycosylation 327
A. Glycogen and Starch 327
B. The Amylopectin Type of Polysaccharides from Sucrose . . 332
C. Amylose from Maltose 333
D. Cyclic Amylosaccharides 334
E. Dextran from Sucrose 334
F. Dextran from Amylodextrin 335
G. Levan from Sucrose 335
IX. Synthesis of Polysaccharides from Sugar Nucleotides . . . 336
A. Glycogen 336
B. Chitin 339
C. Starch 340
D. Cellulose 342
E. D-Gluco-D-mannan 345
F. D-Xylan 346
G. Pectin 347
H. $(1{\rightarrow}3)$-β-D-Glucans 348
I. Alginic Acid 349
J. DL-Galactan 350
K. Hyaluronic Acid 351
L. Capsular Pneumococcal Polysaccharides 352
M. Bacterial Cell-Wall Lipopolysaccharides 353
N. Teichoic Acids 356
O. Glycoproteins 358
P. Sulfated Mucopolysaccharides 361
References 362

35. Introduction to Polysaccharide Chemistry

I. Danishefsky, Roy L. Whistler, and F. A. Bettelheim

I. Occurrence, Types, and Functions by Roy L. Whistler . . . 375
A. Classification and Nomenclature 375
B. Structural Characteristics 377
C. Functions of Polysaccharides 382
II. Physical Methods of Characterization of Polysaccharides by F. A.
Bettelheim 383
A. Electron Microscopy 383
B. X-Ray Diffraction 384
C. Infrared Spectra and Infrared Dichroism 385
D. Optical Rotatory Dispersion 386
E. Osmotic Pressure and Other Colligative Properties . . . 387
F. Light Scattering 388
G. Ultracentrifugation 390
H. Viscosity 391

III. Characterization by Chemical, Enzymic, and Immunological Methods
by I. Danishefsky 394
   A. Isolation and Purification 394
   B. Identification of the Constituents of the Polysaccharide . . 396
   C. Linkage and Sequence 397
   References 410

36. Cellulose, Lichenan, and Chitin

K. Ward, Jr. and P. A. Seib

I. Introduction and Chemical Structure 413
II. Occurrence 415
III. Enzymic Synthesis 415
IV. Isolation 417
V. Molecular Weight and Fine Structure 418
VI. Sorption, Swelling, and Solution 420
VII. Degradation 422
   A. Enzymolysis. 422
   B. Hydrolysis 426
   C. Oxidation 426
   D. Miscellaneous Degradations 428
VIII. Substitution Reactions 429
   A. Reactivity in General 429
   B. Esters 430
   C. Ethers. 432
   D. Cross-linking 434
IX. Graft Polymers 434
X. Lichenan and Isolichenan 435
XI. Chitin 435
   References 438

37. Hemicelluloses

Roy L. Whistler and E. L. Richards

I. Introduction 447
II. Isolation 449
   A. Extraction 449
   B. Chemical Modification during Isolation 450
   C. Purification 450
III. Properties 451
   A. Molecular Weight 451
   B. Crystal Structure 452
IV. Hemicelluloses 452
   A. D-Xylans 452
   B. D-Mannans 458

C. D-Gluco-D-mannans 459

D. D-Galacto-D-gluco-D-mannans 461

E. L-Arabino-D-galactans 462

V. Enzymic Hydrolysis of Hemicelluloses 464

A. D-Xylans 464

B. D-Mannans, D-Gluco-D-mannans, and D-Galacto-D-gluco-D-
mannans 466

References 467

AUTHOR INDEX (VOLUME IIA) *1*

SUBJECT INDEX (VOLUME IIA) *41*

LIST OF CONTRIBUTORS

Numbers in parentheses indicate the pages on which the authors' contributions begin.

F. A. Bettelheim (375), Adelphi University, Garden City, Long Island, New York

Jean Émile Courtois ·(213), Laboratoire de Chimie Biologique, Faculté de Pharmacie, Paris, France

I. Danishefsky (375), Department of Biochemistry, New York Medical College, New York, New York

C. A. Dekker (1), Department of Biochemistry, University of California, Berkeley, California

L. Goodman (1), Bio-Organic Chemistry Department, Stanford Research Institute, Menlo Park, California

Stephen Hanessian* (139), Research Laboratories, Parke, Davis and Co., Ann Arbor, Michigan

Y. Hashimoto (241), Department of Biochemistry, Saitama University, Urawa, Japan

Theodore H. Haskell (139), Research Laboratories, Parke, Davis and Co., Ann Arbor, Michigan

W. Z. Hassid (301), Department of Biochemistry, University of California, Berkeley, California

K. Nisizawa (241), Botanical Institute, Tokyo Kyoiku University, Otsuka, Bunkyo-ku, Tokyo, Japan

John H. Pazur (69), Department of Biochemistry, The Pennsylvania State University, University Park, Pennsylvania

François Percheron (213), Laboratoire de Chimie Biologique, Faculté de Pharmacie, Paris, France

* Present address: Department of Chemistry, University of Montreal, Montreal, Quebec, Canada.

xv

E. L. Richards* (447), Department of Biochemistry, Purdue University, Lafayette, Indiana

P. A. Seib (413), The Institute of Paper Chemistry, Appleton, Wisconsin

K. Ward, Jr. (413), The Institute of Paper Chemistry, Appleton, Wisconsin

Roy L. Whistler (375, 447), Department of Biochemistry, Purdue University, Lafayette, Indiana

* Present address: Chemistry and Biochemistry Department, Massey University, Palmerston North, New Zealand.

PREFACE

This edition of "The Carbohydrates" is a complete revision of the 1957 work which was based on "The Chemistry of the Carbohydrates" (1948). Because of its size, it is divided into two volumes, each in two separate parts. Its length of approximately 2500 pages, including indexes, is a reflection of the rapid growth in research in this field. The statement has been made that scientific knowledge has recently been doubling every 10–15 years; the present edition illustrates this for the 1957 edition had 900 pages, despite the brief treatment of polysaccharides.

In retrospect, the previous edition has very little that needs correction, but new fields of knowledge have developed. Thus, conformational analysis has made spectacular advances with the development of nuclear magnetic resonance methods. Amino sugars and uronic acids have attained great importance because of their widespread occurrence in important biological substances. Unsaturated sugars have been especially studied recently, partly because of the utility of nuclear magnetic resonance spectroscopy in determining their structures and stereochemistry. A new chapter has been added on the effects of ionizing radiations and of autoxidation reactions. Newly developed physical methods and separation methods have been described in additional chapters. The literature on nucleosides and antibiotics has expanded to the extent that these subjects have necessitated full chapters. With the discovery of transglycosylation reactions, the number of known oligosaccharides and enzymes acting on carbohydrates has greatly increased. A new chapter on the biosynthesis of sugars and complex saccharides was required to cover this rapidly growing field.

In the previous edition, the discussion of polysaccharides was reduced to two chapters because of the prior appearance of "The Polysaccharides" by Whistler and Smart. In this new edition, the original practice of having separate chapters for the main types of polysaccharides has been restored. Chapters on the rapidly growing fields of glycolipids and glycoproteins have been introduced.

The two final chapters cover the official nomenclature rules for carbohydrates and for enzymes having carbohydrates as substrates. The latter were extracted from the official report, but the names have been modified to conform as much as possible to official carbohydrate nomenclature. In the other chapters, official carbohydrate nomenclature has been used, but both old and new enzyme names are given.

xvii

Despite constant efforts to minimize their number, approximately 10,000 references are cited as compared with 4500 in the earlier edition. This is again a reflection of the rapid growth of the field and of the problems faced by the individual authors. The book is an international collaborative effort, and sixty-three authors were involved in the writing of the various chapters; they reside in Australia, British Isles, Canada, France, Germany, Japan, and the United States.

Most of the chapters were read by other workers in the field. We thank the following for their assistance in this way: Drs. I. Danishefsky, H. El Khadem, J. Fox, S. Hannessian, Michael Harris, R. Hems, M. Horowitz, K. L. Loening, R. H. McCluer, D. J. Manners, Fred Parrish, N. K. Richtmyer, R. Schaffer, C. Szymanski, and the late M. L. Wolfrom.

We owe special appreciation for the help of Drs. Anthony Herp, Hewitt Fletcher, Jr., and L. T. Capell. Dr. Herp acted as a co-editor and translated or rewrote several chapters in Volume II, Dr. Fletcher read all of the galley proofs, and Dr. Capell was responsible for the indexes, both important and onerous tasks.

Our own institutions, New York Medical College and The Ohio State University, gave important support and encouragement to us in the preparation of these volumes. Academic Press gave the expected hearty cooperation.

THE CARBOHYDRATES

Chemistry and Biochemistry

SECOND EDITION

VOLUME IIA

29. NUCLEOSIDES

C. A. DEKKER AND L. GOODMAN

I. Introduction BY C. A. DEKKER 1
   A. Definition of Term 1
   B. Structures and Nomenclature 2
II. Isolation and Characterization BY C. A. DEKKER . 3
   A. Isolation and Fractionation 3
   B. Characterization 5
III. Synthesis of Nucleosides BY L. GOODMAN. . . 9
   A. Direct Methods with Protected Sugar Derivatives. 9
   B. Direct Methods with Unprotected Sugars . . 15
   C. Synthesis by Construction of a Heterocyclic Base
      after Glycosylation 16
   D. Developments in Conventional Synthetic Methods 17
   E. *C*-Nucleosides 20
IV. Chemical Reactions of the Sugar Portion of Nucleosides
   BY L. GOODMAN 22
   A. Blocking Groups 22
   B. Cyclonucleosides and Their Reactions . . 28
   C. Neighboring-Group Reactions 35
   D. Oxidation 38
   E. Transglycosylation 39
   F. Substitution and Elimination Reactions . . 41
   G. Hydrolysis BY C. A. DEKKER 43
V. Physical Properties of Nucleosides BY L. GOODMAN . 52
   A. Proton Magnetic Resonance Spectra . . . 52
   B. Optical Rotatory Dispersion Measurements. . 55
   C. Ultraviolet Spectra 57
   D. Miscellaneous Physical Data 58
   References 59

I. INTRODUCTION*

A. DEFINITION OF TERM

While the term "nucleoside," in the strict sense, is reserved for those glycosyl purines and pyrimidines derived from nucleic acids, it has gradually come to include both natural and synthetic glycosylamines in which the

* This section was prepared by C. A. Dekker.

1

amine is a heterocyclic base. In addition, compounds in which C-1 of the sugar residue is linked to an oxygen or carbon atom of the heterocyclic base are frequently referred to as nucleosides—for example, vicine (2,4-diamino-5-*O*-β-D-glucopyranosyl-5,6-dihydroxypyrimidine) and pseudouridine (5-β-D-ribofuranosyluracil). In this chapter the term will be applied to *C*-glycosyl and *N*-glycosyl derivatives but not to *O*-glycosyl derivatives. Because of the nature of this treatise, the primary emphasis will be on the carbohydrate portion of the nucleosides. However, digressions will be made on occasion to note properties of the heterocyclic bases which may affect reactions of the sugar moiety.

B. Structures and Nomenclature

The structures of the most common nucleosides are shown in the accompanying formulas together with their systematic and trivial names.

NUCLEOSIDES DERIVED FROM RIBONUCLEIC ACID (RNA)

1-β-D-Ribofuranosyluracil
(uridine)

1-β-D-Ribofuranosylcytosine
(cytidine)

9-β-D-Ribofuranosyladenine
(adenosine)

9-β-D-Ribofuranosylguanine
(guanosine)

NUCLEOSIDES DERIVED FROM DEOXYRIBONUCLEIC ACID (DNA)

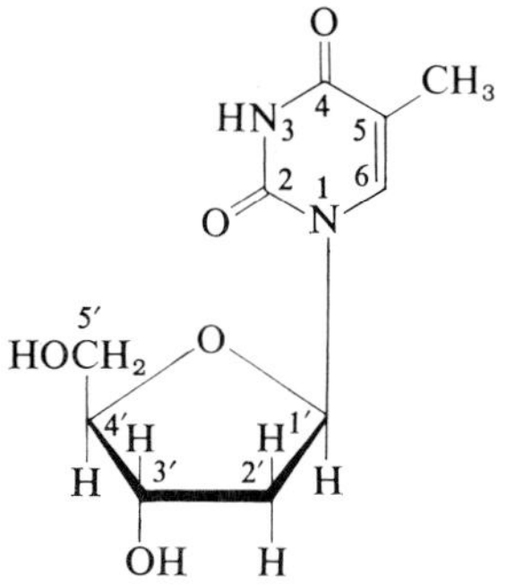

1-(2-Deoxy-β-D-*erythro*-
pentofuranosyl)thymine

(thymidine)

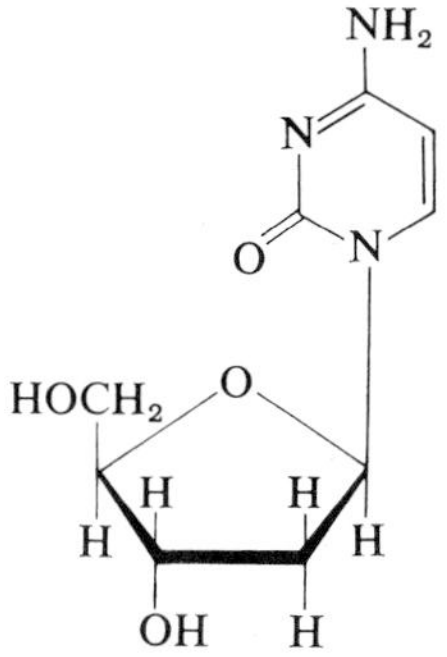

1-(2-Deoxy-β-D-*erythro*-
pentofuranosyl)cytosine

(2′-deoxycytidine)

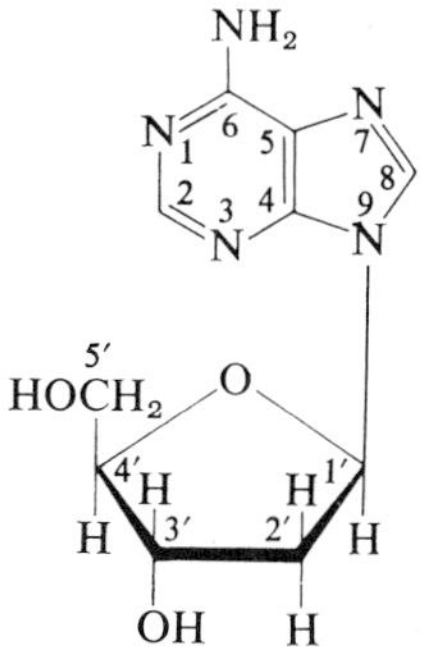

9-(2-Deoxy-β-D-*erythro*-
pentofuranosyl)adenine

(2′-deoxyadenosine)

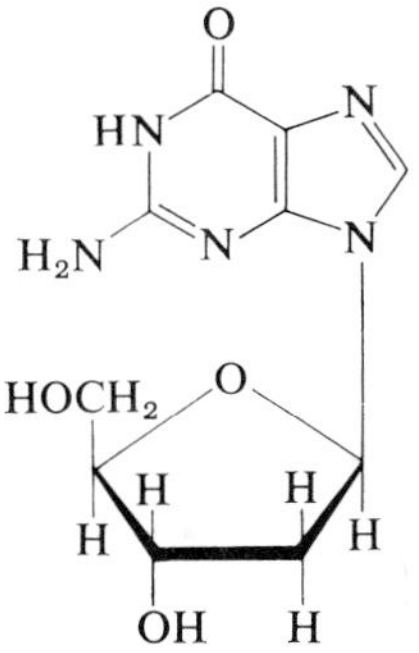

9-(2-Deoxy-β-D-*erythro*-
pentofuranosyl)guanine

(2′-deoxyguanosine)

II. ISOLATION AND CHARACTERIZATION†

A. ISOLATION AND FRACTIONATION

The preparation of nucleosides from nucleic acids requires the hydrolysis of both the internucleotide (phosphate diester) linkage and the mono-phosphate ester group of the resultant nucleotide. Both steps can be accom-plished chemically with aqueous pyridine[1] or lead oxide[2] (and, less conveniently, with other catalysts). Enzymic hydrolysis, which requires the combined action of a phosphate diesterase and phosphate monoesterase provides the mildest method of preparation.[3] Deoxyribonucleosides, which

* *References start on p. 59.*
† This section was prepared by C. A. Dekker.

are in most cases more susceptible to both acid and alkaline degradation than their 2′-hydroxylated counterparts, can be obtained from deoxyribonucleic acid (DNA) by either lead oxide-[4] or enzyme-catalyzed hydrolysis.[5]

Isolation of individual nucleosides from the hydrolyzates is almost universally accomplished by a combination of fractional crystallization and

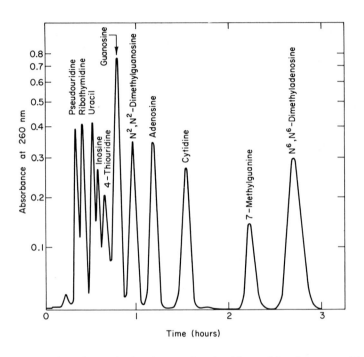

FIG. 1. Separation of a synthetic mixture of nucleosides and free bases on a Bio-Rad A-6 (NH$_4^+$) column, 0.6 × 23 cm; eluant, 0.4M ammonium formate, pH 4.65, 48°, 0.26 ml/minute (24 lb.in^{-2}). M. Uziel, C. Koh, and W. E. Cohn, *Anal. Biochem.*, **25**, 77 (1968).

chromatographic techniques when large-scale preparations of high purity are required. Ion-exchange[6,7] and partition chromatography[8] have been most widely applied. In the former method, variations in the pK_a values of both the base[6,7] and the sugar[9] moieties can be exploited for separation. For smaller-scale separation, paper and gel chromatography,[10] paper electrophoresis,[11,12] and thin-layer chromatography[13] have been applied successfully. Examples of separations that have been achieved by ion-exchange chromatography are shown in Figs. 1 and 2.

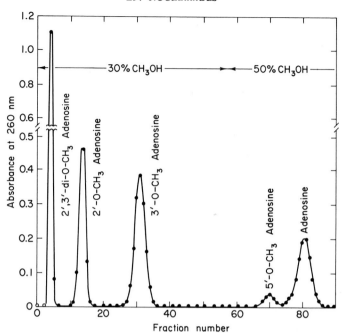

FIG. 2. Separation of a synthetic mixture of adenosine and O-methylated adenosine derivatives on a Bio-Rad AG 1-2x (OH⁻) column, 200–400 mesh, 1.6 × 22 cm; eluant, 30% methanol for fractions 1 to 56 at 1 ml/minute and 50% methanol for fractions 57 to 100 at 1.5 ml/minute. J. B. Gin and C. A. Dekker, *Biochemistry*, 7, 1413 (1968).

B. CHARACTERIZATION

Complete structural characterization of the common nucleosides was not achieved until 80 years after Miescher's initial isolation of nucleic acid from pus cells obtained from surgical bandages.[14] The identification of the heterocyclic-base components obtained upon acid hydrolysis was achieved in the late 1800's by Kossel, Schulze, Fischer, and others,[15] but the nature of the sugar residues was not established until 1910–30 when Levene and co-workers undertook a systematic study.[16] The early structural work on the nucleic acids and their component residues has been reviewed in detail.[17,18]

After developing methods for the hydrolysis and isolation of nucleosides from yeast ribonucleic acid (RNA), Levene was able to identify adenosine, guanosine, cytidine, and uridine on the basis of the heterocyclic bases produced on acid treatment, which led to adenine, guanine, cytosine, and uracil,

* *References start on p. 59.*

respectively. The sugar residue of the purine nucleosides could be isolated after hydrolysis with mild acid, and it was ultimately shown to be D-ribose. The pyrimidine ribonucleosides, which require more drastic conditions for rupture of the N-ribosyl linkage, had to be converted into their corresponding 4,5-dihydropyrimidine derivatives, which could be hydrolyzed under conditions sufficiently mild to preserve the D-ribose liberated. The deoxynucleosides, obtained from DNA, presented a special problem because of the ready conversion of 2-deoxy-D-*erythro*-pentose ("2-deoxy-D-ribose") into levulinic acid under acidic conditions. Eventually it was found that deoxyguanosine or its deamination product, deoxyxanthosine, could be hydrolyzed under very mild conditions, as could the reduced pyrimidine deoxynucleosides. In this way 2-deoxy-D-*erythro*-pentose was shown to be the sugar component of deoxynucleosides and thus of DNA. Levene assumed, because of his failure to find other sugars in RNA and DNA, that D-ribose was the sole sugar moiety to be found in RNA and 2-deoxy-D-*erythro*-pentose in DNA. It is now known that both transfer RNA and ribosomal RNA have 2-O-methyl-D-ribose replacing D-ribose in some of the ribonucleoside residues.[19,20] It is possible that other sugars (most probably derivatives of D-ribose) will be found to occur in small proportion in RNA (1 residue per several hundred, or 1 per thousand sugar residues).[8] Thus far, no other sugar has been found replacing 2-deoxy-D-*erythro*-pentose in the sugar phosphate backbone of DNA, and, because of the unique structural requirements of double-stranded DNA it seems unlikely that such will be found.

The characterization of the sugar residues has been carefully documented elsewhere[18,21] and no further details will be given. It should be noted, however, that nucleoside hydrolases and phosphorylases are now available[22] for the purpose of cleaving sugar from base, and the identification of the sugar portion of a newly discovered nucleoside is a simple operation.

Ascertaining the position of attachment of sugar to heterocyclic base[23,24] in nucleosides was originally quite a formidable problem, and several misassignments were made for nucleosides not derived from nucleic acids (such as vicine and the D-ribosyl derivative of uric acid).[24a] Early methods were based on a process of elimination: The position on the base which failed to react with specific reagents (nitrous acid, methyl iodide, bromine, for example) and which was exposed after splitting off the sugar residue, was assumed to be the point of attachment. Later, comparison of the ultraviolet spectrum of the nucleoside with those of selected methylated bases was used to locate the site of substitution on the heterocyclic base. Today degradative techniques[25,26] are available which facilitate solution of this problem. Degradations of purine and pyrimidine nucleosides are shown in the accompanying equations. Moreover, since a large collection of methylated bases of known spectral characteristics is now available for comparison, the problem has almost ceased to exist. Nevertheless, the position of the base–sugar linkage in pseudouridine, which possesses a *C*-glycosyl bond, was a most difficult problem, finally solved by nuclear magnetic resonance (n.m.r.) spectroscopy.[27]

R = glycosyl residue

$H_3\overset{+}{N}CH_2CH_2CO_2^- + sugar$

The determination of the ring forms (furanose or pyranose) of the sugars has been undertaken by a variety of techniques, all satisfactory but of variable ease of accomplishment. The following techniques are commonly used.

1. Methylation, hydrolysis, and identification of the methylated sugar. For example, 2,3,5-tri-*O*-methyl-D-ribose is produced from a D-ribofuranosyl derivative, and 2,3,4-tri-*O*-methyl-D-ribose from a D-ribopyranosyl derivative.

2. Periodate consumption. In the case of D-ribose derivatives, D-ribofuranosyl compounds take up 1 mole per mole of periodate; D-ribopyranosyl, 2 moles per mole; 2'-deoxy-D-*erythro*-pentofuranosyl, 0 mole; and 2'-deoxy-D-*erythro*-pentopyranosyl, 1 mole per mole.

3. Tritylation. Formation of a monotrityl derivative under mild conditions indicates the presence of an unsubstituted primary hydroxyl group, as in a pentofuranosyl derivative.

4. Borate complexing. The stability of the borate complex has been related

* *References start on p. 59.*

to the oxygen-to-oxygen distance in the vicinal-glycol component of glycosides and nucleosides.[27a]

To complete the characterization of a nucleoside, it is necessary to establish the configuration at the anomeric carbon atom.[24] In the past this has been accomplished by the following procedures.

1. Periodate oxidation. The "dialdehyde" formed is compared with that obtained from a nucleoside of known anomeric configuration.

2. Formation of an anhydronucleoside. The formation of 2,5'-anhydro derivatives from pyrimidine nucleosides and 3,5'-anhydro derivatives from purine nucleosides is possible only for the β-D anomers, in which the hydroxymethyl group and the heterocyclic base are *cis* to one another. For ribonucleosides, 2,2'-anhydronucleoside formation by displacement of a 2'-*p*-tolylsulfonyloxy group is also a satisfactory indication of the β-D configuration.

3. Nuclear magnetic resonance (n.m.r.) spectroscopy. Determination of the H-1–H-2 dihedral angle by application of the Karplus curve will generally permit assignment of anomeric configuration when the conformation of the sugar ring is known with some certainty.[28] Lemieux has discussed the method and its limitations.[28a]

4. Optical rotation and optical rotatory dispersion (o.r.d.). Difficulties are encountered in utilizing the sign and magnitude of the optical rotation for anomeric assignments because of the failure of Hudson's rules of isorotation to apply. However, o.r.d. is more promising, and new generalizations are being worked out.[29] These have not yet been applied widely.

5. Order of elution from Dowex-1 (OH⁻) columns. For the pentofuranosyl derivatives which can be eluted with water or methanol–water, the order of elution is determined by the configuration at the anomeric center, all other structural features being equal. The α-D-compound precedes the β-D-compound in the cases of the *ribo*-, 2'-deoxy-*erythro*-pento, *arabino*-, and *xylo*-compounds.[29a]

Ultimate proof of structure involves synthesis by unambiguous means. This was achieved for the common, naturally occurring nucleosides by Todd and colleagues,[23,29b] thus confirming the structural features that had been deduced by less direct means. Tipson[30] provided an explanation of the factors determining the stereochemistry at C-1 during synthesis by the Fischer–Helferich procedure. The "*trans* rule," which has facilitated subsequent synthetic studies on nucleosides,[30a] states that "condensation of a heavy metal salt of a purine or pyrimidine with an acylated glycosyl halide will form a nucleoside with a C-1–C-2-*trans* configuration in the sugar moiety regardless of the original configuration at C-1–C-2." Anchimeric assistance by the 2-acyloxy group was invoked to explain the stereochemical control.

III. SYNTHESIS OF NUCLEOSIDES†

Two reviews have covered in detail the conventional synthesis of purine nucleosides[31] by using mercury derivatives of the base and of pyrimidine nucleosides[32] by using either the mercuri method or the Hilbert–Johnson technique. More recently, new and, frequently, more convenient methods of synthesis for both types of nucleosides have been reported.

A. DIRECT METHODS WITH PROTECTED SUGAR DERIVATIVES

1. *Fusion Methods*

In a series of papers starting in 1960, Shimadate and co-workers reported the fusion, with acid catalysts, of a number of polyacylated sugars with substituted purines to give the blocked nucleosides. As an example, the fusion of 6-chloropurine (**1**) with 1,2,3,5-tetra-*O*-acetyl-β-D-ribofuranose (**2**) and

p-toluenesulfonic acid (TsOH) gave the blocked nucleoside (**3**), which could be ammonolyzed to adenosine.[33] A variety of acid catalysts have been employed, including zinc chloride,[34] aluminum chloride,[34] sulfur trioxide,[34] manganese sulfate,[34] and bis-(*p*-nitrophenyl) hydrogen phosphate.[35] Use of the latter catalyst is especially noteworthy in the total synthesis of toyocamycin and sangivamycin.[36]

The early papers that utilized tetra-*O*-acetyl-β- (**2**) and α-D-ribofuranose in the fusion synthesis reported that only β-D-nucleosides were formed.[37] More recently the fusion of **2** with 6-acetamidopurine and TsOH was reported[38] to give, after deblocking, the β and α anomers of 9-D-ribofuranosyladenine in a ratio of 58:42, while the similar reaction of tetra-*O*-acetyl-D-xylofuranose with 6-nonanamidopurine gave approximately equal amounts

---

* *References start on p. 59.*

† This section was prepared by L. Goodman.

of the β and α anomers of 9-D-xylofuranosyladenine.[39] The fusion synthesis of purine nucleosides appears generally to be much less stereospecific than the conventional mercuri salt method.[31] Onodera et al.[40] studied the condensation reaction of 6-benzamidopurine (4) with a variety of 1-substituted

BzNH

AcOCH$_2$

NH$_2$

HOCH$_2$

4 + 5 $\longrightarrow$ 6

2,3,4,6-tetra-O-acetyl-D-glucopyranoses (5), using a variety of catalysts, and noted that the nature of both the catalyst and the C-1 substituent of the sugar could influence the anomeric composition of the nucleoside (6) that resulted after deblocking. For example, with TsOH as catalyst, mixtures of α and β forms of 6 were formed with 5 when X = Ac or CCl_3CO, whereas only the β form of 6 resulted with 5 when X = Cl, Br, or OH. With mercuric chloride or stannous chloride as catalyst only the β isomer of 6 was formed from 5 when X = Ac or CCl_3CO. The complexity of the fusion reaction is reemphasized in another study in which iodine was used as a catalyst for the condensation of the tetraacetate 2 with various purine derivatives.[41] Only the β-D-nucleoside was reported to be formed from 2,6-dichloropurine, while $N^6,N^{9(7)}$-diacetyladenine and N^9-acetylhypoxanthine gave both α-D- and β-D-nucleosides. The use of $N^2,N^{9(7)}$-diacetylguanine gave both 9-β- and 9-α- as well as 7-β- and 7-α-D-ribofuranosylguanine.[41]

The noncatalyzed fusion of β-D-ribofuranose tetraacetate (2) with a variety of purines has been reported to give only the β-D-nucleosides,[42,43] while the similar reaction of 4 and 5 (X = CCl_3CO) gave[40] a mixture of the α-D- and β-D anomers of 6. The yields in the fusions utilizing β-D-ribofuranose tetraacetate[42] (2) were strongly influenced by the fusibility of the purine bases, ranging from 10% with 6-chloropurine to 67% with 2(6)-chloro-6(2)-iodopurine. Other papers have stressed the importance of obtaining a homogeneous melt in these reactions, such as the use of a relatively low-melting purine derivative, in obtaining good yields of condensation products.[44] This requirement probably accounts for the small number of pyrimidine nucleosides that have been prepared by the fusion technique. The uncatalyzed reaction of 2,4-dimethoxypyrimidine with the 1-O-trichloroacetyl derivatives of tetra-*O*-acetyl-D-glucopyranose and tri-*O*-acetyl-D-xylopyranose gave, after deblocking, low yields of 1-(β-D-glucopyranosyl)uracil and 1-(β-D-xylo-pyranosyl)uracil, while the fusion of 4-ethoxy-2(1*H*)-pyrimidinone with 2,3,4-tri-*O*-acetyl-1-*O*-trichloroacetyl-α-D-ribopyranose and TsOH gave, as final product, 1-(β-D-ribopyranosyl)uracil.[45] Similarly, the reaction of β-D-ribofuranose tetraacetate (2) with 4-hydroxypyrimidine and TsOH

7

yielded 4-oxo-3-(2,3,5-tri-*O*-acetyl-β-D-ribofuranosyl)dihydropyrimidine[46] (7).

One of the most elegant applications of the fusion nucleoside synthesis is the use of 1,3,5-tri-*O*-acetyl-2-deoxy-D-*erythro*-pentofuranose (9) with a variety of purines (8) to give, after deblocking, both anomers of a number of 2-deoxy-D-*erythro*-pentofuranose nucleosides.[47] The catalyst was the weaker

8   9   10

acid, chloroacetic acid. Reasonable yields of 10 were obtained with 8 (X = H, Y = Cl), (X = Y = H), (X = H, Y = Me), and (X = H, Y = BzNH), while an excellent yield was obtained with 8 (X = Y = Cl). The condensation of 6-chloropurine (1) with 1,3,4-tri-*O*-acetyl-2-deoxy-β-D-*erythro*-pentopyranose with TsOH led[48] to the mixture of anomers of the pyranose analog of 10

* *References start on p. 59.*

(X = H, Y = Cl), also formed by the fusion of 6-chloropurine with 3,4-di-*O*-acetyl-D-arabinal and TsOH.[48,49] The use of another unsaturated sugar, 3,4,6-tri-*O*-acetyl-D-glucal (12), in a fusion reaction with theophylline (11) and TsOH, however, gave the unsaturated nucleoside[50] (13).

2. Use of Trimethylsilyl Derivatives

The use of trialkylsilyl groups to activate heterocyclic bases for condensation with glycosyl halides was first reported by Birkhofer *et al.*[51] in their synthesis of 3-ribofuranosyluric acid (16) by deblocking the product from the condensation of the tetrakis(triethylsilylated) uric acid (14) and 2,3,5-tri-*O*-benzoyl-D-ribosyl bromide (15). Nishimura and co-workers[52] extended this method, with trimethylsilyl activating groups, to provide a new synthesis of purine and pyrimidine nucleosides which gives especially good yields with the pyrimidine bases and thus complements the fusion method (Section III,A,1), which is more useful for the synthesis of purine nucleosides. Thus, uracil was converted into the 2,4-O-bis(trimethylsilyl) derivative (17) which, fused with

the chloro sugar (18) at 180° to 190°, followed by deblocking, gave a 35% overall yield of uridine. Originally only β-D-nucleosides were isolated by this method, but in a later paper the same authors reported[53] the formation of both α-D and β-D anomers in the condensation of 17 and 18. Significantly, the condensation of 17 with 2,3,5-tri-O-benzoyl-α-D-arabinofuranosyl bromide gave nearly equal amounts of the α-D and β-D anomers of 1-(D-arabino-furanosyl)uracil.[54] Extension of the trimethylsilyl technique to 6-benzamido-purine and to hypoxanthine gave low yields of adenosine and of 2',3',5'-tri-O-acetylinosine.[53]

Wittenburg prepared a number of pyrimidine nucleosides using the tri-methylsilyl derivatives of the bases with poly-O-acylated sugar halides (including 2-deoxy-3,5-di-O-p-toluoyl-α-D-erythro-pentofuranosyl chloride) by heating the molten compounds at 90° to 100°, by heating the reactants in dry solvents (benzene, nitromethane, or N,N-dimethylformamide), or by reaction in solvents at room temperature in the presence of silver perchlorate or mercuric acetate.[55] The third technique was utilized to prepare both anomers of 2'-deoxy-5-(trifluoromethyl)uridine,[56] while the former method was used to prepare 5-allyl-2'-deoxyuridine.[57] A significant effect of tempera-ture as well as the nature of the base on the anomeric composition of the product was noted in the condensation of the 2,4-O-bis(trimethylsilyl) derivatives of 5-acetylthio-, 5-methyl-, and 5-fluorouracil with 2-deoxy-3,5-di-O-p-chlorobenzoyl-α-D-erythro-pentofuranosyl chloride.[58] In all three cases when the reactants were fused at 100° to 110°, only β-D-nucleosides were formed. When the reactants were refluxed in benzene, the silyl derivatives of 5-acetylthio- and 5-fluorouracil gave essentially only the β-D-nucleosides, while that of thymine gave 26% of β-D-nucleoside and 8% of α-D-nucleoside.

* References start on p. 59.

When the reactants were kept at 37° for 90 hours in benzene as a solvent, the silyl derivative of 5-acetylthiouracil gave an $\alpha:\beta$ ratio of 2.3, and that of thymine 5.5, whereas 5-fluorouracil gave essentially pure α-D-nucleoside. The high-temperature, short reaction periods that gave largely β-D-nucleosides were postulated to occur by SN2 attack on the α-D-glycosyl halide, whereas the source of the α-D-nucleosides appeared to be connected with a slow anomerization of the glycosyl halide in the benzene solutions.

3. Miscellaneous Methods

Several techniques for direct coupling of acylamido purines and pyrimidines with poly-O-acylglycosyl halides have been reported. These methods avoid the sometimes difficult preparation of a metal derivative of the heterocyclic base. For example, the condensation of 6-benzamidopurine with 2,3,4,6-tetra-O-acetyl-α-D-glucopyranosyl bromide in hot nitromethane with mercuric cyanide gave a good yield of the blocked nucleoside.[59] Similarly 6,8-dichloropurine, 2,6-dichloropurine, theophylline, and benzimidazole could be used to prepare nucleosides.[59] The syntheses, by acid-catalyzed and autocatalytic reactions of fully acetylated sugars with purine derivatives, similar to the fusion reactions noted in Section III,A,1, have also been carried out in refluxing nitromethane.[60] In the rather rare example of the formation of a β-nucleoside from a D-arabinosyl halide, the condensation of 2,3,5-tri-O-benzyl-D-arabinofuranosyl chloride with 6-benzamidopurine in dichloromethane at room temperature gave, after deblocking, 9-(β-D-arabinofuranosyl)adenine.[61] This appears to be an SN2 reaction, since the sugar halide is known to be essentially the pure α-D anomer. The use of 2,6-dichloropurine with the arabinosyl halide, similarly, provided a route to 9-(β-D-arabinofuranosyl)-2-chloroadenine.[62]

The direct alkylation of adenine with 2,3,5-tri-O-benzoyl-D-ribosyl bromide (15) in acetonitrile at 50° gave 25% of 3-(2,3,5-tri-O-benzoyl-β-D-ribofuranosyl)adenine, accompanied by 18% of the 9 isomer, 2′,3′,5′-tri-O-benzoyladenosine.[63] By a similar technique, 7-(pivaloyloxymethyl)adenine was alkylated with 2-deoxy-3,5-di-p-chlorobenzoyl-α-D-erythro-pentofuranosyl chloride to yield, after removal of the blocking groups, the α-D and β-D anomers of 3-(2-deoxy-D-erythro-pentofuranosyl)adenine.[64] Alkylation of N^2-acetylguanine with 15 in N,N-dimethylformamide followed by deacylation gave a mixture of 7- and 9-(β-D-ribofuranosyl)guanine, indicating that these 7- and 9-substituted derivatives of N^2-acetylguanine are thermodynamically more stable than the 3 isomer.[65] It is interesting that in the Fletcher procedure for preparation of 9-(β-D-arabinofuranosyl)adenine,[61] which resembles the procedure for direct ribosylation of adenine,[63] a by-product was isolated which, after deblocking, has been identified as 3-(β-D-arabinofuranosyl)-adenine.[66]

B. Direct Methods with Unprotected Sugars

The direct condensation of adenine with D-ribose and 2-deoxy-D-*erythro*-pentose has been reported by several investigators. The reaction in the presence of "polyphosphoric ester" was claimed by Schramm[67] to give a good yield of adenosine and 30% yield of 2'-deoxyadenosine in a stereospecific synthesis. Carbon[68] repeated the reaction with 2-deoxy-D-*erythro*-pentose, however, and reported that six major compounds were formed in the condensation in a total yield of about 13 to 19%. Both 2'-deoxyadenosine and its α-D anomer were present in the product, with the α-D-nucleoside in greater amount. Condensation of 3,5-di-O-benzoyl-2-deoxy-D-*erythro*-pentofuranose with adenine under Schramm's conditions also gave a mixture of products, again containing, after deblocking, 2'-deoxyadenosine and a somewhat greater amount of its α-D anomer.[68] The condensation of D-arabinose and adenine in the presence of ethyl polyphosphate has been reported as a method for synthesis of 9-(β-D-arabinofuranosyl)adenine;[69] a more precise evaluation of the method showed that both α and β anomers were produced with the α anomer preponderating.[70] Schramm *et al.*[71] have described the reaction of adenine, D-ribose, and phenyl polyphosphate at 100° which gives adenosine and its α anomer in 20% yield accompanied by small proportions of pyranose nucleosides; the similar reaction with 2-deoxy-D-*erythro*-pentose gave 40% of a mixture of 2'-deoxyadenosine and its α anomer.

The reaction of adenine with 2-deoxy-D-*erythro*-pentose in aqueous solution, or in N,N-dimethylformamide or methyl sulfoxide, at 100° gave a mixture of two alkali-sensitive, 9-substituted nucleosides whose structures were shown to be 20 and 21; similar results were obtained with 6-dimethyl-

aminopurine and 6-methylthiopurine. The products were postulated to arise by Michael addition of the bases to the unsaturated aldehyde (19) formed in the reaction by dehydration of the *aldehydo* form of 2-deoxy-D-*erythro*-pentose.[72]

* *References start on p. 59.*

C. SYNTHESIS BY CONSTRUCTION OF A HETEROCYCLIC BASE AFTER GLYCOSYLATION

Construction of the heterocyclic base of a nucleoside after formation of a carbon to nitrogen bond at C-1 of the sugar often provides a useful synthetic method for specific nucleosides. Thus, the blocked imidazole derivative[73] (22) was converted into 2′,3′-O-isopropylideneguanosine by treatment with benzoyl isothiocyanate, then methylation, ammonolysis, and alkaline ring closure,[74] and into 2′,3′-O-isopropylideneinosine by treatment with ethyl formate and base.[75] These procedures should be useful for introducing

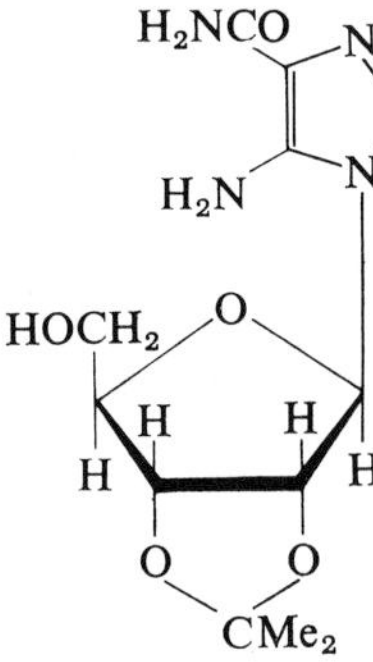

22

labeled atoms into the natural nucleosides. The similar use of an imidazole intermediate provided what is probably the best available synthesis of 7-glycosylpurines.[76] Acid-catalyzed fusion of D-ribofuranose tetraacetate (2) with 4(5)-bromo-5(4)-nitroimidazole gave a good yield of the ribosyl derivative (23) which, with potassium cyanide, was converted into 24. Reduction gave 25, which was treated with ethyl orthoformate and acetic anhydride, and then ammonolyzed to yield 7-(β-D-ribofuranosyl)adenine (26).

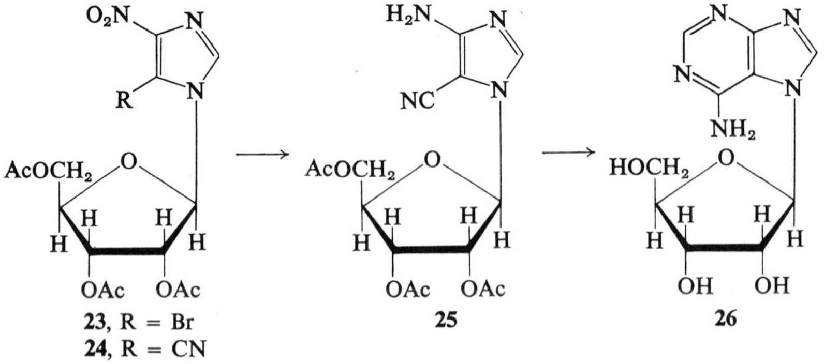

23, R = Br
24, R = CN

25

26

The first synthesis of a 9-glycosyl derivative of theophylline[77] was accomplished by the condensation of tetra-*O*-acetyl-β-D-glucopyranosylamine with 4-chloro-1,3-dimethyl-5-nitrouracil, followed by reduction to give 5-amino-4-(tetra-*O*-acetyl-β-D-glucopyranosylamino)-1,3-dimethyluracil. Ring closure of this diaminopyrimidine with carbon disulfide led to 9-(β-D-glucopyranosyl)theophylline. Conventional coupling methods give only theophylline nucleosides substituted at C-7.

In the field of pyrimidine nucleosides, condensation of 3,5-di-*O*-benzyl-2-deoxy-D-*erythro*-pentofuranose with urea gave a mixture of the anomeric 2-deoxy-D-*erythro*-pentofuranosylureas from which the β anomer could be separated. Reaction of the urea with 3-ethoxyacryloyl chloride, treatment of the product with ammonia, and subsequent hydrogenolysis afforded a 25% yield of 2′-deoxyuridine.[78] The biologically interesting compound 5-azacytidine (**30**) was prepared by a somewhat related route.[79] Treatment of 2,3,5-tri-*O*-acetyl-D-ribofuranosyl chloride with silver cyanate gave the isocyanate (**27**) which, with *O*-methylpseudourea, was converted into the derivative (**28**). Ring closure with ethyl orthoformate yielded the methoxytriazine

29, R′ = OMe, R = Ac
30, R′ = NH₂, R = H

(**29**), which was ammonolyzed to the product (**30**). A number of sugars, including 2-deoxy-3,5-di-*O*-*p*-toluoyl-α-D-*erythro*-pentofuranosyl chloride,[80] have been used successfully in this sequence.

D. DEVELOPMENTS IN CONVENTIONAL SYNTHETIC METHODS

1. The Mercuri Salt Method

In spite of the advent of the new coupling methods described in Section III, A, mercury derivatives of purines and pyrimidines continue to be used for many nucleoside syntheses, probably because more confidence can be

---

* *References start on p. 59.*

placed in the positional and anomeric structures of the products as compared with the products from the newer methods. However, a number of reports[80a–80c] now emphasize the formation of both anomers in the preparation of nucleosides from polyacylglycosyl halides in spite of the presence of a directing group at C-2'. Montgomery and Thomas[81] examined the structures of a number of mercury derivatives of purines by spectral means and showed that, in general, the acylglycosyl halides attack the nitrogen atoms that bear the mercury groups. Thus, bis(theophylline)-mercury has the metal at N^7, in agreement with the sole formation of 7-substituted nucleosides from theophylline, and 3-benzylhypoxanthine forms an N^7 chloromercuri derivative that leads to 7-substituted nucleosides. These considerations were used in preparing 7-α and 7-(β-D-ribofuranosyl)adenine utilizing the benzyl as the N^3 protecting group.[82]

A number of novel adenine nucleosides have been prepared by using 9-chloromercuri-6-benzamidopurine or the corresponding 6-acetamido derivative. The use of 2,3,5-tri-O-benzoyl-3-C-methyl-D-ribofuranosyl bromide and 2,3,5-tri-O-benzoyl-2-C-methyl-D-ribofuranosyl chloride led to the unique branched nucleosides, 3'-C-methyl- and 2'-C-methyladenosine, respectively.[83] The tetrahydrothiophene nucleoside, 4'-thioadenosine,[84] and the pyrrolidine nucleoside, 4'-acetamidoadenosine,[85] were prepared by using the appropriate sugars. The conversion of 2,3:5,6-di-O-isopropylidene-D-mannofuranosyl chloride into 9-(α-D-mannofuranosyl)adenine constitutes the first report of isopropylidene-blocked sugars in this type of nucleoside synthesis.[86] The preparation of the alkali- and acid-sensitive 9-(1-deoxy-β-D-psicofuranosyl)adenine from 1-bromo-1-deoxy-3,4,6-tri-O-p-toluoyl-D-ribo-hexulofuranosyl bromide and 9-chloromercuri-6-benzamidopurine represents an elegant example of the method.[87] Murray and co-workers have made extensive use of the coupling of a 1-O-acetyl-blocked sugar with the appropriate mercury derivative in the presence of titanium chloride to give the adenine nucleosides derived from D- and L-erythrose and D- and L-threose,[88] 3-deoxy-D-ribo-hexose,[89] and 3-deoxy-D-xylo-hexose.[90] This convenient method avoids the sometimes troublesome preparation of the glycosyl halide.

Although adenine nucleosides are prepared conveniently from the 6-acylamidopurine derivatives as in the examples described above, guanine nucleosides have been prepared by less direct and much more laborious routes, usually involving 2,6-diaminopurine derivatives. Recently the use of a mercury derivative of 2-acetamido-6-chloropurine has permitted a more direct synthesis of 9-substituted guanine nucleosides including 2'-deoxyguanosine.[91] 3'-deoxyguanosine,[92] and 9-(3-O-methyl-β-D-ribofuranosyl)-guanine.[93] The synthesis of 3'-deoxyguanosine from chloromercuri-2-acetamidohypoxanthine was also described, but in this case equal amounts of the 9 and 7 isomers were obtained.[94]

In the pyrimidine area, mercury derivatives of thymine and *N*-acetylcytosine have been employed successfully in preparation of the 1-substituted nucleosides, but no such direct preparation of a uracil nucleoside has been reported. It is interesting that the silver salt of uracil has now been successfully condensed with tri-*O*-benzoyl-D-ribofuranosyl chloride (**18**) and with tetra-*O*-acetyl-α-D-glucopyranosyl bromide to give the respective 1-substituted blocked nucleosides of uracil in useful yields.[95] Reaction of the mercury derivatives of acetyl-2-thiouracil and acetyl-4-acetamido-2-thiopyrimidine with (**18**) represents a good method for preparation of 2-thiouridine and 2-thiocytidine.[95a]

2. *The Hilbert–Johnson Method*

The Hilbert–Johnson synthesis of pyrimidine nucleosides via 2,4-dialkoxypyrimidines and poly-*O*-acylglycosyl halides offers a convenient route to both uracil and cytosine nucleosides and continues to be widely used in spite of the introduction of the more direct methods for pyrimidine nucleosides. Prystaš and Šorm have studied intensively the condensation of both blocked D-ribofuranosyl halides and 2-deoxy-D-*erythro*-pentofuranosyl halides with various dialkoxypyrimidines in relation to the stereochemistry of the products and the reactivity of the reagents. Treatment of tri-*O*-benzoyl-D-ribofuranosyl chloride (**18**) with a wide variety of 2,4-dialkoxypyrimidines[96] and 5-halo-2,4-dimethoxypyrimidines[97] in acetonitrile gave only β anomers. The reactivity of the dialkoxypyrimidines decreased in the order benzyl > methyl > ethyl > isopropyl.[96] In reactions of 2,4-dimethoxypyrimidine with 2-deoxy-3,5-di-*O*-(*p*-substituted benzoyl)-D-*erythro*-pentofuranosyl chlorides, the highest yields were reported with the *p*-methyl-substituted derivative in acetonitrile as solvent.[98] These conditions also gave the highest ratio of α to β anomers. The reaction in benzene gave a lower yield, but a lower α:β ratio and the inclusion of mercuric bromide in the reaction medium lowered the α:β ratio still more but had a deleterious effect on the yield. The use of dialkoxypyrimidines having negative substituents at C-5, such as fluoro, gave definitely poorer yields of products, still predominantly the α anomers, in reactions with the blocked 2-deoxy-D-*erythro*-pentofuranosyl halides; 2,4-dimethoxy-5-nitropyrimidine gave no coupling product.[99] The lack of steric control in the formation of pyrimidine 3′-deoxynucleosides from dialkoxypyrimidines and 2,5-di-*O*-*p*-nitrobenzoyl-3-deoxy-β-D-*erythro*-pentofuranosyl bromide has been reported.[100]

The presence of intermediate glycosylpyridinium salts in the Hilbert–Johnson procedure has been noted and these intermediates have been used to prepare 2-thio- and 2-aminopyrimidine nucleosides in an extension of the general method.[100a]

* *References start on p. 59.*

E. C-Nucleosides

The discovery that a novel nucleoside isolated from ribonucleic acids derived from several sources was 5-ribofuranosyluracil[101] (**31**) coupled with the structural elucidation of two antibiotics, formycin[102] (**32**) and showdomycin[103] (**33**), as compounds containing a sugar joined to a heterocyclic base

31　　　　　　**32**　　　　　　**33**

by a carbon to carbon bond, has been the impetus for an intense effort to develop synthetic methods for these types of C-nucleosides.

Pseudouridine (**31**) and related compounds have been prepared by attaching sugars to preformed pyrimidines; the alternative method of constructing the pyrimidine from a suitable sugar derivative has not been reported. The reaction of 2,4-dimethoxy-5-lithiopyrimidine with ribosyl chloride (**18**) gave a mixture of the anomers of 2,4-dimethoxy-5-D-ribofuranosylpyrimidine which, after hydrolysis with dichloroacetic acid, afforded four isomers of pseudouridine, one of which was identical to the natural material[104] (**31**). Brown et al.[105] improved the yield of **31** by using 5-lithio-2,4-di-*tert*-butoxypyrimidine and 2,4:3,5-di-O-benzylidene-D-ribose as reactants. The latter sugar, condensed with 2′,3′-O-isopropylidene-5-lithiouridine, afforded after deblocking, 5-(β-D-ribofuranosyl)uracil, which could be converted into **31** with yeast nucleosidase.[105] By similar procedures, 2,4-dibenzyloxy-5-lithiopyrimidine with the 2,3:4,5-di-O-isopropylidene-*aldehydo* derivatives of D-arabinose, D-xylose, and D-ribose gave 5-(1-deoxy-"α"-D-arabinit-1-yl)-uracil, 5-("β"-D-xylofuranosyl)uracil, and 5-(1-deoxy-"α"-D-ribit-1-yl)-uracil.[106] Treatment of 5-O-acetyl-2,3-O-isopropylidene-D-ribonolactone with 2,4-dibenzyloxy-5-lithiopyrimidine followed by reduction with sodium borohydride and deblocking yielded 10% of pseudouridine (**31**); use of the appropriate sugar lactones similarly afforded 5-(1-deoxy-"α"-D-mannit-1-yl)uracil and 5-(1-deoxy-"β"-D-gulit-1-yl)uracil.[107] Essentially the same method was used to prepare the unstable compounds 1-deazauridine and 2′-deoxy-1-deazauridine.[108] The configurations "α" and "β" of the 1-substituted alditol derivatives refer to the disposition of the C-1 hydroxyl group when the heterocyclic substituent is at the top of the chain in the Fischer projection; these configurations were assigned by comparative optical rotatory dispersion measurements.

The synthesis of 5-(D-ribofuranosyl)-6-azauracil (37) illustrates the synthesis of a *C*-nucleoside by construction of the heterocyclic base onto a sugar derivative.[109] The thiosemicarbazone of D-*altro*-heptulosonic acid (34) was cyclized to yield the 6-azathiouracil derivative (35). Methylation and treatment with mild acid gave the 1-deoxy-D-ribit-1-yl derivative (36). More vigorous acid treatment cyclized 36 to give 37. Similarly, the thiosemi-

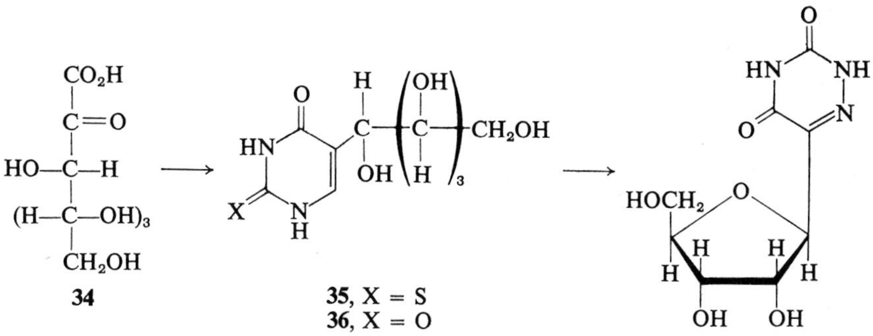

34

35, X = S
36, X = O

37

carbazones of L-*xylo*-, D-*arabino*-, and D-*ribo*-hexulosonic acids were converted into 1,4-anhydrotetrose derivatives of 6-azauracil.[110] As a potential entree into the formycin (32) area, the reaction of diazosugars with acetylenic compounds has been reported.[111] For example, 3,4,5,6-tetra-*O*-acetyl-1-diazo-1-deoxy-D-*ribo*-hexulose reacted with dimethyl acetylenedicarboxylate

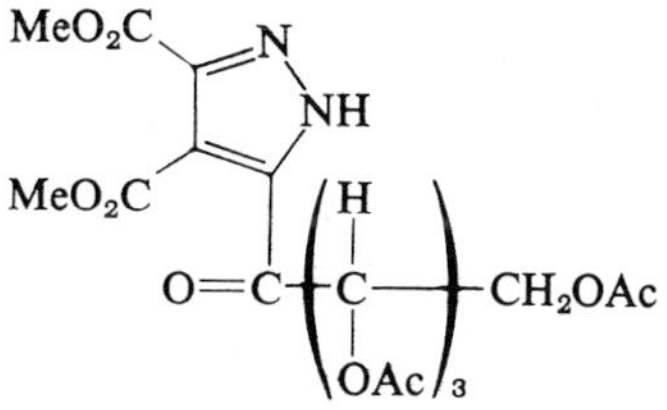

38

to give the pyrazole (38), which has the necessary functionality for cyclization of the sugar ring and construction of the pyrimidine moiety of formycin (32).

* References start on p. 59.

IV. CHEMICAL REACTIONS OF THE SUGAR PORTION
OF NUCLEOSIDES*

A. BLOCKING GROUPS

Interest in the synthesis of polynucleotides has resulted in a large number of papers on blocking groups for nucleosides whose introduction and removal are compatible with the reaction conditions necessary to prepare the polynucleotides.

1. Trityl Ethers

Historically, the most reliable selective blocking group for the primary hydroxyl group of a sugar or a nucleoside has been the trityl (triphenylmethyl) group introduced with chlorotriphenylmethane and pyridine (see Vol. IA, Chap. 12). However, in the preparation of certain nucleotides and deoxynucleotides the acid conditions necessary to remove a 5'-O-trityl group are frequently too severe to permit the retention of certain other blocking groups or, in some cases, to allow survival of the C–N glycosyl linkage. To provide a more acid-labile blocking group, the (p-methoxyphenyl) diphenylmethyl (monomethoxytrityl), the di-(p-methoxyphenyl)phenylmethyl (dimethoxytrityl), and the tri-(p-methoxyphenyl)methyl derivatives of nucleosides have been used.[112] Introduction of each p-methoxy group increases the rate of hydrolysis in 80% aqueous acetic acid at room temperature by a factor of approximately 10. Thus, the times required for complete hydrolysis to uridine were: 5'-O-trityluridine, 48 hours; 5'-O-(monomethoxytrityl)uridine, 2 hours; 5'-O-(dimethoxytrityl)uridine, 15 minutes; 5'-O-(trimethoxytrityl)-uridine, 1 minute.[112] Unfortunately, the ease of acid removal is paralleled by an increased reactivity toward secondary sugar hydroxyl groups, and more care is necessary with the p-methoxy derivatives to ensure selective substitution. Good procedures for the direct preparation of the 5'-O-monomethoxy-trityl derivatives of cytidine, adenosine, and guanosine have been reported, however.[113] The incorporation of β-D-arabinofuranosyl nucleosides into oligonucleotides by use of the dimethoxytrityl group has been employed to advantage.[114]

Although trityl ethers are usually regarded as specific for primary hydroxyl groups, it has been noted that uridine with trityl chloride under fairly vigorous conditions gives both 2',5'-di-O-trityluridine and 3',5'-di-O-trityluridine.[115] Subsequently, the formation, in this reaction, of some 2',3',5'-tri-O-trityluridine was reported by two groups.[116,117] The nature of the base obviously affects the course of these tritylations, since 6-azauridine under the same conditions gives only 6-aza-5'-O-trityluridine,[118] while 3-benzyluridine

* This section was prepared by L. Goodman.

gives 98.5% of 3-benzyl-2′,3′,5′-tri-*O*-trityluridine.[116] The reaction of thymidine 5′-*O*-(isobutylcarbonate) and monomethoxytrityl chloride has been reported to give an excellent yield of the 3′-monomethoxytrityl ether.[119]

2. *Acetals*

The preparation of polyribonucleotides usually requires a ribonucleoside having the 2,3-*cis*-diol group blocked at some stage of the synthesis. The 2,3-isopropylidene acetal, formed with acetone and an acid catalyst, is commonly used (see Vol. IA, Chap. 11). In a few instances where nucleosides of D-xylofuranose have been studied, the 3′,5′-isopropylidene acetal has also been useful,[120,121] and this derivative, as well as the 2′,3′-*O*-isopropylidene analog, has been prepared from 9-(β-D-lyxofuranosyl)adenine.[122] As was the case with the trityl group, however, the 2′,3′-*O*-isopropylidene group is frequently too acid-stable to permit its removal without damage to other groups. Smith *et al.*[112] noted that 2′,3′-*O*-anisylideneuridine was about ten times as labile as the isopropylidene- or the benzylidene-blocked compound, and this observation has been extended by Cramer *et al.*,[123] who prepared the acetals (**39**) and showed that their hydrolysis in 80% aqueous acetic acid

39

at 25° followed the order: R = 2,4-(MeO)$_2$C$_6$H$_3$; R′ = H > R = 4-Me$_2$NC$_6$H$_4$; R′ = H > R = 4-MeOC$_6$H$_4$; R′ = H > R = Ph; R′ = H ∼ R = R′ = Me > R = 4-ClC$_6$H$_4$; R′ = H. The 2′,3′-(4-dimethylaminobenzylidene)ribonucleosides are especially interesting, since they are stable in 90% trifluoroacetic acid because the dimethylamino group is protonated, but they are readily hydrolyzed in dilute aqueous acid because the functional group acts simply as an electron donor; the uridine, adenosine, cytidine, and guanosine derivatives were prepared from *p*-dimethylaminobenzaldehyde by using either trifluoroacetic or trichloroacetic acid as catalyst.[124] The preparation of 2′,3′-*O*-alkylidene derivatives of ribonucleosides by treatment with an acetal of the corresponding carbonyl compound in *N,N*-dimethylformamide with hydrogen chloride catalysis has been described.[125] For a given heterocyclic base, the rate of hydrolysis in 50% aqueous acetic acid decreases in the

* References start on p. 59.

order: p-anisylidene > cyclopentylidene > isopropylidene > cyclohexyli-
dene. Preparation of a variety of 2′,3′-cyclic acetals of uridine (and, in some
cases, guanosine) by treatment at 25° with the ketone, its dimethyl acetal (as
a dehydrating agent), and di-p-nitrophenyl phosphate as catalyst gave excellent
yields of the derivatives from acetone, benzophenone, crotonaldehyde,
cycloheptanone, cyclooctanone, cyclopentanone, methyl tert-butyl ketone,
and 3-pentanone.[126] The acid lability at pH 2 and 26° decreased in the order
listed, ranging from a half-life of 0.5 hour for the crotonylidene derivative
to more than 40 hours for the diphenylmethylidene derivatives. A related
acid-labile derivative, the 2′,3′-benzeneboronate, has been prepared from
adenosine and benzeneboronic acid.[127]

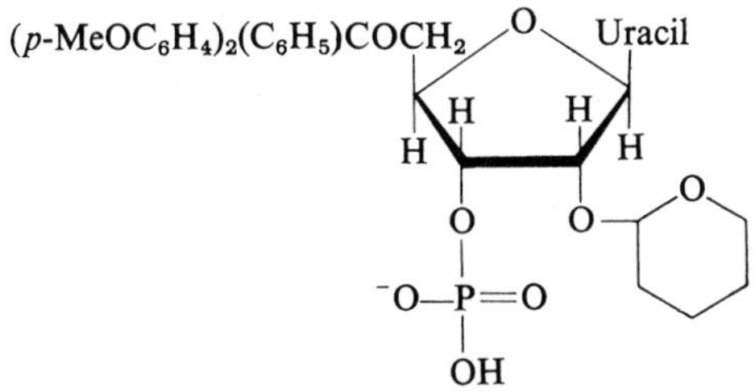

40

Protection of a single hydroxyl group by acetal derivatives has also been
important in syntheses of nucleotides. The reaction of uridine 3′,5′-cyclic
phosphate, as the free acid, with dihydropyran gave the 2′-tetrahydropyranyl
ether, which could be converted in two steps into the blocked 3′-phosphate
(**40**), suitable for preparation of uridine-containing nucleotides.[112] Ordinarily
the addition of an acid catalyst is needed in preparing tetrahydropyranyl
ethers, as in the conversion of 3′,5′-di-O-acetyladenosine with dihydropyran
and TsOH into the 2′-tetrahydropyranyl ether, which could then be deacylated with methanolic ammonia to 2′-O-tetrahydropyranyladenosine.[128]
To provide a group even more acid-labile than the tetrahydropyranyl ether,
5′-O-acetyluridine 3′-phosphate with ethyl vinyl ether gave the 2′-O-(1-
ethoxyethyl) derivative, easily hydrolyzed with 5% aqueous acetic acid at
room temperature.[129] The tetrahydropyranyl group possesses an asymmetric

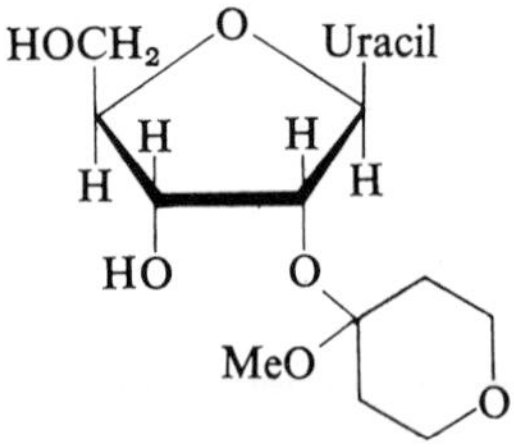

41

center which leads to the formation of mixtures of diastereoisomers. To overcome this difficulty, the preparation of symmetrical acetals by protection of a free hydroxyl group with 4-methoxy-5,6-dihydro-2H-pyran and acid has been reported.[130] Thus, 3',5'-diacetyluridine was converted into the 2'-acetal and then deacylated to **41**, for which the acid lability was of the same order of magnitude as that of the corresponding 2'-tetrahydropyranyl ether. The selective conversion of the primary hydroxyl group of uridine, by treatment with 2,2-dimethoxypropane and di-p-nitrophenyl phosphate, into the crystalline, very acid-labile, 5'-O-(1''-methoxyisopropyl)uridine was reported by Hampton.[131]

3. *Ortho Esters*

As a substitute for the 2',3'-alkylidene acetals as blocking groups for ribonucleosides, the use of 2',3'-O-alkoxymethylidene derivatives was reported by two groups at about the same time.[132,133] These very acid-labile ortho esters are prepared by the acid-catalyzed condensation of nucleosides with trialkyl ortho esters. The 2',3'-O-methoxymethylidene derivatives are rapidly hydrolyzed by 98% formic acid, 0.01N hydrochloric acid, or 5% aqueous acetic acid, to give a mixture of 2'- and 3'-formates, but are stable to alkaline conditions.[133] Similarly, the use of methyl orthoacetate or methyl ortho-benzoate in reaction with uridine gave cyclic ortho esters that could be hydrolyzed with acid to 2'(3')-acetates and 2'(3')-benzoates, respectively.[134] A particularly interesting synthetic application of the ortho-ester method has been reported. Treatment of uridine, cytidine, or adenosine with N-benzyl-oxycarbonylglycine ethyl ortho ester, with catalysis by methanesulfonic acid, gave high yields of the 2',3'-cyclic ortho esters (**43**), which were hydrolyzed with aqueous acetic acid to a mixture of the 2'- and 3'-esters (**44**) with the 3'-ester preponderating.[135] Hydrogenolysis of **44** gave the O-aminoacyl derivatives (**45**) useful as models for studying transfer RNA. Hydrogenolysis of **43** (base = adenine) gave the ortho ester (**42**), which is remarkably stable to both acid and alkali, since protonation of the amino group impedes further protonation of oxygen in the ester.[136]

4. *Acyl Derivatives*

Simple acyl esters constitute the principal acid-stable, alkali-labile blocking groups used for the hydroxyl groups of nucleosides. These are usually prepared by treatment of the nucleoside in pyridine with an acid chloride or anhydride. Use of these groups for selective blocking is complicated by their ability to migrate under certain conditions (see Vol. IA, Chap. 6). Thus, both 3'-*O*-acetyl- and 2'-*O*-acetyl-β-D-xylofuranosyluracil, when heated in refluxing pyridine, give 5'-*O*-acetyl-β-D-xylofuranosyluracil.[120] This tendency for acyl migration from secondary to primary hydroxyl groups is quite general. Under acidic conditions, no acyl migration was noted. The interconversion of 2',5'-di-*O*-acetyl- and 3',5'-di-*O*-acetyluridine occurs in refluxing alcohol or in dry pyridine at 20°, giving rise to an equilibrium mixture of the two.[137] Formyl groups also migrate in this system. However, benzoyl, and, even more so, anisoyl groups migrate from the 2'-OH to the 3'-OH group very slowly in dry pyridine, thus permitting the synthesis of pure 2'-*O*-acyl ribonucleosides.[137a]

Some cyclohydroxamic acids have been used to effect selective acetylations of nucleosides.[138] For example, compound **46** with uridine in hot pyridine

46

gave a good yield of 3'-*O*-acetyluridine; with inosine under similar conditions a lower yield of what is either 2'- or 3'-*O*-acetylinosine was obtained. With thymidine or 1-(β-D-xylofuranosyl)uracil, however, only the 5'-*O*-acetylnucleoside was formed, emphasizing the importance of the *cis*-2',3'-diol system in these selective acylations. The selective acylation of thymidine to give, among other compounds, 5'-*O*-(chloroacetyl)thymidine has been reported.[139] *N*-Formylimidazole has been used to prepare *O*-formylnucleosides.[123] The introduction of *O*-acyl groups by acid hydrolysis of cyclic orthoesters was mentioned in Section IV, A, 3.

The hydrolysis of *O*-acylnucleosides has been studied as a function of pH by Cramer *et al.*[123] At pH ∼11 and 20°, the half-lives of the 5'-*O*-acyl-2',3'-*O*-isopropylideneuridines were: chloroacetate, 1 to 2 minutes; formate, 1.75 minutes; acetate, 100 minutes; and benzoate, 680 minutes. At pH 10, the 5'-formate had a half-life of 22 minutes. The 5'-trifluoroacetate at pH greater than 7 was the most labile of the esters studied. Selective solvolysis of *O*-acyl groups has been used to prepare certain acylated nucleosides. For instance, 2',3',5'-tri-*O*-formyluridine and 6-aza-2',3',5'-tri-*O*-formyluridine in refluxing methanol for 2 hours gave the corresponding 5'-*O*-formyl derivatives.[140] With a longer period of reflux, 2',3'-di-*O*-acetyl-5'-*O*-formyluridine was converted into the 2',3'-di-*O*-acetylnucleoside. Thus, the lability to boiling methanol of formates combined with the stability of acetates under these conditions is useful in selective blocking of nucleosides. When either 3',5'-di-*O*-acetyladenosine or 3',5'-di-*O*-acetyluridine, in ethanol solution, was heated with morpholine, the 5'-*O*-acetylnucleosides were formed.[141] Under the same conditions, 3',5'-di-*O*-acetyl-2'-deoxyadenosine and 2',5'-di-*O*-acetyl-3'-deoxyadenosine were unchanged, again emphasizing the importance of the *cis*-vicinal OH groups in this selective solvolysis, just as its presence is important in selective acylation.[138]

p-Nitrophenyl chloroformate has been recommended for selective acylation of nucleosides.[142] This reagent gives *p*-nitrophenoxycarbonic esters with isolated OH groups, which can be removed with imidazole in aqueous organic solvents. With *cis*-glycols the reagent gives cyclic carbonates, which can be removed in hot aqueous pyridine or in dilute aqueous alkali.

5. *Miscellaneous Blocking Groups*

A few other substituents have been useful as blocking groups for nucleosides. Treatment of nucleosides with 2,4-dinitrobenzenesulfenyl chloride in pyridine yields acid-stable, base-labile sulfenic esters[143,144] that are especially useful with uracil and thymine derivatives. The sulfenate group can be removed at room temperature with cyanide, thiosulfate ion, or thiophenol and, under neutral conditions, with Raney nickel.

Benzyl ethers have been of some utility because of their ease of removal by hydrogenolysis. Specifically, substituted uridines have been prepared via the *O*-benzyl blocking group.[145,146] Sodium and liquid ammonia was shown to be a useful agent for *O*-debenzylation with an adenine nucleoside.[147] Methyl ethers, although not useful blocking groups, have been selectively introduced into the carbohydrate positions of nucleosides. Diazomethane and adenosine in aqueous 2-methoxyethanol gave 2'-*O*-methyladenosine,[148] also prepared by ammonolysis of the product from diazomethane and 6-chloro-9-β-D-ribofuranosylpurine.[149] Similar methylation of 2-amino-6-chloro-9-β-D-ribofuranosylpurine provided a route to 2'-*O*-methylguanosine.[149] It was shown subsequently that methylation of adenosine gives the 3'- and 5'-methyl ethers in addition to the major 2'-methyl ether.[150] Preparation of the 2'- and 3'-*O*-methyl derivatives of both adenosine and cytidine via the diazomethane route has been described.[151]

Nitric esters have been advantageously employed as blocking groups in transformations of (2-deoxy-D-*erythro*-pentofuranosyl)pyrimidines.[152] The direct nitration of 2'-deoxy-5-fluorouridine under anhydrous conditions gave the 3',5'-dinitrate. The 5'-nitrate reacted with sodium iodide in acetone to afford 2',5'-dideoxy-5-fluoro-5'-iodouridine 3'-nitrate, and the 3'-ester group was intramolecularly displaced under mildly alkaline conditions to give 2,3'-anhydro-2'-deoxy-5-fluorouridine 5'-nitrate. The nitric esters could be removed by catalytic hydrogenolysis without affecting the fluoro substituent or the 5,6-double bond.

B. Cyclonucleosides and Their Reactions

The compounds formed by bonding of an atom of the heterocyclic base with a carbon of the sugar portion of a nucleoside are called cyclonucleosides. They are important synthetically and for structure determinations in both the purine and pyrimidine nucleosides.*

1. *Purine Derivatives*

Chemical evidence for the β-D configuration at the anomeric center of adenosine and 2'-deoxyadenosine was provided by the ready formation of the cyclonucleosides[153] **47** and[154] **48**. Guanosine has been converted into similar 3,5'-anhydro-2',3'-O-isopropylideneguanosine salts[155,156] and thence into the free cyclonucleoside.[157] The β-D (or α-L) configuration of other synthetic purine or purine-like nucleosides has been established by formation of cyclonucleosides, accompanied by characteristic ultraviolet spectral

47   **48**

---

* These compounds have been called both cyclonucleosides and anhydronucleosides. The term cyclonucleoside will be used here as a general name, but the term anhydro-nucleoside will be used in systematic names such as 8,3'-anhydro-9-(β-D-xylopyranosyl)-8-mercaptoadenine and O^2,2'-anhydro-1-β-D-xylofuranosyluracil. The term epoxide has been used by some authors for anhydro.

changes, in the case of tubercidin[158] (**49**), 9-(3-deoxy-3-*C*-hydroxymethyl-α-L-*threo*-pentofuranosyl)adenine[159] (**50**), and 3-β-D-ribofuranosyl-3*H*-imidazo[4,5-*b*]pyridine[160] (**51**). In the case of **51**, formation of a quaternized

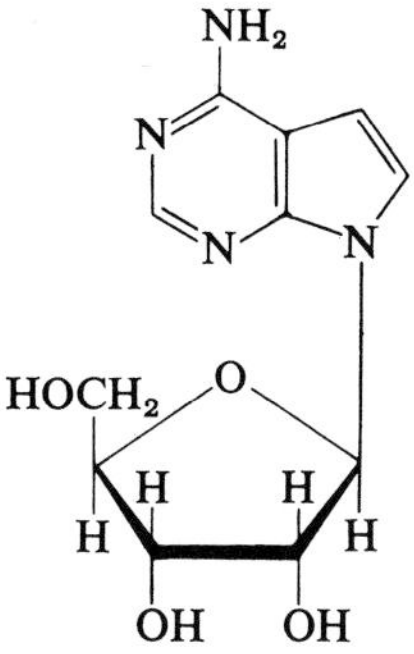

49 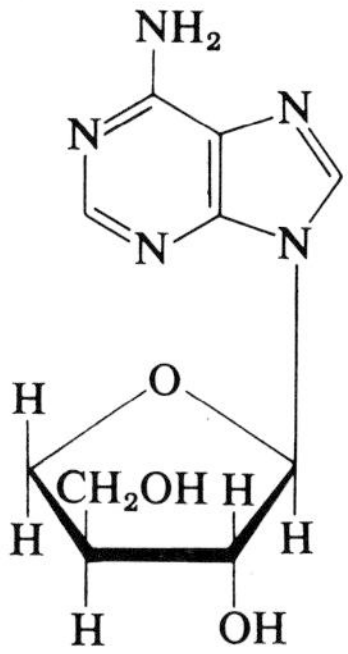   50 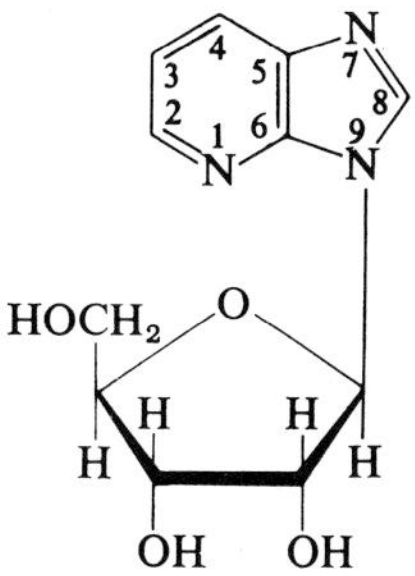   51

derivative proved not only that a β-D-nucleoside had been prepared but that substitution of the sugar had occurred at N-3, since the N-1 nucleoside cannot form a cyclonucleoside. Evidence for 3,3′-cyclonucleoside formation from purine nucleosides has been noted. Thus, the reaction of the benzylthio-nucleoside (**52**) with sodium acetate in aqueous 2-methoxyethanol gave, in addition to other products, a salt whose analysis and spectral properties were in accord with the cyclonucleoside structure (**54**), rationalized by attack of the purine base on the intermediate episulfonium ion[161] (**53**). Similarly, in

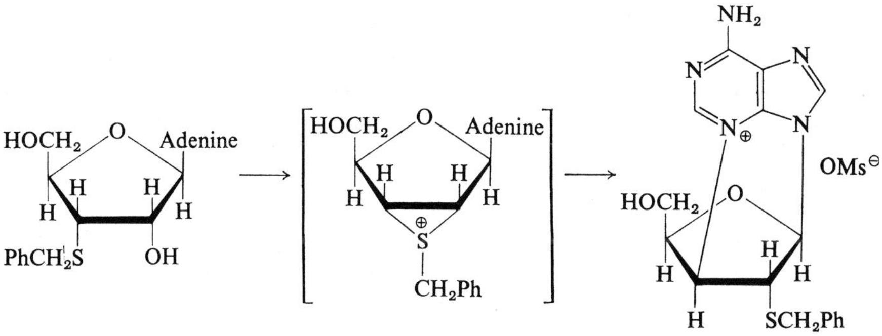

52   53   54

attempts to open the epoxide ring of 9-(2,3-anhydro-5-deoxy-β-D-ribofurano-syl)adenine with certain nucleophiles, there was spectral evidence that the 3,3′-cyclonucleoside was an important product.[162]

* *References start on p. 59.*

Some synthetic applications of purine 3,5'-cyclonucleosides have been reported. Acid hydrolysis of these compounds gave 3-substituted purines, **55** being derived from the salt[157] (**47**). The conversion of 3,5'-cyclonucleoside salts into their N^6-acyl derivatives facilitates nucleophilic attack at C-5'. Treatment of **47** with acetic anhydride and pyridine followed by reaction

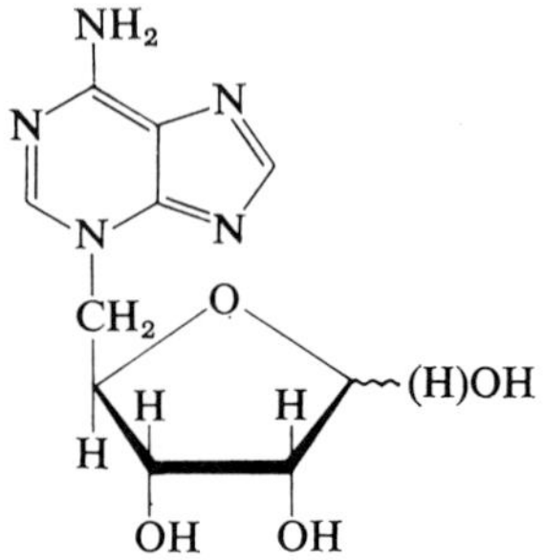

55

with dibenzyl hydrogen phosphate gave a good yield of 5'-O-(dibenzyl-phosphono)-2',3'-O-isopropylideneadenosine; dinucleotides were formed similarly by reaction with uridine 2'(3')-phosphate and adenosine 3'-phosphate.[163] These reactions were suggested by Jahn's finding[164] that N^6-acylation of 5'-O-(p-tolylsulfonyl)adenosines hindered formation of 3,5'-cyclonucleosides.

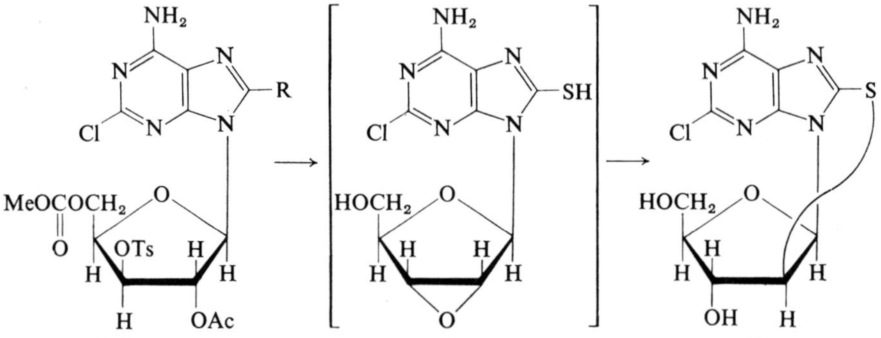

56, R = Cl
57, R = SH

58

59

The use of 8-substituted purine nucleosides has led to other purine cyclo-nucleosides that have utility in synthetic work. Treatment of nucleoside (**56**) with thiourea gave the mercaptonucleoside (**57**) which, with sodium methoxide, yielded the 8,2'-cyclonucleoside (**59**), presumably by way of the

2′,3′-anhydro compound[165] (58). The conversion of 59 into 2′-deoxyadenosine served as a proof of structure. From 8-bromoguanosine, via an 8-mercaptonucleoside, an 8,5′-nucleoside was obtained that could be converted into 5′-deoxyguanosine.[166] By methanesulfonylation of 5′-O-acetyl-8-bromoadenosine followed by treatment with thiourea, both 8,2′-anhydro-9-(β-D-arabinofuranosyl)-8-mercaptoadenine, convertible into 2′-deoxyadenosine, and 8,3′-anhydro-9-(β-D-xylofuranosyl)-8-mercaptoadenine, convertible into 3′-deoxyadenosine, could be obtained by appropriate treatment.[167] Conversion of 5′-O-acetyl-8-bromoadenosine into 5′-O-acetyl-8-hydroxy-2′(3′)-O-(p-tolylsulfonyl)adenosine followed by treatment with sodium benzoate in N,N-dimethylformamide yielded 8,2′-anhydro-8-hydroxy-9-(β-D-arabinofuranosyl)adenine, which could be hydrolyzed to 9-(β-D-arabinofuranosyl)-8-hydroxyadenine with mild acid or converted into 9-(2′-O-benzoyl-β-D-ribofuranosyl)-8-hydroxyadenine with sodium benzoate and benzoic acid.[168] The same 8,2′-anhydroarabinose nucleoside was formed when 8-bromo- or 8-azido-9-(2,3,5-tri-O-acetyl-β-D-arabinofuranosyl)adenine was treated at 0° with methanolic sodium methoxide.[169] Treatment of 8-bromo-2′,3′-O-isopropylideneadenosine with sodium hydride in dioxane gave 8,5′-anhydro-2′,3′-O-isopropylideneadenosine.[170,171] The O-isopropylidene group of this 8,5′-cyclonucleoside could be removed with relatively mild acid hydrolysis, whereas, under more vigorous conditions, the glycosidic bond was cleaved to give 5-(adenine-8-yl)-5-deoxy-D-ribose.[170]

2. Pyrimidine Derivatives

Cyclonucleoside formation in pyrimidine nucleosides, as in the purine derivatives, has been useful in structure proofs and has been even more useful as a synthetic tool. A wide variety of pyrimidine cyclonucleosides has been prepared.

The β-D configuration of pseudouridine C (31) (Section III, E) was established by conversion into its 2′,3′-isopropylidene acetal and then into the 5′-O-(p-tolylsulfonyl) nucleoside. Treatment of the latter compound with sodium tert-butoxide gave O⁴,5′-anhydro-2′,3′-O-isopropylidenepseudouridine.[172]

While O²,2′- and O²,5′-cyclonucleosides have been prepared from derivatives of both uridine and cytidine under relatively mild conditions, the O²,3′ derivatives of these compounds have been much less accessible. Treatment of 3′-O-(p-tolylsulfonyl)uridine with sodium tert-butoxide gave O²,3′-cyclouridine as a major product;[173] the conditions were much more stringent than those required to prepare the O²,2′ analog. Similarly, the reaction of 3′-O-mesylcytidine with sodium tert-butoxide at 100° afforded

$O^2,3'$-cyclocytidine.[174] This greater tendency for attack of the pyrimidine carbonyl on the C-2' of the sugar is emphasized by the report that p-toluenesulfonylation of a mixture of $N^4,O^{3'},O^{5'}$-tri-acetyl- and $N^4,O^{2'},O^{5'}$-tri-acetylcytidine gave, after treatment with water at 20°, $N^4,O^{3'},O^{5'}$-tri-acetyl-β-D-arabinofuranosylcytosine, the reaction probably proceeding by hydrolysis of the $O^2,2'$-cyclocytidine derivative[175] whose formation is not interfered with by the N^4-acetyl group. There was no evidence for either an $O^2,3'$-anhydrocytidine derivative or the D-xylofuranosylcytosine nucleoside that would have resulted from its hydrolysis. Formation and facile hydrolysis of an $O^2,2'$-cyclocytidine derivative is involved also in a simple synthesis of 1-(β-D-arabinofuranosyl)cytosine by treatment of cytidine with polyphosphoric acid.[176]

A simple preparation of $O^2,2'$-cyclouridine and its 5'-O-trityl derivative is available by treatment of uridine or 5'-O-trityluridine with thiocarbonylbis(imidazole);[177] the intermediate 2',3'-thionocarbonate has been isolated.[178] The same method was used to prepare 6-aza-$O^2,2'$-cyclouridine.[179] Synthesis of 1-(β-D-arabinofuranosyl)thymine and 1-(β-D-arabinofuranosyl)-5-fluorouracil by a method involving the conversion of a 2',3'- anhydro derivative into an $O^2,2'$-cyclonucleoside has been reported;[180] the process is related to the conversion of **58** into **59**. Two examples of cyclonucleosides derived from pyrimidine pyranose nucleosides have been recorded. Treatment of the 3'-acetamidonucleoside (**60**) with sodium ethoxide

gave the $O^2,2'$-anhydride (**61**); there was no evidence for participation of the 3'-acetamido group to give an oxazolino sugar nucleoside.[181] By heating at reflux the 3',4'-cyclic carbonate of 1-(2-deoxy-β-D-*erythro*-pentopyranosyl)-thymine, the cyclonucleoside (**62**) was formed in good yield. The 3',4'-di-methanesulfonate of the parent nucleoside was transformed[182] into **63** with potassium *tert*-butoxide at 100°. The similar derivatives of 1-(2-deoxy-α-D-

erythro-pentopyranosyl)thymine could not be converted into cyclonucleosides, as is predictable from structural considerations.

Cyclonucleosides can also be formed by addition of functional groups of a sugar to the 5,6-double bonds of pyrimidine bases. Treatment of 1-(β-D-arabinofuranosyl)-5-fluorouracil or -cytosine with hot 0.1N alkali gave **64** by addition of the "up" 2'-hydroxyl group, a required structural feature, to

64   **65, R = I**   **67**
                         66, R = H

the 5,6-double bond, followed by hydrolysis.[183] When 1-(β-D-arabino-furanosyl)cytosine was iodinated in the presence of periodic acid, the expected 5-iodo derivative was a minor product. Treatment of the residues with alkali afforded the 6-hydroxy-O^6,2'-cyclonucleoside (**65**) which could be hydrogenated to the cyclouridine derivative[184] (**66**). The precursor to **65** was 6-hydro-5,6-diiodo (or 5-hydro-5,6-diiodo)-6-hydroxy-O^6,2'-cyclouridine, which could also be formed by iodination of 1-(β-D-arabinofuranosyl)-uracil in the presence of nitric acid. The addition of the 5'-thiol group to the 5,6-double bond of a pyrimidine nucleoside was first noted in the formation of **67** by treatment of 5'-deoxy-5'-iodo-2',3'-O-isopropylideneuridine with sodium hydrogen sulfide.[185] Such 6-thio-S^6,5'-cyclonucleosides are in equilibrium with the 5'-mercaptonucleosides, and, because of the ready oxidation of the mercaptide ion to disulfide, the situation is quite complex. The position of equilibrium of the cyclonucleoside with the 5'-mercapto-nucleoside depends on the substituent at C-5 in the base.[186] Intermediate formation of the cyclonucleoside formed by addition of the 5'-hydroxyl group of 2',3'-O-isopropylidene-5-bromouridine to the 5,6-double bond was postulated in conversion of the parent nucleoside into 1-β-D-ribofuranosyl-barbituric acid and the isobarbituric acid nucleoside.[186a] Loss of hydrogen

* *References start on p. 59.*

bromide from the intermediate gave a 6,5′-anhydronucleoside that could be hydrolyzed with acid to 1-β-D-ribofuranosylbarbituric acid.

The intermediate (**64**) is useful for the transformation of the pyrimidine residues into nucleosides of 2-oxoimidazoline-4-carboxylic acid and ultimately into nucleosides of barbituric acid.[186a]

The $O^2,2'$-, $O^2,3'$-, and $O^2,5'$-cyclonucleosides of pyrimidines are exceedingly useful synthetic intermediates which undergo two main types of reaction. Nucleophiles can attack the C-2 position of the base, or they can attack the carbon atom of the sugar at the anhydride junction; there are many examples of each type of reaction.[32]

The alkaline hydrolysis of $O^2,2'$-cyclo-uridines and -cytidines to give the 1-(β-D-arabinofuranosyl)nucleosides is probably the best example of the first type of reaction.[176,177] The reaction of hydrogen sulfide in alkali with 1-(5-O-trityl-β-D-arabinofuranosyl)-$O^2,2'$-anhydrouracil to give, after detritylation, 1-(β-D-arabinofuranosyl)-2-thiouracil proceeds similarly.[187] Attack of ammonia at C-2 leads to isocytosine nucleosides, as with **61**, which gives a 1-(β-D-mannopyranosyl)isocytosine derivative[181] and with 6-aza-2′,3′-O-isopropylidene-5′-O-(methylsulfonyl)uridine, which yields 6-aza-2′,3′-O-isopropylideneisocytidine by way of the $O^2,5'$-cyclonucleoside.[188] As a result of the intervention of isocytosine nucleosides in opening of cyclonucleosides with ammonia, a synthesis of $N^2,3'$-cyclonucleosides has been developed.[189] The $O^2,5'$-cyclonucleoside (**68**), prepared from 5′-deoxy-5′iodo-

3′-O-(methylsulfonyl)thymidine and silver acetate, with liquid ammonia gave **69** and with methylamine yielded **70**.

The reaction of $O^2,2'$-anhydro-1-(β-D-arabinofuranosyl)-uracil and -thymine with anhydrous hydrogen fluoride, hydrogen chloride, and hydrogen bromide gave the 2′-fluoro, 2′-chloro, and 2′-bromo analogs of 2′-deoxyuridine and thymidine;[190] this is an example of nucleophilic attack on a pyrimidine cyclonucleoside at the sugar residue of the anhydro bridge. Similarly 2'-O-(p-tolylsulfonyl)uridine and 2'-O-(p-tolylsulfonyl)-5'-O-tritylυridine with alkali bromides or chlorides gave the 2'-chloro- or 2'-bromonucleosides by way of the O^2,2'-cyclonucleosides;[191] alkali iodides gave a mixture of the 2'-iodo derivative and the cyclonucleoside. The reaction of 3'-O-(methylsulfonyl)-5'-O-tritylthymidine with potassium thiolbenzoate or potassium phthalimide yielded, after detritylation, 3'-S-benzoyl-3'-thiothymidine and 3'-deoxy-3'-phthalimidothymidine, respectively, the products of opening of the intermediate O^2,3'-cyclonucleoside.[192]

Some reactions of cyclonucleosides proceed at carbon atoms at both ends of the bridge. The 2',3'-isopropylidene acetal of O^2,5'-cyclouridine with hydrogen sulfide gave 2',3'-O-isopropylidene-2-thiouridine and its disulfide as well as the S^6,5'-epithio derivative[193] (67).

The reaction of O^2,3'-[174,194] and O^2,5'-cyclonucleosides[195,196] with nucleoside 3'- and 5'-phosphates to give dinucleotides is a recent synthetic application of these compounds.

The complexities of reactions involving pyrimidine cyclonucleosides are well illustrated in a series of papers by Fox and co-workers.[197-199] The reaction of 2',3',5'-tri-O-(methylsulfonyl)uridine with sodium benzoate in N,N-dimethylformamide gave O^2,2'-anhydro-(3,5-di-O-benzoyl-β-D-arabinofuranosyl)uracil and 1-(2,3,5-tri-O-benzoyl-β-D-xylofuranosyl)uracil as direct products.[197] The residue from the isolation of the two nucleosides, after alkaline treatment, gave all four 1-(β-D-aldopentofuranosyl)uracils, the formation of which was rationalized by intervention of the 2',3'-anhydronucleoside and the 2',3'-orthobenzoate ion.[197]

C. NEIGHBORING-GROUP REACTIONS

Direct substitution of the sugar hydroxyl groups in nucleosides, especially the secondary hydroxyl groups, is difficult, and many such reactions are accomplished by utilizing suitable neighboring groups. The formation of cyclonucleosides represents such a neighboring-group reaction, and especially in the pyrimidine area (Section IV, B, 2) many of the substitution reactions require the intervention of a cyclonucleoside. A relatively small number of purine nucleosides containing "unnatural" sugars have been prepared by reactions on a preformed nucleoside; usually it is simpler to synthesize the required sugar, often by neighboring-group reactions, and to couple it with the purine base.

Because of the operation of the *trans* rule[30,30a] in nucleoside couplings, the first preparation of 9-(β-D-arabinofuranosyl)adenine was accomplished by way of a neighboring-group reaction on a preformed β-D-nucleoside;[200] the same method was used to prepare 9-(β-D-arabinofuranosyl)guanine.[121] The same considerations suggested that the preparation of 9-(β-D-lyxofuranosyl)-adenine (74) would require a suitable β-D-nucleoside as the starting material. The fully blocked derivative (71 or 72) was prepared from 9-(β-D-xylo-furanosyl)adenine by suitable reactions and was converted into the lyxoside (74) by way of the orthobenzoate ion[122] (73). The product (74) was accompanied by various proportions, depending on the reaction conditions, of

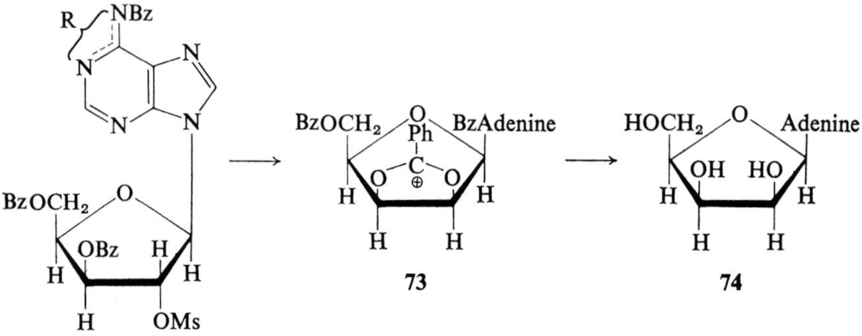

71, R = Bz
72, R = H

73

74

the D-arabinose and D-xylose nucleosides formed by opening of 73 at C-3′ and C-2′, respectively. Another example of complex neighboring-group substitution, this time in the conversion of a *trans*-1,2-amino alcohol into a *cis*-1,2-aminothiol, is represented by the synthesis of 3′-amino-3′-deoxy-

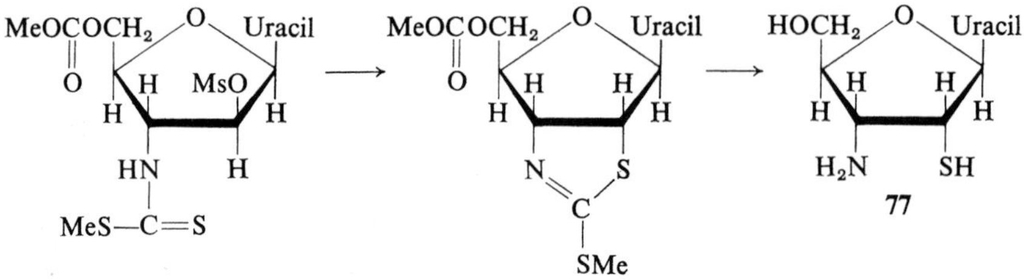

75

76

77

2′-thiouridine[201] (77). Compound 75, prepared from 1-(3-amino-3-deoxy-β-D-arabinofuranosyl)uracil, was refluxed in pyridine to give the thiazoline (76) which, by reduction and hydrolysis, was converted into the aminothiol (77). The generally powerful neighboring-group effects of sulfur-containing groups, noted above, was used in the preparation of the thiocyanato nucleoside (80), a precursor of 2′,3′-dideoxyadenosine. Treatment of the chloro-(ethylthio)nucleoside (78) with potassium thiocyanate yielded the 3′-thiocyano derivative (80) as the only isolated product;[202] intervention of the epi-

78   **79**   **80**

sulfonium ion (**79**) is reasonable, based on analogous reactions in the sugar field. The transformation of a nucleoside containing a *trans*-vicinal amino alcohol group into one containing a *cis*-1,2-amino alcohol group by means of the acetamido neighboring group has been thoroughly investigated by Baker and co-workers; an example occurs in the synthesis of 9-(2-amino-2-deoxy-β-D-allopyranosyl)-6-dimethylaminopurine.[203]

81   **82**, R = H, R′ = Ts   **84**
       83, R = Ts, R′ = H

The conversion of a *trans*-1,2-hydroxysulfonate into an epoxide represents one of the simplest neighboring-group transformations. Nucleoside epoxides are usually formed in this way and have proved to be very useful synthetic intermediates. This is illustrated by the two mono(*p*-toluenesulfonates) (**82** and **83**) which, by treatment with base, give the 2′,3′-anhydro-D-lyxosyl (**81**)

85, R = PhCH₂   **87**, R = Ms
86, R = H   **88**, R = H

89, R = H   **91**
90, R = Ms

and the 2',3'-anhydro-D-ribosyl (**84**) derivatives.[204] Since, in common with all 9-(2',3'-anhydro-β-D-pentofuranosyl)adenines studied to date, **81** and **84** are opened by nucleophiles predominantly at C-3',[162] such intermediates provide a wide variety of substituted adenine nucleosides. In the pyrimidine nucleosides the conversion of a 2',3'-anhydro- into a 2',5'-anhydro- and into a 3',5'-anhydronucleoside has been reported.[198] The epoxide (**87**) with sodium benzyloxide yielded **85**, in which the 2'-anionoid oxygen atom formed in the epoxide opening gave the 2',5'-anhydride by neighboring-group attack on the 5'-methanesulfonate. Catalytic hydrogenation converted **85** into **86**. When epoxide (**88**) was hydrolyzed in aqueous acid the D-xylosyl derivative (**89**) was the minor product. The methanesulfonate (**90**) from **89** by treatment with base gave the 3',5'-anhydronucleoside (**91**). Intervention of an epoxide intermediate seems probable in the hydrolysis of **63** with hot water containing calcium carbonate to give 1-(2-deoxy-α-L-*threo*-pentopyranosyl)thymine.[182]

D. OXIDATION

The direct oxidation of nucleosides to the uronic acid nucleosides has been achieved with several reagents. Reaction with oxygen in the presence of platinum at pH 9 was first used to prepare the 5'-carboxylic acids from uridine, thymidine, and adenosine[205] and, subsequently, from 2'-deoxyuridine and its 5-bromo- and 5-iodo derivatives, from 2'-deoxy-5-iodocytidine, and from the 1-(β-D-arabinofuranosyl) derivatives of uracil and cytosine.[206] The direct nitration of uridine gave 5-nitro-1-(β-D-ribosyluronic acid)uracil.[207] The conversion of 2',3'-O-isopropylideneadenosine to the 5'-carboxylic acid with aqueous potassium permanganate provides the most attractive synthesis of adenosine-5'-carboxylic acid available.[164a] Oxidation, with pyridine–chromium trioxide, of thymidine, 2'-deoxyadenosine, 2'-deoxyguanosine, and 2'-deoxycytidine to the respective uronic acids has been reported.[208] The products in these reactions were accompanied by the heterocyclic bases, apparently formed by the accompanying oxidation of the 3'-hydroxyl group of these nucleosides followed by base-catalyzed elimination of the base from the 2'-deoxy-3'-pentulosyl nucleosides. The synthesis of nucleoside 5'-aldehydes and 2'- and 3'-ketones has been accomplished with the methyl sulfoxide oxidation systems. Treatment of 3'-O-acetylthymidine with methyl sulfoxide, N,N'-dicyclohexylcarbodiimide, and anhydrous orthophosphoric acid gave an excellent yield of 3'-O-acetylthymidine-5'-carboxaldehyde;[209] similar reactions with 2',3'-O-isopropylidene-uridine and -adenosine afforded, after deblocking with acid, the respective 5'-aldehydes. When this reaction was applied to a nucleoside containing a free 3'-hydroxyl group, the glycosidic bond was cleaved by spontaneous β elimination of the base from the resultant 3'-pentulosylnucleoside. By proper choice of compound, however, this latter problem has been solved. The oxidation of 2′,5′-di-*O*-trityluridine with the methyl sulfoxide–*N*,*N*′-dicyclohexylcarbodiimide system (or with the methyl sulfoxide and acetic anhydride or phosphorus pentaoxide) gave 1-(2,5-di-*O*-trityl-D-*erythro*-3-pentulofuranosyl)uracil, converted into the alkali-labile "3′-ketouridine" with acid;[210] similarly 3′,5′-di-*O*-trityluridine afforded "2′-ketouridine."[210]

Another valuable oxidation technique in the nucleoside area has been the use of periodate, which originally was used to determine the ring size of the sugar moiety or the anomeric configuration at C-1′, but more recently has found utility in synthesis. The periodate oxidation of a nucleoside gives a "dialdehyde" (see Vol. IB, Chap. 25) that can be condensed with an aliphatic nitro compound (usually nitromethane) to give a mixture of *C*′-nitro sugars; these can be hydrogenated to the corresponding *C*′-amino sugars. This is illustrated for the case of cytidine.[211] The "dialdehyde" (**92**) (probably existing as a cyclic hemialdal) with nitromethane gave **93** which, without isolation, was converted into the mixture of the hydrochlorides of **94** and **95**

($R = O$). Hydrogenation with Raney nickel gave the mixed amino nucleosides (**94** and **95**, $R = H$) with the ratio of the D-*gluco* (**95**) to the D-*manno* (**94**) isomers being 20:1. The same sequence applied to 7-(β-D-ribofuranosyl)- or 7-(β-D-glucopyranosyl)theophylline afforded 7-(3-amino-3-deoxy-β-D-mannopyranosyl, -β-D-galactopyranosyl, and -β-D-glucopyranosyl)theophyllines.[212] It has been applied to a wide variety of nucleosides to give similar mixtures of products.[213,214]

E. TRANSGLYCOSYLATION

The transfer of the sugar moiety from one position of the heterocyclic base to another has been noted in both the purine and pyrimidine nucleosides.

---

* *References start on p. 59.*

These glycosyl migrations give important clues about the mechanism of certain methods of synthesis of nucleosides.

When the hydrochloride of 3-(2,3,5-tri-O-benzoyl-β-D-ribofuranosyl)-N-benzoyladenine was heated in acetonitrile solution, it rearranged to 9-(2,3,5-tri-O-benzoyl-β-D-ribofuranosyl)-N-benzoyladenine.[215] The D-ribosyl migration could also be accomplished with mercuric halides in other organic solvents. When the 3-substituted derivative was heated with labeled N-benzoyladenine in a mixture of N,N-dimethylformamide and xylene containing mercuric bromide, the 9-substituted nucleoside obtained contained 51% of the radioactivity, establishing that the migration was an intermolecular reaction. It was also observed that the reaction of N-benzoyladenine (or its chloromercuri salt) and tri-O-benzoyl-D-ribofuranosyl bromide (**15**) in hot acetonitrile gave a mixture of the 3 and the 9 isomers with the proportion of the 9 isomer increasing with increased reaction time. These observations suggest that the reactions of glycosyl halides with N-acyladenines or their chloromercuri salts give the 3-glycosyl derivatives as the initial product, which then rearrange to the more stable 9-glycosyl derivatives.[215] Transglycosylation from pyrimidines to purines has been noted. Treatment of 2',3',5'-tri-O-acetyl-N^4-acetylcytidine with N-benzoyladenine and mercuric bromide at 150° in xylene-N,N-dimethylacetamide gave, after removal of protecting groups, adenosine and its α-anomer.[215a] Many other purine bases could be used in this sugar transfer.

The O → N migration of sugars in pyrimidine nucleosides has been studied in some detail. The reaction of the silver salts of cytosine and N-acetylcytosine with 2,3,4,6-tetra-O-acetyl-α-D-glucosyl bromide gave good yields of the glycosides, which could be rearranged to the nucleosides with mercuric bromide in toluene.[216] Similarly, the acylated ribofuranosides and 2'-deoxy-D-*threo*-pentofuranosides could be prepared; the D-riboside could be rearranged with mercuric bromide, but the 2'-deoxy-D-*erythro*-pentofuranoside decomposed under these conditions.[217] The O → N rearrangements were shown to be intermolecular and to give only the β-D-nucleosides; this result suggested that a free polyacylglycosyl cation was not a reaction intermediate. The reaction of thymine with 2,3,4,6-tetra-O-acetyl-α-D-glucosyl bromide and 2,3,4-tri-O-acetyl-β-D-ribopyranosyl bromide in the presence of silver salts gave good yields of the O^2,O^4-bis(poly-O-acetyl-β-D-glycosyl)thymines, which could be rearranged with heavy metals in acetonitrile to the N^1-(poly-O-acetyl-β-D-glycosyl)thymines;[218] in later work it was shown that the rearrangement was specifically initiated by a proton source and that traces of acid were necessary.[219] The transglycosylation of other heterocyclic bases has been studied in detail. Perhaps the observation most pertinent to the mechanism of nucleoside formation concerns the reaction of 2-chloromercuroxypyrimidine (derived from 2-pyridone) with tri-O-benzoyl-D-ribofuranosyl chloride (18) which gave the O-β-D-glycoside, converted readily into 1-(2,3,5-tri-O-benzoyl-β-D-ribofuranosyl)-2-pyridone with mercuric bromide in refluxing xylene;[220] the presence of mercuric bromide in the condensation mixture gave the nucleoside directly. The acetylated O-β-D-glucoside of 5-nitro-2-pyridone, however, with mercuric bromide in toluene, anomerized to the O-α-D-glucoside concomitantly with rearrangement to the N-β-nucleoside;[221] the O-α-D-glucoside rearranged to the N-β-nucleoside with mercuric chloride or stannic chloride. These results suggest that different mechanisms are involved in the anomerization and the O → N rearrangement.

F. SUBSTITUTION AND ELIMINATION REACTIONS

Few examples exist of SN2 displacements of the sulfonic esters (or analogous leaving groups) of sugar hydroxyl groups in purine nucleosides. Esters of the secondary hydroxyl groups are ordinarily resistant to such displacements, while those of the primary hydroxyl groups ordinarily give cyclonucleosides (which, in some cases, then react with nucleophiles to give the same end products as the SN2 reaction). With sufficiently powerful nucleophiles, direct substitutions can be observed, as has been noted in the reaction of 2',3'-O-isopropylidene-5'-O-(p-tolylsulfonyl)guanosine[156] and 9-(2,3-di-O-acetyl-5-O-(p-tolylsulfonyl)-β-D-ribofuranosyl)-2-methylthioadenine[222] with alkali salts of alkanethiols to give the respective 5'-S-alkyl-5'-thionucleosides. The conversion of 2'-deoxy-3'-O-(p-tolylsulfonyl)adenosine into the presumed and probable 9-(2-deoxy-3-S-ethyl-3-thio-β-D-$threo$-pentofuranosyl)adenine with sodium ethanethioxide represents an SN2 reaction on a secondary sugar sulfonate;[223] apparently the 2'-deoxy group results in an enhanced reactivity of the 3'-p-toluenesulfonate. Since N-acyladenine nucleosides resist formation of 3,5'-cyclonucleosides, substitutions of the 5'-sulfonates of these compounds are possible.[164] Thus, 2',3'-O-isopropylidene-5'-O-(p-tolylsulfonyl)adenosine, treated in acetic anhydride with sodium iodide or sodium bromide, and then deblocked, gave the respective 5'-deoxy-5'-haloadenosines.[164] The N-formyl derivative of the parent 5'-p-toluenesulfonate was similarly converted into 5'-chloro-5'-deoxyadenosine and 5'-azido-5'-deoxyadenosine. More recently conditions have been found for the direct conversion of 5'-O-(p-tolylsulfonyl)-adenosine into 5'-amino-5'-deoxyadenosine and into 5'-piperidino-5'-deoxy-adenosine.[164a] The pyrimidine nucleosides provide an example of what is best explained as a direct SN2 reaction on a secondary sulfonate—again, as in the purine nucleoside case,[223] on a 2'-deoxy derivative. The reaction of the thymine nucleoside (96) with azide ion or with iodide ion yielded the respective inverted derivatives[224] (97 and 98), while the isomer (99) with iodide ion also gave **98**, intervention of the $O^2,3'$-cyclonucleoside permitting retention of configuration at C-3'. In other examples of SN2 reactions in pyrimidine nucleosides, 3'-O-(methylsulfonyl)-2',5'-di-O-trityluridine was treated with a large excess of sodium ethanethioxide in hot N,N-dimethylformamide to give, after acid treatment, 11% of 1-(3-S-ethyl-3-thio-β-D-xylofuranosyl)uracil which could be desulfurized to 3'-deoxyuridine.[224a] When 5'-O-acetyl-3'-O-(p-tolylsulfonyl)-6-azauridine was treated with sodium iodide at elevated temperatures a 3'-iodo derivative was obtained that was assigned as the 3-deoxy-3-iodoxylofuranosyl derivative on the basis of its p.m.r. spectrum and its reduction to 3'-deoxy-6-azauridine.[224b] Intervention of the 2',3'-cyclonucleoside in this reaction, which would lead to the 3-deoxy-3-iodoribofuranosyl derivative, cannot be ruled out on the basis of the spectral data. A number of other substitution reactions in the pyrimidine nucleoside area, such as the reactions of methyltriphenoxyphosphonium iodide with 5'-O-acetylthymidine to give 5'-O-acetyl-3'-deoxy-3'-iodothymidine, with retention,[225] and the substitution reactions of 3'-O-acetyl-2'-deoxy-5'-O-p-tolylsulfonyl-N-benzoyl-cytidine,[226] can also be explained as occurring by way of cyclonucleosides.

The preparation of unsaturated nucleosides, either for comparison with natural products or as reaction intermediates, has received considerable attention recently. The reactions of $O^2,3'$-cyclonucleosides and of 3',5'-anhydro derivatives with potassium *tert*-butoxide in N,N-dimethylformamide have been used to prepare 2',3'-didehydro-2',3'-dideoxyuridine[227] and 2',3'-didehydro-2',3'-dideoxycytidine.[228] The E2 reaction of 2'-deoxy-3'-O-(p-tolylsulfonyl)adenosine with sodium methoxide in N,N-dimethylformamide gave 2',3'-didehydro-2',3'-dideoxyadenosine;[229] the elimination reaction in ethanol gave the unsaturated nucleoside as well as some 9-(3,5-anhydro-2-deoxy-β-D-*threo*-pentofuranosyl)adenine.[230] The reaction of a *trans*-1,2-iodosulfonate system with sodium iodide in acetone has also been used to generate a 2',3'-didehydro-2',3'-dideoxynucleoside from

1-[5-O-benzoyl-3-deoxy-3-iodo-2-O-(methylsulfonyl)-β-D-arabinofuranosyl]-uracil[231] and from 1-[4,6-O-benzylidene-3-deoxy-3-iodo-2-O-(methylsulfonyl)-β-D-altropyranosyl]-4-ethoxy-2(1H)-pyrimidinone.[232] All these unsaturated nucleosides could be hydrogenated to the 2′,3′-dideoxynucleosides.

Some 4′,5′-unsaturated nucleosides have also been prepared. Treatment of 2′,3′-di-O-acetyl-5′-deoxy-5′-iodouracil with silver fluoride in pyridine afforded, after treatment with base, the unsaturated nucleoside[233] (**100**).

100, B = uracil
101, B = adenine

The 2′,3′-O-isopropylidene derivative of **100** was obtained by treating either 2′,3′-O-isopropylidene-5′-O-(p-tolylsulfonyl)uridine or O^2,3′-anhydro-2′,3′-O-isopropylideneuridine with potassium *tert*-butoxide.[234] Similarly, the reaction of 2′,3′-O-ethoxymethylidene-5′-O-(p-tolylsulfonyl)adenosine with potassium *tert*-butoxide at room temperature gave, after the mild acid treatment required to remove the cyclic orthoester, the adenine derivative[235] (**101**), which is closely related to the antibiotic decoyinine.[236]

G. HYDROLYSIS†

Because of the paucity of good kinetic data, little progress has been made toward an understanding of the mechanism of hydrolysis of nucleosides. However, numerous qualitative or semiquantitative observations have been made which permit certain generalizations: (1) Most nucleosides are hydrolyzed by aqueous acid but are remarkably stable in aqueous base. (2) Under conditions of acid catalysis, purine nucleosides are more easily hydrolyzed than pyrimidine nucleosides. (3) Deoxyribonucleosides are more labile to acid than ribonucleosides.

As with simple glycosylamines[237] (see Vol. IB, Chap. 20), the most likely mechanism would appear to be protonation of the ring oxygen atom followed

* *References start on p. 59.*
† This section was prepared by C. A. Dekker.

by ring opening and attack by water.[238] The alternative mechanism involving protonation of the aglycon followed by loss of the base and attack by water is not favored because of the stability of tri-*N*-methylglycosylammonium

salts.[239] According to the first mechanism, hydrolysis of the quaternary ammonium compound would be difficult because of the repulsion of the incoming proton by the charge on the nitrogen. (The author does not mean to imply that the mechanism involving the oxonium–carbonium intermediate has been ruled out for all nucleosides.)

If the first mechanism is accepted, it is possible to list the various factors[238,240] that can affect the ease of protonation of the sugar ring-oxygen atom of nucleosides and, hence, their rate of hydrolysis:

1. The basic strength of the aglycon. Mitts and Hixon[241] noted that, for most simple glycosylamines, the ease of hydrolysis appears to parallel the K_b value of the amine. It is important to note that these studies were made in aqueous solution, and no acid was added.

2. The site of protonation of the heterocyclic base and the degree of charge dispersal (through mesomerism). Obviously, a fixed charge near the ring-oxygen atom will impede its protonation, whereas a remote charge or a delocalized charge will have a lesser effect.

3. The ease of direct intramolecular transfer of a proton from the base to the sugar ring-oxygen atom. Intramolecular catalysis of this type would appear to be possible in many nucleosides wherein a hetero atom of the heterocyclic base is localized proximal to the sugar ring-oxygen atom in one of the possible conformations of the nucleoside. All cationic tautomeric forms should be considered as possible intermediates in this process.

If the above factors are considered, it is found that factor 1 has little relevancy when the amine is a heterocyclic base, presumably because the nitrogen atom bonded to C-1 of the sugar is seldom, and possibly never, the site of protonation. The Mitts and Hixon effect can probably be invoked to explain the more rapid hydrolysis of 5,6-dihydrouridine than of uridine. Reduction of the 5,6-ethylenic bond of uridine results in a loss in aromaticity and a gain in basicity of the aglycon. By the same token one might expect that 2'-deoxy-5-bromouridine would be more difficultly hydrolyzed than 2'-deoxyuridine or 2'-deoxythymidine. Wacker and Träger,[242] however, have found the opposite to be true when hydrolysis is catalyzed by 5% trichloro-acetic acid at 100°. No satisfactory explanation exists for the 4- to 5-fold faster rate of the 5-bromo compound. Similar results have been obtained[243] by using 1.1N hydrochloric acid at 60° to 80°.

It has also been observed that 2′-deoxycytidine is hydrolyzed more rapidly than 2′-deoxyuridine in dilute acid, whereas uridine is hydrolyzed more rapidly than cytidine in strong acid.[242] The latter result was obtained by using the orcinol reaction as a measure of the sugar produced, and the possibility of specific catalysis by the reagent cannot be eliminated. Nevertheless, a simple explanation can be offered for the result in dilute acid if a substantial contribution to the rate is made by a protonated form of the nucleoside via intramolecular transfer of a proton. In 5% trichloroacetic acid, only 2′-deoxycytidine would be protonated, whereas in 6N hydrochloric acid both cytidine and uridine would be protonated. If all other factors were equal, the protonated uracil residue would be expected to be a more effective catalyst than the protonated cytosine residue because of its stronger acidity[244] ($pK_a \sim 0$ vs $pK_a \sim 4$).

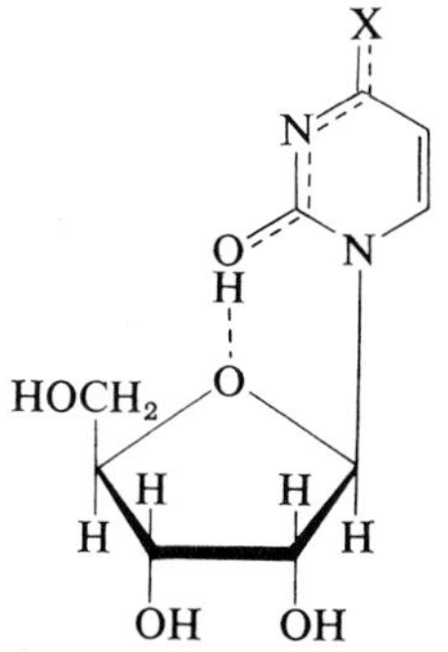

Cytidine, X = NH$_2$

Uridine, X = OH

It has also been observed[244] that cytosine "trialcohol" (the compound obtained from cytidine by successive treatment with sodium metaperiodate and sodium borohydride) is more stable than uracil "trialcohol" to hydrolysis by 6N hydrochloric acid. Thus, the explanation for the different rates of hydrolysis of the two pyrimidine nucleosides in 6N hydrochloric acid does not lie in some subtle conformational difference.

Similar arguments may be offered to explain the more rapid hydrolysis of "isocytidine" (102) than of cytidine[245] and of 5-amino-1-(β-D-ribofuranosyl)-imidazole (103) than of 1-(β-D-ribofuranosyl)imidazole.[246] In the case of "isocytidine" the catalytically active form is also the favored tautomer; thus, the faster rate could be due to a higher concentration of the reactive species.

* References start on p. 59.

102   103

Internal catalysis also provides a possible explanation for the observation that, in dilute acid, 2′-deoxyadenosine and 2′-deoxyguanosine undergo hydrolysis somewhat more rapidly than 2′-deoxycytidine, which, in turn, is hydrolyzed much more rapidly than 2′-deoxyuridine (or thymidine).[242,247] (Conformational differences between purine and pyrimidine nucleosides are discussed later in this section.) Intramolecular protonation through the species 104 and 105 possibly is involved. The species illustrated are not the favored

104   105

tautomers in acid solution but undoubtedly are present in low concentration. This type of explanation surmounts the obstacle presented by consideration of only the favored tautomers, such as 106. By analogy with the ribosylbenzimidazoles, compound 106 would be expected to show high stability to acid because of resistance to approach of a second proton as a result of the charge on the imidazole moiety. However, guanosine is completely hydrolyzed under conditions in which ribosylbenzimidazole is stable.

It is difficult to determine how important factor 3, charge delocalization, is to the mechanism of acid hydrolysis. It has been suggested[238] to be relevant to the case of glycosyladenines for which the charge resulting from initial protonation at N-1 could be shared by three nitrogen atoms through mesomerism. Thus, it is argued, the approach of a second proton would not be repulsed so strongly. Some evidence for a contribution to the rate constant from such a charged species comes from the careful study of Garrett[248] on psicofuranine (9-β-D-psicofuranosyladenine). Garrett is one of the few investigators who have applied exact kinetic methods to the study of nucleoside hydrolysis, and he has recorded useful data for the acid-catalyzed solvolysis of pyrimidine nucleosides.[243,249]

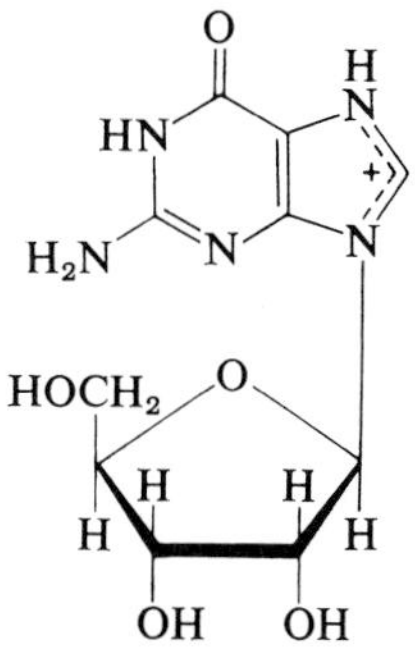

106

Modification of the glycosyl portion of a nucleoside can conceivably affect the hydrolysis rate by either steric or polar effects. In the first case, there is the possibility of an increased rate to relieve the strain of an unfavorable conformation forced on a molecule by a bulky substituent. Little is known about such effects in the nucleoside field. Polar effects, however, are well known and are best exemplified by the series shown below in which the stability increases in the order:[250]

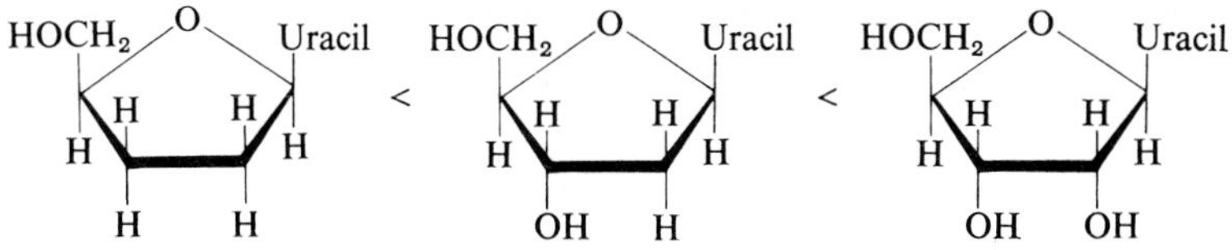

107   **108**   **109**

Progressive replacement of the hydroxyl groups at C-2′ (**108**) and C-3′ (**107**) of uridine (**109**) with hydrogen atoms, which are less strongly electron-withdrawing, leads to an increase in the basicity of the oxygen of the tetrahydrofuran ring and hence in its ability to accept a proton. A similar increase in stability has been observed for a series of ethyl glycosides:[251]

2,3-Dideoxy-D-*erythro*-hexopyranoside < 2-Deoxy-D-*arabino*-hexopyranoside <
D-Glucopyranoside

* *References start on p. 59.*

In aldopentofuranosyl derivatives, each of the three hydroxyl oxygen atoms is two carbon atoms removed from the ring-oxygen atom. It is of interest that replacement of any of these hydroxyl groups by hydrogen has a destabilizing effect on the glycosyl bond with respect to acid-catalyzed hydrolysis.[252] That inductive effects alone cannot adequately account for acid stability can be seen by a comparison of adenosine and psicofuranine (1'-*C*-hydroxymethyl-adenosine). Although the latter compound has a fourth hydroxyl group within two carbon atoms of the ring-oxygen atom, the stability of its glycosyl bond is considerably less than that of adenosine.[248] This is presumably related to the known lower stability of ketofuranosides relative to aldo-furanosides.[253]

An alternative suggestion has been made by Richards[254] to account for the greater stability to acid of normal glycosides compared with 2-deoxy-glycosides—namely, that competitive protonation of the hydroxyl at C-2 electrostatically hinders protonation of one or the other of the acetal oxygen atoms. This hypothesis seems unlikely, since the rate constant for hydrolysis of acetaldehyde diethyl acetal is greater than that for glyceraldehyde diethyl acetal by a factor[255] of ~300. Competitive protonation of the hydroxyl group of the latter compound cannot be invoked in this case, since intra-molecular transfer of the proton to an acetal oxygen atom is possible. It is noteworthy that the strong inductive effect of the 3'-*p*-tolylsulfonyloxy group permits the acid-catalyzed hydrolysis of the 5'-trityl group from 2'-deoxy-3'-*O*-(*p*-tolylsulfonyl)-5'-*O*-trityladenosine without cleavage of the glycosyl linkage.[256]

Nucleosides having furanoid rings are more labile than those having pyranoid rings, all other structural features being similar. This has been extensively studied for comparable glycosides (see Vol. IA, Chap. 9), and the effect is ascribed to (1) strain due to distortion of tetrahedral valency angles and (2) the presence of eclipsed bonds in the case of the five-membered ring.[253]

Finally, to emphasize the complexity of this problem, the data of Smith *et al.*[257] on the hydrolysis of the 3',5'-cyclic phosphates of ribonucleosides (**110**) can be cited.

R = adenine, guanine, cytosine, or uracil

110

Conversion of the 5'-phosphate into the 3',5'-cyclic phosphate has the unusual effect of stabilizing the guanosine and adenosine derivatives and destabilizing the uridine derivative. The effect of the conversion of cytidine 5'-phosphate into the corresponding 3',5'-cyclic phosphate on the stability of the glycosyl bond cannot be measured, since hydrolysis of the phosphate diester bond is more facile. This is probably also true in the case of the purine nucleoside 3',5'-cyclic phosphates;[257] thus the values given are minimum values.

TABLE I

ACID HYDROLYSIS OF RIBONUCLEOSIDE 5'-PHOSPHATES AND 3',5'- CYCLIC PHOSPHATES[a] IN N HCl AT 100°

Compound	Half-life (minutes)
Adenosine 3',5'-cyclic phosphate	> 30
Adenosine 5'-phosphate	~ 3
Cytidine 3',5'-cyclic phosphate	—[b]
Cytidine 5'-phosphate	> 120
Guanosine 3',5'-cyclic phosphate	> 28
Guanosine 5'-phosphate	~ 3
Uridine 3',5'-cyclic phosphate	8
Uridine 5'-phosphate	> 120

[a] M. Smith, G. I. Drummond, and H. G. Khorana, J. Amer. Chem. Soc., **83**, 698 (1961).

[b] The cyclic phosphate is hydrolyzed more readily than the glycosyl linkage.

Although no reliable comparative kinetic data are available for acid hydrolysis of nucleosides and their corresponding 5'-phosphates, it is known that the general order of stability of the four common nucleoside 5'-phosphates is similar to that of the parent nucleosides; that is, the pyrimidine nucleotides are much more stable than the purine nucleotides. An alternative to inductive effects must be sought to explain the observations of Table I. The possibility that steric changes are involved comes from the n.m.r. studies of Jardetzky[258] on adenosine 3',5'-cyclic phosphate. She has proposed marked changes in the conformation of the furanoid ring resulting from introduction of the 3',5'-cyclic phosphate structure. Subsequently, X-ray diffraction studies[259] of adenosine 3',5'-cyclic phosphate confirmed the unusual puckered form of the furanoid ring, where C-4' is out of the best four-atom plane. Furthermore, it was shown that in the *syn* form (having the base located approximately over the sugar ring), the bond angle (C-4–N-9–C-1') is 137°, or 10° larger than the corresponding bond angle in adenosine

* References start on p. 59.

5'-phosphate.[260] If this tilting of the base away from the plane of the sugar is assumed to be common to all four ribonucleoside 3',5'-cyclic phosphates, an explanation can be offered for the results of Table I. In the unsubstituted purine nucleoside (or the corresponding 5'-phosphate), one finds a better initial steric arrangement for intramolecular catalysis, involving proton transfer from N-3 to O-4', than one does in the case of pyrimidine ribo-nucleosides (or the corresponding 5'-phosphates). By using the data obtained by crystal-structure analysis of nucleosides and nucleotides, it has been calculated that there is a greater restriction to rotation about the glycosyl bond of pyrimidine nucleosides than of purine nucleosides. In particular, models based on the bond angles and distances of Furberg[261] show that O-2 of the pyrimidine moiety cannot easily be rotated past O-4' of the sugar.

111   112

Enlarging the angle "a" in each case, by formation of a 3',5'-cyclic phosphate, will allow an O-2–H–O-4' bond of 111 to become more linear (hence permitting improved catalysis) while the N-3–O-4' distance of 112 becomes too large to support a hydrogen bond. The charge on the adenine ring now repels the incoming proton, hindering its approach to O-4', with the consequence that hydrolysis is considerably slower. Alternatively, the data of M. Smith *et al.* might be interpreted as indicating (1) different conformations for the pyrimidine and purine ribonucleoside 3',5'-cyclic phosphates or (2) basically different mechanisms for the hydrolysis of pyrimidine and purine ribonucleotides.

Although most nucleosides are stable to alkali, exceptions to this generalization occur. As might be expected, nucleosides which generate a negative charge on the aglycon in alkaline solution show high stability to base. This situation includes glycosyl derivatives of uracil, guanine, hypoxanthine, and xanthine. Prevention of this dissociation by alkylation reduces alkali stability. Thus, 3-methyluridine is readily hydrolyzed by dilute alkali.[262] Reduction of uridine to 5,6-dihydrouridine also results in labilization to alkali. In both cases cleavage occurs in the heterocyclic ring. Similar alkaline lability of the aglycon moiety is seen in nebularine (9-β-D-ribofuranosylpurine)[263] and in numerous nucleosides having quaternary nitrogen atoms in the heterocyclic ring by virtue of alkylation.[264,264a]

Of the common purine ribonucleosides, only adenosine shows extensive degradation when treated with N alkali at 100° for 1 hour.[265] Three types of reaction occur: (1) opening of the imidazole ring by attack at C-8, (2) hydrolysis of the glycosyl bond by attack at C-1', and (3) deamination by attack at C-6. About 55 to 60% of the adenosine is destroyed, and slightly less than half of this is converted into adenine by hydrolysis of the glycosidic bond. It has also been observed[265a] that cytidine undergoes some deamination (2 to 3%) when subjected to the conditions used for alkaline hydrolysis of RNA—for example, N sodium hydroxide at 37° for 18 hours. Attack at C-1' is negligible under these conditions.

Four classes of nucleosides which show sensitivity to alkali with loss of the aglycon are recognized: (1) Glycosyl derivatives of quaternary ammonium compounds; the best-known example is nicotinamide nucleoside (1-β-D-ribofuranosylnicotinamide)[266] (113). For these compounds, strong electron withdrawal by the quaternary nitrogen atom facilitates attack at C-1'. (2) Glycosides in which the sugar is linked to the heterocyclic base via an exocyclic oxygen—for example, vicine (114). These purine or pyrimidine glycosides frequently behave like phenyl glycosides.[266a] (3) Ketofuranosyladenines. Both psicofuranine (9-β-D-psicofuranosyladenine)[248] (115a) and 1'-deoxypsicofuranine (C-1'-methyladenosine)[267] (115b) are labile to 0.1N

113   114   115a, X = OH
                          115b, X = H

sodium hydroxide at 20° to 50°. These compounds are also considerably more labile to acid than adenosine, a fact which suggests an inherent instability in the ketofuranoid ring. (4) Pentofuranosyl derivatives having strong electron-withdrawing substituents at C-5'. These compounds undergo

* References start on p. 59.

base-catalyzed β elimination rather than hydrolysis as shown in **116** for *S*-(5′-deoxyadenosin-5′-yl)methionine ("*S*-adenosyl methionine").[268] Similar elimination reactions have been observed for the vitamin B_{12} coenzyme[269] and for dialkyl esters of 5′-deoxynucleoside phosphonic acids.[269a]

116

V. PHYSICAL PROPERTIES OF NUCLEOSIDES*

A. PROTON MAGNETIC RESONANCE SPECTRA

Proton magnetic resonance (p.m.r.) spectra of nucleosides have been extremely useful in characterizations, especially in establishing the anomeric nature of the compounds and the conformations of the sugar rings (see Vol. IB, Chap. 27). Jardetzky[270] studied the natural purine and pyrimidine ribonucleosides in aqueous solution and noted, simply on the basis of the first-order H-1′–H-2′ coupling constant, that the furanose rings of the two types of nucleosides had to have different conformations. In the purine nucleosides, C-2′ was considered to be out of the plane defined either by C-1′, O-4′, and C-4′, or by C-1′, O-4′, C-3′, and C-4′, and to be pointing on the same side as the C-4′–C-5′ bond; in the pyrimidine nucleosides C-3′ was out of the plane defined by either C-1′, O-4′, and C-4′, or by C-1′, C-2′, O-4′, and C-4′. (Subsequent X-ray data—see Section V, D—suggest that this is an over-simplification.) In the natural (β-D)-2′-deoxyribonucleosides, both the purine and the pyrimidine compounds show similar p.m.r. spectra for the sugar protons, with the H-1′ signal appearing as a triplet, which strongly suggests that the ring-oxygen atom is twisted out of the plane of the furanose ring.[271] The anomeric proton (H-1′) in α-thymidine gives rise to a quartet,[272] in contrast to the triplet of the β anomer, and this difference in the appearance of the H-1′ signal in the p.m.r. spectra of a number of anomeric pairs of

* This section was prepared by L. Goodman.

2′-deoxy-D-ribofuranosyl nucleosides has been used for anomeric assignment of such compounds.[47] This empirical correlation should be used with some caution, however, since the spectra of 3′-amino-2′,3′-dideoxyadenosine[273] and 1-(2-deoxy-β-D-*erythro*-pentofuranosyl)-5-(trifluoromethyl)uracil[56] show the H-1′ signal as quartets in which the coupling constants with the C-2′ protons are slightly different. A similar observation has been made for 6,8-diamino-9-(2-deoxy-β-D-*erythro*-pentofuranosyl)purine.[274]

Difficulties arise in assignment of anomeric configuration to ribofuranose nucleosides simply on the basis of appearance of the H-1′ signal. Thus, adenosine, its α anomer, and 3-(β-D-ribofuranosyl)adenine have $J_{1',2'}$ values ranging from 4.5 to 5.2 Hz.[63] However, the 2′,3′-O-isopropylidene derivatives of adenosine and 3-(β-D-ribofuranosyl)adenine have $J_{1',2'}$ values of 2.9 Hz and 2.2 Hz, sufficiently small to permit their assignment as β-nucleosides on the basis of conformational considerations.[63] Similarly, two anomeric pairs of pyrimidine ribonucleosides have been examined. In the free nucleosides both anomers had a $J_{1',2'}$ value of 4.0 Hz, but on formation of the 2′,3′-isopropylidene acetals the α anomers showed no change in $J_{1',2'}$ whereas the value for the β anomers dropped to 2.7 Hz.[53]

The β-D-anomeric nature of certain 3′-deoxy-D-*erythro*-pentofuranose nucleosides was established[100] on the basis of a small (<1.8 Hz) value of $J_{1',2'}$. Anomeric configuration has also been assigned on the basis of the chemical shift of the H-1′ proton. In the anomeric pairs of 1-(D-ribofuranosyl)-uracil and -thymine, 1-(D-arabinofuranosyl)-uracil and -thymine, and 1-(D-lyxofuranosyl)-uracil and -thymine, it was noted that the H-1′ signal of the 1′,2′-*cis*-nucleosides always appeared at lower field than the H-1′ signal of the corresponding 1′,2′-*trans*-nucleoside, for which the anomeric proton is shielded by a C-2′ hydroxyl group.[54] This consideration was used to assign anomeric configurations to the 7-(D-ribofuranosyl)guanines[41] and to 3-(β-D-arabinofuranosyl)adenine.[66]

In determining the nature of the 3′-amino-3′-deoxy sugars formed via the nitromethane–nucleoside "dialdehyde" reaction (Section IV, D), the chemical shifts of the O- and N-acetyl methyl resonances were assigned to axial or equatorial acetyl groups on an empirical basis, and these values were used to deduce the structure of the sugar.[211] This method, however, has been shown to be unreliable. For the acetylated pyrimidine nucleosides formed by the periodate-cleavage method, Fox et al.[275] have shown that hydrogenation of the 5,6-double bond of the base causes a diamagnetic (upfield) shift of the 2′-acetoxy methyl resonance in the 1′,2′-*cis*-nucleosides (the D-*manno* compounds); a similar shift was noted with some cis-1′,2′-pentofuranosyl nucleosides containing acetoxy groups at C-2′. Removal of the 5,6-double bond in the 3'-acetamido-3'-deoxynucleosides (D-*gluco* and D-*galacto* derivatives) and in some pentofuranosyl pyrimidine nucleosides having a 1',2'-*trans* relationship causes a paramagnetic (downfield) shift in the 2'-acetoxy methyl resonance. These considerations permit anomeric assignment when only one of the two anomers is available.

In the pyranosyl nucleosides, the $J_{1',2'}$ value can be used to determine anomeric configuration and sugar conformation. In the case of 9-(β-D-ribo-pyranosyl)adenine and a number of purine-substituted derivatives of this compound the $J_{1',2'}$ value was 9 Hz, in complete agreement with the β-anomeric assignment and the $C1$(D) sugar conformation of **117**, since only in this representation is the *trans*-diaxial arrangement of H-1' and H-2'

117

possible that leads to such a large J value.[276] On the basis of their p.m.r. spectra, anomeric assignments were made for the pair of nucleosides prepared by the fusion of 6-chloropurine with 3,4-di-*O*-acetyl-D-arabinal or with 1,3,4-tri-*O*-acetyl-2-deoxy-β-D-*threo*-pentopyranose; the spectra showed that, in the α-nucleoside, the sugar ring existed in the $1C$(D) conformation while in the β derivative the sugar was the $C1$(D) form.[48] Similarly the nucleosides, 1-(2-deoxy-D-*arabino*-hexopyranosyl)-6-azauracil and its 6-azathymine analog, were assigned the β configuration, since the H-1' signal appeared as a doublet of doublets, with $J_{1',2'} = 9$ Hz (diaxial protons) and $J_{1',2'} = 3$ to 4 Hz (axial–equatorial protons) indicative of an axial H-1' if the C1 (D) conformation of the sugar is assumed.[277] P.m.r. data were used to assign the conformation of the furanose sugar in 3'-*C*-methyladenosine,[277a] in 2'-deoxy-2'-fluoro-uridine[277b] and in 3'-deoxy-3'-fluoro-β-D-arabinofuranosyluracil.[277b]

The chemical shift of H-1' and the $J_{1',2'}$ splitting have been used to distinguish between the isomeric 2'- and 3'-substituted ribonucleosides for which the sugar substituents were phosphates, sulfonic esters, alkyl and aryl esters, and benzyl and trityl ethers.[278] For a pair of 2'- and 3'-substituted isomers, the H-1' resonance is at a lower field for the 2'- than for the 3'-substituted isomer, and the $J_{1',2'}$ value is greater for the 3'- than for the 2'-substituted isomer, providing the spectra are determined in the same solvent and at the same concentration. For some adenine nucleosides substituted in the sugar moiety with benzylthio groups, but for which the sugar is not a D-ribose derivative, this chemical shift rule is obeyed.[161]

Proton magnetic resonance spectra have been useful for differentiation of isomerically substituted adenines and purines on the basis of the chemical shifts of the C-2 and C-8 protons.[279] Thus, the 3-′ and the 1-substituted adenine derivatives show a much greater difference ($\Delta\delta$) between the signals for the 2- and 8-protons than do the corresponding 7- and 9-substituted derivatives; this consideration was important in assigning the structure of 3-(β-D-arabinofuranosyl)adenine.[66]

The association of purine nucleosides in solution has been studied with the aid of p.m.r. spectrometry. Intramolecular hydrogen bonding of the 2′-OH group of the ribose moiety to N^3 of the base in adenosine was indicated by these measurements.[280]

B. Optical Rotatory Dispersion Measurements

Many nucleosides, especially pyrimidine derivatives, violate Hudson's rules of isorotation, and it has become obvious that optical rotations, taken at a single wavelength, of a pair of anomers cannot be used to assign configuration; optical rotatory dispersion (o.r.d.) methods (see Vol. IB, Chap. 27) have been found very useful in overcoming this difficulty.[281,282]

In general, the o.r.d. curves of α-D-pyrimidine nucleosides give negative Cotton effects, whereas those of the β anomers give positive effects;[283] with purine nucleosides, the β-D anomers give negative Cotton effects and the α anomers give positive effects. O.r.d. data for a large number of anomeric nucleosides and nucleotides have been summarized.[283a] These simple rules were used to assign configurations to the anomers formed in a 1:1 ratio, from the reaction of monomercurithymine and 3,4-di-O-acetyl-D-ribopyranosyl chloride,[284] and to the anomeric pyrimidine 3′-deoxy-D-erythro-pentofuranose nucleosides.[100]

In a more detailed sense, Ulbricht et al.[29] proposed a rule for predicting the sign of the Cotton effect in pyrimidine furanose nucleosides. The sign of the Cotton effect will be positive if (1) the nucleoside has a favored conformation owing to restricted rotation about the N–C-1′ bond; and (2) a line from the C^6=O group passing through the C^2=O group (as in **118**) passes from above (the same side of the furanose ring as C-5′) to below the plane of the furanose ring, provided that the chromophore is not twisted to such an extent that the line passes through C-5′. The rule is obeyed by cyclonucleosides; thus $O^2,2'$- and $O^2,3'$-cyclouridine fulfil the conditions of the rule and give positive Cotton effects, whereas in $O^2,5'$-cyclouridine the line passes through C-5′ and the Cotton effect is negative. In α-D anomers and in pseudouridine (and in the xylo and arabino isomers[106]) the line passes from below to above the plane of the sugar ring, giving rise to negative Cotton effects. The magnitude of the Cotton effect depends on the configuration at C-2′, and thus arabinosides give larger Cotton effects than ribosides.[285] The substitution of N for CH in the heterocyclic base (as in 6-azauracil nucleosides) causes a reversal in the sign of the Cotton effect.[285] The Cotton effect of 1-(α-D-glucopyranosyl)thymine, contrary to expectations, shows a positive Cotton effect, probably because the bulky pyrimidine substituent is oriented equatorially, giving a $1C(D)$ conformation for the pyranose ring;[286] thus assignment of anomeric configuration seems to demand a reference compound having both identical chromophore and sugar-ring conformation.

118

In purine nucleosides it has been noted that the magnitude of the Cotton effect is noticeably lower than with pyrimidine nucleosides because of the smaller restriction of rotation about the glycosidic bond.[287] For the purine compounds, the sign of the Cotton effect is not influenced by presence of O-acetyl or O-isopropylidene groups in the sugar, by substitution of N for CH in the heterocyclic base, or by replacement of the 2′-OH group by hydrogen or chlorine, and the magnitude of the effect is not significantly affected by the nature of the sugar residue.[287] Although the authors reported that the cyclonucleoside (**47**, as the iodide salt), which is fixed in the *syn* conformation, showed a positive Cotton effect contrary to the general pattern, this observation has been disputed by Hampton *et al.*[288] They determined the o.r.d. curves for a variety of purine cyclonucleosides and found that all showed negative Cotton effects except for 3,5′-cycloxanthosine, which showed a small positive effect at pH 7.6. In the purine nucleosides, as expected, the β-L form of adenosine has a positive Cotton effect.[287] Subsequently, it was noted that adenine β-D-nucleosides containing a sulfur substituent at C-5′ have a positive Cotton effect, suggesting that the bulky substituent causes a change in the orientation of the base with respect to the sugar moiety.[289] Efforts have been made to correlate the sign of the Cotton effect in purine nucleosides on the basis of the *syn* or *anti* orientation of the base with respect to the sugar, by using various cyclonucleosides as model *syn* or *anti* systems. However, a number of the observations reported[287–290] are in conflict with the correlations, and further studies are required to clarify the situation.

C. ULTRAVIOLET SPECTRA

Ultraviolet spectra of nucleosides have been used mainly in determining the position of substitution of a sugar on the heterocyclic base or in establishing formation of cyclonucleosides.

Certain generalities about the ultraviolet spectra of *N*-alkylated adenines have been formulated and used in establishing the correct structures of 3-glycosylpurines previously reported as the N^7 structures.[279] Thus, 1-alkyladenines show a hyperchromic shift in base, an isosbestic point on the short-wavelength side of the maxima, and a hypsochromic shift in acid. 3-Alkyladenines show a hyperchromic shift in acid and an isosbestic point on the long-wavelength side of the maxima. The basic and neutral spectra show very little difference. 7-Alkyladenines exhibit a hyperchromic shift and possibly a bathochromic shift in acid, with an isosbestic point on the short-wavelength side of the maxima. 9-Alkyladenines give spectra that are relatively insensitive to pH. A slight hypsochromic shift in acid and an isosbestic point near and on the short-wavelength side of the maxima are generally observed. These generalizations seem to apply also to the 6-methylamino- and 6-dimethylaminopurines. Such comparisons of ultraviolet spectra of nucleosides with those of the simpler alkyl heterocycles have been used in assigning the structure of 7-(β-D-ribofuranosyl)purine, obtained from the fusion of tetra-*O*-acetyl-D-ribofuranose (2) with purine,[35] and of the 7-β-D-ribofuranosylpyrrolo[2,3-*d*]pyrimidines obtained by way of the fusion of 2 and 4-chloro-5-cyano-6-methylthiopyrrolo[2,3-*d*]pyrimidine.[291]

Ultraviolet spectra have been especially helpful in establishing the formation of cyclonucleosides, with both the purines and pyrimidines. The characteristic shift of the maxima to longer wavelengths on cyclization has been noted in a branched-chain adenine nucleoside[159] and in the formation of a 3,3'-cycloadenosine derivative.[161] The loss of selective ultraviolet absorption in the cyclonucleosides of pyrimidine nucleosides that form by addition to the 5,6-double bond is characteristic.[183,185,186] The O^2,2'-, O^2,3'-, and O^2,5'-cyclonucleosides of a given pyrimidine base show characteristic differences in their ultraviolet spectra.

Ultraviolet spectra are widely used for the determination of the pK values for nucleosides. The subject has been reviewed by Fox and Wempen.[32]

* *References start on p. 59.*

Such spectra have had considerable value for the demonstration of inter-actions between the sugar and base residues of purines and pyrimidines. All aldopentofuranosyl nucleosides form a hydrogen bond from the 2-carbonyl group to hydroxyl groups of the sugar (mainly at C-2). This was first shown for 1-β-D-aldopentosylpyrimidines.[291a]

The spectrum of 3'-O-p-nitrophenylsulfonyladenosine shows a hypso-chromic shift of 2 nm from that of adenosine and is accompanied by a strong hypochromic effect.[292] The spectral evidence for interaction correlates with chemical evidence and p.m.r. evidence.

D. Miscellaneous Physical Data

Infrared spectra have not been very useful in the study of the sugar portion of nucleosides. The technique has been helpful in demonstrating the tauto-meric nature of the heterocyclic bases—for example, showing that cytidine and deoxycytidine exist as the 4-aminonucleosides[293] as does 5-azacytidine.[294]

The determination of pK values has been helpful in determining position of substitution of a sugar on a heterocyclic base and in studying interactions of hydroxyl groups in the sugar ring. A substantial difference in pK_a values exists for the different N-alkyl-substituted adenines with 9- < 7- < 3- < 1-methyladenine;[295] these same differences are maintained in the substituted D-ribofuranosyladenines that have been reported. These pK_a values have been used as corroborative evidence in assigning the structures of 3-(β-D-ribofuranosyl)adenine[63] and 3-(β-D-arabinofuranosyl)adenine.[66] Thermo-metric titrations have been carried out to determine the pK_a values for ionization of the sugar hydroxyl groups of some adenosine analogs. The heat effects associated with these titrations are markedly less for 2'-deoxyadeno-sine, 3'-deoxyadenosine, and 2'-O-methyladenosine than for adenosine[296] or 9-(β-D-xylofuranosyl)adenine;[297] these differences indicate that both OH groups at C-2' and C-3' (cis or trans) are necessary for the enhanced acidity noted with the latter compounds. The differences in pK_a have been used to develop an ion-exchange separation of purine nucleosides.[10]

The methodology for separating mixtures of nucleosides, as their O-tri-methylsilyl derivatives, by gas chromatography has been described.[298] Separation of the fully silylated nucleosides (for example, pentakis(trimethyl-silyl)guanosine) was better than with the derivatives in which only the sugar hydroxyl groups were derivatized.

Structure determinations by X-ray diffraction have been especially helpful with some of the nucleoside antibiotics such as showdomycin[299] (33) and aristeromycin[300] (119) for which absolute configurations of all the sugar carbon atoms have been assigned. The puckering of the furanose-sugar ring in nucleosides and its consequences with respect to rotation about the glycosidic bond[301] have been measured by this technique, and interesting differences have been noted. For instance, in 2'-deoxyadenosine C-3' is displaced to the side of the ring opposite ("exo") from C-5', with respect to the plane defined by the other four ring atoms, whereas in 2'-deoxy-5-fluoro-uridine[302] C-2' is out of the plane but is "endo" to C-5'. These differences in ring puckering have important effects on intramolecular interactions of the sugar and the base.

119

Specific conductance measurements of mixtures of nucleosides in methanol solution have been used to identify interactions by hydrogen bonding of certain pairs (for example, thymidine–2'-deoxyadenosine, 2'-deoxycytidine–2'-deoxyguanosine) as well as the absence of interaction in other pairs (for example, thymidine–guanosine).[303]

REFERENCES

1. H. Bredereck, A. Martin, and F. Richter, *Ber.*, **74**, 694 (1941).
2. K. Dimroth, L. Jaenicke, and D. Heinzel, *Ann.*, **566**, 206 (1950).
3. R. H. Hall, *J. Biol. Chem.*, **237**, 2283 (1962).
4. F. Weygand, A. Wacker, and H. Dellweg, *Z. Naturforsch.*, **6b**, 130 (1951).
5. W. Klein and S. J. Thannhauser, *Hoppe-Seyler's Z. Physiol. Chem.*, **231**, 96 (1935); W. Anderson, C. A. Dekker, and A. R. Todd, *J. Chem. Soc.*, 2721 (1952).
6. W. E. Cohn, in "The Nucleic Acids," E. Chargaff and J. N. Davidson, Eds., Vol. I, Academic Press, New York, 1955, Chapter 6.
7. D. T. Elmore, *J. Chem. Soc.*, 2084 (1950).
8. R. H. Hall, *Biochemistry*, **4**, 661 (1965).
9. C. A. Dekker, *J. Amer. Chem. Soc.*, **87**, 4027 (1965).
10. G. R. Wyatt, in "The Nucleic Acids," E. Chargaff and J. N. Davidson, Eds., Vol. I, Academic Press, New York, 1955, Chapter 7; R. Brown, *Biochim. Biophys. Acta*, **149**, 601 (1967).
11. J. D. Smith, in "The Nucleic Acids," E. Chargaff and J. N. Davidson, Eds., Vol. I, Academic Press, New York, 1955, Chapter 8.

12. L. Jaenicke and I. Vollbrechtshausen, *Naturwissenschaften*, **39**, 86 (1952).
13. K. Randerath, "Thin-Layer Chromatography," Academic Press, New York, 1963, p. 186.
14. F. Miescher, *Hoppe-Seyler's Med.-Chem. Unters.*, 441, 502 (1871).
15. A. Bendich, in "The Nucleic Acids," E. Chargaff and J. N. Davidson, Eds., Vol. I, Academic Press, New York, 1955, Chapter 3.
16. P. A. Levene and L. E. Bass, "Nucleic Acids," Chemical Catalog Co., New York, 1931.
17. R. S. Tipson, *Advan. Carbohyd. Chem.*, **1**, 193 (1945).
18. G. R. Barker, *Advan. Carbohyd. Chem.*, **11**, 285 (1956).
19. J. D. Smith and D. B. Dunn, *Biochim. Biophys. Acta*, **31**, 573 (1959).
20. R. H. Hall, *Biochemistry*, **3**, 876 (1964).
21. W. G. Overend and M. Stacey, in "The Nucleic Acids," E. Chargaff and J. N. Davidson, Eds., Vol. I, Academic Press, New York, 1955 Chapter 2.
22. M. Friedkin and H. Kalckar in "The Enzymes," P. D. Boyer, H. Lardy, and K. Myrbäck, Eds., Vol. 5, Academic Press, New York, 1961, pp. 237–255.
23. J. Baddiley, in "The Nucleic Acids," E. Chargaff and J. N. Davidson, Eds., Vol. I, Academic Press, New York, 1955, Chapter 4.
24. A. M. Michelson, "The Chemistry of Nucleosides and Nucleotides," Academic Press, New York, 1963, pp. 7–14.
24a. For correct structures see: A. Bendich and G. C. Clements, *Biochim. Biophys. Acta*, **12**, 462 (1953); and H. S. Forrest, D. L. Hatfield, and J. M. Lagowski, *J. Chem. Soc.*, 963 (1961).
25. W. E. Cohn and D. G. Doherty, *J. Amer. Chem. Soc.*, **78**, 2863 (1965).
26. E. Shaw, *J. Amer. Chem. Soc.*, **81**, 6021 (1959).
27. W. E. Cohn, *Biochim. Biophys. Acta*, **32**, 569 (1959); *J. Biol. Chem.*, **235**, 1488 (1960).
27a. J. Böeseken, *Advan. Carbohyd. Chem.*, **4**, 189 (1949).
28. E. Walton, F. W. Holly, G. E. Boxer, and R. F. Nutt, *J. Org. Chem.*, **31**, 1163 (1966).
28a. R. U. Lemieux and D. R. Lineback, *Ann. Rev. Biochem.*, **32**, 155 (1963).
29. T. Ulbricht, T. R. Emerson, and R. J. Swan, *Tetrahedron Lett.*, 1561 (1966).
29a. C. A. Dekker and J. Gin, *Abstracts Papers Amer. Chem. Soc. Meeting*, **153**, C37 (1967).
29b. J. Davoll, B. Lythgoe, and A. R. Todd, *J. Chem. Soc.*, 967, 1685 (1948); for a comprehensive review of the synthetic studies by Todd and colleagues see refs. 23 and 24a.
30. R. S. Tipson, *J. Biol. Chem.*, **130**, 55 (1939).
30a. B. R. Baker, in "Ciba Foundation Symposium on The Chemistry and Biology of Purines," G. E. W. Wolstenholme and C. M. O'Connor, Eds., Little, Brown, Boston, Massachusetts, 1957, pp. 120–133.
31. J. A. Montgomery and H. J. Thomas, *Advan. Carbohyd. Chem.*, **17**, 301 (1962).
32. J. J. Fox and I. Wempen, *Advan. Carbohyd. Chem.*, **14**, 283 (1959).
33. T. Shimadate, Y. Ishido, and T. Sato, *Nippon Kagaku Zasshi*, **82**, 938 (1961); *Chem. Abstr.*, **57**, 15216 (1962).
34. Y. Ishido, A. Hosono, K. Fujii, Y. Kikuchi, and T. Sato, *Nippon Kagaku Zasshi*, **87**, 752 (1966); *Chem. Abstr.*, **65**, 17034 (1966).
35. T. Hashizume and H. Iwamura, *Tetrahedron Lett.*, 643 (1966).
36. R. L. Tolman, R. K. Robins, and L. B. Townsend, *J. Amer. Chem. Soc.*, **90**, 524 (1968).

37. T. Shimadate, *Nippon Kagaku Zasshi*, **82**, 1268 (1961); *Chem. Abstr.*, **57**, 16726 (1962).
38. L. Pichat, P. Dufay, and Y. Lamorre, *C.R. Acad. Sci.*, **259**, 2453 (1964).
39. W. W. Lee, A. P. Martinez, G. L. Tong, and L. Goodman, *Chem. Ind. (London)*, 2007 (1963).
40. K. Onodera, S. Hirano, H. Fukumi, and F. Masuda, *Carbohyd. Res.*, **1**, 254 (1965).
41. K. Imai, A. Nohara, and M. Honjo, *Chem. Pharm. Bull (Tokyo)*, **14**, 1377 (1966).
42. Y. Ishido, T. Matsuba, A. Hosono, K. Fujii, H. Tanaka, K. Iwabuchi, S. Isome, A. Maruyama, Y. Kikuchi, and T. Sato, *Bull. Chem. Soc. Jap.*, **38**, 2019 (1965).
43. Y. Ishido, T. Matsuba, A. Hosono, K. Fujii, T. Sato, S. Isome, A. Maruyama, and Y. Kikuchi, *Bull. Chem. Soc. Jap.*, **40**, 1007 (1967).
44. Y. Ishido, A. Hosono, S. Isome, A. Maruyama, and T. Sato, *Bull. Chem. Soc. Jap.*, **37**, 1389 (1964).
45. K. Onodera and H. Fukumi, *Agr. Biol. Chem. (Tokyo)*, **27**, 526 (1963).
46. W. Pfleiderer and R. K. Robins, *Ber.*, **98**, 1511 (1965).
47. M. J. Robins and R. K. Robins, *J. Amer. Chem. Soc.*, **87**, 4934 (1965).
48. E. E. Leutzinger, W. A. Bowles, R. K. Robins, and L. B. Townsend, *J. Amer. Chem. Soc.*, **90**, 127 (1968).
49. N. Nagasawa, I. Kumashiro, and T. Takenishi, *J. Org. Chem.*, **32**, 251 (1967), reported that adenine and 3,4-di-*O*-acetyl-D-arabinal in the presence of hydrogen chloride gave 9-(3,4-di-*O*-acetyl-2-deoxy-β-D-*erythro*-pentopyranosyl)adenine.
50. W. A. Bowles and R. K. Robins, *J. Amer. Chem. Soc.*, **86**, 1252 (1964).
51. L. Birkhofer, A. Ritter, and H. P. Kuehlthau, *Angew. Chem.*, **75**, 209 (1963).
51a. M. J. Covill, H. G. Garg, and T. L. V. Ulbricht, *Tetrahedron Lett.*, 1033 (1968) suggest that a reexamination of the reaction products from the silylated xanthine derivatives may be in order.
52. T. Nishimura, B. Shimizu, and I. Iwai, *Chem. Pharm. Bull. (Tokyo)*, **11**, 1470 (1963).
53. T. Nishimura, B. Shimizu, and I. Iwai, *Chem. Pharm. Bull. (Tokyo)*, **12**, 1471 (1964).
54. T. Nishimura and B. Shimizu, *Chem. Pharm. Bull. (Tokyo)*, **13**, 803 (1965).
55. E. Wittenburg, *Z. Chem.*, **4**, 303 (1964).
56. K. J. Ryan, E. M. Acton, and L. Goodman, *J. Org. Chem.*, **31**, 1181 (1966).
57. J. Montgomery and K. Hewson, *J. Heterocycl. Chem.*, **2**, 313 (1965).
58. T. J. Bardos, M. P. Kotick, and C. Szantay, *Tetrahedron Lett.*, 1759 (1966).
59. N. Yamaoka, K. Aso, and K. Matsuda, *J. Org. Chem.*, **30**, 149 (1965).
60. Y. Ishido, H. Tanaka, T. Yoshino, M. Sekiya, K. Iwabuchi, and T. Sato, *Tetrahedron Lett.*, 5245 (1967).
61. C. P. J. Glaudemans and H. G. Fletcher, Jr., *J. Org. Chem.*, **28**, 3004 (1963).
62. F. Keller, I. J. Botvinick, and J. E. Bunker, *J. Org. Chem.*, **32**, 1644 (1967).
63. N. J. Leonard and R. A. Laursen, *J. Amer. Chem. Soc.*, **85**, 2026 (1963).
64. M. Rasmussen and N. J. Leonard, *J. Amer. Chem. Soc.*, **89**, 5439 (1967).
65. B. Shimizu and M. Miyaki, *Chem. Pharm. Bull. (Tokyo)*, **15**, 1066 (1967).
66. K. R. Darnall and L. B. Townsend, *J. Heterocycl. Chem.*, **3**, 371 (1966).
67. G. Schramm, H. Groetsch, and W. Pollman, *Angew. Chem.*, **73**, 619 (1961).
68. J. A. Carbon, *Chem. Ind. (London)*, 529 (1963).
69. M. P. de Garilhe and J. de Rudder, *C. R. Acad. Sci.*, **259**, 2725 (1964).
70. S. S. Cohen, in "Progress in Nucleic Acid Research and Molecular Biology," J. N. Davidson and W. E. Cohn, Eds., Academic Press, New York, 1966, p. 24.

71. G. Schramm, G. Lünzmann, and F. Bechmann, *Biochim. Biophys. Acta*, **145**, 221 (1967).
72. J. A. Carbon, *J. Amer. Chem. Soc.*, **86**, 720 (1964).
73. Ajinomoto Co., Inc., *Netherlands Patent* 6,409,133 (Feb. 8, 1965); *Chem. Abstr.*, **63**, 3029 (1965).
74. A. Yamazaki, I. Kumashiro, and T. Takenishi, *J. Org. Chem.*, **32**, 1825 (1967).
75. A. Yamazaki, I. Kumashiro, and T. Takenishi, *J. Org. Chem.*, **32**, 3258 (1967).
76. R. J. Rousseau, L. B. Townsend, and R. K. Robins, *Chem. Commun.*, 265 (1966).
77. E. Buehler and W. Pfleiderer, *Angew. Chem.*, **76**, 713 (1964).
78. J. Šmejkal, J. Farkaš, and F. Šorm, *Collect. Czech. Chem. Commun.*, **31**, 291 (1966).
79. A. Pískala and F. Šorm, *Collect. Czech. Chem. Commun.*, **29**, 2060 (1964).
80. J. Pliml and F. Šorm, *Collect. Czech. Chem. Commun.*, **29**, 2576 (1964).
80a. G. T. Rogers and T. L. V. Ulbricht, *Tetrahedron Lett.*, 1025 (1968).
80b. J. D. Stevens, R. K. Ness, and H. G. Fletcher, Jr., *J. Org. Chem.*, **33**, 1806 (1968).
80c. A. P. Martinez, W. W. Lee, and L. Goodman, *J. Org. Chem.*, **34**, 92 (1969).
81. J. A. Montgomery and H. J. Thomas, *J. Org. Chem.*, **31**, 1411 (1966).
82. J. A. Montgomery and H. J. Thomas, *J. Amer. Chem. Soc.*, **87**, 5442 (1965).
83. E. Walton, S. R. Jenkins, R. F. Nutt, M. Zimmerman, and F. W. Holly, *J. Amer. Chem. Soc.*, **88**, 4524 (1966).
84. E. J. Reist, D. E. Gueffroy, and L. Goodman, *J. Amer. Chem. Soc.*, **86**, 5658 (1964).
85. E. J. Reist, D. E. Gueffroy, R. W. Blackford, and L. Goodman, *J. Org. Chem.*, **31**, 4025 (1966).
86. L. M. Lerner and P. Kohn, *J. Org. Chem.*, **31**, 399 (1966).
87. J. Farkaš and F. Šorm, *Collect. Czech. Chem. Commun.*, **32**, 2663 (1967).
88. D. H. Murray and J. Prokop, *J. Pharm. Sci.*, **56**, 865 (1967).
89. D. H. Murray and J. Prokop, *J. Pharm. Sci.*, **54**, 1468 (1965).
90. J. Prokop and D. H. Murray, *J. Pharm. Sci.*, **54**, 359 (1965).
91. R. H. Iwamoto, E. M. Acton, and L. Goodman, *J. Med. Chem.*, **6**, 684 (1963).
92. G. L. Tong, K. J. Ryan, W. W. Lee, E. M. Acton, and L. Goodman, *J. Org. Chem.*, **32**, 859 (1967).
93. G. L. Tong, W. W. Lee, and L. Goodman, *J. Org. Chem.*, **32**, 1984 (1967).
94. S. R. Jenkins, F. W. Holly, and E. Walton, *J. Org. Chem.*, **30**, 2851 (1965).
95. I. A. Mikhailopulo, V. I. Gunar, and S. I. Zav'yalov, *Izv. Akad. Nauk SSSR, Ser. Khim.*, 470 (1967); *Chem. Abstr.*, **67**, 32913 (1967).
95a. H. J. Lee and P. W. Wigler, *Biochemistry*, **7**, 1427 (1968).
96. M. Prystaš and F. Šorm, *Collect. Czech. Chem. Commun.*, **31**, 1035 (1966).
97. M. Prystaš and F. Šorm, *Collect. Czech. Chem. Commun.*, **29**, 2956 (1964).
98. M. Prystaš and F. Šorm, *Collect. Czech. Chem. Commun.*, **30**, 2960 (1965).
99. M. Prystaš and F. Šorm, *Collect. Czech. Chem. Commun.*, **30**, 1900 (1965).
100. E. Walton, F. W. Holly, G. E. Boxer, and R. F. Nutt, *J. Org. Chem.*, **31**, 1163 (1966).
100a. T. Ueda and H. Nishino, *J. Amer. Chem. Soc.*, **90**, 1678 (1968).
101. W. E. Cohn, *Biochim. Biophys. Acta*, **32**, 569 (1959).
102. R. K. Robins, L. B. Townsend, F. Cassidy, J. F. Gerster, A. F. Lewis, and R. L. Miller, *J. Heterocycl. Chem.*, **3**, 110 (1966).
103. K. R. Darnall, L. B. Townsend, and R. K. Robins, *Proc. Nat. Acad. Sci. U.S.*, **57**, 548 (1967).
104. R. Shapiro and R. W. Chambers, *J. Amer. Chem. Soc.*, **83**, 3920 (1961).
105. D. M. Brown, M. G. Burdon, and R. P. Slatcher, *Chem. Commun.*, 77 (1965).

106. W. Asbun and S. B. Binkley, *J. Org. Chem.*, **31**, 2215 (1966).
107. W. Asbun and S. B. Binkley, *J. Org. Chem.*, **33**, 140 (1968).
108. M. P. Mertes, J. Zielinski, and C. Pillar, *J. Med. Chem.*, **10**, 320 (1967).
109. M. Bobek, J. Farkaš, and F. Šorm, *Tetrahedron Lett.*, 3115 (1966).
109a. Later work[109b] showed that the product thought to be 37 is actually 5-(2,5-anhydro-D-*altro*-pentahydroxypentyl)-6-azauracil. Compound 37 was prepared by ozonolysis of 2′,3′,5′-tri-O-acetylpseudouridine, reduction of the ozonide with methyl sulfide, treatment of the resulting ketoformamide with thiosemicarbazide to give the 3-thioxo-1,2,4-triazine-5-one, methylation, and hydrolysis to 6-azapseudouridine (**37**).
109b. M. Bobek, J. Farkaš, and F. Šorm, *Tetrahedron Lett.*, 1543 (1968).
110. M. Bobek, J. Farkaš, and F. Šorm, *Collect. Czech. Chem. Commun.*, **32**, 3572 (1967).
111. M. Sprinzl and J. Farkaš, *Collect. Czech. Chem. Commun.*, **30**, 3787 (1967).
112. M. Smith, D. H. Rammler, I. H. Goldberg, and H. G. Khorana, *J. Amer. Chem. Soc.*, **84**, 430 (1962).
113. R. Lohrmann and H. G. Khorana, *J. Amer. Chem. Soc.*, **86**, 4188 (1964).
114. J. Smrt and F. Šorm, *Collect. Czech. Chem. Commun.*, **32**, 3169 (1967).
115. N. C. Yung and J. J. Fox, *J. Amer. Chem. Soc.*, **83**, 3060 (1961).
116. H. U. Blank and W. Pfleiderer, *Tetrahedron Lett.*, 869 (1967).
117. J. F. Codington and J. J. Fox, *Carbohyd. Res.*, **3**, 124 (1966–7).
118. J. Žemlička, *Collect. Czech. Chem. Commun.*, **29**, 1734 (1964).
119. K. K. Ogilvie and R. L. Letsinger, *J. Org. Chem.*, **32**, 2365 (1967).
120. J. A. R. Johnston, *Tetrahedron Lett.*, 2679 (1967).
121. E. J. Reist and L. Goodman, *Biochemistry*, **3**, 15 (1964).
122. .E. J. Reist, D. F. Calkins, and L. Goodman, *J. Org. Chem.*, **32**, 169 (1967).
123. F. Cramer, H. P. Bär, H. J. Rhaese, W. Saenger, K. H. Scheit, G. Schneider, and J. Tennigkeit, *Tetrahedron Lett.*, 1039 (1963).
124. F. Cramer, W. Saenger, K. Scheit, and J. Tennigkeit, *Justus Liebigs Ann. Chem.*, **679**, 156 (1964).
125. S. Chládek and J. Smrt, *Collect. Czech. Chem. Commun.*, **28**, 1301 (1963).
126. A. Hampton, J. C. Fratantoni, P. M. Carroll, and S. Wang, *J. Amer. Chem. Soc.*, **87**, 5481 (1965).
127. A. M. Yurkevich, I. I. Kolodkina, and N. A. Preobrazhenskii, *Dokl. Akad. Nauk SSSR*, **164**, 828 (1965); *Chem. Abstr.*, **64**, 3661 (1966).
128. B. E. Griffin and C. B. Reese, *Tetrahedron Lett.*, 2925 (1964).
129. S. Chládek and J. Smrt, *Chem. Ind. (London)*, 271 (1964).
130. C. B. Reese, R. Saffhill, and J. E. Sulston, *J. Amer. Chem. Soc.*, **89**, 3366 (1967).
131. A. Hampton, *J. Amer. Chem. Soc.*, **87**, 4654 (1965).
132. J. Žemlička, *Chem. Ind. (London)*, 581 (1964).
133. M. Jarman and C. B. Reese, *Chem. Ind. (London)*, 1493 (1964).
134. C. B. Reese and J. E. Sulston, *Proc. Chem. Soc. (London)*, 214 (1964).
135. J. Žemlička and S. Chládek, *Collect. Czech. Chem. Commun.*, **31**, 3775 (1966).
136. J. Žemlička and S. Chládek, *Tetrahedron Lett.*, 3057 (1965).
137. C. B. Reese and D. R. Trentham, *Tetrahedron Lett.*, 2467 (1965).
137a. H. P. M. Fromageot, C. B. Reese, and J. E. Sulston, *Tetrahedron*, **24**, 3533 (1968).
138. Y. Mizuno, T. Itoh, and H. Tagawa, *Chem. Ind. (London)*, 1498 (1965).
139. K. L. Agarival and M. M. Dhar, *Experientia*, **21**, 432 (1965).
140. J. Žemlička, J. Beránek, and J. Smrt, *Collect. Czech. Chem. Commun.*, **27**, 2784 (1962).

141. B. E. Griffin and C. B. Reese, *Proc. Nat. Acad. Sci. U.S.*, **51**, 440 (1964).
142. R. L. Letsinger and K. K. Ogilvie, *J. Org. Chem.*, **32**, 296 (1967).
143. R. L. Letsinger, J. Fontaine, V. Mahadevan, D. A. Schexnayder, and R. E. Leone, *J. Org. Chem.*, **29**, 2615 (1964).
144. E. Eckstein, *Tetrahedron Lett.*, 531 (1965).
145. C. B. Reese and D. R. Trentham, *Tetrahedron Lett.*, 2459 (1965).
146. B. E. Griffin, C. B. Reese, G. F. Stephenson, and D. R. Trentham, *Tetrahedron Lett.*, 4349 (1966).
147. E. J. Reist, V. J. Bartuska, and L. Goodman, *J. Org. Chem.*, **29**, 3725 (1964).
148. A. D. Broom and R. K. Robins, *J. Amer. Chem. Soc.*, **87**, 1145 (1965).
149. T. A. Khwaja and R. K. Robins, *J. Amer. Chem. Soc.*, **88**, 3640 (1966).
150. J. Gin and C. A. Dekker, *Biochemistry*, **7**, 1413 (1968).
151. D. M. G. Martin, C. B. Reese, and G. F. Stephenson, *Biochemistry*, **7**, 1406 (1968).
152. R. Duschinsky and U. Eppenberger, *Tetrahedron Lett.*, 5103 (1967).
153. V. M. Clark, A. R. Todd, and J. Zussman, *J. Chem. Soc.*, 2952 (1951).
154. W. Anderson, D. H. Hayes, A. M. Michelson, and A. R. Todd, *J. Chem. Soc.*, 1882 (1954).
155. R. W. Chambers, J. G. Moffatt, and H. G. Khorana, *J. Amer. Chem. Soc.*, **79**, 3747 (1957).
156. E. Reist, P. A. Hart, L. Goodman, and B. R. Baker, *J. Org. Chem.*, **26**, 1557 (1961).
157. R. E. Holmes and R. K. Robins, *J. Org. Chem.*, **28**, 3483 (1963).
158. Y. Mizuno, M. Ikehara, K. A. Watanabe, S. Suzaki, and T. Itoh, *J. Org. Chem.*, **28**, 3329 (1963).
159. E. J. Reist, *Chem. Ind. (London)*, 1957 (1967).
160. Y. Mizuno, M. Ikehara, T. Itoh, and K. Saito, *Chem. Pharm. Bull. (Tokyo)*, **11**, 265 (1963).
161. A. P. Martinez, W. W. Lee, and L. Goodman, *J. Org. Chem.*, **31**, 3263 (1966).
162. E. J. Reist, D. F. Calkins, and L. Goodman, *J. Org. Chem.*, **32**, 2538 (1967).
163. Y. Mizuno and T. Sasaki, *J. Amer. Chem. Soc.*, **88**, 863 (1966).
164. W. Jahn, *Ber.*, **98**, 1705 (1965).
164a. R. R. Schmidt, U. Schloz, and D. Schwille, *Ber.*, **101**, 590 (1968).
165. M. Ikehara and H. Tada, *J. Amer. Chem. Soc.*, **87**, 606 (1965).
166. M. Ikehara, H. Tada, and K. Muneyama, *Chem. Pharm. Bull. (Tokyo)*, **13**, 639 (1965).
167. M. Ikehara and H. Tada, *Chem. Pharm. Bull. (Tokyo)*, **15**, 94 (1967).
168. M. Ikehara, H. Tada, K. Muneyama, and M. Kaneko, *J. Amer. Chem. Soc.*, **88**, 3165 (1966).
169. E. J. Reist, D. F. Calkins, L. V. Fisher, and L. Goodman, *J. Org. Chem.*, **33**, 1600 (1968).
170. M. Ikehara and M. Kaneko, *J. Amer. Chem. Soc.*, **90**, 497 (1968).
171. K. L. Nagpal and M. M. Dhar, *Tetrahedron Lett.*, 47 (1968).
172. A. M. Michelson and W. E. Cohn, *Biochemistry*, **1**, 490 (1962).
173. R. Letters and A. M. Michelson, *J. Chem. Soc.*, 1410 (1961).
174. Y. Mizuno and T. Sasaki, *Tetrahedron Lett.*, 4579 (1965).
175. H. P. N. Fromageot and C. B. Reese, *Tetrahedron Lett.*, 3499 (1966).
176. W. K. Roberts and C. A. Dekker, *J. Org. Chem.*, **32**, 816 (1967).
177. J. J. Fox and I. Wempen, *Tetrahedron Lett.*, 643 (1965).
178. W. V. Ruyle, T. Y. Shen, and A. A. Patchett, *J. Org. Chem.*, **30**, 4353 (1965).
179. J. Farkaš, J. Beránek, and F. Šorm, *Collect. Czech. Chem. Commun.*, **31**, 4002 (1966).

180. E. J. Reist, J. H. Osiecki, L. Goodman, and B. R. Baker, *J. Amer. Chem. Soc.*, **83**, 2208 (1961).
181. K. A. Watanabe and J. J. Fox, *J. Org. Chem.*, **31**, 211 (1966).
182. G. Etzold, R. Hintsche, and P. Langen, *Tetrahedron Lett.*, 4827 (1967).
183. J. J. Fox, N. C. Miller, and R. J. Cushley, *Tetrahedron Lett.*, 4927 (1966).
184. M. Honjo, Y. Furukawa, M. Nishikawa, K. Kamiya, and Y. Yoshioka, *Chem. Pharm. Bull. (Tokyo)*, **15**, 1076 (1967).
185. B. Bannister and F. Kagan, *J. Amer. Chem. Soc.*, **82**, 3363 (1960).
186. E. J. Reist, A. Benitez, and L. Goodman, *J. Org. Chem.*, **29**, 554 (1964).
186a. B. A. Otter and J. J. Fox, *J. Amer. Chem. Soc.*, **89**, 3663 (1967).
187. W. V. Ruyle and T. Y. Shen, *J. Med. Chem.*, **10**, 331 (1967).
188. J. Žemlička and F. Šorm, *Collect. Czech. Chem. Commun.*, **32**, 576 (1967).
189. I. L. Doerr and J. J. Fox, *J. Amer. Chem. Soc.*, **89**, 1760 (1967).
190. J. F. Codington, I. L. Doerr, and J. J. Fox, *J. Org. Chem.*, **29**, 558 (1964).
191. T. Naito, M. Hirata, Y. Nakai, T. Kobayashi, and M. Kaneo, *Chem. Pharm. Bull. (Tokyo)*, **13**, 1258 (1965).
192. N. Miller and J. J. Fox, *J. Org. Chem.*, **29**, 1772 (1964).
193. N. K. Kochetkov, E. I. Budovskii, and V. N. Shibaev, *Khim. Prir. Soedin., Akad. Nauk Uz. SSR*, 409 (1965); *Chem. Abstr.*, **64**, 19742 (1966).
194. K. L. Agarival and M. M. Dhar, *Tetrahedron Lett.*, 2451 (1965).
195. J. Žemlička and J. Smrt, *Tetrahedron Lett.*, 2081 (1964).
196. J. Nagyvary and J. S. Roth, *Tetrahedron Lett.*, 617 (1965).
197. J. F. Codington, R. Fecher, and J. J. Fox, *J. Amer. Chem. Soc.*, **82**, 2794 (1960).
198. I. L. Doerr, J. F. Codington, and J. J. Fox, *J. Org. Chem.*, **30**, 467 (1965).
199. J. F. Codington, I. L. Doerr, and J. J. Fox, *J. Org. Chem.*, **30**, 476 (1965).
200. W. W. Lee, A. Benitez, L. Goodman, and B. R. Baker, *J. Amer. Chem. Soc.*, **82**, 2648 (1960).
201. T. Sekiya and T. Ukita, *Chem. Pharm. Bull. (Tokyo)*, **15**, 503 (1967).
202. G. L. Tong, W. W. Lee, and L. Goodman, *J. Org. Chem.*, **30**, 2854 (1965).
203. F. J. McEvoy, M. J. Weiss, and B. R. Baker, *J. Amer. Chem. Soc.*, **82**, 205 (1960).
204. E. J. Reist, V. J. Bartuska, D. F. Calkins, and L. Goodman, *J. Org. Chem.*, **30**, 3401 (1965).
205. C. B. Reese, K. Schofield, R. Shapiro, and A. Todd, *Proc. Chem. Soc. (London)*, 290 (1960).
206. K. Imai and M. Honjo, *Chem. Pharm. Bull. (Tokyo)*, **13**, 7 (1965).
207. I. Wempen, I. L. Doerr, L. Kaplan, and J. J. Fox, *J. Amer. Chem. Soc.*, **82**, 1624 (1960).
208. A. S. Jones, A. R. Williamson, and M. Winkley, *Carbohyd. Res.*, **1**, 187 (1965).
209. K. E. Pfitzner and J. G. Moffatt, *J. Amer. Chem. Soc.*, **85**, 3027 (1963).
210. A. F. Cook and J. G. Moffatt, *J. Amer. Chem. Soc.*, **89**, 2697 (1967).
211. H. A. Friedman, K. A. Watanabe, and J. J. Fox, *J. Org. Chem.*, **32**, 3775 (1967).
212. F. W. Lichtenthaler and T. Nakagawa, *Ber.*, **100**, 1833 (1967).
213. J. Beránek, H. A. Friedman, K. A. Watanabe, and J. J. Fox, *J. Heterocycl. Chem.*, **3**, 188 (1965).
214. F. W. Lichtenthaler and H. P. Albrecht, *Ber.*, **99**, 575 (1966).
215. B. Shimizu and M. Miyaki, *Chem. Ind. (London)*, 664 (1966).
215a. B. Shimizu and M. Miyaki, *Tetrahedron Lett.*, 855 (1968).
216. T. L. V. Ulbricht and G. T. Rogers, *J. Chem. Soc.*, 6125 (1965).
217. T. L. V. Ulbricht and G. T. Rogers, *J. Chem. Soc.*, 6130 (1965).

218. G. Schmidt and J. Farkaš, *Collect. Czech. Chem. Commun.*, **31**, 4442 (1966).
219. G. Schmidt and J. Farkaš, *Tetrahedron Lett.*, 4251 (1967).
220. T. Ukita, R. Funakoshi, and Y. Hirose, *Chem. Pharm. Bull.* (*Tokyo*), **12**, 828 (1964).
221. D. Thacker and T. L. V. Ulbricht, *Chem. Commun.*, 122 (1967).
222. Y. Ishikawa, T. Kanazawa, and T. Sato, *Bull. Chem. Soc. Jap.*, **35**, 731 (1962).
223. M. J. Robins, J. R. McCarthy, Jr., and R. K. Robins, *Biochemistry*, **5**, 224 (1966).
224. J. P. Horwitz, J. Chua, and M. Noel, *J. Org. Chem.*, **29**, 2076 (1964).
224a. G. Kowollik and P. Langen, *Chem. Ber.*, **101**, 235 (1968).
224b. J. Beránek and F. Šorm, *Collect. Czech. Chem. Commun.*, **33**, 901 (1968).
225. J. P. H. Verheyden and J. G. Moffatt, *J. Amer. Chem. Soc.*, **86**, 2093 (1964).
226. E. Benz, N. F. Elmore, and L. Goldman, *J. Org. Chem.*, **30**, 3067 (1965).
227. J. P. Horwitz, J. Chua, M. A. Da Rooge, and M. Noel, *Tetrahedron Lett.*, 2725 (1964).
228. J. P. Horwitz, J. Chua, M. Noel, and J. T. Donatti, *J. Org. Chem.*, **32**, 817 (1967).
229. J. R. McCarthy, Jr., M. J. Robins, L. B. Townsend, and R. K. Robins, *J. Amer. Chem. Soc.*, **88**, 1549 (1966).
230. J. P. Horwitz, J. Chua, and M. Noel, *Tetrahedron Lett.*, 1343 (1966).
231. J. P. Horwitz, J. Chua, I. L. Klundt, M. A. Da Rooge, and M. Noel, *J. Amer. Chem. Soc.*, **86**, 1896 (1964).
232. C. L. Stevens, N. A. Nielsen, and P. Blumbergs, *J. Amer. Chem. Soc.*, **86**, 1894 (1964).
233. J. P. H. Verheyden and J. G. Moffatt, *J. Amer. Chem. Soc.*, **88**, 5684 (1966).
234. M. J. Robins, J. R. McCarthy, Jr., and R. K. Robins, *J. Heterocycl. Chem.*, **4**, 313 (1967).
235. J. R. McCarthy, Jr., M. J. Robins, and R. K. Robins, *Chem. Commun.*, 536 (1967).
236. H. Hoeksema, G. Slomp, and E. E. van Tamelen, *Tetrahedron Lett.*, 1787 (1964).
237. H. S. Isbell and H. L. Frush, *J. Res. Nat. Bur. Stand.*, **46**, 132 (1951); H. L. Frush and H. S. Isbell, *ibid.*, **47**, 239 (1951).
238. G. W. Kenner in "Ciba Foundation Symposium on the Chemistry and Biology of Purines," G. E. W. Wolstenholme and C. M. O'Connor, Eds., Little, Brown, Boston, Massachusetts, 1957, pp. 312–313.
239. P. Karrer and J. ter Kuile, *Helv. Chim. Acta*, **5**, 870 (1922).
240. C. A. Dekker, *Ann. Rev. Biochem.*, **29**, 453 (1960).
241. E. Mitts and R. M. Hixon, *J. Amer. Chem. Soc.*, **66**, 483 (1944).
242. A. Wacker and L. Träger, *Z. Naturforsch.*, **18b**, 13 (1963).
243. E. R. Garrett, J. K. Seydel, and Alan J. Sharpen, *J. Org. Chem.*, **31**, 2219 (1966).
244. J. X. Khym and W. E. Cohn, *J. Amer. Chem. Soc.*, **82**, 6380 (1960).
245. D. M. Brown, personal communication, 1959.
246. S. G. A. Alivisatos, L. LaMantia, and B. L. Matijevitch, *Biochim. Biophys. Acta*, **58**, 209 (1962).
247. C. Tamm, M. E. Hodes, and E. Chargaff, *J. Biol. Chem.*, **195**, 49 (1952).
248. E. R. Garrett, *J. Amer. Chem. Soc.*, **82**, 827 (1960).
249. E. R. Garrett, T. Suzuki, and D. J. Weber, *J. Amer. Chem. Soc.*, **86**, 4460 (1964).
250. K. E. Pfitzner and J. G. Moffatt, *J. Org. Chem.*, **29**, 1508 (1964).
251. K. Butler, S. Laland, W. G. Overend, and M. Stacey, *J. Chem. Soc.*, 1433 (1950).
252. Independent observations of L. Goodman and J. Gin, 1968.
253. F. Shafizadeh, *Advan. Carbohyd. Chem.*, **13**, 24 (1958).
254. G. N. Richards, *Chem. Ind.* (*London*), 228 (1944).
255. M. M. Krevoy and R. W. Taft, Jr., *J. Amer. Chem. Soc.*, **77**, 5590 (1955).
256. M. J. Robins and R. K. Robins, *J. Amer. Chem. Soc.*, **86**, 3585 (1964).

257. M. Smith, G. I. Drummond, and H. G. Khorana, *J. Amer. Chem. Soc.*, **83**, 698 (1961).
258. C. D. Jardetzky, *J. Amer. Chem. Soc.*, **84**, 62 (1962).
259. K. Watenpaugh, J. Dow, L. H. Jensen, and S. Furberg, *Science*, **159**, 206 (1968).
260. J. Kraut and L. H. Jensen, *Acta Cryst.*, **16**, 79 (1963); M. Sundaralingam, *ibid.*, **21**, 495 (1966).
261. S. Furberg, *Acta Cryst.*, **3**, 325 (1950); S. Furberg, C. S. Petersen, and C. H. R. Romming, *ibid.*, **18**, 313 (1965).
262. W. Szer and D. Shugar, *Acta Biochim. Pol.*, **7**, 491 (1960).
263. M. P. Gordon, V. S. Weliky, and G. B. Brown, *J. Amer. Chem. Soc.*, **79**, 3245 (1957).
264. E. Shaw, *J. Amer. Chem. Soc.*, **80**, 3899 (1958).
264a. P. D. Lawley, *Biochim. Biophys. Acta*, **26**, 450 (1957).
265. A. S. Jones, A. M. Mian, and R. T. Walker, *J. Chem. Soc.*, (*C*), 692 (1966).
265a. C. A. Dekker, unpublished results, 1965.
266. N. O. Kaplan, S. P. Colowick, and C. C. Barnes, *J. Biol. Chem.*, **191**, 461 (1951).
266a. C. E. Ballou, *Advan. Carbohyd. Chem.*, **9**, 59 (1954).
267. J. Farkaš and F. Šorm, *Collect. Czech. Chem. Commun.*, **32**, 2663 (1967).
268. J. Baddiley, W. Frank, N. A. Hughes, and J. Wieczorkowski, *J. Chem. Soc.*, 1999 (1962).
269. A. W. Johnson and N. Shaw, *Proc. Chem. Soc.* (*London*), 447 (1961).
269a. D. H. Rammler, L. Yengoyan, A. V. Paul, and P. C. Bax, *Biochemistry*, **6**, 1828 (1967).
270. C. D. Jardetzky, *J. Amer. Chem. Soc.*, **82**, 229 (1960).
271. C. D. Jardetzky, *J. Amer. Chem. Soc.*, **83**, 2919 (1961).
272. R. U. Lemieux, *Can. J. Chem.*, **39**, 116 (1961).
273. W. W. Lee, A. Benitez, C. D. Anderson, L. Goodman, and B. R. Baker, *J. Amer. Chem. Soc.*, **83**, 1906 (1961).
274. R. A. Long, R. K. Robins, and L. B. Townsend, *J. Org. Chem.*, **32**, 2751 (1967).
275. R. J. Cushley, K. A. Watanabe, and J. J. Fox, *J. Amer. Chem. Soc.*, **89**, 394 (1967).
276. Y. H. Pan, R. K. Robins, and L. B. Townsend, *J. Heterocycl. Chem.*, **4**, 246 (1967).
277. G J. Durr, J. F. Keiser, and P. A. Ierardi, III, *J. Heterocycl. Chem.*, **4**, 291 (1967).
277a. R. F. Nutt, M. J. Dickinson, F. W. Holly, and E. Walton, *J. Org. Chem.*, **33**, 1789 (1968).
277b. R. J. Cushley, J. F. Codington, and J. J. Fox, *Can. J. Chem.*, **46**, 1131 (1968).
278. H. P. M. Fromageot, B. E. Griffin, C. B. Reese, J. E. Sulston, and D. R. Trentham, *Tetrahedron*, **22**, 705 (1966).
279. L. B. Townsend, R. K. Robins, R. N. Loeppky, and N. J. Leonard, *J. Amer. Chem. Soc.*, **86**, 5320 (1964).
280. A. D. Broom, M. P. Schweizer, and P. O. Ts'o, *J. Amer. Chem. Soc.*, **89**, 3612 (1967).
281. R. U. Lemieux and M. Hoffer, *Can. J. Chem.*, **39**, 110 (1961).
282. T. R. Emerson and T. L. V. Ulbricht, *Chem. Ind.* (*London*), 2129 (1964).
283. T. L. V. Ulbricht, J. P. Jennings, P. M. Scopes, and W. Klyne, *Tetrahedron Lett.*, 695 (1964).
283a. T. Nishimura, B. Shimizu, and I. Iwai, *Biochim. Biophys. Acta*, **157**, 221 (1968).
284. G. Etzold and P. Langen, *Naturwissenschaften*, **53**, 178 (1966).
285. T. L. V. Ulbricht, T. R. Emerson, and R. J. Swan, *Biochem. Biophys. Res. Commun.*, **19**, 643 (1965).
286. I. Frič, J. Šmejkal, and J. Farkaš, *Tetrahedron Lett.*, 75 (1966).

287. T. R. Emerson, R. J. Swan, and T. L. V. Ulbricht, *Biochem. Biophys. Res. Commun.*, **22**, 505 (1966).
288. A. Hampton and A. W. Nichol, *J. Org. Chem.*, **32**, 1688 (1967).
289. W. A. Klee and S. H. Mudd, *Biochemistry*, **6**, 988 (1967).
290. M. Ikehara, M. Kaneko, K. Mineyama, and H. Tanaka, *Tetrahedron Lett.*, 3977 (1967).
291. R. L. Tolman, R. K. Robins, and L. B. Townsend, *J. Heterocycl. Chem.*, **4**, 230 (1967).
291a. J. J. Fox, L. F. Cavalieri, and N. Chang, *J. Amer. Chem. Soc.*, **75**, 4315 (1953).
292. M. Ikehara and H. Tada, *Chem. Pharm. Bull.* (*Tokyo*), **14**, 197 (1966).
293. T. L. V. Ulbricht, *Tetrahedron Lett.*, 1027 (1963).
294. P. Pithkova, A. Piskala, J. Pitha, and F. Šorm, *Collect. Czech. Chem. Commun.*, **30**, 1626 (1965).
295. N. J. Leonard and J. A. Deyrup, *J. Amer. Chem. Soc.*, **84**, 2148 (1962).
296. R. M. Izatt, J. H. Rytting, L. D. Hansen, and J. J. Christensen, *J. Amer. Chem. Soc.*, **88**, 2641 (1966).
297. J. J. Christensen, J. H. Rytting, and R. M. Izatt, *J. Amer. Chem. Soc.*, **88**, 5105 (1966).
298. Y. Sasaki and T. Hashizume, *Anal. Biochem.*, **16**, 1 (1966).
299. Y. Nakagawa, H. Kano, Y. Tsukuda, and H. Koyama, *Tetrahedron Lett.*, 4105 (1967).
300. T. Kishi, M. Muroi, T. Kusaka, M. Nishikawa, K. Kamiya, and K. Mizuno, *Chem. Commun.*, 852 (1967).
301. A. E. V. Haschemeyer and A. Rich, *J. Mol. Biol.*, **27**, 369 (1967).
301a. P. Tollin, H. R.ʹWilson, and D. W. Young, *Nature*, **217**, 1148 (1968).
302. M. Sundaralingam, *J. Amer. Chem. Soc.*, **87**, 599 (1965).
303. T. Okano, T. Iwaguchi, and S. Mizuno, *Chem. Pharm. Bull.* (*Tokyo*), **15**, 373 (1967).

30. OLIGOSACCHARIDES

JOHN H. PAZUR

I. Introduction 69
II. Classification and Nomenclature 70
  A. Classification 70
  B. Nomenclature 71
III. Methods of Synthesis 73
  A. General Remarks 73
  B. Biosynthesis 74
  C. Hydrolysis of Polymers 74
  D. Chemical Synthesis 79
IV. Methods of Isolation 83
  A. Types of Impurities 83
  B. Precipitation and Extraction 84
  C. Chromatography 85
  D. Concentration and Crystallization . . . 86
V. Determination of Structure 88
VI. General Properties 91
  A. Physical Properties 91
  B. Organoleptic Properties 94
  C. Susceptibility to Hydrolysis by Enzymes . . 95
  D. Susceptibility to Hydrolysis by Acids . . . 96
  E. Chemical Reactions 99
VII. Important Oligosaccharides of Biological Origin . 101
  A. Disaccharides 101
  B. Trisaccharides 121
  C. Other Oligosaccharides 127
  References 129

I. INTRODUCTION

Oligosaccharides comprise a large and important class of polymeric carbohydrates which are found either free or in combined form in virtually all living entities.[1,2] In structure, the oligosaccharides are composed of relatively few monosaccharide residues joined through glycosidic bonds which are readily hydrolyzed in acid solution to yield the constituent monosaccharides. It is generally agreed that a carbohydrate consisting of two to ten monomeric

* References start on p. 129.

residues and with a chemically defined structure is appropriately classed as an oligosaccharide. Oligosaccharides that occur free in Nature often possess some unique structural feature or unusual property which is of special value to the biological entity. The most familiar examples of this group are sucrose and lactose, with the former being universally distributed in plants and the latter being the principal carbohydrate constituent of mammary secretions. Oligosaccharides that occur as structural units of glycosides, polysaccharides, or other oligosaccharides are numerous and have been prepared for the most part by fragmentation methods which leave the oligosaccharide intact in the final reaction mixture in sufficient concentration for subsequent isolation. Well-known examples of compounds of the latter group are maltose, iso-maltose, and cellobiose, which are derived from starch, dextran, and cellulose, respectively. More recent examples are the novel types of hetero-oligosaccharides consisting of various combinations of pentoses, hexoses, hexosamines, deoxyhexoses, and uronic acids. These oligosaccharides have been obtained from the various types of heteropolysaccharides found in plant, microbial, and animal tissues.

In the past decade, there has been much emphasis in biochemistry on the nature and functioning of enzymes including carbohydrases, and many new oligosaccharides have been synthesized by the use of enzymes. Generally, these oligosaccharides are products of action of enzymes having glycosyl transferase activity on suitable substrates and cosubstrates. A variety of different types of oligosaccharides can be produced, depending on the substrate and co-substrate specificities of the individual transferase. A number of chemical methods are available for synthesizing oligosaccharides. The principal chemical methods include coupling of properly substituted monosaccharide derivatives in the presence of catalysts, condensation of unsubstituted monosaccharides under dehydration conditions, and alterations of structure of available oligosaccharides by appropriate chemical reactions. Whereas the structure of most known oligosaccharides has been established, the relation of the structure to the properties of the compound requires further study. The biological function of some oligosaccharides is well understood, but that of others remains obscure. Many challenging areas for research on the chemistry, the enzymology, and the biological functions of oligosaccharides remain to be explored.

II. CLASSIFICATION AND NOMENCLATURE

A. CLASSIFICATION

Oligosaccharides have been classified according to a number of criteria including the type of functional groups, the number of monomeric (monosaccharide) residues, and the types of monomeric residues in the compound. Since at least one of the monosaccharide residues of an oligosaccharide is combined through an oxygen bridge of its hemiacetal group to a hydroxyl group of a second residue, in a manner analogous to the combination of D-glucose to an alcohol in glucosides, the oligosaccharides are indeed true glycosides. The oligosaccharides represent a unique type of glycoside in which the aglycon is also a carbohydrate residue. As a result, these compounds have been classed as holosides, as distinct from glycosides having alcoholic or phenolic aglycons, which are classified as heterosides.

A classification scheme for oligosaccharides which is commonly used is based on the number of monomeric residues in the compound. Thus, disaccharides are composed of two monosaccharide residues, trisaccharides of three, tetrasaccharides of four, and so forth. Each of these groups can be further subdivided according to several types of criteria. A subdivision into homo-oligosaccharides and hetero-oligosaccharides is based on the types of constituent monomeric residues in the oligosaccharide, with the homo group consisting of only one type of monosaccharide, and the hetero group of two or more types of monosaccharides. A subdivision into reducing and non-reducing oligosaccharides is based on the presence or absence of a free hemiacetal hydroxyl group in the compound. A reducing oligosaccharide has a free hemiacetal group and therefore reacts with alkaline copper(II) reagents, exhibits mutarotation, and forms glycosides and osazones in a manner analogous to that of the monosaccharides. An oligosaccharide in which all the hemiacetal groups of the monomeric residues are involved in the formation of the glycosidic bond does not give the above reactions and is accordingly classified as nonreducing. The reducing property of oligosaccharides is important, for it provides a test for the existence of a monosaccharide residue having an unsubstituted hemiacetal hydroxyl group. The reducing property, along with solubility and other properties, has been used to devise schemes for the systematic identification of carbohydrates.[3] A subdivision on the basis of the chemical nature of the constituent residues of oligosaccharides has been employed in a recent monograph on oligosaccharides.[1] This subdivision has merit, since oligosaccharides containing common structural residues are easily identifiable.

B. Nomenclature

In recent years much attention has been given to the establishment of a systematic nomenclature for carbohydrates, including oligosaccharides. Some oligosaccharides, as for example maltose, sucrose, and lactose, were known before their chemical constitution had been fully elucidated and were

* *References start on p. 129.*

given trivial names which, in the interest of brevity, are still useful. As new oligosaccharides are prepared, trivial names are not always apparent, and, since such names give little information on the structure of the compound, their use is to be discouraged. Systematic rules for the nomenclature of carbohydrates have been developed by committees of professional societies. The most recent rules have been officially adopted by the British Chemical Society and by the American Chemical Society and appear in published form[4,5] (see Vol. IIB, Chap. 46).

In brief, the rules state that reducing oligosaccharides are named as glycosyl aldoses or ketoses, and nonreducing oligosaccharides as glycosyl aldosides or ketosides. Numbers and arrows are utilized to indicate the carbon atoms at which substitution has occurred and to indicate glycosidic bonds. Appropriate symbols, α or β, D or L, and O, are inserted to indicate the stereochemistry of the glycosidic bonds, the configuration of the constituent monosaccharide residues, and a substitution at the oxygen atom. A few examples using well-known compounds will serve to illustrate the use of this system. Maltose is 4-O-α-D-glucopyranosyl-D-glucopyranose, the 4 indicating substitution at C-4 of the D-glucose residue containing the reducing group, the O indicating that substitution is on the oxygen at position 4, the α indicating the relative stereochemistry of the glycosidic linkage, and the D indicating the chirality of the monosaccharide residues. Cellobiose is 4-O-β-D-glucopyranosyl-D-glucopyranose, lactose is 4-O-β-D-galactopyranosyl-D-glucopyranose, and sucrose is β-D-fructofuranosyl α-D-glucopyranoside. As the number of residues increases in an oligosaccharide and two or more residues are attached to a common residue, the systematic names become increasingly complicated. Maltotriose is O-α-D-glucopyranosyl-(1 → 4)-O-α-D-glucopyranosyl-(1 → 4)-D-glucopyranose, and a D-glucosyl oligosaccharide from starch or glycogen having two D-glucosyl residues attached to a common D-glucose moiety is O-α-D-glucopyranosyl-(1 → 6)-O-[α-D-glucopyranosyl-(1 → 4)]-D-glucopyranose.

Systematic names, although informative from the standpoint of conveying the structure for the oligosaccharides, are cumbersome when used repeatedly in scientific writing. Two systems have been proposed which shorten the name of the oligosaccharide and yet indicate at least a partial structure for the compound. First, a system that utilizes abbreviations for the monosaccharide residues is in use in scientific journals.[6] The names of the monosaccharide residues are shortened to the first three letters of the name, as for example Rib, Gal, and Fru for ribose, galactose, and fructose, except for glucose, which is shortened to Glc. Pyranose and furanose ring structures are indicated p and f, respectively. Abbreviations of other structural residues are those recommended in recent issues of biochemical journals,[6] as, for example, GlcN for 2-amino-2-deoxyglucose (glucosamine), GalNAc for 2-acetamido-

2-deoxygalactose (*N*-acetylgalactosamine), and ManA for mannuronic acid. In this abbreviated system, maltose is 4-*O*-α-D-Glc*p*-D-Glc*p*, cellobiose is 4-*O*-β-D-Glc*p*-D-Glc*p*, lactose is 4-*O*-β-D-Gal*p*-D-Glc*p*, sucrose is β-D-Fru*f*-α-D-Glc*p*, maltotriose is *O*-α-D-Glc*p*-(1 → 4)-*O*-α-D-Glc*p*-(1 → 4)-D-Glc*p*, and the branched D-glucosyl trisaccharide is *O*-α-D-Glc*p*-(1 → 6)-*O*-[α-D-Glc*p*-(1 → 4)]-D-Glc*p*.

The second system is based on a series of rules formulated for conveying the structure of the oligosaccharide in a manner similar to the rules governing the naming of hydrocarbons.[7] The name of an oligosaccharide is divided into names corresponding to linear segments of the compound. Trivial names are used for the linear segments such as the disaccharides and higher homologs. A number system is used to indicate points of substitution on specific residues of the compound. A superscript numbering system is used to indicate the residue of the compound containing a substitutent group, with the residue containing the reducing group being number 1. Names utilizing this system, as well as trivial and complete names for most of the known oligosaccharides, are used in a recent monograph on oligosaccharides.[1] To illustrate the second system with the examples cited above, maltotriose is 4^2-α-D-glucopyranosylmaltose, where the 4^2 indicates that substitution is at C-4 of the glycosyl residue which is residue 2 of maltose. Maltotriose may also be named 4-*O*-α-maltosyl-D-glucose, where the 4 indicates that a maltosyl unit is at C-4 of the D-glucose residue. The other trisaccharide is 4^1, 6^1-di-α-D-glucopyranosyl-D-glucose. In this system more than one name is possible for a single oligosaccharide, and a name can be selected to emphasize some particular structural feature of the compound under consideration.

III. METHODS OF SYNTHESIS

A. GENERAL REMARKS

Methods for the synthesis of oligosaccharides may be divided into three major types: biosynthesis, fragmentation of biological polymers, and chemical synthesis. While some oligosaccharides have been prepared by all three methods, others may have been obtained by only one of these procedures. Many factors must be considered in the selection of a method for the preparation of a specific oligosaccharide. One very important consideration is the availability of a suitable method for the isolation of the oligosaccharide from the final reaction mixture. The successful synthesis of a compound may require a major modification of the basic procedures that are available.

---

* *References start on p. 129.*

B. Biosynthesis

The biosynthesis of oligosaccharides is covered in this volume, Chap. 34, and transglycosylation reactions are discussed in this volume, Chap. 33.

C. Hydrolysis of Polymers

Oligosaccharides have been obtained from biological polymers by hydrolysis under conditions that leave the desired oligosaccharide intact for subsequent isolation. This procedure is also useful for obtaining information on the structure of biological polymers. Enzymes and acids are used as hydrolytic agents and often yield a different series of oligosaccharides from the same polymer. Sources of specific oligosaccharides include polysaccharides, other oligosaccharides, and glycosides, with the polysaccharides as the most common source.

1. Hydrolysis by Enzymes

The use of enzymic procedures for hydrolyzing polymers is dependent on the availability of suitable enzyme preparations. In many cases impure enzyme preparations are adequate, but for the hydrolysis of specific linkages, pure enzymes having a high degree of specificity are needed. Enzymes that hydrolyze polysaccharides are numerous and for convenience have been divided into two groups, the exo-hydrolases and the endo-hydrolases. The exo-hydrolases cleave a polysaccharide by successive removal of residues from one end of the polymeric chain. If the polymer is of uniform structure, a high percentage of the polymer is converted into a specific oligosaccharide. A familiar example of this type of enzyme is maltohydrolase (β-amylase), which converts amylose, the linear fraction of starch, almost quantitatively into maltose.[8] Other examples of exo-hydrolases are a fungal hydrolase[9] used to prepare isomaltose from dextran,[10] a yeast galacturonanase used to obtain a digalacturonic acid from pectic acid,[11] and a protozoan hydrolase used to obtain xylobiose from xylan.[12]

The enzymes of the endo-hydrolase type are more common in biological materials, and many types of endo-hydrolases have been used for preparing oligosaccharides. These hydrolases effect random fragmentation of a homopolysaccharide to give a homologous series of oligosaccharides, as for example, malto-oligosaccharides from amylose[13] and xylo-oligosaccharides from xylan.[14] With polysaccharides of more complicated structures, endo-hydrolases cleave the various glycosidic bonds at different rates and in some cases may be without action on certain bonds. Accordingly, a wide array of oligosaccharides can be obtained from these polysaccharides, depending on the specificity of the enzyme employed. Glucosyl oligosaccharides having α-D-$(1 \rightarrow 4)$ and α-D-$(1 \rightarrow 6)$ linkages, representing the branch point of starch and glycogen, have been prepared by use of the endo-hydrolase from saliva.[15,16] Oligosaccharides representing branch points of dextran were prepared by use of an endo-hydrolase from *Lactobacillus bifidus*.[17] Hetero-oligosaccharides composed of D-xylose and L-arabinose residues were prepared from arabinoxylans by use of enzymes specific for certain glycosidic bonds[18] or by use of inhibitors to suppress hydrolysis of some of the glycosidic linkages.[19] An increase in the yield of a specific oligosaccharide, as well as in the number of oligosaccharides that can be produced from a polymer, results from use of dialysis techniques.[20,21] In this method, the reaction products are continually removed from the domain of the enzyme by dialysis. Oligosaccharides dialyze at different rates, depending on the molecular weight, the extent of agitation of the digestion mixture, and the pore size of the dialysis tubing. Data on the relative rate of dialysis of oligosaccharides under certain standard conditions are available,[22] and these data can be utilized for the selection of the conditions for obtaining a specific oligosaccharide.

Several interesting oligosaccharides have been prepared from naturally occurring oligosaccharides or glycosides. Specific bonds of the latter compounds are hydrolyzed by enzymic procedures, and the resulting oligosaccharides are isolated by appropriate methods. Melibiose, gentiobiose, planteobiose, and manninotriose have been obtained from raffinose, gentianose, planteose, and stachyose, respectively. Rutinose (6-*O*-α-L-rhamnosyl-D-glucose) has been isolated from an enzymic hydrolyzate of flavonoid glycosides. Enzymes (glycoside glycohydrolases) which have been used for hydrolyzing oligosaccharides and glycosides include α- and β-glucosidases, β-fructosidases, and α- and β-galactosidases (Chapter 33). Often these enzymes, if available in pure form, are valuable in a structural investigation of the oligosaccharide, particularly the assignment of configuration to the glycosidic linkages. The sources of these enzymes have been yeasts,[23] plant seeds,[24,25] fungi, [26,27] and bacteria.[28,29]

2. *Hydrolysis by Acids*

Controlled acid hydrolysis of polysaccharides, oligosaccharides, and glycosides results in reaction mixtures from which numerous oligosaccharides have been obtained. The conditions for effecting hydrolysis of these polymers must be established largely by trial and error. As shown in Table I,[30] differences in the rate of hydrolysis of different glycosidic bonds do exist. If hydrolysis constants are known, these can be utilized in selecting the proper conditions for hydrolyzing the polymer. The procedures that have been employed can be grouped into the following types: autohydrolysis, hydrolysis in dilute acids, hydrolysis in concentrated acids, acetolysis, and hydrolysis by acidic resins.

* *References start on p. 129.*

The autohydrolysis procedure is applicable with those polysaccharides that are composed of monomeric residues having acidic groups. Many of these acidic polysaccharides are sufficiently ionized in solution for auto-hydrolysis to occur when a solution of the polymer is heated for a sufficient period of time. Disaccharides and other uronic acid-containing oligosaccharides have been prepared from plant gums[31,32] and plant residues[33] by this procedure. Solutions of neutral polysaccharides containing labile linkages may undergo autohydrolysis at elevated temperatures, to give oligosaccharides.[34]

TABLE I

RELATIVE RATES OF ACID HYDROLYSIS OF SOME OLIGOSACCHARIDES

Oligosaccharide	$k/a_{H^+} \times 10^6$ (sec^{-1})	Activation energy $(cal/mole)$
β-D-Fructofuranosyl α-D-glucopyranoside (sucrose)	14,600	25,830
α-D-Glucopyranosyl α-D-glucopyranoside (trehalose)	0.864	40,180
6-O-β-D-Glucopyranosyl-D-glucopyranose (gentiobiose)	1.24	33,390
4-O-β-D-Glucopyranosyl-D-glucopyranose (cellobiose)	5.89	30,710
3-O-α-D-Glucopyranosyl-D-fructofuranose (turanose)	11.9	32,450
4-O-α-D-Glucopyranosyl-D-glucopyranose (maltose)	16.8	30,970
4-O-β-D-Galactopyranosyl-D-glucopyranose (lactose)	16.6	26,900
6-O-α-D-Galactopyranosyl-D-glucopyranose (melibiose)	15.5	38,590

Hydrolysis with dilute acids is generally achieved with hydrochloric or sulfuric acid in the concentration range from $0.01N$ to $2N$. Often the hydrolysis is effected at high temperatures, but in some cases hydrolysis of acid-labile linkages may be effected at room temperature. Table II contains data on the conditions that have been used to hydrolyze some representative polymers.

If prolonged periods of hydrolysis are employed, dehydration of the oligosaccharide may occur. In such a case, procedures based on methanolysis[35] are more satisfactory for hydrolyzing the polymers. Such procedures yield the methyl glycosides, which may then be converted into the free oligosaccharides by enzymic or controlled acid hydrolysis.

When all the linkages of the polysaccharide are hydrolyzed at approximately the same rate, random hydrolysis leads to a homologous series of oligosaccharides. Malto-oligosaccharides from amylose,[36,37] and isomalto-oligosaccharides from dextran,[38] are two examples of this type. Hydrolysis conditions for such polymers can be altered depending on whether a higher- or lower-molecular weight homolog of the series is desired. The use of hydrolysis constants for establishing the proper conditions for hydrolyzing a

TABLE II

REACTION CONDITIONS FOR HYDROLYSIS OF POLYSACCHARIDES TO OLIGOSACCHARIDES

Polymer	Type of oligosaccharide produced	Acid concentration	Temperature(°C) and time (hr)	Ref.
Amylose	Malto	0.33 N H_2SO_4	100°, 2	a
	Malto	0.1 N HCl	100°, 4	b
Dextran	Isomalto	0.33 N H_2SO_4	100°, 10	c
Laminaran	Laminaro	0.25 N H_2SO_4	100°, 5	d
	Laminaro	0.33 N H_2SO_4	100°, 2	e
(1 → 2)-β-D-Glucan	Sophoro	1 N H_2SO_4	70°, 18	f
Nigeran	Nigero	1 N H_2SO_4	85°, 3	g
Pustulan	Gentio	1 N H_2SO_4	100°, 2	h
Glucomannan	Gluco-manno	30% HCOOH	100°, 4	i
Aspen hemicellulose	Xylo	1 N H_2SO_4	90°, 8	j
Inulin	Inulo	0.01 N HCl	70°, 0.5	k
Anogeissus schimperi gum	Arabino	0.01 N H_2SO_4	100°, 4	l
	Galacto	0.05 N H_2SO_4	100°, 1	l
	Uronic acid	1 N H_2SO_4	100°, 5	l
Capsular polysaccharides	Uronic acid	2 N H_2SO_4	100°, 3	m
Blood-group substance (glycoprotein)	Hexosamine	1 N HCl	100°, 0.5	n
Plasma glycoprotein	Hexosamine	0.3 N H_2SO_4	95°, 2	o

[a] W. J. Whelan, J. M. Bailey, and P. J. P. Roberts, *J. Chem. Soc.*, 1293 (1953). [b] J. H. Pazur and T. Budovich, *J. Biol. Chem.*, **220**, 25 (1956). [c] J. R. Turvey and W. J. Whelan, *Biochem. J.*, **67**, 49 (1957). [d] V. C. Barry and J. E. McCormick, *Methods Carbohyd. Chem.*, **1**, 328 (1962). [e] S. Peat, W. J. Whelan, and H. G. Lawley, *J. Chem. Soc.*, 724 (1958). [f] P. A. J. Gorin, J. F. T. Spencer, and D. W. S. Westlake, *Can. J. Chem.*, **39**, 1067 (1961). [g] S. A. Barker, E. J. Bourne, and M. Stacey, *J. Chem. Soc.*, 3084 (1953). [h] B. Lindberg and J. McPherson, *Acta Chem. Scand.*, **8**, 985 (1954). [i] A. Tyminski and T. E. Timell, *J. Amer. Chem. Soc.*, **82**, 2823 (1960). [j] J. K. N. Jones and L. E. Wise, *J. Chem. Soc.*, 2750 (1952). [k] J. H. Pazur and A. L. Gordon, *J. Amer. Chem. Soc.*, **75**, 3458 (1953). [l] G. O. Aspinall and T. B. Christensen, *J. Chem. Soc.*, 3461 (1961). [m] S. A. Barker, A. B. Foster, I. R. Siddiqui, and M. Stacey, *J. Chem. Soc.*, 2358 (1958). [n] R. H. Côté and W. T. J. Morgan, *Nature*, **178**, 1171 (1956). [o] E. H. Eylar and R. W. Jeanloz, *J. Biol. Chem.*, **237**, 622 (1962).

* References start on p. 129.

polymer is well illustrated in the isolation of isomaltose from glycogen.[39] With heteropolysaccharides it may be advantageous to use a multistage hydrolysis. In this method a preliminary acid hydrolysis is used to cleave the acid-labile linkages, and the large fragments are isolated from the reaction mixture. The individual fragments are then subjected to a second acid hydrolysis under conditions which favor the accumulation of the desired oligosaccharide. Polysaccharides from plants[40,41] and blood-group glycoproteins[42] have been fragmented by the multistage procedure.

Glycosidic linkages involving furanose residues are extremely acid-labile, and very mild hydrolytic conditions must be employed to obtain oligosaccharides having such linkages.[34,43] It is often advantageous to use enzymic procedures with polymers containing acid-labile linkages.[44] Another approach is to alter the structure of certain residues of the polymer by chemical means. Thus, catalytic oxidation of primary alcoholic groups of the L-arabinofuranose residues in arabinoxylans converts these residues into uronic acid residues.[45] Oligosaccharides having furanose residues can be obtained from acid hydrolyzates of the oxidized polymer. In general, glycosidic linkages involving an ionizable group, such as an uronic acid or hexosamine, are extremely acid-stable, and compounds having these residues accumulate in the reaction mixture.

When the glycosidic linkages of a polymer are resistant to hydrolysis in dilute acids, hydrolysis can be effected in concentrated acids at refluxing temperatures for prolonged periods of time. Oligosaccharides so prepared include oligosaccharides of 2-amino-2-deoxy-D-glucose from chitin,[46] oligosaccharides of D-glucose from cellulose,[47] and oligosaccharides of D-xylose from xylan.[48] Acetolysis, involving the simultaneous use of sulfuric acid and acetic anhydride, is another method for hydrolyzing such resistant polymers. The oligosaccharides from cellulose were prepared initially by acetolysis.[49] In the latter procedure, acetylation and hydrolysis occur simultaneously, and a mixture of oligosaccharide acetates is obtained. One advantage of acetolysis is that the acetylated oligosaccharides are often separated more easily than the free oligosaccharides. Further, it is also possible to alter the order of cleavage of glycosidic bonds in the polymer because of different rates of acetylation of the monomeric residues, so that different types of oligosaccharides can be obtained from the same polymer, according to whether acetolysis or hydrolysis by concentrated acid is employed.[50,51]

A novel method has been described for partial acid hydrolysis of polysaccharides, based on the use of water-soluble polystyrenesulfonic acid resin.[52] This method was used to obtain oligosaccharides from glycoproteins[53] in yields several times as great as was possible by conventional hydrolysis with dilute acid. This method, coupled with dialysis techniques, should be ideal for obtaining oligosaccharides in increased yields from other polymers.

D. Chemical Synthesis

Chemical procedures have been used, not only as a means of synthesizing new oligosaccharides, but also as a method of providing proof of structure for naturally occurring oligosaccharides and polysaccharides. Chemical methods, though generally applicable, nevertheless suffer from the disadvantage that mixtures of isomeric oligosaccharides are often produced. Further, it is at times difficult to prepare monosaccharide derivatives having proper substitution for use in the synthesis of specific compounds. The reactions for chemical synthesis can be divided into three principal types: (1) formation of new glycosidic linkages, (2) alteration of one of the residues of available oligosaccharides, and (3) alteration of glycosidic linkages of available oligosaccharides.

The Koenigs–Knorr reaction for the preparation of glycosides, described in Vol. IA, Chap. 9, is probably the most widely applicable and important chemical method for the synthesis of oligosaccharides. In this reaction a per-*O*-acylated glycosyl halide, usually the bromide, is coupled in the presence of silver carbonate or silver oxide with a second monosaccharide derivative containing an unsubstituted hydroxyl group. Elimination of hydrogen bromide and formation of a glycosidic bond result. The synthesis of α-gentiobiose octaacetate[54] by this reaction is illustrated in formula **1**.

The reaction generally occurs with inversion of configuration at C-1 and usually proceeds in moderate to high yield when the halogen and acetyl groups on C-1 and C-2, respectively, have a *cis* relationship.[54] If these groups

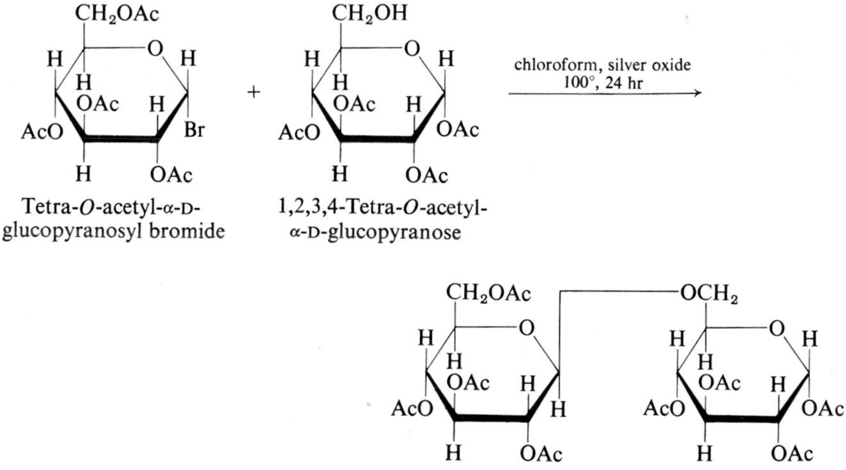

Tetra-*O*-acetyl-α-D-glucopyranosyl bromide

1,2,3,4-Tetra-*O*-acetyl-α-D-glucopyranose

chloroform, silver oxide 100°, 24 hr

α-Gentiobiose octaacetate

1

are *trans*, however, ortho esters may be formed, or reaction without net inversion of configuration may occur because of the participation of the *trans*-related acetoxy group at C-2 in the elimination process.[55] Disaccharides having D-galactose, D-glucose, D-arabinose, or D-xylose as the glycosyl portion of the molecule are easily prepared by this reaction. The glycosyl residues may be attached to the same type of monosaccharide residue or to other monosaccharides, including ketoses,[56,57] hexosamines,[58,59] and uronic acids.[60]

The synthesis of oligosaccharides having an α-D-glycosidic linkage has been achieved by use of mercuric acetate, pyridine, or other organic base as catalyst rather than silver salts. 6-*O*-α-D-Galactopyranosyl-D-glucose (melibiose) was synthesized from tetra-*O*-acetyl-α-D-galactopyranosyl bromide and 1,2,3,4-tetra-*O*-acetyl-D-glucose in the presence of quinoline.[61] Condensation of tri-*O*-acetyl-α-D-xylopyranosyl bromide and a D-glucose derivative having an unsubstituted hydroxyl group at C-6 takes place in the presence of mercuric acetate to yield the two disaccharides having the α-D- and the β-D-glycosidic linkage.[62] The yield of the α-D-linked disaccharide in such reactions is low and may be attributed to the fact that the acetoxy group at C-2 participates in the elimination reaction and promotes reaction with inversion of configuration. The synthesis of α-D-linked oligosaccharides is greatly facilitated[63] by use of derivatives having a nonparticipating group at C-2. Thus, 3,4,6-tri-*O*-acetyl-2-*O*-nitro-β-D-glucopyranosyl chloride and 1,2,3,4-tetra-*O*-acetyl-β-D-glucopyranose react in the presence of silver carbonate and silver perchlorate to form a product which, on reductive removal of the nitro group and acetylation, gives β-isomaltose octaacetate in 60% overall yield.[63] An extensive study of the factors which affect the Koenigs–Knorr reaction has been reported.[64]

Methods for the preparation of trisaccharides by utilizing acetylated glycosyl halides of disaccharides in place of tetra-*O*-acetyl-α-D-glucosyl bromide in the Koenigs–Knorr reaction have been described. *O*-β-Cellobiosyl-D-glucose and 6-*O*-β-maltosyl-D-glucose are two compounds prepared in this manner.[65] Trisaccharides having the α-D configuration can also be prepared by use of this general procedure and selection of the proper catalysts.[66] Panose, the α-D-linked trisaccharide representing the branch point in amylopectin, has been synthesized.[66a]

It is difficult to obtain monosaccharide derivatives having single free hydroxyl groups other than glycosidic or primary hydroxyl groups. However, valuable derivatives of this type are the diisopropylidene acetals and the *O*-isopropylidene anhydro derivatives. By condensation of 1,6-anhydro-2,3-*O*-isopropylidene-D-mannopyranose with tetra-*O*-acetyl-α-D-galactopyranosyl bromide, subsequent hydrolysis of the *O*-isopropylidene and *O*-acetyl groups, and cleavage of the anhydro ring with acids, 4-*O*-β-D-galactopyranosyl-D-mannose has been obtained.[67] This compound was then converted by means of the glycal synthesis into lactose. Cellobiose has been synthesized through a similar series of reactions.[68] These reaction sequences have been valuable for establishing the chemical structure of lactose and cellobiose.

Glycosidic bonds have also been formed by elimination of sodium bromide when a tetra-O-acetylhexose having a free primary or secondary hydroxyl group is heated with sodium and then treated with a tetra-O-acetylglycosyl bromide. The octaacetates of gentiobiose and cellobiose have been prepared by this method.[69] Under different reaction conditions, 3,4,6-tri-O-acetyl-β-D-glucopyranosyl chloride reacted with 1,2:5,6-di-O-isopropylidene-α-D-glucofuranose to yield a derivative from which 3-O-α-D-glucopyranosyl-D-glucose was obtained.[70]

As indicated earlier, chemical synthesis has been used to obtain proof of structure for oligosaccharides or oligosaccharide fragments isolated from biological materials. Assignment of configuration of the glycosidic linkage in the oligosaccharides on the basis of the type of catalyst used in a Koenigs–Knorr reaction has at times led to postulation of erroneous structures for some naturally occurring compounds. In this connection, the original assignment of the β-L linkage in 6-O-L-rhamnosyl-D-glucose[71] and the more recent evidence for the α-L linkage in this compound[72] may be cited.

Anhydro sugar derivatives have been very useful for the synthesis of oligosaccharides. When 5,6-anhydro-1,2-O-isopropylidene-α-D-glucofuranose was treated with tetra-O-acetyl-α-D-glucopyranosyl bromide, condensation occurred, with simultaneous opening of the anhydro ring, addition of the glucosyl group to the oxygen attached to C-5, and addition of bromine to C-6. Catalytic reduction replaced the bromine by a hydrogen atom, and there was obtained an unusual disaccharide derivative, 6-deoxy-5-O-β-D-glucosyl-D-glucose.[73] When 3,4,6-tri-O-acetyl-1,2-anhydro-α-D-glucopyranose was treated with alcohols,[74] β-D-glucosides of the alcohol were obtained. The latter type of reaction was used for the first true chemical synthesis of sucrose. 1,3,4,6-Tetra-O-acetyl-β-D-fructose was condensed with 3,4,6-tri-O-acetyl-1,2-anhydro-α-D-glucopyranose to yield a hepta-O-acetyl derivative of sucrose. After acetylation of the reaction mixture, sucrose octaacetate (2) was isolated by paper and column chromatographic methods, in a 5% overall yield.[75] Deacetylation of the acetate in methanolic sodium methoxide yielded sucrose, obtained crystalline from ethanol. The anhydro derivatives are of general applicability for the synthesis of oligosaccharides. Maltose[76] and 2-O-α-D-glucopyranosyl-D-glucose (kojibiose)[77] are two other examples of oligosaccharides that have been synthesized by this reaction route.

When monosaccharides are refluxed in dilute acid solution, water is eliminated between the hemiacetal group of one residue and a hydroxyl group

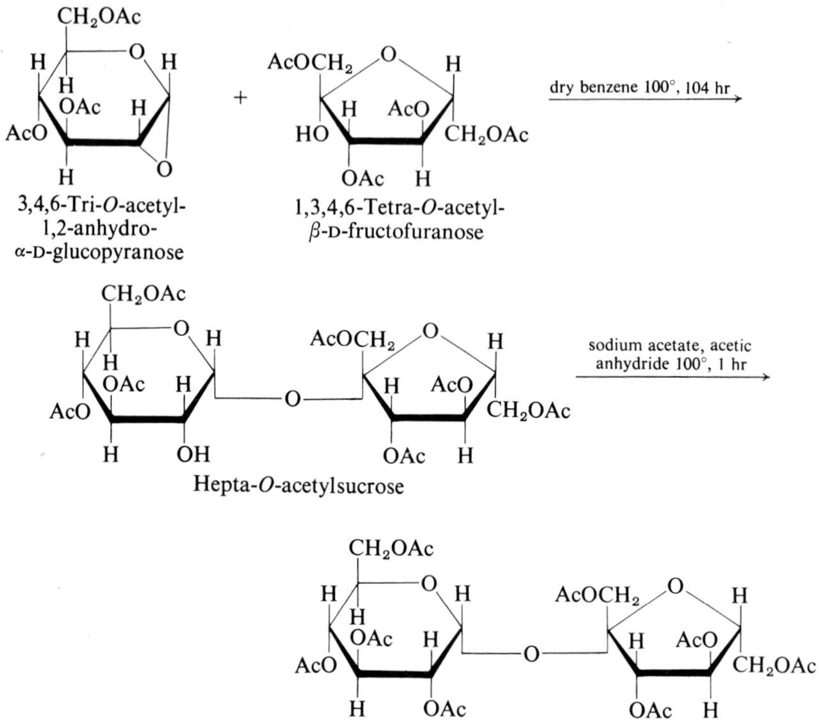

JOHN H. PAZUR

CH$_2$OAc

H   O   H
   H
   OAc   H
AcO
H   O

3,4,6-Tri-O-acetyl-
1,2-anhydro-
α-D-glucopyranose

+

AcOCH$_2$ O   H
   H   AcO
HO   CH$_2$OAc
   OAc   H

1,3,4,6-Tetra-O-acetyl-
β-D-fructofuranose

$\xrightarrow{\text{dry benzene } 100°,\ 104\ hr}$

CH$_2$OAc

H   O   H
   H
   OAc   H
AcO
H   OH

AcOCH$_2$ O   H
   O   H   AcO
          CH$_2$OAc
   OAc   H

Hepta-O-acetylsucrose

$\xrightarrow{\text{sodium acetate, acetic} \atop \text{anhydride } 100°,\ 1\ hr}$

CH$_2$OAc

H   O   H
   H
   OAc   H
AcO
H   OAc

AcOCH$_2$ O   H
   O   H   AcO
         CH$_2$OAc
   OAc   H

Octa-O-acetylsucrose

2

of a second residue in the process known as "reversion." The result is the formation of a new glycosidic bond and thus the synthesis of a new oligosaccharide. Since many alternative possibilities for elimination are available, a complex mixture of oligosaccharides is produced by this route. Acetylation of a reaction mixture obtained by heating D-glucose in acid solution followed by separation of the acetylated compounds by column chromatography has resulted in the isolation of octa-O-acetyl derivatives of nine disaccharides of D-glucose.[78] Deacetylation of the octaacetates yields the pure disaccharides. Charcoal-column chromatography of an unacetylated reaction mixture has also been employed to obtain the same series of compounds.[79] A modification of this type of reaction involves heating of the appropriate monosaccharides *in vacuo* in the dry state.[80]

New oligosaccharides can be made by altering the structure of existing oligosaccharides by application of suitable chemical reactions. Through isomerization reactions in dilute ammonia or lime water (see Vol. IA, Chap. 4), terminal aldose residues of oligosaccharides can be converted into ketoses, and oligosaccharides having ketose moieties are produced. Lactulose,[81] maltulose,[82] cellobiulose,[83] and maltotriulose[84] are among some of the compounds that have been prepared by this method. Oxidation of oligosaccharides with mild reagents such as hypochlorite and hypoiodite converts the hemiacetal group into a carboxyl group. The preparation of maltobionic acid[85] illustrates an application of this reaction. Reduction of the aldehyde group yields oligosaccharides containing alditol residues.[86] Reduction of uronic acid-containing oligosaccharides through the ester derivatives results in new oligosaccharides.[87]

Reactions used for shortening carbon chains of monosaccharides (Vol. IA, Chap. 3) are applicable to oligosaccharides and yield new types of oligosaccharides. Reactions of this group are the Ruff degradation, employing oxidation with hydrogen peroxide and cleavage with ferric salts,[88] the Wohl-Zemplén method for degrading nitriles with sodium methoxide,[89] ozonolysis of glycals,[90] lead tetraacetate oxidation,[91] and ninhydrin oxidation.[92] Reactions of this type, coupled with reactions by which the carbon chain is lengthened, such as the cyanohydrin or the nitromethane synthesis, are especially useful for the preparation of 1-^{14}C-labeled oligosaccharides.[93]

Transformation of glycosidic linkages of certain glycosides from the β-D to the α-D configuration is possible by use of stannic chloride or titanium tetrachloride.[94] The method has been applied for the isomerization of cellobiose octaacetate to maltose octaacetate,[95] and the isomerization of methyl hepta-O-acetyl-α-cellobioside to methyl hepta-O-acetyl-α-maltoside.[96] Such methods have not been widely used and are apparently of limited value for the synthesis of oligosaccharides.

IV. METHODS OF ISOLATION

A. Types of Impurities

Naturally occurring oligosaccharides are found in solution in the fluids and the cells of plants, animals, and microorganisms. These fluids contain, in addition to oligosaccharides, the complex mixture of organic and inorganic compounds characteristic of living systems. Whereas some naturally occurring oligosaccharides can be isolated with relative ease, others are more difficult to obtain and special techniques must be devised. The reaction mixtures resulting from the hydrolysis of polysaccharides are composed of oligosaccharides, monosaccharides, and polymeric fragments of high molecular weight that are resistant to hydrolysis. In addition, these mixtures contain the hydrolytic agent, either enzyme or acid, and after neutralization of the latter,

* *References start on p. 129.*

other ions are introduced. Reaction mixtures resulting from chemical synthesis of oligosaccharides contain the reaction products, unreacted starting
materials, products of side reactions, and inorganic salts employed as catalysts.
For the isolation of the desired oligosaccharide, methods must be devised for
removing the impurities from such mixtures.

B. PRECIPITATION AND EXTRACTION

Removal of large unreacted fragments of the starting polymer is often easy
to achieve by precipitation of the fragment with a suitable organic solvent.
Filtration or centrifugation of the reaction mixture yields a solution from
which the oligosaccharide is obtained by fractional precipitation or crystallization methods. Dialysis or gel filtration (Vol. IB, Chap. 28) is a useful
method for separating oligosaccharides from high-molecular weight contaminants such as proteins, enzymes, lipids, and other compounds. Proteins and
enzymes can also be removed by first denaturing the protein or enzyme with
heat or other agents, followed by filtration or centrifugation. The latter procedure is utilized in the preparation of lactose from whey, and sucrose from
sugar cane and sugar beet extracts.

The removal of monosaccharides and structural homologs of an oligosaccharide can sometimes be achieved by the use of organisms or enzymes
which convert the undesired compounds into products having different
solubility and physical properties. Turanose can be obtained from dilute acid
hydrolyzates of melezitose after removal of D-glucose by fermentation.[97]
Homologs of maltotriose and maltotetraose are removed from enzymic
hydrolyzates of amylose by yeast fermentation.[98] Specific enzymes also have
merit in some isolation procedures, as, for example, the use of glucose
oxidase for removing D-glucose from the reaction mixture from which
laminarabiose was isolated.[99] Such procedures are dependent on the availability of an organism or enzyme which selectively alters a contaminant and
thereby facilitates its removal from the reaction mixture.

Inorganic salts in reaction mixtures can be removed by selective precipitation methods whereby either the inorganic salt or the desired oligosaccharide
is preferentially precipitated. Several precipitations may be required, and such
manipulations are time-consuming and can result in a considerable loss of
the oligosaccharide.[100] In recent years ion-exchange resins have become the
method of choice for removal of inorganic ions.[101]

Solvent extraction procedures are used in several ways for the purification
of oligosaccharides. First, lipophilic aglycons that are present in acid or
enzyme hydrolyzates of glycosides can be extracted with an immiscible
organic solvent. The water layer will contain the desired oligosaccharide.
Second, fully substituted oligosaccharides obtained by chemical methods of

synthesis can often be extracted from the reaction mixture with solvents such as chloroform or benzene in which the derivative is soluble but the catalyst and other impurities are insoluble. Details of solvent extraction procedures can be found in literature describing the preparation of individual oligosaccharides.[8,102]

C. CHROMATOGRAPHY

In recent years chromatography has become the principal method for the isolation of oligosaccharides and is responsible to a large extent for the increase in the number of oligosaccharides that are now available. A description of the chromatographic methods for the isolation of oligosaccharides is the subject of a recent review,[103] and the general topic of chromatography is the subject of several monographs[104,105] (see also Vol. IB, Chap. 28). Many modifications of the basic techniques of chromatography have been devised for the isolation of specific oligosaccharides, and some of these are indicated in this section.

From the standpoint of preparative methods, column chromatography is the most useful. Column-chromatographic methods that were used in earlier studies were based on the use of selective adsorbents such as magnesium acid silicates and activated carbons. The former is especially useful for separation of oligosaccharide derivatives,[106] and the latter for the separation of free oligosaccharides.[107] In the original method for chromatography on activated carbon, Celite was added to improve the flow rate, but it has been claimed that cellulose is superior for this purpose.[108] Pretreatment of the charcoal with stearic acid[109] or hydrochloric acid[110] and gradient elution of the oligosaccharides[111,112] are other improvements in the carbon chromatographic method that have been devised.

Column procedures have also been developed with ion-exchange resins, powdered cellulose, and cross-linked dextrans. The resins are particularly useful for the separation of oligosaccharides having ionizable groups,[113–115] but separation of neutral oligosaccharides as borate complexes is possible on resin columns.[112] The cross-linked dextrans are the basis of the gel-filtration procedures.[116,117] Separations by the latter method are based on molecular size of the oligosaccharides. The gel-filtration procedure is especially useful for a preliminary separation into fractions from which the individual oligosaccharides are obtained by other procedures.

The separation of oligosaccharides on columns of powdered cellulose is possible because of the differences in the partition coefficients of the oligosaccharides in the various solvent systems used. Compounds whose partition coefficients are similar will be poorly resolved, whereas compounds having

References start on p. 129.

large differences in partition coefficients will be easily separated by this technique. Partition coefficients also determine the extent of separation of compounds on paper chromatograms. A detailed theoretical discussion on the selection of a solvent for separating oligosaccharides is available.[118] Compilations of R_f values for oligosaccharides in different solvent systems are also available.[119]

In theory, a particular solvent system should, for a specific mixture of compounds, give the same pattern of separation on a cellulose column as on a paper chromatogram. However, it is difficult at times to achieve identical conditions with the two procedures. A successful use of a column procedure for separating oligosaccharides whose R_f values differ only to a small extent has been reported.[120] Partition chromatography on Celite has been proposed and is claimed to be superior to cellulose-column chromatography with respect to speed, ease of packing, and operation of the column.[121] The rate of formation of methyl furanosides with some oligosaccharides is high in comparison to that of the corresponding methyl pyranosides. The differences in chromatographic behavior of these glycosides have also been utilized to effect separation of oligosaccharides.[122]

Isolation of oligosaccharides by paper chromatography has at times been the only available means of obtaining structurally related compounds in pure form. The partition principle has been utilized for thin-layer chromatography of oligosaccharides. Many materials have been recommended as the support media for this type of chromatography.[123] The separation of oligosaccharides by thin-layer chromatography can be effected in much shorter time than by paper or column methods. The method is finding increasing use as a qualitative technique for identification of oligosaccharides and as a preparative procedure.[124]

D. Concentration and Crystallization

Reaction mixtures containing oligosaccharides often contain large volumes of solvent, especially water. Chromatographic procedures that are utilized for resolution of these mixtures lead to still larger volumes of solvent, so that concentration is necessary if the oligosaccharide is to be crystallized or precipitated. Concentration must be effected under conditions mild enough not to damage the oligosaccharide, such as evaporation under vacuum at low temperatures or freeze drying. The latter technique is especially valuable for concentrating solutions of biological substances that are sensitive to elevated temperatures or are easily hydrolyzed in water solutions.

The final purification and isolation of oligosaccharides, like other organic compounds of low molecular weight, is effected usually by crystallization.[125] Although use of the procedures outlined in the preceding sections, individually or in combination, results in a product of fairly high purity, small proportions of impurities may be carried along with the oligosaccharides. In concentrated solution, these impurities may exist in sufficiently high concentration to interfere with crystallization of the desired oligosaccharide. The crystallization of reducing oligosaccharides is further complicated by the presence of anomers which, though closely related, are nevertheless distinct molecular species.

The principle that crystal growth proceeds by addition of molecular particles from the environment implies the existence of an initiating crystal center. Induction of the crystal center of oligosaccharides is at times a serious experimental problem. A unit cell of a crystal comes into existence when a minimum number of molecules falls into proper arrangement. The formation of the initial crystal is dependent on such factors as the size and shape of the molecules, the relative energy levels of the amorphous and crystalline products, the temperature, the solvent, the types of impurities, and probability statistics.

The temperature of attempted crystallization of oligosaccharides is an important factor which can be controlled. The belief that low temperature aids in crystallization is only partially true and is based on the fact that the degree of saturation increases as the temperature is decreased. However, solutions of oligosaccharides are often supersaturated at room temperatures so that crystallization from solution may occur at ambient or elevated temperatures.[126]

Selection of the solvent for crystallization of various oligosaccharides is largely empirical. Solvents that are frequently suitable include low-molecular-weight alcohols, such as methanol, ethanol, and propyl alcohol; ketones, such as acetone and butanone; and miscellaneous solvents, such as acetic acid, mono- and dimethyl ethers of ethylene glycol, pyridine, methyl sulfoxide, and p-dioxane. For acetylated derivatives of oligosaccharides, acetone, chloroform, ether, benzene, and low-molecular-weight alcohols are commonly used. Mixed solvents may produce better crystallizing conditions than a single solvent. In the latter case, the oligosaccharide is dissolved in one solvent and the other solvent, in which the oligosaccharide is poorly soluble, is added to incipient turbidity of the solution. Solvent systems that are suitable for the initial crystallization of a substance are not necessarily those to be preferred for recrystallization.

The following procedure has been recommended for inducing crystallization of carbohydrates.[127] Small amounts of the purest material available are dissolved in several solvents kept at several temperatures (0°, 25°, and 70°) with the hope that solvent systems and conditions have been selected which result in crystallization. In some cases a slow evaporation of the solvent may

* *References start on p. 129.*

be conducive to crystallization. In other cases, the principle of crystal iso-
morphism may also be exploited. A solution of the compound is seeded with
a variety of crystals, with the expectation that one crystal may be an isomorph
of the sugar, so that it acts as a nucleus for crystallization. Many months may
be required for the crystallization of some oligosaccharides. However, on
many occasions it has not been possible to effect a crystallization, and such
oligosaccharides are available only in amorphous form.

V. DETERMINATION OF STRUCTURE

The following information is required for total elucidation of the structure
of an oligosaccharide: (1) molecular weight, (2) identity of the monomeric
residues, (3) sequence of the monomeric residues, (4) the positions of the
monomeric residues involved in the glycosidic bond, (5) the type of ring
structures, and (6) the stereochemical configuration of the glycosidic bond.
Many methods are in use for determining the molecular weights of oligosaccharides. One of the early methods is the quantitative measurement of a
functional group, such as the hemiacetal hydroxyl group [128,129] and esterifiable
hydroxyl groups.[130] Often the reaction for measuring a functional group is
not stoichiometric,[131,132] and consequently calibration curves are required.
If the molecular weights of a few oligomers of a series of oligosaccharides are
known, the molecular weights of other compounds of the series can be calculated from the molecular rotation of the compound by using appropriate
empirical relationships.[131,133] The effect of oligosaccharides in solution on
colligative properties, such as osmotic pressure, freezing point, and vapor
pressure, can be measured, and the molecular weight of the compound can
be calculated. The availability of new instruments, as, for example, osmometers for measuring small differences in vapor pressures of solutions of the
oligosaccharide, have facilitated accurate determination of molecular
weight by use of colligative properties.[134,135]

Identification of monosaccharide residues after acid hydrolysis of an
oligosaccharide has been greatly facilitated with the aid of chromatographic
techniques [104,105] (see also Vol. IB, Chap. 28).

The sequence of monosaccharides in reducing hetero-oligosaccharides
must be determined in order that the residue having an unsubstituted hemiacetal group may be identified. Chemical modification of the hemiacetal
group, followed by hydrolysis and identification of the hydrolytic fragments,
is utilized for this purpose. Chemical reactions commonly employed include
oxidation to the aldonic acid, reduction to the alditol, and the formation of
osazone, flavazole, or similar derivatives. In hydrolyzates of disaccharides
that have been so modified, the nonreducing moiety of the disaccharide and

the modified reducing group can be identified, and the sequence of the monomeric residues in the disaccharide can be thus established. The sequence of
monosaccharide residues in trisaccharides can be deduced from the nature of
the products of complete and partial hydrolysis by acid or enzymes, as
illustrated for planteose in the accompanying reactions:[24]

$$\text{Planteose} \xrightarrow{\text{H}^+} \text{D-Galactose} + \text{D-fructose} + \text{D-glucose}$$

$$\text{Planteose} \xrightarrow{\text{dilute H}^+} \text{Planteobiose} + \text{D-glucose}$$

$$\text{Planteose} \xrightarrow{\text{galactosidase}} \text{D-Galactose} + \text{sucrose}$$

These results establish that the order of the monosaccharide residues is
D-galactose–D-fructose–D-glucose. Since the trisaccharide is nonreducing, the
arrangement of the residues is fully established by the above results. With
reducing trisaccharides the residue containing the reducing unit must be
identified by one of the methods described for the disaccharides. Oligosaccharides of higher molecular weight may present a more difficult problem in
structural analysis. For a tetrasaccharide, the two trisaccharide fragments and
a sufficient number of disaccharide and monosaccharide fragments must be
isolated and identified. From such information the sequence of the monomeric residues in the compound can be deduced.

The classical method for determining the sequence of monosaccharide
residues in oligosaccharides is by use of methylation techniques followed by
identification of the hydrolytic products from the methylated oligosaccharide
(see Vol. IA, Chapters 9 and 12, and under individual sugars in this chapter).
Information from such studies also leads to establishment of the positions of
the glycosidic linkages. The procedure used for establishing the structure of
lactose[136] illustrates the essential principles of methylation techniques.
Methyl hepta-O-methyl-lactoside was prepared by methylation of lactose
with methyl sulfate. From an acid hydrolyzate of the methylated lactose,
2,3,6-tri-O-methyl-D-glucose and 2,3,4,6-tetra-O-methyl-D-galactose were
obtained. Since the disaccharide is reducing, C-1 of either the D-glucose or
the D-galactose moiety must exist as an unsubstituted hemiacetal. The reducing moiety cannot be D-galactose, since the hydroxyl group at C-1 is the
only available group for the glycosidic linkage. D-Galactose, therefore,
constitutes the nonreducing moiety. If the D-galactose residue were linked to
the hydroxyl group on C-4 of D-glucose, the D-glucosyl residue would have
the pyranose ring structure. If the D-galactose residue were linked to the
hydroxyl group at C-5, the D-glucose moiety would have the furanose ring
structure. In either situation, 2,3,6-tri-O-methyl-D-glucose would result from
hydrolysis of the methylated lactose. Oxidation of lactose followed by

* References start on p. 129.

methylation and hydrolysis yielded 2,3,4,6-tetra-O-methyl-D-galactose and 2,3,5,6-tetra-O-methyl-D-gluconic acid. Both moieties of the disaccharide must be pyranoid to account for these results.

With increasing number of residues in the oligosaccharide, the methylation method must be coupled with fragmentation methods to establish the structure for the compound. Sometimes it is possible to assign a definitive structure on the basis of results of fragmentation experiments only. For example, in partial acid hydrolyzates of a trisaccharide, D-glucose, maltose, and isomaltose were identified by paper chromatographic methods. In partial acid hydrolyzates of the aldonic acid or the alditol of the trisaccharide, D-glucose and isomaltose were the only reducing products. The structure for the compound was therefore established as O-α-D-glucopyranosyl-(1 → 6)-O-α-D-glucopyranosyl-(1 → 4)-D-glucose.[137,138]

Other procedures for determining the position of linkages in disaccharides are periodate oxidation, lead tetraacetate oxidation, and the formation of imidazole derivatives. Under standardized conditions the (1 → 2)-linked disaccharides are not oxidized by lead tetraacetate, the (1 → 3)-linked disaccharides require one mole of lead tetraacetate for oxidation, the (1 → 4)-linked disaccharides require two moles, and the (1 → 6)-linked disaccharides require three moles of oxidant.[139] Lactose requires two moles of lead tetraacetate for oxidation, as expected for a compound having a (1 → 4) linkage. Methods based on periodate oxidation are applicable directly with non-reducing oligosaccharides[140] but are not necessarily applicable with reducing oligosaccharides. The latter compounds are first converted into their glycosides, to stabilize the pyranose structure and prevent complete oxidation of the compound. Assignment of the positions of linkages in the compound is based on the number of moles of periodate used and moles of formic acid and formaldehyde produced during oxidation of the glycoside[141] (see Vol. IB, Chap. 23). The rate of formation of imidazole derivatives of disaccharides has been related to the type of linkage in the molecule.[142] However, this method needs further exploration.

Criteria that have been used for establishing the stereochemistry of the glycosidic linkage include the magnitude and sign of the specific rotation of the oligosaccharide and the susceptibility of the oligosaccharide to hydrolysis by specific enzymes. The latter is dependent on the availability of enzymes that are specific for α-D- and β-D-glycosidic linkages.[143–145] Pure enzymes having the proper specificity have, however, not always been available. The specific rotation of an oligosaccharide includes an algebraic contribution from the glycosidic bond. Oligosaccharides having high specific rotations are generally α-D-linked, whereas those having low specific rotations may be β-D-linked.[146] Assignments of configuration of the glycosidic link, by use of Hudson's rules, are more reliable with the disaccharide alditols than with the free sugars.

Specific rotation values of products from periodate and lead tetraacetate oxidation have also been correlated with the configuration of the glycosidic linkage of oligosaccharides (see Vol. IB, Chap. 25).[147-149] More recently, spectral data from infrared studies,[150] and more especially from nuclear magnetic resonance (n.m.r.) studies,[151] have been used for establishing the configuration of glycosidic linkages of some oligosaccharides (see Vol. IB, Chap. 24). With the advent of newer spectrometers of high resolving power, the n.m.r. technique may become one of the most useful methods for the assignment of structure and configuration of the linkage in oligosaccharides.[152] Mass spectrometry of suitable derivatives is also useful.[152a]

Much attention is currently focused on the conformation of the monomeric residues in oligosaccharides.[152,153] Conformational aspects of structure are important in determining the rates of chemical reactions and the ease with which enzyme–substrate complexes can form. The complexing ability, in turn, determines whether an oligosaccharide can function as a substrate or as an inhibitor for a specific enzyme. Information of this type can also indicate which functional groups of the substrate participate in the formation of the enzyme–substrate complex. Conformational aspects of structure of carbohydrates and the significance of such considerations are discussed elsewhere (Vol. IA, Chap. 5).

VI. GENERAL PROPERTIES

A. PHYSICAL PROPERTIES

Properties of oligosaccharides such as specific rotation, melting point, solubility, and chromatographic behavior are of importance in their isolation and characterization. The availability of homologous series of oligosaccharides has permitted detailed studies of certain of these properties, and several linear relationships for such series have been established. A relationship between the degree of polymerization and molecular rotation is shown in Fig. 1 for a series of acetylated cello-oligosaccharides[133] and between degree of polymerization and melting points in Fig. 2 for a series of xylo-oligosaccharides.[154] Such data are of value for estimating the degree of polymerization of new compounds of the series and for aiding in the identification of members of the series prepared by new methods.

A very useful relationship has been established between the degree of polymerization and the mobility of the oligosaccharides on paper chromatograms.[155] Figure 3 gives data for several series of homologous oligosaccharides. The relationship may permit the estimation of molecular weight for an

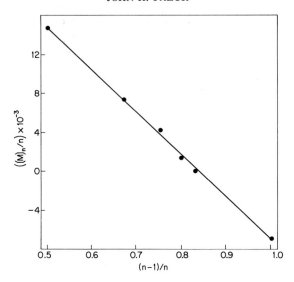

FIG. 1. Relation between molecular rotation (M) and degree of polymerization (n) for the α-peracetates of cello-oligosaccharides. From E. E. Dickey and M. L. Wolfrom, *J. Amer. Chem. Soc.*, **71**, 825 (1949).

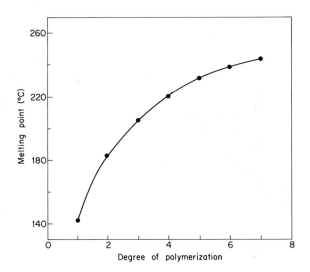

FIG. 2. Relation between melting point and degree of polymerization for xylo-oligosaccharides. From R. L. Whistler and C. C. Tu, *J. Amer. Chem. Soc.*, **75**, 645 (1953).

oligosaccharide and yield information on the structure of a new oligosaccharide. Thus, on the basis of comparison of R_f values, it was suggested that a new D-glucosyl oligosaccharide detected in enzymic hydrolyzates of maltose probably contained the α-D-$(1 \rightarrow 3)$-glucosidic linkage. This suggestion was later verified when the oligosaccharide was isolated in pure form, and its structure established by chemical methods.[156]

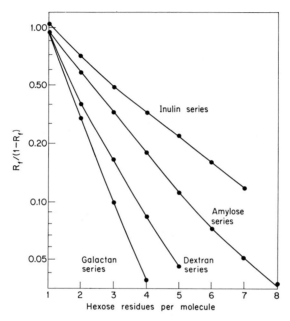

FIG. 3. Relation between chromatographic mobilities and degree of polymerization for several oligosaccharide series; the R_f value is the quotient of the distance the compound has moved to the distance the solvent front has moved. From D. French and G. M. Wild, *J. Amer. Chem. Soc.*, **75**, 2612 (1953).

Oligosaccharides are as a rule very soluble in water and other polar solvents. Addition of polar organic solvents to solutions of oligosaccharides results in precipitation of an amorphous or crystalline form of the oligosaccharide. The solubilities of maltose and D-glucose in various concentrations of methanol and ethanol are illustrated in Fig. 4.[131]

Although the solubility of maltose decreases rapidly with increasing concentration of alcohol, the oligosaccharide is slightly soluble even in absolute ethanol. This solubility behavior of oligosaccharides has been utilized to effect a separation of the oligosaccharide from inorganic salts and is of

* *References start on p. 129.*

particular importance in chromatographic methods based on partition coefficients. For a given pair of immiscible solvents, the differences in the partition coefficients of oligosaccharides between these solvents are responsible
for the different migration rates of the oligosaccharides on chromatograms
developed with that solvent system. The solubility of oligosaccharides in
water-immiscible solvents is extremely low. Oligosaccharides can often be
precipitated quantitatively from water solution by addition of immiscible
solvents such as hydrocarbons, ethers, and high-molecular-weight alcohols.

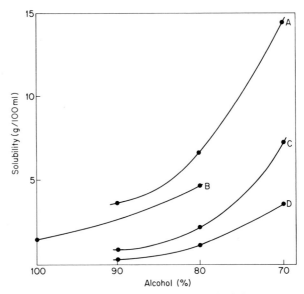

FIG. 4. Solubility of maltose and D-glucose in aqueous alcohol: *A*, maltose in methanol
at 40°; *B*, D-glucose in ethanol at 20°; *C*, maltose in methanol at 25°; *D*, maltose in
ethanol at 25°. From M. Levine, J. F. Foster, and R. M. Hixon, *J. Amer. Chem. Soc.*,
64, 2331 (1942).

Reducing oligosaccharides, like monosaccharides, undergo mutarotation
when dissolved in water. The increase or decrease in rotation is due to a conversion of the anomer constituting the crystalline compound into other
tautomeric forms (see Vol. IA, Chap. 4). Such conversions occur by ring opening and closure in the terminal, reducing residue, and the various intermediates
in this process may contribute to the observed mutarotation.

B. ORGANOLEPTIC PROPERTIES[156a]

Most oligosaccharides are sweet to the taste, a property which makes them
useful in many food and confectionary products. A list of relative sweetness

of oligosaccharides, based on an arbitrary value of 100 for sucrose, is compiled in Table III.[157] Values for some of the common monosaccharides are included for comparative purposes. The acceptability of many foods is markedly improved by the use of sweetening agents. However, in some food products, lack of sweetness is a desirable property, and the moisture-retention property of the compound is the more important consideration.

TABLE III

RELATIVE SWEETNESS OF SOME OLIGOSACCHARIDES
AND MONOSACCHARIDES

Carbohydrate	Relative sweetness[a]
Sucrose	100
Lactose	16
Raffinose	22
Maltose	32
D-Galactose	32
D-Xylose	40
D-Glucose	74
D-Fructose	173

[a] Sucrose is assigned an arbitrary value of 100.

Oligosaccharides in the pure state are odorless. Preparations that possess an odor have undergone decomposition by oxidation or dehydration. Dehydration products have a characteristic caramel-like odor. Precautions must be used to minimize dehydration reactions in preparative procedures that employ high temperatures.

C. SUSCEPTIBILITY TO HYDROLYSIS BY ENZYMES

Oligosaccharides in the diet of animals are a source of readily available energy. However, for this energy to become available, the oligosaccharide must be metabolized in the animal tissues, in which case hydrolysis of the oligosaccharide to its constituent residues will first occur. Enzymes (glycan glycohydrolases, see Vol. IIB, Chap. 47) that hydrolyze different types of oligosaccharides are present in the digestive tracts of animals. The monosaccharides liberated by the action of these enzymes are absorbed and transported to various tissues and organs. In these tissues the energy produced by oxidation of the sugars is "trapped" in the form of adenosine triphosphate by the complex series of metabolic reactions characteristic of living systems. The

* References start on p. 129.

principal enzymes that hydrolyze oligosaccharides in the digestive tract include sucrase, lactase, and maltase. Some of these enzymes have been purified,
and their modes of action have been extensively investigated (see Vol. IIA,
Chap. 33). In general, these enzymes have rather broad substrate specificity.
For example, the maltase isolated from intestinal mucosa is an α-glucosidase
that hydrolyzes many different types of α-D-glucosides.[158] Some of the
intestinal carbohydrases (glycoside glycohydrolases) exist in isoenzyme
forms[159] for which the biological significance is not yet established.
Enzymes that hydrolyze oligosaccharides are universally distributed in
biological materials and are found in high concentrations in many
plant seeds, and in secretions from numerous bacteria and fungi.

D. Susceptibility to Hydrolysis by Acids

The glycosidic linkage of oligosaccharides is readily hydrolyzed by acids
but is relatively stable in alkaline media. The mechanism of the acid-catalyzed
hydrolysis of pyranosides has been the subject of many investigations. A
detailed discussion of hydrolysis of pyranosides is presented in Vol. IA, Chap.
9. The hydrolysis of oligosaccharides occurs, as with other glycosides, with
fission of the bond between C-1 of the glycosyl residue and the glycosidic
oxygen atom.[160] Two possible mechanisms consistent with these and other
observations are shown[161] in formulas 3 to 10. One mechanism involves rapid,
reversible protonation of the glycosidic oxygen atom to yield the protonated
oligosaccharide (3) which undergoes a slow unimolecular decomposition to a
stable monosaccharide (R'OH) and a cyclic carbonium ion (4). The cyclic
ion is presumed to be stabilized by resonance with the oxonium ion (5).
Nucleophilic addition of water to the electron-deficient carbon center yields a
protonated reducing sugar (6), and subsequent loss of a proton yields the
expected hydrolytic products (7). A possible, but not widely accepted,
alternative mechanism would involve protonation of the ring oxygen atom of
the glycosyl moiety to yield a protonated oligosaccharide (8), followed by the
cleavage of the pyranose ring to give an acyclic carbonium ion (9). Nucleophilic addition of water on the acyclic ion would yield an unstable intermediate (10), which would predictably eliminate a monosaccharide residue
(R'OH) to yield an *aldehydo*-monosaccharide. Rearrangement of the latter
into the pyranose structure would yield the final hydrolytic product (7).
Indirect evidence from experiments on the rates of hydrolysis of D-xylose
derivatives[162] supports the first mechanism as being operative in the hydrolysis of these compounds. Conformational effects and intramolecular steric
interactions have also been used to explain the relative rates of hydrolysis of
pyranosides in terms of the first mechanism.[163] It has been postulated that the

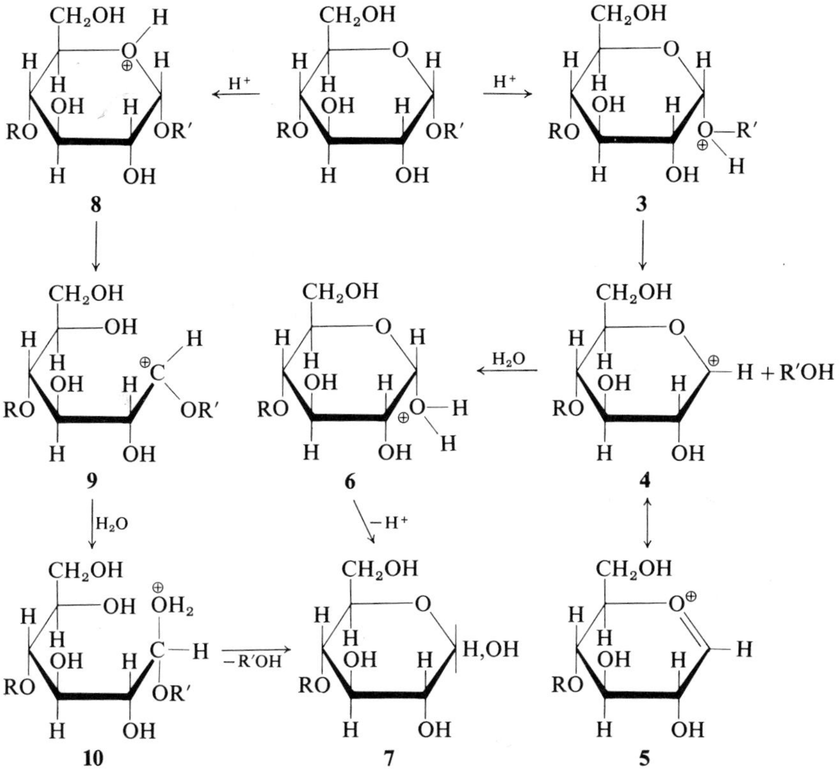

CH$_2$OH   H

H   O   H

H

OH   H

RO   OR'

H   OH

8

←H$^+$

CH$_2$OH

H   O   H

H

OH   H

RO   OR'

H   OH

H$^+$ →

CH$_2$OH

H   O   H

H

OH   H

RO   O—R'

H   OH   H

3

CH$_2$OH

OH

H   H

H   ⊕C

OH   H   OR'

RO   OR'

H   OH

9

CH$_2$OH

H   O   H

H

OH   H

RO   O—H

H   OH   H

6

←H$_2$O

CH$_2$OH

H   O

H

OH   H   ⊕   H + R'OH

RO

H   OH

4

↓ H$_2$O

CH$_2$OH

OH   ⊕OH$_2$

H

H

OH   H   C—H

RO   OR'

H   OH

10

−R'OH →

CH$_2$OH

H   O

H

OH   H   H,OH

RO

H   OH

7

↖ −H$^+$

CH$_2$OH

H   O⊕

H

OH   H   H

RO

H   OH

5

TABLE IV

HYDROLYTIC RATE DATA FOR SOME DISACCHARIDES OF D-GLUCOSE IN 0.1N HYDROCHLORIC ACID

Substance	Linkage	$k \times 10^4$ min^{-1}		Molar activation energy (cal/mole)
		80°	99.5°	
Cellobiose	β-D-(1 → 4)	0.66	6.6	30,800
Gentiobiose	β-D-(1 → 6)	0.58	7.2	33,750
Isomaltose	α-D-(1 → 6)	0.40	5.0	33,800
Kojibiose	α-D-(1 → 2)	1.46	17.3	33,000
Laminarabiose	β-D-(1 → 3)	0.99	9.3	30,000
Maltose	α-D-(1 → 4)	1.55	16.3	31,500
Nigerose	α-D-(1 → 3)	1.78	14.1	27,200
Sophorose	β-D-(1 → 2)	1.17	10.1	28,900

* References start on p. 129.

rate-determining step is the formation of the carbonium ion, which is considered to exist in a half-chair conformation. Formation of the half-chair form involves a small rotation about the C-2 to C-3 and C-4 to C-5 bonds. It is believed that the rate of hydrolysis is dependent primarily on the extent of interaction of equatorial substituents on C-2 relative to C-3 and on C-4 relative to C-5, since these groups must eclipse each other in reaching the transition state for the reaction (see Vol. IA, Chap. 9).

Considerable variation in the rate of hydrolysis of glycosidic bonds in hetero-oligosaccharides can be expected. Relative rates of hydrolysis for some oligosaccharides are presented in Table I. It will be noted that considerable variations do indeed exist in these values. By proper selection of hydrolytic conditions it is possible to effect hydrolysis of one type of interglycosidic linkage while another remains essentially unhydrolyzed. Not only are variations in the rates of hydrolysis of hetero-oligosaccharides observed, but differences are also found between homo-oligosaccharides. Listed in Table IV are the hydrolytic rate constants for the various isomeric disaccharides containing D-glucopyranosyl residues.[164]

The terminal residue of an oligosaccharide can be removed selectively by hydrolysis if the oligosaccharide is first converted into the phenylosotriazole and the hydrolysis is conducted with water-soluble polystyrenesulfonic acid.[164a]

In homologous series of oligosaccharides, such as the malto- or the cellooligosaccharides, not all the glycosidic bonds are hydrolyzed at the same rate (Table V).[49,165] Factors responsible for these differences of rate need yet to be determined.[166]

TABLE V

HYDROLYTIC RATE CONSTANTS OF CELLULOSE AND CELLO-
OLIGOSACCHARIDES IN 51% SULFURIC ACID

| | | $k \times 10^4 \ min^{-1}$ | |
Oligosaccharide	d.p.[a]	18°	30°
Cellobiose	2	1.07	6.9
Cellotriose	3	0.64	4.5
Cellotetraose	4	0.51	3.7
Cellopentaose	5		3.5
Cellohexaose	6		3.2
Cellodextrin	15–20	0.39	
Cellulose	200–250	0.30, 0.39	2.3

[a] d.p. = degree of polymerization.

E. Chemical Reactions

Reducing oligosaccharides contain the same types of functional groups present in monosaccharides—namely, hemiacetals, primary, and secondary hydroxyl groups. Many reactions characteristic of the hemiacetal groups of monosaccharides readily occur with reducing oligosaccharides. Conditions for effecting these reactions are similar to those employed with monosaccharides. Thus, hydrazone, osazone, triazole, and flavazole derivatives of oligosaccharides can be made by conventional methods (Vol. IB, Chap. 21). These derivatives are often useful for identifying the residue of an oligosaccharide containing the reducing group. For example, hydrolysis of lactose 1-phenylflavazole in dilute acid yields D-glucose 1-phenylflavazole and D-galactose, thereby establishing that the D-glucose moiety contains the unsubstituted hemiacetal group.

Other reactions associated with the hemiacetal group include degradation of the reducing moiety, as by oxidation and decarboxylation reactions, and chain-extension reactions as by the cyanohydrin or the nitromethane synthesis. The hemiacetal group of reducing oligosaccharides is reduced to the alditol by sodium borohydride or other reducing agents, and oxidized to the aldonic acid by many of the commonly employed oxidizing agents (Vol. IB, Chap. 22–24). Nonreducing oligosaccharides do not, of course, undergo such oxidation or reduction reactions.

Many procedures used in monosaccharide chemistry, involving high concentrations of acid or alkali, cannot be applied directly with oligosaccharides. In acidic solutions, oligosaccharides tend to undergo hydrolysis, whereas in alkali the reducing oligosaccharides undergo isomerization and fragmentation reactions. The preparation of alkyl or aryl glycosides of oligosaccharides cannot be effected by treating the oligosaccharide and alcohol (or phenol) with an acid catalyst. A several-step reaction sequence, by way of the acetylated glycosyl halide, which adequately protects the glycosidic linkages and other reactive groups of the molecule can be employed, as illustrated by the synthesis of methyl β-maltoside[167] (**11**).

The synthesis of glycosides of oligosaccharides can sometimes be achieved by use of enzymes. In such reactions a readily available glycoside of the monosaccharide is made to serve as the acceptor molecule in an enzymic transfer reaction involving a suitable donor of monosaccharide residues. Thus, methyl α- or β-maltoside can be made from methyl α-D- or β-D-glucopyranoside as the acceptor molecule and a cyclodextrin as the donor of D-glucosyl residues, for *Bacillus macerans* amylase.[168] Dextran sucrase from *Leuconostoc mesenteroides*[169] has also been used for the synthesis of the methyl glycosides of the isomaltose series of oligosaccharides. In some of these reactions,

* *References start on p. 129.*

glycosides of the higher-molecular-weight homologs of the series are also synthesized, such as methyl α-maltotrioside and methyl α-maltotetraoside.

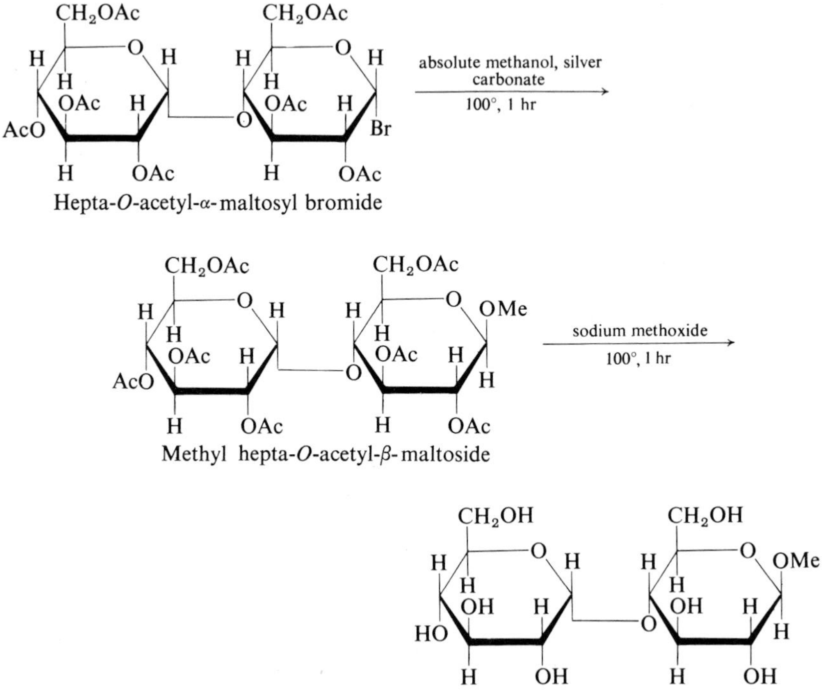

Hepta-O-acetyl-α-maltosyl bromide

absolute methanol, silver carbonate
100°, 1 hr

Methyl hepta-O-acetyl-β-maltoside

sodium methoxide
100°, 1 hr

Methyl β-maltoside

11

The glycosides of oligosaccharides are especially valuable as substrates for studies on the nature of the enzyme–substrate complex, and have been widely used to aid in the elucidation of the mechanism of action of carbohydrases.

In the case of reactions under highly alkaline conditions, methods must be selected to minimize isomerization or fragmentation of the oligosaccharides. Reagents that react at rates considerably higher than the rates of isomerization are sometimes available. For example, methylation of oligosaccharides may be effected by use of methyl sulfate in alkali, since methylation of the reducing group is rapid with this reagent.

The hydroxyl groups in oligosaccharides undergo reactions typical of hydroxyl groups of monosaccharides, and acetylation, benzylation, tritylation, and p-toluenesulfonation are effected by conventional methods of monosaccharide chemistry. Methods minimizing hydrolytic reactions must be selected for oligosaccharides that are particularly acid-labile. For example, sucrose can be acetylated by acetic anhydride in pyridine, but not by procedures involving the use of acidic catalysts. Oligosaccharides undergo oxidation with periodate and similar reagents, to yield formic acid and formaldehyde. Adequate protection of the reactive groups is necessary to oxidize or reduce oligosaccharides selectively at specific positions of the molecule. Oligosaccharides having altered structure at specific carbon atoms are especially useful for studies on the nature of enzyme–substrate complexes in biological reactions.

VII. IMPORTANT OLIGOSACCHARIDES OF BIOLOGICAL ORIGIN

A. DISACCHARIDES

Sucrose

β-D-Fructofuranosyl α-D-Glucopyranoside

Properties. Melting point 160° to 186°, depending on the medium used for purification; $[\alpha]_{20}^{D} = +66.53°$ (c 26, H_2O).[170,171] Nonreducing. Fermentable by yeasts. Diazouracil test is a specific test for the glycosidic linkage of sucrose[172] and is also positive with sucrose-containing oligosaccharides.[24] Another specific test is based on serological identification of a dextran produced by a sucrose-specific dextransucrase.[173] Sucrose octaacetate; m.p. 89°; $[\alpha]_D = +59°$ (c 0.8, $CHCl_3$).[174] Many other derivatives are known.

Occurrence. Sucrose occurs almost universally throughout the plant kingdom in seeds, leaves, fruits, flowers, and roots. It functions as the energy source for the metabolic processes of the various tissues. It also functions as a source of carbon chain for the biosynthesis of the various cellular components. It is synthesized primarily in the leaves from products of photosynthetic reactions and is transported to the various parts of the plant. Honey consists principally of sucrose and its hydrolysis products,

* *References start on p. 129.*

D-glucose and D-fructose (invert sugar). The principal sources for commercial production of sucrose are sugar cane, sugar beets, and the sap of maple trees.

Preparation. Sugar cane (*Saccharum officinarum* L.) is harvested at a stage of maturity when maximal synthesis of sucrose has occurred. To minimize the hydrolysis of sucrose by the invertase in the plant, the initial steps in the isolation of sucrose are performed as rapidly as possible after harvesting. The stalks are passed first through cutting machines to facilitate handling and then through roller crushers to force out the plant fluids containing the sucrose. The juice is collected and made slightly alkaline by addition of lime to prevent hydrolysis of the acid-labile glycosidic linkage of sucrose. This solution is heated, and many of the nonsugar impurities, including invertase, are coagulated and float on top of the solution. After separation of these impurities, the plant extract is drawn into large evaporation pans in which preliminary evaporation to about 50% solids is achieved. The concentrated solution is then transferred to other evaporating pans in which the temperature and the evaporation rates are carefully controlled and in which crystallization of the sucrose occurs. When the desired amount of crystal growth has been obtained, the crystals are separated by centrifugation, washed, and finally dried. These crystals constitute the raw sugar of commerce. Several preparations of crystalline sucrose of decreasing purity can be obtained from the mother liquors by further concentration of the solution. The final mother liquor (blackstrap molasses) is a dark, viscous liquid which is used primarily in animal feeds or in the production of industrial and beverage alcohols. The crystalline sucrose is shipped to refineries, where the sucrose is subjected to further purification. The purification steps include treatment with a variety of adsorbents and decolorizing materials to remove impurities, evaporation to concentrate the solution, nucleation, and crystallization in evaporating pans under controlled temperatures. The resulting crystals are collected by centrifugation, dried in air, and screened to predetermined particle sizes. The final crystalline sucrose meets high standards of chemical purity.

In other than tropical and subtropical countries, the sugar beet (*Beta vulgaris*) is the principal source of sucrose. After harvesting, the beets are washed and cut into slices called "cosettes." The cosettes are delivered from a central spout into a series of diffusion vessels in which the sucrose is extracted with hot water, by utilizing the countercurrent principle. Fresh water passes first into the diffusion vessel having the most extracted cosettes; the resulting solution then goes through the diffusion vessels in order of increasing sucrose content, and finally it passes through the fresh sample of cosettes. The dark diffusion juice, containing about 12% of sucrose, is agitated with lime for several hours. Carbon dioxide is passed into the solution, and the precipitate, which contains most of the impurities, is separated by filtration.

The light-yellow filtrate is decolorized by treatment with sulfur dioxide, and after a final filtration is concentrated in vacuum pans. Crystallization is effected during the evaporation in the same manner as for cane sugar. Evaporation and crystallization are carried out repeatedly with the mother liquors as long as enough sucrose is obtained to make the process economical.

Additional quantities of sucrose may be obtained from the molasses by diluting it to a concentration of about 7% of sugar, cooling to 12°, and adding lime. A sparingly soluble compound of sucrose with three moles of calcium hydroxide, tricalcium saccharate, crystallizes. The tricalcium saccharate, after separation from the final molasses, serves in the place of lime for the purification of sucrose from the diffusion chambers. Some sucrose in the final molasses may be recovered by treatment with barium hydroxide, which forms a sparingly soluble barium saccharate. This precipitate is collected and is decomposed with carbon dioxide. The insoluble barium carbonate is separated from the sucrose and reconverted into barium hydroxide.

Synthesis. The *in vivo* synthesis of sucrose in plants occurs primarily by the uridine 5'-(D-glucosyl pyrophosphate) pathway as discussed[175] in this volume, Chap. 34. The first successful *in vitro* synthesis of sucrose was accomplished by use of a phosphorylase from *Pseudomonas saccharophila*, which catalyzes reversibly the phosphorolysis of sucrose.[176] *In vitro* synthesis has also been achieved by the action of levan sucrase on levan and D-glucose[177] and by the action of fungal invertase on raffinose and D-glucose.[178] The synthesis of sucrose by reversal of the normal hydrolytic action of invertase has not been achieved, however. (For more details on biosynthesis, see this volume, Chap. 34.)

A chemical synthesis of sucrose by condensation of tetra-*O*-acetyl-D-fructofuranose and tetra-*O*-acetyl-D-glucopyranose was claimed[179] in 1928. However, other investigators were not able to synthesize sucrose from the two acetylated monosaccharides.[180,181] It was not until a quarter of a century later that an authentic chemical synthesis was accomplished.[75] 3,4,6-Tri-*O*-acetyl-1,2-anhydro-α-D-glucopyranose and 1,3,4,6-tetra-*O*-acetyl-D-fructofuranose were heated in a sealed tube at 100° for 104 hours. The reaction mixture was acetylated, and the products were then separated by paper- and column-chromatographic methods. Sucrose octaacetate was isolated in an overall yield of 5%. Deacetylation of the product yielded free sucrose, which was crystallized from ethanol.

Structure. Sucrose is hydrolyzed by acids and by enzymes to a mixture of equal amounts of D-fructose and D-glucose. The process is called inversion because the optical rotation changes from dextro to levo on account of the high levorotation of the D-fructose. The mixture formed is called invert sugar. The structure of sucrose was established primarily by methylation

* *References start on p. 129.*

techniques; a detailed account of this work is given in a review.[182] In brief, hydrolysis of octa-O-methylsucrose yields two dextrorotatory tetra-O-methyl-hexoses, one of which was easily identified as 2,3,4,6-tetra-O-methyl-D-glucose. Through a series of oxidation and degradation reactions the other tetramethyl ether was identified as 1,3,4,6-tetra-O-methyl-D-fructofuranose. If the assumption is made that the residues of sucrose have the same ring structures in the methylated hydrolytic products, the glycosidic bond of sucrose is at the anomeric carbons of the two hexose residues. The configuration of the glycosidic linkages of sucrose is α for the D-glucose component and β for the D-fructose component. The hydrolysis of sucrose by yeast α-D-glucosidase and not by the β-D-glucosidase of almond emulsin supports the α-D-glucoside configuration. Similarly, the hydrolysis of the sugar by yeast invertase, an enzyme which hydrolyzes β-D- but not α-D-fructofurano-sides, supplies evidence for the β-D-fructofuranoside configuration. Comparisons of the measured optical rotation with that calculated from isorotation rules agree with the above evidence for an α,β configuration for the glycosidic linkage of sucrose.[182a] Unequivocal proof of the total structure of sucrose, and its conformation in the crystalline state, is provided by crystallographic studies by X-ray and neutron diffraction (see Vol. IB, Chap. 27).

Lactose

4-O-β-D-Galactopyranosyl-D-glucopyranose

Properties. Monohydrate of α-lactose; m.p. 202°; $[\alpha]_D = +52.6°$ (c 8, H_2O). Anhydrous β-lactose; m.p. 252°; $[\alpha]_D = +55.4°$ (c 4, H_2O).[183] Reducing. Not fermentable by ordinary yeasts, but fermentable by yeasts adapted to lactose. α-Lactose octaacetate; m.p. 152°; $[\alpha]_D = +53°$ (c 10, $CHCl_3$). β-Lactose octaacetate; m.p. 90°; $[\alpha]_D = -4.4°$ (c 10, $CHCl_3$).[184]

Occurrence. Lactose is present in the milk of mammals in approximately 5% concentration. It is a constituent of several oligosaccharides that contain neuraminic acids, which also occur in milk and in mammary tissues in small amounts.[185] In the plant kingdom the presence of lactose has been reported in methanolic extracts of the longstyled pollen of forsythia flowers[186] and in the fruit of *Achras sapota.*[187] A report that lactose occurs in digests containing phenyl β-D-galactoside, D-glucose, and enzymes from *Escherichia coli* and other organisms[188] was refuted.[186a]

Preparation. Lactose is prepared from whey, a by-product in the manufacture of cheese. The monohydrate of α-lactose crystallizes when whey is evaporated. This material is collected by filtration and is recrystallized from water to yield lactose of high purity. Different conditions of crystallization may result in other crystalline forms of lactose. For example, at high temperatures (95°) the β anomer is obtained.[188]

Synthesis. The biochemical reactions for the *in vivo* synthesis of lactose in mammary glands have been extensively investigated. The overall reaction sequence has been indicated in the section on biosynthesis of oligosaccharides (this volume, Chap. 34).

The synthesis of a D-galactosyl-D-glucose disaccharide by condensation of tetra-*O*-acetyl-α-D-galactosyl chloride and D-glucose in the presence of sodium ethoxide has been reported.[189] This compound was probably lactose.[190] A synthesis of more importance, because it also provides information on the structure of lactose, involves, as the first step, the condensation of tetra-*O*-acetyl-α-D-galactopyranosyl bromide and 2,3-*O*-isopropylidene-1,6-anhydro-β-D-mannopyranose in an organic solvent and in the presence of silver oxide.[67] After removal of the *O*-isopropylidene group by mild acid hydrolysis, the product was acetylated, and the anhydro ring was opened by the action of sulfuric acid in acetic acid–acetic anhydride. The resulting substance, the octaacetate of a disaccharide epimeric with lactose, is called epilactose. This disaccharide derivative was then converted into hexa-*O*-acetyl-lactose by means of the glycal synthesis, with oxidation of the lactal with peroxybenzoic acid. In all probability, the initial condensation produces a β-D-galactosidic linkage, since tetra-*O*-acetyl-D-galactosyl bromide normally condenses with alcohols with the formation of β-D-galactosides.

Structure. Acidic hydrolysis, or enzymic hydrolysis with β-D-galactosidase, yields one mole of D-galactose and one mole of D-glucose. If the disaccharide is first oxidized with bromine to lactobionic acid and then hydrolyzed, D-gluconic acid and D-galactose are the products obtained. This evidence established lactose as being a D-galactosyl-D-glucose. The determination of the structure of lactose by methylation techniques has been discussed. On the basis of the results of methylation studies, the glycosidic linkage was established to be from C-1 of the D-galactose moiety to C-4 of the D-glucose moiety. The principal evidence for a β-D configuration of the glycosidic linkage rests on the known specificity of the galactosidases of almond emulsin. The structure of lactose is, therefore, 4-*O*-β-D-galactopyranosyl-D-glucose. Additional support for this structure is given through the synthesis of lactose by the chemical methods outlined above.

* *References start on p. 129.*

α,α-Trehalose

α-D-Glucopyranosyl α-D-Glucopyranoside

Properties. Dihydrate of α,α-trehalose; m.p. 97°; $[\alpha]_D = +178°$ (c 2, H_2O).[191] Nonreducing. Fermentable by most yeasts. Trehalose octaacetate; m.p. 98°; $[\alpha]_D = +162°$ (c 10, $CHCl_3$).[192]

Occurrence. The widespread occurrence of trehalose in the lower orders of the plant kingdom has been known for many years. The sugar is particularly common in fungi where the spores and fruiting bodies of some fungi contain over 7% of trehalose on a dry-weight basis.[193,194] The disaccharide is found in young mushrooms, but, as the plants develop, the trehalose is replaced largely by D-mannitol.[195] Yeasts are a rich source of trehalose, although the concentration is dependent on the age of the cells.[196] Many species of lichens and algae have been found to contain small amounts of trehalose.[197] The resurrection plant (*Selaginella lepidophylla*) appears to be the only higher plant reported to contain free trehalose. Trehala manna, which is now considered to be of insect origin, is another good source of trehalose.[198]

In the animal kingdom, the role of trehalose as an energy source in insects has only recently become apparent. Thus, trehalose has been found to be the major circulatory sugar in the fluids of insects, where it appears to function much like D-glucose in the mammalian circulatory system.[199] It is also an important carbohydrate reserve in the eggs, larvae, and pupae of insects.[200] In some cases, trehalose contributes as much as 10% of the dry weight of these structures.

Preparation. The disaccharide is extracted from trehala manna with hot, 75% alcohol.[198] After concentration, the extracts are purified by treatment with basic lead acetate, and the excess lead is removed with hydrogen sulfide. The filtered solution, after concentration, deposits crystals of trehalose. Essentially the same process is employed in obtaining the sugar from coarsely ground *Selaginella*, but the extraction may be carried out with water rather than alcohol. Yeast may be used as the raw material for obtaining the disaccharide.[196]

Synthesis. The enzymic synthesis of trehalose *in vitro* can be effected by the reaction[201]

Uridine 5'-(D-glucopyranosyl pyrophosphate) + D-glucose 6-phosphate ⟶
Trehalose phosphate + uridine pyrophosphate

It is likely that this route is operative in the biosynthesis of trehalose in the intact cells. (For further details, see this volume, Chap. 34.)

The chemical synthesis of trehalose has been accomplished[202] by heating a mixture of 2,3,4,6-tetra-*O*-acetyl-α,β-D-glucose and 3,4,6-tri-*O*-acetyl-1,2-anhydro-D-glucose at 100°. Chromatographic separation of the deacetylated products gives α,α-trehalose, as well as α,β-trehalose. Earlier attempts at the synthesis had given the α,β isomer.[203]

Maltose

4-*O*-α-D-Glucopyranosyl-D-glucopyranose

Properties. Monohydrate of the β-maltose; m.p. 103°; $[\alpha]_D = +130°$, (c 4, H_2O).[102,204] Reducing. Fermentable by yeasts in the presence of D-glucose. α-Maltose octaacetate; m.p. 125°; $[\alpha]_D = +122°$, (c 5, $CHCl_3$). β-Maltose octaacetate; m.p. 159°; $[\alpha]_D = +63°$, (c 10, $CHCl_3$).[205] Many other derivatives have been prepared and their constants recorded.[206]

Occurrence. Maltose occurs as a structural component of two important D-glucans—starch and glycogen. Maltose occasionally has been recorded as present in the free state in plant tissues. However, since it is a product of the enzymic hydrolysis of starch and since starch and a hydrolytic enzyme are found in the same plants, it may be a secondary product formed during the extraction process.

Preparation. Soluble starch, made from commercial starch by a mild treatment with acid, is hydrolyzed by the enzymes of barley flour to a mixture of maltose and other D-glucosyl oligosaccharides. The maltose is precipitated with ethanol, and the crude maltose is then crystallized from aqueous ethanol. Commercial maltose contains considerable quantities of other reducing

* *References start on p. 129.*

oligosaccharides, which can be removed by fractional crystallization of the β-maltose octaacetate.[207] By use of a single enzyme such as sweet potato β-amylase [(1 → 4)-α-D-glucan maltohydrolase] rather than a mixture of enzymes, maltose free from other oligosaccharides can be obtained.[208]

Synthesis. The *in vivo* synthesis of glycogen and starch, and in turn the indirect synthesis of maltose, is discussed by Hassid (this volume, Chap. 34) and by Greenwood (Vol. IIB, Chap. 38). Direct biosynthesis of maltose can be achieved with enzymes such as maltose phosphorylase,[209] *Bacillus macerans* amylase,[210] and amylo-maltase,[211] which catalyze the following reactions, respectively. (For the systematic names for these enzymes, see Vol. IIB, Chap. 47.)

D-Glucosyl phosphate + D-glucose ⟶ Maltose + phosphate
Cyclohexaamylose + D-glucose ⟶ Maltose + malto-oligosaccharides
(1 → 4)-α-D-Glucan + D-glucose ⟶ Maltose + malto-oligosaccharides

The latter two reactions result in the formation of a homologous series of malto-oligosaccharides, including maltose. Accordingly, these reactions can also be used for the preparation of other oligosaccharides of the series. The enzymic reactions are of particular value for the synthesis of maltose labeled with ^{14}C at specific positions of the molecule.[85,212] Maltose can be detected in reaction mixtures in which D-glucose has been incubated with oligosaccharide hydrolases for a long period.[213] However, such reactions are of little importance for the preparation of maltose.

Structure. Methylation of maltose leads to a methyl hepta-*O*-methyl-maltoside, which by acid hydrolysis is converted into 2,3,4,6-tetra-*O*-methyl-D-glucose and a tri-*O*-methyl-D-glucose.[214] The tri-*O*-methyl-D-glucose does not form an osazone and on methylation gives the well-known methyl 2,3,4,6-tetra-*O*-methyl-D-glucopyranoside. Of the many possible tri-*O*-methyl-D-glucoses, only three conform to these specifications. These are the 2,3,4-, 2,3,6-, and 2,4,6-trimethyl ethers. The synthetic 2,3,4-tri-*O*-methyl-D-glucose differs from the product of the hydrolysis of the methylated maltose. Since the tri-*O*-methyl-D-glucose from the methylated maltose is oxidized by nitric acid to di-*O*-methyl-L-threaric acid, it must be the 2,3,6-trimethyl ether. Oxidation of authentic 2,3,6-tri-*O*-methyl-β-D-glucose with nitric acid yields di-*O*-methyl-L-threaric acid, but 2,4,6-tri-*O*-methyl-D-glucose does not.

The position of the linkage in maltose is shown by oxidation of maltose to maltobionic acid with bromine, followed by methylation and hydrolysis. This sequence yields 2,3,5,6-tetra-*O*-methyl-D-gluconic acid, in addition to 2,3,4,6-tetra-*O*-methyl-D-glucose.[215] The unsubstituted hydroxyl group at C-4 of the D-gluconic acid derivative represents the position of the glycosidic linkage in the original disaccharide.

The above evidence proves that maltose consists of two D-glucose residues connected between C-1 and C-4 by an oxygen bridge, but the configuration of the glucosidic linkage is not established by these experiments.[216] Another important disaccharide, cellobiose, gives exactly the same final products as those outlined above for maltose. The best proof of the configurations of the glucosidic carbon of these two disaccharides is obtained by use of enzymes (glycoside glycohydrolases; see Vol. IIB, Chap. 47). Maltose is hydrolyzed by the same yeast enzyme (α-D-glucosidase) as that which hydrolyzes methyl α-D-glucopyranoside. The β-D-glucosidase of almond emulsin causes no significant cleavage of maltose. Cellobiose, however, is hydrolyzed by the same enzyme (β-D-glucosidase) as that acting on β-D-glucosides. From this evidence, maltose is assigned the formula 4-O-α-D-glucopyranosyl-D-glucose, and cellobiose the formula 4-O-β-D-glucopyranosyl-D-glucose. These formulas are confirmed by the high dextro rotation of maltose and the low rotation of cellobiose, and by synthetic studies.

Cellobiose

4-O-β-D-Glucopyranosyl-D-glucopyranose

Properties. β-Cellobiose; m.p. 225°; $[\alpha]_D = +35°$, (c 8, H_2O).[102,217] Reducing. Not fermentable by yeasts. α-Cellobiose octaacetate; m.p. 229°; $[\alpha]_D = +41°$ (c 5, $CHCl_3$). β-Cellobiose octaacetate; m.p. 202°; $[\alpha]_D = -15°$ (c 5, $CHCl_3$).[205]

Occurrence. The compound is not known to exist in the free state in products of biological origin but is the basic repeating unit of cellulose, which is a skeletal element of virtually all plants. It occurs in repeating units of lichenan, laminaran, and other related polysaccharides. The disaccharide also occurs as the carbohydrate constituent of many plant glycosides.

Preparation. Cellulose in the form of cotton or filter paper is simultaneously acetylated and acetolyzed by the action of acetic anhydride and sulfuric acid at low temperatures.[218] Cellobiose octaacetate crystallizes from the reaction mixture and after separation is recrystallized from alcohol. The acetyl groups are removed with barium methoxide in methanol solution, and the cellobiose is crystallized from ethanol. Crystalline cellobiose has been obtained from hydrolyzates of cellulose obtained with a cell-free enzyme preparation from *Aspergillus niger.*[219]

* *References start on p. 129.*

Synthesis. The biosynthesis of polymers containing cellobiose units proceeds via nucleotide pathways[220] which are discussed in this volume, Chap. 34. Although cellobiose has been detected in products of enzymic reversion with β-glucosidases,[213] this route of biosynthesis is not of major importance. Chemical synthesis of cellobiose was achieved by condensation of 1,6-anhydro-β-D-glucopyranose and tetra-*O*-acetyl-D-glucopyranosyl bromide, with subsequent hydrolysis of the anhydro ring by sulfuric acid.[221] The stereospecificity of the method provides evidence for the β configuration of the D-glucosidic linkage. The exact position of the glucosidic linkage is not defined, however, as three unsubstituted hydroxyl groups are present in the 1,6-anhydro-β-D-glucose. A structurally definitive synthesis, by the condensation of 1,6-anhydro-2,3-*O*-isopropylidene-β-D-mannopyranose (which contains only one free hydroxyl group, at C-4) and tetra-*O*-acetyl-D-glucopyranosyl bromide, has been described.[68] After rupture of the anhydro ring, a derivative of the 2-epimer of cellobiose is obtained, which is converted into cellobiose through a glycal intermediate. The method is similar to that used for the synthesis of lactose.

Structure. In the proof of structure for cellobiose, the principles outlined for proof of structure for maltose were utilized. The position of the glycosidic bond was established by methylation studies, and the β-D configuration of the linkage was established by the susceptibility of cellobiose to hydrolysis by β-glucosidases.

Isomaltose

6-*O*-α-D-Glucopyranosyl-D-glucopyranose

Properties. Amorphous; $[\alpha]_D = +122°$ (*c* 2, H_2O).[38] Reducing. Not fermentable by yeasts. β-Isomaltose octaacetate; m.p. 144°; $[\alpha]_D = +97°$ (*c* 2.7, $CHCl_3$).[222,223] Isomaltose 1-phenylflavazole; m.p. 234° (decomp.).[224]

Occurrence. This disaccharide occurs as a structural component in the polysaccharides amylopectin and glycogen and represents the branch point of these compounds. It is the major repeating unit of many bacterial polysaccharides of the dextran type. The occurrence of isomaltose in the free state in biological materials has not been established. Some reports have appeared on the presence of isomaltose in honey and in snake, where the compound is probably a product of an enzymic reaction.

Preparation. Isomaltose has been obtained as the crystalline octaacetate derivative in 5% yields from partial acid hydrolyzates of bacterial dextran,[222] from enzymic hydrolyzates of amylopectin,[225] and from the acid reversion mixture of D-glucose.[226] It has been prepared in high yields from enzymic hydrolyzates of the dextran from *Leuconostoc mesenteroides.*[10] The compound can also be prepared from maltose by fungal glucosyl transferases.[224]

Synthesis. The first chemical synthesis that probably led to isomaltose as one of the end products was carried out by allowing D-glucose to revert in the presence of acids.[227] Recently, definitive methods of chemical synthesis of isomaltose have been achieved by two means: first by condensation of 1,2,3,4-tetra-*O*-acetyl-D-glucose and 3,4,6-tri-*O*-acetyl-2-*O*-nitro-β-D-glucopyranosyl chloride,[63] and the other by catalytic rearrangement of gentiobiose octaacetate to isomaltose octaacetate.[95] The free oligosaccharide is liberated from the derivatives obtained in the above reactions.

Gentiobiose

6-*O*-β-D-Glucopyranosyl-D-glucopyranose

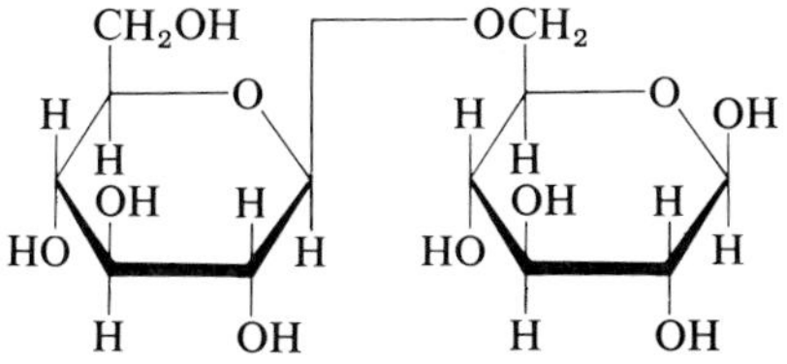

Properties. α-Gentiobiose crystallizes with two moles of methanol; m.p. 86°; $[\alpha]_D = +8.7°$ (*c* 5, H_2O).[102] β-Gentiobiose (solvent-free); m.p. 190°, $[\alpha]_D = +10$ (*c* 4, H_2O).[126] Reducing. Not fermentable by yeasts. α-Gentiobiose octaacetate; m.p. 189°; $[\alpha]_D = +52°$ (*c* 20, $CHCl_3$).[228] β-Gentiobiose octaacetate; m.p. 193°; $[\alpha]_D = -5°$ (*c* 1, $CHCl_3$).[229]

Occurrence. The disaccharide is the carbohydrate constituent of a large number of glycosides, the most important of which are amygdalin and crocin. The two D-glucose residues of the trisaccharide gentianose, which is present in the roots of plants of the *Gentian* species, constitute a gentiobiose moiety.[230,231] Gentiobiose is found in mixtures obtained by the action of acids[78] and enzymes[232] on D-glucose, as a result of reversion-type reactions. The disaccharide is a structural unit of a number of polysaccharides, among which are the yeast β-D-glucans,[233] the seaweed pustulans,[234] and the luteic acids of *Penicillium.*[235] It occurs in the products of enzymic disproportionation of cellobiose.[235,236] Free gentiobiose is rare in Nature, although a report on its presence in xylem sap has appeared.[237]

* *References start on p. 129.*

Preparation. Gentiobiose is obtained from partial acid or enzymic hydrolysis of gentianose after removal of the fructose by fermentation with yeast. It is usually separated as the octaacetate.[228] The hydrogenation of hepta-*O*-acetyl-amygdalin[238] and acetylation of the mother liquors from the preparation of D-glucose from starch[226] are two other procedures that have been recommended. Isolation of the disaccharide from enzymic digests of D-glucose with almond β-glucosidase has also been described.[239]

Synthesis. Chemical synthesis from 1,2,3,4-tetra-*O*-acetyl-β-D-glucopyranose and tetra-*O*-acetyl-α-D-glucopyranosyl bromide, yielding β-gentiobiose octaacetate, is readily achieved and is a good method for the preparation of gentiobiose.[54]

Laminarabiose

3-*O*-β-D-Glucopyranosyl-D-glucopyranose

Properties. Melting point 205°; $[\alpha]_D = +18°$ (*c* 2, H_2O).[240] Reducing. Nonfermentable by yeasts. β-Laminarabiose octaacetate; m.p. 161°; $[\alpha]_D = -29°$ (*c* 2.5, $CHCl_3$).[241]

Occurrence. Laminarabiose in the free state is rare in Nature, but it has been reported that a compound having the properties of laminarabiose was detected in pine needles[242] and in seeds of *Cycas revoluta*.[243] β-Laminarabiosyl-azoxymethane, which occurs in the latter seeds, represents the only known occurrence of laminarabiose as a glycoside. In Nature the disaccharide is a structural unit of many types of polysaccharides, some of which, like laminaran[240] and pachyman,[244] contain laminarabiose as the major repeating unit. In others, like the yeast glucans[233] and seaweed lichenans,[245] the disaccharide alternates with gentiobiose as the repeating unit. Laminarabiose has been detected as a product of enzymic disproportionation of cellobiose[235] and of reversion reactions of D-glucose.[78]

Preparation. Laminarabiose was first isolated as an amorphous powder from a partial acid hydrolyzate of laminaran.[240] A fungal polysaccharide, pachyman, which contains exclusively the β-D-(1 → 3) linkages, is a better source of the disaccharide.[244] Techniques utilizing partial acid fragmentation of the latter polysaccharide, and isolation of the disaccharide by carbon–Celite chromatographic methods, are recommended. Enzymes have been used also to hydrolyze these polysaccharides, resulting in digests from which the disaccharide may be isolated.[246]

Synthesis. Little is known about the pathways for the biosynthesis of the polysaccharides having the laminarabiose unit in their structure. A role for nucleotide sugars such as uridine 5'-(α-D-glucosyl pyrophosphate) has been suggested, however.[247] Chemical synthesis of laminarabiose by the Koenigs–Knorr reaction has been achieved from 2,3,4,6-tetra-*O*-acetyl-α-D-gluco-pyranosyl bromide and 1,2:5,6-di-*O*-isopropylidene-α-D-glucofuranose.[241]

Chitobiose

2-Amino-2-deoxy-4-*O*-(2-amino-2-deoxy-β-D-glucopyranosyl)-D-glucopyranose

Properties. *N,N'*-Diacetyl-chitobiose; m.p. 247°; $[\alpha]_D = +18°$ (*c* 1, H_2O).[248] Octaacetyl derivative; m.p. 296°; $[\alpha]_D = +55°$ (*c* 1, CH_3COOH).[249]

Occurrence. *N,N'*-Diacetyl-chitobiose is the repeating unit of the poly-saccharide chitin, which is widely distributed in both the animal and plant kingdoms and may well be one of the most widely distributed biological compounds in Nature. In the animal kingdom it occurs in invertebrates, particularly shellfish and insects. In the plant kingdom it occurs in the cell walls of many molds and fungi.

Preparation. Chitobiose was first obtained as the octaacetate from chitin by acetolysis procedures[250] or by hydrolysis of chitin in concentrated acids followed by acetylation.[251] *N,N'*-Diacetyl-chitobiose was isolated by chro-matographic methods from saponification mixtures of fully acetylated chitobiose.[248] Free chitobiose may be isolated from a partial acid hydrolyzate of chitosan (*N*-deacetylated chitin), by ion-exchange chromatographic methods.[115] The free oligosaccharide has not yet been obtained in crystalline form. Although enzymes are present in molds and invertebrates which hydro-lyze chitin to oligosaccharides,[252] the latter have not been used for the prepara-tion of the oligosaccharides.

* *References start on p. 129.*

Xylobiose

4-O-β-D-Xylopyranosyl-D-xylopyranose

Properties. Melting point 186°; $[\alpha]_D = -25°$ (c 2, H_2O).[48] Reducing. Nonfermentable by yeasts. β-Xylobiose hexaacetate; m.p. 155°; $[\alpha]_D = -75°$ (c 0.9, $CHCl_3$).[253] Several other derivatives are known.

Occurrence. The disaccharide is the major repeating unit of many plant polysaccharides, particularly D-xylans, L-arabino-D-xylans, and D-glucurono-D-xylans. These polysaccharides are important structural elements in mature cell walls of grasses, cereal grains, weeds, trees, and virtually all other higher plants.

Preparation. Xylobiose has been isolated from acid hydrolyzates of xylans from corn cobs,[48] from aspen wood,[254] and from wheat straw.[33] It can also be isolated from enzymic digests obtained by the action of various endo-xylanases on xylans.[255]

Synthesis. Many studies have been reported on the biosynthesis of xylan in plant materials. On the basis of labeling patterns of the D-xylose from xylan, nucleotide pathways have been implicated in the biosynthesis of xylan.[256] More recently the guanosine 5'-(glycosyl pyrophosphate) pathway has been suggested as the possible route of synthesis of polymers of the xylan and cellulose types (see this volume, Chap. 34).[220] Procedures for the chemical synthesis of xylobiose have been described, by condensation of 2,3,4-tri-O-acetyl-α-D-xylopyranosyl bromide and benzyl 2,3-anhydro-β-D-ribopyranoside. Treatment of the resulting disaccharide epoxide with sodium hydroxide led to deacetylation and opening of the epoxide ring to give benzyl 4-O-β-D-xylopyranosyl-β-D-xylopyranoside which on reduction yielded the corresponding free disaccharide.[257]

3-O-β-L-Arabinofuranosyl-L-arabinofuranose

Properties. Amorphous; $[\alpha]_D = +94°$ (c 0.5, H_2O).[258,259] Phenylosazone; m.p. 200°.[258]

Occurrence. The disaccharide is a structural unit in many heteropoly-saccharides from plants. The glycosidic linkage between the arabinose residues is readily hydrolyzed in dilute acids. The lability of the bond may be of special importance to the plant.

Preparation. The compound is isolated from partial acid hydrolyzates of arabinans from sugar beets[258] and *Acacia pycnantha.*[259]

3-*O*-β-L-Arabinopyranosyl-L-arabinopyranose

Properties. Amorphous; $[\alpha]_D = +220°$, (*c* 3, H_2O). Phenylosazone; m.p. 235°.[260] Degraded to 2-*O*-β-L-arabinopyranosyl-glycerol; m.p. 155°; $[\alpha]_D = +204°$ (*c* 1, H_2O).[261] The pentabenzoate of the latter; m.p. 50°; $[\alpha]_D = +164°$ (*c* 0.8, 2,4-lutidine).[261]

Occurrence. The compound is the major repeating unit of L-arabinans and related heteropolymers in plants. It is especially abundant in the polymers from various trees and shrubs.

Preparation. Partial acid hydrolyzates of gums from various fruit trees[262,263] and from arabinogalactans from larch[260] contain the disaccharide. Chromatographic methods of isolation were employed to obtain the compound. The compound is also present in acid-reversion mixtures of L-arabinose.[264]

Inulobiose

1-*O*-β-D-Fructofuranosyl-D-fructofuranose

Properties. Amorphous; $[\alpha]_D = -32°$ (*c* 2, H_2O)[43] and $-72°$ (*c* 2.7, H_2O).[265] β-Inulobiose octaacetate; $[\alpha]_D = -6°$ (*c* 1.5, $CHCl_3$)[43] and $-14°$ (*c* 1.6, $CHCl_3$).[265]

* *References start on p. 129.*

Occurrence. Inulobiose has been reported to occur free in the Jerusalem artichoke.[266] The disaccharide may well be present in other plant species in which fructans are a major carbohydrate reserve. Fructans having inulobiose as a repeating structural unit include inulin, graminan, and irisan, from artichoke, cereal grains, and iris, respectively.

Preparation. Inulobiose was isolated initially by a paper-chromatographic procedure from an acid hydrolyzate of inulin.[43] The compound is present in digests of sucrose with the enzyme levansucrase [(1 → 2)-β-D-fructan: D-glucose 6-fructosyltransferase], particularly in the presence of D-fructose.[267,268] The isolation of the oligosaccharides from such mixtures is difficult to achieve but is possible by use of chromatographic methods.

Mannobiose

4-*O*-β-D-Mannopyranosyl-D-mannopyranose

Properties. Melting point 194°; $[\alpha]_D = -2°$ (*c* 0.9, H_2O).[269] Mannobiose phenylhydrazone; m.p. 199°.[270]

Occurrence. Mannobiose is the primary repeating unit in the polymeric chain of plant mannans, galactomannans, and glucomannans. The polysaccharides are widely distributed in various species of trees and grasses, particularly those in the tropical and subtropical regions.

Preparations. The disaccharide was first isolated from an enzymic hydrolyzate of the polysaccharide guaran by chromatographic methods and was crystallized from methanol.[269] Mannobiose has also been isolated from partial acid hydrolyzates of other plant polysaccharides, including ivory-nut mannan[271] and lucerne (alfalfa) galactomannan.[272]

3-*O*-β-D-Galactopyranosyl-D-galactopyranose

Properties. Monohydrate; m.p. 160°; $[\alpha]_D = +62°$ (*c* 0.5, H_2O).[273] Crystallized from methanol; m.p. 203°.[274] The compound has been degraded by oxidation with lead tetraacetate to 2-*O*-β-D-galactopyranosyl-glycerol. The hexabenzoate of the latter has been prepared; m.p. 56°–59°; $[\alpha]_D = +36°$ (*c* 0.7, 2,4-lutidine).[261]

Occurrence. The disaccharide occurs as the structural unit of many plant polysaccharides. Thus it is the major structural unit of agars from various seaweeds and an important structural unit of gum arabic and other plant gums. Galactans from snail eggs contain this disaccharide in their structure.[275] The compound appears in reaction mixtures of *p*-nitrophenyl β-D-galactoside and D-galactose incubated with yeast β-galactosidase[120] and of D-galactose incubated with almond emulsin β-galactosidase.[274]

Preparation. A number of plant polysaccharides have been used for the preparation of the disaccharide. The polysaccharides are subjected to hydrolysis with concentrated acid, and the oligosaccharide is isolated by chromatographic methods. Polysaccharides from many plant species have been used.[276,277] Chemically, the compound has been prepared by condensing 2,3,4,6-tetra-*O*-acetyl-α-D-galactopyranosyl bromide with 4,6-*O*-ethylidene-1,2-*O*-isopropylidene-α-D-galactopyranose in the presence of silver oxide as catalyst.[278]

2-Amino-2-deoxy-4-*O*-(α-D-glucopyranosyl)-D-glucopyranose

Properties. Hydrochloride; m.p. 183° to 187° (decomp.); $[\alpha]_D = +81°$ (*c*. 0.5 H_2O).[279]

Occurrence. The *N*-sulfated, partially *O*-sulfated derivative of the D-glucuronosyl analog of this disaccharide,[280] occurs as a major repeating structural unit of heparin, a polysaccharide distributed widely in mammalian tissues.

Preparation. The disaccharide can be isolated by hydrolysis of *O*,*N*-desulfated, carboxyl-reduced heparin,[280a] and it has been synthesized from maltose.[279]

* *References start on p. 129.*

Hyalobiouronic Acid

2-Amino-2-deoxy-3-O-(β-D-glucopyranuronosyl)-D-glucopyranose

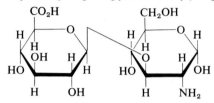

Properties. Melting point 190° (decomp.); $[\alpha]_D = +30°$ (c 1, 0.1N HCl).[281] Methyl α-glycoside of the compound; m.p. 210°; $[\alpha]_D = +31°$ (c 0.7, CH_3OH).[282,283]

Occurrence. The N-acetyl derivative of hyalobiouronic acid is a structural unit in the polysaccharide, hyaluronic acid. The latter compound is widely distributed in mammalian tissues in which it has a multifold function.

Preparation. The hyalobiouronic acid is liberated by combined enzymic and acid hydrolysis of hyaluronic acid.[283] A more recent method is a two-stage enzymic procedure involving the use of hyaluronidase and β-glucosid-uronase.[284] Chemical synthesis of the free compound has been achieved from methyl 2,3,4-tri-O-acetyl-1-bromo-1-deoxy-α-D-glucopyranuronate and ben-zyl 2-(benzyloxycarbonyl)amino-2-deoxy-4,6-O-ethylidene-α-D-glucopyrano-side by the Koenigs–Knorr reaction followed by removal of the blocking groups by catalytic hydrogenation and acid hydrolysis.[281] The fully acetylated methyl ester of the methyl glycoside of hyalobiouronic acid was obtained by degradative methanolysis of hyaluronic acid followed by acetylation with acetic anhydride.[282]

Chondrosine

2-Amino-2-deoxy-3-O-(β-D-glucopyranuronosyl)-D-galactopyranose

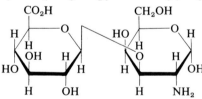

Properties. Amorphous; $[\alpha]_D = +42°$ (c 2, H_2O). Methyl ester hydro-chloride; m.p. 159°; $[\alpha]_D = +42°$ (c 2, H_2O).[285] Methyl ester of hepta-O-acetyl-N-acetyl-chondrosinitol; m.p. 123°; $[\alpha]_D = -23°$ (c 1.8, C_2H_5OH).[286,287]

Occurrence. The N-acetyl derivative, O-sulfated at C-4 or C-6 of the hexosamine moiety, is the major repeating unit of the chondroitin 4- and 6-sulfates.

Preparation. Chondrosine has been isolated as the ethyl ester hydrochloride from partial acid hydrolysis of chondroitin sulfate.[288] Chondrosine hydrochloride is also readily obtained from acid hydrolyzates of the polysaccharide.[286,287] The *N*-acetyl derivative and a sulfate ester of chondrosine have been isolated by enzymic hydrolysis, employing hyaluronidase and microbial enzymes.[289] Chemical synthesis from appropriately blocked monosaccharide residues has also been effected.[281]

6-*O*-β-D-Glucopyranuronosyl-D-galactopyranose

Properties. Melting point 116°; $[\alpha]_D = -8.3°$ (*c* 5, H_2O).[290] Methyl ester heptaacetate; m.p. 203°; $[\alpha]_D = -17.5°$ (*c* 3, $CHCl_3$).[60]

Occurrence. This disaccharide is probably the most common aldobiouronic acid present in Nature. It occurs primarily as a structural unit of plant gums and mucilages. It also occurs as a structural component of polysaccharides in wood and woody plants. The 4-*O*-methyl derivative of the disaccharide is also a constituent of many of these polysaccharides.

Preparation. The disaccharide was isolated originally from partial acid hydrolyzates of gum arabic.[290] However, many other plant polysaccharides may be fragmented into oligosaccharide mixtures, from which this disaccharide may be isolated by ion-exchange chromatography. The chemical synthesis involves condensation of 1,2:3,4-di-*O*-isopropylidene-α-D-galactopyranose and the methyl ester of tri-*O*-acetyl-α-D-glucopyranuronosyl bromide in the presence of a silver oxide catalyst. Deacetylation of the product with barium hydroxide and removal of the *O*-isopropylidene groups by acid yields the crystalline disaccharide.[60]

Cellobiouronic Acid

4-*O*-β-D-Glucopyranuronosyl-D-glucopyranose

---

* References start on p. 129.

Properties. $[\alpha]_D = +7°$ (c 2, H_2O).[291] Heptaacetate; m.p. 230°; $[\alpha]_D = +32.9°$. Methyl ester of heptaacetate; m.p. 251°; $[\alpha]_D = +42°$ (c 0.7, $CHCl_3$).[291,292]

Occurrence. The disaccharide occurs as a structural component of capsular polysaccharides from *Pneumococcus* type III and type IV and as a minor structural unit of some plant polysaccharides.

Preparation. Cellobiouronic acid is prepared by oxidation of cellobiose derivatives, such as penta-*O*-acetyl-1,6-anhydrocellobiose.[293] It was initially isolated from partial acid hydrolyzates of *Pneumococcus* capsular polysaccharides.[294] This disaccharide, as well as other members of the series, was obtained in large amounts from acid hydrolyzates of the latter polysaccharides by chromatographic methods.[291]

2-*O*-(4-*O*-Methyl-α-D-glucopyranuronosyl)-D-xylopyranose

Properties. Amorphous compound; $[\alpha]_D = +110°$ (c 1, H_2O). Methyl ester methyl glycoside tetraacetate; m.p. 201°; $[\alpha]_D = +100°$ (c 1.5, $CHCl_3$).[295,296]

Occurrence. The disaccharide is an important aldobiouronic acid that occurs as a structural unit of polysaccharides widely distributed in plant materials. It has been isolated from hydrolyzates of polysaccharides from some thirty different plant species. The compound is especially abundant in the woody tissues of plants.

Preparation. Wood meal or shavings are hydrolyzed in concentrated sulfuric acid. The oligosaccharide is separated from the resulting reaction mixture by chromatography on ion-exchange resins.[296]

4-*O*-α-D-Galactopyranuronosyl-D-galactopyranuronic Acid

Properties. Calcium salt pentahydrate; m.p. 130°–140° (decomp.); $[\alpha]_D = +119°$ (*c* 1.3, *N* HCl).[297,298] Dimethyl ester methyl glycoside; m.p. 122°; $[\alpha]_D = +163°$ (*c* 1, H$_2$O).[299]

Occurrence. This acid is the major repeating unit in polysaccharides of the pectin class. These polymers function as the cementing material in the cell wall of virtually every plant.

Preparation. The disaccharide acid is isolated from an enzymic digest of suitable D-galacturonans and fungal bacterial galacturonanase.[300,301] Other methods are based on partial acid hydrolysis of the pectins. However, the enzymic method appears to be the preferred method, by which large quantities of the acid have been prepared.[302]

2-Acetamido-2-deoxy-3-*O*-(D-1-carboxyethyl)-
6-*O*-(2-acetamido-2-deoxy-β-D-glucopyranosyl)-D-glucopyranose
("6-β-*N*-Acetylglucosaminyl-*N*-acetylmuramic Acid")

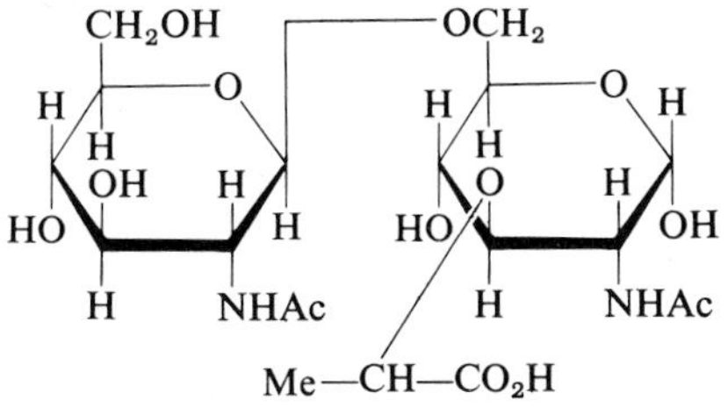

Occurrence. The disaccharide occurs as a repeating unit of the glycopeptide in the cell wall of gram-positive microorganisms.[303–305] A uridyl glycosyl-peptide which accumulates in Penicillin-treated organisms also contains a disaccharide moiety of this general type[306] (see Vol. IIB, Chap. 41).

Preparation. Cell walls of microorganisms are hydrolyzed by lysozyme or other enzyme preparations having similar activity, and the disaccharide is isolated by chromatographic procedures.[304,305]

B. TRISACCHARIDES
Raffinose
β-D-Fructofuranosyl *O*-α-D-Galactopyranosyl-(1 → 6)-α-D-glucopyranoside

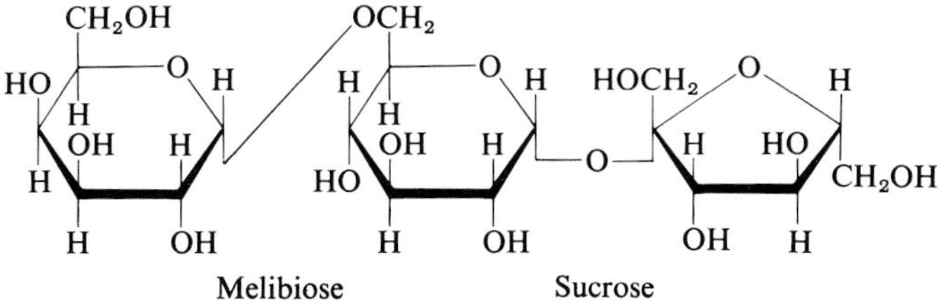

Melibiose   Sucrose

* References start on p. 129.

Properties. Raffinose pentahydrate; m.p. 78°; $[\alpha]_D = +105°$ (*c* 4, water).[307,308] Nonreducing. Partially fermented by top yeast (Bakers' yeast) with formation of melibiose; completely fermented by bottom yeast. Raffinose hendecaacetate; m.p. 101°; $[\alpha]_D = +100°$ (*c* 8, EtOH).[309]

Occurrence. Raffinose is distributed almost as widely in the plant kingdom as sucrose, but is present only in low concentration as, for example, less than 0.05% in sugar beets. The trisaccharide is concentrated in the mother liquors during the preparation of sucrose from sugar beets and sugar cane.

Preparation. Raffinose is available as a by-product of the barium process for the recovery of sucrose from beet molasses and can be crystallized directly from the final molasses.[308] Cottonseed meal may also be utilized for the preparation of raffinose by extracting the sugar with water and precipitating the raffinose as a slightly soluble compound with calcium or barium hydroxide.[310] (For biosynthesis, see this volume, Chap. 34.)

Planteose

O-α-D-Galactopyranosyl-(1 → 6)-β-D-fructofuranosyl
α-D-Glucopyranoside

Properties. Melting point 124°; $[\alpha]_D = +130°$ (*c* 5, H_2O).[24,311] Nonreducing. Planteose hendecaacetate; m.p. 135°; $[\alpha]_D = +97°$ (*c* 1, $CHCl_3$).[24]

Occurrence. Planteose was discovered in plants of the *Plantago* genus and at one time was considered to be a compound characteristic of this genus.[311] However, many other plants of the *Teucrium* and *Fraxinus* also contain the compound.[24]

Preparation. The sugar may be obtained by methanol extraction of the defatted seeds of *P. ovata*, followed by yeast fermentation of the extract to remove sucrose, and final isolation by the technique of charcoal-column chromatography.[24]

Melezitose

O-α-D-Glucopyranosyl-(1 → 3)-β-D-fructofuranosyl
α-D-Glucopyranoside

Properties. Melezitose dihydrate; m.p. 148°; $[\alpha]_D = +88.2°$ (*c* 4, H_2O).[312,313] Not fermented by top (Bakers') yeast. Nonreducing. Melezitose hendècaacetate; m.p. 117°; $[\alpha]_D = +104°$ (*c* 1, $CHCl_3$).[313]

Occurrence. Melezitose is present in the sweet exudations of many plants such as the "honeydew" of limes and poplars, and the manna exuded from insect-produced wounds of the Douglas fir, Virginia pine, larch, etc. The question as to whether the melezitose is of plant or insect origin appears to have been resolved with the discovery of a glycosyl transferase enzyme in the bodies of the insects, which effect the synthesis of melezitose from sucrose.[314] In dry seasons when the supply of flower nectar is insufficient, bees may collect the mannas of honeydews, and the honey may contain considerable quantities of melezitose.[313] When the percentage of the trisaccharide increases in the honey, crystallization may occur in the honey combs. Honey that contains this trisaccharide will not serve as food for bees, probably because of the resistance of the melezitose to hydrolysis by invertase.

Preparation. Melezitose-rich honey provides the best source of the compound, since the crystallized melezitose is easily separated by dilution of the honey with alcohol followed by centrifugation.[313] Mannas from various sources may be utilized by first extracting the impurities with aqueous alcohol and then extracting the trisaccharide with water. The melezitose is crystallized from the aqueous extracts after the addition of alcohol.

* *References start on p. 129.*

Gentianose

β-D-Fructofuranosyl O-β-D-Glucopyranosyl-
$(1 \rightarrow 6)$-α-D-glucopyranoside

Gentiobiose　　　　　　Sucrose

Properties. Melting point 211°; $[\alpha]_D = +33°$ (c 2, H_2O).[315] Nonreducing.

Occurrence. Gentianose is found in the rhizomes of many plants of the *Gentian* genus. It undoubtedly serves as a carbohydrate reserve in these rhizomes.

Preparation. Powdered gentian root is extracted with 90% alcohol, and the sugar is isolated from the extracts.[315] The trisaccharide can be synthesized by the action of levansucrase on sucrose and gentiobiose.[316]

"Neuramin-Lactose"

O-(N-Acetylneuraminyl)-$(2 \rightarrow 3)$-O-β-D-galactopyranosyl-$(1 \rightarrow 4)$-D-glucopyranose

←――――――― Lactose ―――――――→

Occurrence. "Neuramin-lactose" is a unique trisaccharide composed of neuraminic acid and lactose. It was originally isolated from mammary tissue homogenates[185] and is present in various amounts in mammary gland tissue and secretions from most animals.[317] A number of other nitrogen-containing oligosaccharides in which 2-acetamido-2-deoxy-D-glucose replaces neuraminic acid often accompany the "neuramin-lactose" in these biological materials.[318] Some of these oligosaccharides contain additional carbohydrate residues, such as L-fucose and D-galactose, attached to the amino sugar. Since the oligosaccharides are potent inhibitors of the precipitin reactions of glycoproteins with their antibodies,[318] it is probable that the glycoproteins contain these oligosaccharides as structural components, most likely at the terminal end of the carbohydrate portion of the glycoproteins. In many glycoproteins the carbohydrate component is the immunologically determinant group of the antigen, and even the nature and the configuration of the glycosidic linkages are of great importance. This class of oligosaccharides is very useful not only in immunological studies but also as reference compounds in studies on the structure of glycoproteins.

Preparation. "Neuramin-lactose" has been isolated from rat mammary glands[185] and colostrum.[317] The trisaccharide is very acid-labile at the bond linking the neuraminic acid to the lactose. The neuraminic acid can also be liberated from the oligosaccharide by neuraminidase.[319] Nucleotide pathways are apparently involved in the biosynthesis of the compound.[320] Biosynthetic pathways for "neuramin-lactose" and the related oligosaccharides are discussed in this volume, Chap. 34. Methods for the preparation and isolation of the oligosaccharides follow the principles outlined in an earlier section of this chapter, and directions for the preparation of the individual compounds may be found in references cited in various reviews.[318]

Maltotriose

O-α-D-Glucopyranosyl-(1 → 4)-O-α-D-glucopyranosyl-
(1 → 4)-D-glucopyranose

Maltose   Maltose

Properties. Amorphous; $[\alpha]_D = +160°$ (c 1, H_2O).[36,98] Reducing. Slowly fermented by yeasts.[98] β-Maltotriose hendecaacetate; m.p. 136°; $[\alpha]_D = +86°$ (c 1.6, $CHCl_3$).[321] Maltotriose 1-phenylflavazole; m.p. 248° (decomp.).[98]

Occurrence. The trisaccharide is present as a structural unit in starch and glycogen. It is also present as a product of enzymic disproportionation of maltose by transferases from several sources.[210,211] The compound does not occur naturally in biological material, although it may accumulate in some animal tissues, presumably owing to action of amylase on glycogen.[322]

Preparation. Acid hydrolysis of starch, amylopectin, and amylose have yielded mixtures from which the oligosaccharide was isolated initially in an acetylated form[321] and later as the free compound.[36] Chromatographic procedures were utilized to separate the trisaccharide in the derivatives from

* *References start on p. 129.*

other glucosyl oligosaccharides in the reaction mixture. Enzymic digestion of amylose by salivary amylase yields initially, maltose, maltotriose, and malto-tetraose, with the latter being hydrolyzed to maltose. By stopping the reaction at an appropriate stage the amylose is converted largely into maltose and maltotriose. The maltose is removed by fermentation, and the maltotriose is isolated by adsorption on charcoal and elution with alcohol.[98]

Cellotriose

O-β-D-Glucopyranosyl-(1 → 4)-O-β-D-glucopyranosyl-
(1 → 4)-D-glucopyranose

Cellobiose   Cellobiose

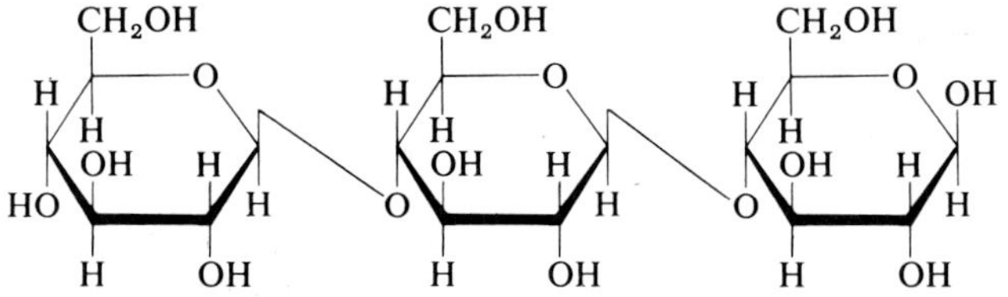

Properties. Melting point 209°; $[\alpha]_D = +22°$ (c 4, H_2O).[49] β-Cellotriose hendecaacetate; m.p. 223°; $[\alpha]_D = +23°$ (c 5, $CHCl_3$).[49,133]

Occurrence. Cellotriose is a structural unit of cellulose.

Preparation. The trisaccharide is produced as the hendecaacetate by acetolysis of cellulose.[133] The acetylated cellotriose is obtained from the reaction mixture by chromatography on Magnesol and fractional crystallization. Recrystallization of the derivative yields a product of high purity from which cellotriose is regenerated by saponification procedures. Cellotriose may also be obtained by partial acid hydrolysis of cellulose with fuming hydrochloric acid[323] or from enzymic hydrolyzates of cellulose.[324] The trisaccharide is a product of transferase action on cellobiose.[235] It has been synthesized by the Koenigs–Knorr reaction from hepta-O-acetyl-α-cellobiosyl bromide and 1,2,3,6-tetra-O-acetyl-D-glucose.[325]

Panose

O-α-D-Glucopyranosyl-(1 → 6)-O-α-D-glucopyranosyl-(1 → 4)-
D-glucopyranose

Isomaltose   Maltose

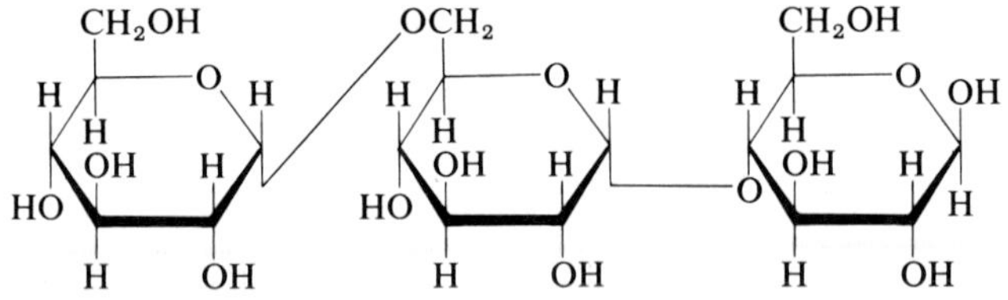

Properties. Melting point 213° (decomp.); [α]$_D$ = +154° (*c* 2, H$_2$O).[326]
Panitol dodecaacetate; m.p. 148°–150°; [α]$_D$ = +120° (*c* 4, CHCl$_3$).[138,327]
Panose 1-phenylflavazole; m.p. 210° (decomp.).[328]

Occurrence. The compound is a structural unit of amylopectin and glycogen. It is also present in the enzymic digests of maltose with various α-
glucoside glucohydrolases.

Preparation. Enzymic synthesis from maltose by use of D-glucosyl transferase is the principal method of synthesis of this compound.[326,328] Chromatographic methods and fractional precipitation methods, coupled with a
fermentation step, are utilized in the isolation of the compound. The compound has also been isolated from partial acid hydrolyzates of glycogen[329]
and other D-glucans.[327] It has been synthesized by condensation of tri-*O*-
acetyl-2-*O*-nitro-β-D-glucopyranosyl chloride and a suitably protected
derivative of maltose, followed by removal of the protecting groups.[66a]

C. OTHER OLIGOSACCHARIDES

Stachyose

β-D-Fructofuranosyl *O*-α-D-Galactopyranosyl-(1 → 6)-
O-α-D-galactopyranosyl-(1 → 6)-α-D-glucopyranoside

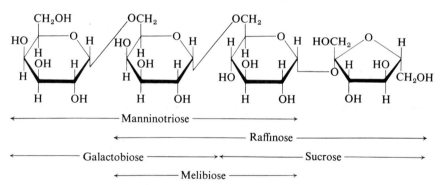

Properties. Stachyose hydrate; m.p. 101°; [α]$_D$ = +133° (*c* 4.5, H$_2$O).[8,330]
Partially fermentable by yeasts. Nonreducing.

Occurrence. The tetrasaccharide has been isolated from about forty
different plant species, and is usually found associated with sucrose and
raffinose. Thus, it has been reported in the roots of *Stachys* species, in the
twigs of white jasmine, in the seeds of yellow lupine (*Lupinus luteus*), in
soybeans (*Soja max*), in lentils (*Ervum lens*), and in ash manna (*Fraxinus
ornus*).

* *References start on p. 129.*

128   JOHN H. PAZUR

Preparation. Ash manna, soybeans, and rhizomes of *Stachys tuberifera* have been utilized as sources of the sugar. Methods of preparation are critically evaluated in a review.[331] Specific directions for the preparation of stachyose from *Stachys tuberifera* are recorded.[332] (For biosynthesis, see this volume, Chap. 34.)

Verbascose

β-D-Fructofuranosyl *O*-α-D-Galactopyranosyl-(1 → 6)-[*O*-α-D-galacto-pyranosyl-(1 → 6)]$_2$-α-D-glucopyranoside

Properties. Melting point 220°; $[\alpha]_D$ = +169° (*c* 5, H_2O)[333] Nonreducing.

Occurrence. Verbascose is present in storage organs, tubers, rhizomes, and seeds of some plants. The oligosaccharide accumulates in the seeds of the plants during maturation and disappears during germination.

Preparation. The mullein (*Verbascum thapsus*) root is the commonest source of verbascose.[333] Legume seeds are also a source of the oligosaccharide.[334]

Cyclohexaamylose (Schardinger α-dextrin)

Properties. Melting point 200°; $[\alpha]_D$ = +150° (*c* 1, H_2O). Cyclohexa-amylose peracetate; m.p. 242°; $[\alpha]_D$ = +105° (*c* 1, $CHCl_3$).[335,336]

Occurrence. Cyclohexaamylose and several other cyclic dextrins (Schardinger dextrins)[337,338] are present in digests of starch with enzyme preparations that have been produced by *Bacillus macerans*. The cyclic dextrins are beautifully crystalline compounds and represent the highest molecular oligosaccharides that have thus far been obtained in crystalline form.

Preparation. Cyclohexaamylose can be isolated from enzymic hydro-lyzates of starch with precipitation agents such as toluene and trichloro-ethylene and crystallized from water or propyl alcohol–water solutions.[336] Several reviews of the methods of preparation and other properties of cyclohexaamylose and other members of the series are available.[339,340]

Maltoheptaose

O-α-D-Glucopyranosyl-(1 → 4)-[*O*-α-D-glucopyranosyl-(1 → 4)-]$_5$-D-glucopyranose

Properties. $[\alpha]_D$ = +175° (*c* 2, H_2O)[341] and +186° (H_2O).[36]

Occurrence. This oligosaccharide occurs as a structural unit of amylose and of the linear portions of amylopectin and glycogen molecules.

Preparation. Maltoheptaose was initially prepared from cyclohepta-amylose by partial acid hydrolysis of the cyclic compound, removal of unhydrolyzed cycloheptaamylose with a complexing agent, and fractional precipitation of the oligosaccharide with ethyl alcohol.[341] The compound is also produced by partial acid hydrolysis of amylose, from which it is isolated by a chromatographic method.[36,342] The oligosaccharide represents one of the first high-molecular weight oligosaccharides that is available in chemically pure form.

REFERENCES

1. R. W. Bailey, "Oligosaccharides," Macmillan (Pergamon), New York, 1965; S. Tsuiki, Y. Hashimoto, and W. Pigman, in "Comprehensive Biochemistry," M. Florkin and E. H. Stotz, Eds., Vol. 5, Elsevier, Amsterdam, 1963, pp. 153.
2. J. Staněk, M. Černý, and J. Pacák, "The Oligosaccharides," Academic Press, New York, 1965.
3. W. E. Militzer, *J. Chem. Educ.*, **18**, 25 (1941).
4. Nomenclature of Carbohydrates, *J. Chem. Soc.*, 5307 (1962); see Vol. IIB, Chap. 46.
5. Rules of Carbohydrate Nomenclature, *J. Org. Chem.*, **28**, 281 (1963); (see Vol. IIB, Chap. 46).
6. Biochemical Nomenclature, *Biochemistry*, **5**, 1445 (1966).
7. W. J. Whelan, *Ann. Rev. Biochem.*, **29**, 105 (1960).
8. R. L. Whistler and M. L. Wolfrom, Eds., "Methods in Carbohydrate Chemistry," Vol. I, Academic Press, New York, 1962.
9. H. M. Tsuchiya, A. Jeanes, H. M. Bricker, and C. A. Wilham, *J. Bacteriol.*, **64**, 513 (1952).
10. A. Jeanes, C. A. Wilham, R. W. Jones, H. M. Tsuchiya, and C. E. Rist, *J. Amer. Chem. Soc.*, **75**, 5911 (1953).
11. A. L. Demain and H. J. Phaff, *J. Biol. Chem.*, **210**, 381 (1954).
12. R. W. Bailey, R. T. J. Clarke, and D. E. Wright, *Biochem. J.*, **83**, 517 (1962).
13. S. K. Dube and P. Nordin, *Arch. Biochem. Biophys.*, **99**, 105 (1962).
14. R. L. Whistler and E. Masak, Jr., *J. Amer. Chem. Soc.*, **77**, 1241 (1955).
15. D. French, *Baker's Dig.*, **32**, 50 (1957).
16. B. J. Bines and W. J. Whelan, *Biochem. J.*, **76**, 253 (1960).
17. R. W. Bailey, D. H. Hutson, and H. Weigel, *Biochem. J.*, **80**, 514 (1961).
18. C. T. Bishop and D. R. Whitaker, *Chem. Ind. (London)*, 119 (1955).
19. G. O. Aspinall, I. M. Cairncross, R. J. Sturgeon, and K. C. B. Wilkie, *J. Chem. Soc.*, 3881 (1960).
20. T. E. Timell, *Chem. Ind. (London)*, 999 (1959).
21. O. Perila and C. T. Bishop, *Can. J. Chem.*, **39**, 815 (1961).
22. R. W. Bailey and B. D. E. Gaillard, *Nature*, **207**, 292 (1965).
23. C. S. Hudson and T. S. Harding, *J. Amer. Chem. Soc.*, **37**, 2734 (1915).
24. D. French, G. M. Wild, B. Young, and W. J. James, *J. Amer. Chem. Soc.*, **75**, 709 (1953).
25. H. Suzuki, *Arch. Biochem. Biophys.*, **99**, 476 (1962).
26. G. W. Hay, D. W. S. Westlake, and F. J. Simpson, *Can. J. Microbiol.*, **7**, 921 (1961).
27. Y. Li and M. R. Shetlar, *Arch. Biochem. Biophys.*, **108**, 523 (1964).
28. K. Wallenfels, M. L. Zarnitz, G. Laule, H. Bender, and M. Keser, *Biochem. Z.*, **331**, 459 (1959).
29. R. W. Bailey, *Biochem. J.*, **86**, 509 (1963).
30. E. A. Moelwyn-Hughes, *Trans. Faraday Soc.*, **25**, 503 (1929).

31. A. J. Charlson, J. R. Nunn, and A. M. Stephen, *J. Chem. Soc.*, 1428 (1955).
32. P. Andrews and J. K. N. Jones, *J. Chem. Soc.*, 583 (1955).
33. C. T. Bishop, *Can. J. Chem.*, **33**, 1073 (1955).
34. G. O. Aspinall and R. G. J. Telfer, *J. Chem. Soc.*, 1106 (1955).
35. C. C. Sweeley and B. Walker, *Anal. Chem.*, **36**, 1461 (1964).
36. W. J. Whelan, J. M. Bailey, and P. J. P. Roberts, *J. Chem. Soc.*, 1293 (1953).
37. J. H Pazur and T. Budovich, *J. Biol. Chem.*, **220**, 25 (1956).
38. J. R. Turvey and W. J. Whelan, *Biochem. J.*, **67**, 49 (1957).
39. M. L. Wolfrom, E. N. Lassettre, and A. N. O'Neill, *J. Amer. Chem. Soc.*, **73**, 595 (1951).
40. F. Smith and A. M. Stephen, *J. Chem. Soc.*, 4892 (1961).
41. S. Haq and G. A. Adams, *Can. J. Chem.*, **39**, 1563 (1961).
42. G. Schiffman, E. A. Kabat, and S. Leskowitz, *J. Amer. Chem. Soc.*, **84**, 73 (1962).
43. J. H. Pazur and A. L. Gordon, *J. Amer. Chem. Soc.*, **75**, 3458 (1953).
44. R. Zelikson and S. Hestrin, *Biochem. J.*, **79**, 71 (1961).
45. G. O. Aspinall and I. M. Cairncross, *J. Chem. Soc.*, 3998 (1960).
46. S. A. Barker, A. B. Foster, M. Stacey, and J. M. Webber, *J. Chem. Soc.*, 2218 (1958).
47. G. L. Miller, J. Dean, and R. Blum, *Arch. Biochem. Biophys.*, **91**, 21 (1960).
48. R. L. Whistler and C. C. Tu, *J. Amer. Chem. Soc.*, **74**, 3609 (1952).
49. M. L. Wolfrom and J. C. Dacons, *J. Amer. Chem. Soc.*, **74**, 5331 (1952).
50. P. A. J. Gorin and A. S. Perlin, *Can. J. Chem.*, **34**, 1796 (1956).
51. S. Peat, W. J. Whelan, and T. E. Edwards, *J. Chem. Soc.*, 29 (1961).
52. T. J. Painter, *Chem. Ind. (London)*, 1214 (1960).
53. T. J. Painter, W. M. Watkins, and W. T. J. Morgan, *Nature*, **193**, 1042 (1962).
54. D. D. Reynolds and W. L. Evans, *J. Amer. Chem. Soc.*, **60**, 2559 (1938).
55. R. U. Lemieux, *Advan. Carbohyd. Chem.*, **9**, 1 (1954).
56. C. W. Klingensmith and W. L. Evans, *J. Amer. Chem. Soc.*, **61**, 3012 (1939).
57. R. Kuhn and H. H. Baer, *Ber.*, **87**, 1560 (1954).
58. R. Kuhn, H. H. Baer, and A. Gauhe, *Ber.*, **88**, 1713 (1955).
59. T. Osawa and R. W. Jeanloz, *Carbohyd. Res.*, **1**, 181 (1965).
60. R. D. Hotchkiss and W. F. Goebel, *J. Biol. Chem.*, **115**, 285 (1936).
61. B. Helferich and H. Bredereck, *Ann. Chem.*, **465**, 166 (1928).
62. G. Zemplén and R. Bognár, *Ber.*, **72**, 1160 (1939).
63. M. L. Wolfrom, A. O. Pittet, and I. C. Gillam, *Proc. Nat. Acad. Sci. U.S.*, **47**, 700 (1961); M. L. Wolfrom and D. R. Lineback, *Methods Carbohyd. Chem.*, **2**, 341 (1963); cf. R. U. Lemieux *et al.*, *Can. J. Chem.*, **46**, 405 (1968).
64. H. R. Goldschmid and A. S. Perlin, *Can. J. Chem.*, **39**, 2025 (1961).
65. S. H. Nichols, Jr., W. L. Evans, and H. D. McDowell, *J. Amer. Chem. Soc.*, **62**, 1754 (1940).
66. R. de Souza and I. J. Goldstein, *Tetrahedron Lett.*, 1215 (1964).
66a. M. L. Wolfrom and K. Koizumi, *J. Org. Chem.*, **32**, 656 (1967).
67. W. T. Haskins, R. M. Hann, and C. S. Hudson, *J. Amer. Chem. Soc.*, **64**, 1852 (1942).
68. W. T. Haskins, R. M. Hann, and C. S. Hudson, *J. Amer. Chem. Soc.*, **64**, 1289 (1942).
69. V. E. Gilbert, F. Smith, and M. Stacey, *J. Chem. Soc.*, 622 (1946).
70. S. Haq and W. J. Whelan, *J. Chem. Soc.*, 1342 (1958).
71. G. Zemplén and A. Gerecs, *Ber.*, **68**, 1318 (1935).
72. P. A. J. Gorin and A. S. Perlin, *Can. J. Chem.*, **37**, 1930 (1959).

73. K. Freudenberg, H. Eich, C. Knoevenagel, and W. Westphal, *Ber.*, **73**, 441 (1940).

74. P. Brigl, *Hoppe-Seyler's Z. Physiol. Chem.*, **122**, 245 (1922).

75. R. U. Lemieux and G. Huber, *J. Amer. Chem. Soc.*, **78**, 4117 (1956).

76. R. U. Lemieux, *Can. J. Chem.*, **31**, 949 (1953).

77. S. Haq and W. J. Whelan, *Nature*, **178**, 1222 (1956).

78. A. Thompson, K. Anno, M. L. Wolfrom, and M. Inatome, *J. Amer. Chem. Soc.*, **76**, 1309 (1954).

79. S. Peat, W. J. Whelan, T. E. Edwards, and O. Owen, *J. Chem. Soc.*, **586** (1958).

80. A. Pictet and H. Vogel, *Helv. Chim. Acta.*, **11**, 209 (1928).

81. E. M. Montgomery and C. S. Hudson, *J. Amer. Chem. Soc.*, **52**, 2101 (1930).

82. J. H. Pazur and K. Kleppe, *J. Biol. Chem.*, **237**, 1002 (1962).

83. W. M. Corbett and J. Kenner, *J. Chem. Soc.*, 1431 (1955).

84. G. Avigad, *Biochem. J.*, **73**, 587 (1959).

85. C. Fitting and E. W. Putman, *J. Biol. Chem.*, **199**, 573 (1952).

86. M. Abdel-Akher, J. K. Hamilton, and F. Smith, *J. Amer. Chem. Soc.*, **73**, 4691 (1951).

87. S. A. Barker, A. Gómez-Sánchez, and M. Stacey, *J. Chem. Soc.*, 3264 (1959).

88. O. Ruff and G. Ollendorff, *Ber.*, **33**, 1798 (1900).

89. G. Zemplén, *Ber.*, **60**, 1309 (1927).

90. A. M. Gakhokidze, *J. Gen. Chem.*, **16**, !914 (1946).

91. A. J. Charlson, P. A. J. Gorin, and A. S. Perlin, *Can. J. Chem.*, **34**, 1811 (1954).

92. P. J. Stoffyn and R. W. Jeanloz, *Arch. Biochem. Biophys.*, **52**, 373 (1954).

93. R. Schaffer and H. S. Isbell, *J. Res. Nat. Bur. Stand.*, **57**, 333 (1956).

94. E. Pacsu, *J. Amer. Chem. Soc.*, **52**, 2571 (1930).

95. B. Lindberg, *Acta Chem. Scand.*, **3**, 1355 (1949).

96. Y. Inouye, K. Onodera, I. Karasawa, and Y. Nishisawa, *J. Agr. Chem. Soc. Jap.*, **26**, 631 (1952).

97. C. S. Hudson and E. Pacsu, *J. Amer. Chem. Soc.*, **52**, 2519 (1930).

98. J. H. Pazur, *J. Biol. Chem.*, **205**, 75 (1953).

99. J. J. Connell, E. L. Hirst, and E. G. V. Percival, *J. Chem. Soc.*, 3494 (1950).

100. Reference 8, page 3.

101. O. Samuelson, "Ion-exchangers in Analytical Chemistry," Wiley, New York, 1953, p. 262.

102. F. J. Bates, "Polarimetry, Saccharimetry and the Sugars," *Nat. Bur. Stand. Circ.* C 440, Washington D.C., 1942.

103. R. W. Bailey and J. B. Pridham, *Chromatogr. Rev.*, **4**, 114 (1962).

104. R. J. Block, E. L. Durrum, and G. Zweig, "Paper Chromatography and Paper Electrophoresis," Academic Press, New York, 1958.

105. E. Lederer and M. Lederer, "Chromatography," Elsevier, Amsterdam, 1957.

106. W. H. McNeely, W. W. Binkley, and M. L. Wolfrom, *J. Amer. Chem. Soc.*, **67**, 527 (1945).

107. R. L. Whistler and D. F. Durso, *J. Amer. Chem. Soc.*, **72**, 677 (1950).

108. M. A. Jermyn, *Aust. J. Chem.*, **10**, 55 (1957).

109. N. Hoban and J. W. White, *Anal. Chem.*, **30**, 1294 (1958).

110. P. M. Taylor and W. J. Whelan, *Chem. Ind. (London)*, 44 (1962).

111. J. S. D. Bacon and D. J. Bell, *J. Chem. Soc.*, 2528 (1953).

112. S. A. Barker, E. J. Bourne, and O. Theander, *J. Chem. Soc.*, 4276 (1955).

113. B. Weissmann, K. Meyer, P. Sampson, and A. Linker, *J. Biol. Chem.*, **208**, 417 (1954).

114. G. O. Aspinall, E. L. Hirst, and N. K. Matheson, *J. Chem. Soc.*, 989 (1956).

115. S. T. Horowitz, S. Roseman, and H. J. Blumenthal, *J. Amer. Chem. Soc.*, **79**, 5046 (1957).
116. J. K. N. Jones, R. A. Wall, and A. O. Pittet, *Can. J. Chem.*, **38**, 2285 (1960).
117. P. Flodin, J. D. Gregory, and L. Roden, *Anal. Biochem.*, **8**, 424 (1964).
118. J. A. Thoma and D. French, *Anal. Chem.*, **29**, 1645 (1957).
119. H. M. Rauen, "Biochemisches Taschenbuch," Springer Verlag, Berlin, 1956, p. 1134.
120. J. H. Pazur, M. Shadaksharaswamy, and A. Cepure, *Arch. Biochem. Biophys.*, **94**, 142 (1961).
121. R. U. Lemieux, C. T. Bishop, and G. E. Pelletier, *Can. J. Chem.*, **34**, 1365 (1956).
122. L. Hough, J. K. N. Jones, and W. H. Wadman, *J. Chem. Soc.*, 1702 (1950).
123. J. M. Bobbitt, "Thin-Layer Chromatography," Reinhold, New York, 1963.
124. E. Stahl and U. Kaltenbach, in "Thin-Layer Chromatography," E. Stahl, Ed., Academic Press, New York, 1965, p. 461.
125. R. S. Tipson, "Technique of Organic Chemistry," A. Weissberger, Ed., Vol. III, Part I, Wiley (Interscience), New York, 1956, p. 395.
126. A. Thompson and M. L. Wolfrom, *J. Amer. Chem. Soc.*, **75**, 3605 (1953).
127. Reference 8, p. 8.
128. F. F. Farley and R. M. Hixon, *Ind. Eng. Chem. Anal. Ed.*, **13**, 616 (1941).
129. N. Nelson, *J. Biol. Chem.*, **153**, 375 (1944).
130. A. G. Perkin, *J. Chem. Soc.*, **87**, 107 (1905).
131. M. Levine, J. F. Foster, and R. M. Hixon, *J. Amer. Chem. Soc.*, **64**, 2331 (1942).
132. J. F. Robyt and W. J. Whelan, *Biochem. J.*, **95**, 10P (1965).
133. E. E. Dickey and M. L. Wolfrom, *J. Amer. Chem. Soc.*, **71**, 825 (1949); *cf.* M. L. Wolfrom, J. C. Dacons, and D. L. Fields, *Tappi*, **39**, 803 (1956).
134. W. Simon and C. Tomlinson, *Chimia*, **14**, 301 (1960).
135. J. van Dam and W. Prins, *Methods Carbohyd. Chem.*, **5**, 253 (1965).
136. W. N. Haworth and C. W. Long *J. Chem. Soc.,* 544 (1927).
137. D. French, *Science*, **113**, 352 (1951).
138. M. L. Wolfrom, A. Thompson, and T. T. Galkowski, *J. Amer. Chem. Soc.*, **73**, 4093 (1951).
139. A. S. Perlin, *Anal. Chem.*, **27**, 396 (1955).
140. P. Fleury and J. Courtois, *Bull. Soc. Chim. Fr.*, **10**, 245 (1943).
141. E. L. Hirst and J. K. N. Jones, *J. Chem. Soc.*, 1659 (1949).
142. M. R. Grimmett, R. W. Bailey, and E. L. Richards, *Chem. Ind. (London)*, 651 (1965).
143. B. Helferich and W. W. Pigman, *Hoppe-Seyler's Z. Physiol. Chem.*, **259**, 253 (1939).
144. M. Adams and C. S. Hudson, *J. Amer. Chem. Soc.*, **65**, 1359 (1943).
145. W. W. Pigman, *J. Amer. Chem. Soc.*, **62**, 1371 (1940).
146. C. S. Hudson, *J. Amer. Chem. Soc.*, **52**, 1707 (1930).
147. J. M. Bobbitt, *Advan. Carbohyd. Chem.*, **11**, 1 (1956).
148. A. J. Charlson and A. S. Perlin, *Can. J. Chem.*, **34**, 1804 (1956).
149. P. A. J. Gorin and A. S. Perlin, *Can. J. Chem.*, **39**, 2474 (1961).
150. S. A. Barker, E. J. Bourne, M. Stacey, and D. H. Whiffen, *J. Chem. Soc.*, 171 (1954).
151. J. M. van der Veen, *J. Org. Chem.*, **28**, 564 (1963); C. A. Glass, *Can. J. Chem.*, **43**, 2652 (1965).
152. W. W. Binkley, D. Horton, and N. S. Bhacca, *Carbohyd. Res.*, **10**, 245 (1969).
152a. O. S. Chizhov, N. V. Molodtsov, and N. K. Kochetkov, *Carbohyd. Res.*, **4**, 273 (1967); *ibid.*, *Tetrahedron*, **24**, 5587 (1968).
153. R. E. Reeves, *Advan. Carbohyd. Chem.*, **6**, 123 (1951); R. U. Lemieux and R. Nagarajan, *Can. J. Chem.*, **42**, 1270 (1964).

154. R. L. Whistler and C. C. Tu, *J. Amer. Chem. Soc.*, **75**, 645 (1953).
155. D. French and G. M. Wild, *J. Amer. Chem. Soc.*, **75**, 2612 (1953).
156. J. H. Pazur, T. Budovich, and C. L. Tipton, *J. Amer. Chem. Soc.*, **79**, 625 (1957).
156a. R. S. Shallenberger and T. E. Acree, "Handbook of Physiology," Vol. 4: The Chemical Sensors, Springer-Verlag, Berlin, 1970.
157. B. L. Oser, "Hawk's Physiological Chemistry," 14th ed., McGraw-Hill, New York, 1965, p. 62.
158. J. Larner and C. M. McNickle, *J. Biol. Chem.*, **215**, 723 (1955).
159. A. Dahlqvist, *Acta Chem. Scand.*, **14**, 1 (1960).
160. C. A. Bunton, T. A. Lewis, D. R. Llewellyn, and C. A. Vernon, *J. Chem. Soc.*, 4419 (1955).
161. F. Shafizadeh, *Advan. Carbohyd. Chem.*, **13**, 9 (1958).
162. R. L. Whistler and R. M. Rowell, *J. Org. Chem.*, **29**, 3290 (1964).
163. M. S. Feather and J. F. Harris, *J. Org. Chem.*, **30**, 153 (1965).
164. M. L. Wolfrom, A. Thompson, and C. E. Timberlake, *Cereal Chem.*, **40**, 82 (1963).
164a. J. N. BeMiller and D. R. Smith, *Carbohyd. Res.*, **6**, 118 (1968).
165. K. Freudenberg and G. Blomqvist, *Ber.*, **68**, 2070 (1935).
166. R. W. Jones, R. J. Dimler, and C. E. Rist, *J. Amer. Chem. Soc.*, **77**, 1659 (1955).
167. T. J. Schoch, E. J. Wilson, Jr., and C. S. Hudson, *J. Amer. Chem. Soc.*, **64**, 2871 (1942).
168. J. H. Pazur, J. M. Marsh, and T. Ando, *J. Amer. Chem. Soc.*, **81**, 2170 (1959).
169. R. W. Jones, A. Jeanes, C. S. Stringer, and H. M. Tsuchiya, *J. Amer. Chem. Soc.*, **78**, 2499 (1956).
170. F. J. Bates and R. F. Jackson, *Bull. Bur. Stand.*, **13**, 125 (1916).
171. A. Pictet and H. Vogel, *Helv. Chim. Acta*, **11**, 901 (1928).
172. H. W. Raybin, *J. Amer. Chem. Soc.*, **59**, 1402 (1937).
173. J. Y. Sugg and E. J. Hehre, *J. Immunol.*, **43**, 119 (1942).
174. R. P. Linstead, A. Rutenberg, W. G. Dauben, and W. L. Evans, *J. Amer. Chem. Soc.*, **62**, 3260 (1940).
175. L. F. Leloir and C. E. Cardini, *J. Biol. Chem.*, **214**, 157 (1955).
176. W. Z. Hassid, M. Doudoroff, and H. A. Barker, *J. Amer. Chem. Soc.*, **66**, 1416 (1944).
177. C. Péaud-Lenoël and R. Dedonder, *C. R. Acad. Sci.*, **241**, 1418 (1955).
178. H. J. Breuer and J. S. D. Bacon, *Biochem. J.*, **66**, 462 (1957).
179. A. Pictet and H. Vogel, *Helv. Chim. Acta*, **11**, 436 (1928).
180. G. Zemplén and A. Gerecs, *Ber.*, **62**, 984 (1929).
181. J. C. Irvine and J. W. H. Oldham, *J. Amer. Chem. Soc.*, **51**, 3609 (1929).
182. I. Levi and C. B. Purves, *Advan. Carbohyd. Chem.*, **4**, 1 (1949).
182a. M. L. Wolfrom and F. Shafizadeh, *J. Org. Chem.*, **21**, 88 (1956).
183. C. S. Hudson, *Z. Phys. Chem.*, **44**, 487 (1903).
184. C. S. Hudson and J. M. Johnson, *J. Amer. Chem. Soc.*, **37**, 1270 (1915).
185. R. E. Trucco and R. Caputto, *J. Biol. Chem.*, **206**, 901 (1954).
186. R. Kuhn and I. Löw, *Ber.*, **82**, 479 (1949).
186a. R. Kuhn and I. Löw, *Ber.*, **93**, 1009 (1960).
187. R. Venkataraman and F. J. Reithel, *Arch. Biochem. Biophys.*, **75**, 443 (1958).
188. K. Wallenfels and O. P. Malhotra, *Advan. Carbohyd. Chem.*, **16**, 239 (1961).
189. E. Fischer and E. F. Armstrong, *Ber.*, **35**, 3144 (1902).
190. H. H. Schlubach and W. Rauchenberger, *Ber.*, **59**, 2102 (1926).
191. J. C. Sowden and A. S. Spriggs, *J. Amer. Chem. Soc.*, **78**, 2503 (1956).
192. C. S. Hudson and J. M. Johnson, *J. Amer. Chem. Soc.*, **37**, 2748 (1915).
193. A. S. Sussman and B. T. Lingappa, *Science*, **130**, 1343 (1959).
194. J. S. Clegg and M. F. Filosa, *Nature*, **192**, 1077 (1961).

195. E. Bourquelot, *C. R. Acad. Sci.*, **111**, 578 (1890).
196. L. C. Stewart, N. K. Richtmyer, and C. S. Hudson, *J. Amer. Chem. Soc.*, **72**, 2059 (1950).
197. B. Lindberg, *Acta. Chem. Scand.*, **9**, 917 (1955).
198. T. S. Harding, *Sugar*, **25**, 476 (1923): *Chem. Abstr.*, **18**, 78 (1924).
199. G. F. Kalf and S. V. Rieder, *J. Biol. Chem.*, **230**, 691 (1958).
200. J. Dutrieu, *C. R. Acad. Sci.*, **252**, 347 (1961).
201. E. Cabib and L. F. Leloir, *J. Biol. Chem.*, **231**, 259 (1958).
202. R. U. Lemieux and H. F. Bauer, *Can. J. Chem.*, **32**, 340 (1954).
203. W. N. Haworth and W. J. Hickinbottom, *J. Chem. Soc.*, 2847 (1931).
204. H. S. Isbell and W. W. Pigman, *J. Res. Nat. Bur. Stand.*, **18**, 141 (1937).
205. C. S. Hudson and J. M. Johnson, *J. Amer. Chem. Soc.*, **37**, 1276 (1915).
206. J. Conchie, G. A. Levvy, and C. A. Marsh, *Advan. Carbohyd. Chem.*, **12**, 157 (1957).
207. Reference 8, p. 334.
208. Reference 8, p. 335.
209. C. Fitting and M. Doudoroff, *J. Biol. Chem.*, **199**, 153 (1952).
210. D. French, M. L. Levine, E. Norberg, P. Nordin, J. H. Pazur, and G. M. Wild, *J. Amer. Chem. Soc.*, **76**, 2387 (1954).
211. H. Wiesmeyer and M. Cohn, *Biochim. Biophys. Acta*, **39**, 427 (1960).
212. J. H. Pazur, *J. Amer. Chem. Soc.*, **77**, 1015 (1955).
213. S. Peat, W. J. Whelan, and K. A. Hinson, *Chem. Ind. (London)*, 385 (1955).
214. W. N. Haworth, J. V. Loach, and C. W. Long, *J. Chem. Soc.*, 3146 (1927).
215. W. N. Haworth and S. Peat, *J. Chem. Soc.*, 3094 (1926).
216. W. N. Haworth, C. W. Long, and J. H. G. Plant, *J. Chem. Soc.*, 2809 (1927).
217. F. C. Peterson and C. C. Spencer, *J. Amer. Chem. Soc.*, **49**, 2822 (1927).
218. G. Braun, *Org. Syntheses*, **17**, 36 (1937).
219. R. A. Whistler and C. L. Smart, *J. Amer. Chem. Soc.*, **75**, 1916 (1953).
220. G. L. Barber, A. D. Elbein, and W. Z. Hassid, *J. Biol. Chem.*, **239**, 4056 (1964).
221. K. Freudenberg and W. Nagai, *Ber.*, **66**, 27 (1933).
222. M. L. Wolfrom, L. W. Georges, and I. L. Miller, *J. Amer. Chem. Soc.*, **71**, 125 (1949).
223. E. E. Bacon and J. S. D. Bacon, *Biochem. J.*, **58**, 396 (1954).
224. Reference 8, p. 319.
225. E. M. Montgomery, F. B. Weakley, and G. E. Hilbert, *J. Amer. Chem. Soc.*, **69**, 2249 (1947).
226. M. L. Wolfrom, A. Thompson, and A. M. Brownstein, *J. Amer. Chem. Soc.*, **80**, 2015 (1958).
227. E. Fischer, *Ber.*, **28**, 3024 (1895).
228. C. S. Hudson and J. M. Johnson, *J. Amer. Chem. Soc.*, **39**, 1272 (1917).
229. G. Zemplén, *Hoppe-Seyler's Z. Physiol. Chem.*, **85**, 399 (1913).
230. E. Bourquelot and H. Herissey, *C. R. Acad. Sci.*, **135**, 290 (1902).
231. W. N. Haworth and B. Wylam, *J. Chem. Soc.*, **123**, 3120 (1923).
232. S. Peat, W. J. Whelan, and K. A. Hinson, *Nature*, **170**, 1056 (1952).
233. S. Peat, W. J. Whelan, and T. E. Edwards, *J. Chem. Soc.*, 3862 (1958).
234. B. Lindberg and J. McPherson, *Acta Chem. Scand.*, **8**, 985 (1954).
235. E. M. Crook and B. A. Stone, *Biochem. J.*, **65**, 1 (1957).
236. K. V. Giri, V. N. Nigam, and K. S. Srinivasan, *Nature*, **173**, 953 (1954).
237. S. Haq and G. A. Adams, *Can. J. Biochem. Physiol.*, **40**, 989 (1962).
238. R. Kuhn and A. Kolb, *Ber.*, **91**, 2408 (1958).
239. Reference 8, p. 313.

240. V. C. Barry, *Sci. Proc. Roy Dublin Soc.*, **22**, 423 (1941): *Chem. Abstr.*, **35**, 7985 (1941).
241. P. Bachli and E. G. V. Percival, *J. Chem. Soc.*, 1243 (1952).
242. A. Assarsson and O. Theander, *Acta Chem. Scand.*, **12**, 1319 (1958).
243. T. Nagahama, K. Nishida, and T. Numata, *Mem. Fac. Agr.*, *Kagoshima Univ.*, **4**, 9 (1960): *Chem. Abstr.*, **55**, 703 (1961).
244. S. A. Warsi and W. J. Whelan, *Chem. Ind. (London)*, 1573 (1957).
245. S. Peat, W. J. Whelan, and J. G. Roberts, *J. Chem. Soc.*, 3916 (1957).
246. R. A. Aitken, B. P. Eddy, M. Ingram, and C. Weurman, *Biochem. J.*, **64**, 63 (1956).
247. D. S. Feingold, E. F. Neufeld, and W. Z. Hassid, *J. Biol. Chem.*, **233**, 783 (1958).
248. F. Zilliken, G. A. Braun, C. S. Rose, and P. György, *J. Amer. Chem. Soc.*, **77**, 1296 (1955).
249. Reference 8, p. 305.
250. M. Bergmann, L. Zervas, and E. Silberkweit, *Ber.*, **64**, 2436 (1931).
251. L. Zechmeister and G. Tóth, *Ber.*, **64**, 2028 (1931).
252. L. Dierickx and J. M. Ghuysen, *Biochim. Biophys. Acta*, **58**, 7 (1962).
253. R. L. Whistler and C. C. Tu, *J. Amer. Chem. Soc.*, **74**, 4334 (1952).
254. J. K. N. Jones and L. E. Wise, *J. Chem. Soc.*, 2750 (1952).
255. B. H. Howard, G. Jones, and M. R. Purdom, *Biochem. J.*, **74**, 173 (1960).
256. H. A. Altermatt and A. C. Neish, *Can. J. Biochem. Physiol.*, **34**, 405 (1956).
257. G. O. Aspinall and K. M. Ross, *J. Chem. Soc.*, 3674 (1961).
258. P. Andrews, L. Hough, and D. B. Powell, *Chem. Ind. (London)*, 658 (1956).
259. G. O. Aspinall, E. L. Hirst, and A. Nicolson, *J. Chem. Soc.*, 1697 (1959).
260. J. K. N. Jones, *J. Chem. Soc.*, 1672 (1953).
261. A. J. Charlson, P. A. J. Gorin, and A. S. Perlin, *Can. J. Chem.*, **35**, 365 (1957).
262. P. Andrews and J. K. N. Jones, *J. Chem. Soc.*, 583 (1955).
263. P. Andrews, D. H. Ball, and J. K. N. Jones, *J. Chem. Soc.*, 4090 (1953).
264. J. K. N. Jones and W. H. Nicholson, *J. Chem. Soc.*, 27 (1958).
265. H. H. Schlubach and A. Scheffler, *Ann. Chem.*, **588**, 192 (1954).
266. S. M. Strepkov, *Dokl.*, *Akad. Nauk SSSR*, **124**, 1344 (1959); *Chem. Abstr.*, **53**, 21686 (1959).
267. D. J. Bell and J. Edelman, *J. Chem. Soc.*, 4652 (1954).
268. D. S. Feingold, G. Avigad, and S. Hestrin, *Biochem. J.*, **64**, 351 (1956).
269. R. L. Whistler and J. Z. Stein, *J. Amer. Chem. Soc.*, **73**, 4187 (1951).
270. A. K. Mukherjee, D. Choudhury, and P. Bagchi, *Can. J. Chem.*, **39**, 1408 (1961).
271. G. O. Aspinall, R. B. Rashbrook, and G. Kessler, *J. Chem. Soc.*, 215 (1958).
272. M. E. Henderson, L. Hough, and T. J. Painter, *J. Chem. Soc.*, 3519 (1958).
273. E. L. Hirst and A. S. Perlin, *J. Chem. Soc.*, 2622 (1954).
274. A. M. Stephen, S. Kirkwood, and F. Smith, *Can. J. Chem.*, **40**, 151 (1962).
275. H. Weinland, *Hoppe-Seyler's Z. Physiol. Chem.*, **305**, 87 (1956).
276. B. O. Lindgren, *Acta Chem. Scand.*, **11**, 1365 (1957).
277. J. K. N. Jones and G. H. S. Thomas, *Can. J. Chem.*, **39**, 192 (1961).
278. D. H. Ball and J. K. N. Jones, *J. Chem. Soc.*, 905 (1958).
279. M. L. Wolfrom, H. El Khadem, and J. R. Vercellotti, *J. Org. Chem.*, **29**, 3284 (1964).
280. M. L. Wolfrom, R. Montgomery, J. V. Karabinos, and P. Rathgeb, *J. Amer. Chem. Soc.*, **72**, 5796 (1950); H. Masamune, T. Ishikawa, and Y. Katabira, *Tohoku J. Exp. Med.*, **55**, 29 (1951); *Chem. Abstr.*, **46**, 5631 (1952).
280a. M. L. Wolfrom, J. R. Vercellotti, and D. Horton, *J. Org. Chem.*, **29**, 540 (1964).
281. S. Takanashi, Y. Hirasaka, and M. Kawada, *J. Amer. Chem. Soc.*, **84**, 3029 (1962).

282. R. W. Jeanloz and H. M. Flowers, *J. Amer. Chem. Soc.*, **84**, 3030 (1962).
283. B. Weissmann and K. Meyer, *J. Amer. Chem. Soc.*, **76**, 1753 (1954).
284. A. Linker, K. Meyer, and P. Hoffman, *J. Biol. Chem.*, **235**, 924 (1960).
285. E. A. Davidson and K. Meyer, *J. Amer. Chem. Soc.*, **76**, 5686 (1954).
286. P. A. Levene, *J. Biol. Chem.*, **140**, 267 (1941).
287. M. L. Wolfrom, R. K. Madison, and M. J. Cron, *J. Amer. Chem. Soc.*, **74**, 1491 (1952).
288. J. Hebting, *Biochem. Z.*, **63**, 353 (1914).
289. S. Suzuki and J. L. Strominger, *J. Biol. Chem.*, **235**, 274 (1960).
290. M. Heidelberger and F. E. Kendall, *J. Biol. Chem.*, **84**, 639 (1929).
291. J. K. N. Jones and M. B. Perry, *J. Amer. Chem. Soc.*, **79**, 2787 (1957).
292. W. F. Goebel, *J. Biol. Chem.*, **110**, 391 (1935).
293. B. Lindberg and L. Selleby, *Acta Chem. Scand.*, **14**, 1051 (1960).
294. M. Heidelberger and W. F. Goebel, *J. Biol. Chem.*, **74**, 613 (1927).
295. T. E. Timell, *Can. J. Chem.*, **37**, 827 (1959).
296. Reference 8, p. 301.
297. H. J. Phaff and B. S. Luh, *Arch. Biochem. Biophys.*, **36**, 231 (1952).
298. J. K. N. Jones and W. W. Reid, *J. Chem. Soc.*, 1361 (1954).
299. M. Gee, F. T. Jones, and R. M. McCready, *J. Org. Chem.*, **23**, 620 (1958).
300. E. F. Jansen and L. R. MacDonnell, *Arch. Biochem. Biophys.*, **8**, 97 (1945).
301. R. M. McCready and E. A. McComb, *J. Agr. Food Chem.*, **1**, 1165 (1953).
302. Reference 8, p. 309.
303. M. R. J. Salton and J. M. Ghuysen, *Biochim. Biophys. Acta*, **45**, 355 (1960).
304. H. R. Perkins, *Biochem. J.*, **74**, 182 (1960).
305. R. W. Jeanloz, N. Sharon, and H. M. Flowers, *Biochem. Biophys. Res. Commun.*, **13**, 20 (1963).
306. P. Mandelstam, R. Loercher, and J. L. Strominger, *J. Biol. Chem.*, **237**, 2683 (1962).
307. C. Scheibler, *Ber.*, **19**, 2868 (1886).
308. E. H. Hungerford and A. R. Nees, *Ind. Eng. Chem.*, **26**, 462 (1934).
309. G. Tanret, *Bull. Soc. Chim. Fr.*, **13**, 261 (1895).
310. E. P. Clark, *J. Amer. Chem. Soc.*, **44**, 210 (1922).
311. N. Wattiez and M. Hans, *Bull. Acad. Roy. Med. Belg.*, **8**, 386 (1943); *Chem. Abstr.*, **39**, 4849 (1945).
312. G. Tanret, *Bull. Soc. Chim. Fr.*, **35**, 816 (1906).
313. C. S. Hudson and S. F. Sherwood, *J. Amer. Chem. Soc.*, **42**, 116 (1920).
314. J. S. D. Bacon and B. Dickinson, *Biochem. J.*, **66**, 289 (1957).
315. M. Bridel and M. Desmarest, *J. Pharm. Chim.*, **9**, 465 (1929).
316. S. Hestrin and G. Avigad, *Biochem. J.*, **69**, 388 (1958).
317. L. W. Mayron and Z. A. Tokes, *Biochim. Biophys. Acta*, **45**, 601 (1960).
318. P. György, in "Chemistry and Biology of Mucopolysaccharides," G. E. W. Wolstenholme and M. O'Connor, Eds., J. and A. Churchill Ltd., London, 1958; R. Kuhn, *Proc. 4th Intern. Congr. Biochem., Vienna, 1958*, Pergamon, New York, 1959, pp. 67–79; W. M. Watkins and W. T. J. Morgan, *Vox Sanguinis*, **7**, 129 (1962); J. K. Huttunen, *Ann. Med. Exp. Biol. Fenniae, Helsinki Suppl.*, **12**, 44 (1966).
319. A. Gottschalk, *Biochim. Biophys. Acta*, **23**, 645 (1957).
320. S. Roseman, *Federation Proc.*, **21**, 1075 (1962).
321. M. L. Wolfrom, L. W. Georges, A. Thompson, and I. L. Miller, *J. Amer. Chem. Soc.*, **71**, 2873 (1949).

322. W. H. Fishman and H. G. Sie, *J. Amer. Chem. Soc.*, **80**, 121 (1958).
323. G. L. Miller, *Anal. Biochem.*, **1**, 133 (1960).
324. E. T. Reese, E. Smakula, and A. S. Perlin, *Arch. Biochem. Biophys.*, **85**, 171 (1959).
325. K. Freudenberg and W. Nagai, *Ann. Chem.*, **494**, 63 (1932).
326. S. C. Pan, L. W. Nicholson, and P. Kolachov, *J. Amer. Chem. Soc.*, **73**, 2547 (1951).
327. S. Peat, W. J. Whelan, and T. E. Edwards, *J. Chem. Soc.*, 355 (1955).
328. J. H. Pazur and D. French, *J. Biol. Chem.*, **196**, 265 (1952).
329. M. L. Wolfrom and A. Thompson, *J. Amer. Chem. Soc.*, **79**, 4212 (1957).
330. G. Tanret, *Bull. Soc. Chim. Fr.*, **13**, 176 (1913).
331. D. French, *Advan. Carbohyd. Chem.*, **9**, 149 (1954).
332. Reference 8, p. 368.
333. E. Bourquelot and M. Bridel, *C. R. Acad. Sci.*, **151**, 760 (1910).
334. V. N. Nigam and K. V. Giri, *Can. J. Biochem. Physiol.*, **39**, 1847 (1961).
335. K. Freudenberg and R. Jacobi, *Ann. Chem.*, **518**, 102 (1935).
336. D. French, M. L. Levine, J. H. Pazur, and E. Norberg, *J. Amer. Chem. Soc.*, **71**, 353 (1949).
337. F. Schardinger, *Zentr. Bakt. Parasitenk. Abt.* II, **29**, 188 (1911): *Chem. Abstr.*, **5**, 2575 (1911).
338. D. French, A. O. Pulley, J. A. Effenberger, M. A. Rougvie, and M. Abdullah, *Arch. Biochem. Biophys.*, **111**, 153 (1965).
339. D. French, *Advan. Carbohyd. Chem.*, **12**, 189 (1957).
340. J. A. Thoma and L. Stewart, in "Starch: Chemistry and Technology," R. L. Whistler and E. F. Paschall, Eds., Vol. 1, Academic Press, New York, 1965, p. 209.
341. D. French, M. L. Levine, and J. H. Pazur, *J. Amer. Chem. Soc.*, **71**, 356 (1949).
342. R. L. Whistler and J. L. Hickson, *Anal. Chem.*, **27**, 1514 (1955).

31. ANTIBIOTICS CONTAINING SUGARS

Stephen Hanessian and Theodore H. Haskell

I. Introduction 139
II. Classification and Chemistry of Antibiotics Containing
Sugars 140
  A. Macrolide Antibiotics 140
  B. Cyclitol Antibiotics 159
  C. Nucleoside Antibiotics 172
  D. Antibiotics of the Glycosylamine Type . . 180
  E. Antibiotics Containing Aromatic Groups . . 183
  F. Miscellaneous Antibiotics 190
III. Characterization of Antibiotics and Their Sugar Components by Physical Means 194
  A. Nuclear Magnetic Resonance Spectroscopy . . 194
  B. Mass Spectrometry 196
IV. Biological Concepts 198
  References 200

I. INTRODUCTION

An antibiotic substance is a chemical (antibiotic) compound, elaborated by microorganisms, that inhibits the growth of other microorganisms and animal and plant tumors. The mechanism of action often involves interference with DNA, RNA, or protein synthesis. From the viewpoint of clinical efficacy the most prolific producers of antibiotics are the actinomycetes[1] (genus *Actinomyces*). The age of antibiotics began in the early 1940's, when the first patient was successfully treated with penicillin to combat bacterial infections. Many books and review articles have covered the development of the entire field of antibiotics. The occurrence of unique and unusual amino, deoxy, and branched-chain sugars in some of the antibiotics has stimulated increased interest in the distribution of unusual carbohydrates in Nature, and an extensive list of unusual sugars has resulted from chemical investigations on bacterial cell walls, capsular materials, and other naturally occurring macromolecules.[1a]

* *References start on p. 200.*

The chemistry of the amino sugars[2,3] and deoxy sugars[4] derived from antibiotic substances has been reviewed in detail. It is the purpose of this chapter to classify some of the sugar-containing antibiotics and to discuss the chemistry of those members whose complete structures are known. Discussions are preceded by the complete structure of the antibiotic whenever it is known. The important degradative reactions that led to isolation of the sugar component(s) and also shed some light on the gross structure of the intact molecule are discussed. Degradative reactions and other physical and chemical procedures used to elucidate the structure of the sugar portion(s), as well as the methods employed for establishing configurations and for synthesis, are given. A further aspect is the position of attachment of the sugar portion to the remainder of the molecule. For reasons of brevity, only general and some of the more significant references, especially those containing final structures, have been cited for a given antibiotic. The mechanism of action of certain antibiotics is discussed. The biosynthesis[4a] of antibiotics is not included in this chapter. Additional data on the chemistry and biochemistry of all classes of antibiotics reported up to the end of 1963 can be found in the excellent book by Umezawa.[5]

II. CLASSIFICATION AND CHEMISTRY OF ANTIBIOTICS CONTAINING SUGARS

For the purposes of this chapter the classification of sugar-containing antibiotics will be based on overall common structural relationships rather than on the similarity of sugars present. This system has been adopted because common structural similarities have usually resulted in analogous types of biological activities and mechanisms of action. A list of sugar-containing antibiotics along with the individual sugars present is compiled in Table I. The antibiotics are listed in the order in which their respective classes are discussed, and the various names that have been used for the same or presumably identical antibiotics are also included.

The antibiotics to be discussed in detail have been divided into six classes. Those antibiotics that have been shown to be of clinical value and are commercially available are indicated in Table I (p. 142).

A. MACROLIDE ANTIBIOTICS

The macrolide antibiotics are organic-soluble substances containing a large-membered lactone ring to which amino and/or other sugars are attached glycosidically. This class is divided into the polyene types and a group containing highly branched, polyfunctional macrocyclic rings. The latter group is further subdivided into those antibiotics containing nitrogen and those that are nitrogen-free. The chemistry and biogenesis of this class of antibiotics have been reviewed.[6]

1. Polyene Antifungal Antibiotics

These substances, which are elaborated by *Actinomyces*, exhibit complex ultraviolet spectra because of the conjugated polyene system present. They usually have pronounced activity in inhibiting the growth of fungi, but are not as effective in their action against bacteria. Those antibiotics containing sugars in this class are divided into two general types—namely, the conjugated tetraenes and the conjugated heptaenes.

a. Conjugated Tetraenes—Nystatin ($C_{47}H_{75}NO_{18}$) has been assigned[7,7a] a partial structure (**1**). The sugar portion (mycosamine), which is attached

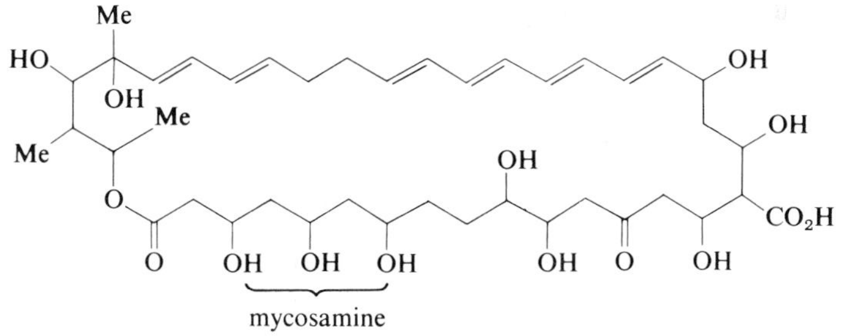

mycosamine

Nystatin (**1**)

glycosidically to the aglycon nystatinolide[7] ($C_{41}H_{64}O_{15}$), has been shown[8] to be 3-amino-3,6-dideoxy-D-mannose (**2**). This assignment has been verified by degradation[9] and also by synthesis[10] from D-glucose. The configuration of **2** at C-2 and C-3 was established by the following sequence of reactions.[9]

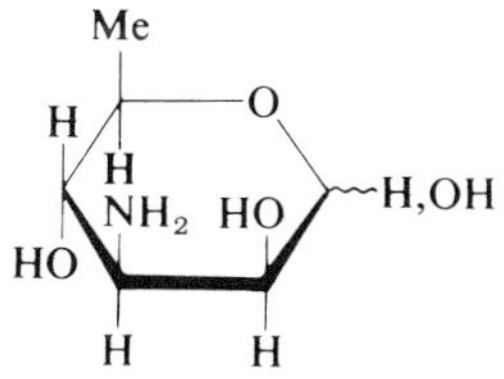

Mycosamine (**2**)

The configuration at C-5 was determined[11] by periodate oxidation of methyl *N*-ethylmycosaminide and the isolation of the known D′-methoxy-D-methyl-

* References start on p. 200.

TABLE I

Antibiotics Containing Sugars[a,d]

Antibiotic	Sugar component(s)	Method of isolation[b]	References
Macrolides			
Nystatin[c] (**1**)	Mycosamine (**2**)	A, Ac, H, M	7
Pimaricin (**4**)	2	A, Ac	13
Lucensomycin	2		16
Rimocidin	2	A	297
Amphotericin B[c]	2	Ac	298
Trichomycin[c]	2	A	19
Perimycin	Perosamine (**4a**)		20a
Levorin	2		299
Methymycin (**6a**)	Desosamine (**5**)	H	23, 24
Neomethymycin (**6b**)	5	H	25
Picromycin (**7**)	5	H	22, 23, 23a
Narbomycin (**7a**)	5	H	31, 32
Erythromycin[c] (**9**)	5	H	42
	Cladinose (**9**)		
Oleandomycin[c] (**10**)	5	H	47
	Oleandrose (**11**)	A	
Spiramycin[c] (**12**)	Mycarose (**13**)	H	49
	Forosamine (**14**)	H	54
	Mycaminose (**15**)	H	56
Magnamycin (**16**)	13, 15	H	52, 59
Niddamycin	13, 15		61
Tylosin	8, 15		68
	Mycinose (**17**)		
Leucomycins A3 (**18a**)	13, 15	H	70a, 300, 300a
Angolamycin	L-Mycarose (**13**)	H	70b, c
	Mycinose (**17**)		
	Angolosamine (**18b**)		
Chalcomycin (**19**)	17	A	71
	Chalcose (**22**)		
Lankamycin (**24**)	Arcanose (**23**)	A	18, 80, 81
Aldgamycin E	Aldgarose (**27**)	A	85, 85a
Neutramycin	17, 22	A	86
Cyclitol Antibiotics			
Kasugamycin[c] (**29**)	Kasugamine (**31**)	A	87, 92b
Hygromycin A (**33**)	*neo*-Inosamine (**34**)	H	93, 94
	6-Deoxy-D-*arabino*-hexos-5-ulose	H	
Bluensomycin (**35**)	Bluensidine (**37**)	H	36
	2-Methylamino-2-deoxy-L-glucose (**39**)		
	Dihydrostreptose		

TABLE I

ANTIBIOTICS CONTAINING SUGARS[a, d]—*continued*

Antibiotic	Sugar component(s)	Method of isolation[b]	References
Streptomycin[c] (38)	39		100
	Streptose (40)		
	Streptidine (41)		
Actinospectacin (44)	Actinamine (45)	H	114
	4,6-Dideoxy-hexos-2,3-diulose (46)		
Neomycin B[c] (48a)	Neosamine B (52)	A, H	119, 127
	Deoxystreptamine (51)	H	123
	D-Ribose	H	124
	Neosamine C (53)		119, 127
Neomycin C (48b)	D-Ribose	A, H	119, 123
	51, 53		127
Paromomycin I[c] (48c)	Paromose (52)	A, H, M	138
	D-Ribose	H	137
	2-Amino-2-deoxy-D-glucose	H	136
	51		
Paromomycin II (48d)	51, 53		5, 118
	D-Ribose, 2-amino-2-deoxy-D-glucose		
Kanamycin A[c] (54)	51, Kanosamine (55)	H	143
	6-Amino-6-deoxy-D-glucose (56)		
Kanamycin B	51, 53, 55		156
Kanamycin C	2-Amino-2-deoxy-D-glucose	H	155
	51, 53		
Gentamycin[c] (56a)	Garosamine (3-deoxy-3-methylaminopentose)	A	157, 157a, b
Hygromycin B$_2$ (58)	Hyosamine (57)		150
	D-Talose		
Destomycin A (59)	N-Methyldeoxystreptamine	H	161
	D-Talose		
Nucleoside Antibiotics			
Nebularine (62)	D-Ribose	H	165
Nucleocidin (63)	D-Ribose		170
Cordycepin (64)	D-Ribose		171, 174
Psicofuranine (65)	D-Psicose		177, 178
Decoyinine (66)	6-Deoxy-D-*erythro*-hexos-2,5-diulose		181
Puromycin (67)	3-Amino-3-deoxy-D-ribose		183
Septacidin (68)	4-Amino-4-deoxy-L-*glycero*-L-*gluco*-heptose (69)		186

(*continued*)

* *References start on p. 200.*

TABLE I

ANTIBIOTICS CONTAINING SUGARS[a, d]—*continued*

Antibiotic	Sugar component(s)	Method of isolation[b]	References
Amicetin (**70**)	Amosamine (**71**)	A	192, 193
	Amicetose (**72**)	A	192, 193
Bamicetin	Bamosamine		196a
Plicacetin	**71, 72**		196
Blasticidin S (**77**)	4-Amino-4-deoxy-D-*erythro*-hex-2-enopyranuronic acid (**78**)		202
Gougerotin (**86**)	4-Amino-4-deoxy-D-gluco-pyranuronamide		209
Polyoxin C (**86a**)	5-Amino-5-deoxy-D-alluronic acid		209a
Tubercidin (**87**)	D-Ribose		211
Toyokamycin (**88a**)	D-Ribose		214
Sangivamycin (**88b**)	D-Ribose		214, 214a, 214b
Formycin (**89**)	D-Ribose		215
Formycin B (**90**)	D-Ribose		215
Showdomycin (**90a**)	D-Ribose		217a
Glycosylamine Types			
Streptothricin (**92**)	2-Amino-2-deoxy-D-gulose	H	222
Streptolin (**93**)	2-Amino-2-deoxy-D-glucose	H	222
Racemomycin A (**94a**)	2-Amino-2-deoxy-D-gulose		225, 225a
Racemomycin O (**96**)	2-Amino-2-deoxy-D-glucose	H	226
Streptolydigin (**97**)	2,3,4-Trideoxy-L-*threo*-hexose (**115**)	H	195
Aromatic Types			
Novobiocin[c] (**100**)	Noviose (**102**)	A	232
Coumermycin A-1, A-2[c] (**103**)	Coumerose (**104**)	A	239
Olivomycins ⎫ (**105**) Chromomycins[e] ⎬	Olivose (chromose C) (**106**) Olivomose (chromose A) (**107**) Acetyloliose (chromose D) (**108**) Acetylolivomycose (chromose B) (**109**)	A	241, 242
Pyrromycin (**110**)	Rhodosamine (**112**)	H	250, 251
Rhodomycins	**112**		248
Cinerubins	**112**		248
	2,6-Dideoxy-L-*lyxo*-hexose (**114**)		

TABLE I

ANTIBIOTICS CONTAINING SUGARS[a,d]—*continued*

Antibiotic	Sugar component(s)	Method of isolation[b]	References
γ-Rhodomycins	**112, 114, 115** (rhodinose)	H	248, 301
Daunomycin	Daunosamine (**116**)	H	255
Chartreusin (**117**)	2,6-Dideoxy-D-*ribo*-hexose	H	258
	D-Fucose		
Curamycin	L-Lyxose	H	302
	4-*O*-Methyl-D-fucose (curacose)		
Everninomycins	Evermicose	H	303, 258b
	Evernitrose	H	258a
Tolypomycin (**117a**)	Tolyposamine (**117b**)	H	258c
Miscellaneous			
α,α-Trehalosamine (**118**)	D-Glucose		259, 260
	2-Amino-2-deoxy-D-glucose		
Lincomycin[c] (**119**)	Lincosamine (**120**)	Z	261
Celesticetin (**123**)	7-*O*-Methyl-lincosamine		262
Streptozotocin (**123a**)	2-Amino-2-deoxy-D-glucose		263a
Nojirimycin (**123b**)	5-Amino-5-deoxy-D-glucose		263b, 263c
Labilomycin (**124**)	Labilose (**125**)	A	264
"γ-Activity X"	Trideoxyoctose (**126**)	H	266
Ristocetins[c]	Glucose, mannose, arabinose, rhamnose	H	267, 268
Vancomycin[c]	Glucose	H	269
Actinoidin	Glucose, mannose	H	270
Statolon	Arabinose, xylose, galactose, 2-amino-2-deoxy-D-galactose, galacturonic acid, glucose, rhamnose	H	271
Moenomycin	D-Glucose, 2-amino-2-deoxy-D-glucose, 2-amino-2,6-di-deoxy-D-glucose, inositol	H	304, 304a

[a] Only those antibiotics whose sugar components have been definitely identified are listed. Several[5] have as yet unidentified sugars in addition to the ones listed. [b] A = alcoholysis; Ac, acetolysis; H, acid hydrolysis; M, mercaptolysis; Z, hydrazinolysis. [c] Clinically important. [d] Bold face numbers refer to formulas.

* References start on p. 200.

diglycolic aldehyde. The configuration at C-4 was rigorously established by converting the phenylosazone derived from **2** into the 1-phenylflavazole derivative (**3**) (obtained in only 1% yield) and comparison with derivatives from L-rhamnose and L-fucose.

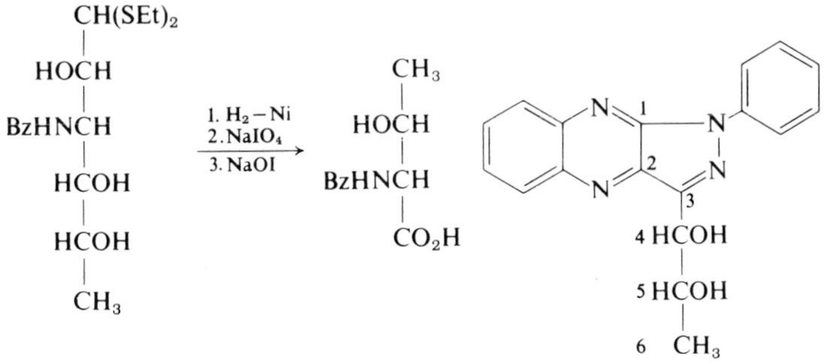

3

Pimaricin (tennecetin) is a crystalline antibiotic ($C_{33}H_{47}NO_{13}$) which also contains mycosamine (**2**) as its sole sugar constituent. A structure proposed[12] in 1958 and revised[13] in 1964 has been modified[14] and is shown as **4**.

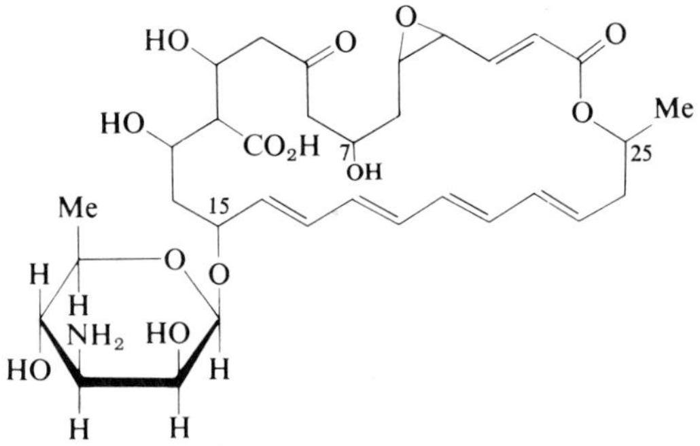

Pimaricin (**4**)

In a series of papers,[13] the determination of the carbon skeleton, its size, and the position of attachment of mycosamine were demonstrated by using nuclear magnetic resonance (n.m.r.) and mass spectrometry on hydrogenated and similarly modified derivatives of pimaricin. The ring form of mycosamine in pimaricin has been established, based on methylation studies.[13a] The configuration of the anomeric center in the intact antibiotic has been designated[15] as β-D on the basis of optical rotatory data.

Lucensomycin[16] appears to be the same as **4**, except that the methyl group at C-25 is replaced by a butyl group.

Many of the tetraene antibiotics have been assigned an all-*trans* arrangement of groups on the basis of ultraviolet spectral studies.

b. Conjugated Heptaenes—Interpretation of the ultraviolet spectra of this class of polyene antibiotics reveals that many do not have the all-*trans* arrangement found in the tetraenes. Whereas the heptaenes are supposed to have the strongest antifungal activities,[17] the complexity of their structures and their general insolubility and instability have impeded complete characterization. Among the better characterized and purified members are the trichomycins.[18] The polyenic structure, the presence of mycosamine (2), a *p*-aminobenzoyl group, and other structural features of this antibiotic were demonstrated. Ethanolysis of the *N*-2,4-dinitrophenyl derivative of perhydrotrichomycin A afforded a crystalline ethyl 3,6-dideoxy-3-(2,4-dinitroanilino)-D-mannopyranoside.[19] A tentative partial structure for the trichomycins has been proposed.[20] Perimycin has been shown[20a] to contain perosamine (4a) (4-amino-4,6-dideoxy-D-mannose).

Perosamine (4a)

c. General Aspects Concerning the Structure Elucidation of Polyene Antibiotics Containing Sugars—The polyene antibiotics investigated so far invariably have mycosamine (2) attached glycosidically to the macrocyclic ring, which accounts for the single nitrogen atom present in the molecule. The ultraviolet spectra of the various antibiotics usually show four or five discernible peaks, which are of the same general shape but differ in their wavelengths. The infrared spectra usually show a carbonyl-stretching peak at about 5.85 μm because of an ester or more probably a lactone group. Determination of the number of hydroxyl groups in these molecules is conveniently achieved by quantitative n.m.r. spectroscopy of the corresponding *O*-acetylated derivatives. The fission of some carbon–carbon bonds in these molecules under mild alkaline conditions (*N* sodium hydroxide, 3 days at room temperature) is a useful degradative tool and reflects the aldol type of structure in certain locations. The fission can be regarded as a reverse aldol cleavage; it produces a carbonyl compound as one of the fragments, as illustrated[13] in the partial structure of **4**.

* *References start on p. 200.*

Another reaction takes advantage of the vicinal hydroxyl groups, as in the oxidative fission[7a] of **1** with the formation of tiglic aldehyde.

2. Macrocyclic Lactone Antibiotics

This extensive group of antibiotics is characterized by the presence of a large lactone ring from which the name "macrolide" is derived.[21] Several members of this family of antibiotics are active primarily against gram-positive organisms and are important chemotherapeutic substances. Despite the common feature of a macrocyclic lactone in these compounds, subtle differences are apparent in their structures. These include the size of the ring, the types and number of functional groups present, and the nature and position of attachment of the sugar components.

a. Macrocyclic Lactone Antibiotics Containing Nitrogen—The first macrolide antibiotic to be discovered was picromycin.[22,23] Methymycin[23,24] and neomethymycin[25] have in common with picromycin the amino sugar component desosamine (**5**) (3-dimethylamino-3,4,6-trideoxy-β-D-*xylo*-hexose). The structures of methymycin (**6a**) and neomethymycin (**6b**) are shown on p. 149. A revised structure[23a] of picromycin (**7**) shows its similarity to narbomycin (**7a**), rather than methymycin and neomethymycin (see p. 150).

Desosamine (5)

6a, Methymycin: $R_1 = 5$; $R_2, R_4 = H$; $R_3 = OH$
6b, Neomethymycin: $R_1 = 5$; $R_2, R_3 = H$; $R_4 = OH$

The structure of **6a** was the first of the series to be elucidated completely.[26] A key degradation reaction was mild acid hydrolysis, which removed desosamine (**5**) without causing structural changes in the aglycon, methynolide ($C_{17}H_{23}O_5$). Since acetylation of methynolide led to a monoacetate, and oxidation with chromium trioxide afforded a ketone, dehydromethynolide, the presence of a reactive hydroxyl group was indicated, presumably the point of attachment of desosamine (**5**) to the aglycon. This hydroxyl group was shown to be β to the carbonyl group of the lactone because the ketone resulting from oxidation yielded carbon dioxide upon saponification and acidification. The structure of **6b**, which differs from **6a** in the location of a hydroxyl group, has also been elucidated.[27,28] Mild acid hydrolysis liberated desosamine, but the aglycon neomethynolide ($C_{17}H_{28}O_5$) was accompanied by an isomer, cycloneomethynolide. The latter proved to be a five-membered anhydro derivative formed by the addition of the secondary hydroxyl group on the ethyl side chain to the double bond at C-9 of the lactone ring.

The position of attachment of desosamine in **7** was for a long time debatable.[29,30] The differences in properties between **6a** and **7**, including totally different optical rotatory dispersion (o.r.d.) curves,[28] can now be understood in the light of a new structural assignment to picromycin.[23a] It is interesting that mild acid treatment of **7** affords the aglycon, kromycin which, in contrast to methynolide, is altered and contains a C-5–C-6 double bond. All three antibiotics **6a**, **6b**, and **7** on oxidation with permanganate yield a 7-carbon lactonic acid containing the three methyl side chains.

A related antibiotic of the macrocyclic lactone type, which also contains desosamine as its sole sugar component, is narbomycin[31,32] (7a) ($C_{28}H_{49}NO_7$). A series of oxidation reactions gave the above C-7 lactone acid (from C-3 to C-9), while reduction with various catalysts gave dihydro and tetrahydro derivatives of narbomycin, which were degraded further. The attachment of desosamine (5) is indicated to be β-D from n.m.r. data.[15]

7, Picromycin: R = OH
7a, Narbomycin: R = H

The gross structure of desosamine (5) (also called picrocin), obtained from 7, was derived mainly from periodate oxidation experiments[33] according to established procedures.[34] It is noteworthy that the hydrochloride of desosamine (m.p. 189° to 191°, dec.) can be sublimed under high vacuum

without decomposition and that the free base is ether-soluble.[33]

The structure, stereochemistry, and synthesis of desosamine (5) have been the subject of numerous reports. Alkaline degradation[34] of 5, followed by the steps shown, gave (−)-pentane-1,4-diol, which was correlated with D-glyceraldehyde, thereby indicating that desosamine is a D-hexose. Chemical evidence[35,36] for the *trans* arrangement of the C-2 and C-3 substituents is based on the *trans* elimination of the methiodide obtained from desosamine. The D-*xylo* configuration has been assigned by n.m.r. spectral studies[37,38] on the free sugar and its diacetate. The degradation of ethyl desosaminide to a 3,4-epoxy-2-ethoxy-6-methyltetrahydropyran and the reconversion of the latter into desosamine is the basis of one synthesis[39] of desosamine. A synthesis[40] from methyl 3-amino-3-deoxy-D-glucopyranoside, as well as several nonstereospecific syntheses[39,41] of the racemic compound, is also available.

Erythromycin ($C_{37}H_{67}NO_{13}$), which is one of the most widely used antibiotics and medicinally the most important of the macrolides, has been the subject of extensive chemical and biogenetic studies. This antibiotic contains

not only desosamine (5) but also a branched-chain, neutral deoxy sugar named cladinose (8) (2,6-dideoxy-3-C-methyl-3-O-methyl-L-ribo-hexose). The complete structure[42] of erythromycin (9) has been corroborated by X-ray crystallographic data.[42a]

Cladinose (8)   Erythromycin (9)

Initial degradative studies on erythromycin were facilitated by operating with the reduced antibiotic, dihydroerythromycin (the C-9 methylene analog of 9). Acid hydrolysis of this material afforded desosamine, cladinose, and the modified aglycon, dihydroerythronolide. The gross structure was established[43] for cladinose by functional group analysis, by its resistance to periodate oxidation, and by its negative iodoform reaction. The position of attachment of cladinose to the aglycon was clarified from the behavior of

* References start on p. 200.

erythromycin on treatment with acid under very mild conditions (pH 2.5, a few minutes). This treatment caused the loss of water between the hydroxyl groups at C-12 and C-6 and afforded an anhydro derivative of erythromycin. The unfavorable location of the C-3 hydroxyl group excludes its involvement in acetal formation, and consequently indicates the point of attachment of cladinose in the antibiotic. Treatment of erythromycin or its anhydro form with acid causes the loss of two molecules of water and frees the cladinose moiety. Vigorous alkaline hydrolysis of the remaining fragment produces dimethylamine, which must arise by scission of the desosamine residue (5) and subsequent decomposition of desosamine. The elimination of desosamine from C-5 of the macrocycle has been rationalized[42] as proceeding by way of an activated allylic intermediate. The configuration of the anomeric linkage of cladinose and of desosamine in erythromycin has been proposed to be β-L and β-D, respectively, on the basis of n.m.r. spectral studies.[44] However, from optical rotatory data and independent n.m.r. spectral studies, an α-L configuration has been suggested[15] for the cladinose moiety. The synthesis of cladinose has been reported.[45,45a]

Erythromycin B, which is more stable than erythromycin, differs from it in the absence of the C-12 tertiary hydroxyl group.

Oleandomycin was assigned the formula $C_{35}H_{61}NO_{12}$ on the basis of chemical characterization and preliminary degradation studies.[46] Its complete structure[47] is shown as 10. Acid hydrolysis of oleandomycin was shown to give desosamine (5), accounting for the nitrogen atom in the molecule. Mild methanolysis gave the aglycon desoleandomycin ($C_{28}H_{49}NO_9$) and a methyl glycoside of the dideoxy sugar, L-oleandrose (11) (2,6-dideoxy-3-O-methyl-α-L-*arabino*-hexose).[48]

Oleandomycin (10)   L-Oleandrose (11)

Mild alkaline treatment of oleandomycin caused the elimination of the C-11 hydroxyl group to give an anhydro derivative ($C_{35}H_{59}NO_{11}$). Methanolysis of this substance caused the loss of L-oleandrose and the addition of methanol across the epoxide ring. This product and related derivatives were investigated by n.m.r. spectroscopy and by further degradative sequences.

A key degradative fragment shown above, which did not lactonize, still contained the desosamine residue but could be lactonized after removal of the amino sugar by acid hydrolysis. This sequence established the position

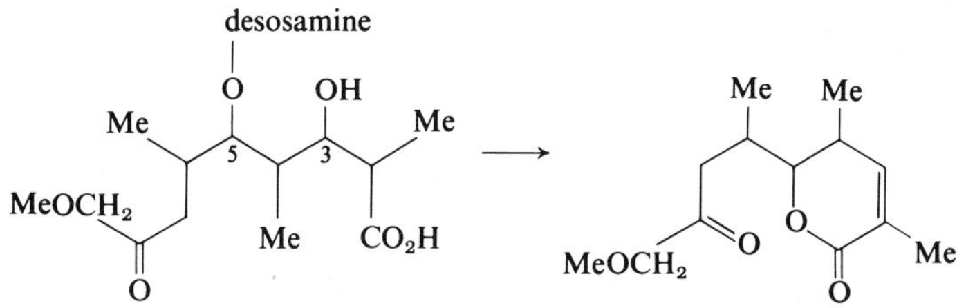

of attachment of desosamine (**5**) at C-5 and of L-oleandrose (**11**) at C-3 in the molecule. The α-L anomeric linkage of **11** and the β-D linkage of **5** in oleandomycin have been deduced from n.m.r. and optical rotatory data.[15]

The spiramycin complex[49] (foromacidins)[50] has been separated into components A (an alcohol or formate ester) and components B and C (as the corresponding acetate and propionate, respectively). A structure[51] proposed initially for the spiramycins has been revised[52,52a] and is shown as **12**.

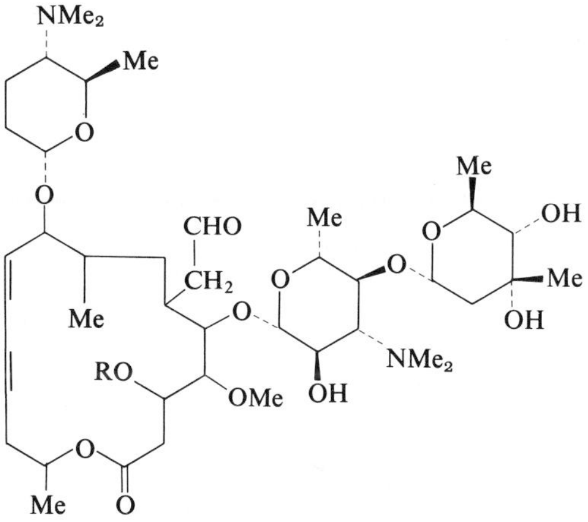

(**12**) Spiramycin A, R = H
Spiramycin B, R = Ac
Spiramycin C, R = COEt

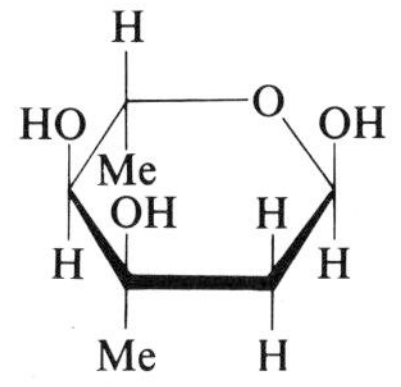

Mycarose (**13**)

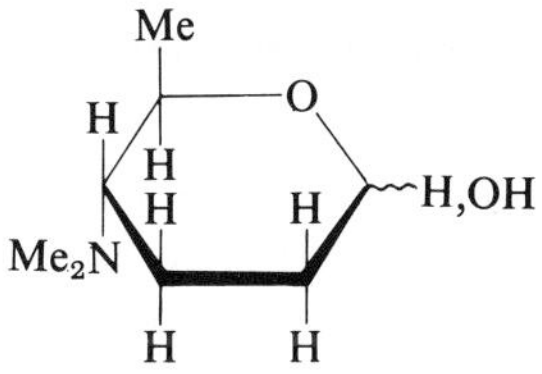

Forosamine (**14**)

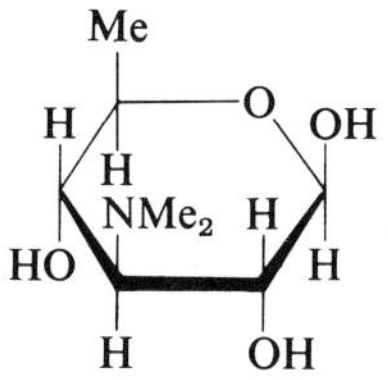

Mycaminose (**15**)

* *References start on p. 200.*

Graded acid hydrolysis of the spiramycins gives mycarose[49] (13) (2,6-di-deoxy-3-C-methyl-L-*ribo*-hexose),[45,45a,53] forosamine[54] (14) (4-dimethyl-amino-2,3,4,6-tetradeoxy-D-*erythro*-hexose),[55] and mycaminose[56] (15) (3,6-dideoxy-3-dimethylamino-β-D-glucose).[57,58] Degradation of 12 and of the various products obtained by selective cleavage of the sugars demonstrated that mycarose was linked to mycaminose and the latter was attached to the lactone ring, as was the forosamine residue. Considerations based on molec-ular rotational calculations[15] allowed the assignment of an α-L linkage of mycarose to mycaminose, and of a β-D linkage of mycaminose to the macro-cyclic lactone ring.

The structure[21] of magnamycin (carbomycin), which contains mycarose and mycaminose, has also been revised[52,52a,59] and is shown as 16.

Magnamycin (16)

The presence of 4-O-isovaleroyl-mycarose (carimbose) is noteworthy, as well as the presence of an epoxide group in the lactone ring. Magnamycin B[60] (carbomycin B) contains a double bond instead of the epoxide group and can be obtained chemically from magnamycin. Niddamycin has properties similar to those of magnamycin B and has been shown[61] to be the 3-desacetyl analog of magnamycin B.

The structure[62] of mycarose (13) ($C_7H_{14}O_4$) was deduced from functional-group analysis and periodate-oxidation data. Oxidation of 13 by hypo-bromite to give a lactone ($C_7H_{12}O_4$) was in agreement with the aldose structure assigned. The close relationship between mycarose (13) and cladinose (8) was demonstrated through methylation[45,63] and demethylation[53,63] experi-ments. By independent oxidative methods, the configuration at C-5 in mycarose was correlated with that in L-rhamnose[64] and with L-lactic acid,[45] thus establishing the L configuration of mycarose. Detailed n.m.r. spectral analysis[53] of mycarose diacetate, as well as syntheses of DL-mycarose[45,65] and of the natural L enantiomorph,[45a] have been reported.

The structure[66] of mycaminose (15) was likewise established from periodate oxidation data, formation of a triacetate, and rapid liberation of dimethyl-amine when 15 was treated with base (indicating a β-dimethylamino aldehyde structure). Studies on the stereochemistry of mycaminose led to the assign-ment[67] of an *erythro* relationship at C-4 and C-5. After the *galacto* and *altro* structures had been excluded,[67] the synthesis of 15 by standard procedures was reported.[57,58] The glycosidic attachment of mycaminose in magnamycin is shown to be β-D on the basis of n.m.r. spectral studies.[44] Optical rotatory data have corroborated this assignment and indicate[15] the α-L linkage for the mycarose–mycaminose portion.

Tylosin ($C_{45}H_{77}NO_{17}$) has been shown[68] to contain the sugars cladinose, mycaminose, and mycinose (17) (6-deoxy-2,3-di-O-methyl-D-allose).[69,70] A degradation product, which was assigned the structure 18, indicates the nature and position of attachment of mycinose and the identity of the lactone ring-oxygen atom. Nuclear magnetic resonance spectral data[15] suggest an α-L linkage to cladinose and a β-D linkage to mycaminose and mycinose.

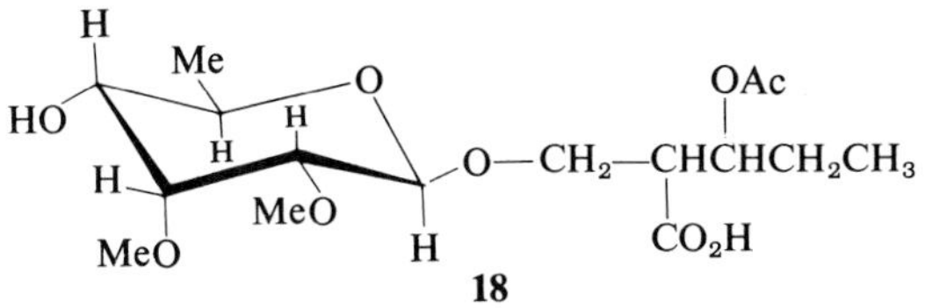

18

The complete structure[52a,70a] of leucomycin A3, a component of the leuco-mycins, has been elucidated (18a).

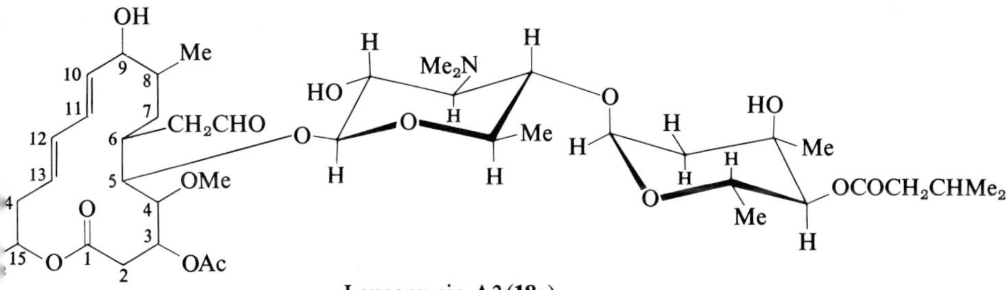

Leucomycin A3 (18a)

Angolamycin[70b] affords on acid hydrolysis the sugars L-mycarose, mycinose, and angolosamine (18b) (3-dimethylamino-2,3,6-trideoxy-D-*xylo*-hexose).[70c] The stereochemistry of angolosamine was derived from n.m.r. spectroscopic studies.

* References start on p. 200.

A configurational model for macrolide antibiotics has been proposed,[15] based on the established absolute configuration of oleandomycin (10). The predicted configuration in various macrolide antibiotics could have interesting implications in the areas of molecular structures, biogenesis, and mode of action.

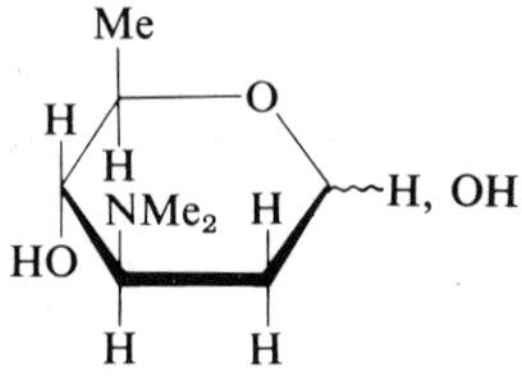

Angolosamine (18b)

b. *Macrocyclic Lactone Antibiotics Not Containing Nitrogen*—This group of antibiotics contains large, unsaturated lactone rings but contains unusual types of deoxy sugar instead of amino sugars.

The structure of chalcomycin (19), a complex macrocylic antibiotic ($C_{35}H_{56}O_{14}$), has been totally elucidated[71] by the elegant application of degradation reactions and n.m.r. spectroscopy.

Two fragments, each containing one of the sugar residues, could be obtained by degrading chalcomycin. Thus, treatment of hexahydro chalcomycin with sodium hydroxide opened the lactone ring, and subsequent periodate oxidation effected cleavage between C-8 and C-9 to give fragments 20 and 21.

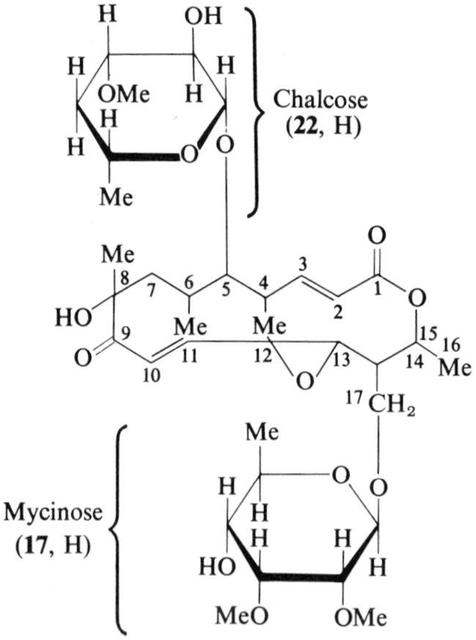

Chalcose (22, H)

Mycinose (17)

Mycinose (17, H)

Chalcomycin (19)

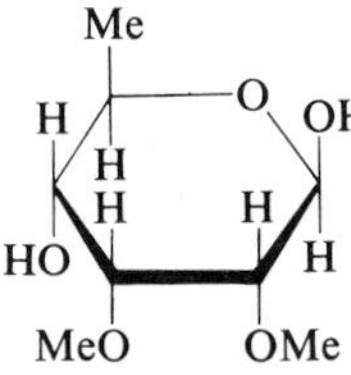

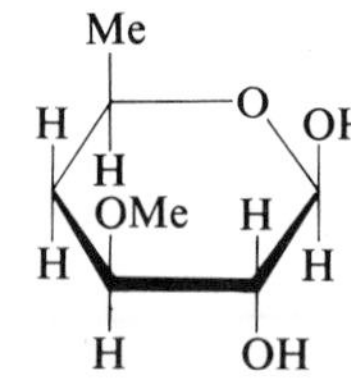

Chalcose (22)

Reduction of chalcomycin with selected catalysts, followed by periodate oxidation, afforded other fragments which confirmed the overall structure of the antibiotic. The presence of unsaturated and epoxide groups and methyl side chains was established by n.m.r. spectral studies on chalcomycin and its transformation products.

Methanolysis of chalcomycin followed by acid hydrolysis gives crystalline chalcose[72] (22) (4,6-dideoxy-3-O-methyl-β-D-xylo-hexose).[73] Its structure was deduced from periodate oxidation, borohydride reduction, and demethylation experiments. Periodate–permanganate oxidation of chalcose afforded a 5-carbon methoxylactone, which was synthesized independently. The complete structure and stereochemistry of chalcose were elucidated by n.m.r. spectroscopy[73] and by a correlation[74] with desosamine (5); several syntheses of chalcose have been reported.[58,75,76] The point of attachment of chalcose in chalcomycin was shown to be at C-5, since hypoiodite oxidation of fragment 21 and hydrolysis of the resulting diacid afforded a γ-lactonic acid. The β-D-glycosidic attachment was confirmed by n.m.r. spectroscopy. The other deoxy sugar, mycinose (17), was also isolated from chalcomycin by methanolysis and acid hydrolysis.[72] Its structure was established by the isolation of meso-2,3-dimethoxysuccinic acid (C-1 to C-4) by oxidative reactions and by demethylation with boron trichloride, which afforded crystalline 6-deoxy-D-allose.[77] Examination of 17 and its methyl glycoside by n.m.r. spectroscopy revealed a β-D configuration for the anomeric center, which could also be predicted from mutarotation studies. A synthesis of 17 from L-rhamnose by established routes has been described.[70]

* References start on p. 200.

The antibiotic lankamycin[78] ($C_{42}H_{72}O_{16}$) has also been shown[79] to contain chalcose (22) (also termed lankavose)[80] and another branched-chain deoxy sugar, arcanose (23) (2,6-dideoxy-3-C-methyl-3-O-methyl-L-*xylo*-hexose),[81,81a] occurring as the 4-acetate in the antibiotic.[80] The structure[80] of lankamycin, which was arrived at by degradative and n.m.r. studies, is represented as 24.

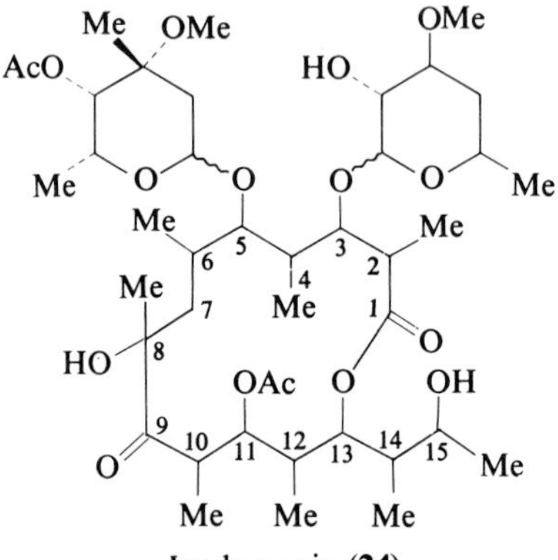

Lankamycin (24)

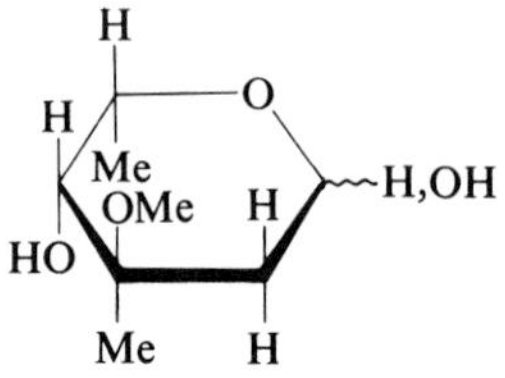

Arcanose (23)

Mild methanolysis of lankamycin affords 4-O-acetylarcanose and the aglycon, darcanolide, containing chalcose. Total methanolysis of lankamycin gives the methyl glycoside of chalcose and the aglycon, monoacetyllankanol- ide. The position of attachment of the sugars was arrived at by the isolation and further degradation of fragments 25 and 26.

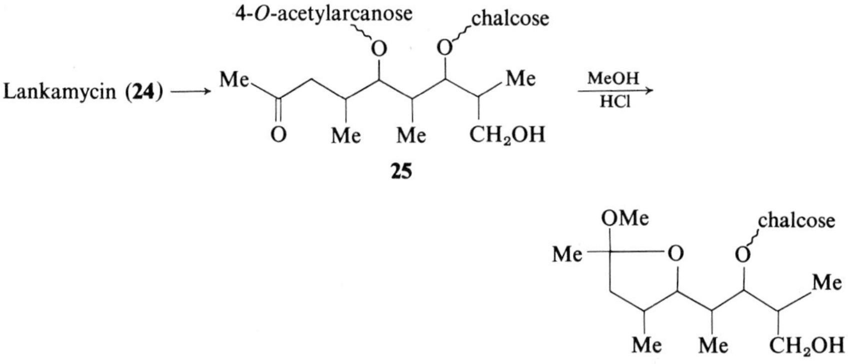

25

26

Although anomeric assignments have not been finally made, a β-D linkage can be predicted[15] for chalcose and an α-L linkage for arcanose on the basis of optical rotatory data. The degradation of arcanose and cladinose to the same 3-methoxy-3'-methyl-1,4-butanediol indicated the structural identity at

C-1 to C-4 and a possible difference in the configuration at C-4 or C-5. The L-*xylo* configuration was ultimately established[81] for arcanose by its transformation into cladinose through the reduction of a 4-keto derivative with lithium aluminum hydride. Thus, arcanose is the C-4 epimer of cladinose.

The antibiotics aldgamycin E,[82] megacidin,[83] and bandamycin A[84] show an absorption band at 1800 cm^{-1} in their infrared spectra. The origin of this band in the former antibiotic has been ascribed[85] to the presence of a sugar containing a cyclic carbonate group. Methanolysis of aldgamycin E produces the methyl glycoside of mycinose (17) and two new sugars, methyl aldgarosides A (27) and B (28). Structures 27 and 28 were assigned[85a] on the basis

27   28

of n.m.r. and mass spectral data. Treatment of the glycoside 27 with dilute barium hydroxide removed the carbonate group, and subsequent periodate oxidation revealed the branched-chain structure. It is interesting to note that the free sugar aldgarose has the dideoxyhexose structure encountered in chalcose (22), one of the sugar components of chalcomycin (19) and neutramycin.[86] An added feature of biogenetic interest is that mycinose (17) is also present in chalcomycin, aldgamycin E, and neutramycin.

B. CYCLITOL ANTIBIOTICS

Grouped under this section are many important antibiotics which have previously been classified under the heading "aminoglycosidic antibiotics." This general name is in a way appropriate, since most of the members are composed solely of carbohydrates attached to each other by glycosidic linkages. A common feature in these substances is the presence of a cyclitol ring, either as inositol or in a substituted form, usually containing one or two amino groups. Classification according to the type of cyclitol provides a more coherent system, because it includes all antibiotics that contain a cyclitol ring in their structures.

* *References start on p. 200.*

1. Inositol Antibiotics

The antibiotic kasugamycin ($C_{15}H_{17}N_3O_{10}$), which inhibits the growth of various bacteria, has been reported[87,87a] to have structure **29**. A unique feature

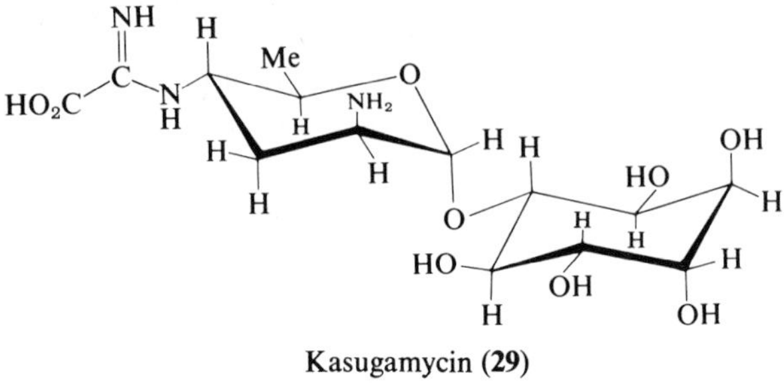

Kasugamycin (**29**)

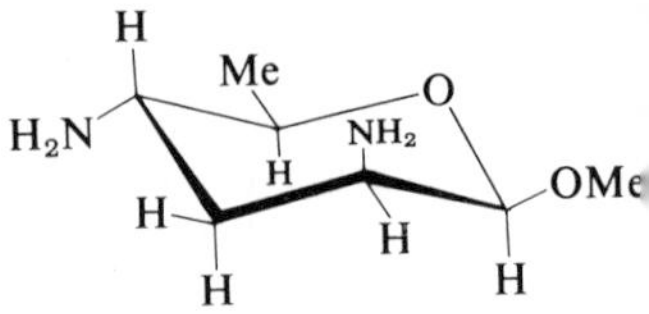

30

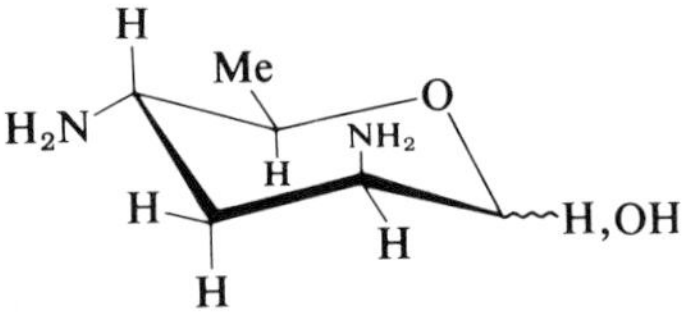

31

in its structure is the presence of a 2,4-diamino-2,3,4,6-tetradeoxyhexose, containing an unusual C-4 amidine group. Acid hydrolysis of kasugamycin affords ($+$)-inositol, but the remainder of the molecule is destroyed. Methanolysis,[88] on the other hand, replaces the inositol group in kasugamycin by a methoxyl group and affords a glycoside, $C_9H_{17}N_3O_4$. Alkaline hydrolysis of this product gives oxalate, ammonia, and methyl kasugaminide (**30**), which is resistant to periodate oxidation. Its structure and stereochemistry were established[89] mainly by n.m.r. spectroscopy, which indicated that the free sugar kasugamine (**31**) is 2,4-diamino-2,3,4,6-tetradeoxy-D-*arabino*-hexose. The only other related example in biological products is the 4-acetamido-2-amino-2,4,6-trideoxyhexose[90] isolated from a polysaccharide from *Bacillus subtilis*. Alkaline hydrolysis[91] of kasugamycin itself affords kasuganobiosamine ($C_{12}H_{24}N_2O_7$) (**32a**, R = H) and kasugamycinic acid ($C_{14}H_{24}N_2O_{10}$) (**32b**), together with oxalate and ammonia.

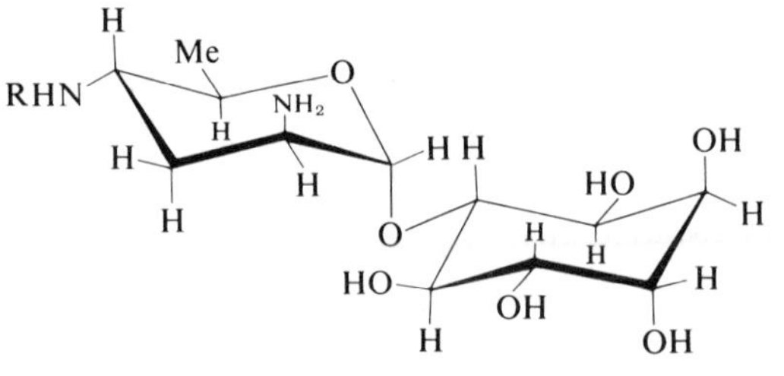

32a, R = H

32b, R = CCO$_2$H
           ‖
           O

Conclusive evidence for the amidine type of structure in kasugamycin was achieved through a partial synthesis.[87] Thus, treatment of **32a** with the di-ethyl ester of oxalimidic acid and subsequent mild hydrolysis of the product afforded a product identical with natural kasugamycin. The α-D anomeric configuration in kasugamycin was suggested from n.m.r. spectral data.[91] The position of attachment of the (+)-inositol residue, as well as the complete structure of kasugamycin, has been confirmed by an X-ray crystallographic analysis[92] of kasugamycin hydrobromide. A synthesis of methyl kasug-aminide[92a] as well as the total synthesis of kasugamycin have been reported.[92b]

2. *Aminocyclohexanepentol Antibiotics*

Hygromycin A (homomycin) $(C_{23}H_{29}NO_{12})$, a metabolic product of *Actinomyces* sp., has been shown[93,94] to have structure **33**.

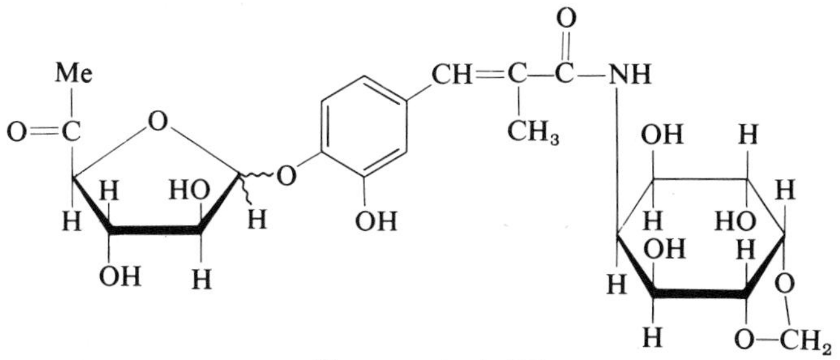

Hygromycin A (**33**)

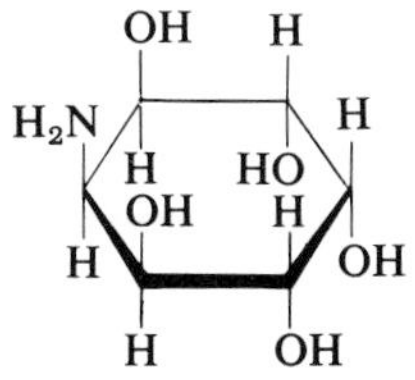

34

Of particular interest is the presence of the 6-deoxy-D-*arabino*-hexos-5-ulose structure and the methylene acetal grouping in the inosamine moiety. Furthermore, the phenolic portion, and not the inosamine portion, forms the aglycon. Acid hydrolysis[93–95] of hygromycin A afforded *neo*-inosamine-2 (**34**) and formaldehyde. Alkaline hydrolysis of hygromycin A gave 3,4-dihydroxy-α-methylcinnamic acid, arising from the central portion of the molecule. The intact methylene *neo*-inosamine could also be isolated from alkaline hydrolyzates, and its structure was established[93] by periodate oxidation. Methylation and hydrolysis of hygromycin A afforded 4-hydroxy-3-methoxy-α-methylcinnamic acid, thus establishing the position of the

* *References start on p. 200.*

glycosidic linkage. Assigning the position of the carbonyl group was accomplished by treatment of the antibiotic with ethanethiol followed by desulfurization of the resulting dithioacetal. Mercaptolysis of this modified product gave a dideoxyhexose diethyl dithioacetal which produced propionaldehyde upon periodate oxidation (from the C-4–C-6 fragment). When the position of the carbonyl group had been established, hygromycin A was reduced with sodium borohydride, and the reduced product was subjected to mercaptolysis. L-Fucose diethyl dithioacetal was isolated, which indicated the D-*arabino* stereochemistry of the sugar moiety. The synthesis of the sugar component of hygromycin A has been accomplished.[95a]

Except for the assignment of the nature of the R and R′ groups, the antibiotic bluensomycin[96] (glebomycin)[97] has been given structure **35**. Methanolysis[98] gave methyl dihydrostreptobiosaminide (**36**) and bluensidine (**37**) (glebidine) ($C_8H_{16}N_4O_6$), which is 1-deoxy-3-*O*-carbamoyl-1-guanidino-*scyllo*-inositol.

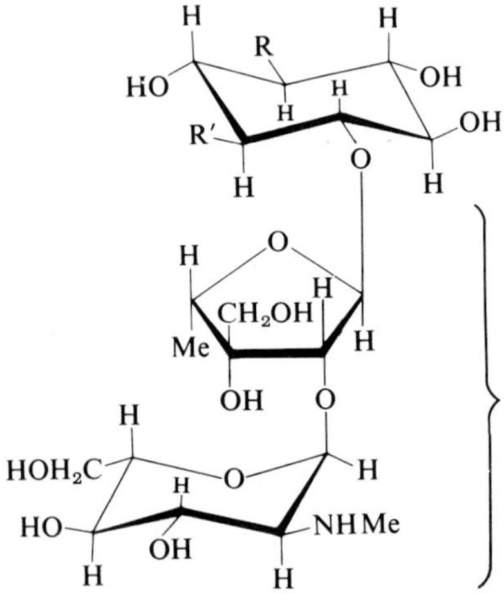

Bluensomycin (**35**)

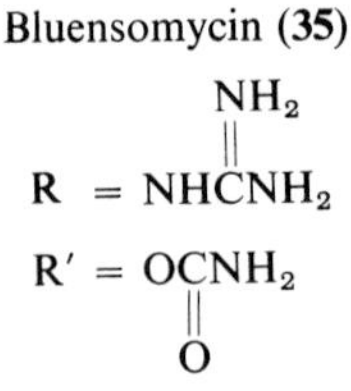

$$R = NHCNH_2$$
$$\overset{\parallel}{}NH_2$$

$$R' = OCNH_2$$
$$\overset{\parallel}{}O$$

(or vice versa)

36, OMe

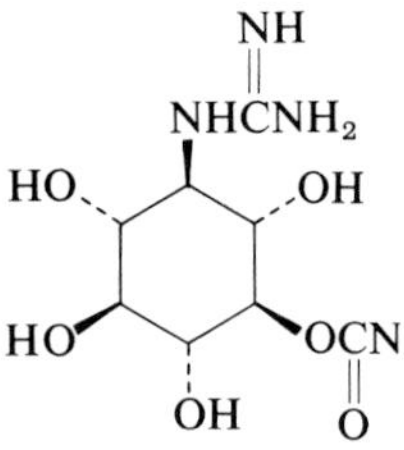

Bluensidine (**37**)

Structure **37** was established by alkaline and acid hydrolysis, by comparison with model compounds, by periodate-oxidation data, and by spectral observations. The identity of the methyl dihydrostreptobiosaminide fragment (**36**) was confirmed by comparison with authentic material available from dihydrostreptomycin. Consideration of the n.m.r. spectral characteristics of the anomeric hydrogen atoms in the sulfate of bluensomycin allowed the assignment[99] of the α-L configuration for both anomeric linkages. Further experiments[96,98] established that bluensomycin is the same as dihydrostreptomycin except that one of the guanidino groups is replaced by the biogenetically related carbamate group.

3. Diaminocyclohexanetetrol Antibiotics

The gross structural features of streptomycin have been reviewed.[2,3,100] As a result of the application of modern instrumental techniques in the past few years, a complete structure, including absolute stereochemistry, can now be written for streptomycin; it is shown as **38**. The configuration of the glycosidic linkage between 2-deoxy-2-methylamino-L-glucose (**39**) and streptose (**40**) was previously shown[101] to be α-L; that between streptose and streptidine (**41**) was suggested[101] to be β-L, on the basis of rotational data for perbenzoylated derivatives. The latter anomeric configuration has been revised to that of α-L on the basis of n.m.r. spectral data, whereas the α-L assignment to the linkage between **39** and **40** was confirmed.[99] However, an α-L

Streptomycin (**38**)

assignment made earlier[102] for the linkage between **40** and **41** is in agreement with the n.m.r. data. With the available information on the structure of streptomycin, only the assignment of absolute stereochemistry to the substituted streptidine fragment (**41**) remained to be established. Although

* *References start on p. 200.*

streptidine itself is a *meso* form (all *trans*), it is asymmetrically substituted by streptobiosamine (**42**) in the intact antibiotic. The absolute configuration of the streptidine moiety (as shown in **38**) was determined[103] on N,N^1-diacetyl-4-deoxystreptamine[104] (obtained from bluensomycin (**35**) by a series of reactions) by the application of Reeves' cuprammonium method.[105] The absolute configuration of the streptidine moiety in dihydrostreptomycin (**43**) (CH_2OH in place of CHO in **38**) has been determined[106,107] by using Reeves' method[105] on di-*O*-methylstreptamine, obtained by acid hydrolysis of *N*-acetylated, permethylated dihydrostreptomycin. A partial synthesis[108] of dihydrostreptose (CH_2OH in place of CHO in **40**), originally isolated[109] by hydrolysis of methyl *N*-acetyldihydrostreptobiosaminide with mild acid, has been reported. Synthesis[110] of streptose (**40**) has also been achieved.

Two forms of salts are known[111] for streptomycin, and their structures have been rationalized[112] on the basis of a reversible cyclization between the aldehyde group and the *N*-methyl group. The crystal and molecular structure of streptomycin oxime selenate has been reported.[112a]

Dihydrostreptomycin (**43**)

Hydroxystreptomycin[113] is the same as **38** except that the C-5 methyl group in the streptose moiety is replaced by a CH_2OH group.

A proposed structure,[114] exclusive of stereochemistry, of the antibiotic actinospectacin ($C_{14}H_{24}N_2O_7$) is shown as **44**. Acid hydrolysis of **44** gave actinamine[115,115a] (**45**), which, from optical rotational data (*meso*, inactive) and periodate oxidation data, was formulated as 1,3-di-*N*-methylamino-

2,3,4,6-tetrahydroxycyclohexane. Both the structure and the conformation of **45** were established by n.m.r. spectral studies.[116] Syntheses of **45** as well as the parent diamine have been reported.[116a–c] Various hydrolytic procedures[114] on actinospectacin or its dihydro form ($C_{14}H_{26}N_2O_7$) failed to provide the remaining 6-carbon fragment and afforded only actinamine. The

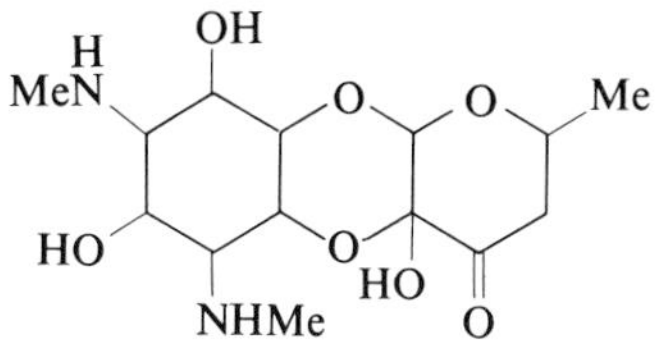

Actinospectacin (**44**) 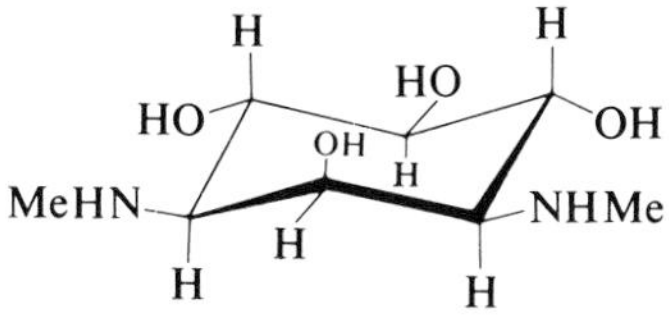 Actinamine (**45**)

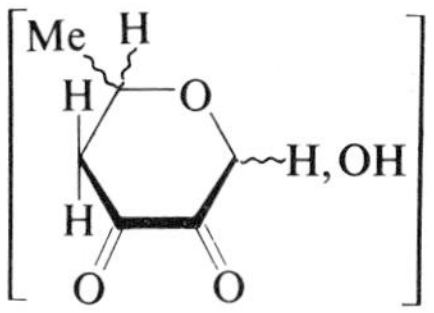

Actinospectose (**46**)

nature of the sugar portion, called actinospectose (**46**) (not isolated) was suggested from periodate oxidation data of various derivatives. Mild base converts actinospectacin into actinospectoic acid[114] (**47**) ($C_{14}H_{28}N_2O_8$) by a tertiary ketol rearrangement, and acid hydrolysis of **47** affords a 5-carbon fragment by the following sequence of reactions.

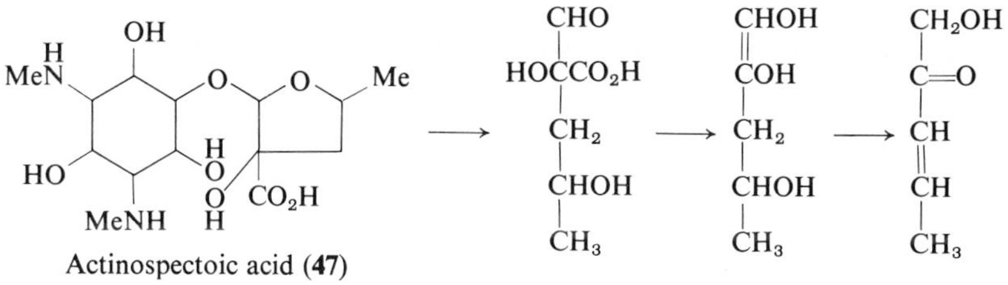

Actinospectoic acid (**47**)

$$
\begin{array}{ccc}
\text{CHO} & \text{CHOH} & \text{CH}_2\text{OH} \\
| & \| & | \\
\text{HOCCO}_2\text{H} & \text{COH} & \text{C}=\text{O} \\
| & | & | \\
\text{CH}_2 & \text{CH}_2 & \text{CH} \\
| & | & \| \\
\text{CHOH} & \text{CHOH} & \text{CH} \\
| & | & | \\
\text{CH}_3 & \text{CH}_3 & \text{CH}_3
\end{array}
$$

The structure of the unusual sugar actinospectose can thus be represented as 4,6-dideoxy-D- or L-*glycero*-hexos-2,3-diulose.

4. *Diaminocyclohexanetriol Antibiotics*

Neomycin[117] was the first to be discovered among the group of cyclitol-containing antibiotics. Two important components of this antibiotic complex are neomycin B (**48a**) and neomycin C (**48b**). The physical and chemical

* *References start on p. 200.*

properties of these as well as degradative sequences leading to the structure elucidation are described in several reviews.[2,3,118]

A key step in the degradation of the neomycins is methanolysis,[119] which cleaves the molecule at the site remotest from an amino group—that is, the

49, H, a, b: R″ = CH$_2$NH$_2$
  c: R″ = CH$_2$OH

50, H, a: R = H; R′ = CH$_2$NH$_2$
  b: R = CH$_2$NH$_2$; R′ = H

48a, Neomycin B: R′,R″ = CH$_2$NH$_2$; R = H

48b, Neomycin C: R,R″ = CH$_2$NH$_2$; R′ = H

48c, Paromomycin I: R′ = CH$_2$NH$_2$; R = H; R″ = CH$_2$OH

48d, Paromomycin II: R = CH$_2$NH$_2$; R′ = H; R″ = CH$_2$OH

Deoxystreptamine (51)

bond between D-ribose and deoxystreptamine (**51**). The products, neamine[120] (**49b**) [2-(2,6-diamino-2,6-dideoxy-D-glucopyranosyl)deoxystreptamine][121,121a] and methyl neobiosaminides B and C (**50a** and **50b**), are obtained from **48a** and **48b**, respectively.

Informative data regarding the point of linkage and ring size have been secured from methylation and hydrolytic studies.[122] Besides deoxystreptamine[123] and D-ribose,[124] the neomycins contain neosamine B (**52**) (2,6-diamino-2,6-dideoxy-L-idose)[125] and neosamine C (**53**) (2,6-diamino-2,6-dideoxy-D-glucose).[126]

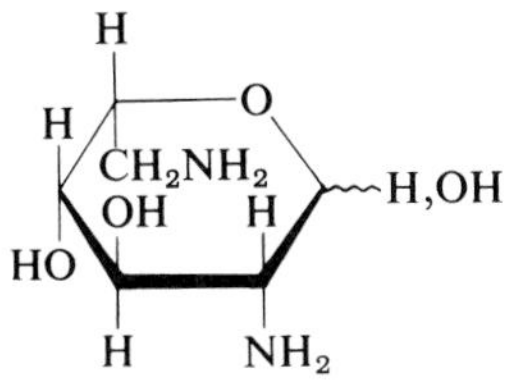

Neosamine B (Paromose) (**52**) 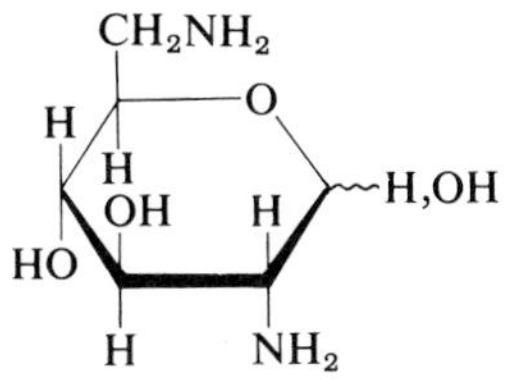   Neosamine C (**53**)

The structures of these rare sugars were established by various oxidative degradations of suitable derivatives; an example in the case of neosamine C (**53**) is shown below.

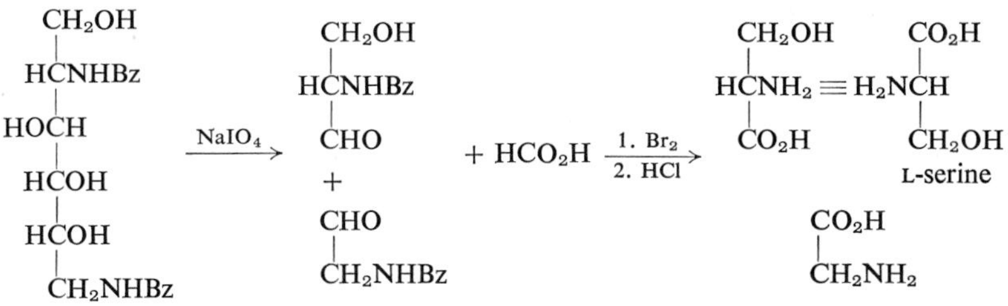

The total stereochemistry of neosamine C was established[127] mainly from optical rotatory considerations and by comparisons with model compounds, and ultimately by synthesis.[128,129] The stereochemistry of neosamine B (**52**)

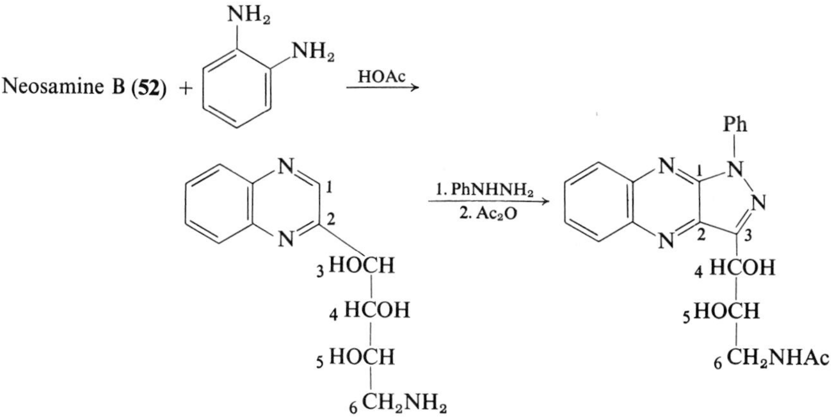

Neosamine B (**52**) +

* *References start on p. 200.*

was first suggested[122] on the basis of conformational arguments and on biogenetic considerations. The conclusions were substantiated by synthesis[125] and by degradation;[118] the latter procedure indicated the stereochemistry at C-3, C-4, and C-5, as illustrated.

The attachment of the neosamines (52 and 53) to C-3 of the D-ribose moiety, and that between the neobiosamines and neamine,[130] were established by methylation and hydrolysis studies.[122] Definitive evidence for the all-*trans* stereochemistry of deoxystreptamine (51) was obtained from n.m.r. spectra.[131] The absolute stereochemistry was derived[132] by application of Reeves' cuprammonium method[105] to appropriate derivatives. Evidence for the configuration of the anomeric linkages in neomycins B and C has been obtained from n.m.r. spectral studies.[133] The stereochemistry of deoxystreptamine in neomycins B and C has been established chemically.[134]

The structure of the paromomycins (catenulin, hydroxymycin, aminosidin, zygomycin, monomycin) has been established independently[135] as 48c and 48d through degradative reactions similar to those discussed for the neomycins. Methanolysis[136] of paromomycin I (48c) gives paromamine (49c) [2-(2-amino-2-deoxy-α-D-glucopyranosyl)deoxystreptamine] and methyl paromobiosaminide (50a).[137] The structure of the diaminohexose 52 (paromose,[138] identical with neosamine B), as well as its configuration at C-2 (D-*glycero*), was established[135] by the isolation of L-alanine and glycine from oxidative reactions on 1-deoxyparomitol derivatives. Mercaptolysis of paromomycin I gave crystalline paromose dibenzyl dithioacetal. Through a sequence of reactions outlined below, this derivative was used to establish the stereo-

HC(SCH$_2$Ph)$_2$ HC(SO$_2$CH$_2$Ph)$_2$
|
HCNHAc   HCNHAc
|
HOCH   CH$_3$CH$_2$CO$_3$H   HOCH   NH$_4$OH
| →
HCOH   HCOH
|
HOCH   HOCH
|
CH$_2$NHAc   CH$_2$NHAc

+ CH$_2$(SO$_2$CH$_2$Ph)$_2$
+ AcNH$_2$

CH$_2$NHAc

chemistry at C-3, C-4, and C-5, thus characterizing[139] paromose as 2,6-diamino-2,6-dideoxy-L-idose.

The resulting 5-acetamido-5-deoxy-L-xylose existed as two separate ring forms and was compared with synthetic samples.[140,141] Paromomycin II (48d), a minor component in the paromomycin complex, has been shown[118] to contain neobiosamine C (50b), rather than paromobiosamine (50a). The configurations of the glycosidic centers and the absolute stereochemistry of deoxystreptamine (51) in paromomycins I and II have been determined by n.m.r.[133] and by optical rotational measurements,[132] respectively. The zygomycin A_1 and A_2 structures, which are identical with 48c and 48b, respectively,[106,142] and the hybrimycins[142a] (neomycin analogs), have been investigated.

Kanamycin, discovered by Umezawa and co-workers,[143] also contains deoxystreptamine (51). The major component kanamycin A, has structure[5] 54, which has been corroborated by X-ray crystallographic analysis.[143a]

CH₂OH structure — 3-Amino-3-deoxy-D-glucose (kanosamine) (55)

Kanamycin A (54)

6-Amino-6-deoxy-D-glucose (56)

Proof of the total structure of kanamycin A was reported by three groups of workers.[144-146] The pertinent degradative reactions are essentially those described previously. Because of the proximity of the amino groups in the deoxystreptamine moiety to the glycosidic linkages, methanolysis of kanamycin A required relatively vigorous conditions. Acid hydrolysis produced, in addition to deoxystreptamine (51), two amino sugars—kanosamine[147] (3-amino-3-deoxy-D-glucose)[145,148,148a] (55) and 6-amino-6-deoxy-D-glucose[147,149,150] (56). Graded acid hydrolysis[151] afforded two disaccharides ($C_{12}H_{25}N_3O_7$), each containing deoxystreptamine and one of the sugars, thus indicating the type of glycosidic linkage in the intact antibiotic.

* References start on p. 200.

The linkage between compounds **51** and **55** is more resistant to acid hydrolysis than the linkage between **51** and **56** because of the proximity of the amino group in **55** to the glycosidic bond in the former disaccharide. The proof of structure and ring size of **56** was based on periodate oxidation data[149,150] of various derivatives, n.m.r. data,[145] deamination experiments,[145] including its conversion into D-glucose pentaacetate, and direct comparison with model compounds.[150] Treatment of **56** with acid gave 5-(aminomethyl)-furfural.[149] The structure of **55**, including its ring size, was also deduced from periodate-oxidation data[150,152,153] of its glycoside and the *N*-acetyl derivative. Evidence regarding both structure and stereochemistry was secured from comparison with synthetic analogs[145] and from the agreement of molecular rotations with those of D-glucose derivatives. Independent proof was obtained[148] by converting **55** into 1-phenyl-(D-*erythro*-trihydroxypropyl)-flavazole, demonstrating the D-*erythro* configuration at C-4 and C-5.

The position of attachment of the amino sugars **55** and **56** to deoxystrept-amine (**51**) was established[154] by methylation and hydrolysis studies, which gave an optically inactive derivative of **51**, and also by partial hydrolysis studies. Periodate oxidation[153] of kanamycin A followed by acid hydrolysis produced only deoxystreptamine, thus indicating the 4,6 substitution pattern in the deoxystreptamine moiety in the intact antibiotic. The same substitu-tion pattern was arrived at by hydrolysis of methylated *N*-acetylated kanamy-cin A, which afforded di-*N*-acetyl-2-deoxy-5-*O*-methylstreptamine.[154] Finally, the absolute stereochemistry of the deoxystreptamine residue in kanamycin A was deduced from optical rotatory data.[132] That both anomeric linkages in kanamycin A are α-D was deduced[145,146,154] from infrared absorption data and from the molecular rotations of appropriate degradation products.

Kanamycin C resembles the A component and has equal antibacterial activity; it differs only in that the 6-amino-6-deoxy-D-glucose moiety is replaced by 2-amino-2-deoxy-D-glucose. On acid hydrolysis[155] it gives paromamine (**49c**). Kanamycin B, which is more active against staphylococci than kanamycin A, has been shown[156] to contain 2,6-diamino-2,6-dideoxy-D-glucose (**53**) in place of 6-amino-6-deoxy-D-glucose (**56**). It is interesting, from a biogenetic standpoint, that the minor components B and C are re-placed by biogenetically possible amino sugars. The occurrence of the diamino-sugar **53** in the kanamycin complex is also noteworthy, since this sugar has so far been encountered only in another family of antibiotics, comprising the neomycins and paromomycins. The total synthesis of Kana-mycin A,[156a,156b] B,[156c] and C[156d,156e] has been accomplished.

The gentamycin complex,[157] isolated from cultures of microorganisms other than *Streptomyces*, contains deoxystreptamine (**51**), 2-amino-2-deoxy-D-glucose, and a 3-deoxy-3-methylaminopentose (garosamine).[157a] The gross structure[157b] of gentamycin is shown as **56a**.

Gentamycin (56a)

The antibiotic hygromycin B_2[158] isolated from culture filtrates of *Streptomyces* producing the A component (33), was shown[159] to have the formula $C_{15}H_{30}N_2O_{10}$. Acid hydrolysis produced D-talose and hyosamine (57) (*N*-methyl-2-deoxystreptamine).

Hyosamine (57)   Hygromycin B_2 (58)

Mild acid hydrolysis produced hygromycin B_2 ($C_{13}H_{26}N_2O_8$), for which structure 58 was suggested[159] exclusive of the anomeric configuration. The remaining $C_2H_4O_2$ residue in hygromycin B_2 has been placed on the talose portion on the basis of periodate-oxidation data. A synthesis of deoxystreptamine (51) and hyosamine (57) has been reported.[159a]

A closely related antibiotic, destomycin A ($C_{20}H_{37}N_3O_{13}$), was assigned structure 59.[160,161] It contains D-talose, *N*-methyldeoxystreptamine (having an optical rotation opposite in sign to that of 57), and a polyhydroxyamino acid, destomic acid[162] (60), which forms a seven-membered lactam ring. From considerations of optical rotatory data, periodate oxidation, and correlation of degradation products from the cyclitol with *N*-methyl-L-aspartic acid, structure 61 was assigned to the main acid degradation product of destomycin A.

An unusual ortho ester type of attachment at the C-2 and C-3 hydroxyl groups of the D-talose moiety was suggested[161] on the basis of methylation and hydrolysis studies.[163] Conclusive evidence for this linkage must await further experiments.

* *References start on p. 200.*

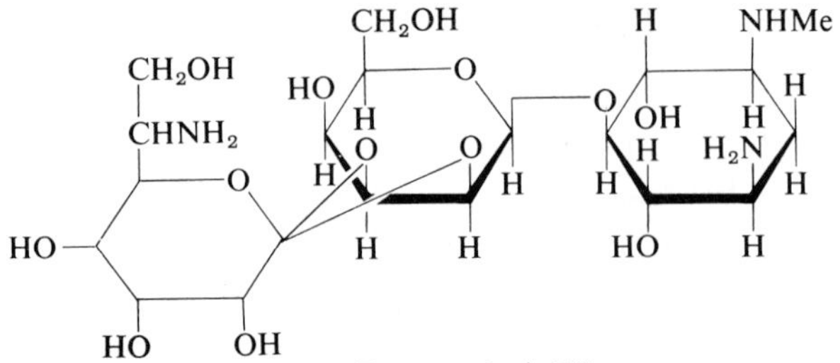

Destomycin A (59)

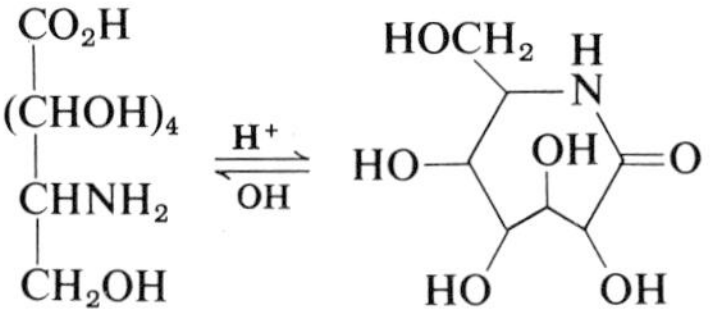

Destomic acid
(60)

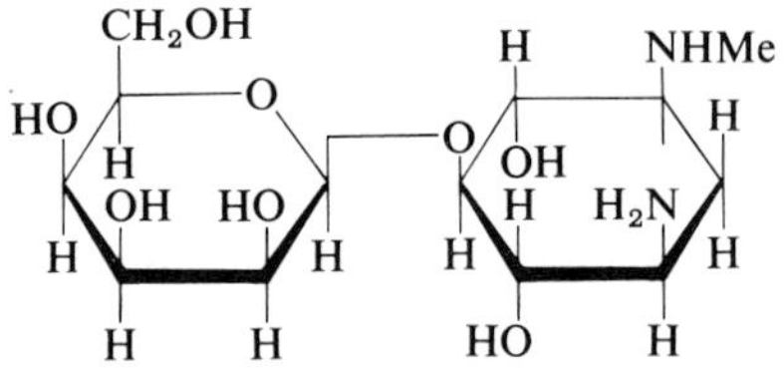

61

C. Nucleoside Antibiotics[163a]

1. Purine Nucleosides[164]

Nebularine, which has strong inhibitory action against microbacteria, has been shown by degradation[165] and synthesis[166–168] to be 9-(β-D-ribofuranosyl)purine (**62**). An improved synthesis, utilizing the fusion procedure catalyzed by bis(p-nitrophenyl) hydrogen phosphate, has been reported.[169] Nucleocidin has been assigned structure **63**, in which a fluoro sugar and a sulfamic acid residue are present.[170,170a]

Cordycepin (**64**), long considered to be a branched-chain sugar nucleoside,[171] has been shown to be 9-(3-deoxy-β-D-*erythro*-pentofuranosyl)-adenine[172,173] by comparison[174] with an authentic sample. The revised structure has been corroborated by ^{14}C-labeling experiments[175] and by mass-spectrometric examination.[176]

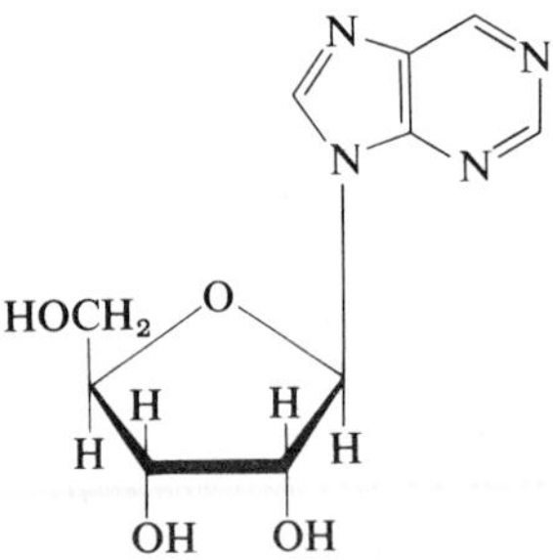

Nebularine (**62**)

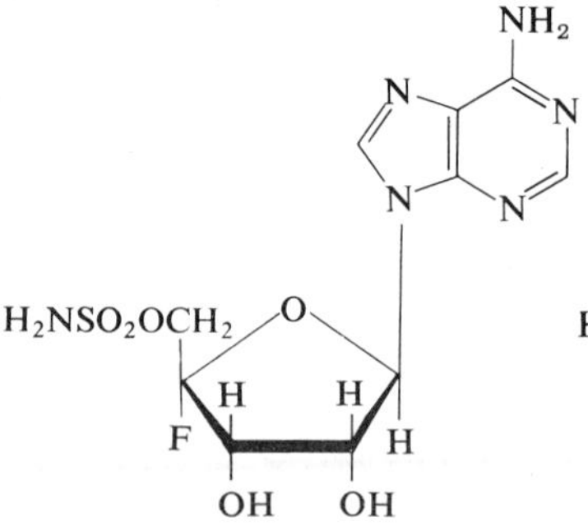

Nucleocidin (**63**)

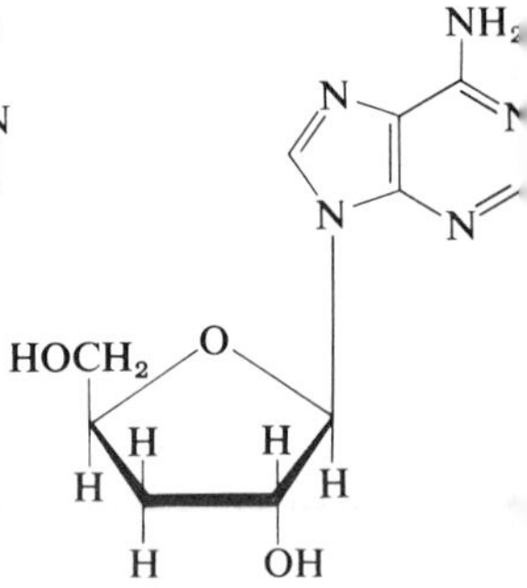

Cordycepin (**64**)

Psicofuranine[177] (angustmycin C),[178] produced by a *Streptomyces* species, was shown to be 9-(D-psicofuranosyl)adenine[179] (**65**). Periodate oxidation data, together with the recognition of the ketose structure, indicated that the nucleoside sugar moiety was psicose.

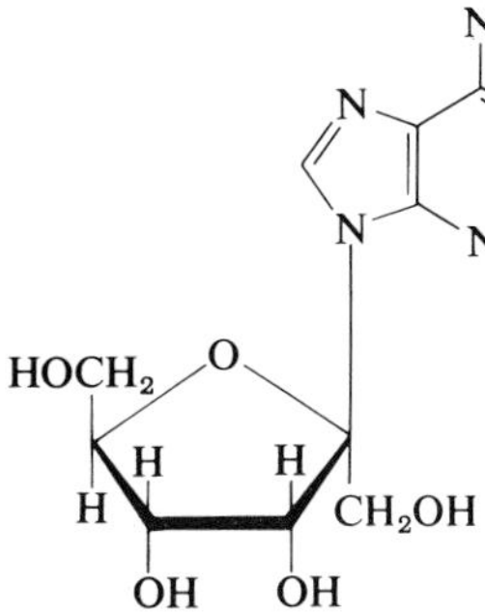

Psicofuranine (**65**)

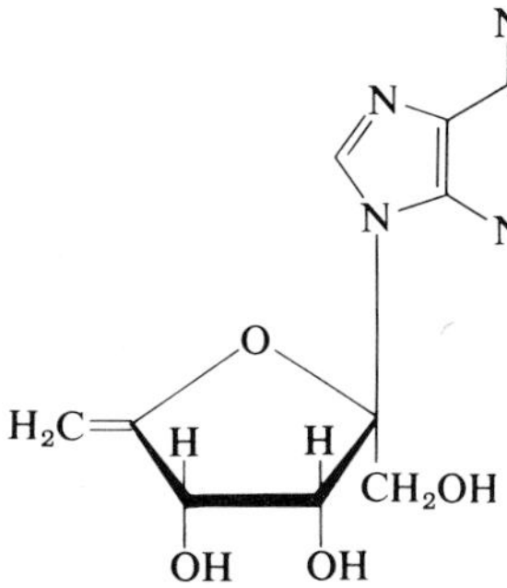

Angustmycin A (**66**)
Decoyinine

Another nucleoside-like product, named angustmycin A, was initially assigned an erroneous structure.[180] The same product has been given the name decoyinine by other workers,[181] and its structure was presented as **66**.

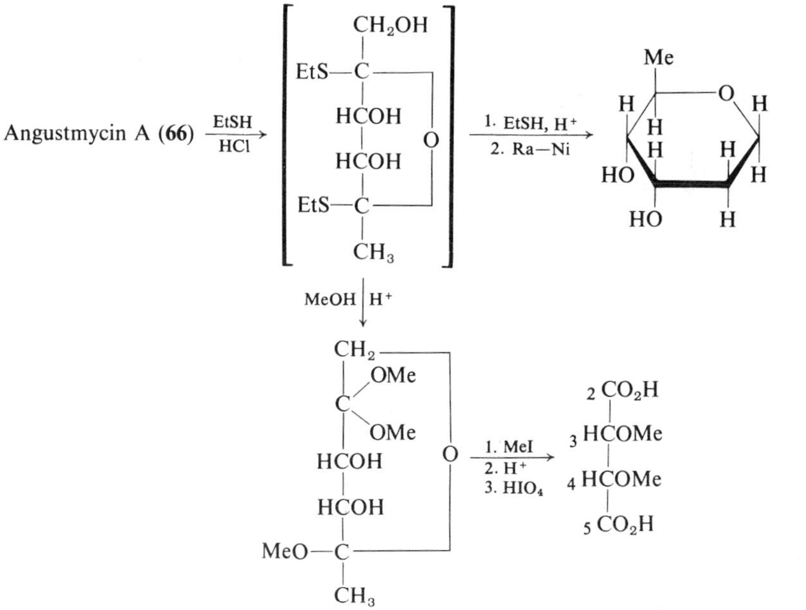

* *References start on p. 200.*

The structural proof[181] of this antibiotic is based on the degradative scheme summarized below. Other evidence is provided from n.m.r. studies on the acetate of angustmycin A and its reduced derivatives, a mixture of C-5 epimeric, 6-deoxyhexulosyl nucleosides.

Puromycin (**67**) was the first member of the nucleoside antibiotic group to be discovered.[182] Elucidation of its structure was achieved through elegant synthetic studies.[3,183] A related nucleoside, 3'-amino-3'-deoxyadenosine, has been isolated[184] from a fungus and has been shown to possess antitumor activity. It is thus the second example of the occurrence of a 3-amino-3-deoxypentose in antibiotic products.

Septacidin (**68**), a cytotoxic and antifungal agent, has been reported[185] to give adenine, an amino sugar, and an insoluble acid compound, on acid hydrolysis. The amino sugar has been identified by n.m.r. spectral studies[186] as a 4-amino-4-deoxyaldoheptose ($C_7H_{15}NO_6$). Acid hydrolysis of septacidin produced glycine and isopalmitic acid. In contrast, methanolysis affords a

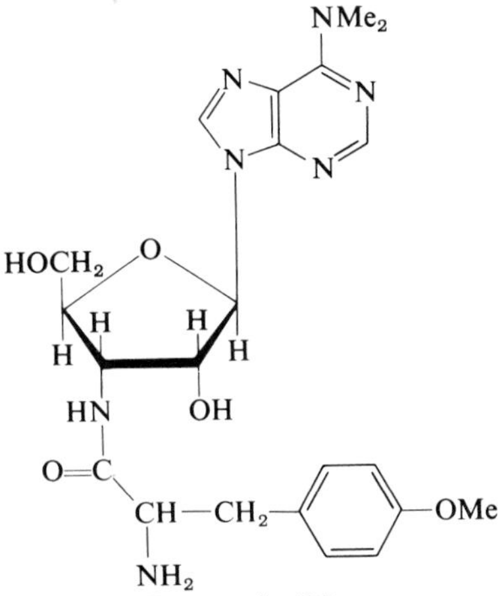

Puromycin (**67**)

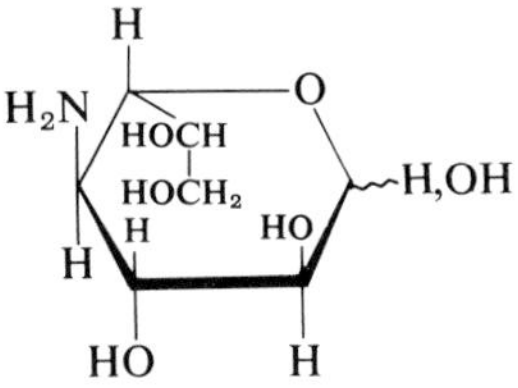

(**69**)

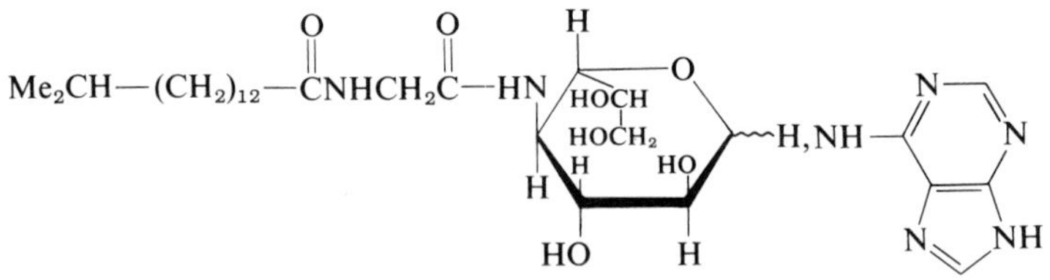

Septacidin (**68**)

"nucleoside" containing the amino sugar and adenine. Controlled periodate oxidation of the amino sugar derivative, and subsequent borohydride reduction and acid hydrolysis, afford 4-amino-4-deoxy-L-glucose.[187] This observation, coupled with the L-*glycero* configuration established for C-6, identifies the aminodeoxyheptose as 4-amino-4-deoxy-L-*glycero*-L-*gluco*-heptose (**69**).

2. Pyrimidine Nucleosides[188]

A feature common to this class of antibiotics isolated from culture filtrates of *Streptomyces* is the presence of cytosine and very unusual amino[2] and deoxy[4] sugars.

Amicetin ($C_{29}H_{42}N_6O_9$) is a metabolic product of several *Streptomyces*.[189-191] A complete structure including total stereochemistry is shown as **70**. This complex molecule contains two unusual sugars, amosamine[192,193]

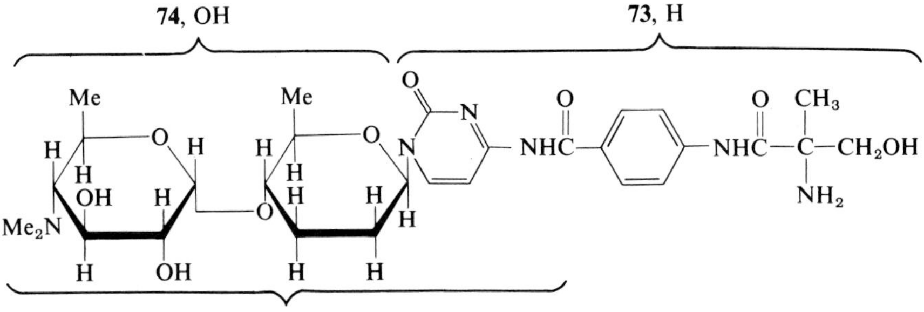

74, OH   73, H

75, H
Amicetin (**70**)

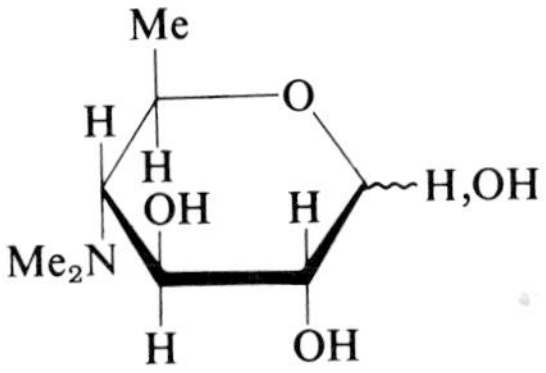

Amosamine (**71**) 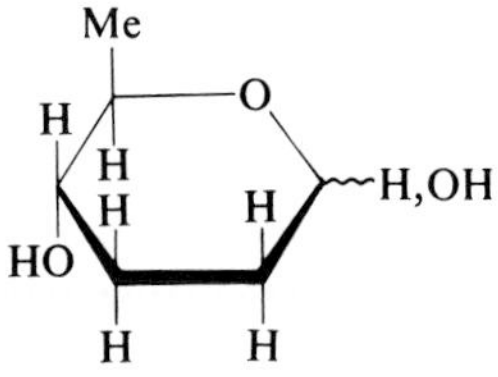   Amicetose (**72**)

(**71**) (4,6-dideoxy-4-dimethylamino-D-glucose)[194] and amicetose[192,193] (**72**) (2,3,6-trideoxy-D-*erythro*-hexose).[195] Hydrolysis of amicetin with mild acid affords cytimidine[190] (**73**) ($C_{15}H_{17}N_5O_4$) and the disaccharide portion, amicetamine[192,196] (**74**). Treatment of amicetin with alkali cleaved the cytosine–*p*-aminobenzoic acid bond and afforded crystalline cytosamine[190,196] (**75**) ($C_{18}H_{30}N_4O_6$). The neutral sugar, amicetose (**72**), isolated as the methyl glycoside by methanolysis of amicetin, was shown by periodate oxidation data to possess the 2,3,6-trideoxyhexose structure.[193] The D-*erythro* stereochemistry was assigned by synthesis[195] of amicetose from D-glucose. The amino sugar portion, also isolated by methanolysis, had erroneously been assigned[193] a 3,6-dideoxy-3-dimethylaminohexose structure on the basis of

* *References start on p. 200.*

periodate oxidation and comparative pK_a data. Its identity, as shown in **71**, was ascertained by synthesis.[194] The configuration of the anomeric linkages in amicetin was established from n.m.r. spectral studies[197] of appropriate derivatives. The n.m.r. spectra of **75** and amicetaminol,[193] obtained from **74** by borohydride reduction, clearly indicated an α-D linkage for amosamine. A β-D linkage and a *C1* (D) conformation for amicetose (**72**) in the intact antibiotic were assigned by investigating the n.m.r. spectra of the crystalline

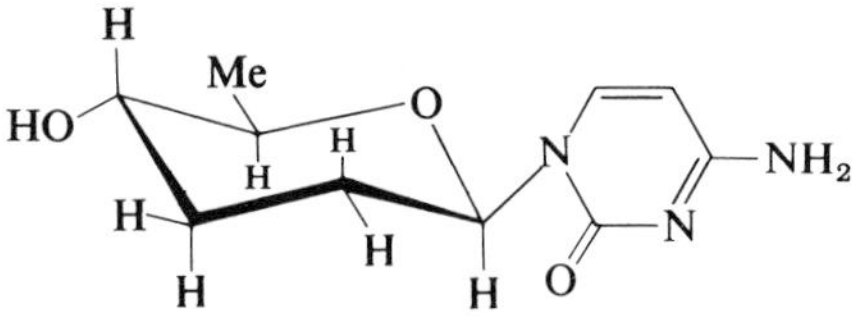

(76)

amicetose nucleoside **76** and the corresponding acetate. The nucleoside **76**, obtained[197] by periodate oxidation of **75** followed by mild acid hydrolysis, has also been synthesized.[198]

Two minor components were also isolated[191] from cultures producing amicetin and were termed bamicetin and plicacetin. The latter was shown[196] to be the same as amicetin except for the lack of the terminal serine moiety. Bamicetin ($C_{28}H_{40}N_6O_9$) on acid hydrolysis gave **73** and bamicetamine ($C_{13}H_{25}NO_6$),[196] which was shown[196a] to be presumably the *N*-methylamino disaccharide corresponding to **74**.

Blasticidin S ($C_{18}H_{24}N_8O_5$), isolated[199] from *Streptomyces* broths, is a powerful inhibitor of a fungus disease in rice plants. Degradation studies[200,201] led to the assignment of structure **77**. The stereochemistry of the sugar at C-4 and C-5, initially designated as L-*erythro*,[202] has been revised to D-*erythro* (as shown in the structure), based on a reinterpretation of the

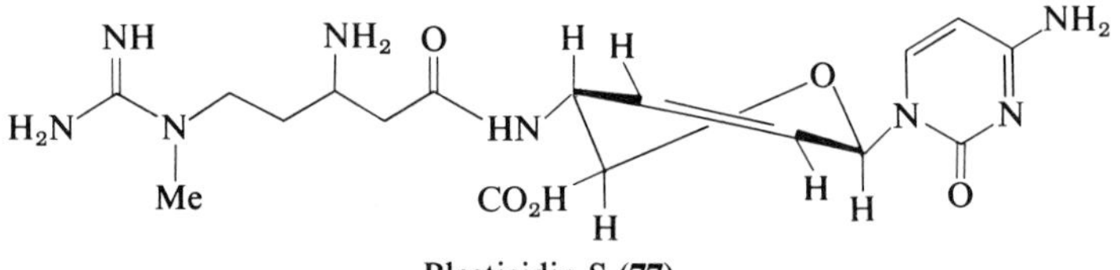

Blasticidin S (77)

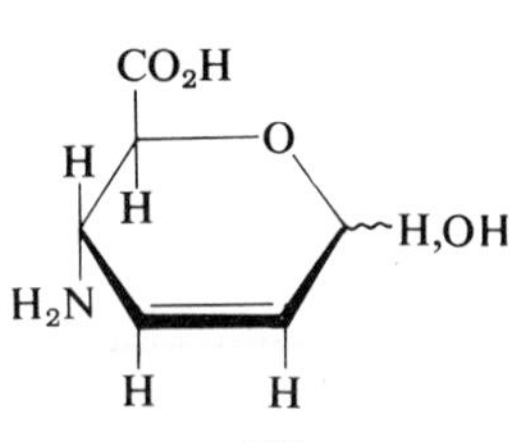

(78)

data.[203,204] The sugar component **78** is 4-amino-4-deoxy-D-*erythro*-hex-2-enopyranuronic acid, a rather remarkable structure. Controlled acid hydrolysis[205] of blasticidin gives cytosinine (**79**) and blastidic acid (**80**) ($C_7H_{16}N_4O_2$). The structure of **79** ($C_{10}H_{12}N_4O_4$) was proved by ultraviolet (cytidine-like) and n.m.r. spectral analysis, and by hydrogenation. Thus, in the presence of platinum oxide and acetic acid three hydrogenation products were formed—dihydrocytosinine, cytosine, and 3-amino-2-carboxypyran (**81**). The last two fragments arise from β elimination of the C-1 substituent through an allylic type of activation. Ozonolysis of *N*-acetylcytosinine

Cytosinine (**79**)

Blastidic acid (**80**)   **81**

followed by oxidation with hydrogen peroxide and acid hydrolysis affords *erythro*-3-hydroxy-D-aspartic acid. The latter fragment establishes the position of the double bond, and the amino and carboxy substituents. Alkaline hydrolysis[205] of blasticidin S, on the other hand, affords cytomycin (**82**)

Cytomycin (**82**)   Pseudoblastidone (**83**)

* *References start on p. 200.*

($C_{17}H_{23}N_7O_5$), which can be degraded further by acid hydrolysis to cytosinine and pseudoblastidone (83) ($C_7H_{13}N_3O_2$). The latter could also be obtained from blastidic acid (80) by alkaline treatment (internal cyclization). That the side chain was attached to the C-4 amino group, and not to the carboxyl group, was ascertained from pK_a measurements and from degradation of the hydrogenolysis product (84) of blasticidin S. Finally, the anomeric con-

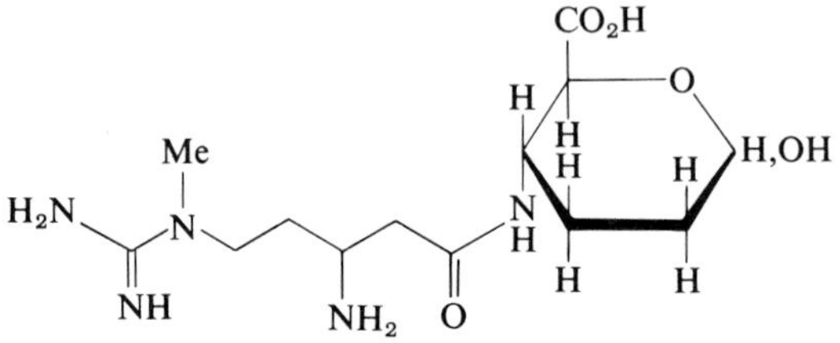

84

figuration was assumed[204] to be β-D, and a method to verify this assignment has been proposed.[203] An X-ray analysis of blasticidin S monohydrobromide has been reported.[205a]

Gougerotin[206] ($C_{16}H_{25}N_7O_8$) has been reported[207] to inhibit the transfer of amino acids from s-RNA to proteins. A structure proposed initially[208] as 1-(N-sarcosylcytosin-1-yl)-1,3-dideoxy-3-(D-serylamino)-β-D-allopyranuronamine (85) has been revised[209] to the structure 86, namely 1-(cytosinyl)-1,4-dideoxy-4-(sarcosyl-D-serylamido)-β-D-glucopyranuronamide. The revised structure 86 is in accord with the physical, chemical, and spectral

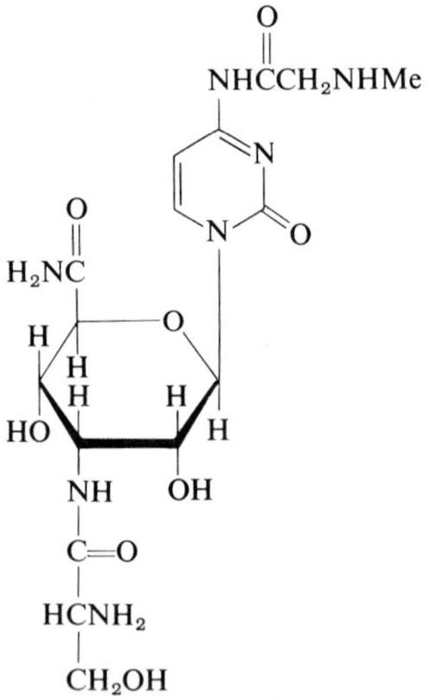

85

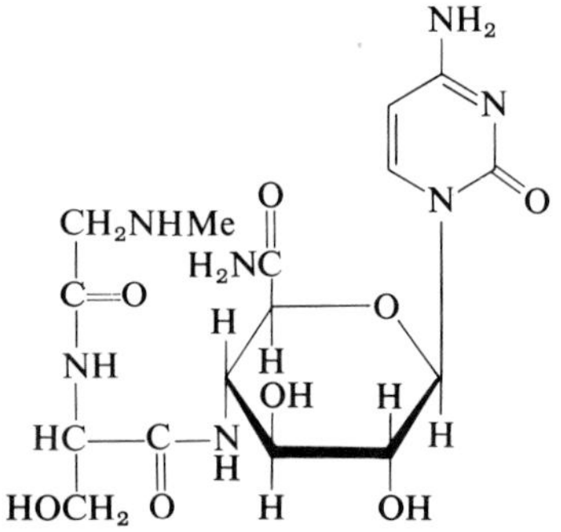

Gougerotin (86)

properties expected of its various degradation products. It is interesting to note that the amino sugar again belongs to the 4-amino-4-deoxyhexose series. Polyoxin C[209a] (**86a**) is the only example of a nucleoside antibiotic containing the 5-hydroxymethyluracil moiety. The sugar portion has been identified as 5-amino-5-deoxy-D-allofuranuronic acid. Thus all the other known antibiotics of the pyrimidine nucleoside type are elaborated by *Streptomyces* species, and all contain, among other components, cytosine and a 4-amino-4-deoxyhexose in their structures.

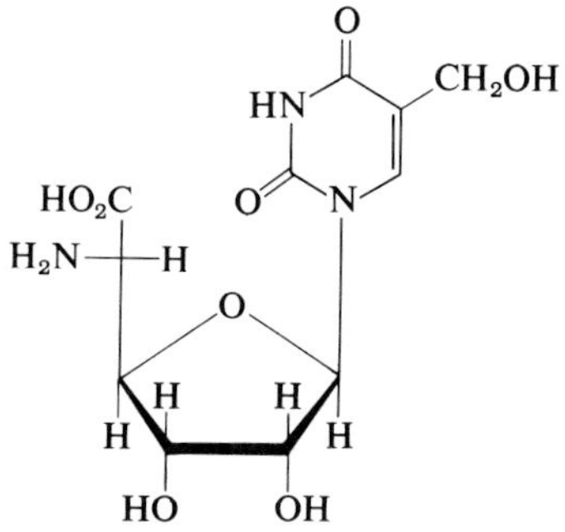

Polyoxin C (**86a**)

3. *Miscellaneous*

Tubercidin[210,211] (**87**) (4-amino-7-(β-D-ribofuranosyl)pyrrolo(2,3-*d*)pyrimidine) is similar to adenosine except that the N-7 atom is replaced by CH; it has been synthesized.[212,213]

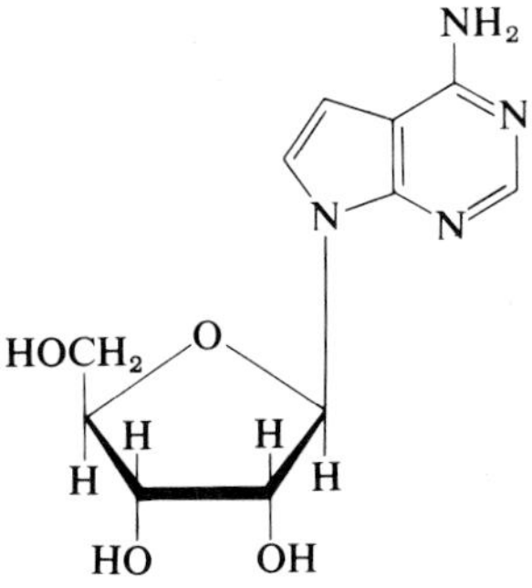

Tubercidin (**87**)

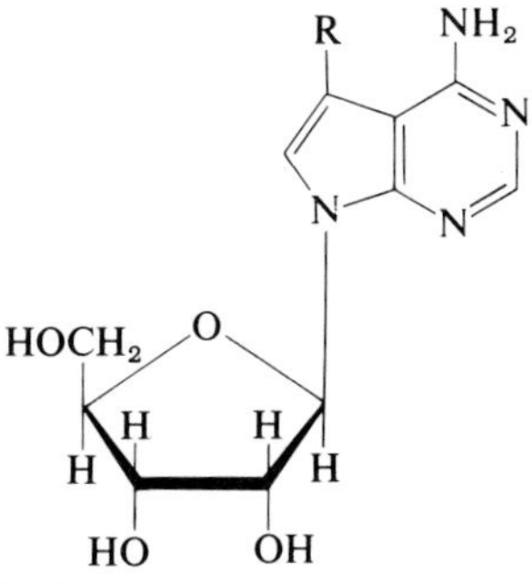

88a, Toyocamycin: R = CN
88b, Sangivamycin: R = CONH₂

* References start on p. 200.

Formycin (89)   Formycin B (90)

Toyocamycin[214] (88a) and sangivamycin (88b), which are structurally related to tubercidin, have been synthesized.[214a] Formycin and formycin B[215] belong to the class of C-nucleosides. They have been shown by X-ray analysis[216] and by degradation[217] to have structures 89 and 90, respectively. Showdomycin[217a] has been identified[217b,c] as 3-(β-D-ribofuranosyl)maleimide (90a) by chemical, spectroscopic, and crystallographic analyses.

Showdomycin (90a)

It is interesting to note that the D-ribose portion is common to all the above-mentioned nucleosides and that changes in the base structure account for the unusual characteristics of some of their members.

D. ANTIBIOTICS OF THE GLYCOSYLAMINE TYPE

The few members of this class of antibiotics are characterized by the presence of aglycons having unique structures, which are attached to the sugar portion through a nitrogen atom.

The streptothricin family of antibiotics has complex molecules which contain 2-amino-2-deoxy-D-gulose as the sole sugar component. Besides streptolidine (91), which is common to these antibiotics, one or more residues of 3,6-diaminohexanoic acid (β-lysine) are usually present. The structures of streptothricin and streptolin were investigated by several workers,[218–220] and their structures can be written as **92** and **93**, respectively.[221]

Acid hydrolysis of **92** and **93** afforded the amino sugar, which was identified[222] through the phenylosazone derivative and from optical rotatory data.

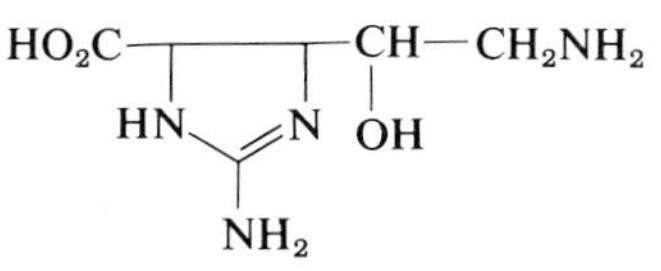

Streptolidine (91)

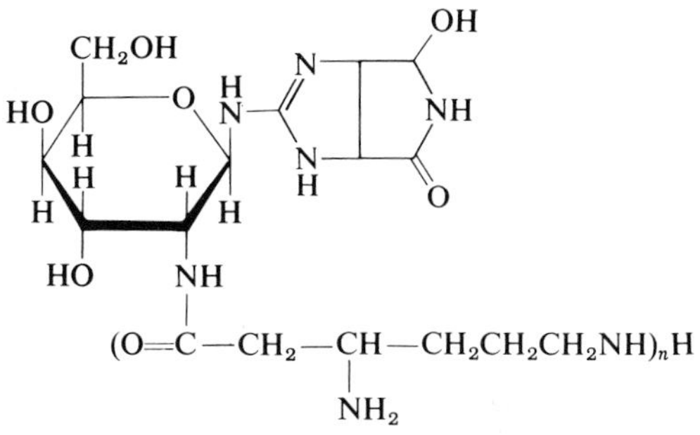

Streptothricin (**92**, $n = 1$)
Streptolin (**93**, $n = 2$)

Other hydrolysis products were identified as carbon dioxide, β-lysine, and a basic amino acid, streptolidine[220] (**91**), which is 4-(2-amino-1-hydroxyethyl)-2-aminoimidazoline-5-carboxylic acid. The structure of streptolidine was elucidated by periodate oxidation followed by borohydride reduction. An insight into the manner of linkage of the amino sugar in the molecules was achieved[223] by the isolation and characterization of N-(2-amino-2-deoxy-D-gulopyranosyl)guanstreptolidine (**94**).

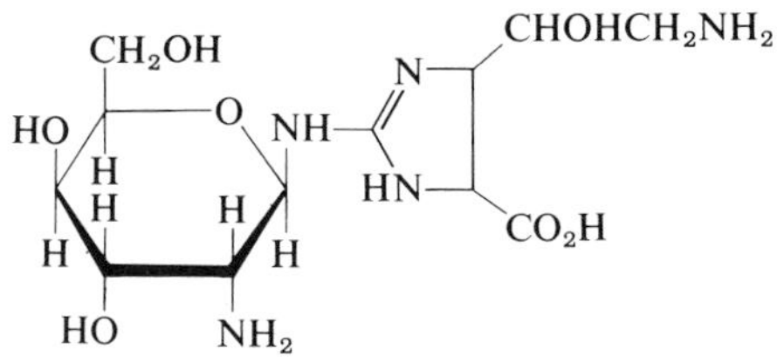

94

A β-D-glycosidic linkage has been proposed[223] on the basis of optical rotatory data. The assignments of the position of the carboxamido function as well as the size of the lactam ring (which would be determined by the actual position of a hydroxyl and amino group) have not been substantiated.

The racemomycins have also been isolated from *Streptomyces*.[224] Racemomycin A[225] has been shown[225a] to have structure **94a**. Racemomycin O

* *References start on p. 200.*

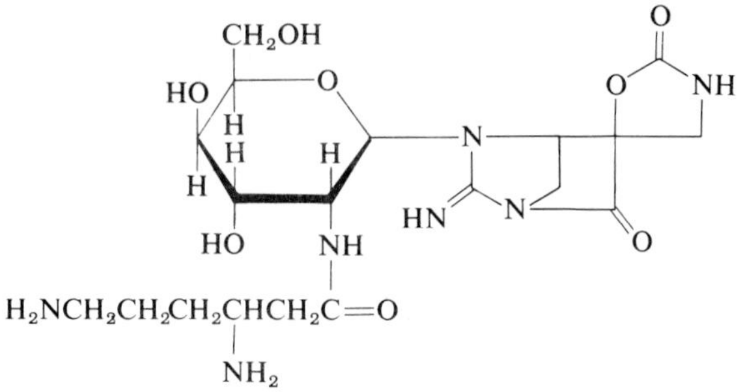

Racemomycin A (**94a**)

has been shown[226] to give, on acid hydrolysis, β-lysine, streptolidine (**91**) (also termed roseonine), racemonic aldehyde (**95**), carbon dioxide, and ammonia, thus resembling streptothricin. The amino sugar component, however, is 2-amino-2-deoxy-D-glucose and not the D-*gulo* isomer. The proposed structure for racemomycin O, originally formulated with an incorrect aglycon portion, is shown here in its correct form as **96**.

$$CH_3 — CHCH_2OCH_2CH_2CHO$$
$$|$$
$$OH$$

Racemonic aldehyde (**95**)

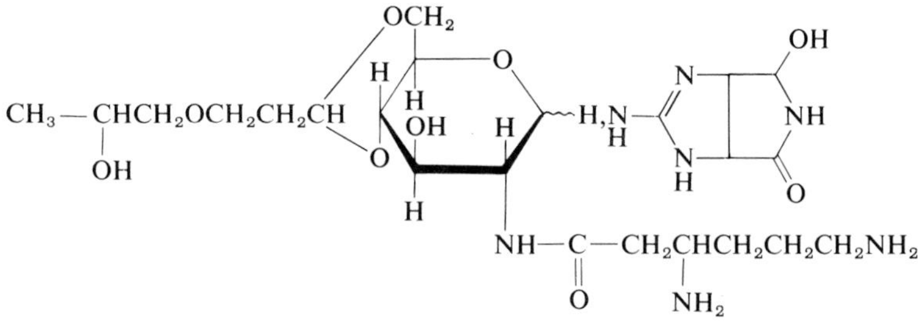

Racemomycin O (**96**)

Streptolydigin ($C_{32}H_{44}N_2O_9$) is another *Streptomyces* antibiotic; it has been assigned[227] structure **97**.

The 2,3,6-trideoxyhexose component was shown by synthesis of its enantiomorph to have the L-*threo* stereochemistry.[195] Acid hydrolysis of streptolydigin gives streptolic acid[228] (**98**) ($C_{18}H_{24}O_5$), the structure of which was elucidated by hydrogenation studies, n.m.r. (including spin-decoupling experiments), and mass-spectral data. Ozonolysis of streptolydigin gives a water-soluble product, ydiginic acid[229] (99) ($C_{14}H_{20}N_2O_7$), which with streptolic acid (98) accounts for all the carbon atoms in the molecule. Structure 99 was assigned to ydiginic acid mainly from mass-spectral and n.m.r. data. Alkaline hydrolysis of 99 gave L-*threo*-3-methylaspartic acid.

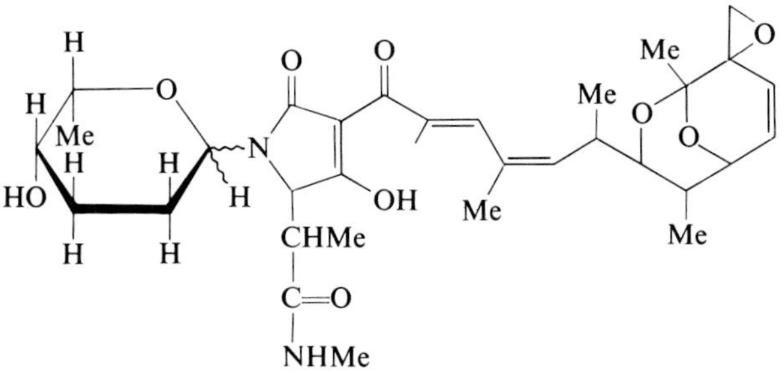

Streptolydigin (97)

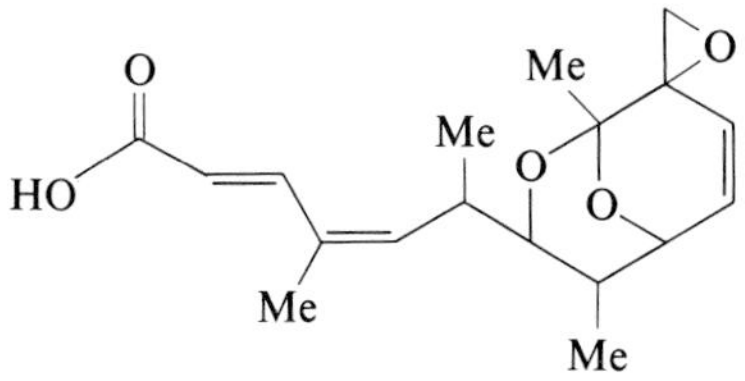

Streptolic acid (98)

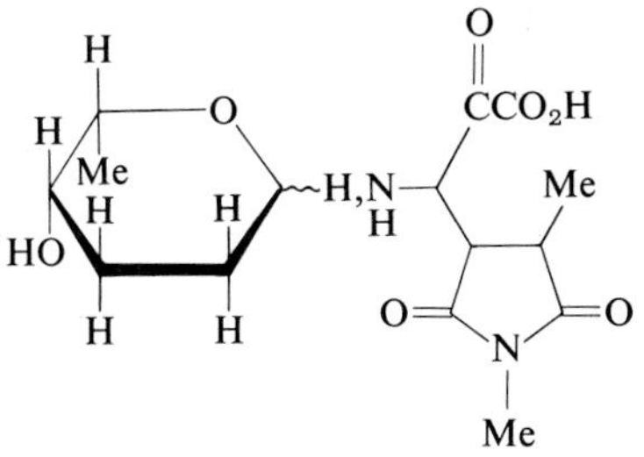

Ydiginic acid (99)

E. ANTIBIOTICS CONTAINING AROMATIC GROUPS

Novobiocin ($C_{31}H_{36}N_2O_{11}$) is an antibiotic isolated from culture filtrates of several *Streptomyces* species and is known under many names.[2] Structural studies have been reviewed;[2,230] the complete structure,[231] including the configuration of the sugar moiety, is depicted in formula 100.

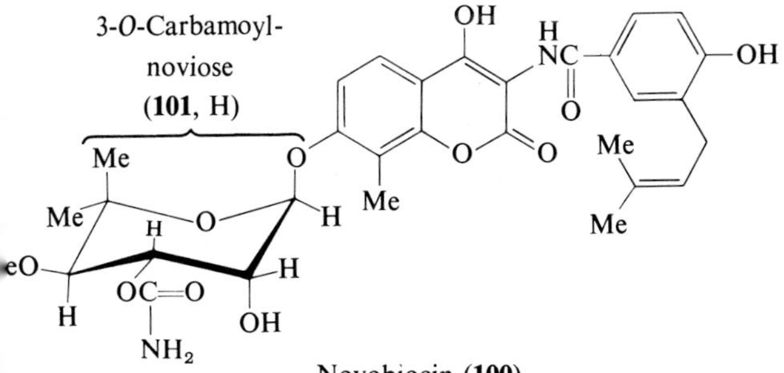

3-*O*-Carbamoyl-
noviose
(101, H)

Novobiocin (100)

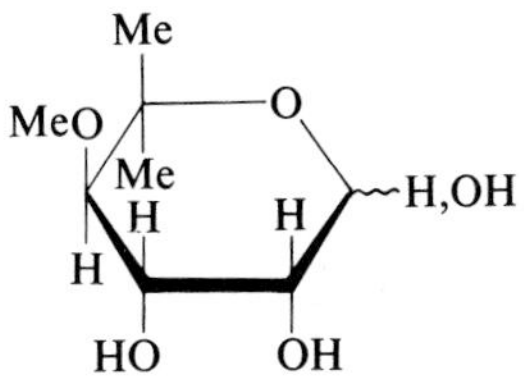

Noviose (102)

*References start on p. 200.

An unexpected reaction occurred during attempted acetylation[231] of novobiocin. Instead of the acetylated antibiotic, the two cleavage products shown below were obtained. This reaction facilitated degradative studies on novobiocin.

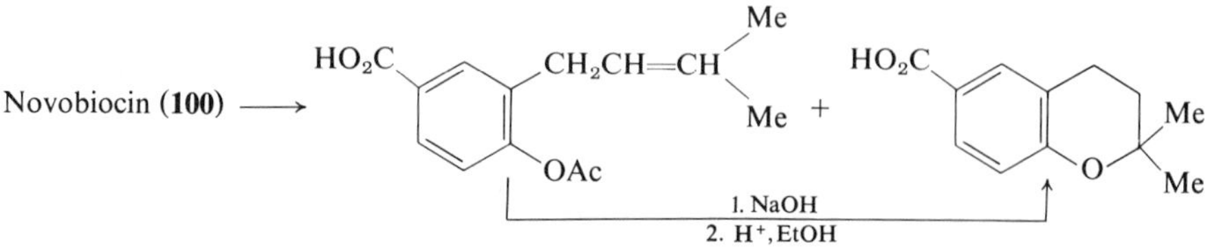

Novobiocin (100) ⟶

1. NaOH
2. H⁺, EtOH

Methanolysis[232] of novobiocin affords methyl 3-O-carbamoylnovioside ($C_{10}H_{19}NO_6$), the structure of which was determined by periodate oxidation studies[231] before and after complete hydrolysis of the 3 substituent. Furthermore, alkaline hydrolysis afforded ammonia and carbon dioxide, as would be expected from the presence of the urethane group. The position of the latter (at C-3) was verified by periodate oxidation of the alkali-treated glycoside, which now consumed one mole of periodate and produced (−)-3-hydroxy-2-methoxy-3-methylbutyric acid (from C-3 to C-6) after hydrolysis and oxidation with bromine. The L-*lyxo* configuration for 3-O-carbamoylnoviose (101) was assigned[233] in part on the basis of the empirical rules of optical rotation. Thus the benzyl-(*p*-methoxyphenyl)hydrazone derivative had a rotation expected for an aldose having the C-2 hydroxyl group on the right-hand side in the Fischer projection. Moreover, 2,3-O-isopropylidene-5-O-methyl-novionic acid was found to be identical to 6-deoxy-2,3-O-isopropylidene-5-C-methyl-4,5-di-O-methyl-L-*lyxo*-hexonic acid, prepared from L-rhamnose, thus confirming the configuration assigned. The synthesis of noviose (102) (6-deoxy-5-C-methyl-4-O-methyl-L-*lyxo*-hexose) has been reported[234] by a route starting from a D-glucose derivative; a related sugar, epi-noviose (L-*xylo* configuration) has been epimerized[235] to noviose. The conformation of noviose as depicted in the parent antibiotic (100), has been proposed on the basis of n.m.r. data.[236] Finally, the synthesis[237] of the aglycon, novobionic acid, and the total synthesis[238] of novobiocin have confirmed previous structural and stereochemical findings.

The noviose structure (102) has also been encountered in the antibiotics coumermycin A-1 (103) ($C_{55}H_{59}N_5O_{20}$)[239] and coumermycin A-2.[240]

The latter has been shown to have the same structure as 103, except for the absence of the 2-methyl group on the pyrrole rings. This feature seems to be responsible for the diminished biological activity of coumermycin A-2 as compared with coumermycin A-1.

Methanolysis of coumermycin A-1 gave a methyl glycoside and the aglycon, coumermic acid ($C_{27}H_{21}N_3O_{10}$), the central portion of the molecule. Aqueous acidic hydrolysis of the glycoside gave amorphous coumerose (104),

Me

HN

Coumerose

104, H

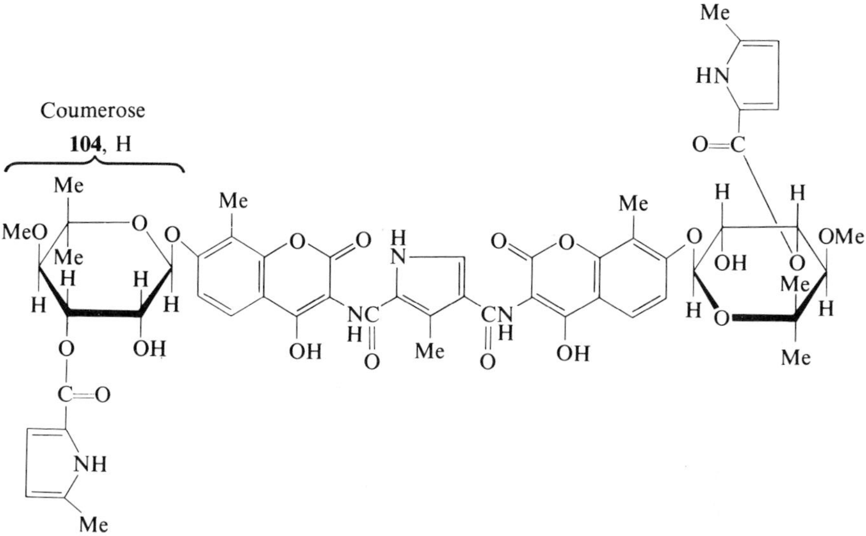

O=C

Me   Me   Me   H   H

MeO ⟨   O   O   O   H   O   O   O   OMe
  Me   N   OH   O
  H   H   NC   Me   CN   Me
H   H   H‖   ‖H   H   H
  O   OH   OH   O   Me   O   OH   Me
  C=O

NH

Me

Coumermycin A-1 (**103**)

which is 6-deoxy-5-*C*-methyl-4-*O*-methyl-3-*O*-(5-methyl-2-pyrrolyl)-L-*lyxo*-hexose.

The mixtures of antibiotics called olivomycin and chromomycin have distinct similarities.[4,240a] Chromomycin A_3 is the principal constituent of the chromomycins, a group of cancerostatic antibiotics. The general structural features of these substances are expressed in formulas **105a–d**.

Structural studies[241] on the chromomycins led to the above formulation except for the nature of the anomeric linkages. A detailed comparison of degradation reactions of the olivomycins and chromomycins has suggested[242] the presence of two molecules of chromose C (**106**) (olivose) in the molecule, instead of the one as originally postulated by the Japanese workers. According to these comparative data, it appears that the olivomycins differ from the chromomycins essentially in the absence of the C-7 methyl group (R) in the central aglycon moiety. The aglycons olivin[243] and chromomycinone[244] are formed by hydrolysis of the respective antibiotics with mild acid. Their structures were established mainly by n.m.r. spectral data on the corresponding acetates. The position of attachment of the sugar residues to the aglycons, and the disaccharide linkages, were deduced by hydrolytic studies (methanolysis) of suitably protected derivatives, followed by isolation of the partially blocked fragments. Graded acid hydrolysis led to fragments in which one of the terminal sugar residues was cleaved. Periodate oxidation data as well as the formation of an isopropylidene acetal indicated the

* References start on p. 200.

R = Me, chromomycins
R = H, olivomycins

Olivomycin A
Chromomycin A_2 } (105a) R′ = isobutyryl; R″ = acetyl

Olivomycin B
Chromomycin A_3 } (105b) R′, R″ = acetyl

Olivomycin C (105c) R′ = isobutyryl; R″ = H

Olivomycin D
Chromomycin A_4 (105d) R″ = acetyl; chromose B portion replaced by H.

presence of a pair of vicinal, unsubstituted hydroxyl groups, while ultraviolet absorption and pK_a data established the involvement of a phenolic hydroxyl group in glycoside formation. Component 106, chromose C (olivose), is 2,6-dideoxy-D-*arabino*-hexose.[245] The synthesis[246,247] of chromose A (olivomose) (107) and of chromose D (acetyl oliose) (108, R″ = acetyl) established their identities as 2,6-dideoxy-4-O-methyl-D-*lyxo*-hexose and 3-O-acetyl-2,6-dideoxy-D-*lyxo*-hexose, respectively. Chromose B (acetyl olivomycose) (109, R′ = acetyl), is 4-O-acetyl-2,6-dideoxy-3-C-methyl-L-*arabino*-hexose (epimeric at C-3 with mycarose, 13).[247a] A related antibiotic, chromocyclomycin ($C_{48}H_{64}O_{21}$), contains mycarose, olivose, and desacetyl chromose D (oliose).[247b]

The cinerubins, pyrromycins, and rhodomycins constitute another family[248] of pigmented aromatic antibiotics (anthracycline type) containing sugars. The red antibiotic pyrromycin[249] ($C_{30}H_{35}NO_{11}$) has structure 110. Hydrolysis of pyrromycin with mild acid affords the aglycon, ε-pyrromycinone (111) ($C_{22}H_{20}O_9$), which is partially transformed into η-pyrromycinone ($C_{22}H_{16}O_7$)

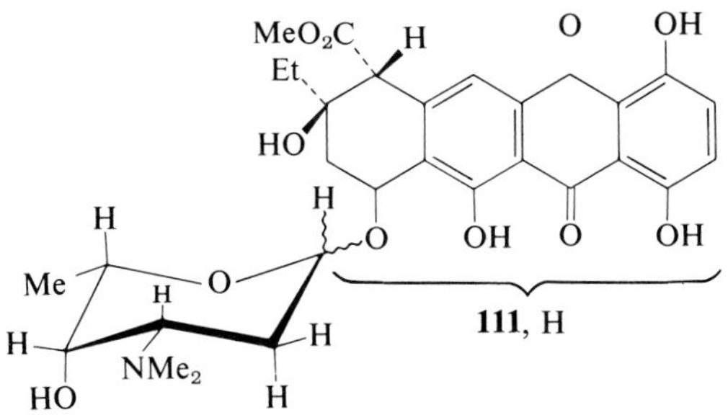

Pyrromycin (110)

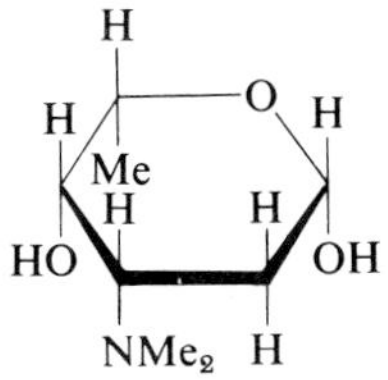

Rhodosamine (112)

as a result of the elimination of two molecules of water during the acid hydrolysis. Although the amino sugar rhodosamine (112) ($C_8H_{17}NO_3$) (3-dimethylamino-2,3,6-trideoxy-L-*lyxo*-hexose)[248] was isolated[250] in crystalline form during the early structure studies on this type of antibiotic, its actual structure and configuration[251] were not established until 1963. Periodate oxidation data, analytical data, color tests, and the liberation of dimethylamine on alkaline treatment suggested the gross structure and the presence of a 3-dimethylamino or 4-dimethylamino group in rhodosamine. The position of the dimethylamino group and the configuration at C-4 and C-5 were determined by treating rhodomycin sulfate with alkali, whereby two neutral sugars, 2,6-dideoxy-L-*lyxo*-hexose and 2,6-dideoxy-L-*xylo*-hexose, were formed, presumably according to the following mechanism.

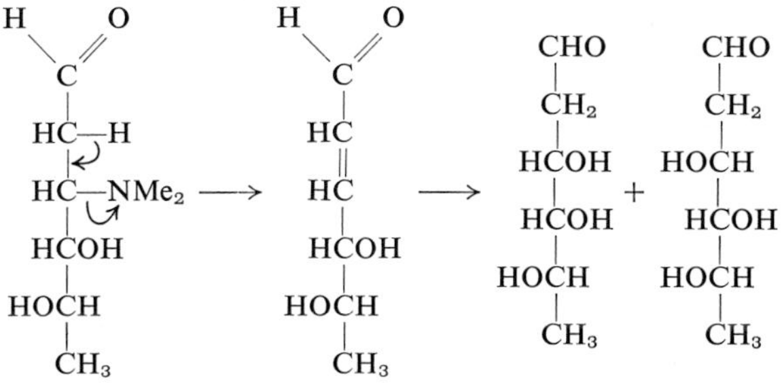

The configuration at C-3 of rhodomycin was elucidated by investigation of the n.m.r. spectrum of its diacetate.[248,251] This amino sugar (112) has been obtained from other colored antibiotics of related structure,[248] such as the rhodomycins A and B, the cinerubins A and B, and the γ-rhodomycins. The latter have been separated into four components that differ in the type of sugars that they contain. The aglycon produced by acid hydrolysis of these red substances is γ-rhodomycinone (113).

* References start on p. 200.

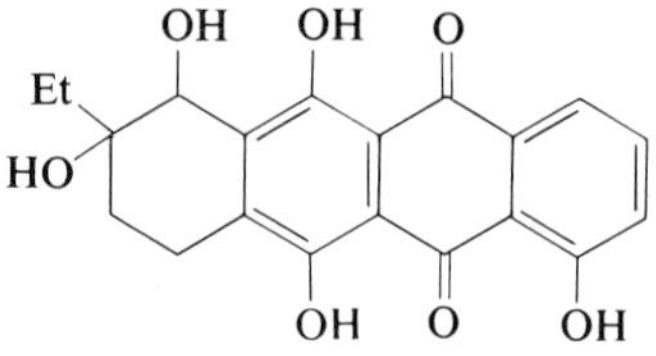

113

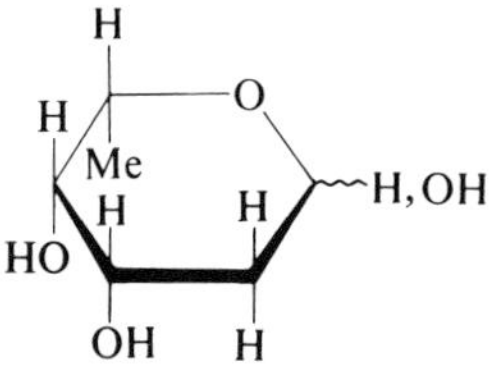

114

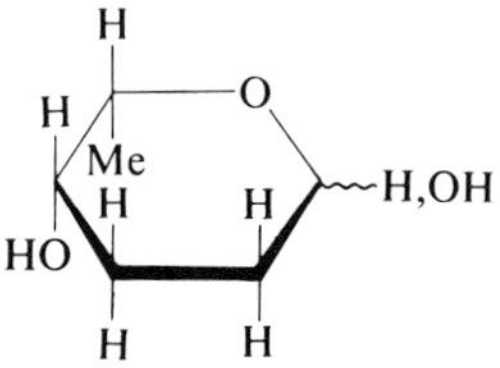

115

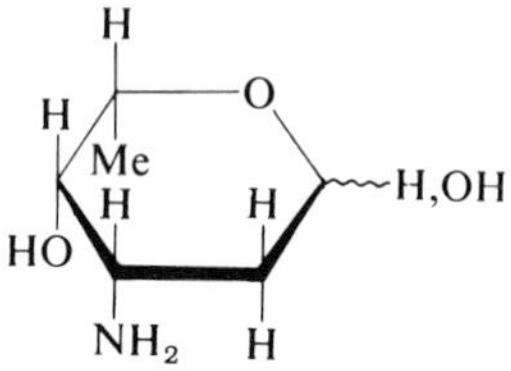

116

Hydrolysis of rhodomycin II gave in addition to **112** and **113**, a neutral sugar, 2,6-dideoxy-L-*lyxo*-hexose **114**. Rhodomycin III produced the same components but a new trideoxyhexose called rhodinose **115** $(C_6H_{12}O_3)$[248,252] was also formed. The structure of this unusual sugar was derived from periodate oxidation studies. Its stereochemistry was elucidated by synthesis[195] of its enantiomer. It was thus shown to be 2,3,6-trideoxy-L-*threo*-hexose and was found[195] to be identical with the sugar component in the antibiotic streptolydigin,[227] **97**.

A closely related antibiotic, daunomycin[253] has been shown[254] to have the tetracyclic structure of the rhodomycins. Hydrolysis of this antibiotic gave an amino sugar, daunosamine[255] **(116)** $(C_6H_{13}NO_3)$. Periodate oxidation studies, optical rotation, and n.m.r. data on appropriate derivatives established the identity of **116** as 3-amino-2,3,6-trideoxy-L-*lyxo*-hexose, the *N*-demethyl analog of **112**. A synthesis of *N*-benzoyl-D-daunosamine has been reported.[256,256a] The total structure[254a,b] and the absolute configuration of daunomycin are depicted in formula **116a**.

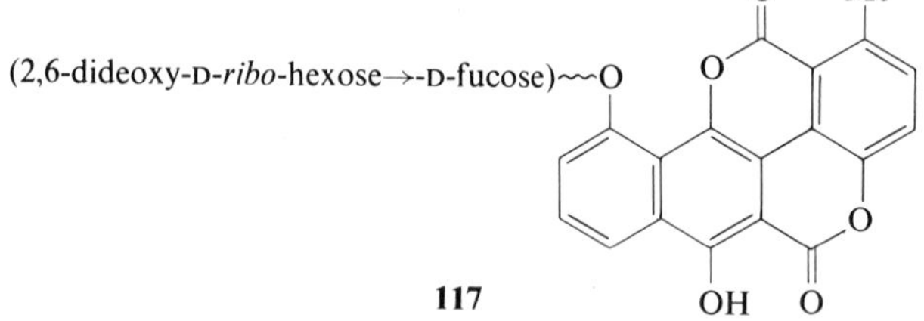

(2,6-dideoxy-D-*ribo*-hexose—→-D-fucose)—O

117

The crystalline antibiotic chartreusin[257,258] has structure **117**. Acid hydrolysis gave D-fucose and 2,6-dideoxy-D-*ribo*-hexose (digitoxose), and the aglycon $(C_{19}H_{10}O_6)$ containing a phenolic hydroxyl group and two lactone rings. Dehydrogenation of the aglycon gave 2-(3-methylphenyl)naphthalene

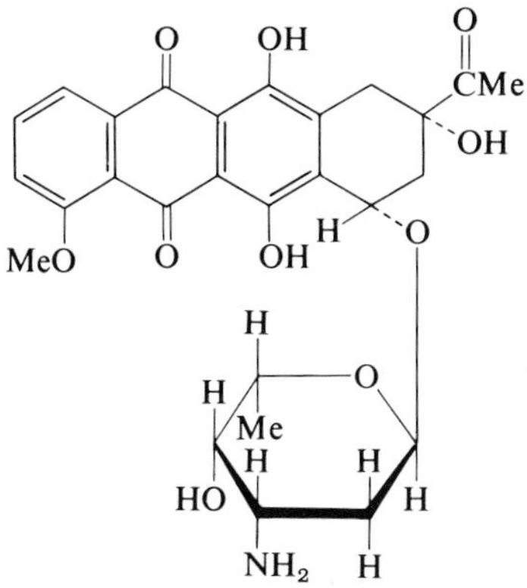

Daunomycin (**116a**)

and a methyl 2,3-benzofluorene; other sequences gave 1,4-dimethoxypyromellitic anhydride from the central aromatic ring. Hydrolysis of chartreusin with mild acid produced a fragment which did not contain the dideoxy sugar portion. It follows that the disaccharide is attached to a phenolic hydroxyl group through the D-fucose moiety.

Evernitrose, from a hydrolyzate of everninomycins B and D, is 4-*O*-methyl-3-*C*-methyl-3-nitro-2,3,6-trideoxy-L-*ribo*-(or-*arabino*)-hexose.[258a] Another hydrolysis product, evermicose, is 2,6-dideoxy-3-*C*-methyl-D-*arabino*-hexose.[258b]

Tolypomycin Y has been reported to have structure **117a**.[258c] Mild acid hydrolysis gives tolyposamine (**117b**) (4-amino-2,3,4,6-tetradeoxy-L-*erythro*-hexose). A related amino sugar, ossamine, isolated from a fungal metabolite, is 2,3,4,6-tetradeoxy-4-dimethylamino-D-*threo*-hexose.[258d]

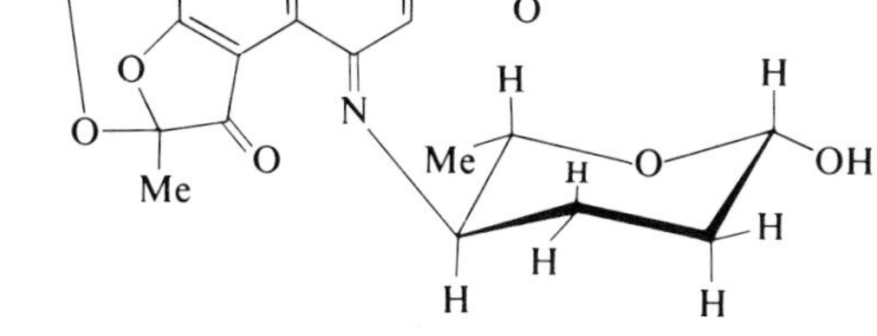

Tolypomycin γ (**117a**)

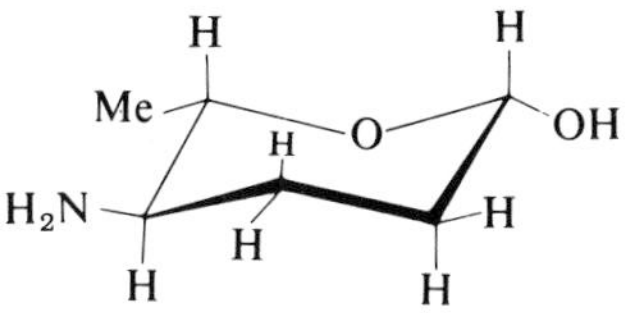

Tolyposamine (**117b**)

* *References start on p. 200.*

F. Miscellaneous Antibiotics

Listed under this heading are several antibiotics that do not fit into any of the previous general classes.

Trehalosamine[259,260] (**118**) (2-amino-2-deoxy-α-D-glucopyranosyl α-D-glucopyranoside) is a *Streptomyces* product having weak antibiotic activity against some bacteria and fungi. It has been synthesized.[260a]

Trehalosamine (**118**)

The antibiotic lincomycin has been reported[261–261d] to have structure **119**. It is unique in that it provides the first example of the occurrence of an aminodeoxyoctose, lincosamine (**120**) (6-amino-6,8-dideoxy-D-*erythro*-D-*galacto*-octose), in a biologically synthesized material. Vigorous acid hydrolysis of lincomycin gave *n*-propylhygric acid (**121**), which was identified by synthesis. The formation of an isopropylidene acetal and the behavior of the molecule toward periodate oxidation indicated a pyranose structure and a pair of *cis*-hydroxyl groups. Hydrolysis with mild acid liberated methanethiol, and the remainder of the molecule could be isolated as a crystalline phenylosazone that still contained the amino acid portion. The most successful degradative procedure for the cleavage of the amide bond in lincomycin proved to be hydrazinolysis, which provided the intact thioglycoside (**122**). The structure and stereochemistry of **122** were determined by periodate oxidation studies of appropriate derivatives, followed by further degradations of the fragments formed. The isolation of galactaric acid through a series of such reactions established the stereochemistry at C-2 to C-5. The D-*galacto* configuration in the ring system and the α-D anomeric configuration were confirmed by n.m.r. spectral studies and by the isolation of D-galactose (2-methyl-2-phenyl)hydrazone from the thioglycoside **122** by another sequence of reactions. The nature of the side chain was established by selective oxidation of the C-7 hydroxyl group and observation of the resultant terminal CH_3CO group by n.m.r. spectroscopy. The antibiotic gives a negative iodoform test, suggesting, quite ambiguously, that a terminal methyl group is not present. The stereochemistry of the side chain (C-6 and C-7) was

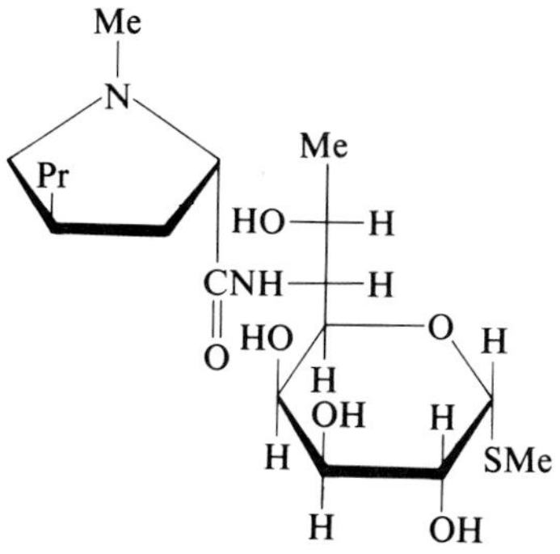

Lincomycin (119) 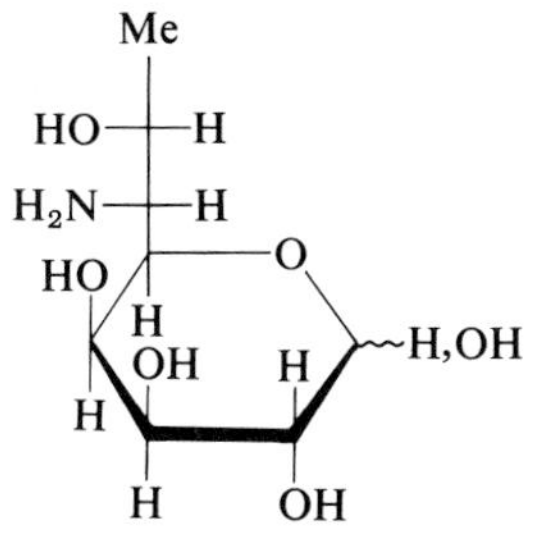   120

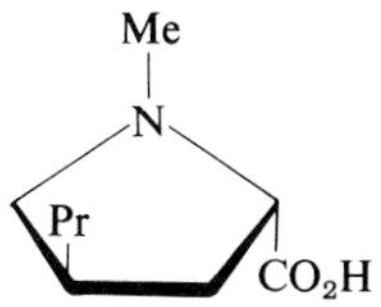

121 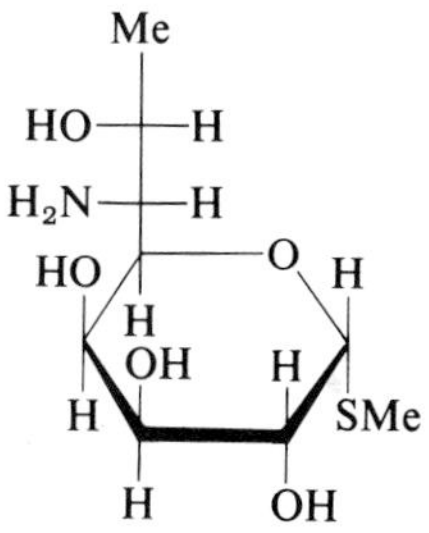   122

determined by the isolation of N-(2,4-dinitrophenyl)-D-allothreonine (C-5 to C-8) after oxidation of N-(2,4-dinitrophenyl)lincosamine dimethyl dithioacetal with periodate–permanganate. N-Acetyl-lincosamine has been synthesized.[261e]

Celesticetin[262,262a] (123) is closely related in structure to lincomycin but is produced by a different actinomycete.

Thus celesticetin has the same carbon chain, order of substitution, and stereochemistry as lincomycin. The main differences are the nature of the aglycon, the absence of the n-propyl group in the amino acid portion, and the presence of a methyl ether group at C-7 of the amino sugar. By essentially the same series of degradation reactions used with lincomycin, the amino sugar moiety in celesticetin was found to be 7-O-methyl-lincosamine.

* *References start on p. 200.*

New lincomycin-related antibiotics have been isolated[263] by the addition of DL-ethionine and alkyl α-thiolincosaminides to the fermentation media producing lincomycin. The structures of these substances are similar to that of lincomycin itself, but differ from it and from each other in the nature of the alkyl groups at C-1 and in the proline portion.

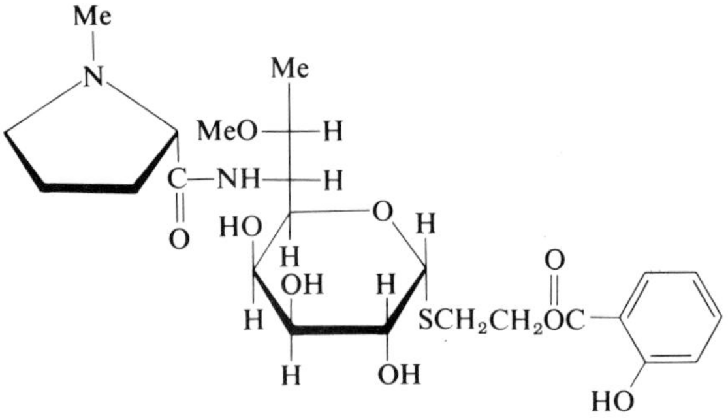

Celesticetin (**123**)

Streptozotocin is one of the rare examples of antibiotics that are composed of a single amino sugar moiety. Its structure[263a] (**123a**) has been deduced from degradation studies and by synthesis.

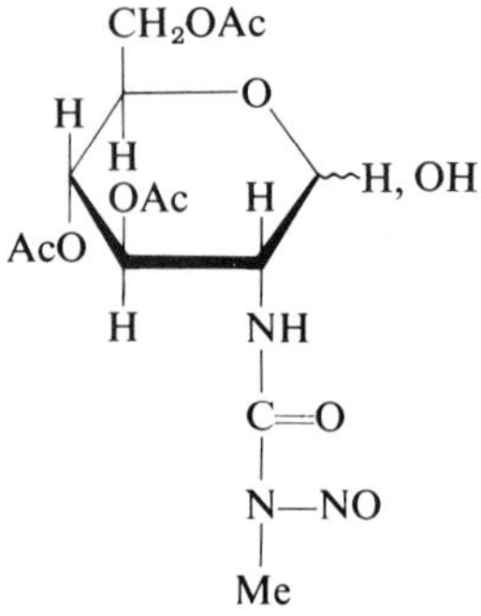

Streptozotocin (**123a**)

Nojirimycin[263b] has been shown by appropriate transformations and synthesis[263c] to be 5-amino-5-deoxy-D-glucopyranose (**123b**). It is unstable

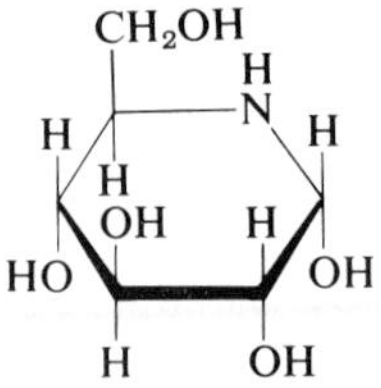

Nojirimycin (**123b**)

in neutral and acidic media, but relatively stable at pH 7 to 9. Nojirimycin is the sole representative of a sugar containing a nitrogen atom in the ring, in the group of biologically derived antibiotics.

Labilomycin $(C_{23}H_{34}O_8)$ is an antimycobacterial antibiotic and has structure[264] **124**.

Labilomycin (**124**)   Labilose (**125**)

Hydrogenation gives a perhydro derivative $(C_{23}H_{24}O_8)$, and alkaline hydrolysis of this substance gives a deacyl derivative $(C_{20}H_{28}O_7)$. Acid hydrolysis of labilomycin gives the aglycon, a dihydroxyketone $(C_{12}H_{24}O_3)$, and labilose[265] (**125**) $(C_8H_{16}O_5)$ (6-deoxy-2,4-di-*O*-methyl-D-galactose). Methyl labiloside, produced by methanolysis of labilomycin, was transformed into methyl 6-deoxy-2,3,4-tri-*O*-methyl-D-galactopyranoside. The positions of substitution in labilose were deduced from the formation of a 1,5-lactone from the derived aldonic acid and the failure of the derived alditol to react with periodate. A β-D-anomeric linkage was assigned on the basis of n.m.r. spectral data.

An octose believed to have the unusual structure[266] **126** has been isolated as the anhydro derivative **127** by hydrolysis with mild acid of an antibiotic termed "γ-activity X."

126   **127**

128

Structure **126** has been suggested on the basis of periodate oxidation of **127**, and isolation of a dicarbonyl compound as its 2,4-dinitrophenyl derivative (**128**) (arising from cleavage of the C-3–C-4 bond). Detailed n.m.r. spectral investigation suggested the L-*gluco* stereochemistry for **126**.

A series of antibiotics previously listed as antibiotics of high molecular weight are not included here because of the lack of information regarding their actual structural complexity.

Ristocetin A[267] and B[268] are closely related, crystalline, amphoteric antibiotics having gram-positive activity. Components A and B, which have different microbiological activities, have been separated by chromatography on a carbon column. Acid hydrolytic studies showed the presence of glucose, mannose, arabinose, and rhamnose in addition to a phenolic moiety.

Vancomycin,[269] another amphoteric, crystalline antibiotic effective against gram-positive bacteria, contains organically combined chlorine, glucose, aspartic acid, and other unidentified components. Actinoidin[270] is an antibiotic complex similar to ristocetin or vancomycin and contains glucose and mannose.

The only example of a polysaccharide antibiotic is statolon[271] which has antiviral properties. It was reported to be composed of arabinose, xylose, galactose, 2-amino-2-deoxy-galactose, galacturonic acid, glucose, and rhamnose.

III. CHARACTERIZATION OF ANTIBIOTICS AND THEIR SUGAR COMPONENTS BY PHYSICAL MEANS

The advent of modern instrumentation, especially nuclear magnetic resonance spectroscopy and mass spectrometry, has greatly facilitated structural and stereochemical studies on antibiotic substances.

A. Nuclear Magnetic Resonance Spectroscopy

This section demonstrates briefly some important applications of nuclear magnetic resonance spectroscopy (see Vol. IB, Chap. 27) in deciphering both structural and conformational aspects of complex sugars derived from antibiotics. The following are some examples in which both the stereochemistry and conformation of some sugars have been independently established by n.m.r. spectroscopy. The references cited pertain to the n.m.r. work only. The abbreviations are as follows: s, singlet; d, doublet; q, quartet; m, multiplet.

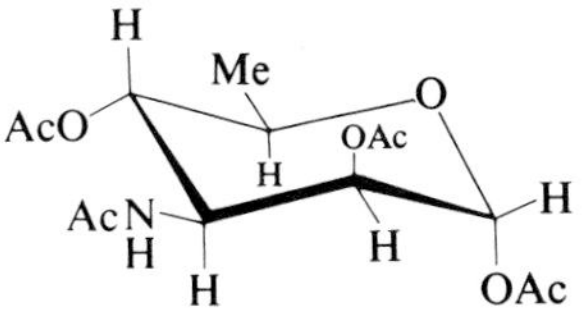

α-Mycosamine (2) tetraacetate[13] (CDCl₃)

α-Mycosamine (**2**) tetraacetate[13] (CDCl$_3$)
H–1 τ 3.99 (d); $J_{1,2} = 1.5$ Hz
H–5 τ 6.02 (m); $J_{5,6} = 6.0$ Hz
$\qquad\qquad\qquad J_{5,4} = 8.5$ Hz
H–6 τ 8.80 (d)
OAc τ 7.88 (axial)
$\quad$ τ 7.91 (equat.)
NAc τ 8.02

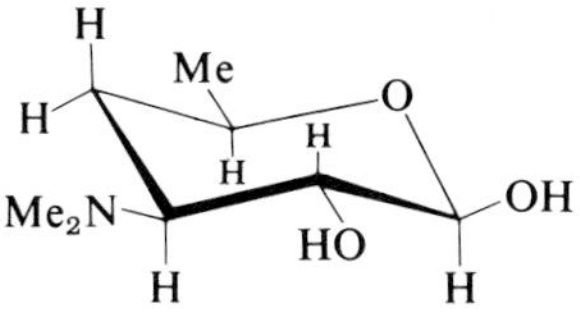

Desosamine [38](**5**) (D$_2$O)
H–1 τ 5.03 (d); $\quad J_{1,2} = 7.8$ Hz
H–2 τ 6.25 (q) $\quad J_{2a,3} = 10.2$ Hz
H–3 τ 6.66 (m); $J_{3a,4e} = 3.9$ Hz
$\qquad\qquad\qquad J_{3a,4a} = 12.0$ Hz
H–4 τ 7.86; $\quad J_{4a,4e} = 13.1$ Hz
$\quad$ τ 7.53 $\qquad J_{4a,5a} = 11.1$ Hz
$\qquad\qquad\qquad J_{4e,5a} = 1.8$ Hz
NMe$_2$ τ 7.26 (s)

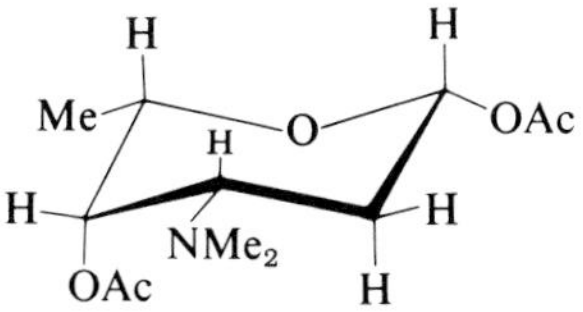

β-Rhodosamine (**112**) diacetate[251]
(CDCl$_3$)
H–1 τ 4.23; $\quad J_{1,2e} = 4.0$ Hz
$\qquad\qquad\quad J_{1,2a} = 8.0$ Hz
H–2 τ 8.3–7.9; $J_{2a,3} = 12.0$ Hz
$\qquad\qquad\quad J_{2e,3} = 5.0$ Hz
H–3 τ 7.57; $\quad J_{3,4} = 2.0$ Hz
H–4 τ 4.79; $\quad J_{4,5} = 2.0$ Hz
H–5 τ 6.24
H–6 τ 8.82; $\quad J_{5,6} = 6.5$
OAc τ 7.87, 7.85
NMe$_2$ τ 7.72 (s)

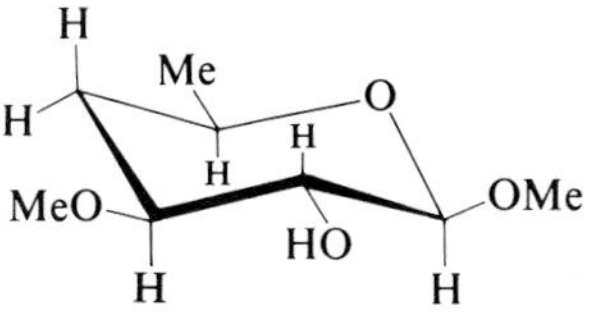

Chalcose (**22**) methyl glycoside[73]
(pyridine)
H–1 τ 5.60 $\quad J_{1,2} = \sim 7.0$ Hz
H–4 τ 7.95; $\quad J_{4a,4e} = \sim 12.5$ Hz
$\quad$ τ 9–8.42

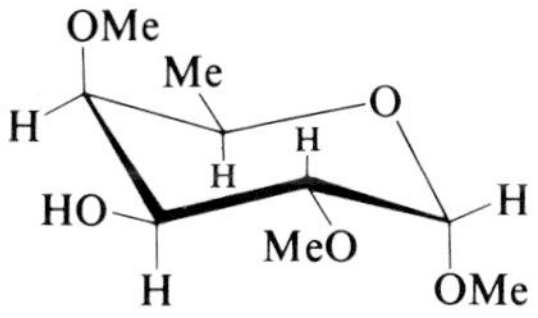

Labilose (**125**) methyl glycoside[265]
(CDCl$_3$)
H–1 $\quad \tau$ 5.13; $J_{1,2} = 3.0$ Hz
H–6 $\quad \tau$ 8.71 (d)

OMe τ 6.37, 6.50, 6.61

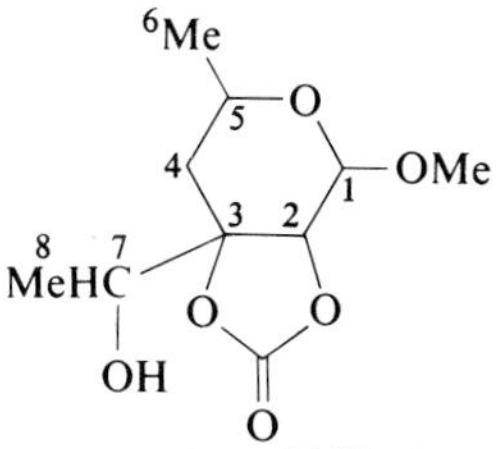

Methyl aldgaroside A[85,85a] (**28**) (CDCl$_3$)

H–1 τ 5.52; $J_{1,2} = 5.5$ Hz
H–2 τ 5.73; $J_{1,2} = 5.5$ Hz
H–4 τ 8.40⎫
$\quad$ τ 8.10⎭ $J_{4a,4e} = 12.0$ Hz
H–7 τ 6.55; $J_{7,8} = 7$ Hz
C–Me τ 8.48, 8.77
OMe τ 6.25

References start on p. 200.

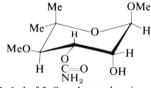

Anhydrooctose[266] (127) (D₂O)

Anhydrooctose²⁶⁶ (127) (D₂O)

H–1 τ 6.81; $J_{1,2}$ = 2.4 Hz

$J_{1,2}$ = 1.8 Hz

H–2 τ 7.48; $J_{2,2}$ = 14.9 Hz

τ 8.14 $J_{2,3}$ = 3.6 Hz

$J_{2,3}$ = 9.8 Hz

H–3 τ 5.69

H–5,H–7, τ 5.45, 5.68

Me τ 8.77, 8.79, J = 6.7

Mycarose (13) diacetate⁵³ (CDCl₃)

H–1 τ 4.12; $J_{1,2}$ = 9.0 Hz

$J_{1,2}$ = 3.1 Hz

H–4 τ 5.50 $J_{4,5}$ = 9.6 Hz

H–5 τ 6.10; $J_{5,6}$ = 6.0 Hz

Me τ 8.87, 8.90

OAc τ 7.92

N,N-dimethylactinamine¹¹⁶ (45) (D₂O)

H–1, H–3 τ 6.70

H–2 τ 4.99 (d); J = 3.0 Hz

H–4, H–6 τ 5.93; J = 11.0 Hz

H–5 τ 6.51; J = 9.0 Hz

NMe₂ τ 7.00

Methyl of 3-O-carbamoylnoviose (101)glycoside²³⁶ (pyridine)

H–1 τ 4.97; $J_{1,2}$ = 2.3 Hz

H–2 τ 5.32; $J_{2,3}$ = 3.0 Hz

H–3 τ 4.20; $J_{3,4}$ = 10.0 Hz

OMe τ 6.43, 6.61

Nuclear magnetic resonance spectroscopy has also been very useful in assigning the configuration of the anomeric linkage in many antibiotics containing sugars.

B. Mass Spectrometry

Mass spectrometry of intact antibiotics containing sugars has been little exploited, but sugar derivatives obtained by degradation have frequently provided useful spectra²⁷² (see Vol. IB, Chap. 27) furnishing information on the molecular weight and structure (but not configuration) with samples of less than a milligram.

Sugar components are commonly isolated from antibiotics by alcoholysis or mercaptolysis. The resulting glycosides and dithioacetals are well suited for mass spectrometric studies. Study of the spectra of related model compounds usually facilitates the interpretation of mass spectra.

The mass spectra of the sugar dithioacetals are relatively easy to interpret.[273] For example,[274] N,N-diacetylparomose diethyl dithioacetal, obtained from paromomycin[135] (48c) by mercaptolysis and N-acetylation, clearly shows in its mass spectrum the successive cleavage of certain C—C bonds, and fragments derived from the 2-acetamido and 6-acetamido groups are observed. Classical proof[138] of structure by degradation is much more tedious. The mass spectra[274] of 3-acetamido-3-deoxy-D-ribose and 6-acetamido-6-deoxy-D-glucose diethyl dithioacetals, which can be obtained from puromycin and kanamycin, respectively, are also indicative of their specific structures.

Deoxy and dideoxy sugars derived from antibiotics can also be characterized by mass spectrometry of their dithioacetals.[4,275] The mass spectra[275] of the diethyl dithioacetals of chalcose (22) and mycinose (17), sugar components in the antibiotic chalcomycin[71] (19), give peaks for the molecular ion $(M \cdot^+)$ and indicate the positions of methoxyl and deoxy groups. Other classes of deoxy sugars derived from antibiotics, such as the 2-, 3-, and 6-deoxyhexoses and 2,6-, and 3,6-dideoxyhexoses, in the form of their dialkyl dithioacetals, also provide informative mass spectra.[275] The MacDonald–Fischer degradation[276] can be used with the sugar dialkyl dithioacetals to obtain the lower aldoses by way of the corresponding disulfones. The complete structure and partial stereochemistry (C-3–C-5) of paromose (52) can be deduced from its mass spectrum,[274] in conjunction with this useful degradative reaction.[139]

The mass spectra of glycosides are usually more complex than those of the corresponding dialkyl dithioacetals. Published data[273,273a] on the mass spectra of methyl glycosides (including amino and deoxy sugars), their methyl ethers, and acetates may facilitate elucidation of structure for sugars that have been isolated as glycosides from the parent antibiotics.

There are some examples of the use of mass spectrometry for elucidation of structure of glycosides isolated directly from the parent antibiotics by alcoholysis, as in the fragmentation of methyl aldgaroside B (28), which affords peaks consistent with the structure proposed.[85]

Mass spectrometry of antibiotics containing sugars[276a] can be used to determine accurate molecular weights. Thus N-acetylpimaricin, in the form of its hexakis(trimethylsilyl) ether, shows a molecular-ion peak at m/e 1139 and corroborates the revised formula for pimaricin.[14] The complete carbon skeleton and the oxygenation pattern of nystatinolide, the aglycon portion of nyastatin, has been established by mass spectral studies.[7] Macrolide antibiotics not containing nitrogen all give molecular-ion peaks. Aldgamycin E shows the M^+ at m/e 742, and several peaks corresponding to loss of various combinations of its sugar components are also present.[85] Applications of mass spectrometry in the field of nucleosides[277] and nucleoside antibiotics[276a]

* References start on p. 200.

involve the characterization of cordycepin,[176] puromycin,[278] components of amicetin,[276a] and the C-nucleosides, formycin, formycin B, and showdomycin.[278b] Aminocyclitol antibiotics such as neomycin, paromomycin, and kanamycin, in the form of their N-acetyl-O-trimethylsilyl derivatives, are adaptable to mass spectral studies.[276a,278a] Information regarding molecular weight, the gross structures of the units of which the saccharides are composed, and their sequential arrangement can be obtained from fragmentation patterns of the intact antibiotics and of appropriate model compounds.

IV. BIOLOGICAL CONCEPTS

Within recent years major advances have been made in our understanding of the molecular events involved in synthesis of nucleic acids and proteins. The use of antibiotics that selectively inhibit specific biosynthetic reactions has contributed greatly to the development of some of the present concepts of molecular biology. The fact that antibiotics act by specifically blocking certain essential enzymic syntheses has enabled biochemists to study the hitherto inaccessible chemical intermediates which accumulate therefrom. Studies on mechanisms of action[278c,d] at this molecular level have also demonstrated a relationship between the degree of specificity of action of an antibiotic and its toxicity toward living cells. For example, those substances that interfere with synthesis of DNA, RNA, and protein are normally too toxic to be of clinical value, whereas those that selectively inhibit only protein synthesis are frequently nontoxic and therapeutically effective. The effect of antibiotics on molecular biological syntheses in cell-free systems is an active field of present research. Reviews[279] on this subject have appeared.

Studies on mode of action with the macrolide antibiotics have been made with erythromycin (9). This antibiotic has been shown to inhibit cell-free protein synthesis[280–282] and to precipitate RNA.[281] It has been suggested that erythromycin interacts with a 50S ribosomal subunit and inhibits protein synthesis induced by m-RNA, although the exact mechanism is unclear.[280] The polycationic cyclitol antibiotics appear to have similar mechanisms of action and inhibit protein synthesis at the ribosomal level, while allowing continued synthesis of RNA and DNA.[283]

It has been shown that streptomycin (38) binds irreversibly to sensitive ribosomes in *Escherichia coli* but does not prevent s-RNA or aminoacyl-RNA from binding to the ribosomal complex.[284–286] By combining with the 30S ribosomal subunit it alters the function but not the formation of the ternary complex. Other members of this class of antibiotics also cause extensive and specific alterations in the coding properties of synthetic and natural polynucleotides *in vitro*.[283,287] Streptomycin, for example, inhibits the poly-U-stimulated incorporation of phenylalanine into protein, but stimulates the incorporation of other amino acids (isoleucine, serine, and leucine) not normally coded for by poly-U.[288] Kanamycin (**54**), neomycin (**48a,b**), paromomycin (**48c,d**), gentamycin, and hygromycin B also disturb the fidelity of the reading of polynucleotides, and the effect is found with all four ribohomopolymers: poly-U, poly-C, poly-A, and poly-I.[284,287,288] This effect is supposedly caused by alteration of the triplet code and the configuration of the 30S ribosome subunit. Streptomycin is bactericidal and kills only rapidly growing cells. During cell growth the m-RNA is alternately bound to, and then freed from, the ribosomes. When the ribosomes are free of messenger, the cells are susceptible to the antibiotic.

The nucleotide antibiotics, by virtue of their close structural similarity to normal nucleosides, interfere with synthesis of RNA and DNA and are thus highly toxic to living cells. Cordycepin (**64**), which is 3'-deoxyadenosine, acts on Ehrlich ascites cells and causes accumulation of corresponding mono-, di-, and triphosphate derivatives of the antibiotic, and these are the active substances inhibiting DNA and RNA synthesis.[289] The antibiotic is incorporated into RNA by the RNA polymerase, and, since the resulting polynucleotide lacks a 3'-hydroxyl group to which the next incoming nucleotide can be linked, chain growth ceases at this position.[290] Tubercidin (**87**) (7-deazaadenosine) also becomes incorporated into nucleic acids and leads to the formation of biologically defective units. The antibiotic most widely studied in this class is puromycin (**67**). Its close structural resemblance to the amino acid-bearing end of t-RNA led to the proposal that it inhibited protein synthesis specifically at the site involving t-RNA.[291] It was subsequently found to prevent the transfer of aminoacyl residues from t-RNA into polypeptides. This was shown to occur on the 50S ribosomal subunit, where the amino group of puromycin competes with the incoming aminoacyl t-RNA for the activated carboxyl group on the growing end of the peptidyl t-RNA.[292]

Antibiotics of the aromatic type whose modes of action have been studied most extensively can be subdivided into the anthracycline and chromomycin (**105**) types. The anthracyclines inhibit synthesis of RNA by reacting strongly with native DNA. They resemble the acridine dyes in that they increase the viscosity and decrease the sedimentation of native DNA by intercalation.[293] This binding persists at high ionic strength and stabilizes the native DNA toward thermal denaturation. Daunomycin stabilizes double-stranded DNA by binding to both strands by way of the hydroxyl groups on the chromophore and the amino group on the sugar.[294] The amino group appears to make the greater contribution to the stability of the complex. The chromophore intercalates between adjacent base pairs of helical DNA, and the sugar side chain probably projects into the minor grooves of DNA, causing steric interference with the RNA polymerase.[294] The chromomycin types also bind to DNA but do not intercalate and do not stabilize native

* *References start on p. 200.*

DNA to thermal denaturation. They all sediment with DNA and cause a decrease in the buoyant density of DNA in cesium chloride gradients.[293] Novobiocin (100), a substituted coumarin type of antibiotic, causes immediate inhibition of DNA synthesis, a lesser inhibition of RNA synthesis, and a delayed effect on protein synthesis, growth, and membrane permeability.[295] It appears to interfere with DNA polymerization, possibly by binding to or reacting with DNA polymerase. The carbamoyl group on the C-3 hydroxyl of the sugar is essential for its biological activity.

Lincomycin (119), classified under the miscellaneous group, interferes with protein synthesis in gram-positive organisms by inhibiting the binding of s-RNA to the ribosome–messenger complex. The binding of phenylalanine–s-RNA to the 50*S* subunit of the ribosome was inhibited in gram-positive organisms, but not in gram-negative organisms, which are insensitive to lincomycin.[296]

REFERENCES

1. H. Umezawa, Ed., "Index of Antibiotics from Actinomycetes," University Park Press, State College, Pennsylvania, 1967.
1a. M. R. J. Salton, *Ann. Rev. Biochem.*, **34**, 143 (1965).
2. J. D. Dutcher, *Advan. Carbohyd. Chem.*, **18**, 259 (1963); V. I. Veksler, *Russian Chem. Rev.*, **33**, 424 (1964).
3. D. Horton, in "The Amino Sugars," R. W. Jeanloz, Ed., Vol. IA, Academic Press, New York, 1969, Chap. 1.
4. S. Hanessian, *Advan. Carbohyd. Chem.*, **21**, 143 (1966).
4a. D. Gottlieb and P. D. Shaw, Eds., "Antibiotics-Biosynthesis," Vol. II, Springer-Verlag, New York, New York, 1967.
5. H. Umezawa, "Recent Advances in the Chemistry and Biochemistry of Antibiotics," edited by The Microbiol. Chem. Res. Found., Tokyo, Japan, 1964.
6. S. Lewak, *Postepy Biochem.*, **6**, 487 (1960); *Chem. Abstr.*, **55**, 6605 (1961); R. Huetter, W. Keller-Schierlein, and H. Zähner, *Arch. Microbiol.*, **39**, 158 (1961); W. Oroshnik and A. D. Mebane, *Fortschr. Chem. Org. Naturstoffe*, **21**, 17 (1963); M. Berry, *Quart. Rev.*, p. 343 (1964).
7. M. Ikeda, M. Suzuki, and C. Djerassi, *Tetrahedron Lett.*, 3745 (1967).
7a. A. J. Birch et al. *Tetrahedron Lett.*, 1491 (1964).
8. J. D. Dutcher, D. R. Walters, and O. Wintersteiner, *J. Org. Chem.*, **28**, 995 (1963).
9. M. H. von Saltza, J. Reid, J. D. Dutcher, and O. Wintersteiner, *J. Org. Chem.*, **28**, 999 (1963).
10. M. H. von Saltza, J. Reid, J. D. Dutcher, and O. Wintersteiner, *J. Amer. Chem. Soc.*, **83**, 2785 (1961).
11. D. R. Walters, J. D. Dutcher, and O. Wintersteiner, *J. Amer. Chem. Soc.*, **79**, 5076 (1957).
12. J. B. Patrick, R. P. Williams, and J. S. Webb, *J. Amer. Chem. Soc.*, **80**, 6689 (1958).
13. O. Ceder, *Acta Chem. Scand.*, **18**, 126 (1964), and previous papers.
13a. W. E. Meyer, *Chem. Commun.*, 470 (1968).
14. B. T. Golding, R. W. Rickards, W. E. Meyer, J. B. Patrick, and M. Barber, *Tetrahedron Lett.*, p. 3551 (1966).
15. W. D. Celmer, *J. Amer. Chem. Soc.*, **87**, 1799 (1965).

16. G. Guadiano, P. Bravo, A. Quilico, B. T. Golding, and R. W. Rickards, *Tetrahedron Lett.*, 3567 (1966); G. Guadiano, P. Bravo, and A. Quilico, *ibid.*, 3559 (1966).
17. S. Ball, C. J. Bessell, and A. Mortimer, *J. Gen. Microbiol.*, 17, 96 (1959).
18. H. Nakano, K. Hattori, M. Seki and Y. Hirata, *J. Antibiot. (Tokyo)*, A9, 172 (1956).
19. H. Nakano, *J. Antibiot. (Tokyo)*, A14, 72 (1961).
20. K. Hattori, *J. Antibiot. (Tokyo)*, B15, 37 (1962); *Chem. Abstr.*, 58, 5532 (1963).
20a. C. H. Lee and C. P. Schaffner, *Tetrahedron Lett.*, 5837 (1966).
21. R. B. Woodward, *Angew. Chem.*, 69, 50 (1957).
22. H. Brockmann and W. Henkel, *Ber.*, 84, 284 (1951).
23. S. E. DeVoe, H. B. Renfroe, and W. K. Hausmann, *Antimicrobial Agents Chemotherapy*, 125 (1963).
23a. H. Muxfeldt, S. Shrader, P. Hansen, and H. Brockmann, *J. Amer. Chem. Soc.*, 90 4748 (1968).
24. M. N. Donin, J. Pagano, J. D. Dutcher, and C. M. McKee, *Antibiot. Ann.*, 179 (1953-1954).
25. R. Anliker, D. Dvornik, K. Gubler. H. Heusser, and V. Prelog, *Helv. Chim. Acta*, 39, 1785 (1956).
26. C. Djerassi and J. A. Zderic, *J. Amer. Chem. Soc.*, 78, 6390 (1956).
27. C. Djerassi and O. Halpern, *J. Amer. Chem. Soc.*, 79, 2022 (1957).
28. C. Djerassi and O. Halpern, *Tetrahedron*, 3, 255 (1958).
29. H. Brockmann and R. Oster, *Ber.*, 90, 605 (1957).
30. R. Anliker and K. Gubler, *Helv. Chim. Acta*, 40, 119, 1768 (1957).
31. R. Corbaz, L. Ettlinger, E. Gäumann, W. Keller-Schierlein, F. Kradolfer, E. Kyburz, L. Neipp, V. Prelog, R. Reusser, and H. Zähner, *Helv. Chim. Acta*, 38, 935 (1955).
32. V. Prelog, A. M. Gold, G. Talbot, and A. Zamojski, *Helv. Chim. Acta*, 45, 4 (1962).
33. H. Brockmann, H. B. König, and R. Oster, *Ber.*, 87, 856 (1954).
34. R. K. Clark, Jr., *Antibiot. Chemotherapy*, 3, 663 (1953).
35. C. H. Bolton, A. B. Foster, M. Stacey, and J. M. Webber, *J. Chem. Soc.*, 4831 (1961); *Chem. Ind. (London)*, 1945 (1962).
36. H. Newman, *Chem. Ind. (London)*, 372 (1963).
37. W. Hofheintz and H. Grisebach, *Tetrahedron Lett.*, 377 (1962).
38. P. W. K. Woo, H. W. Dion, L. Durham and H. S. Mosher, *Tetrahedron Lett.*, 735 (1962).
39. H. Newman, *J. Org. Chem.*, 29, 1461 (1964).
40. A. C. Richardson, *Proc. Chem. Soc.*, 131 (1963).
41. F. Korte, A. Bilow, and R. Heinz, *Tetrahedron*, 18, 657 (1962).
42. P. F. Wiley, K. Gerzon, E. H. Flynn, M. V. Sigal, Jr., O. Weaver, J. C. Quarck, R. R. Chauvette, and R. Monahan, *J. Amer. Chem. Soc.*, 79, 6062 (1957), and previous papers.
42a. D. R. Harris, S. G. McGeachin, and H. H. Mills, *Tetrahedron Lett.*, 679 (1965).
43. E. H. Flynn, M. V. Sigal, Jr., P. F. Wiley, and K. Gerzon, *J. Amer. Chem. Soc.*, 76, 3121 (1954).
44. W. Hofheintz and H. Grisebach, *Ber.*, 96, 2867 (1963).
45. D. M. Lemal, P. D. Pacht, and R. B. Woodward, *Tetrahedron*, 18, 1275 (1962).
45a. G. B. Haworth and J. K. N. Jones, *Can. J. Chem.*, 45, 2253 (1967).
46. H. Els, W. D. Celmer, and K. Murai, *J. Amer. Chem. Soc.*, 80, 3777 (1958).
47. F. A. Hochstein, H. Els, W. D. Celmer, B. L. Shapiro, and R. B. Woodward, *J. Amer. Chem. Soc.*, 82, 3225 (1960).
48. F. Blindenbacher and T. Reichstein, *Helv. Chim. Acta*, 31, 2061 (1948).

49. R. Paul and S. Tchelitcheff, *Bull. Soc. Chim. Fr.*, 443 (1957).
50. R. Corbaz, L. Ettlinger, E. Gäumann, W. Keller-Schierlein, F. Kradolfer, E. Kyburz, L. Neipp, V. Prelog, A. Wettstein, and H. Zähner, *Helv. Chim. Acta*, **39**, 304 (1956).
51. R. Paul and S. Tchelitcheff, *Bull. Soc. Chim. Fr.*, p. 650 (1965).
52. M. E. Kuehne and B. W. Benson, *J. Amer. Chem. Soc.*, **87**, 4660 (1965).
52a. S. Ōmura, *et al.*, *J. Amer. Chem. Soc.*, **91**, 3401 (1969).
53. W. Hofheintz, H. Grisebach, and H. Friebolin, *Tetrahedron*, **18**, 1265 (1962).
54. R. Paul and S. Tchelitcheff, *Bull. Soc. Chim. Fr.*, 734 (1957).
55. C. L. Stevens, R. P. Glinski, G. E. Gutowski, and J. P. Dickerson, *Tetrahedron Lett.*, p. 649 (1967); E. L. Albano and D. Horton, *Carbohyd. Res.*, **11**, 485 (1969).
56. R. Paul and S. Tchelitcheff, *Bull. Soc. Chim. Fr.*, 1059 (1957).
57. A. C. Richardson, *Proc. Chem. Soc.*, 430 (1961); *J. Chem. Soc.*, 2758 (1962).
58. A. B. Foster, T. D. Inch, J. Lehmann, M. Stacey, and J. M. Webber, *Chem. Ind. (London)*, 142 (1962); *J. Chem. Soc.*, 2116 (1962).
59. R. B. Woodward, L. S. Weiler, and P. C. Dutta, *J. Amer. Chem. Soc.*, **87**, 4662 (1965).
60. F. A. Hochstein and K. Murai, *J. Amer. Chem. Soc.*, **76**, 5080 (1954).
61. F. Lidner, *Arzneimittel-Forsch.*, **12**, 1191 (1962).
62. P. P. Regna, F. A. Hochstein, R. L. Wagner, Jr., and R. B. Woodward, *J. Amer. Chem. Soc.*, **75**, 4625 (1953).
63. A. B. Foster, T. D. Inch, J. Lehmann, and J. M. Webber, *Chem. Ind. (London)*, 1619 (1962).
64. A. B. Foster, T. D. Inch, J. Lehmann, L. F. Thomas, J. M. Webber, and J. A. Wyer, *Proc. Chem. Soc.*, 254 (1962).
65. F. Korte, U. Claussen, and K. Göhring, *Tetrahedron*, **18**, 1257 (1962).
66. F. A. Hochstein and P. P. Regna, *J. Amer. Chem. Soc.*, **77**, 3353 (1955).
67. A. B. Foster, J. Lehmann, and M. Stacey, *J. Chem. Soc.*, 1396 (1962).
68. R. B. Morin and M. Gorman, *Tetrahedron Lett.*, 2339 (1964).
69. H. W. Dion, P. W. K. Woo, and Q. R. Bartz, *J. Amer. Chem. Soc.*, **84**, 880 (1962).
70. J. S. Brimacombe, M. Stacey, and L. C. N. Tucker, *Proc. Chem. Soc.*, 83 (1964); *J. Chem. Soc.*, 5391 (1964).
70a. S. Ōmura, H. Ogura, and T. Hata, *Tetrahedron Lett.*, 1267 (1963).
70b. R. Corbaz, L. Ettlinger, E. Gäumann, W. Keller-Schierlein, L. Neipp, V. Prelog, R. Reusser, and H. Zähner, *Helv. Chim. Acta*, **38**, 1202 (1955).
70c. M. Brufani and W. Keller-Schierlein, *Helv. Chim. Acta*, **49**, 1962 (1966).
71. P. W. K. Woo, H. W. Dion, and Q. R. Bartz, *J. Amer. Chem. Soc.*, **86**, 2724 (1964).
72. P. W. K. Woo, H. W. Dion, and Q. R. Bartz, *J. Amer. Chem. Soc.*, **83**, 3352 (1961).
73. P. W. K. Woo, H. W. Dion, and L. F. Johnson, *J. Amer. Chem. Soc.*, **84**, 1066 (1962).
74. A. B. Foster, M. Stacey, J. M. Webber, and J. H. Westwood, *Proc. Chem. Soc.*, 279 (1963).
75. N. K. Kochetkov and A. I. Usov, *Tetrahedron Lett.*, 519 (1963).
76. S. McNally and W. G. Overend, *Chem. Ind. (London)*, 2021 (1964).
77. F. Micheel, *Ber.*, **63**, 347 (1930).
78. E. Gäumann, R. Heutter, W. Keller-Schierlein, L. Neipp, V. Perlog, and H. Zähner, *Helv. Chim. Acta*, **43**, 601 (1960).
79. W. Keller-Schierlein and G. Roncari, *Helv. Chim. Acta*, **45**, 138 (1962).
80. W. Keller-Schierlein and G. Roncari, *Helv. Chim. Acta*, **47**, 78 (1964).
81. G. Roncari and W. Keller-Schlierlein, *Helv. Chim. Acta*, **49**, 705 (1966).
81a. G. B. Haworth, W. A. Szarek, and J. K. N. Jones, *Chem. Commun.*, 62 (1968).

82. M. P. Kunstmann, L. A. Mitscher and E. L. Patterson, *Antimicrobial Agents Chemotherapy*, 87 (1964).

83. L. Ettlinger, E. Gäumann, R. Heutter, W. Keller-Schierlein, F. Kradolfer, L. Neipp, V. Prelog, R. Reusser, and H. Zähner, *Monatsh. Chem.*, **88**, 989 (1957).

84. S. Kondo, J. M. Sakamoto, and H. Yumoto, *J. Antibiot. (Tokyo)*, **A14**, 365 (1961).

85. M. P. Kunstmann, L. A. Mitscher, and N. Bohonos, *Tetrahedron Lett.*, 389 (1966).

85a. G. A. Ellestad, M. P. Kunstmann, J. E. Lancaster, L. A. Mitscher, and G. Morton, *Tetrahedron*, **23**, 3893 (1967).

86. M. P. Kunstmann and L. A. Mitscher, *Experientia* **21**, 372 (1965).

87. Y. Suhara, K. Maeda, H. Umezawa, and M. Ohno, *Tetrahedron Lett.*, 1239 (1966).

87a. Y. Suhara, K. Maeda, H. Umezawa, and M. Ohno, *Advan. Chem.*, No. 74, 15 (1968).

88. Y. Suhara, K. Maeda, and H. Umezawa, *J. Antibiot. (Tokyo)*, **A18**, 182 (1965).

89. Y. Suhara, K. Maeda, H. Umezawa, and M. Ohno, *J. Antibiot. (Tokyo)*, **A18**, 184 (1965).

90. N. Sharon and R. W. Jeanloz, *J. Biol. Chem.*, **235**, 1 (1960).

91. Y. Suhara, K. Maeda, and H. Umezawa, *J. Antibiot. (Tokyo)*, **A18**, 187 (1965).

92. T. Ikekawa, H. Umezawa, and Y. Iitaka, *J. Antibiot. (Tokyo)*, **A19**, 49 (1966).

92a. K. Kitahara *et al.*, *Agr. Biol. Chem. (Tokyo)*, **33**, 748 (1969).

92b. Y. Suhara, F. Sasaki, K. Maeda, and H. Umezawa, *J. Am. Chem. Soc.*, **90**, 6559 (1968).

93. R. L. Mann and D. O. Woolf, *J. Amer. Chem. Soc.*, **79**, 120 (1957).

94. M. Namiki, K. Isono, and S. Suzuki, *J. Antibiot. (Tokyo)*, **A10**, 160 (1957).

95. J. B. Patrick, R. P. Williams, C. W. Walker, and B. L. Hutchings, *J. Amer. Chem. Soc.*, **78**, 2652 (1956).

95a. S. Takahashi and M. Nakajima, *Tetrahedron Lett.*, 2285 (1967).

96. B. Bannister and A. D. Argoudelis, *J. Amer. Chem. Soc.*, **85**, 234 (1963).

97. T. Miyaki, H. Tsukiura, M. Wakae, and H. Kawaguchi, *J. Antibiot. (Tokyo)*, **A15**, 15 (1962); **B15**, 373 (1962).

98. B. Bannister and A. D. Argoudelis, *J. Amer. Chem. Soc.*, **85**, 119 (1963).

99. I. J. McGilveray and K. L. Rinehart, Jr., *J. Amer. Chem. Soc.*, **87**, 4003 (1965).

100. R. U. Lemieux and M. L. Wolfrom, *Advan. Carbohyd. Chem.*, **3**, 337 (1948).

101. M. L. Wolfrom, M. J. Cron, C. W. DeWalt, and R. M. Husband, *J. Amer. Chem. Soc.*, **76**, 3675 (1954).

102. R. U. Lemieux, C. W. DeWalt, and M. L. Wolfrom, *J. Amer. Chem. Soc.*, **69**, 1838 (1947).

103. J. R. Dyer and A. W. Todd, *J. Amer. Chem. Soc.*, **85**, 3896 (1963).

104. F. A. Kuehl, Jr., R. L. Peck, C. E. Hoffhine, Jr., and K. Folkers, *J. Amer. Chem. Soc.*, **70**, 2325 (1948).

105. R. E. Reeves, *Advan. Carbohyd. Chem.*, **6**, 107 (1951).

106. S. Tatsuoka and S. Horii, *Proc. Jap. Acad.*, **39**, 314 (1963).

107. S. Tatsuoka, S. Horii, K. L. Rinehart, Jr., and T. Nakabayashi, *J. Antibiot. (Tokyo)*, **A17**, 88 (1964).

108. Y. Wang, L. Loh, W. Lin, and C. Chang, *Acta Chim. Sinica*, **25**, 257 (1959); *Chem. Abstr.*, **54**, 18371 (1960).

109. I. J. McGilveray and J. B. Stenlake, *Chem. Ind. (London)*, 238 (1964).

110. J. R. Dyer, W. E. McGonigal, and K. C. Rice, *J. Amer. Chem. Soc.*, **87**, 654 (1965).

111. L. J. Heusen, M. A. Dolliven, and E. T. Stiller, *J. Amer. Chem. Soc.*, **75**, 4013 (1953).

112. D. P. Young, *J. Chem. Soc.*, 1337 (1961).

112a. S. Neidl, D. Rogers, and M. B. Hursthouse, *Tetrahedron Lett.*, 4725 (1968).

113. F. H. Stodola, D. L. Shotwell, A. M. Borud, R. G. Benedict, and A. C. Riley, Jr., *J. Amer. Chem. Soc.*, **73**, 2290 (1951).

114. H. Hoeksema, A. D. Argoudelis, and P. F. Wiley, *J. Amer. Chem. Soc.*, **84**, 3212 (1962); *ibid.*, **85**, 2652 (1963).

115. P. F. Wiley, *J. Amer. Chem. Soc.*, **84**, 1514 (1962).

115a. A. L. Johnson, R. H. Gourlay, D. S. Tarbell, and R. L. Autrey, *J. Org. Chem.*, **28**, 300 (1963).

116. G. Slomp and F. A. MacKellar, *Tetrahedron Lett.*, 521 (1962).

116a. M. Nakajima, N. Kurihara, A. Hasegawa, and T. Kurokawa, *Ann.*, **689**, 243 (1965).

116b. T. Suami and S. Ogawa, *Bull. Chem. Soc. Jap.*, **38**, 2026 (1965); F. Lichtenthaler, H. Leinert, and T. Suami, *Ber.*, **100**, 2383 (1967).

116c. S. Ogawa, T. Abe, H. Sano, K. Kotera, and T. Suami, *Bull. Chem. Soc. Jap.*, **40**, 2405 (1967).

117. S. A. Waksman and H. A. Lechevalier, *Science*, **109**, 305 (1949).

118. K. L. Rinehart, Jr., "The Neomycins and Related Antibiotics," Wiley, New York, 1964.

119. J. D. Dutcher, N. Hosansky, M. N. Donin, and O. Wintersteiner, *J. Amer. Chem. Soc.*, **73**, 1384 (1951).

120. B. E. Leach and C. M. Teeters, *J. Amer. Chem. Soc.*, **73**, 2794 (1951).

121. K. L. Rinehart, Jr. and P. W. K. Woo, *J. Amer. Chem. Soc.*, **83**, 643 (1961) and previous papers.

121a. K. Tatsuta, E. Kitazawa, and S. Umezawa, *Bull. Chem. Soc. Jap.*, **40**, 2371 (1967).

122. K. L. Rinehart, Jr., M. Hichens, A. D. Argoudelis, W. S. Chilton, H. E. Carter, M. P. Georgiadis, C. P. Schaffner, and R. T. Schillings, *J. Amer. Chem. Soc.*, **84**, 3218 (1962).

123. F. A. Kuehl, Jr., M. N. Bishop, and K. Folkers, *J. Amer. Chem. Soc.*, **73**, 881 (1951).

124. K. L. Rinehart, Jr., P. W. K. Woo, and A. D. Argoudelis, *J. Amer. Chem. Soc.*, **79**, 4568 (1957).

125. W. Meyer zu Reckendorf, *Angew. Chem.*, **75**, 573 (1963); *Tetrahedron*, **19**, 2033 (1963).

126. H. Weidmann and H. K. Zimmerman, Jr., *Ann.*, **644**, 127 (1961).

127. K. L. Rinehart, Jr., P. W. K. Woo, and A. D. Argoudelis, *J. Amer. Chem. Soc.*, **80**, 6461 (1958).

128. K. L. Rinehart, Jr., M. Hichens, K. Striegler, K. R. Rover, T. P. Culbertson, S. Tatsuoka, S. Horii, T. Yamaguchi, H. Hitomi, and A. Miyake, *J. Amer. Chem. Soc.*, **83**, 2964 (1961).

129. W. Meyer zu Reckendorf, *Ber.*, **96**, 2017 (1963).

130. H. E. Carter, J. R. Dyer, P. D. Shaw, K. L. Rinehart, Jr., and M. Hichens, *J. Amer. Chem. Soc.*, **83**, 3723 (1961).

131. R. U. Lemieux and R. J. Cushley, *Can. J. Chem.*, **41**, 858 (1963).

132. M. Hichens and K. L. Rinehart, Jr., *J. Amer. Chem. Soc.*, **85**, 1547 (1963).

133. K. L. Rinehart, Jr., W. S. Chilton, M. Hichens, and W. von Phillipsborn, *J. Amer. Chem. Soc.*, **84**, 3216 (1962).

134. S. Tatsuoka and S. Horii, *Proc. Japan Acad.*, **39**, 314 (1963).

135. T. H. Haskell, J. C. French, and Q. R. Bartz, *J. Amer. Chem. Soc.*, **81**, 3482 (1959).

136. T. H. Haskell, J. C. French, and Q. R. Bartz, *J. Amer. Chem. Soc.*, **81**, 3480 (1959).

137. T. H. Haskell, J. C. French, and Q. R. Bartz, *J. Amer. Chem. Soc.*, **81**, 3481 (1959).
138. T. H. Haskell, J. C. French, and Q. R. Bartz, *J. Amer. Chem. Soc.*, **81**, 3481 (1959).
139. T. H. Haskell and S. Hanessian, *J. Org. Chem.*, **28**, 2598 (1963).
140. S. Hanessian and T. H. Haskell, *J. Org. Chem.*, **28**, 2604 (1963).
141. For a review on sugars containing nitrogen in the ring, see H. Paulsen, *Angew. Chem. Internat. Ed. Engl.*, **5**, 495 (1966).
142. S. Horii, *J. Antibiot. (Tokyo)*, **A15**, 187 (1962) and previous papers.
142a. W. T. Shier *et al.*, *Proc. Nat. Acad. Sci. U.S.*, **63**, 198 (1969).
143. H. Umezawa, M. Ueda, K. Maeda, K. Yagishita, S. Kondo, Y. Okami, R. Utahara, Y. Osato, K. Nitta, and T. Takeuchi, *J. Antibiot. (Tokyo)*, **A10**, 181 (1957).
143a. G. Koyama, Y. Iitaka, K. Maeda, and H. Umezawa, *Tetrahedron Lett.*, 1875 (1968).
144. K. Maeda, M. Murase, H. Mawatari, and H. Umezawa, *J. Antibiot. (Tokyo)*, **A11**, 163 (1958).
145. M. J. Cron, D. L. Evans, F. M. Palermiti, D. F. Whitehead, I. R. Hooper, P. Chu, and R. U. Lemieux, *J. Amer. Chem. Soc.*, **80**, 4741 (1958).
146. H. Ogawa, T. Ito, S. Kondo, and S. Inoue, *J. Antibiot. (Tokyo)*, **A11**, 169 (1958).
147. M. J. Cron, O. B. Fardig, D. L. Johnson, H. Schmitz, D. F. Whitehead, I. R. Hooper, and R. U. Lemieux, *J. Amer. Chem. Soc.*, **80**, 2342 (1958).
148. H. Ogawa, T. Ito, S. Inoue, and S. Kondo, *J. Antibiot. (Tokyo)*, **A11**, 166 (1958).
148a. H. H. Bear, *J. Amer. Chem. Soc.*, **83**, 1882 (1961).
149. H. Ogawa, T. Ito, S. Inoue, and S. Kondo, *J. Antibiot. (Tokyo)*, **A11**, 70 (1958).
150. K. Maeda, M. Murase, H. Mawatari, and H. Umezawa, *J. Antibiot. (Tokyo)*, **A11**, 73 (1958).
151. H. Ogawa, T. Ito, S. Kondo, and S. Inoue, *Bull. Agr. Chem. Soc. Jap.*, **23**, 289 (1959).
152. H. Ogawa, T. Ito, S. Inoue, and S. Kondo, *J. Antibiot. (Tokyo)*, **A11**, 72 (1958).
153. M. J. Cron, O. B. Fardig, D. L. Johnson, D. F. Whitehead, I. R. Hooper, and R. U. Lemieux, *J. Amer. Chem. Soc.*, **80**, 4115 (1958).
154. S. Umezawa, Y. Ito, and S. Fukatsu, *J. Antibiot. (Tokyo)*, **A11**, 162 (1958).
155. M. Murase, *J. Antibiot. (Tokyo)*, **A14**, 367 (1961).
156. T. Ito, M. Nishio, and H. Ogawa, *J. Antibiot. (Tokyo)*, **A17**, 189 (1964).
156a. M. Nakajima, A. Hasegawa, N. Kurihara, H. Shibata, T. Ueno, and D. Nishimura, *Tetrahedron Lett.*, 623 (1968).
156b. S. Umezawa, K. Tatsuta, and S. Koto, *Bull. Chem. Soc. Jap.*, **42**, 533 (1969).
156c. S. Umezawa, S. Koto, K. Tatsuta, H. Hineno, Y. Nishimura, and T. Tsumura, *Bull. Chem. Soc. Jap.*, **42**, 537 (1969).
156d. S. Umezawa, S. Koto, K. Tatsuta, and T. Tsumura, *Bull. Chem. Soc. Jap.*, **42**, 529 (1969).
156e. S. Umezawa, S. Kotō, K. Tatsuta, and T. Tsumura, *J. Antibiot. (Tokyo)*, **21**, 162 (1968).
157. M. J. Weinstein, G. M. Leudemann, E. M. Oden, G. H. Wagman, J. P. Rosselet, J. A. Marquez, C. T. Coniglio, W. Charney, and H. L. Herzog, *J. Med. Chem.*, **6**, 463 (1963).
157a. D. J. Cooper and M. D. Yudis, *Chem. Commun.*, 821 (1967).
157b. H. Maehr and C. P. Schaffner, *J. Amer. Chem. Soc.*, **89**, 6787 (1967).
158. R. L. Mann and W. W. Bromer, *J. Amer. Chem. Soc.*, **80**, 2714 (1958).
159. P. F. Wiley, M. V. Sigal, Jr., and O. Weaver, *J. Org. Chem.*, **27**, 2793 (1962).
159a. M. Nakajima, A. Hasegawa, and N. Kurihara, *Ann.*, **689**, 235 (1965).
160. S. Kondo, M. Sezaki, M. Koike, M. Shimura, E. Akita, K. Satoh, and T. Hara, *J. Antibiot. (Tokyo)*, **A18**, 38 (1965).

161. S. Kondo, E. Akita, and M. Koike, *J. Antibiot.* (*Tokyo*), **A19**, 139 (1966).
162. S. Kondo, E. Akita, and M. Sezaki, *J. Antibiot.* (*Tokyo*), **A19**, 137 (1966).
163. S. Kondo, M. Sezaki, M. Koike, and E. Akita, *J. Antibiot.* (*Tokyo*), **A18**, 192 (1965).
163a. J. J. Fox, K. A. Watanabe, and A. Bloch, *Progr. Nucleic Acid Res. Mol. Biol.*, **5**, 251 (1966).
164. J. A. Montgomery and H. J. Thomas, *Advan. Carbohyd. Chem.*, **17**, 301 (1962).
165. N. Löfgren, B. Lüning, and H. Hedström, *Acta Chem. Scand.*, **8**, 670 (1954).
166. G. B. Brown and V. S. Weliky, *J. Biol. Chem.*, **204**, 1019 (1953).
167. J. J. Fox, I. Wempen, A. Hampton, and I. L. Doerr, *J. Amer. Chem. Soc.*, **80**, 1669 (1958).
168. H. J. Schaeffer and H. J. Thomas, *J. Amer. Chem. Soc.*, **80**, 4896 (1958).
169. T. Hashizume and H. Iwamura, *Tetrahedron Lett.*, 643 (1966).
170. C. W. Waller, J. B. Patrick, W. Fulmor, and W. E. Meyer, *J. Amer. Chem. Soc.*, **79**, 1011 (1957).
170a. G. O. Morton *et al.*, *J. Amer. Chem. Soc.*, **91**, 1535 (1969).
171. H. R. Bentley, K. G. Cunningham, and F. G. Spring, *J. Chem. Soc.*, 2301 (1951).
172. W. W. Lee, A. Benitez, C. D. Anderson, L. Goodman, and B. R. Baker, *J. Amer. Chem. Soc.*, **83**, 1906 (1961).
173. E. Walton, R. F. Nutt, S. R. Jenkins, and F. W. Holly, *J. Amer. Chem. Soc.*, **86**, 2952 (1964).
174. E. A. Kaczka, N. R. Trenner, B. Arison, R. W. Walker, and K. Folkers, *Biochem. Biophys. Res. Commun.*, **14**, 456 (1964).
175. R. J. Suhadolnik and J. G. Cory, *Biochim. Biophys. Acta*, **91**, 661 (1964).
176. S. Hanessian, D. C. DeJongh, and J. A. McCloskey, *Biochim. Biophys. Acta*, **117**, 480 (1966).
177. W. Schroeder and H. Hoeksema, *J. Amer. Chem. Soc.*, **81**, 1767 (1959).
178. H. Yüntsen, *J. Antibiot.* (*Tokyo*), **A11**, 244 (1958).
179. J. Farkaš and F. Šorm, *Tetrahedron Lett.*, 813 (1962).
180. H. Yüntsen, *J. Antibiot.* (*Tokyo*), **A11**, 233 (1958) and earlier papers.
181. H. Hoeksema, G. Slomp, and E. E. Van Tamelen, *Tetrahedron Lett.*, 1787 (1964).
182. J. W. Porter, R. J. Hewitt, C. W. Hesseltine, G. Krupta, J. A. Lowery, W. S. Wallace, N. Bohonos, and J. H. Williams, *Antibiot. Chemotherapy*, **2**, 409 (1952).
183. B. R. Baker, R. E. Schaub, J. P. Joseph, and J. H. Williams, *J. Amer. Chem. Soc.*, **77**, 12 (1955).
184. N. N. Gerber and H. A. Lechevalier, *J. Org. Chem.*, **27**, 1731 (1962).
185. J. D. Dutcher, M. H. von Saltza, and F. E. Pansy, *Antimicrobial Agents Chemotherapy*, 83 (1963).
186. H. Agahigian, G. D. Vickers, M. H. von Saltza, J. Reid, A. I. Cohen, and H. Gauthier, *J. Org. Chem.*, **30**, 1085 (1965).
187. M. H. von Saltza, J. D. Dutcher and J. Reid, *Abstr. Papers Am. Chem. Soc. Meeting*, **148**, 15Q (1964).
188. J. J. Fox and I. Wempen, *Advan. Carbohyd. Chem.*, **14**, 283 (1959).
189. C. DeBoer, E. L. Caron, and J. W. Hinman, *J. Amer. Chem. Soc.*, **75**, 499 (1957).
190. E. H. Flynn, J. W. Hinman, E. L. Caron, and D. O. Woolf, *J. Amer. Chem. Soc.*, **75**, 5887 (1953).
191. T. H. Haskell, A. Ryder, R. D. Frohardt, S. A. Fusari, Z. L. Jakubowski, and Q. R. Bartz, *J. Amer. Chem. Soc.*, **80**, 743 (1958) and references cited therein.
192. C. L. Stevens, R. J. Gasser, T. K. Mukherjee, and T. H. Haskell, *J. Amer. Chem. Soc.*, **78**, 6212 (1956).
193. C. L. Stevens, K. Nagarajan, and T. H. Haskell, *J. Org. Chem.*, **27**, 2991 (1962).

194. C. L. Stevens, P. Blumbergs, and F. A. Daniher, *J. Amer. Chem. Soc.*, **85**, 1552 (1963).
195. C. L. Stevens, P. Blumbergs, and D. L. Wood, *J. Amer. Chem. Soc.*, **86**, 3592 (1964); S. Hanessian and N. R. Plessas, *J. Org. Chem.*, **34**, 1035, 1045 (1969); E. L. Albano and D. Horton, *J. Org. Chem.*, **34**, 3519 (1969).
196. T. H. Haskell, *J. Amer. Chem. Soc.*, **80**, 747 (1958).
196a. C. L. Stevens, P. Blumbergs, F. A. Daniher, D. H. Otterbach, and K. G. Taylor, *J. Org. Chem.*, **31**, 2822 (1966).
197. S. Hanessian and T. H. Haskell, *Tetrahedron Lett.*, 2451 (1964).
198. C. L. Stevens, N. A. Nielsen, P. Blumbergs, and K. G. Taylor, *J. Amer. Chem. Soc.*, **86**, 5695 (1964).
199. S. Takeuchi, K. Hirayama, K. Ueda, H. Sakai, and H. Yonehara, *J. Antibiot. (Tokyo)*, **A11**, 1 (1958).
200. N. Otake, S. Takeuchi, T. Endo, and H. Yonehara, *Agr. Biol. Chem. (Tokyo)*, **30**, 126 (1966).
201. N. Otake, S. Takeuchi, T. Endo, and H. Yonehara, *Agr. Biol. Chem. (Tokyo)*, **30**, 132 (1966).
202. N. Otake, S. Takeuchi, T. Endo, and H. Yonehara, *Tetrahedron Lett.*, 1405 (1965).
203. J. J. Fox and K. A. Watanabe, *Tetrahedron Lett.*, 897 (1966).
204. H. Yonehara and N. Otake, *Tetrahedron Lett.*, 3785 (1966).
205. N. Otake, S. Takeuchi, T. Endo, and H. Yonehara, *Tetrahedron Lett.*, 1411 (1965).
205a. S. Onuma, Y. Nawata, and Y. Saito, *Bull. Chem. Soc. Jap.*, **39**, 1091 (1966).
206. T. Kanzaki, E. Higashide, H. Yamamoto, M. Shibata, K. Nakazawa, and A. Miyake, *J. Antibiot. (Tokyo)*, **A15**, 93 (1962).
207. J. M. Clark and J. K. Gunter, *Biochim. Biophys. Acta*, **76**, 3638 (1963).
208. E. Iwasaki, *Yakugaku Zasshi*, **82**, 1358, 1361, 1365, 1368, 1372, 1376, 1380, 1384, 1390, 1393 (1962).
209. J. J. Fox, Y. Kuwada, and K. A. Watanabe, *Tetrahedron Lett.*, 6029 (1968).
209a. K. Isono and S. Suzuki, *Tetrahedron Lett.*, 203 (1968).
210. K. Anzai, G. Nakamura, S. Suzuki, *J. Antibiot. (Tokyo)*, **A10**, 201 (1957).
211. S. Suzuki and S. Marumo, *J. Antibiot. (Tokyo)*, **A14**, 34 (1961).
212. Y. Mizuno, M. Ikehara, K. Watanabe, and S. Suzaki, *Chem. Pharm. Bull. (Tokyo)*, **11**, 1091 (1963).
213. J. E. Pike, L. Slechta, and P. F. Wiley, *J. Heterocycl. Chem.*, **1**, 159 (1964).
214. K. Ohkuma, *J. Antibiot. (Tokyo)*, **A14**, 343 (1961).
214a. R. L. Tolman, R. K. Robins, and L. B. Townsend, *J. Amer. Chem. Soc.*, **90**, 524 (1968).
214b. K. V. Rao, *J. Med. Chem.*, **11**, 939 (1968).
215. G. Koyama and H. Umezawa, *J. Antibiot. (Tokyo)*, **A18**, 175 (1965) and references cited therein.
216. G. Koyama, K. Maeda, and H. Umezawa, *Tetrahedron Lett.*, 597 (1966).
217. R. K. Robins, L. B. Townsend, F. Cassidy, J. F. Gerster, A. F. Lewis, and R. L. Miller, *J. Heterocycl. Chem.*, **3**, 110 (1966).
217a. H. Nishimura, M. Mayama, Y. Komatsu, H. Kato, N. Shimaoka, and Y. Tanaka, *J. Antibiot. (Tokyo)*, **A17**, 148 (1964).
217b. K. R. Darnall, L. B. Townsend, and R. K. Robins, *Proc. Nat. Acad. Sci. U.S.*, **57**, 548 (1967).
217c. Y. Nakagawa, H. Kano, Y. Tsukuda, and H. Koyama, *Tetrahedron Lett.*, 4105 (1967).

218. K. Nakanishi and M. Ohaski, *Bull. Chem. Soc. Jap.*, **30**, 725 (1957) and previous papers.
219. A. W. Johnson and J. W. Westley, *J. Chem. Soc.*, 1642 (1962).
220. H. E. Carter, C. C. Sweeley, E. E. Daniels, J. E. McNary, C. P. Schaffner, C. A. West, E. E. Van Tamelen, J. R. Dyer, and H. A. Whaley, *J. Amer. Chem. Soc.*, **83**, 4296 (1961), and references cited therein.
221. E. E. Van Tamelen, J. R. Dyer, H. A. Whaley, H. E. Carter, and G. B. Whitefield, *J. Amer. Chem. Soc.*, **83**, 4295 (1961).
222. E. E. Van Tamelen, J. R. Dyer, H. E. Carter, J. V. Pierce, and E. E. Daniels, *J. Amer. Chem. Soc.*, **78**, 4817 (1958).
223. H. E. Carter, J. V. Pierce, G. B. Whitefield, J. E. McNary, E. E. Van Tamelen, J. R. Dyer, and H. A. Whaley, *J. Amer. Chem. Soc.*, **83**, 4287 (1961).
224. H. Taniyama and S. Takemura, *Chem. Pharm. Bull.* (*Tokyo*), **8**, 150 (1960), and previous papers.
225. H. Taniyama and F. Miyoshi, *Chem. Pharm. Bull.* (*Tokyo*), **10**, 156 (1962).
225a. H. Taniyama and F. Miyoshi, *Chem. Pharm. Bull.* (*Tokyo*), **10**, 156 (1962).
226. S. Takemura, *Chem. Pharm. Bull.* (*Tokyo*), **8**, 574, 578 (1960).
227. K. L. Rinehart, Jr., J. R. Beck, D. B. Borders, T. H. Kinstle, and D. Krauss, *J. Amer. Chem. Soc.*, **85**, 4038 (1963).
228. K. L. Rinehart, Jr., J. R. Beck, W. W. Epstein, and L. D. Spicer, *J. Amer. Chem. Soc.*, **85**, 4035 (1963).
229. K. L. Rinehart, Jr. and D. B. Borders, *J. Amer. Chem. Soc.*, **85**, 4037 (1963).
230. H. Hoeksema and C. G. Smith, *Progr. Ind. Microbiol.*, **3**, 93 (1961).
231. J. W. Hinman, E. L. Caron, and H. Hoeksma, *J. Amer. Chem. Soc.*, **79**, 3789 (1957).
232. E. A. Kaczka, C. H. Shunk, J. W. Richter, F. J. Wolf, M. M. Gasser, and K. Folkers, *J. Amer. Chem. Soc.*, **78**, 4125 (1956).
233. E. Walton, J. O. Rodin, C. H. Stammer, F. W. Holly, and K. Folkers, *J. Amer. Chem. Soc.*, **80**, 5168 (1958).
234. B. P. Vaterlaus and H. Spiegelberg, *Helv. Chim. Acta*, **47**, 381 (1964).
235. J. Kiss and H. Spiegelberg, *Helv. Chim. Acta*, **47**, 398 (1964).
236. B. T. Golding and R. W. Rickards, *Chem. Ind.* (*London*), 1081 (1963).
237. C. F. Spencer, J. O. Rodin, E. Walton, F. W. Holly, and K. Folkers, *J. Amer. Chem. Soc.*, **80**, 140 (1958).
238. B. P. Vaterlaus, K. Doebel, J. Kiss, A. I. Rachlin, and H. Spiegelberg, *Helv. Chim. Acta*, **47**, 390 (1964).
239. H. Kawaguchi, T. Naito, and H. Tsukiura, *J. Antibiot.* (*Tokyo*), A**18**, 11, (1965).
240. H. Kawaguchi, T. Miyaki, and H. Tsukiura, *J. Antibiot.* (*Tokyo*), A**18**, 220 (1965).
240a. G. F. Gauze, *Advan. Chemother.*, **2**, 179 (1965).
241. M. Miyamoto, Y. Kawamatsu, K. Kowashima, M. Shinohara, and K. Nakanishi, *Tetrahedron Lett.*, 545 (1966).
242. Y. A. Berlin, S. E. Esipov, M. N. Kolosov, and M. M. Shemyakin, *Tetrahedron Lett.*, 1643 (1966).
243. Y. A. Berlin, O. A. Chuprunova, B. A. Klyashchitskii, M. N. Kolosov, G. Yu Peck, L. A. Piotrovich, M. M. Shemyakin, and I. V. Vasina, *Tetrahedron Lett.*, 1425 (1966).
244. M. Miyamoto, Y. Kawamatsu, M. Shinohara, K. Nakanishi, Y. Nakadaira, and N. S. Bhacca, *Tetrahedron Lett.*, 2371 (1964).
245. W. W. Zorbach and J. P. Ciaudelli, *J. Org. Chem.*, **30**, 451 (1965).
246. J. S. Brimacombe, D. Portsmouth, and M. Stacey, *Chem. Ind.* (*London*), 1758 (1964); *J. Chem. Soc.*, 5614 (1964).

247. J. S. Brimacombe and D. Portsmouth, *Chem. Ind. (London)*, 468 (1965); *Carbohyd. Res.* 1, 128 (1965).

247a. M. Miyamoto, Y. Kawamatsu, M. Shinohara, Y. Nakadaira, and K. Nakanishi, *Tetrahedron*, 22, 2785 (1966).

247b. Yu. A. Berlin *et al.*, *Chem. Commun.*, 762 (1968).

248. H. Brockmann, *Fortschr. Chem. Org. Naturstoffe*, 21, 121 (1963).

249. H. Brockmann and W. Leuk, *Ber.*, 92, 1904 (1959).

250. H. Brockmann and E. Spohler, *Naturwissenschaften*, 42, 154 (1955).

251. H. Brockmann, E. Spohler, and T. Waehneldt, *Ber.*, 96, 2925 (1963).

252. H. Brockmann and T. Waehneldt, *Naturwissenschaften*, 50, 43 (1963).

253. A. Di Marco, M. Gaetani, P. Orezzi, B. M. Scarpinato, R. Silvestrini, M. Soldati, T. Dasdia, and L. Valentini, *Nature*, 201, 706 (1964).

254. F. Arcamone, G. Franceschi, P. Orezzi, G. Cassinelli, W. Barbieri, and R. Mondelli, *J. Amer. Chem. Soc.*, 86, 5334 (1964).

254a. F. Arcamone, G. Franceschi, P. Orezzi, S. Penco, and R. Mondelli, *Tetrahedron Lett.*, 3349 (1968).

254b. F. Arcamone, G. Cassinelli, G. Franceschi, P. Orezzi, and R. Mondelli, *Tetrahedron Lett.*, 3353 (1968).

255. F. Arcamone, G. Cassinelli, P. Orezzi, G. Franceschi, and R. Mondelli, *J. Amer. Chem. Soc.*, 86, 5335 (1964).

256. A. C. Richardson, *Chem. Commun.*, 627 (1965).

256a. J. P. Marsh, C. W. Mosher, E. M. Acton, and L. Goodman, *Chem. Commun.*, 573 (1967).

257. L. H. Sternbach, S. Kaiser, and M. W. Goldberg, *J. Amer. Chem. Soc.*, 80, 1639 (1958) and references cited therein.

258. E. Simonitsch, W. Eisenhuth, O. A. Stamm, and H. Schmid, *Helv. Chim. Acta*, 53, 58 (1960).

258a. A. K. Ganguly, O. Z. Sarre, and H. Reimann, *J. Amer. Chem. Soc.*, 90, 7129 (1968).

258b. A. K. Ganguly and O. Z. Sarre, *Chem. Commun.*, 1149 (1969).

258c. T. Kishi, S. Harada, M. Asai, M. Muroi, and K. Mizuno, *Tetrahedron Lett.*, 97 (1969).

258d. C. L. Stevens *et al.*, *Tetrahedron Lett.*, 1181 (1969).

259. F. Arcamone, G. Canevazzi, and M. Ghione, *Giorn. Microbiol.*, 2, 205 (1956).

260. F. Arcamone and F. Bizioli, *Gazzetta Chimica Italiana*, 87, 896 (1957).

260a. S. Umezawa, K. Tatsuta, and R. Muto, *J. Antibiot. (Tokyo)*, 20, 388 (1967).

261. H. Hoeksema, B. Bannister, R. D. Birkenmeyer, F. Kagan, B. J. Magerlein, F. A. MacKellar, W. Schroeder, G. Slomp, and R. R. Herr, *J. Amer. Chem. Soc.*, 86, 4223 (1964).

261a. R. R. Herr and G. Slomp, *J. Amer. Chem. Soc.*, 89, 2444 (1967).

261b. W. Schroeder, B. Bannister, and H. Hoeksema, *J. Amer. Chem. Soc.*, 89, 2448 (1967).

261c. G. Slomp and F. A. MacKellar, *J. Amer. Chem. Soc.*, 89, 2454 (1967).

261d. B. J. Magerlein, R. D. Birkenmeyer, R. R. Herr, and F. Kagan, *J. Amer. Chem. Soc.*, 89, 2459 (1967).

261e. G. W. Howarth *et al.*, *Chem. Commun.*, 1139 (1969).

262. H. Hoeksema, *J. Amer. Chem. Soc.*, 86, 4224 (1964) and previous papers.

262a. H. Hoeksema, *J. Amer. Chem. Soc.*, 90, 755 (1968).

263. A. D. Argoudelis, J. J. Fox, D. J. Mason, and T. E. Eble, *J. Amer. Chem. Soc.* 86, 5044 (1964).

263a. R. R. Herr, H. K. Jahnke, and A. D. Argoudelis, *J. Amer. Chem. Soc.*, 89, 4808 (1967).

210   STEPHEN HANESSIAN AND THEODORE H. HASKELL

263b. N. Ishida, K. Kumagai, T. Niida, T. Tsuruoka, and H. Yumoto, *J. Antibiot.* (*Tokyo*), **20**, 66 (1967).
263c. S. Inouye, T. Tsuruoka, T. Ito, and T. Niida, *Tetrahedron*, **23**, 2125 (1968).
264. E. Akita, K. Maeda, and H. Umezawa, *J. Antibiot.* (*Tokyo*), **A17**, 200 (1964).
265. E. Akita, K. Maeda, and H. Umezawa, *J. Antibiot.* (*Tokyo*), **A16**, 147 (1964); *ibid.*, **A17**, 37 (1964).
266. J. S. Webb, R. W. Broschard, D. B. Cosulich, J. H. Mowat, and J. E. Lancaster, *J. Amer. Chem. Soc.*, **84**, 3183 (1962).
267. J. E. Philip, J. R. Schenck, and M. P. Hargie, *Antibiot. Ann.*, p. 699 (1956–57).
268. W. E. Grundy, J. C. Holper, E. F. Alford, C. J. Rickher, C. M. Vojtko, and J. Sylvester, *Antibiot. Ann.*, 158 (1957–58).
269. H. M. Higgins, W. H. Harrison, G. M. Wild, H. R. Bungay, and M. H. McCormick, *Antibiot. Ann.*, 906 (1957–58).
270. N. N. Lomakina, M. S. Yurina, M. F. Lavrova, and M. G. Brazhnikova, *Antibiotiki*, 609 (1961).
271. W. J. Kleinschmidt and G. W. Probst, *Antibiot. Chemotherapy*, **12**, 298 (1962).
272. See Chapter 27.
273. D. C. DeJongh, *J. Org. Chem.*, **30**, 1563 (1965) and previous papers.
273a. N. K. Kochetkov and O. S. Chizov, *Advan. Carbohyd. Chem.*, **21**, 39 (1966).
274. D. C. DeJongh and S. Hanessian, *J. Amer. Chem. Soc.*, **87**, 1408, 3744 (1965).
275. D. C. DeJongh and S. Hanessian, *J. Amer. Chem. Soc.*, **88**, 3114 (1966).
276. D. L. MacDonald and H. O. L. Fischer, *J. Amer. Chem. Soc.*, **74**, 2087 (1952).
276a. S. Hanessian, *Methods Biochem. Anal.*, **18** (1969).
277. K. Biemann and J. A. McCloskey, *J. Amer. Chem. Soc.*, **84**, 2005 (1962).
278. S. H. Eggers, S. I. Biedron and A. O. Hawtrey, *Tetrahedron Lett.*, 3271 (1966).
278a. D. C. DeJongh, J. D. Hribar, S. Hanessian, and P. W. K. Woo, *J. Amer. Chem. Soc.*, **89**, 3364 (1967).
278b. L. B. Townsend and R. K. Robins, *J. Heterocyclic Chem.*, **6**, 459 (1969).
278c. D. Gottlieb and P. D. Shaw, Eds., "Antibiotics. Mechanism of Action," Vol. I, Springer-Verlag, New York, 1967.
278d. W. Carter and K. S. McCarty, *Ann. Internal Med.*, **64**, 1087 (1966).
279. I. H. Goldberg, *Amer. J. Med.*, **39**, 722–752 (1965).
280. A. S. Weisberger, S. Wolfe, and S. Armentrout, *J. Exp. Med.*, **120**, 161 (1964).
281. A. D. Wolfe and F. E. Hahnn, *Science*, **143**, 1445 (1964).
282. A. G. So, J. W. Bodley, and E. W. Davie, *Biochemistry*, **3**, 1977 (1964).
283. N. Anand and B. D. Davis, *Nature*, **185**, 22 (1960).
284. S. Pestka, R. Marshall, and M. Nirenberg, *Proc. Nat. Acad. Sci. U.S.*, **53**, 639 (1965).
285. J. E. Davies, *Proc. Nat. Acad. Sci. U.S.*, **51**, 659 (1964).
286. E. C. Cox, J. R. White, and J. G. Flaks, *Proc. Nat. Acad. Sci. U.S.*, **51**, 703 (1964).
287. J. E. Davies, L. Gorini, and B. D. Davis, *J. Mol. Pharmacol.*, **1**, 93 (1965).
288. J. E. Davies, W. Gilbert, and L. Gorini, *Proc. Nat. Acad. Sci. U.S.*, **51**, 883 (1964).
289. H. Klenow, *Biochim. Biophys. Acta*, **76**, 347, 354 (1963).
290. H. T. Shigeura and G. E. Boxer, *Biochem. Biophys. Res. Commun.*, **17**, 758 (1964).
291. M. B. Yarmolinsky and G. L. de la Haba, *Proc. Nat. Acad. Acad. Sci. U.S.*, **45**, 1721 (1959).
292. D. Nathans, *Federation Proc.* **23**, 984 (1964).
293. W. Kersten, H. Kersten, and W. Szybalski, *Biochemistry*, **5**, 236 (1966).
294. E. Calendi, A. DiMarco, M. Reggiani, B. M. Scarpinato, and L. Valentini, *Biochim. Biophys. Acta*, **103**, 25 (1965).

295. D. H. Smith and B. D. Davis, *Biochem. Biophys. Res. Commun.*, **18**, 796 (1965).
296. F. N. Chang, C. J. Sih, and B. Weisblum, *Proc. Nat. Acad. Sci. U.S.*, **55**, 431 (1966).
297. A. C. Cope, E. P. Bunones, M. E. Derieg, S. Moon, and W. D. Wirth, *J. Amer. Chem. Soc.*, **87**, 5452 (1955); A. C. Cope, U. Axen, and E. P. Burrows, *ibid.*, **88**, 4221 (1966).
298. J. D. Dutcher, M. B. Young, J. H. Sherman, W. Hibbits, and D. R. Walters, *Antibiot. Ann.*, 866 (1956–1957); A. C. Cope, U. Axen, E. P. Burrows, and J. Weinlich, *J. Amer. Chem. Soc.*, **88**, 4228 (1966).
299. M. A. Malyshkina, B. G. Belenky, and S. N. Soloviev, *Antibiotiki*, **11**, 1002 (1963).
300. T. Watanabe, *Bull. Chem. Soc. Jap.*, **34**, 15 (1961); T. Watanabe, T. Fujii, and K. Satake, *J. Biochem. (Tokyo)*, **50**, 197 (1961); T. Watanabe, T. Fujii, H. Sakurai, J. Abe, and K. Watanabe, *Abstr. Intern. Symp. Chem. Natural Products*, C17-2, 145 (1964).
300a. S. Ōmura, M. Katagiri, H. Ogura, and T. Hata, *Chem. Pharm. Bull.*, **16**, 1167 (1968).
301. H. Brockmann and T. Waehneldt, *Naturwissenschaften*, **48**, 717 (1961).
302. O. L. Galmarini and V. Deulofeu, *Tetrahedron*, **15**, 76 (1961).
303. H. L. Herzog, E. Meseck, S. DeLorenzo, A. Murawaski, W. Charney, and J. P. Rosselet, *Appl. Microbiol.* **13**, 515 (1966).
304. G. Huber, U. Schacht, H. L. Weidenmüller, J. Schmidt-Thomé, J. Duphorn, and R. Tschesche, *Antimicrobial Agents Chemotherapy*, 737 (1965).
304a. R. Tschesche, F. X. Brook, and J. Duphorn, *Tetrahedron Lett.*, 2905 (1968).

32. COMPLEX GLYCOSIDES*

JEAN ÉMILE COURTOIS AND FRANÇOIS PERCHERON

I. Introduction	. . . .	213
II. Carbohydrate Constituents of Naturally Occurring Glycosides		214
III. Glycosides of Aliphatic Alcohols and of Alditols	.	216
IV. Cyanogenetic Glycosides		217
A. Heterosides Yielding an Aromatic Aldehyde on Hydrolysis		217
B. Heterosides Yielding an Aliphatic Ketone on Hydrolysis		218
V. Glycosides of Phenols		218
VI. Coumarin Glycosides		220
VII. Glycosides of Anthracene Derivatives . .		220
VIII. Phenanthrene Glycosides		221
A. Cardiac Glycosides.		221
B. Steroid Glycosides		222
C. Steroid Saponins		223
D. Triterpenoid Saponins		224
E. Glyco-alkaloids of *Solanum*		225
IX. Glycosides of Natural Pigments		225
A. Indole Glycosides		226
B. Anthocyanidin Glycosides		226
C. Hydroxyflavone Glycosides		227
D. Carotenoid Glycosides		229
X. 1-Thioglycosides		230
XI. Miscellaneous Glycosides		232
XII. *C*-Glycosyl Compounds		232
A. Anthraquinone Derivatives		233
B. *C*-Glycosylflavones and Related Compounds .		234
XIII. Biogenesis and Metabolism		235
A. Biogenesis		235
B. Metabolism		236
References		237

I. INTRODUCTION

A glycoside may be defined as a mixed acetal resulting from the exchange of an alkyl or aryl group for the hydrogen atom of the hemiacetal hydroxyl group of a cyclic aldose or ketose. The aglycon (or genin) is the noncarbohydrate portion attached to the carbohydrate moiety (glycosyl residue) of the glycoside.

* Translated from the French by Anthony Herp.

The general properties and synthesis of simple glycosides are considered in Vol. IA, Chap. 9, but the widely distributed natural glycosides, found principally in the plant kingdom, are discussed in this chapter.

For a long time the terms glycoside and glucoside were used interchangeably. To avoid ambiguity, it is now customary to designate as glycosides (or heterosides) those substances which, on acid hydrolysis, liberate one or several monosaccharides and an aglycon. In the naturally occurring glycosides, which constitute a most heterogeneous group of carbohydrates, a great variety of genins has been identified. The carbohydrate moiety of these glycosides is a mono- or disaccharide, or less frequently a higher oligosaccharide. The genin may contain widely diverse functional groups which markedly influence the chemical, physical, and pharmacological properties of the corresponding glycosides. Numerous glycosides are used as drugs, and only a few synthetic compounds are capable of replacing the natural glycosides employed as therapeutic agents. Most complex glycosides are isolated from plants and often have trivial names designating their origin. Some, such as the flavone derivatives, are widely distributed in higher plants. Others are localized in a small number of botanical species; such examples include glycosides of the anthraquinone or steroid types, and the 1-thioglycosides of the Crucifereae and related families. The distribution of glycosides in Nature has been outlined in numerous works of taxonomy.

In plant tissues, glycosides are principally located in the epidermis, endodermis, medullary radius, and the periphery of the marrow. The methods used for extraction and purification vary according to the structure of the genin. Since glycosides are optically active, their specific rotations are frequently utilized for physical characterization. Other physical methods of value are chromatographic mobility, and visible, ultraviolet, infrared, and n.m.r. spectroscopy.

The majority of complex glycosides are O-glycosides. They are formally derived by condensation between the hemiacetalic hydroxyl group of the carbohydrate portion and the hydroxyl group of an alcohol or a phenol. Thioglycosides are condensation products of a thiol and a sugar. Some natural glycosides are C-glycosides in which C-1 of the carbohydrate is attached directly to a carbon atom of the aglycon. N-Glycosyl derivatives of amines or nitrogen heterocycles, such as the nucleosides, are glycosylamines; these are considered in Vol. IB, Chapter 20 and this volume, Chapter 29.

II. CARBOHYDRATE CONSTITUENTS OF NATURALLY OCCURRING GLYCOSIDES

Among the sugars obtained by complete hydrolysis of the glycosides are many of the common monosaccharides; however, the degree of distribution varies greatly. D-Glucose is the most widely distributed hexose. D-Fructose has been described only in pajaneelin, a glycoside of the bark of *Pajaneelia rheedii*,[1] and in a few saponins. D-Galactose and D-mannose are also encountered rarely in natural glycosides. D-Galactose occurs in hyperin and in a few saponins; glycosides containing D-mannose are found only in algae. However, 6-deoxy-L-mannosides (L-rhamnosides) and 6-deoxy-D-(or L-) galactosides (fucosides) occur more frequently. Among pentoses, L-arabinose is more common than D-xylose.

Mention should be made of a group of deoxy sugars which are constituents of the cardiac glycosides.[2] They all are derivatives of 6-deoxyhexoses; in addition, the C-2 hydroxyl group can be replaced by a hydrogen atom (2,6-dideoxyhexoses), and the OH group at C-3 can be methylated. These sugars are listed in Table I.

D-Apiose (1), found in the glycoside apiin in the leaves of parsley, is a branched-chain sugar, and in apiin it is combined with a flavone aglycon.[3] Other branched-chain sugars are encountered in antibiotics (see Vol. IB, Chap. 17 and this volume, Chap. 31.)

$$\begin{array}{c} CHO \\ | \\ HCOH \\ | \\ HOCH_2-C-CH_2OH \\ | \\ OH \end{array}$$

Apiose (1)

In some glycosides, the carbohydrate moiety is a single sugar, usually D-glucose; in other instances, two sugar molecules are attached to different hydroxyl groups of the aglycon, as in the flavone diglycosides. Often, the glycosidic component is an oligosaccharide. Such constituents are discussed in this volume, Chap. 30. Sometimes the carbohydrate residue is a disaccharide: a glucosylglucose (for example, gentiobiose of amygdalin, sophorose of sophoraflavonoloside), an arabinosylglucose (vicianose in vicianin), a rhamnosylglucose (rutinose in rutin), or a xylosylglucose (primeverose in primeveroside). Among the more complex oligosaccharides are lycotetraose (in tomatin) and the carbohydrate chains of the cardiac heterosides. These oligosaccharides have been reviewed by Staněk.[4]

In plants, the glycosides are often accompanied by enzymes capable of hydrolyzing them. Before extraction, it is, therefore, necessary to stabilize the wet material, as by rapid heating in boiling ethanol in the presence of calcium carbonate as a neutralizing agent. This treatment also inactivates the enzymes that catalyze oxidation of certain genins. Without such precautions, the

* *References start on p. 237.*

conclusions reached concerning the carbohydrate constituents may be erroneous, since one or more carbohydrate components may have been split off by enzymic hydrolysis—for example, during drying of the raw material (as in the case of cardiac glycosides). The free oligosaccharide can be isolated by mild hydrolysis with sulfuric acid, or acetolysis (acetic anhydride–zinc chloride) can be used to yield the per-O-acetylated derivative.

TABLE I

TRIVIAL AND SYSTEMATIC NAMES OF THE DEOXY SUGARS OF THE
CARDIAC GLYCOSIDES

Trivial name	Systematic name
L-Acofriose	6-Deoxy-3-O-methyl-L-mannose
L-Acovenose	6-Deoxy-3-O-methyl-L-talose
D-Allomethylose	6-Deoxy-D-allose
L-Altromethylose	6-Deoxy-L-altrose
D-Antiarose	6-Deoxy-D-gulose
D-Boivinose	2,6-Dideoxy-D-*xylo*-hexose
D-Cymarose	2,6-Dideoxy-3-O-methyl-D-*ribo*-hexose
D-Diginose	2,6-Dideoxy-3-O-methyl-D-*lyxo*-hexose
D-Digitalose	6-Deoxy-3-O-methyl-D-galactose
D-Digitoxose	2,6-Dideoxy-D-*ribo*-hexose
D-(and L-)Fucose	6-Deoxy-D-(and L-)galactose
L-Oleandrose	2,6-Dideoxy-3-O-methyl-L-*arabino*-hexose
D-Quinovose	6-Deoxy-D-glucose
L-Rhamnose	6-Deoxy-L-mannose
D-Sarmentose	2,6-Dideoxy-3-O-methyl-D-*xylo*-hexose
L-Talomethylose	6-Deoxy-L-talose
D-(and L-)Thevetose	6-Deoxy-4-O-methyl-D-(and L-)glucose

The naturally occurring glycosides are generally β-D-linked, levorotatory, and, in the case of β-D-glucosides, hydrolyzable by the enzymic preparation of almond emulsin. However, a few heterosides possess the α-D configuration.

III. GLYCOSIDES OF ALIPHATIC ALCOHOLS AND OF ALDITOLS

The simplest of these glycosides is methyl β-D-glucopyranoside, isolated from the leaves of *Scabiosa succisa*.[5] It seems to exist only in Dipsacaceae and Rhamnaceae.[6] Ethyl α-D-glucopyranoside has been obtained from the phosphatides of lupin.[7]

Several glycosides of alditols also occur in Nature: floridoside (2-O-α-D-galactopyranosylglycerol) in red seaweeds,[8] O-α-D-mannopyranosyl-(1 → 3)-O-α-D-galactopyranosyl-(1 → 2)-glycerol from *Furcellaria*,[9] O-α-D-galactopyranosyl-(1 → 6)-O-β-D-galactopyranosyl-(1 → 1)-glycerol of *Polysiphonia*, and the glycolipids of wheat flour.[10] Umbilicin, of the lichen *Umbilicaria*, is O-β-D-galactofuranosyl-(1 → 3)-D-arabinitol;[11] the lichen *Peltigera* contains a galactoside of D-mannitol.[12]

An α-D-mannoside of a glyceric acid has been extracted from the seaweed of the genus *Polysiphonia*.[13]

Several glycosides of *myo*-inositol are known: galactinol (O-α-D-galacto-pyranosyl-(1 → 1)-*myo*-inositol) of the sugar beet,[14] and di-, tri-, and pentamannosides of *myo*-inositol from the phosphatides of *Mycobacterium tuberculosis*.[15]

IV. CYANOGENETIC GLYCOSIDES[16]

Many plant species contain glycosides that yield hydrogen cyanide upon hydrolysis, but the structures of only a few of these have been established. They all possess a nitrile group attached to the aglycon (with the sole exception of lotusin, which has a flavonoid aglycon). Their hydrolysis may be given schematically as follows:

$$\begin{array}{ccc} R_1 \diagdown \quad \diagup O\text{-glycosyl} & & R_1 \diagdown \quad \diagup OH \\ C & \xrightarrow{\text{H}_2\text{O, H}^+} & C \qquad + \text{ Mono- or oligosaccharides} \\ R_2 \diagup \quad \diagdown CN & & R_2 \diagup \quad \diagdown CN \end{array}$$

A. HETEROSIDES YIELDING AN AROMATIC ALDEHYDE ON HYDROLYSIS

Amygdalin, which was one of the first glycosides to be discovered, was isolated from bitter almond by Robiquet and Boutron-Charlard[17] in 1830. Wöhler and Liebig[18] showed, in an early observation of enzymic hydrolysis, that almond emulsin, like acids, splits amygdalin into benzaldehyde, hydrogen cyanide, and D-glucose. The carbohydrate portion, which contains two moles of D-glucose,[19] was later identified as gentiobiose.[20] Hydrolysis of amygdalin with mild acid yields the nitrile corresponding to levorotatory mandelic acid. Synthesis of amygdalin from hepta-O-acetylgentiobiosyl bromide and ethyl DL-mandelate involved a separation of the resultant diastereoisomers by fractional crystallization.[21] This synthesis confirmed the structure of amygdalin as the β-D-gentiobioside of D-($-$)mandelonitrile.

$$\text{O-}\beta\text{-D-Glucopyranosyl-(1} \rightarrow \text{6)-}\beta\text{-D-glucopyranosyl-O-}\overset{\displaystyle\text{Ph}}{\underset{\displaystyle\text{CN}}{\text{CH}}}$$

Amygdalin

Amygdalin has been isolated only from the seeds of Rosaceae species. The corresponding gentiobioside of (+)-mandelic acid has been synthesized but has not been found in Nature.

In the case of vicianin, extracted from *Vicia angustifolia*,[22] the aglycon is the same, but the carbohydrate component is the disaccharide vicianose (*O*-α-L-arabinopyranosyl-(1 → 6)-β-D-glucose).

Amygdalin and vicianin can be partially hydrolyzed by almond emulsin to give D-glucose or L-arabinose and the β-D-glucoside of mandelonitrile. The latter is subsequently hydrolyzed to D-glucose, hydrogen cyanide, and benzaldehyde. This result suggests that emulsin contains several enzymes of different specificity (see also this volume, Chap. 33). However, some enzymic preparations (such as one from *Geum urbanum*) can liberate vicianose from vicianin without cleavage of the disaccharide moiety.[23]

Three mandelonitrile glucosides have been found in plants: amygdalonitrile glucoside (prunasin),[24] which is D-(−)-mandelonitrile glucoside, from Rosaceae and Scrophulariaceae; sambunigrin (L-(+)-mandelonitrile glucoside), from *Sambucus nigra*;[25] and prulaurasin[26] (DL-mandelonitrile glucoside) from *Prunus laurocerasus*.

Related glycosides, such as dhurrin and phyllanthin, on hydrolysis by almond emulsin yield *p*-hydroxybenzaldehyde; zierin yields *m*-hydroxy-benzaldehyde.

B. HETEROSIDES YIELDING AN ALIPHATIC KETONE ON HYDROLYSIS

The main glycoside of this group is linamarin,[27] the β-D-glucoside of the cyanohydrin of acetone. Others (lotaustralin, acacipetalin) have aglycons of the same type, but are more complex.

V. GLYCOSIDES OF PHENOLS

The aglycons of naturally occurring glycosides are often substituted phenols. The leaves of various Ericaceae (bearberry or *Arctostaphylos uva-ursi*) and of Rosaceae (*Pyrus*) contain arbutin (hydroquinone β-D-gluco-pyranoside) and methylarbutin (*p*-methoxyphenyl β-D-glucopyranoside).[28] In certain varieties of *Pyrus*, arbutin is the predominant glycoside, yielding hydroquinone on enzymic hydrolysis; the oxidation of the latter causes blackening of pear leaves. In the varieties containing methylarbutin as the main glycoside, the leaves turn yellow when they fall. Arbutin has been synthesized from tetra-*O*-acetyl-α-D-glucopyranosyl bromide and hydro-quinone having one hydroxyl group protected as the benzoate.[29]

Another well-known glycoside is salicin (**2**), which is found in the bark of the willow (*Salix purpurea*). It is 2-(hydroxymethyl)phenyl β-D-gluco-pyranoside.[30] This glycoside was used by Rabaté when he made his observation of the existence of glycosyl transferase enzymes. An extract from the

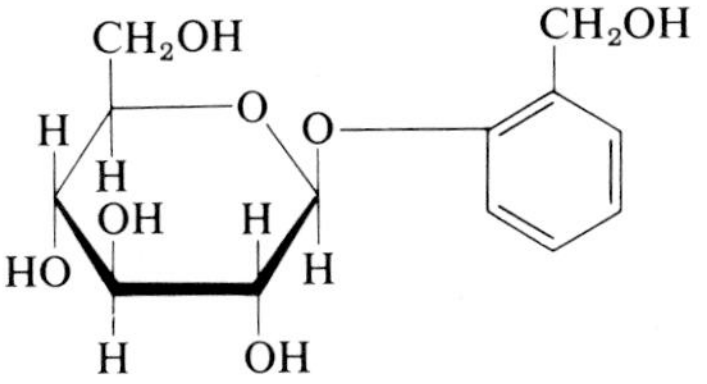

Salicin (**2**) 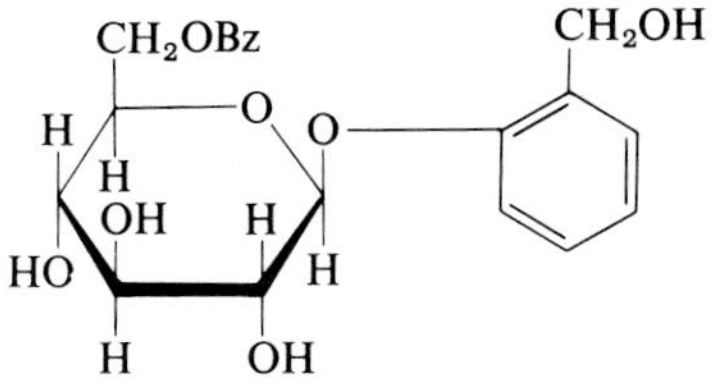   Populin (**3**)

powdered leaves of *Salix* catalyzed the transfer of a D-glucosyl residue from the phenolic hydroxyl group of salicin to the alcoholic hydroxyl group of the aglycon.[31] In poplar bark, salicin is accompanied by populin (**3**), which is the 6-*O*-benzoyl derivative of salicin. The substitution at C-6 prevents the hydrolysis of populin by emulsin,[32] whereas salicin is easily hydrolyzed. Weidenhagen proposed salicin as the standard substrate for evaluation of β-glucosidase activity.[33] Gaultherin (methyl salicylate 2-primeveroside) is identical to monotropitin of *Spiraea*.[34]

It is generally accepted that coniferin (**4**) from the cambium of *Pinus* trees is involved in the lignification process; its aglycon easily polymerizes by enzymic dehydrogenation to lignin-like materials.[35]

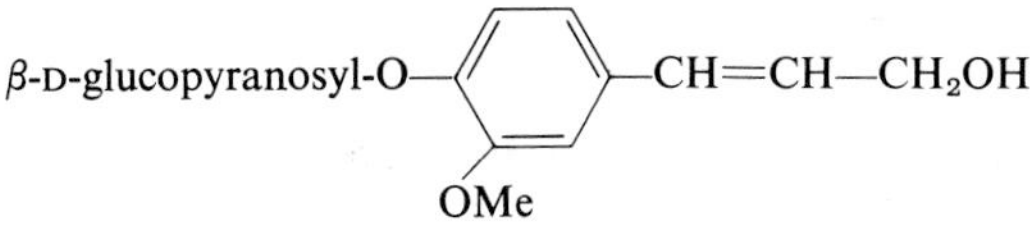

β-D-glucopyranosyl-O—⟨benzene ring⟩—CH=CH—CH$_2$OH

ÓMe

Coniferin (**4**)

Syringin is a methoxyconiferin. On oxidation of the ethylenic linkage by chromic acid, coniferin yields vanillin D-glucoside, which itself occurs naturally in the fruit of *Vanilla planifolia*.

Gein, of *Geum urbanum*, is the vicianoside of eugenol.[36] Phloretin (2',4',6'-trihydroxy-3-[4''-hydroxyphenyl]propiophenone) is a more complex phenol aglycon; its 2'-β-D-glucosyl derivative is phloridzin, found in the bark of the apple tree;[37] it has the property of inhibiting the tubular reabsorption of D-glucose in the kidneys, causing "phloridzin diabetes." Naringin, which is of similar structure, contains rhamnosylglucose instead of D-glucose. The 4-β-D-glucosyl derivative of protocatechuic acid has been isolated from the collateral glands of certain insects (*Periplaneta, Blatta*).[38]

* *References start on p. 237.*

VI. COUMARIN GLYCOSIDES

No glycosides of coumarin itself are known, since this lactone has no free hydroxyl group. It is believed that coumarin is present in plants as the glycoside of o-coumaric acid (*trans*) and o-coumarinic acid (*cis*). On treatment with acids, the glycosides are hydrolyzed with isomerization of the *trans* acid to the *cis* form followed by lactonization; hydrolysis with almond emulsin maintains the initial form intact. These rearrangements have been shown especially for furocoumarin (hydroxy and methoxy) derivatives.[39]

The β-D-glucoside of o-coumaric acid has been isolated from *Melilotus altissima*.[40] Several glycosides of hydroxyl derivatives of coumarin are known. These include aesculin (**5**) of *Aesculus hippocastanum*, which is the 6-β-D-glucosyl derivative of 6,7-dihydroxycoumarin.[41] Daphnin, isolated from *Daphne*, is the 7-β-D-glucosyl derivative of 7,8-dihydroxycoumarin.[42] Fraxin (**6**), from the bark of the ash-tree (*Fraxinus excelsior*), is the 8-β-D-glucosyl-fraxetin, a dihydroxy and methoxy derivative of coumarin.[43]

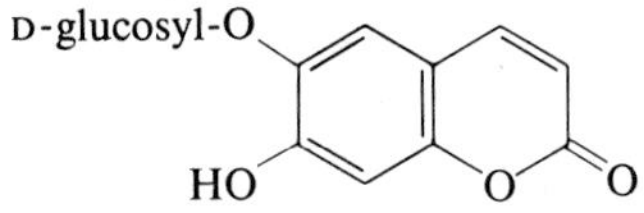

D-glucosyl-O

HO

Aesculin (**5**)

MeO

HO

D-glucosyl-O

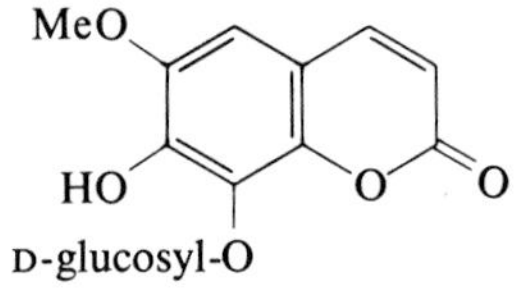

Fraxin (**6**)

VII. GLYCOSIDES OF ANTHRACENE DERIVATIVES[44]

These glycosides can be divided into those possessing a so-called oxidized aglycon (hydroxyl derivatives of anthraquinone, **7**) and those having a reduced aglycon (hydroxy derivatives of anthrone, **8** or its tautomeric form, anthranol, **9**).

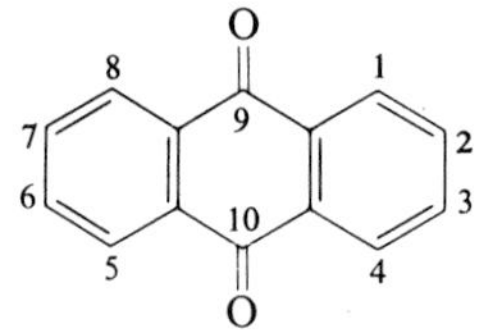

O

8 9 1
7   2
6   10   3
5   4
O

7

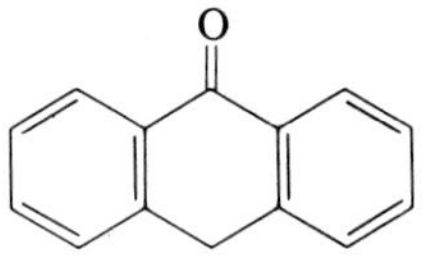

O

8

OH

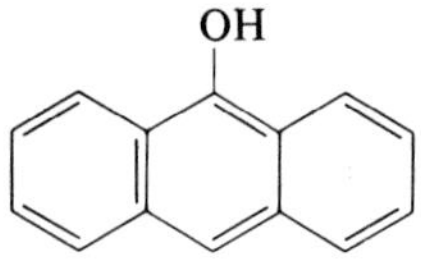

9

Ruberythric acid, a constituent of the roots of *Rubia tinctorum*, is the 2-primeveroside of alizarin (1,2-dihydroxyanthraquinone); it was utilized for a long time to produce alizarin for a textile dyestuff; aglycons of closely related glycosides accompany the natural alizarin, and their presence explains the difference in color of the natural product from that of synthetic alizarin.

Other anthracene glycosides containing yellow aglycons have been widely used as purgatives. The roots of *Rheum officinale* contain glycosides of chrysophanol (1,8-dihydroxy-3-methylanthraquinone), emodol (1,6,8-tri-hydroxy-3-methylanthraquinone), and related derivatives such as rhein and rheochrysidin.

Frangulin (**10**) (a rhamnoside of emodol) is a glycoside found in the bark of *Rhamnus frangula*; glucofrangulin is a more complex glycoside.[45]

Glycosides having reduced aglycons include frangularoside, a rhamnoside of frangularol.[46] Sennoside (**11**), from the leaves of the genus *Cassia*, is a bis(D-glucoside) of dianthrone and is used as a purgative.[47]

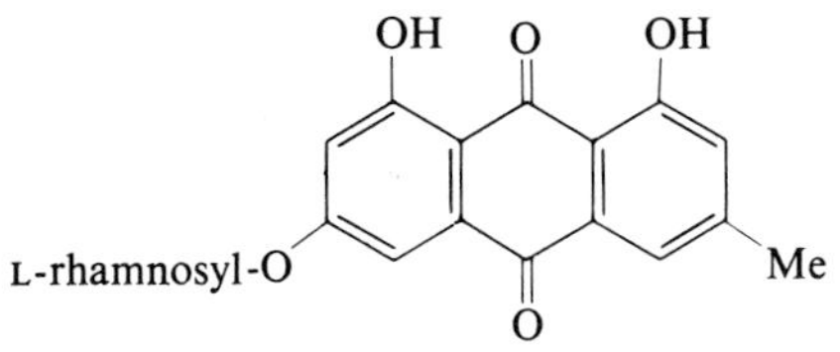

Frangulin (**10**)

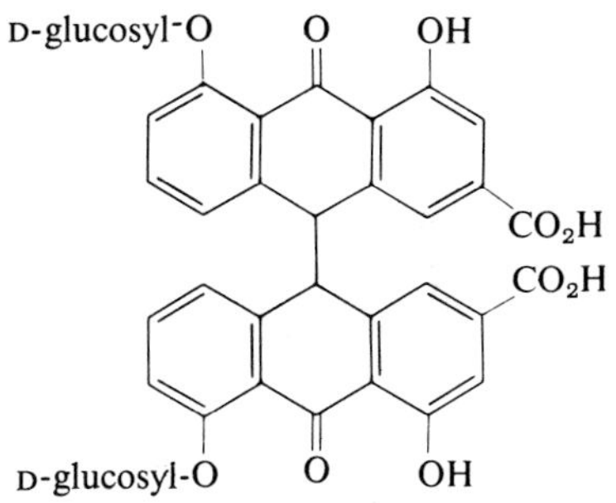

Sennoside (**11**)

VIII. PHENANTHRENE GLYCOSIDES

A. CARDIAC GLYCOSIDES[48]

Several glycosides having a steroid-type aglycon are used in therapy as cardiotonics. The most important ones belong to the group of "cardenolides" which contain C_{23} aglycons having the following characteristics: an unsaturated, lateral lactone chain having four carbon atoms (butenolide); rings C and D *cis*, with a β-oriented hydroxyl group at C-14. They are mainly of the 5β-series and have the C-3 hydroxyl group in the β configuration; hydroxyl groups may also be found at C-5, C-11, C-12, C-16, and C-19.

These glycosides, found mainly in the leaves of Scrophulariaceae, Apocynaceae, and Liliaceae, generally contain deoxy sugars (see Table I) linked directly to the aglycon and to D-glucose according to the scheme

$$\text{Aglycon-(deoxysugar)}_m\text{-(D-glucose)}_n$$

On enzymic hydrolysis, during drying, the plant yields D-glucose, whereas acid hydrolysis liberates all sugar components.

* *References start on p. 237.*

One of the principal cardiac glycosides is digitoxin, which occurs in *Digitalis purpurea* as purpurea glycoside *A*, and in *Digitalis lanata* as acetyl purpurea glycoside *A* or lanatoside *A*.[49] Digitoxin contains a chain of three molecules of digitoxose linked to the hydroxyl group at C-3 of the aglycon.

The seeds of *Strophanthus kombé* contain *k*-strophanthoside[50] whose aglycon has hydroxyl groups at C-3, C-5, and C-14, and an aldehyde group at C-19, a structure rare in Nature. The glycosidic portion is strophantho-triose (*O*-α-D-glucosyl-*O*-β-D-glucosylcymarose).

Another group of aglycons is that of "bufadienolides" (C_{24}); here, the lateral chain is lactonized, with five carbon atoms containing two double bonds; these glycosides are found in scilla (star flower) (*Urginea scilla*) and, in nonglycosidic form, in toad poison. As an example, the structure of scillaren *A* is shown[51](**12**).

The cardiotonic activity of these glycosides (of which the most used remains digitoxin) has been extensively studied. The glycosides reduce the pulse rate, regularize the rhythm, and strengthen the heart beat. A secondary diuretic action is often observed, which can become predominant in the glycosides of *Scilla*. Some glycosides have been used as arrow poisons.

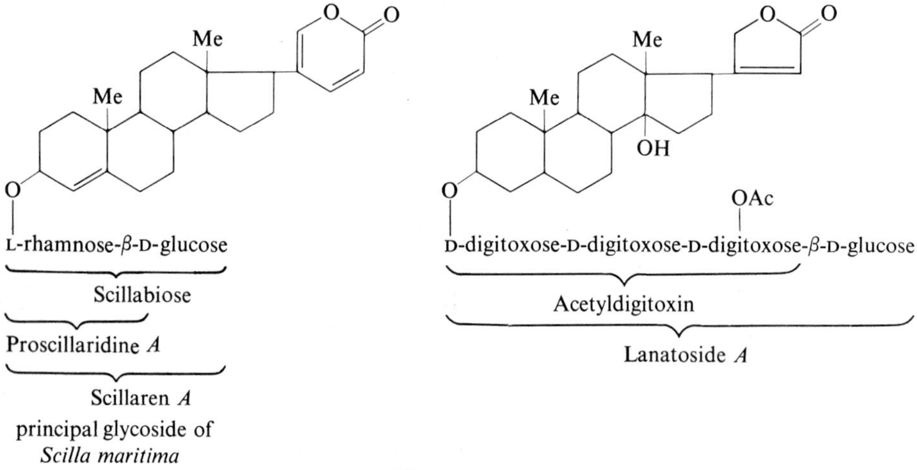

L-rhamnose-β-D-glucose

Scillabiose

Proscillaridine *A*

Scillaren *A*
principal glycoside of
Scilla maritima

D-digitoxose-D-digitoxose-D-digitoxose-β-D-glucose

Acetyldigitoxin

Lanatoside *A*

12

B. STEROID GLYCOSIDES

The phytosterols are derivatives of ergostane (C_{28}) and stigmastane (C_{29}). They all possess a β-oriented hydroxyl group at C-3 and double bonds at one or all of the positions: C-5, C-7, or C-22. One of the most widespread of these sterols is β-sitosterol. These sterols have generally been isolated in the free state from plants, but only a few of their glycosides have been described;

they are known as phytosterolins. This group includes daucosterin, a glyco-side of β-sitosterol,[52] which may be identical to ipuranol,[53] and a glycoside of γ-spinasterol which has been isolated from *Spinacia oleracea*.[54]

C. Steroid Saponins[55]

The best known of these glycosides are mainly distributed in the plant families Scrophulariaceae, Liliaceae, Dioscoreaceae, and Amaryllidaceae. They have some common characteristics, such as strong foaming power in solution. They have hemolytic action when given intravenously but not if taken orally. They are toxic in cold-blooded animals, especially fish, and have been used as fish poisons. The majority of these saponins form insoluble complexes with 3-β-hydroxysteroids.

The first ones discovered were digitonin, isolated by Schmiedeberg in 1875 from *Digitalis purpurea*, gitonin, and tigonin. On acid or enzymic hydrolysis, they yield sugars, together with C_{27} aglycons called sapogenins, which are structurally "spirostanes." They are of three types: 5β (the most frequent), 5α, and $\Delta 5$. They all possess a hydroxyl group in the β configuration at C-3, and two oxygenated heterocycles, E and F. The terms "normal" and "iso," which were previously used to describe the two series of compounds thought to be isomers at C-22, should be avoided, since the isomerism is really[56] at C-25.

The sequence of sugars in the carbohydrate residues has not been com-pletely elucidated in all cases. In sarsasaponin (**13**), the OH group at C-3 of the aglycon is combined to a glycosidic chain containing two molecules

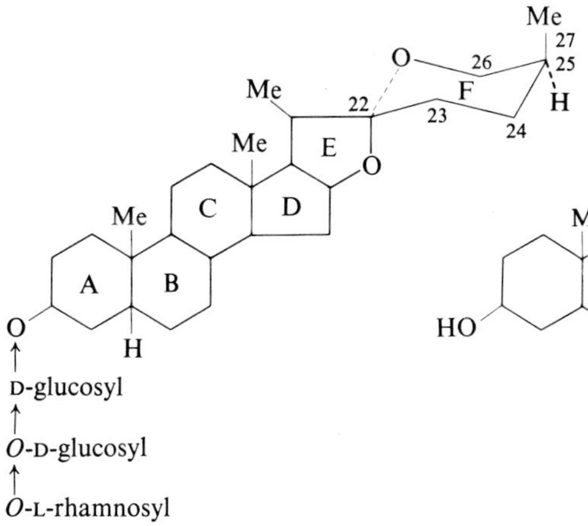

D-glucosyl
↑
O-D-glucosyl
↑
O-L-rhamnosyl

Sarsasaponin (**13**)

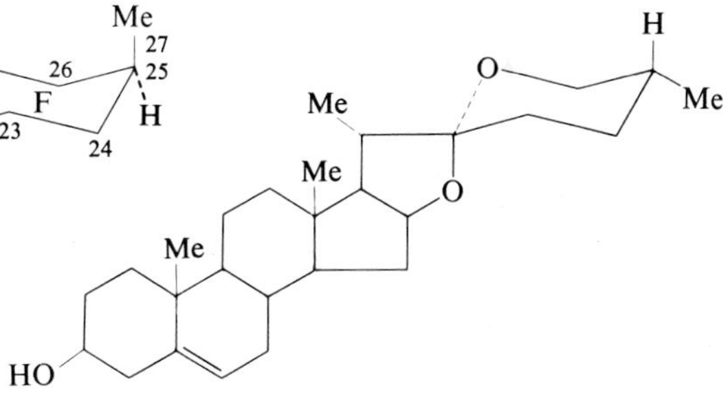

Diosgenin (**14**)

* References start on p. 237.

of D-glucose and one molecule of L-rhamnose. The carbohydrate portion of digitonin is more complex; it contains two molecules of D-glucose, two molecules of D-galactose, and one molecule of D-xylose.

Diosgenin (14), found in numerous Dioscoreaceae, is of the $\Delta 5$ type and is used in the partial synthesis of C_{21} steroid hormones.

Holothurin is an animal saponin extracted from *Actinopyga agassizi*.[57] It is very toxic and irreversibly inhibits cholinesterase; it also exhibits strong hemolytic action. Holothurin has a steroid skeleton carrying a cyclic lactone, an oxygenated heterocycle, and seven methyl groups. The carbohydrate fraction, which is linked to the OH group at C-3, is composed of D-xylose, D-glucose, 3-*O*-methyl-D-glucose, and D-quinovose.

In mammals, most steroidal hormones and their metabolic products are eliminated in the urine, in the form of β-D-glucosiduronic acids.

D. Triterpenoid Saponins[58]

The triterpenes, natural C_{30} compounds sometimes found in animals, are more widespread in many botanical families. These compounds, which exist in great number, are almost all hydroxylated. However, only a few of their glycosides have been isolated in pure state. Many of the triterpenes probably exist in nature in the form of esters or glycosides.

The glycosides of known triterpene saponins belong mainly to the group of pentacyclic triterpenes. Their general properties are comparable to those of the saponins.

The β-amyrin group includes several glycosides. The soya saponins yield various sapogenols on hydrolysis.[59] The saponin of the sugar-beet is a β-D-glucosiduronate of oleanolic acid,[60] and that of alfalfa blossom has the probable formula[61] (15).

Asiaticoside is not a true glycoside. Its carbohydrate chain, composed of *O*-L-rhamnosyl-(1 → 4)-*O*-D-glucosyl-(1 → 6)-D-glucose, is linked to the carboxyl group of asiatic acid.[62]

β-D-glucopyranosyl-O
↑
O-β-D-glucopyranuronosyl
↑
O-β-L-rhamnopyranosyl

15

The cucurbitacins, which are purgatives of the Cucurbitaceae family, are glycosides and have genins that are tetracyclic triterpenes.

E. Glyco-alkaloids of *Solanum*[63]

The alkaloids of the genera *Solanum* and *Lycopersicum* may be divided into two classes, the solanins and the solasonins. The solasonins, for example tomatin (**16**), may be considered as the nitrogenous analogs of the steroid sapogenins, and have an NH group instead of an oxygen atom in the F nucleus. The solanin group (**17**) have aglycons that lack the oxygenated heterocycle but have a tertiary nitrogen atom.

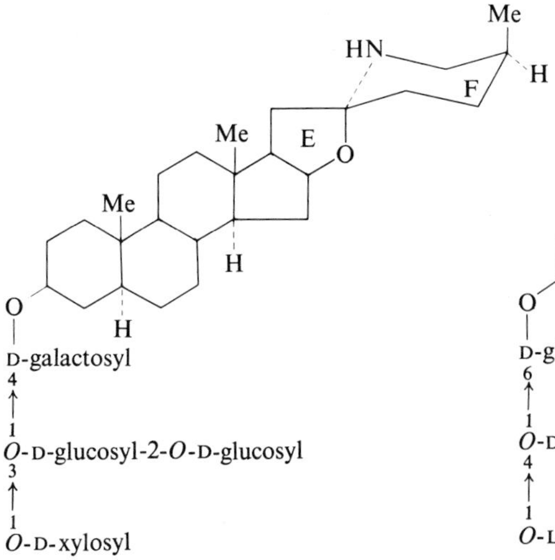

D-galactosyl
4
↑
1
O-D-glucosyl-2-*O*-D-glucosyl
3
↑
1
O-D-xylosyl

Tomatin (**16**)

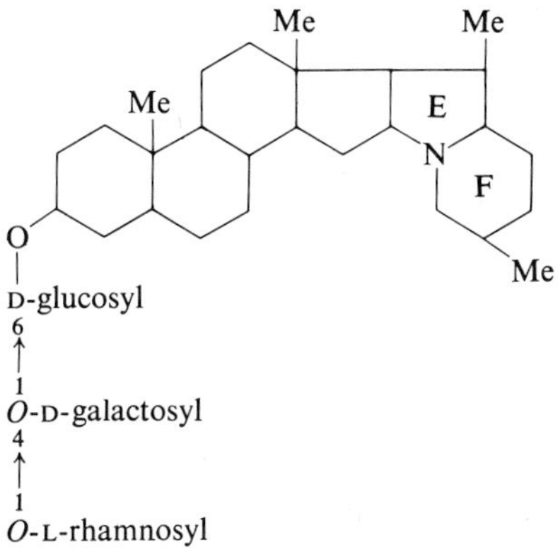

D-glucosyl
6
↑
1
O-D-galactosyl
4
↑
1
O-L-rhamnosyl

Solanin (**17**)

The OH group in the 3-β position is glycosidically linked to an oligosaccharide. In the case of solanin,[64] this is a trisaccharide, and in the case of tomatin and demissin, it is a branched tetrasaccharide called lycotetraose.[65]

The larvae of Doryphore (potato beetle) do not attack *Solanum* species, which contain demissin,[66] but some hybrids from varieties containing solanin are poorly resistant to the insect because of their low content of demissin.

IX. GLYCOSIDES OF NATURAL PIGMENTS

The natural pigments of plants frequently consist of glycosides and are usually classified according to the structure of their aglycons.

* *References start on p. 237.*

A. INDOLE GLYCOSIDES

Indican (3-hydroxyindole β-D-glucoside) occurs mainly in the genera *Indigofera*, *Isatis*, and *Polygonum*. On hydrolysis it yields D-glucose and indoxyl (**18**), which on aerial oxidation becomes the insoluble dye indigo (**19**).

Indoxyl (**18**)   Indigo (**19**)

The structure of indican was established by Robertson.[67] The urine of mammals contains indoxyl sulfate, which is derived from the degradation of tryptophan by intestinal bacteria.

B. ANTHOCYANIDIN GLYCOSIDES[68]

These pigments, localized in the sap of cells, are responsible for the red, violet, and blue coloring of many flowers, fruits, and leaves. They change their color according to the pH of the solution.

Willstätter, who in 1913 isolated cyanin from the cornflower, introduced the term anthocyanins to designate the glycosides, and the term anthocyanidins for the corresponding aglycons.[69]

The anthocyanidins are hydroxy derivatives of flavylium salts (2-phenyl-benzopyrylium) (**20**). On alkaline fusion, they yield phloroglucinol and

Flavylium chloride (**20**)

hydroxy derivatives of benzoic acid. Generally, they are derived from pelargonidin, cyanidin, or delphinidin chlorides. These three types possess the 3,5,7-trihydroxyflavylium system (**21**) in common, and they differ by the number of hydroxyl groups present on the phenyl nucleus in ring 2. Sometimes, one or several of these hydroxyl groups are methylated. These structures may be summarized as follows:

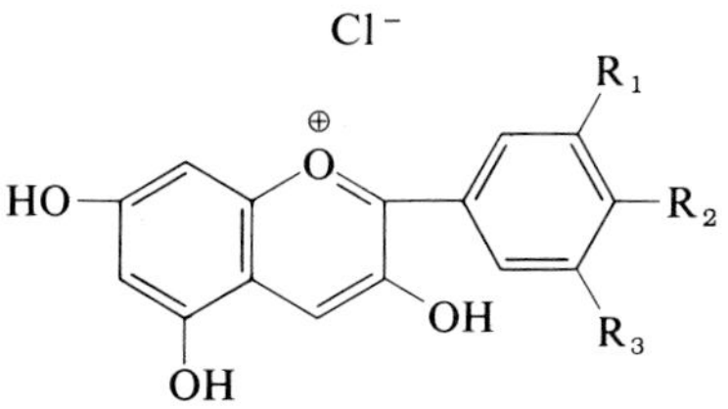

21

	R_1	R_2	R_3
Pelargonidin chloride	H	OH	H
Cyanidin chloride	OH	OH	H
Delphinidin chloride	OH	OH	OH
Peonidin chloride	OMe	OH	H
Malvidin chloride	OMe	OH	OMe

In each group, the anthocyanins also differ by the number and position of the carbohydrate and methoxyl groups. The best-known glycosides are the 3,5-di-D-glucosyl derivatives: pelargonin, cyanin, delphinin, peonin, malvin, and so on. 3-Mono-D-glucosyl, 3-mono-D-galactosyl, and 3-[O-L-rhamnosyl-D-glucosyl] derivatives are also known. The classical procedure for synthesis of anthocyanins is that outlined by Robinson for peonin, starting from phloroglucinaldehyde and 3-methoxy-4-acetoxyacetophenone.[70]

Before the introduction of synthetic dyes, some of these glycosides were widely used as dyestuffs.

Many plant tissues contain colorless materials which, on treatment with mineral acids, are converted into red compounds related to the anthocyanins. Although not comparable to the leuco derivatives of dyes, these colorless substances were given the names of leuco-anthocyanidins by Rosenheim. Some seem to be 3,4-dihydroxyflavans,[71] which relation confirms the similarity of the anthocyanins to the flavone glycosides.

C. Hydroxyflavone Glycosides[72]

These widely distributed glycosides are responsible for the yellow or brown color of many flowers and fruits. Their aglycons are hydroxy derivatives of 2-phenylbenzopyrone (flavone) (**22**) or, more rarely, of 3-phenylbenzopyrone (isoflavone) (**23**).

They may be classified according to the position of the hydroxyl groups and the state of saturation of the oxygenated heterocycle. The aglycons of the true flavone glycosides have hydroxyl groups at C-5, C-7, C-3', or C-4', but not at C-3, for example, apiin and glucoluteolin.

References start on p. 237.

The flavonol glycosides have an aglycon possessing a hydroxyl group at C-3, and one or more hydroxyl groups at either C-5, C-7, C-3′, or C-4′. The most important aglycons of this group are kaempferol (3,5,7,4′-tetrahydroxy-flavone) and quercetin (3,5,7,3′,4′-pentahydroxyflavone), which occurs in the bark of *Quercus tinctoria*. Some quercetin-like glycosides have a carbo-hydrate component represented either by a monosaccharide residue: 3-*O*-β-L-rhamnopyranosyl (quercitrin), 3-*O*-D-galactosyl (hyperin), or by a disaccharide residue: 3-*O*-rutinosyl (rutin) from *Ruta graveolens*, and 3-*O*-sophorosyl of the pollen of *Alnus*.[73] In addition, *O*-methylated derivatives of

Flavone (**22**)   Isoflavone (**23**)

(For numbering system, see flavylium formula **20**)

quercetin and their glycosides are also found in Nature—for example, xanthorhamnin, which on hydrolysis yields two molecules of L-rhamnose, one molecule of D-galactose, and rhamnetin (the 7-methyl ether of quercetin).

In some instances the glycosyl residues are esterified by organic acids. Tiliroside[74] is 3-*O*-D-glucosylkaempferol esterified in the D-glucose moiety by *p*-coumaric acid.

The aglycons (hydroxyflavanones) of the flavanone glycosides are saturated at positions 2 and 3, and an asymmetric center (C-2) is thereby introduced into the molecule. Hesperidin (**24**) is a glycoside of 5,7,3′-trihydroxyflavanone. It is found in the peels of citrus fruits in the form of the 7-rutinoside. Eriodictyol L-rhamnoside has the methoxyl group replaced by a hydroxyl group. When treated with concentrated acid or alkali, the flavanones are converted into chalcones (**25**), through opening of the pyran ring. The chalcones take up hydrogen readily; the reduced forms lose hydrogen easily when exposed to air. The reduced chalcone can also transfer hydrogen to oxidation–reduction coenzymes.

Phloridzin may be considered as a dihydrochalcone.

Chalcones are localized in a limited number of botanical families, including Leguminosae and Compositae.

Several flavone glycosides have been described as the so-called vitamin P factors, which are reputed to have the ability to reduce capillary permeability and to increase resistance to hemorrhages. This activity is exhibited by rutin, and to a lesser extent by eriodictyol glycoside and hesperidin. These glycosides, and some of their water-soluble derivatives, are used as therapeutic agents, usually in combination with ascorbic acid.

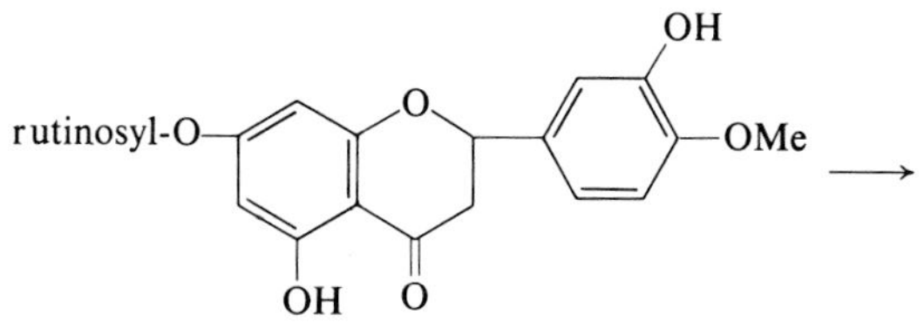

Hesperidin (24)

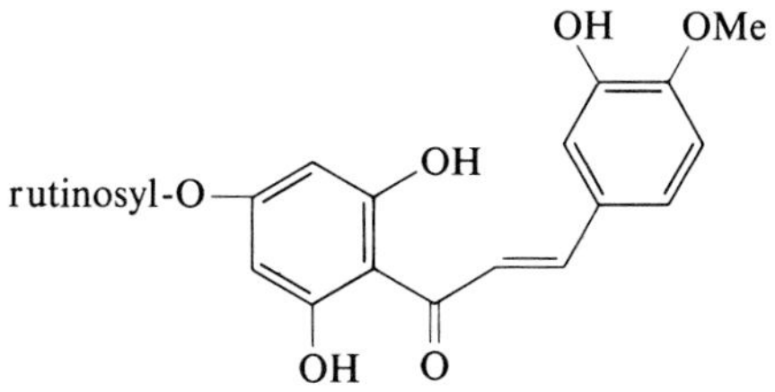

Chalcone (25)

D. CAROTENOID GLYCOSIDES

The stigma of the flower *Crocus sativus* (saffron) contain a pigment called crocin, which is the bis(gentiobiosyl) ester of the carotenoid pigment crocetin. In plants, this ester occurs in combination with the glycoside picrocrocin, a β-D-glucoside of hydroxysafranal. Crocin arises in the plant only upon exposure to light. According to Kuhn,[75] crocin (26) and picrocrocin (27) are derived from the same precursor, procrocin (28), a compound structurally related to β-carotene.

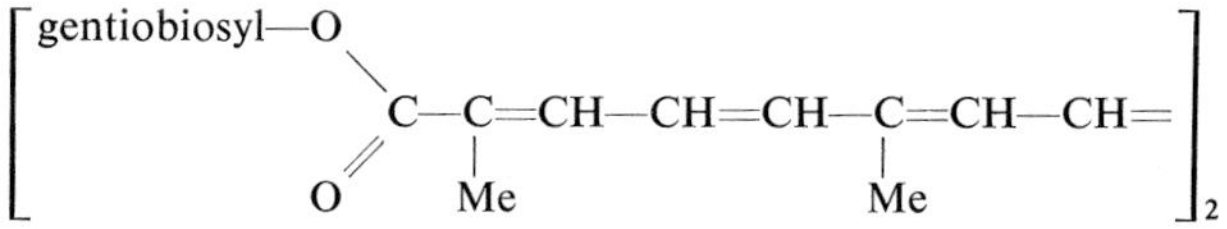

Crocin (26)

A similarity in structure seems to exist between these derivatives and the substances that are responsible for the differentiation of gametes in the alga *Chamydomonas eugametos*.

* References start on p. 237.

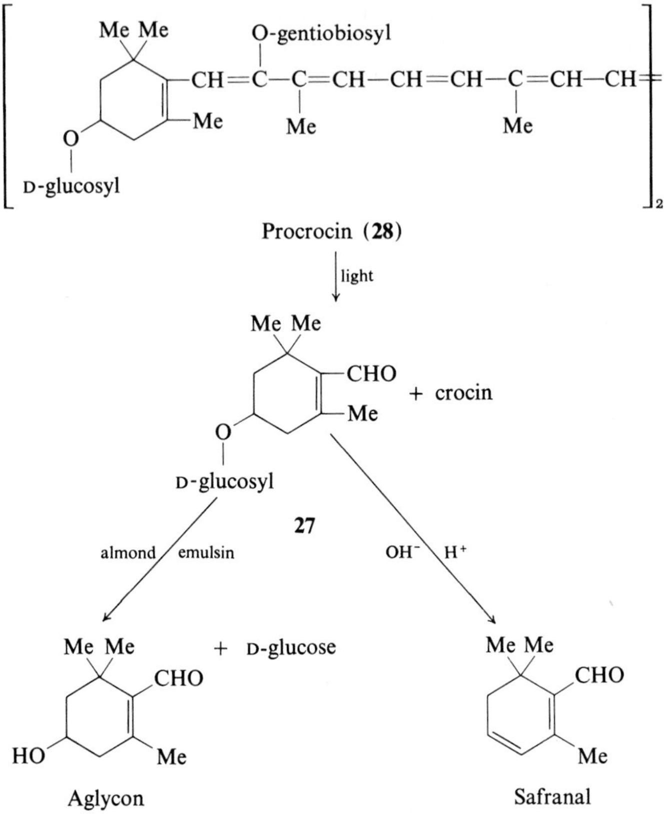

Procrocin (28)

↓ light

27

+ crocin

Aglycon   + D-glucose   Safranal

X. 1-THIOGLYCOSIDES[76]

Several plants of the Cruciferae contain thioglycosides, which upon macera-
tion with water give a vesicatory solution. The thioglycosides are easily
hydrolyzed by a mixture of enzymes called myrosinase, present in the tissues
of these plants.

The thioglycosides have the following general structure:[76a]

$$R-\underset{\underset{N-OSO_3^-}{\|}}{C}-SC_6H_{11}O_5$$

The seeds of black mustard (*Sinapis nigra*) contain sinigrin ($R = CH_2{=}CH{-}
CH_2{-}$), which is hydrolyzed to allyl isothiocyanate, D-glucose, and potassium
hydrogen sulfate.[76] The seeds of the wallflower contain glucocheirolin

(R = $MeSO_2CH_2CH_2-$), and the glycoside of cress (*Nasturtium*) yields 2-phenylethyl isothiocyanate (R = $PhCH_2CH_2-$). Sinalbin, from *Sinapis alba*, is a more complex thioglycoside (R = $HOC_6H_4CH_2-$), and on hydrolysis it yields sinapine sulfate (a derivative of choline) in addition to *p*-hydroxybenzyl isothiocyanate and D-glucose.

Although the D-glucose residues have the β-D configuration, these glycosides are not hydrolyzed by almond emulsin. Alkaline hydrolysis by sodium methoxide leads to 1-thio-D-glucose (see Vol. IB, Chap. 18).

Synthetic thioglycosides have been prepared by the action of thiophenols or mercaptans on *O*-acetylglycosyl bromides in the presence of sodium hydroxide.[77] Several natural 1-thio-D-glucosides have been synthesized.[77a]

The sulfur essence of the seeds of black mustard has been used therapeutically as a rubefacient.

Kjær and co-workers[77b] have elucidated the structure of several glycosides from mustard-oil. The crucifer *Sisymbrium austriacum* Jacq, for example, was found to contain a mixture of glucosides: glucosisymbrin, glucosisaustricin, and their *O*-benzoylated derivatives. The former two glucosides undergo enzymic hydrolysis by myrosin to D-glucose, sulfate, and (+)-4-methyl-2-oxazolidinethione (sisymbrin) and (+)-4-ethyl-2-oxazolidinethione (sisaustricin) (29), respectively. Glucocleomin is a thiocyanate derivative encountered in the seeds of *Cleome spinosa* Jacq which, upon enzymic hydrolysis yields D-glucose, sulfate, and (−)-5-ethyl-5-methyl-2-oxazolidinethione (30); leaves of *Capparis linearis* Jacq contain a glucoside having the structure 3-methyl-3-butenylglucosinolate (31). A thioglucoside of the radish (*Raphanus sativus* L var. esculentus Metzg.) liberates upon enzymic hydrolysis an isothiocyanate which was identified as *trans*-4-methylthiobutenyl isothiocyanate.

Sisaustricin (29) 30 31

XI. MISCELLANEOUS GLYCOSIDES

Several glycosides possess aglycons which are derivatives of furyl-1,2-cyclopentanediol, and are sometimes considered as monoterpene analogs. Aucubin,[78] catalpol and methylcatalpol,[79] gentiopicrin,[80] and verbenalin[81] belong to this group. Bakankoside,[82] which has a nitrogeneous aglycon, and swertiamarin[83] may also be included. It is noteworthy that swertiamarin and gentiopicrin are easily transformed by ammonium hydroxide followed by acid hydrolysis into the alkaloid gentianin in which the oxygen atom of the pyran nucleus is replaced by a nitrogen atom.[84] This conversion confirms that gentianin is not present in the plant, but is an artifact formed during the extraction procedure with ammonium hydroxide.[84]

N-Glycosyl derivatives of heterocycles include the nucleosides; these derivatives are discussed in this volume, Chap. 29. In addition, this group includes the cobalamins (vitamin B_{12}), which are D-ribofuranosyl derivatives of dimethylbenzimidazole, and vicine, an N-β-D-glucosyl derivative of a diaminopyrimidinedione.[85]

Cycadaceae plants contain azoxy glycosides that are of interest in view of the carcinogenic nature of the aglycon. These glycosides have the following general formula

$$ROCH_2N{=}NMe$$
$$\downarrow$$
$$O$$

where R is a glycosyl residue. Neocycasin A, isolated from the seeds of *Cycas revoluta*, was found to be 3-O-β-D-glucopyranosylcycasin.[86a] Cycad emulsin from seed kernels of the plant catalyzes the decomposition of cycasin, liberating nitrogen, D-glucose, formaldehyde, and methanol.[86b]

XII. *C*-GLYCOSYL COMPOUNDS

The number of known *C*-glycosides has steadily increased during recent years, in part because several compounds formerly listed as *O*-glycosides are in reality *C*-glycosides. In some instances, the existence of the C—C bond was confirmed by synthesis of compounds identical to the natural products.[86] The linkage between the genin and C-1 of the carbohydrate is very resistant to dilute acids. Concentrated acids may, however, cleave the bond between C-1 and C-2 of the sugar; the *C*-glycosides of *Aloe*, for example, yield D-arabinose. For a long time, this observation led to the belief that these glycosides are among the few naturally occurring substances containing a D-arabinosyl residue. The isolation of *C*-glycosyl compounds is somewhat simplified by their resistance to dilute acids and to the majority of glycosidases.

They remain intact in hydrolyzed extracts in which normal glycosidic linkages have been split. However, mild oxidants such as ozone and ferric chloride easily disrupt the C—C bond and dissociate the aglycon from the sugar components. Oxidation of the carbohydrate by periodic acid is a useful method for structural studies when the genin is unreactive or can be protected by mild methylation.

Infrared and ultraviolet spectroscopy, and especially nuclear magnetic resonance (n.m.r.) spectroscopy, have considerably facilitated study of the structure of these glycosides.

In all known compounds, D-glucose is the carbohydrate linked to the genin. Sometimes, the sugar moiety is a disaccharide, a O-D-glucosyl- or O-L-rhamnosyl-D-glucosyl group. The only known exception is pseudouridine (5-β-D-ribofuranosyluracil), described by W. E. Cohn.[87] It occurs in some ribonucleic acids and exists in the free form in mammalian urine. (See this volume, Chap. 29.)

The genins of the C-glycosyl compounds are closely related to those of the O-glycosides and for a long time have been classified together. Paradoxically, these genins are generally polyhydroxy derivatives, and the formation of the carbon—carbon bond is not due to the absence of groupings able to react with the hemiacetal hydroxyl group of the carbohydrate.

The principal classes of C-glycosyl compounds are the anthraquinone derivatives and the C-glycosylflavones.

A. ANTHRAQUINONE DERIVATIVES

Several closely related compounds have been extracted from the leaves of different *Aloe* species—for example, barbaloin (**32**) and homonataloin.

Barbaloin (**32**)

* References start on p. 237.

Aloinoside B is both a *C*-glycosyl compound and an *O*-glycoside. The primary alcohol function of barbaloin carries an α-L-rhamnosyl group.

The cascarosides of *Rhamnus purshiana* are *O*- and *C*-glycosyl compounds of the anthraquinone type.

The *C*-glycosides from *Aloe* have the same purgative properties as the anthraquinone α-D-glycosides of *Rheum* and *Cassia*.

The buff-colored substance mangiferin, encountered in the mango tree and in *Iris* flowers, is a *C*-glycosyl compound having the sugar linked to C-2 of a 1,3,6,7-tetrahydroxyanthraquinone.

Carminic acid is a red compound of animal origin which was isolated from cochineal, a hemipter insect. The carmine of cochineal was formerly widely used as natural dyestuff. The structure of carminic acid is not yet definitely established. It is a *C*-glycoside of a trihydroxymethylanthraquinone having a carboxyl group.

B. *C*-GLYCOSYLFLAVONES AND RELATED COMPOUNDS

This group includes a great number of materials having the same genins as the flavonol glycosides. Vitexin (**33**), an 8-*C*-D-glucopyranosylapigenin, is a trihydroxyflavone derivative.

Vitexin (**33**)

Orientin and isoorientin, isolated from various leaves, are the 8- and 6-*C*-β-D-glucopyranosyl derivatives, respectively, of luteolin.[88] Scoparin, of *Sarothamnus scoparia* is the 3′-methyl ether of orientin.[89]

Some compounds have been isolated in which two different molecules of D-glucose are attached to two distinct carbon atoms of the flavone—for example, lucenin, a 6,8-di-*C*-glycosyl derivative of luteolin, and violanthin (6,8-di-*C*-glucosylapigenin).

XIII. BIOGENESIS AND METABOLISM

A. BIOGENESIS

It now seems established that the genin and the sugars are synthesized independently and subsequently combine to form the glycoside. (See this volume, Chap. 34 for more details.) The major, if not exclusive, mode of combination takes place through activitation of the monosaccharide units as the nucleoside sugar pyrophosphates. The work of Leloir,[90] Hassid,[91] and Pridham[92] and their collaborators has shown that the biosynthesis of complex glycosides involves enzymic transfer of the sugar from the nucleoside sugar pyrophosphate according to the scheme

$$\text{XDP-sugar} + \text{genin} \longrightarrow \text{Glycoside} + \text{XDP}$$

where X represents one of the common purine or pyrimidine nucleosides. The mechanism whereby the glycosyl group is transferred seems to be similar to that which takes place in the synthesis of oligo- and polysaccharides.

Uridine 5′-(α-D-glucopyranosyl pyrophosphate) is the principal but not the exclusive donor. Numerous enzyme preparations are able to catalyze the synthesis of various glycosides from sugar moieties linked to nucleoside pyrophosphates. In the case of glycosides containing an oligosaccharide as the carbohydrate portion, a monosaccharidic glycoside is first formed, which in turn serves as an acceptor of a new glycosyl group, which becomes attached to the first sugar residue. In this manner, the phenolic mono-β-D-glucosides can easily form β-gentiobiosides from uridine 5′-(α-D-glucopyranosyl pyrophosphate).[93]

Quercetin first receives a D-glucose residue from thymidine or uridine 5′-(α-D-glucopyranosyl pyrophosphate). This transfer is catalyzed by an enzyme preparation from *Phaseolus aureus* and leads to the formation of quercetin 3-β-D-glucoside. In a second step, the same enzyme preparation transfers L-rhamnose from thymidine 5′-(α-L-rhamnosyl pyrophosphate) to the primary alcohol group of the D-glucose residue of quercetin D-glucoside to yield rutin.[94]

Transglycosylation of glycosides also can occur from a glycosidic donor to a hydroxylic acceptor:

$$\text{X-Glycosyl} + \text{Y-OH} \longrightarrow \text{Y-Glycosyl} + \text{X-OH}$$

where X is the genin and Y is an aryl or acyl compound (see this volume, Chap. 33 for more details).

Transglycosylation reactions on phenolic acceptors are rarely encountered,[95] but such transfer reactions take place easily with acyclic alcohols, especially with methanol and alditols. It is possible that the first step in the

* *References start on p. 237.*

synthesis is initiated by nucleoside pyrophosphates; this reaction would enable the glycosides thus formed to act as donors for other glycosides. Oligosaccharides may also act as carbohydrate donors for alcohol acceptors. Similar mechanisms seem to take place in metabolic pathways. The products of transglycosylation induce the lactose operon in *Escherichia coli*.[96] The best inducer is 1-*O*-β-D-galactosylglycerol, formed by transglycosylation.

The *C*-glycosyl compounds seem to arise by a different biogenetic pathway. It is unlikely that they are formed from nucleotides and very doubtful that they result via transglycosylations. Their mode of formation may be similar to that of the purine nucleotides, in which the sugar becomes attached to a two-carbon fragment, after which the rest of the molecule is built up in subsequent steps. Furthermore, in the case of glycosides, it is not at all necessary for the monosaccharide to become attached directly to the genin. It is conceivable that the sugar combines first with a simpler precursor molecule, before the final genin is formed. The frequent existence in the same plant of several glycosides having similar structures would favor such a concept.

B. Metabolism

Numerous interpretations of the physiological role of glycosides have been advanced. Some may be valid for a limited group of glycosides, but none are acceptable for all the glycosides. This fact is hardly surprising, considering the great structural differences among genins. It is more rational to accept the fact that the glycosides have distinct physiological functions according to the structure of their genin.

The hypothesis which considered the glycosides as intermediates in the synthesis of polysaccharides seems to have been totally abandoned. *O*-Glycosides, as well as oligo- and polysaccharides, are formed from glycosyl derivatives of nucleoside pyrophosphates.

The formation of glycosides as detoxication products is now a certainty for numerous compounds. In plant tissues, absorption of phenolic compounds by the roots or injection of these compounds induces the biosynthesis of the corresponding β-D-glucosides. *o*-Chlorophenol produces not the glucoside, but the β-gentiobioside. Hydroquinone induces the glycosylation of one phenolic function with formation of arbutin, and resorcinol leads to a mono-D-glucoside. Cell-free extracts induce similar glycosylations. Such reactions are understandable if the phenols which result from plant metabolism are automatically blocked by a glycosidic linkage.

Numerous glycosides possess polyhydroxylated genins in which one or several phenolic groups are free. Moreover, salicyl alcohol, when absorbed by the plant, does not form the naturally occurring compound salicin but yields a salicyl D-glucoside. Here, the alcohol group combines with D-glucose, as in the transglucosylation reaction of β-D-glucosides. A possible objection to glycosylation as a mechanism of detoxication is the fact that the glycosides exist in plants *in situ* and are not excreted. However, insects form β-D-glucosides for the blocking of toxic products.[97]

The xylophagic insects contain active β-D-glucosidases at fairly constant levels which do not undergo the fluctuations observed for other glycosidases that are involved in the digestion of polysaccharides.[98]

In mollusks, phenols are glycosylated either by D-glucose or by D-glucuronic acid, according to the particular species.[99] In mammals, birds, fishes, and some amphibia, many toxic, pharmacological, or waste compounds, including degradation products of steroid hormones, are glycosylated by D-glucuronic acid and are eliminated as D-glucosiduronic acids.

Glycosides have also been considered to serve as reserve foods or storage compounds. Indeed, seasonal and diurnal fluctuations in their concentrations have been observed, often with a maximal content occurring before blossoming. It is generally agreed that the α-D-galactoside of red algae is a reserve carbohydrate that replaces sucrose, which is lacking in such alga.[100]

The glycosides of alditols are more closely related to disaccharides than are the glycosides of complex genins.

The concept which considers the cyanogenetic glycosides as a storage source of nitrogen to be used for amino acid synthesis is not convincing. Although the glycosides are usually more soluble in water than the aglycons, the exact physiological role of this property in relation to transport phenomena in plant fluids is difficult to determine.

The presence of five different glycosides of quercetin in the leaves of myrtle, and six glycosides in some apple peels, indicates that the definite physiological function of individual glycosides is far from being elucidated.

REFERENCES

1. A. Kameswaramma and T. R. Sheshadri, *Proc. Indian Acad. Sci. Sect. A*, **25**, 43 (1947).
2. T. Reichstein, in "Carbohydrate Chemistry of Substances of Biological Interest," M. L. Wolfrom, Ed., Vol. I, Pergamon Press, New York, 1959, p. 124.
3. See C. S. Hudson, *Advan. Carbohyd. Chem.*, **4**, 57 (1949).
4. J. Staněk, M. Černý, and J. Pacák, in "The Oligosaccharides," Academic Press, New York, 1965, pp. 214 and 387.
5. N. Wattiez, *Bull. Soc. Chim. Biol.*, **7**, 917 (1925).
6. V. Plouvier, *C. R. Acad. Sci.*, **256**, 1397 (1963).
7. E. Nottbohm and F. Mayer, *Vorratspflege Lebensmittelforsch.*, **1**, 243 (1938).
8. H. Colin, *Bull. Soc. Chim. Fr.*, **4**, 277 (1934); J. Augier and L. du Merac, *C. R. Acad. Sci.*, **258**, 387 (1954); E. W. Putman and W. Z. Hassid, *J. Amer. Chem. Soc.* **76**, 2221 (1954).

9. B. Lindberg, *Acta Chem. Scand.*, **8**, 869 (1954).
10. B. Wickberg, *Acta Chem. Scand.*, **8**, 1183 (1954); H. E. Carter, R. H. McChia, and E. D. Slifer, *J. Amer. Chem. Soc.*, **78**, 3735 (1956).
11. B. Lindberg and B. Wickberg, *Acta Chem. Scand.*, **8**, 821 (1954).
12. G. Puejo, *C. R. Acad. Sci.*, **248**, 2788 (1959).
13. H. Colin and J. Augier, *C. R. Acad. Sci.*, **208**, 1450 (1939).
14. R. J. Brown and R. F. Serro, *J. Amer. Chem. Soc.*, **75**, 1040 (1953); E. A. Kabat, D. L. MacDonald, C. E. Ballou, and H. O. L. Fischer, *J. Amer. Chem. Soc.*, **75**, 4507 (1953).
15. E. Vilkas and E. Lederer, *Bull. Soc. Chim. Biol.*, **42**, 1013 (1960); C. E. Ballou, E. Vilkas, and E. Lederer, *J. Biol. Chem.*, **238**, 69 (1963); Y. C. Lee and C. E. Ballou, *ibid.*, **239**, 1316 (1964).
16. G. Dillemann, in "Handbuch der Pflanzenphysiologie" W. Ruhland, Ed., Vol. 8, Springer Verlag, Berlin, 1958, p. 1050.
17. P. J. Robiquet and A. F. Boutron-Charlard, *Ann. Chim. Phys.* **44**, 352 (1830).
18. F. Wöhler and J. Liebig, *Ann.*, **22**, 1 (1837).
19. J. Giaja, *C. R. Soc. Biol.*, **69**, 235 (1910).
20. R. Kuhn, *Ber.*, **56**, 857 (1923).
21. R. Campbell and W. N. Haworth, *J. Chem. Soc.*, 1337 (1924); G. Zemplén and A. Kunz, *Ber.*, **57**, 1357 (1924); R. Kuhn and H. Sobotka, *ibid.*, **57**, 1767 (1924).
22. G. Bertrand, *C. R. Acad. Sci.*, **143**, 832 (1906); G. Bertrand and L. Rivkind, *ibid.*, **143**, 970 (1906); G. Bertrand and G. Weisweiler, *ibid.*, **147**, 252 (1908).
23. J. E. Courtois, F. Percheron, and Huyhn Nhut Quang, *Bull. Soc. Chim. Biol.*, **46**, 543 (1964).
24. H. E. Armstrong, R. F. Armstrong, and E. Horton, *Proc. Roy. Soc., Ser. B*, **85**, 359 (1912).
25. E. Bourquelot and E. Danjou, *C. R. Acad. Sci.*, **141**, 598 (1905).
26. H. Hérissey, *J. Pharm. Chim.*, **23**, 5 (1906).
27. A. Jorissen and E. Hairs, *Bull. Acad. Roy. Sci. Belg.*, **14**, 923 (1887); **21**, 529 (1891).
28. A. Kawalier, *Ann.*, **82**, 241 (1852); E. Bourquelot and A. Fichtenholz, *C. R. Acad. Sci.*, **151**, 81 (1910); **153**, 468 (1911).
29. A. Robertson and R. B. Waters, *J. Chem. Soc.*, 2729 (1930).
30. J. C. Irvine and R. E. Rose, *J. Chem. Soc.*, **89**, 814 (1906).
31. J. Rabaté, *Bull. Soc. Chim. Biol.*, **17**, 572 (1935).
32. W. W. Pigman and N. K. Richtmyer, *J. Amer. Chem. Soc.*, **64**, 374 (1942).
33. R. Weidenhagen, *Z. Ver. Deut. Zucker-Ind.*, **79**, 591 (1929).
34. W. Procter, *Amer. J. Pharm.*, **15**, 241 (1843); M. Bridel, *C. R. Acad. Sci.*, **177**, 642 (1923); *Bull. Soc. Chim. Biol.*, **6**, 659 (1924).
35. K. Freudenberg, H. Reznik, H. Boesenberg, and D. Rasenack, *Ber.*, **85**, 641 (1952); K. Freudenberg and D. Rasenack, *ibid.*, **86**, 756 (1953).
36. E. Bourquelot and H. Hérissey, *J. Pharm. Chim.*, **21**, 481 (1905); H. Hérissey and J. Cheymol, *C. R. Acad. Sci.*, **180**, 384 (1925); **181**, 505 (1925); **183**, 1307 (1926).
37. L. de Koninck, *Ann.*, **15**, 75, 258 (1835–1836); M. Bridel and A. Kramer, *Bull. Soc. Chim. Biol.*, **15**, 544 (1933).
38. D. Gilmour, Ed., in "Biochemistry of Insects," Academic Press, New York, 1961, p. 191.
39. A. Stoll, A. Perreira, and J. Renz, *Helv. Chim. Acta*, **33**, 1637 (1950).
40. C. Charaux, *Bull. Soc. Chim. Biol.*, **7**, 1056 (1925).
41. Minor, *Arch. Pharm.*, **38**, 130 (1831).
42. F. Wessely and K. Sturm, *Ber.*, **63**, 1299 (1930).

43. F. Wessely and E. Demmer, *Ber.*, **61**, 1270 (1928); **62**, 120 (1929).
44. W. Schmid, in "Modern Methods of Plant Analysis", K. Paech and M. V. Tracey, Eds., Vol. 3, Springer Verlag, Berlin, 1955, p. 549; R. Paris and H. Moyse, *Produits Pharm.*, **14**, 462 (1959).
45. P. Casparis and R. Maeder, *Schweiz. Apotheker-Z.*, **63**, 313 (1925); M. Bridel and C. Charaux, *C. R. Acad. Sci.*, **191**, 1151 (1930); M. Bridel and C. Charaux, *Bull. Soc. Chim. Biol.*, **17**, 793 (1935).
46. M. Bridel and C. Charaux, *C. R. Acad. Sci.*, **191**, 1374 (1930).
47. A. Stoll, B. Becker, and W. Kussmaul, *Helv. Chim. Acta*, **32**, 1892 (1949); A. Stoll, B. Becker, and A. Helfenstein, *ibid.*, **33**, 313 (1950).
48. A. Stoll and E. Jucker, in "Modern Methods of Plant Analysis" K. Paech and M. V. Tracey, Eds., Vol. 3, Springer Verlag, Berlin, 1955, pp. 140–271; T. Reichstein *et al.*, more than 250 articles in *Helv. Chim. Acta* from 1938 to 1969.
49. A. Stoll and W. Kreis, *Helv. Chim. Acta*. **16**, 1049 (1933).
50. A. Stoll, J. Renz, and W. Kreis, *Helv. Chim. Acta*, **20**, 1484 (1937).
51. A. Stoll, A. Hofmann, and W. Kreis, *Helv. Chim. Acta*, **17**, 1334 (1934).
52. H. von Euler and E. Nordenson, *Hoppe-Seyler's Z. Physiol. Chem.*, **56**, 223 (1908).
53. S. Furukawa, *Chem. Zentr.*, **II**, 3901 (1932).
54. F. W. Heyl and D. Larsen, *J. Amer. Pharm. Assoc.*, **22**, 510 (1933).
55. A. Stoll and E. Jucker, in "Modern Methods of Plant Analysis" K. Paech and M. V. Tracey, Eds., Vol. 3, Springer Verlag, Berlin 1955, p. 176; L. F. Fieser and M. Fieser, "Steroids," Reinhold, New York, 1959, p. 810.
56. I. Scheer, R. B. Kostic, and E. Mosettig, *J. Amer. Chem. Soc.*, **75**, 4871 (1953); *ibid.*, **77**, 641 (1955).
57. H. Sobotka, *Bull. Soc. Chim. Biol.*, **47**, 169 (1965).
58. M. Steiner and H. Holtzem, see ref. 44, Vol. 3, pp. 58–140.
59. E. Ochiai, K. Tsuda, and S. Kitagawa, *Ber.*, **70**, 2083, 2093 (1937).
60. K. Rehorst, *Ber.*, **62**, 519 (1929).
61. R. J. Morris and E. W. Hussey, *J. Org. Chem.*, **30**, 166 (1965).
62. J. Polonsky, E. Sach, and E. Lederer, *Bull. Soc. Chim. Fr.*, 880 (1959).
63. T. Reichstein and H. Reich, *Ann. Rev. Biochem.*, **15**, 155 (1946).
64. L. H. Briggs, and L. C. Vining, *J. Chem. Soc.*, 2809 (1953).
65. R. Kuhn, I. Löw, and H. Trischmann, *Ber.*, **90**, 203 (1957).
66. R. Kuhn and I. Löw, *Ber.*, **80**, 406 (1947).
67. A. Robertson, *J. Chem. Soc.*, 1937 (1927).
68. R. Robinson, *Nature*, **135**, 732 (1935); *Endeavour*, **1**, 92 (1942); N. Campbell, in "Chemistry of Carbon Compounds" E. H. Rodd, Ed., Vol. 4*B*, Elsevier, Amsterdam, 1959, p. 855.
69. R. Willstätter and A. E. Everest, *Ann.*, **410**, 189 (1913).
70. R. Robinson and A. R. Todd, *J. Chem. Soc.*, 2488 (1932).
71. R. Robinson and G. M. Robinson, *J. Chem. Soc.*, 744 (1935); F. E. King and W. Bottomley, *J. Chem. Soc.*, 1399 (1954).
72. R. Paris and L. Beauquesne, *Ann. Pharm. Fr.*, **8**, 65, 148, 228, 322 (1950); N. Campbell, in "Chemistry of Carbon Compounds" E. H. Rodd, Ed., Vol. 4*B*, Elsevier, Amsterdam, 1959, pp. 905 and 957; J. B. Harborne, "Comparative Biochemistry of Flavonoids," Academic Press, New York, 1967.
73. F. Sosa and F. Percheron, *C. R. Acad. Sci.*, **261**, 4544 (1965).
74. L. Hörhammer, L. Stichand, and H. Wagner, *Naturwissenschaften*, **46**, 358 (1959); J. B. Harborne, *Phytochemistry*, **3**, 151 (1964).
75. R. Kuhn and A. Winterstein, *Ber.*, **67**, 344 (1934).

76. J. Gadamer, *Arch. Pharm.*, **235**, 44 (1897); A. L. Raymond, *Advan. Carbohyd. Chem.*, **1**, 129 (1945); D. Horton and D. H. Hutson, *ibid.*, **18**, 123 (1963).

76a. M. G. Ettlinger and J. A. Lundeen, *J. Amer. Chem. Soc.*, **78**, 4172 (1956).

77. E. Fischer and K. Delbrück, *Ber.*, **42**, 1476 (1909); C. B. Purves, *J. Amer. Chem. Soc.*, **51**, 3627 (1929); W. Schneider, J. Sepp, and O. Stiehler, *Ber.*, **51**, 220 (1918).

77a. M. H. Benn, *Can. J. Chem.*, **41**, 2836 (1963); **42**, 163 (1964); **43**, 1 (1965); *J. Chem. Soc.*, 4072 (1964); M. H. Benn and D. Neakin, *Can. J. Chem.*, **43**, 1874 (1965); M. H. Benn and L. Yelland, *ibid.*, **45**, 1595 (1967).

77b. A. Kjær and B. W. Christensen, *Acta Chem. Scand.*, **16**, 71, 83, (1962); A. Kjær and H. Thomsen, *ibid.*, **16**, 591 (1962); A. Kjaer and W. Wagnières, *ibid.*, **19**, 1989 (1965); P. Friis and A. Kjær, *ibid.*, **20**, 698 (1966).

78. P. Karrer and H. Schmid, *Helv. Chim. Acta*, **29**, 525 (1946). Y. Nakamura, *J. Chem. Soc. Jap.*, *Pure Chem. Sect.*, **71**, 63, 123, 186 (1950); *Chem. Abstr.*, **45**, 7117 (1951).

79. R. B. Duff, J. S. D. Bacon, C. M. Mundie, V. C. Farmer, J. D. Russell, and A. R. Forrester, *Biochem. J.*, **96**, 1 (1965).

80. F. Korte, *Ber.*, **87**, 512 (1954).

81. L. Bourdier, *J. Pharm. Chim.*, **27**, 49, 101 (1908); J. Cheymol, *Bull. Soc. Chim. Biol.*, **19**, 1609 (1937); M. Cohn, E. Vir, and P. Karrer, *Helv. Chim. Acta*, **37**, 790 (1954).

82. E. Bourquelot and H. Hérissey, *J. Pharm. Chim.*, **25**, 417 (1907).

83. T. Kubota and Y. Tomita, *Tetrahedron Lett.*, **5**, 176 (1961).

84. M. Plat, M. Koch, A. Bouquet, J. Le Men, and M. M. Janot, *Bull. Soc. Chim. Fr.*, 1302 (1963); M. Koch, M. Plat, J. Le Men, and M. M. Janot, *ibid.*, 403 (1964).

85. H. Hérissey and J. Cheymol, *Bull. Soc. Chim. Biol.*, **13**, 29 (1931).

86. L. J. Haynes, *Advan. Carbohyd. Chem.*, **18**, 227 (1963); *ibid.*, **20**, 357 (1965).

86a. K. Nishida, A. Kobayashi, and T. Nagahama, *Bull. Agr. Chem. Soc. Jap.*, **19**, 77 (1955).

86b. K. Nishida, A. Kobayashi, T. Nagahama, and T. Numata, *Bull. Agr. Chem. Soc. Jap.*, **23**, 460 (1959).

87. W. E. Cohn, *Bull. Soc. Chim. Biol.*, **46**, 239 (1964).

88. B. H. Koeppen and D. G. Roux, *Biochem. J.*, **97**, 444 (1965).

89. R. Paris and A. Stambouli, *C. R. Acad. Sci.*, **252**, 1659 (1961).

90. L. F. Leloir, *Biochem. J.*, **91**, 1 (1964).

91. E. F. Neufeld and W. Z. Hassid, *Advan. Carbohyd. Chem.*, **18**, 309 (1963).

92. J. B. Pridham, *Advan. Carbohyd. Chem.*, **20**, 371 (1965).

93. T. Yamaha and C. E. Cardini, *Arch. Biochem. Biophys.*, **86**, 127 (1960); *ibid.*, **86**, 133 (1960).

94. G. A. Barber, *Biochemistry*, **1**, 463 (1962).

95. J. B. Pridham and M. J. Saltmarsh, *Biochem. J.*, **87**, 218 (1963); J. B. Pridham, *Chem. Ind.* (*London*), 1172 (1961); J. B. Pridham and K. Wallenfels, *Nature*, **202**, 488 (1964).

96. C. Burstein, *Bull. Soc. Chim. Biol.*, **47**, 1901 (1965).

97. D. Gilmour, "Biochemistry of Insects," Academic Press, New York, 1961.

98. J. E. Courtois, C. Chararas, M. M. Debris, and H. Laurant-Hubé, *Bull. Soc. Chim. Biol.*, **47**, 2219 (1965).

99. G. J. Dutton, *Biochem J.*, **96**, 36P (1965).

100. S. Peat and D. A. Rees, *Biochem. J.*, **79**, 7 (1961).

33. GLYCOSIDE HYDROLASES AND GLYCOSYL TRANSFERASES

K. Nisizawa and Y. Hashimoto

I. Introduction. 242
II. Classification 242
III. Mechanism of Glycosidase Action 243
   A. Cleavage of Glycosidic Bonds 243
   B. Retention or Inversion of the Configuration of the Potential Reducing Group during Enzyme Action 243
   C. Hydrolysis and Transfer 244
   D. Reversibility of Glycosidase Action . . . 244
   E. Mechanism of Fission of Glycosidic Bonds . . 245
   F. Role of Acceptor and Transglycosylation . . 247
IV. Chemical Nature of Glycoside Hydrolases and Glycosyl Transferases 248
V. Individual Glycosidases and Transglycosylases . . 251
   A. α-D-Glucosidases (α-D-Glucoside Hydrolases and α-D-Glucosyl Transferases) 251
   B. β-D-Glucosidases (β-D-Glucoside Hydrolases and β-D-Glucosyl Transferases) 258
   C. α-D-Galactosidases (α-D-Galactoside Hydrolases and α-D-Galactosyl Transferases) . . . 266
   D. β-D-Galactosidases (β-D-Galactoside Hydrolases and β-D-Galactosyl Transferases) . . . 268
   E. α,β-D-Mannosidases (α,β-D-Mannoside Hydrolases) 274
   F. β-D-Fructofuranosidases (β-D-Fructofuranoside Hydrolases and β-D-Fructofuranosyl Transferases) 275
   G. 2-Acetamido-2-deoxy-α,β-D-hexosidases and Related Enzymes 280
   H. β-D-Glucosiduronases 284
   I. β-D-Xylosidases (β-D-Xyloside Hydrolases) . . 286
   J. Sialidases 287
   K. α,β-D- and L-Fucosidases and α,β-L-Rhamnosidases (α,β-D- and L-Fucoside Hydrolases and α,β-L-Rhamnoside Hydrolases) 288
   L. Cyclodextrin Transglucosylases 288
   M. D-Enzyme 289
   N. Amylomaltase 290
   References 290

I. INTRODUCTION

Glycoside hydrolases or glycosidases are enzymes that catalyze the breakdown of various glycosides and oligosaccharides. Much of our knowledge of their action has been based on studies with crude or partially purified enzyme preparations. However, with advances in methods for protein fractionation, some of the enzymes have been isolated in highly purified form. Purified enzymes that split particular glycosidic linkages or catalyze the transfer of particular glycosyl residues to a suitable sugar have been very useful tools in determination of the structure of complex polysaccharides. They are also of value in studies of the breakdown of carbohydrates and related compounds in animal and plant tissues.

II. CLASSIFICATION

The glycoside hydrolases can be classified according to the nature of the sugar moieties in the glycoside, the anomeric configuration of the glycosidic linkage hydrolyzed, and the type of transferring reactions catalyzed.

Enzymes that hydrolyze common glycosides and oligosaccharides have generally been differentiated from those acting on polysaccharides. However, no sharp delineation can be made between these two groups of enzymes. Some polysaccharide hydrolases cleave disaccharides, and some glycoside hydrolases also split polysaccharides. Glycosyl transferases may be divided into two broad groups: enzymes that are originally hydrolytic but also catalyze the transfer of glycosyl residues to suitable acceptors other than water, and enzymes that are specific only for transfer reactions that lead to synthesis of oligo- and polysaccharides.

The systematic and trivial names of glycosidases that will be dealt with are listed in Vol. IIB, Chap. 47. Many trivial names will be used in this chapter, since they have been used more widely and new names are often not sufficiently discriminating; for example, the same enzyme may be a transferase and also a hydrolase.

The transferring enzymes have been used to produce many new oligosaccharides. Additional information about these enzymes may be found in this volume, Chap. 30. The subject of biosynthesis is covered in this volume, Chap. 34. Since the action of glycosidases is frequently indistinguishable from that of polysaccharidases, additional discussions will be found in the chapters on polysaccharides and in the chapter on bacterial cell walls (Vol. IIB, Chap. 41). Chapter 32 (this volume) contains information on some of the rarer glycosidases.

III. MECHANISM OF GLYCOSIDASE ACTION

Glycosidases have been usually named according to the type of bond hydrolyzed. Some glycosidases also catalyze the transfer of the glycosyl moiety to a suitable acceptor, which becomes the aglycon of a newly formed glycoside.

A. Cleavage of Glycosidic Bonds

An important question concerning the mechanism by which glycosidases act is whether the oxygen atom of the glycosidic bond is retained by the sugar or by the aglycon group. Bunton et al.[1] and Koshland et al.,[2] using $H_2^{18}O$, were able to show that the hydrolysis of methyl α- and β-D-gluco-pyranosides both in acid and by appropriate D-glucosidases proceeds by fission of the hexose–oxygen bond; that is, the linkage broken is that between the C-1 of the sugar and the oxygen bridge, according to Eq. (1):

$$R—O—Me + H_2^{18}O \longrightarrow R—^{18}OH + MeOH \qquad (1)$$

The same type of mechanism was confirmed for the action of α- and β-amylases on starch[3,4] and for β-D-glucosiduronases on benzyl β-D-gluco-pyranosiduronic acid.[5]

B. Retention or Inversion of the Configuration of the Potential Reducing Group during Enzyme Action

It is believed that the configuration of the reducing group of maltose produced by the action of α- or β-amylase on starch is α-D or β-D, respectively, although starches have only α-D linkages. This assumption is based on observed increases or decreases in optical rotation of the starch digest during enzymic hydrolysis. The action of cellulase from *Myrothecium verrucaria* upon cellopentaose revealed that the rotation of the reaction mixture increased progressively, suggesting that the β-D configuration was retained at the reducing terminus in the products.[6] Proton magnetic resonance spectroscopy indicates that D-glucosidases and polysaccharases of the *endo* type, such as α-amylase, yield products in which there is retention of the anomeric configuration of the substrate, whereas polyglycosidases of the *exo* type, such as β-amylase, act with inversion of the anomeric configuration.[6a]

Inversion of configuration was observed during the action of some uridine 5′-(glycosyl pyrophosphate) transglycosylases. Hassid et al.[7] reported that an enzyme from mung beans catalyzed the transfer of a D-glucosyl residue

* References start on p. 290.

from uridine 5'-(D-glucopyranosyl pyrophosphate) to give a β-D-(1 → 3)-linked polymer of D-glucose. The same kind of transfer to cellulose by a particulate enzyme from the seeds of *Lupinus albus* was observed by Brummond and Gibbons.[8] This transfer involves inversion of the configuration at C-1 of the D-glucose residue from α to β. Similar instances are known for maltose[9] and cellobiose phosphorylases,[10] in which β-D-glucopyranosyl phosphate and α-D-glucopyranosyl phosphate are converted in the presence of D-glucose into maltose and cellobiose, respectively.

C. HYDROLYSIS AND TRANSFER[11,12]

It has been established that the same glycosidase can, under suitable conditions, transfer the glycosyl residue of a substrate to various acceptors, and that hydrolysis is a special case in which water serves as the acceptor. Thus, if aliphatic alcohols or sugars are present in suitable concentrations together with the substrate, glycosides having alkyl groups or sugar residues as aglycon groups are formed. Whether the enzymic reaction proceeds toward hydrolysis or to the production of new glycosides depends on experimental conditions as well as on the specificities of the glycosidases for their substrates and acceptors. Since the configuration at the anomeric carbon atom of the new glycoside produced by the action of an α- or β-D-glycosidase is always the same as that in the original substrate, no inversion of configuration at C-1 appears to take place during these transferring reactions.

D. REVERSIBILITY OF GLYCOSIDASE ACTION

The enzymic hydrolysis of most glycosides is essentially reversible, although in most cases it favors hydrolysis. Various β-D-glucosides can be synthesized by almond emulsin in the presence of a minimal amount of water and an excess of D-glucose and aglycon.[13,14] According to Koshland,[15] this reaction arises from a replacement mechanism. He has shown that β-D-glucosidase will catalyze the following reaction, where water acts as both donor and acceptor.

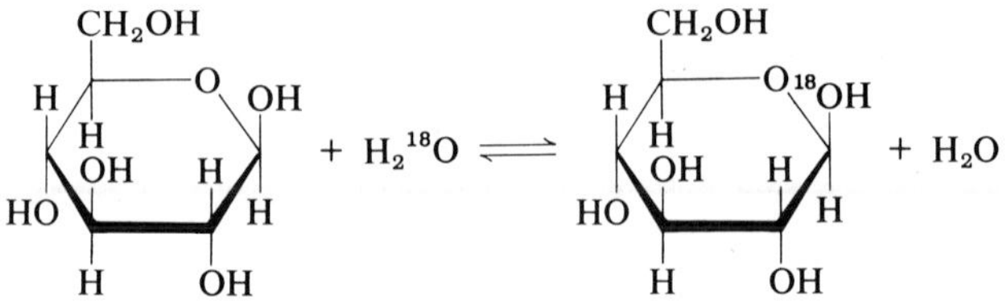

E. MECHANISM OF FISSION OF GLYCOSIDIC BONDS

The problem of interaction between enzyme and glycoside just prior to fission of the glycosidic bond has been discussed by Pigman[16] on the basis of the classic concept of Euler that the enzyme forms a complex with the substrate, involving two areas of the enzyme molecule, illustrated as ovals in Fig. 1. It is assumed that the glycoside is adsorbed on these two areas, the

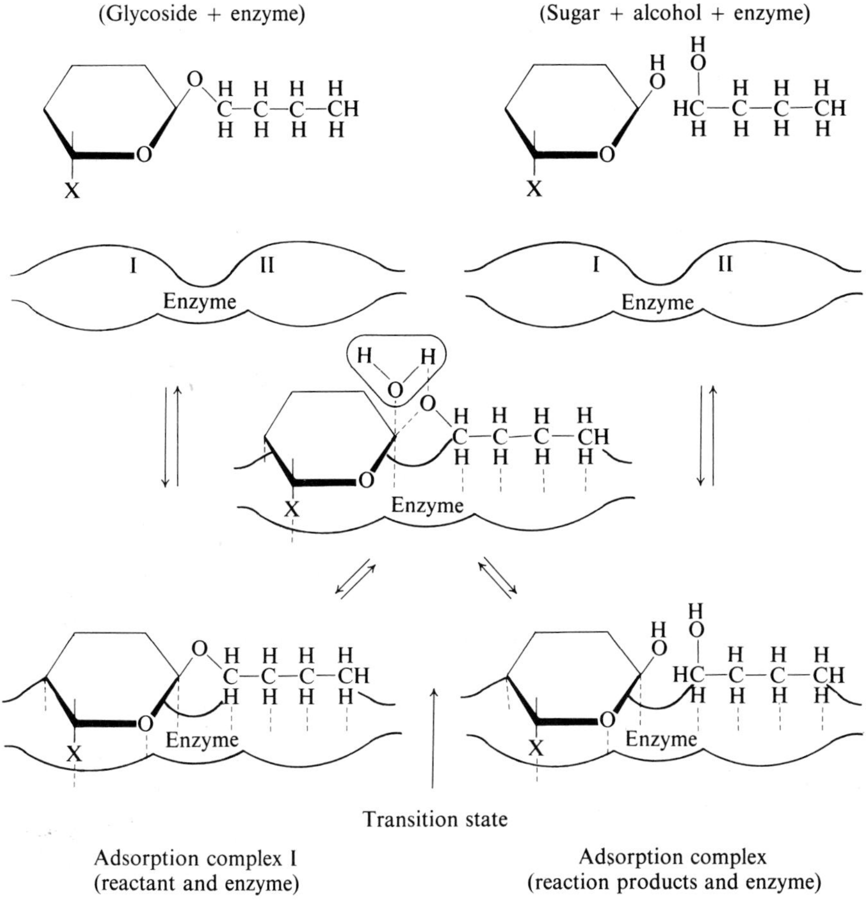

FIG. 1. Possible mechanism for the enzymic hydrolysis of an alkyl glucoside.

aglycon group being taken up by area II and the sugar moiety by area I. Area I exhibits specific adsorption, whereas area II adsorbs many types of groups. After the substrate has been adsorbed, a molecule of water or hydronium ion adds to the glycosidic linkage. Cleavage of the glycosidic

* *References start on p. 290.*

linkage is assumed then to take place, leading to the formation of a complex comprising enzyme, sugar, and aglycon. Dissociation of sugar and aglycon from the enzyme surface constitutes the final stage of hydrolysis.

According to this mechanism, the enzymic hydrolysis is similar to acid hydrolysis, but because of the formation of an intermediate complex, a preliminary activation of the substrate molecule takes place. The activation energy required in the second phase of the reaction, which corresponds to the reaction occurring during acid hydrolysis, is, therefore, lowered. Thus, the activation energy for the acid-catalyzed hydrolysis of methyl β-D-gluco-pyranoside is 32,610 cal mole^{-1} and that of the enzyme-catalyzed reaction is 12,000 cal mole^{-1}.

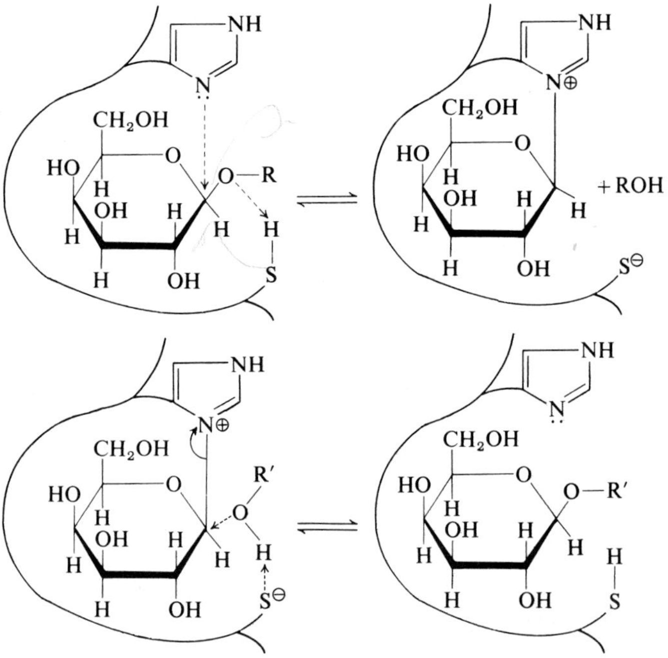

Fig. 2. Possible mechanism for the enzymic transfer of the galactosyl residue from a galactoside to an acceptor (R'OH).

Wallenfels and Malhotra[16a] have confirmed this mechanism for β-D-galacto-sidase of *Escherichia coli* which, on the basis of inhibition and pH–activity studies, appears to have a sulfhydryl and an imidazole group at the active site. This mechanism is shown in Fig. 2. The imidazole group acts as a nucleophilic center, and the sulfhydryl group labilizes the oxygen atom of the glycosidic linkage.

Koshland *et al.*[16b] suggest two roles for the amino acids at or near the active site. One is in adjusting the conformation of the active site to provide a better fit, and a hydrophobic environment. The other provides active groups in the neighborhood of the linkage being cleaved, as in Fig. 2.

F. ROLE OF ACCEPTOR AND TRANSGLYCOSYLATION

The glycoside–enzyme complex in the second and third stages (see Fig. 1) of the reaction mechanisms stated above may combine with a suitable acceptor such as an alcohol that has been added previously to a mixture of carbohydrase and substrate. As a result, the glycosyl moiety is transferred to an alcohol instead of being converted into a free sugar, and a new alkyl glycoside is formed, according to the equation

$$RO\text{-glycosyl} + R'O\text{---}H \overset{E}{\rightleftharpoons} R'O\text{-glycosyl} + RO\text{---}H$$

where E represents the enzyme. This transfer reaction can be divided into two stages (see Fig. 2):

$$EO\text{---}H + RO\text{-glycosyl} \longrightarrow EO\text{-glycosyl} + RO\text{---}H \qquad (2)$$

$$EO\text{-glycosyl} + R'O\text{---}H \longrightarrow EO\text{---}H + R'O\text{-glycosyl} \qquad (3)$$

where R represents the aglycon and E represents the enzyme molecule.

Jermyn[17] assumed that enzyme, substrate, and acceptor are bound in a single, ternary complex. Decomposition to give the products of the enzyme reaction takes place only after formation of this ternary complex has occurred. The enzymic reaction proceeds toward hydrolysis or to formation of a new glycoside according to the nature and concentration of the acceptor and the substrate, since the acceptor specificity is as fundamental a property of glycosidases as is the donor specificity. It has been reported by Suzuki[18] that β-D-glucosidases of higher plants react with a mixture of possible acceptors (water, alcohol, and the D-glucoside) according to a characteristic time pattern. Furthermore, the newly formed transfer products usually possess the same configuration at the glycosidic bond as the original substrate. This net retention may be the result of a double displacement mechanism in which the glycosyl group is covalently bound to a site on the enzyme, with a resulting double Walden inversion at the glycosidic carbon atom. Jermyn's assumption seems to combine Koshland's theory[15] with that of Mayer and Larner.[4] The α-D configuration at the point of cleavage is maintained on hydrolysis of starch by α-amylase, but inversion occurs on cleavage by β-amylase. According to Jermyn, Eqs. (2) and (3) may be applicable to glycoside transferases. An attack on the carbonium-ion intermediate, as assumed by Mayer and Larner, could take place either by an acceptor

* *References start on p. 290.*

molecule bound specifically to the enzyme surface, or by water, or in general by a solvent molecule.

IV. CHEMICAL NATURE OF GLYCOSIDE HYDROLASES AND GLYCOSYL TRANSFERASES

With recent improvements in techniques for the fractionation of proteins, numerous glycosidases from a variety of sources have been isolated in a highly purified state. The first successful crystallization by Wallenfels and associates[19] of a β-D-galactosidase from *Escherichia coli* has greatly increased the understanding of the chemical and physical properties of glycoside hydrolases. Some aspects of recent knowledge concerning the chemical nature of various glycosidases will be summarized here.

TABLE I

CARBOHYDRATE CONTENT OF SOME GLYCOSIDE HYDROLASES OF DIFFERENT ORIGINS

Enzyme	Origin	Carbohydrate content (%)	
Amyloglucosidase	*Aspergillus niger*	20	(D-glucose, D-mannose, D-xylose)[a]
β-D-Fructosidase	Yeast	70	(mannan)[b]
β-D-Fructosidase	*Neurospora crassa*	2.4	(2-amino-2-deoxy-D-hexose)[c]
β-D-Fructosidase	Grape	25[d]	
β-D-Galactosidase	Calf intestine	88	(2-acetamido-2-deoxy-D-glucose, 71%, D-galactose, 17%)[e]
β-D-Glucosidase	*Cucunis duhurica*	2.0[f]	
β-D-Glucosidase	Sweet almond	4-5[g]	
β-D-Glucosiduronase	Snail (*Helix pomatia*)	15[h]	
Taka-amylase A	*Aspergillus oryzae*	3	(2-amino-2-deoxy-D-glucose, D-mannose, D-xylose)[i]
Cellulase III	*Trichederma viride*	16	(D-mannose, D-glucose, 2-amino-2-deoxy-D-glucose)[j]

[a] J. H. Pazur and K. Kleppe, *J. Biol. Chem.*, **237**, 1002 (1962). [b] E. H. Fischer, L. Kohtès, and T. Fellig, *Helv. Chim. Acta*, **34**, 1132 (1951). [c] R. L. Metzenberg, *Arch. Biochem. Biophys.*, **110**, 134 (1965). [d] W. N. Arnold, *Biochim. Biophys. Acta*, **110**, 134 (1965). [e] K. Wallenfels and J. Fischer, *Biochem. Z.*, **321**, 223 (1960). [f] F. J. Joubert, *Arch. Biochem. Biophys.*, **91**, 11 (1960). [g] B. Helferich and T. Kleinschmidt, *Hoppe-Seyler's Z. Physiol. Chem.*, **340**, 31 (1965). [h] A. Alfsen-Blanc and M. F. Jayle, *Bull. Soc. Chim. Biol.*, **40**, 2143 (1958). [i] M. Anai, T. Ikenaka, and Y. Matsushima, *J. Biochem. (Tokyo)*, **59**, 57 (1966). [j] G. Okada, K. Nisizawa, and H. Suzuki, *J. Biochem. (Tokyo)*, **63**, 591 (1968).

It has been known for a long time that even highly purified glycosidases contain an appreciable proportion of carbohydrates. It remains to be elucidated whether or not these carbohydrates are always integral constituents of the enzyme molecule, since in most cases the carbohydrate content changes considerably during purification procedures. Table I summarizes the carbohydrate content of several glycosidase preparations from different sources.

An extreme instance is represented by the β-D-galactosidase of calf intestine, which has 90% of the enzyme composed of carbohydrates.[20a] The carbohydrate composition of the invertases from yeast cells varies widely according to the location of the enzyme with respect to the cell membrane. Whereas the internal invertase has virtually no carbohydrate, the external enzyme contains 50% of D-mannose, linked probably to the protein core through 2-amino-2-

TABLE II

AMINO ACID COMPOSITIONS OF β-D-GALACTOSIDASES OF *Escherichia coli*[a,b]

Amino acid	Strain:		
	ML 35	ML 309	K 12 3300
	(moles/10^5 g protein)		(μmoles, %)
Alanine	68.2	65.0	7.96
Arginine	59.4	56.6	6.30
Aspartic acid	86.9	88.6	10.50
S-(Carboxymethyl)cysteine			1.64
Cysteine	17.0	17.1	
Cystine/2	17.0	17.1	
Glutamic acid	97.7	95.7	12.11
Glycine	59.9	60.6	7.22
Histidine	30.6	31.2	3.11
Isoleucine	35.1	36.2	4.07
Leucine	81.9	82.1	9.40
Lysine	19.5	19.0	2.48
Methionine	20.5	20.1	2.08
Phenylalanine	31.4	31.8	3.08
Proline	41.6	44.1	5.70
Serine	48.6	47.6	5.69
Threonine	44.9	45.9	5.52
Tyrosine	20.7	20.8	3.09
Valine	53.8	53.0	6.35

[a] K. Wallenfels, C. Streiffer, and C. Gölker, *Biochem. Z.*, **342**, 495 (1965).
[b] G. R. Craven, E. Steers, Jr., and C. B. Anfinsen, *J. Biol. Chem.*, **240**, 2468 (1965).

* *References start on p. 290.*

deoxy-D-glucose.[20] In the case of yeast invertase[21] and crude aryl β-D-glucosidase of *Stachybotrys atra*,[22] removal of the bound carbohydrate causes a rapid loss in enzyme activity. Thus, the carbohydrate constituent in glycosidases may contribute to the stabilization of the secondary and tertiary structure of these enzymes.

In some cases, such as *Stachybotrys atra*, the carbohydrates are attached to the enzyme surface through hydrogen bonds or hydrophobic bonds.[22] In Taka-amylase A, the carbohydrate moiety is covalently bound to the protein core through an amide linkage between 2-amino-2-deoxy-D-glucose and aspartic acid.[23] The glucoamylase of *Aspergillus niger* consists of two isoenzymes, and each appears to have a carbohydrate moiety linked glycosidically through a D-mannose residue to the hydroxyl group of a serine or threonine residue in the protein.[23a] A highly purified cellulase preparation from *Trichoderma viride* was found to contain 10 to 15% of carbohydrate, mainly D-mannose and D-glucose, with small proportions of D-galactose, D-xylose, and hexosamines.[24] The carbohydrate component seems to affect the liberation of these enzymes from the cell.[25,26]

The amino acid composition of *E. coli* β-D-galactosidases is shown in Table II. No significant differences are found between normal and constitutive strains and mutants. In general, a characteristic of the amino acid composition of glycosidases, especially exo-enzymes, is the high content of aspartic and glutamic acids, and a low proportion of sulfur-containing amino acids.[24–32]

The molecular weights of some glycosidases, determined by the sedimentation equilibrium method, are shown in Table III. The β-D-galactosidase preparations from strains of *E. coli* have molecular weights ranging from 520,000 to 540,000, and a sedimentation constant of about $16S$. On the basis of hydrodynamic data, the shape of the molecule is assumed to be an oblate ellipsoid with an axial ratio of about 3 when the hydration is 30%. Its length and width are about 150 and 50 Å, respectively.[33]

As observed with many proteins of high molecular weight, the molecule of *E. coli* β-D-galactosidase dissociates reversibly in the presence of urea or guanidine hydrochloride[33–36] from its biologically active, polymeric form into four inactive subunits of identical molecular weight (135,000). Electron microscopy indicated that the four subunits of the parent enzyme are arranged at the corners of a square.[37] Treatment with dodecyl sulfate, 70% formic acid, cyanogen bromide, exposure to pH 12, or partial tryptic digestion produced fragments having average molecular weights ranging from 50,000 to 140,000. These results indicate that *E. coli* β-D-galactosidase consists of several different polypetide chains.[38]

In contrast to the inactive subunits of the *E. coli* enzyme, the β-D-fructofuranosidase of *Neurospora crassa* was found to dissociate into enzymically

TABLE III

MOLECULAR WEIGHTS OF GLYCOSIDASES

Enzyme	Source	Molecular weight
Amyloglucosidase	*Aspergillus niger*	97,000[a]
Amylomaltase	*Escherichia coli*	124,000[b]
β-D-Galactosidase	*E. coli* ML 309	516,000[c]
β-D-Galactosidase	*E. coli* K 12	540,000[d]
α-D-Glucosidase	*Saccharomyces italicus*	85,000 + 30,000[e]
β-D-Glucosidase	Sweet almond	Two components, 117,000 and 66,500[f]
β-D-Glucosidase	*Rhodotrula minute*	300,000[g]
β-D-Glucosidase	*S. dobzhanskii* × *S. fraglis*	315,000[h]
α-D-Transglucosylase	*A. niger*	300,000[i]

[a] J. H. Pazur and K. Kleppe, *J. Biol. Chem.*, **237**, 1002 (1962). [b] H. Wiesmeyer and M. Cohn, *Biochim. Biophys. Acta*, **39**, 417 (1960). [c] H. Sund and K. Weber, *Angew. Chem.*, **78**, 217 (1966). [d] G. R. Craven, E. Steers, Jr., and C. B. Anfinsen, *J. Biol. Chem.*, **240**, 2468 (1965). [e] H. Halvorson and L. Ellias, *Biochim. Biophys. Acta*, **30**, 28 (1958). [f] B. Helferich and T. Kleinschmidt, *Hoppe-Seyler's Z. Physiol. Chem.*, **324**, 211 (1961). [g] J. D. Duerksen and H. Halvorson, *J. Biol. Chem.*, **233**, 1113 (1958). [h] A. S. L. Hu, R. Epstein, H. O. Halvorson, and R. M. Bock, *Arch. Biochem. Biophys.*, **91**, 210 (1960). [i] J. H. Pazur and T. Ando, *Arch. Biochem. Biophys.*, **93**, 43 (1961).

active subunits when the enzyme solution was adjusted to high or low pH at high salt concentrations. Dissociation was reversible, and the resulting subunits having a sedimentation constant of 5.2S retained the same order of activity[39] as the polymerized form of 10.3S. Experiments with protoplasts from conidia showed that the polymeric form of invertase predominated *in vivo*, but it was converted into a monomeric form outside the cells.[40] A similar conversion into subunits by dodecyl sulfate or by a change in pH at high salt concentration was noted for the amyloglucosidase of *Aspergillus niger*.[41]

V. INDIVIDUAL GLYCOSIDASES AND TRANSGLYCOSYLASES

A. α-D-GLUCOSIDASES (α-D-GLUCOSIDE HYDROLASES AND α-D-GLUCOSYL TRANSFERASES)

1. α-D-*Glucosidases of Plant Origin*

The α-D-glucosidases are a family of enzymes catalyzing the hydrolysis and/or transfer of the α-D-glucosyl residue of α-D-glucosidically linked derivatives. They may be classified into three groups according to their modes of action and their substrate specificities. The enzymes of the first

* *References start on p. 290.*

group are strictly hydrolytic, as exemplified by the trehalases and amylo-glucosidases. The second group is composed of α-D-glucosidases, as represented by maltase. They catalyze hydrolysis but are also able to transfer the α-D-glucosyl residue of maltose and α-D-glucosides to suitable acceptors. The third group includes the α-D-glucosyl transferases, which are primarily responsible for the α-D-glucosyl or oligosaccharidyl transfer reactions from certain substrates to produce oligo- or polysaccharides; their acceptor specificities are limited. Enzymes of this type include amylomaltase, dextransucrase, and D-enzyme.

Specificities of the D-glucosidases for the aglycon group have been reviewed by Pigman[16] and by Gottschalk.[42] Helferich and Johannis[43] showed that the relative rates of hydrolysis of aryl α-D-glucosides by yeast α-D-glucosidase parallel the acidity of the substituted phenolic aglycons. With alkyl α-D-glucosides, the rate of hydrolysis depends upon the reactivity of the hydrogen atom of the hydroxyl group of the alcohol. Hydrolysis of phenyl α-D-glucosides by the yeast enzyme is facilitated by electron-attracting substituents.[44] A similar effect of substitution in the phenolic aglycon was reported for Taka-amylase A, which splits phenyl α-maltoside in addition to its typical α-amylolytic activity toward starch.[45,46]

Highly purified preparations of α-D-glucosidase from the *Saccharomyces* genus were found to hydrolyze a number of disaccharides having α-D-glucosidic linkages, as well as various synthetic α-D-glucosides.[47–49] Hutson and Manners[50] found that alfalfa seeds and tomato fruits contained an α-D-glucosidase that acted on maltose, isomaltose, and nigerose. Several components were present in the alfalfa-seed enzyme, as shown by electrophoresis. An α-D-glucosidase from barley malt was active on isomaltose, maltose, and panose, to a lesser extent on their reduced derivatives, and on isomaltotriose.[51] The enzyme also catalyzed the hydrolysis of starch and dextrin to D-glucose by endwise attack.[52] Cell extracts of a rumen strain of *Lactobacillus bifidus*, grown on dextran, contain an intracellular α-D-(1 → 6)-glucosidase specific for the hydrolysis of α-D-(1 → 6) linkages in isomaltodextrins and panose; the rate of hydrolysis decreased with increasing degree of polymerization. Nigerose and a variety of β-D-linked sugars were also substrates for the enzyme.[53]

Transglucosylations catalyzed by α-D-glucosidases depend upon the donor and acceptor specificities of individual enzymes. Mold α-D-glucosidases, in general, transfer the D-glucosyl residue predominantly to the primary alcohol groups of the acceptor saccharides, leading to α-D-(1 → 6)-glucosidic linkages. In studies with *Aspergillus niger* (strain NRRL 337), isomaltose, panose, and O-α-D-glucopyranosyl-(1 → 6)-O-α-D-glucopyranosyl-(1 → 4)-D-glucose were isolated and characterized as the main transfer products from maltose digests.[54,55] Further, under appropriate conditions, successive transfers resulted in the formation of 6-O-isomaltosyl-D-glucose (dextrantriose) and 4-O-α-isomaltotriosyl-D-glucose in addition to isomaltose and panose.[56,57] Reversal of the transfer reaction has been demonstrated by the synthesis of labeled isomaltotriose by the transferring enzyme of *A. oryzae*.[58]

The identity of the enzyme catalyzing both hydrolysis and transfer reactions has been shown by Pazur and Ando, who isolated an α-D-glucosyl transferase in a homogeneous state from cultures of *A. niger*.[59] The specificity of this enzyme was surprisingly broad; maltose, isomaltose, maltotriose, methyl α-D-glucopyranoside, methyl β-maltoside, and starch could serve as either donors or acceptors.

A potato enzyme, named T-enzyme by Abdullah and Whelan,[60] catalyzes the synthesis of isomaltose from maltose and the redistribution of α-D-(1 → 6)-glucosidic linkages between isomaltose and higher isomaltodextrins.

Nigerose (3-O-α-D-glucopyranosyl-α-D-glucose) was synthesized from maltose in the presence of an excess of D-glucose by an enzyme from *A. oryzae*. On further incubation, a trisaccharide was formed, which was identified as O-α-D-glucopyranosyl-(1 → 6)-O-α-D-glucopyranosyl-(1 → 3)-D-glucose.[61] A similar α-D-glucosyl transfer, to C-2 of D-glucuronic acid, afforded 2-O-α-D-glucopyranosyl-D-glucuronic acid, and a further α-D-glucosyl transfer yielded O-α-D-glucopyranosyl-(1 → 6)-O-α-D-glucopyranosyl-(1 → 2)-D-glucuronic acid.[62] With D-xylose as acceptor, 3-O-α-D-glucopyranosyl-D-xylose was synthesized by a transglycosylase from *Penicillium lilacinum*.[63]

The maltases of *Tetrahymena pyriformis* and *Cladophora rupestris* and of some rumen protozoa are able to transfer the α-D-glucosyl group of maltose to positions C-4 and C-6 of other sugars.[64-66] A transglucosylase from gram seeds (*Phaseolus radiatus*) catalyzing α-D-glucosyl transfer to C-4 in the nonreducing D-glucose residue of maltose, to yield maltotriose and maltotetraose, has been partially purified by Nigam and Giri.[67] Edelman and Keys,[68] working with wheat-germ extracts, showed that an enzymic reaction occurs between maltose and labeled D-glucose, resulting in the synthesis of labeled maltose.

A further example of α-D-transglucosylation is the enzymic synthesis of O-α-D-glucosylriboflavin and its isomalto-oligosaccharides from maltose and riboflavin. Enzymes utilizing riboflavin and certain other isoalloxazine derivatives have been found in rat liver, *E. coli*, and *A. oryzae*. Maltose, maltulose, or turanose may function as the donor substrate.[69,70]

Like other glycosidases, the α-D-glucosidases from brewers' yeast and barley bran can catalyze the transfer of the α-D-glucosyl group of aryl α-D-glucosides to various aliphatic alcohols to form new α-D-glucosides.[71]

* References start on p. 290.

2. Mammalian α-D-Glucosidases

Interest in the metabolism of disaccharides and glycogen has resulted in many studies of mammalian α-D-glucosidases and maltase. Lieberman and Eto[72] obtained a crude maltase from equine serum that readily hydrolyzes maltose and methyl α-maltoside but which shows only weak activity for the hydrolysis of phenyl α-D-glucoside, turanose, isomaltose, starch, and glycogen. A bovine blood α-D-transglucosylase, an α_1-globulin, catalyzes the transfer of α-D-glucosyl groups from maltose to give maltodextrins, in addition to its hydrolytic activity for amylose.[73]

Borgstrøm and Dahlqvist[74] developed several methods by using deoxycholate, trypsin, and ethanol for the solubilization of several specific α-D-glucosidases associated with a microsomal fraction from homogenates of hog intestinal mucosa. Chromatographic techniques resolved the solubilized fraction into three components, designated maltases I, II, and III. These components differed in their specificities, K_m values, and thermal stabilities.[75] A trehalase was also separated.[76] Maltase I was heat-labile and had a maltase/invertase activity quotient of 0.6. Maltase II was also thermally labile but had no invertase activity. Maltase III was a heat-stable component and had no invertase activity. The invertase activity of maltase I was illustrated by its activity on sucrose, its very weak activity on melezitose, and its ability to transfer α-D-glucosyl groups from sucrose to form α-maltosylsucrose [O-α-D-glucopyranosyl-(1 → 4)-α-D-glucopyranosyl β-D-fructofuranoside] as the main product. Inhibition of maltase III by various α-D-glucosidic substrates in the presence of maltose was competitive and indicated the existence of a single enzyme.[77]

Kinetic studies of intestinal glycosidases revealed that the marked inhibition of hydrolysis at high substrate concentrations was largely due to transglycosylation reactions. Inhibition was strong in the case of maltases and other α-D-glycosidases, but none was observed for trehalase, presumably because of its lack of transglucosylase activity.[77]

Homogenates of hog intestinal mucosa hydrolyze palatinose (6-O-α-D-glucopyranosyl-D-fructofuranose). The palatinase (isomaltulase) activity was attributed to a mixture of three separate enzymes—a specific isomaltase, maltase II, and maltase III; the latter was responsible for half of the total palatinase activity.[78]

Particle-bound glycosidases of rabbit intestinal mucosa have been solubilized and fractionated by methods similar to those used for the preparation of the hog enzymes. Rabbit intestinal invertase was identified as an α-D-glucosidase showing both hydrolytic and transferring activities for sucrose.[79] With maltose as donor, maltotriose and a tetrasaccharide were formed. In contrast to the hog enzyme, the rabbit maltase was capable of resynthesizing sucrose by an α-D-glucosyl transfer reaction.[80]

Comparison between the enzyme activities in homogenates of the intestinal tract of conventional rats and germ-free rats, as measured with 6-bromo-2-naphthyl α-D-glucopyranoside, maltose, and sucrose, revealed that these enzymes occur mainly in the small intestine.[81] Some activity was also found in the pancreas. Hydrolysis of the 6-bromo-2-naphthyl α-D-glucopyranoside was brought about, apparently, by two maltase components present in the homogenates of the small intestine.[82]

Human intestinal disaccharidases are different from those of hog or rabbit intestine, since they could be solubilized only when treated with papain.[83] Maltase, isomaltase, and invertase were completely recovered in a soluble form after treatment with papain, but trehalase remained insoluble. Five maltases were separated from the soluble fraction. They differed in heat stabilities, substrate specificities, optimum pH's, and modes of activation by sodium ions. Of these enzymes, two maltases (I and II) hydrolyzed only maltose, but two others (III and IV) attacked sucrose as well as maltose. The fifth enzyme, maltase V, split isomaltose and isomaltulose as well as maltose. The congenital absence of maltases IV, V, and probably III in the intestinal mucosa has been demonstrated in human subjects showing sucrose malabsorption.[84]

Glycosidases, or more specifically the disaccharidases, are receiving much clinical attention, especially in pediatrics.[85] Their malfunction may be fatal if it is not diagnosed in time. Such disaccharidases are particulate or insoluble enzymes located in the brush border of the intestinal mucosa and may not be secreted into the small intestine. These disaccharidases are specific for cellobiose, isomaltose, lactose, maltose, palatinose, sucrose, and trehalose.[85a] When enzyme activity is absent, primary deficiences are found for isomaltose, lactose, palatinose, and sucrose, and in the disturbance of active transport of D-galactose and D-glucose through the intestinal wall. Secondary deficiencies due to diseases, drugs, and malnutrition have also been detected. Most non-Caucasian adults lack intestinal lactase, and therefore a majority of the adult population of the world probably cannot properly digest lactose or milk.[85b]

Two α-D-glucosidases differing in optimal pH, cellular location, and specificity were found in rat liver homogenates: The acidic, lysosomal α-D-glucosidase hydrolyzed diverse kinds of α-D-glucosidic substrates including glycogen, whereas the neutral, cytoplasmic enzyme was specific for maltose and maltodextrins.[86,87] Both enzymes exhibited transferring activity, but the neutral enzyme was more effective than the acidic one in producing maltodextrins from maltose.

Human heart, liver, and skeletal muscle contain an acid-active α-D-glucosidase catalyzing the hydrolysis of maltose and glycogen, and the transfer of α-D-glucosyl residues from the disaccharide to glycogen.[88] Lack of this

* References start on p. 290.

enzyme in the tissues of children causes a glycogen-storage disease (Pomp's disease). Shibko and Tappel[89] separated two α-D-glucosidases from rat kidney. One, a microsomal enzyme active at pH 6, represented most of the activity. The other, a lysosomal enzyme having an optimum pH at 5, accounted for only 10% of the total activity.

3. α-D-*Glucosyl Transfer from Sucrose*

Honey invertase was shown to be an α-D-glucosidase that produces six oligosaccharides from sucrose.[90] Its main transfer products were identified as α-maltosyl-β-D-fructose and maltose. This enzyme also produced maltodextrins from maltose. A fructosylmaltose was also obtained after treatment of sucrose with enzymes from the nectar of *Robinia pseudocacia* and other flowers,[91] and from the honeydew of certain insects fed on sucrose.[92,93] Melezitose is formed by an α-D-glucosyl transfer reaction from one sucrose molecule to C-3 of the D-glucosyl residue of another.[94]

Avigad[95] found that the α-D-glucosyl transfer from sucrose by yeast α-D-glucosidase resulted in the synthesis of isomaltulose (6-*O*-α-D-glucopyranosyl-D-fructofuranose) and 1-*O*-α-D-glucopyranosyl-D-fructose in addition to isomaltose (6-*O*-α-D-glucopyranosyl-D-glucose), maltulose (4-*O*-α-D-glucopyranosyl-D-fructose), isomaltotriose, and isomaltotriulose (*O*-α-D-glucopyranosyl-(1 → 6)-*O*-α-D-glucopyranosyl-(1 → 6)-D-fructose). Weidenhagen and Lorenz[96] reported nearly 90% conversion of sucrose into isomaltulose by an α-D-glucosyl transfer catalyzed by an enzyme from *Enterobacterium*.

Dextransucrases are extracellular enzymes produced by *Leuconostoc mesenteroides*,[97] *Betacoccus arabinosaceus*,[98] and *Streptococcus bovis*;[99,100] they catalyze the synthesis of dextran by consecutive transfers of α-D-glucosyl groups from sucrose:

$$n(\text{Sucrose}) + \text{acceptor} \longrightarrow (\text{D-Glucose})_n + n(\text{D-fructose})$$

In this reaction, the usual acceptor is sucrose; isomaltose or panose may also function as donor. The addition of mono- and disaccharides to the reaction system decreases the yield of dextran, but increases the reaction rate, probably by competition by various acceptors.

An analogous enzyme, amylosucrase, has been obtained from *Neisseria perflava*. This enzyme differs from dextransucrase in that it transfers the D-glucosyl group to C-4 of the acceptor, to give a starch-like polysaccharide.[101]

Various oligosaccharides have been isolated as by-products during the synthesis of dextrans through transglycosylation reactions of dextransucrase from *L. mesenteroides*.[102] The amount of such oligosaccharides depends not only on the nature of the acceptor molecules, but also on their relative concentrations. D-Fructose is a poor acceptor, but, when present in large concentration, it is converted in relatively large yield into leucrose (5-*O*-α-D-glucopyranosyl-D-fructopyranose) and isomaltulose. The reaction is of interest, since D-fructose functions as an acceptor in both furanose and pyranose forms, the latter being more effective. Further consecutive transfers led to the formation of 5-O-α-isomaltosylfructose and isomaltotriulose.[103] Incubation of 3-O-methyl-D-glucose or methyl α-D-glucopyranoside as acceptor and sucrose as donor with an enzyme of *B. arabinosaceus* yielded a series of α-D-(1 → 6)-linked D-glucosyloligosaccharides by successive transfer of D-glucosyl residues from sucrose, with the acceptors as end groups.[104]

Two novel branched trisaccharides containing lactose or cellobiose substituted with an α-D-glucosyl group at C-2 of the reducing D-glucose residue have been isolated from the mixtures of an enzyme from *B. arabinosaceus* acting on sucrose and these disaccharides as substrates.[105] In a similar transfer reaction, a disaccharide was obtained that was characterized as α-D-glucopyranosyl D-galactofuranoside.[106] It was further shown that an *A. niger* enzyme and dextran sucrases from *L. mesenteroides* produce a trisaccharide, O-α-D-glucopyranosyl-(1 → 6)-O-α-D-glucopyranosyl β-D-fructofuranoside,[107] and a branched tetrasaccharide, O-α-D-galactopyranosyl-(1 → 6)-[O-α-D-glucopyranosyl-(1 → 2)-]-α-D-glucopyranosyl β-D-fructofuranoside.[108] Sucrose was the donor in both instances; maltose was the acceptor in the former case and raffinose in the latter.

Suzuki *et al.*[109,110] reported that various strains of dextran-producing *L. mesenteroides* yielded large amounts of O-α-D-glucosylriboflavin. Both extracellular and intracellular enzymes, probably dextransucrases, were shown to catalyze α-D-glucosyl transfer from sucrose to riboflavin.

4. *Amyloglucosidase* (see also this volume, Chap. 34)

The amyloglucosidase from *Aspergillus niger* is an α-D-glucosidase that hydrolyzes α-D-glucosidic linkages [α-D-(1 → 6); α-D-(1 → 4); α-D-(1 → 3)] in a variety of D-glucosyloligosaccharides. The enzyme hydrolyzes starch, amylose, amylopectin, and dextrins to D-glucose from the nonreducing end of the molecules. The α-D-(1 → 4)-glucosidic linkage was hydrolyzed approximately 15 to 30 times as fast as the α-D-(1 → 3) and α-D-(1 → 6) linkages.[111,112] Phenyl α-D-glucopyranoside and methyl α-D-glucopyranoside were split at much lower rates. The relative hydrolytic rates of maltodextrins increased with increasing chain lengths, whereas a reverse relation was shown for the isomaltodextrin series.[111] The highly purified enzyme has a molecular weight of 97,000, and it is a glycoprotein containing D-mannose, D-galactose, and D-glucose,[41] with covalent linkages between the D-mannosyl residues and hydroxyl groups in the protein.[23a] Purification of glucoamylase from *A. niger* was achieved by density-gradient centrifugation.[113]

5. Trehalases

Trehalase is an α-D-glucosidase that specifically hydrolyzes α,α-trehalose. It is found widely in bacteria, fungi, higher plants, insects, and vertebrates.[114-118] In organisms that store trehalose as a reserve carbohydrate, trehalases have been assumed to play a physiological role in the breakdown of the disaccharide to D-glucose.[117]

Highly purified trehalases have been prepared from various organisms such as *Neurospora crassa*,[117] silkworm,[119] insects,[120-122] yeast,[123,124] slime mold,[125] and hog intestinal mucosa. About 85% of the total trehalase activity in hog intestine was localized in the microsomal fraction, and 6% in the mitochondrial fraction.[76] Hansen,[126] studying the intracellular localization of trehalase in the flight muscle of the blowfly, found that the enzyme was distributed between cytoplasm and mitochondria, but no significant differences were observed in the enzymes from either source.

The appreciable trehalase activity present in the plasma of human adults has been shown to be associated with the α$_2$-globulin fraction. In a few cases, pathological variations in the concentration of serum trehalase have been reported.[116]

Courtois and Demelier,[127] who studied the occurrence of trehalase in various organs of man and other mammals, detected its presence in kidney, intestine, liver, bile, and urine. In man, trehalase activity was highest in the kidney. Trehalase is present in the kidneys of various mammals except in those of rats.

The only known substrate of the enzyme is α,α-trehalose, which indicates that the enzyme is different from ordinary α-D-glucosidases, which are unable to split α,α-trehalose. α,α-Trehalose 6-phosphate and trehalosamine act as inhibitors.[128] Purified trehalases have no transglucosylation activity. It is assumed that the failure of these enzymes to effect glycosyl transfer is because of the symmetrical structure of the trehalose molecule.

B. β-D-GLUCOSIDASES (β-D-GLUCOSIDE HYDROLASES AND β-D-GLUCOSYL TRANSFERASES)

β-D-Glucosidases are widely distributed, particularly in plant seeds, molds, and bacteria. The classical source of β-D-glucosidase is sweet-almond emulsin, which has been extensively used for the study of specificity and mechanism of transfer reactions.

1. Almond and Apricot β-D-Glucosidases

Almond emulsin is prepared from defatted, crushed almonds by extraction with water and precipitation of the extract with alcohol.[129] The dried powder is known as almond emulsin. A preparation of considerably higher β-D-

TABLE IV

COMPARISON OF THE AGLYCON SPECIFICITIES OF APRICOT β-D-GLUCOSIDASE
AND β-D-GALACTOSIDASE[a]

	Relative activities for 0.012 M solutions of:	
Aglycon	β-D-Glucosides[b]	β-D-Galactosides[c]
o-Aminomethylphenol (o-aminocresol)	0.1	
Butyl alcohol	1.7	0.6
m-Cresol	2.5	3.6
o-Cresol	24.2	16
p-Cresol	0.5	0.6
Ethanol	0.2	
Guaiacol (o-methoxyphenol)	16.9	2.2
Isoeugenol (4-hydroxy-3-methoxy-1-propenylbenzene)	4.9	0.4
Methanol	0.14	0.06
Methyl salicylate	4.4	0.4
β-Naphthol	1.2	
p-Nitrophenol	22.1[d]	25
Phenol	1	1
Propyl alcohol	1	0.7
Salicyl alcohol (o-hydroxybenzyl alcohol)	12.5	
Salicylaldehyde (o-hydroxybenzaldehyde)	51	8.1
Vanillin (4-hydroxy-3-methoxybenzaldehyde)	66	46
m-Xylenol (2,4-dimethylphenol)	9.9	6.7

[a] All figures in the table are average values relative to the phenyl D-glycosides as unity, calculated from the first-order reaction constant under standard conditions. [b] T. Miwa and A. Miwa, *Medicine* (Tokyo), **1**, 229 (1942). [c] K. Nisizawa, *Fac. Textiles Sericult., Shinshu Univ., Ser.* **C1**, 1 (1951). [d] Calculated from another series of experiments.

glucosidase activity is obtained by extraction of the almond meal with zinc sulfate followed by precipitation with tannin. Removal of tannin with acetone leaves a powder called "Rohferment" or crude enzyme. The "Rohferment" has a β-D-glucosidase value of 1; it is a mixture of glycosidases, in which β-D-glucosidase and β-D-galactosidase predominate. Helferich and Klein-schmidt,[130] using (diethylamino)ethyl-Sephadex, obtained a highly purified preparation of almond β-D-glucosidase having a β-D-glucosidase value of 33. Ultracentrifugation yielded two active components, having molecular weights of 117,000 and 66,500. Chromatography on (diethylamino)ethyl- and carboxymethyl-Sephadex resulted in the separation of two components, A and B, having β-D-glucosidase values of 35 to 40. The authors[131] succeeded in

* *References start on p. 290.*

crystallizing component B from ammonium sulfate solution; the β-D-glucosidase value of the crystallized product was 46.

Conflicting views have been presented on the identity of the β-D-glucosidase and β-D-galactosidase in almond emulsin.[132–138] Early attempts by Helferich and associates to separate the two enzymes remained unsuccessful despite the application of a number of fractionation methods. In addition, a similar aglycon specificity observed with apricot glycosidases (Table IV) suggested that a single enzyme is responsible for both activities.[132,133] However, contradictory results were obtained from inhibition and kinetic studies. Veibel *et al.*[134] concluded that the two β-D-glycosidases must be different, in view of the failure of D-galactose to inhibit β-D-glucosidase, and the greater inhibitory effect of D-glucose in the hydrolysis of *o*-cresyl β-D-glucopyranoside than for that of β-D-galactopyranoside. Changes in the ratio of velocity

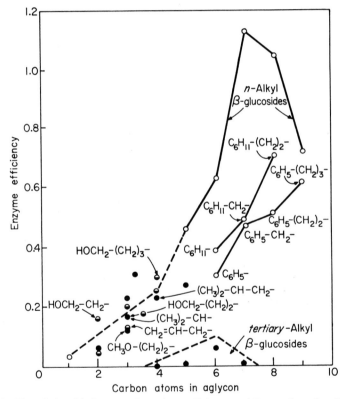

FIG. 3. The relationship between the enzyme efficiency and the number of carbon atoms in the aglycon groups of alkyl β-D-glucosides. Unfilled circles represent data of Pigman and Richtmyer,[141] filled circles are data of Veibel, and half-filled circles are data of Helferich.[140]

constants for both β-D-glycosides resulted when comparisons were made at different pH values. In contrast, Heyworth and Walker[135] insisted on a single enzyme theory, based on the constancy of the rate of hydrolysis of an equimolar mixture of the two hexosides and the competitive inhibition by D-glucose and D-galactose. On the basis of difference in behavior of the two β-hexosidases during adsorption on polystyrene β-D-glucoside,[136] and differences of these enzyme activities during fractionation by ion exchangers, Helferich[130] has concluded that two enzymes are involved in the hydrolysis of the two hexosides. This concept is also supported by the electrophoretic patterns of apricot β-D-galactosidase in $8M$ urea.[137] It is likely, as suggested by Miwa et al.,[138] that the two enzymes possess very similar properties and normally exist as a firm aggregate.

Figure 3 shows the relationship between susceptibility to almond β-D-glucosidase and increase in the size of the aglycon groups of alkyl β-D-glucopyranosides. As may be seen, the rate of hydrolysis increases with initial increase in chain length. The n-alkyl series shows a progressive increase in hydrolysis rate, with increasing chain length up to about seven carbon atoms, and thereafter a rapid decrease. On the basis of the postulated mechanism

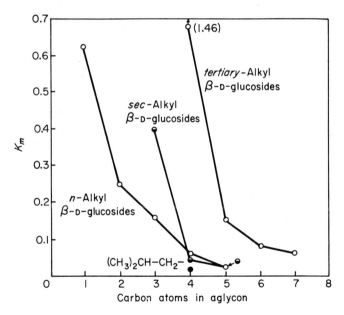

Fig. 4. Relationship between the dissociation constant of the enzyme–substrate complex and the number of carbon atoms in the aglucon groups of alkyl β-D-glucosides.[139]

previously given (Fig. 1), the existence of an optimal chain length for maximal hydrolysis may be ascribed to a counterbalancing of the beneficial effects of an increase in chain length on the formation of the enzyme–substrate complex by a disadvantageous influence of the slow dissociation of the hydrolysis products from the surface of the enzyme. The rate-determining reaction for the nonyl glucoside may be the dissociation of the nonyl alcohol from the surface of the enzyme rather than by cleavage of the D-glucosidic linkage. This is supported by the values of the Michaelis constants for the enzyme–alkyl glucoside complexes as shown in Fig. 4.

Substitution of the hydroxyl groups at C-2, C-3, and C-4 with methoxyl or p-tolylsulfonyloxy groups renders β-D-glucopyranosides unhydrolyzable by almond emulsin.[140] Furthermore, the rates of hydrolysis of β-D-glucosides substituted at C-6 by F, Cl, Br, OMe, or I decrease in this order.[141] The ease of hydrolysis seems to be related to the volume of the substituted groups. It is of interest that the 6-deoxy-β-D-glucopyranosides are hydrolyzed more readily than the unsubstituted β-D-glucopyranosides.[141]

Although aryl 1-thio-β-D-glucopyranosides are usually not hydrolyzed by almond β-D-glucosidase, the introduction of strong electron-attracting groups into the aglycons or the presence of electrophilic, heterocyclic aglycons renders 1-thio- or 1-seleno-β-D-glucopyranosides hydrolyzable by almond emulsin.[142] The degree of hydrolysis increases in the order: β-D-glucopyranosides > 1-thio-β-D-glucopyranosides > 1-seleno-β-D-glucopyranosides.

Alkyl β-D-glucopyranosides were shown to be cleaved by almond emulsin, but furanosides are not.[143] However, Yoshida et al.[144] showed that several aryl β-D-glucofuranosides are hydrolyzed by almond β-D-glucosidase at a rate comparable to that of pyranosides, and that the K_m values are also of the same order for both forms. Involvement of the same enzyme seems probable because of the strong inhibition of the hydrolysis by both D-glucono-1,4- and -1,5-lactones.

2. β-D-Glucosidases from Various Sources

Miwa et al.[145] observed that o-cresyl β-D-glucopyranoside was split much more readily than phenyl β-D-glucopyranoside by enzymes from Rosaceae seeds, whereas fungal β-D-glucosidases were less effective for the o-substituted derivative. Enzymes from other plant sources appear to be much less affected by structural changes in the aglycon groups.

Various strains of yeast have the ability to grow on phenolic β-D-glucopyranosides as well as on cellobiose. Barnett et al.[146] have attempted to classify yeasts according to their differing substrate specificity toward β-D-glucosides. These differences may indicate the existence of different β-D-glucosidases in various types of yeast.

Halvorson and co-workers,[147,148] using strains of *Saccharomyces*, studied induction mechanisms, metabolic controls, and genetic regulation in the synthesis of β-D-glucosidases. From such sources, these authors purified inducible and constitutive β-D-glucosidases, which showed low specificity for the hydrolysis of β-D-glucosides having diverse aglycons. In contrast, for almond emulsin the affinities and rates of hydrolysis increased with increasing chain length of aliphatic aglycons; dodecyl β-D-glucopyranoside was the most efficient substrate.[16] A correct configuration at C-3 and C-4 in the pyranose ring is an absolute requirement for activity of the yeast enzymes, but replacement of the hydroxymethyl group at C-6 by a carboxyl group or hydrogen atom leads only to small changes in dissociation constants.[149] The inducible and constitutive β-D-glucosidases have similar physical and chemical properties, as reflected in the activation energy, optimal pH value, molecular weight, and maximal hydrolytic rate. A small part of the constitutive β-D-glucosidase appears to be bound *in vivo* to ribosomes.[147] A mutant strain of Bakers' yeast, grown on cellobiose and phenyl β-D-glucopyranoside, produced two β-D-glucosidases: an aryl β-D-glucosidase induced by cellobiose, and an unstable cellobiase associated with the cell structure.[150]

An enzyme system comprising two β-D-glucosidases having different specificities and thermal stabilities was found in a wild strain of *Neurospora crassa*. Eberhart *et al.*[151,152] separated several mutants possessing a regulatory gene, *gluc⁻*, that controls production of an aryl β-D-glucosidase that has a lower enzyme activity than wild strains containing the gene *gluc⁺*, responsible for normal enzyme formation. The dominance of the *gluc⁻* gene over the wild type, *gluc⁺*, was demonstrated by the observation that the heterocaryons containing both *gluc⁺* and *gluc⁻* genes in nuclei exhibit low β-D-glucosidase activity characteristic of homocaryon strains having the *gluc⁻* gene.[151] The properties of the β-D-glucosidase from the *gluc⁻* strain resembled those of the enzyme from the wild type except for a significant difference in the K_m values for some β-D-glucopyranosides.

β-D-Glucosidases of cellulolytic microorganisms are, in general, non-specific for β-D-glucosidic substrates. An extracellular β-D-glucosidase of *Stachybotrys atra*, however, is more specific for aryl β-D-glucopyranosides than for cellobiose, and configurational change or substitution in the β-D-glycosyl residue results in a loss of reactivity and leads to compounds that act as competitive inhibitors.[153] For instance, aryl 6-*O*-methyl, 3-*O*-methyl, and 6-*O*-*p*-tolylsulfonyl β-D-glucopyranosides, and *p*-nitrophenyl 6-deoxy-β-D-glucopyranoside,[154] are all suitable substrates, whereas aryl β-D-xylopyranosides are not.

The aryl β-D-glucosidase of *S. atra* is unique in that it can hydrolyze aryl

* *References start on p. 290.*

1-thio-β-D-glucopyranosides. An intracellular β-D-glucosidase isolated from mycelia of *S. atra* differs from aryl β-D-glucosidases in that it catalyzes the hydrolysis of cello-oligosaccharides up to at least eleven D-glucose residues.[155] The production of these enzymes depends upon the type of inducer and the nature of the amino acids in the culture medium.[156]

Enzyme systems responsible for the hydrolysis of β-D-(1 → 3)- or β-D-(1 → 4)-linked D-glucans in *Aspergillus niger*[157] and in extracts of ungerminated barley[158] are mixtures of several β-D-glucosidases, including cellobiase, salicinase, and laminarabiase, which can be separated by fractionation or by using their differential susceptibility toward inhibitors.

Animal β-D-glucosidases are widespread but occur in low concentration. Their presence has been established by fluorescence methods with 4-methylumbelliferone β-D-glucopyranoside as substrate. Appreciable activity was found in the liver, kidney, and duodenum of several animals, the crop fluid of locusts, gut of cockroach, and shellfish. In all cases,[159,160] the optimal pH was between 4.5 and 5.7. A β-D-glucosidase of pig intestinal mucosa, localized in the microsomal fraction, can be solubilized by tryptic digestion.[161] This enzyme appears to be devoid of configurational specificity for C-4, and splits lactose as well as cellobiose and gentiobiose. The relative β-D-glucosidase and β-D-galactosidase activities are not changed by partial heat inactivation or by chromatographic fractionation, which seems to indicate that only one enzyme is involved in the hydrolysis of these β-D-glycopyranosides. Furthermore, a similarity in the activity quotients of cellobiose/lactose and gentiobiose/lactose for intestinal enzyme preparations from adult and newborn pigs supports the concept of a single enzyme. The two enzyme activities were also inhibited similarly by D-galactose, D-glucose, their aldonolactones, aryl β-D-glycopyranosides, and lactose.[160] However, a β-D-glucosidase from beef brain that is involved in the breakdown of glycosphingolipids is different from the coexistent β-D-galactosidase.[161]

3. Transferring Activity of β-D-Glucosidases

Since Rabaté[162] reported the formation of *p*-hydroxybenzyl β-D-glucoside from arbutin, and methyl β-D-glucopyranoside from phenyl β-D-glucopyranoside or picein, by enzyme preparations from *Salix purpurea* and *Gaulteria procumbens*, intensive investigations have been initiated to study the transglycosylation reactions induced by β-D-glucosidases of plant and animal origin. It is believed that this "Umglukosidierung" reaction is catalyzed by an enzyme having both hydrolase and glucosyl transferase activities.[163,164] The relationship between the two enzyme activities can be expressed by the formula $(T - G/T) \times 100$, where T is the amount of the aglycon liberated from the chromogenic substrate such as *p*-nitrophenyl β-D-glucoside, and G is the amount of D-glucose liberated in the presence of a suitable acceptor. With an aryl β-D-glucopyranoside as donor and aliphatic alcohols as acceptor, the transfer reaction proceeds as follows:

Aryl β-D-glucopyranoside + aliphatic alcohol
⟶ Aliphatic β-D-glucopyranoside + phenol

Almond β-D-glucosidase and other plant enzymes have relatively broad acceptor specificities. Among various alcohols tested, primary and secondary alcohols such as cyclohexanol, ethylene glycol, furfuryl alcohol, glycerol, and methanol, are in general efficient acceptors, whereas tertiary alcohols and mono- and disaccharides are poor or inert acceptors. Upon incubation of phenyl β-D-glucopyranoside and sucrose with the almond enzyme, the synthesis of gentianose was observed.[164]

In comparative studies of various plant β-D-glucopyranosides with regard to acceptor specificity, it was found that the enzymes of rice plants, cycad, *Datura tatula*, and plantain are highly active in transferring the β-D-glucosyl group from a β-D-glucopyranoside to D-glucose and D-xylose, in addition to alcohols.

The β-D-glucosidase of rice plants is effective in catalyzing the redistribution of β-D-glucosyl groups of a β-D-glucoside with formation of several aryl β-oligosides, even in the presence of alcohols and phenols. Of the transfer products, two compounds were tentatively identified as *p*-nitrophenyl β-cellobioside and β-gentiobioside. Similarly, arbutin was found to function as both donor and acceptor for almond emulsin and for broad-bean β-D-glucosidase, forming *p*-hydroxyphenyl β-gentiobioside.[165,166]

Cellobiases are involved not only in the degradation of cellulose (along with cellulases) but also in β-D-glucosyl transfer reactions leading to the synthesis of β-D-linked oligosaccharides. Thus, the action of cell-free extracts or cultures of *Aspergillus niger* on cellobiose[167] brings about transfer of β-D-glucosyl groups to C-2, C-3, C-4, and C-6 of the nonreducing D-glucose residue of cellobiose, forming cellotriose, gentiotriose, and other trisaccharides. β-D-Glucosyl transfers to C-2, C-3, and C-6 of D-glucose also occur, giving rise to sophorose, laminarabiose, and gentiobiose. The ease of formation of β-D linkages is: $(1 \rightarrow 6) > (1 \rightarrow 4) > (1 \rightarrow 3) > (1 \rightarrow 2)$. Other monosaccharides, such as D-xylose, L-sorbose, and 2-acetamido-2-deoxy-D-glucose, can serve as acceptors. Such transfer reactions are also catalyzed by enzymes from *Myrothecium verrucaria*, *Aspergillus aureus*, *Aspergillus flavus*, *Penicillium chrysogenum*, snail gut, alfalfa seeds, and some marine algae.[168–170] β-D-Glucosidases from *Aspergillus niger* and from *Cladophora rupestris* were shown to synthesize 3-*O*-β-D-glucosyl-D-xylose[167] and 4-*O*-β-D-glucosyl-D-galactose[171] by transfer of a β-D-glucosyl group to C-3 and C-4 of the acceptor molecules. An intermediate type of transfer was observed with an aryl β-D-glucosidase of *Neurospora crassa* that transfers a D-glucosyl residue from cellobiose to D-glucose and to C-4 or C-6 of the nonreducing D-glucose residue of cellobiose.[172] Extracts of *Chaetomium globosum* contain a transglucosylase that is comparable to amylomaltase of *E. coli* in forming cellotriose, and possibly cellotetraose:

$$\text{2 Cellobiose} \longrightarrow \text{Cellotriose} + \text{D-glucose}$$

Successive β-D-glucosyl transfer or disproportionation of cellodextrins, with formation of higher cello-oligosaccharides, is brought about by the action of cellobiases of some cellulolytic microorganisms.[173-175]

The cells or cell-free extracts of *Sporobolomyces singularis*[176] contain a β-D-transglucosylase capable of transferring the β-D-glucosyl and β-D-galactosyl residues of cellobiose and lactose, respectively, to various sugars. This transfer occurs at a secondary position rather than at the primary hydroxyl group in the acceptor molecule. Transfer of β-D-glucosyl and β-D-galactosyl residues to *myo*-inositol occurs at the 1 and 5 positions of *myo*-inositol, respectively. The minimal structural requirement for the transfer to an acceptor appears to be a free hydroxyl group vicinal to the substituted hydroxyl group.

C. α-D-Galactosidases (α-D-Galactoside Hydrolases and α-D-Galactosyl Transferases)

α-D-Galactosidases are distributed in both animals and plants. Higher plants are able to utilize as substrates naturally occurring oligosaccharides, their degradation products, and some galactolipids containing α-D-galactosyl residues.

Plant α-D-galactosidases that have been studied in detail are those from coffee,[177] alfalfa, plantain,[178] bran,[179] *Vicia faba*,[180] and watermelon.[181] Some strains of yeast,[182] molds, and some Basidiomycetes[183] capable of fermenting melibiose or raffinose are also good enzyme sources. Two distinct α-D-galacto-sidase components were shown to occur in the extracts from coffee beans, *Plantago* seeds, and wheat bran; they differ in K_m values, optimum pH, and transferring activities.[177-179] One of the *Plantago* enzymes forms raffinose and the other gives planteose as transfer products when incubated with phenyl α-D-galactopyranoside and sucrose.

The enzymes of bottom yeast[182] contain several components that have similar transferring activities, but differ from each other in the rates at which they cause hydrolysis. Li and Shetlar[183,184] obtained a partially purified galactosidase from a culture filtrate of *Diplococcus pneumoniae*, and it was shown to be an SH enzyme. Induction of *Aerobacter aerogenes*[185] and *E. coli B*[186] with either α- or β-D-galactopyranosides produced intracellular α-D-galactosidases. These enzymes appear to be firmly associated with the cell structures. Manganous ions and reducing agents are effective in stabilizing the α-D-galactosidase of *E. coli* that is induced by melibiose.[187]

α-D-Galactosidases of almond emulsin and yeast also hydrolyze corresponding L-arabinopyranosides, but α-D-galactosidases of other plants and of bacterial origin are inactive.[188,189]

Most α-D-galactosidases so far studied are similar in their substrate specificity toward alkyl and aryl α-D-galactopyranosides, and toward α-D-galactopyranosyl oligosaccharides including melibiose, manninotriose, stachyose, planteose, and other higher saccharides of the raffinose family. The ease of hydrolysis of α-D-galactopyranosyl residues of the oligosaccharides decreases progressively with increasing size of the oligosaccharide.

Bailey and Howard[190,191] found that α-D-galactosidases from a protozoön (*Epidinium ecaudatum*) and *Streptococcus bovis* catalyzed readily the hydrolysis of a di-*O*-α-D-galactosylglycerol and α-D-(1 → 6)-linked galactosyloligosaccharides. The α-D-galactosyl residues in mono- and di-*O*-α-D-galactosyldilinolenin could be removed by an enzyme occurring in the cell sap of leaves of the runner bean (*Phaseolus multiflorus*).[192]

α-D-Galactosidases of plants and bacteria have been shown to catalyze trans-α-D-galactosylation reactions from aryl α-D-galactopyranosides and oligosaccharides to alcohols and mono- and oligosaccharides.[193–195] In general, with α-D-galactosyl oligosaccharides as substrates, transfer of α-D-galactosyl residues occurs preferentially to the primary alcohol group of the acceptor, resulting in the introduction of one or more D-galactosyl groups into the acceptor molecule. Epimelibiose, or three isomeric mono-*O*-α-D-galactosylcellobiose derivatives, were synthesized by the action of the coffee enzyme on phenyl α-D-galactopyranoside and mannose or cellobiose, respectively.[196] Raffinose is a common transfer product when sucrose is used as acceptor.

An *O*-α-D-galactosyltrehalose was formed in the presence of trehalose under the influence of the α-D-galactosidase from germinating fenugreek seeds.[197] Pneumococcal α-D-galactosidase, acting on melibiose, produced galactobiose, manninotriose, galactotriose, verbascotetraose, and other higher isomeric oligosaccharides.[198] Verbascose and ajugose were found in the transfer products formed by the action of coffee α-D-galactosidase upon raffinose and stachyose.[193]

Incubation of melibiose or phenyl α-D-galactoside and 2-acetamido-2-deoxy-D-galactose with enzymes from *Trichomonas foetus*[199] and *Aspergillus niger*[171] resulted in the synthesis of *O*-α-D-galactosyl-(1 → 6)-2-acetamido-2-deoxy-D-galactose.

---

* References start on p. 290.

The presence of α-D-galactosidases has been demonstrated in homogenates of various mammalian organs.[200,201] α-D-Galactosidase from pig kidney appears to be associated with a particulate fraction. In rat and human tissues, high activities were noted in endocrine organs, such as thyroid, kidney, and spleen.

D. β-D-GALACTOSIDASES (β-D-GALACTOSIDE HYDROLASES AND β-D-GALACTOSYL TRANSFERASES)

The β-D-galactosidases of microorganisms, particularly of *Escherichia coli*, constitute the most extensively purified and characterized glycosidases. This group of enzymes has been valuable not only in the elucidation of the mechanism and specificity of glycosidase action but also for the understanding of genetic regulation of enzyme synthesis. Developments in this field are covered in excellent reviews by Wallenfels and Malhotra.[202,203]

1. Purification and Properties

Much effort has been devoted to the purification and characterization of the β-D-galactosidases in the cells of various organisms, especially in strains of *E. coli*.[204-206] Wallenfels *et al.*[19] crystallized a β-D-galactosidase of *E. coli* ML 309 by extraction of lysed cells, removal of nucleic acids, and fractionation of the enzyme protein with alcohol and ammonium sulfate. The crystallized enzyme[33] was practically homogeneous and had a sedimentation coefficient of 16*S*, corresponding to a molecular weight of 518,000. Anfinsen and co-workers[28,38] isolated a β-D-galactosidase from cells of *E. coli* K 12 3300 in a high state of purity. This enzyme was found to be composed of identical monomer units having a particle weight of approximately 130,000. The sub-units, for which enzymic activity has not been demonstrated, are normally associated in the form of an active tetramer with an aggregate weight of about 540,000. The amino acid compositions were apparently similar for the normal (inducible) strains and for the constitutive mutants (see Table II). Digestion of *E. coli* β-D-galactosidase with trypsin and papain did not significantly alter enzymic activity, antigenicity, or sedimentation coefficient; treatment with chymotrypsin and nagase destroyed all activity.

Immunochemical studies showed that the enzyme produced by different strains of *E. coli* consists of the same type of molecules.[207] Perrin[208] characterized the molecular form of the immunochemically reactive molecule (cross-reactive protein, CRM) produced by mutant cells of *E. coli* that are genetically unable to synthesize active enzymes. The CRM mutant 3310 was found[209] to have a molecular weight of 235,000. The enzyme molecule can be dissociated reversibly in the presence of urea or guanidine hydrochloride into four enzymically inactive subunits,[38] each having a molecular weight of 135,000.

Based on the dissociation and cleavage of the enzyme molecule by peroxyformic acid, cyanogen bromide, and digestion with trypsin, it has been suggested that the monomer comprises several different polypeptide chains.[33,38] End-group analysis by chemical and enzymic methods established that the β-D-galactosidase possesses a molecule of threonine at the amino-terminal end, and a molecule of carboxyl-terminal lysine per monomer unit.[210,211] This result indicates that the monomer may consist of a single long polypeptide. The sequence[210] of amino acids at the carboxyl-terminal end appears to be Tyr-GluNH$_2$-LysCO$_2$H.

According to Appel et al.[36] uninduced strains of E. coli contain a single β-D-galactosidase, but after induction it produces at least seven enzyme components distinguishable by gradient centrifugation and zone electrophoresis. A major component of 34S appears to be an aggregate of the 16S fraction; both components have similar K_m values. The β-D-galactosidase of both inducible and constitutive strains of E. coli is located at the ribosomes, mainly in the 30S and 50S particles. The enzyme protein has the same physical properties as the free enzyme.[212-213]

An extracellular β-D-galactosidase of Diplococcus pneumoniae, which had been purified about 500-fold, not only catalyzed the hydrolysis of aryl β-D-galactopyranosides, β-D-fucopyranosides, and N-acetyllactosamine, but also liberated 80% of the D-galactosyl residues from neuraminidase-treated α_1-acid glycoprotein.[214] This β-D-galactosidase is hydrolytic and lacks transferring activity. Purified enzymes from Bacillus megatherium[215] and B. subtilis[216] differ from each other and from the enzyme of E. coli. The β-D-galactosidase of B. subtilis is relatively thermostable and is not activated by metal ions.

The utilization of lactose by Neurospora crassa has been studied by Landman et al.[217] Studies with purified lactases from lactose-utilizing strains and from mutants indicate that these lactases are closely similar. The mycelial extracts of Neurospora crassa contain two discrete lactases, an acid and a neutral β-D-galactosidase, which differ in thermal stability, in behavior toward inhibitors, and in molecular weight.[218,219] The activity of the neutral enzyme is markedly enhanced by the presence of monosaccharides, disaccharides, and a variety of alcohols. This phenomenon led to the assumption that the enzyme is a trans-β-D-galactosylase capable of utilizing the hydroxyl groups of these compounds as acceptors more readily than water.[219]

Phaseolus vulgaris has been found to be a rich source of several glycosidases. Treatment of its germinating seeds with sodium citrate buffer, fractionation by ammonium sulfate, and chromatography on (diethylamino)ethyl-Sephadex and carboxymethyl-Sephadex yielded α-D-galactosidase, β-D-galactosidase, α-D-mannosidase, β-D-glucosidase, and 2-acetamido-2-deoxy-β-D-glucosidase in very active forms. These enzymes were highly specific for the individual D-glycosyl residues and the anomeric configuration of the glycosidic linkage. The effectiveness of the procedure is ascribed to the initial step of germination which increases the specific activity of the various enzymes in the seeds, whereas nongerminating seeds are much less active.[220]

2. β-D-*Galactosidases in Animals*

The widespread occurrence of β-D-galactosidases in various mammalian organs is probably related to the multiple physiological functions of these enzymes, which catalyze the hydrolysis of the β-D-galactosidic linkage of lactose and of a variety of biologically occurring compounds. They are, thus, related to the nutrition and catabolism of glycoproteins, erythrocyte membranes, and glycolipids. β-D-Galactosidases are localized mainly in subcellular particles, particularly in the lysosomes of the liver,[221,222] kidney, bone tissue,[223] brain,[224,225] testis,[226] and gastrointestinal tract.[226–230] They can be liberated from the latent state by treatment with detergents or cholate, or by osmotic shock. Methods for the determination of enzyme activity of tissue homogenates, leucocytes, platelets, and plasma use β-D-galactosides possessing chromogenic or fluorescent aglycons.[231–234]

In the gastrointestinal tract of rats and humans, β-D-galactosidase activity is highest in the jejunum and ileum, but it is˙absent from the stomach.[228] Significant variations in activity were observed in certain parts of the small intestine of suckling and growing rats. At least two types of β-D-galactosidases seem to exist in rat and human intestinal mucosa. One is a lysosomal enzyme which has a pH optimum of 3.0; it is active against o-nitrophenyl β-D-galactoside, but not against lactose. A second β-D-galactosidase, localized exclusively in the brush border of intestinal mucosa, has an optimum pH value of 6.0, and hydrolyzes lactose readily.[234] The lactase of pig intestinal mucosa is not sensitive to the configuration at C-4 and therefore hydrolyzes both cellobiose and lactose, as well as 4-methylumbelliferyl β-D-glucopyranoside and β-D-galactopyranoside.[160,161]

Calf intestinal β-D-galactosidase, purified some 2100-fold, contained a large amount of carbohydrate.[22]

Semenza et al.[235] isolated two lactases from human jejunum and ileum, designated lactases I and II. Lactase I is present in the particulate fraction and has a lactase/cellobiase quotient of about 5, whereas lactase II has a quotient of 2 and is present in both soluble and particulate fractions. The particulate lactase is solubilized by papain and resolved by gel filtration or disc electrophoresis into two components. The minor component hydrolyzes 6-bromo-2-naphthyl β-D-galactopyranoside readily and seems to be identical with lactase II, which predominates in the soluble fraction. Lactulose, however, is not hydrolyzed.[236] These findings could be important in elucidating the cause of hereditary lactase deficiency such as in malabsorption and intolerance.[233] Enzyme systems also composed of two lactases have been noticed in rat and rabbit intestinal mucosa: a soluble enzyme that is a specific lactase, and a particulate one that is an aryl β-D-galactosidase.[228]

The cytoplasm and lysosomal fractions of liver are rich in β-D-galactosidases. Rat liver β-D-galactosidase is a single enzyme having an optimal pH of 3.0 and a molecular weight of 127,000. The cytoplasmic β-D-galactosidase of beef liver can be separated into two components having optimal pH values of 4.5 and 6.0, and molecular weights of 85,000 and 43,000, respectively. The hepatic lysosomal β-D-galactosidase exists in multiple forms that can be separated into four components by chromatography on (diethylamino)ethyl-cellulose. The elution patterns of β-D-galactosidases of homogenates of kidney, liver, spleen, and urine showed similar multiple components. The activity pattern of the urinary enzymes resembles closely that of kidney homogenates.[237]

β-D-Galactosidases involved in the degradation of cerebrosides occur in a variety of mammalian organs [224–239] and are often called cerebrosidases. Gatt and Rapport[225,239] solubilized the cerebroside-cleaving enzymes from the crude mitochondrial fraction of calf and rat brains by treatment with cholate, and separated the β-D-galactosidase from β-D-glucosidase. The combined action of these glycosidases caused stepwise hydrolysis of p-nitrophenyl β-D-lactoside and affords a model for enzymic breakdown of 1-O-lactosyl-ceramide. Fractionation of a solubilized enzyme preparation from pig brain yielded two β-D-galactosidases: an aryl β-D-galactosidase, and an enzyme specific for the cleavage of 1-O-D-galactosylceramides. A rat intestinal β-D-galactosidase that splits cerebrosides is bound to the particulate fraction; it is similar in its activation behavior to that of the porcine brain enzyme.[238] The enzyme is active on o-nitrophenyl β-D-galactopyranoside as well as on 1-O-D-glucosylceramide and 1-O-D-galactosylceramide.

Mammalian β-galactosidases are inhibited competitively by D-galactono-1,4- and -1,5-lactones, and by 6-deoxy-D-galactono-1,5-lactone. The last two lactones are much the more powerful inhibitors, since the ring size of these compounds is the same as that of the substrates. The β-D-galactosidase of ox liver differs from those of other organs in that it is relatively resistant to inhibition by D-galactono-1,4-lactone but is inhibited strongly by 6-deoxy-D-galactono-1,5-lactone.[240]

The snails *Helix pomatia*, *H. aspersa*, and *Lymnaea stagnalis* contain a potent β-D-galactosidase composed of two components, as shown by immunoelectrophoresis.[241-243] The occurrence of two types of β-D-galactosidase in the hepatopancreas of a Japanese snail, *Eulota peliomphala*, has been reported.[133] One showed the same inhibition pattern toward β-D-galactopyranosides and related substances as the β-D-galactosidase of Taka-diastase; the pattern of the other was similar to that for apricot emulsin.

3. Specificity of β-D-galactosidases

Accounts of the specificity of β-D-galactosidases have been summarized by Veibel.[129] According to Wallenfels and co-workers,[244,245] the galactosidase of *E. coli* requires the D-*galacto* configuration for the hydroxyl groups at C-2, C-3, C-4, and C-5. Oxidation of C-6 to a carboxyl group, reduction of the hydroxyl groups at C-2 and C-6, or substitution of the hydroxyl groups at C-2, C-3, C-4, or C-6 by a methoxyl group stops the action of the enzyme.

Likewise, inversion of the β-D-galactosidic linkage to the α-D configuration or conversion of the pyranose ring into the furanose form prevents enzyme action. Replacement of the CH_2OH group of the galactopyranosyl residue by a hydrogen atom (as in α-L-arabinosides) or its reduction to a methyl group (as in ethyl 6-deoxy-β-D-galactoside) results in a decrease in the rate of hydrolysis. Ethyl 3-deoxy-β-D-*xylo*-hexopyranoside is not hydrolyzed. Methylation at C-2, C-3, C-4, or C-6 leads to a loss of hydrolyzability. Substitution of a sulfur atom for an oxygen atom at C-1 in the D-galactosyl moiety renders the linkage resistant, but the ability to bind the glycoside remains unimpaired. Hence, thioglycosides are competitive inhibitors. β-D-Galactopyranosyl fluoride is a good substrate, yielding D-galactose and hydrogen fluoride.[246]

The specificity of β-D-galactosidases from various plant and animal sources for aglycons has been studied on more than fifteen different β-D-galactosides.[133] Their specificity patterns vary with the source of the enzyme. According to their inhibition pattern toward β-D-galactopyranosides and related substrates, they may be classified into those similar to almond emulsin and those resembling Taka-diastase. Aryl β-D-galactopyranosides are, in general, better substrates than alkyl β-D-galactopyranosides or disaccharides.

A chromogenic substrate, *o*-nitrophenyl β-D-galactopyranoside, is generally used for the routine assay of the activity of *E. coli* β-D-galactosidase. In addition, an ingenious method using 6-hydroxyfluoran β-D-galactopyranoside as a substrate has been devised for the measurement of single molecules of the *E. coli* enzyme.[247]

The type of linkage in *O*-β-D-galactosyl-D-glucoses influences the rates of hydrolysis, and these rates decrease in the order $(1 \rightarrow 6) > (1 \rightarrow 4) > (1 \rightarrow 3) > (1 \rightarrow 1)$. The rate of hydrolysis is the same for *O*-β-D-galactosyl-2-acetamido-2-deoxy-D-glucose. For O-α-L-arabinosyl-D-glucose, the order is $(1 \rightarrow 3) > (1 \rightarrow 4) > (1 \rightarrow 6)$. The ease of hydrolysis of disaccharides containing D-galactose, L-arabinose, or D-fructose as aglycon is lower than for lactose and its isomers.[244]

Calf intestinal β-D-galactosidase of high purity apparently differs in substrate specificity from that of *E. coli*.[20a] The best substrate for the calf enzyme is lactose, which is hydrolyzed more rapidly than o-nitrophenyl β-D-galactopyranoside. The anomeric configuration of lactose influences the hydrolytic rate, since β-lactose is hydrolyzed faster and has a lower K_m value than the α anomer or an equal mixture of both α- and β-lactose. The hydrolysis rates of lactose and its linkage isomers decrease in the order $(1 \rightarrow 3) > (1 \rightarrow 4) > (1 \rightarrow 6)$. The same relation is also observed for the rate of hydrolysis of O-β-D-galactosyl-2-acetamido-2-deoxy-D-glucose and O-α-L-arabinosyl-D-glucoses.

The mold *Sclerotinia fructogena* produces enzymes hydrolyzing aryl β-D-galactopyranosides, α-L-arabinopyranosides, and α-L-arabinofuranosides.[248]

4. β-Galactosyl Transfer

Wallenfels and co-workers[249,250] and Aronson[251] demonstrated the intermediate formation of several oligosaccharides by β-D-galactosyl transfer from lactose by crude enzyme preparations of molds, snail (*Helix pomatia*), *E. coli*, and calf intestine. Pazur[252,253] has reported the enzymic synthesis of four O-β-D-galactosyloligosaccharides by the lactase from *Saccharomyces fragilis*. With these enzymes, β-D-galactosyl transfer occurred preferentially at the primary alcohol group of D-glucose (forming a disaccharide called allolactose or lactobiose, 3-O-β-D-galactopyranosyl-D-galactose), of D-galactose (forming galactobiose, 6-O-β-D-galactopyranosyl-D-galactose), and of the nonreducing D-galactosyl residues of lactose or galactobiose (forming lactotriose or galactotriose). In general, lactobiose and lactotriose are the predominant products.[254] The β-D-galactosidases of *E. coli* ML 309, in either crude or crystalline form, appear to catalyze the same transfer reaction as indicated by the pattern of the transfer products. This may indicate that one and the same enzyme is responsible for both hydrolytic and transferring activities.[19] The enzyme also is able to produce three different disaccharides substituted by β-D-galactosyl groups at C-3, C-4, and C-6 of D-glucose. Quantitative data indicate that the order of formation of these disaccharides is $(1 \rightarrow 6) > (1 \rightarrow 4) > (1 \rightarrow 3)$, and the reverse is true for the order of the hydrolytic rates.[244] In this reaction, a number of mono- and disaccharides and their derivatives, including D-glucose, D-galactose, D- and L-arabinose, D-fructose, 3-deoxy-, 6-deoxy-, and 3,5-dideoxy-D-glucose, D-fucose, 2-amino-2-deoxy-

D-glucose, and sucrose can serve as acceptors of the β-D-galactosyl residues. α-L-Arabinosyl transfer catalyzed by the *E. coli* β-D-galactosidase yields three *O*-α-L-arabinosyl-D-glucoses from *o*-nitrophenyl α-L-arabinopyranoside in the presence of D-glucose. In contrast to the β-D-galactosyl transfer, the synthesis of the disaccharides in this case follows the order (1 → 3) > (1 → 6) > (1 → 4).

Almond emulsin and an enzyme from *Vicia* plants were found to catalyze not only the cleavage of α-L-arabinosyl residues of vicianoside, but also their transfer to methanol.[255] Various alcohols may also function as acceptors for β-D-galactosyl residues of aryl β-D-galactopyranosides in the presence of the plant[253] or mold enzyme.[256] β-D-Galactosyl transfer catalyzed by the enzyme of *E. coli* yielded only 1-*O*-β-D-galactosyl-D-glycerol.[257] Resorcinol is an acceptor for 6-deoxy-β-D-galactosyl or β-D-galactosyl groups of the respective aryl β-D-glycopyranosides.[258]

The extracts and intact cells of a yeast, *Sporobolomyces singularis*,[259] contain a β-D-galactosidase that catalyzes the β-D-galactosyl transfer of lactose to the secondary rather than to the primary hydroxyl group of the acceptors. It appears that three consecutive hydroxyl groups are required, with substitution occurring at the middle hydroxyl group; thus, 2-acetamido-2-deoxy-D-glucose and 1,2-*O*-isopropylidene-α-D-glucofuranose are substituted at the C-4 and C-5 positions, respectively; methyl 2-deoxy-β-D-*arabino*-hexopyranoside does not function as an acceptor. A similar transfer occurs preferentially at C-5 of *myo*-inositol with configurations corresponding to the C-3 of D-glucose, 6-deoxy-D-glucose, and D-xylose.[176]

The synthesis of β-D-linked disaccharides containing a 2-amino-2-deoxy-D-hexose has been achieved by using lactose and β-D-galactosidases from *Lactobacillis bifidus* var. *pennsylvanicus*,[260] yeast, *E. coli*,[261] and mammary glands of the rat.[262] The transfer products isolated from the reaction mixtures of lactose or phenyl β-D-galactopyranoside and 2-acetamido-2-deoxy-D-glucose were characterized as the *O*-β-D-galactosyl-(1 → 3)-, (1 → 4)-, and (1 → 6)-2-acetamido-2-deoxy-D-glucoses. In a similar transfer reaction, an enzyme of *Sporobolomyces* synthesized *N*-acetyllactosamine specifically.[176]

E. α,β-D-MANNOSIDASES (α,β-D-MANNOSIDE HYDROLASES)

α-D-Mannosidases have been found in the particulate fraction of mammalian liver and other organs together with other glycosidases. The general properties of some α-D-mannosidases were studied by Conchie and Hay.[222] The enzyme is commonly encountered in plants, especially in extracts of seeds and kernels—for example, almond emulsin.[129,263,264] According to studies by Pigman,[16] α-D-mannopyranosides, α-D-lyxopyranosides, and heptopyranosides, having the configuration of D-mannose at C-2, C-3, and C-4, are substrates for almond α-D-mannosidase. Mammalian α-D-mannosidases from various sources are inhibited by D-mannono-1,4- and -1,5-lactones. The α-D-mannosidase of *Streptomyces griseus* catalyzes the hydrolysis of phenyl α-D-mannopyranoside and the cleavage of D-mannose from *O*-α-D-mannosyl-streptomycin.[266]

β-D-Mannosidases occur in yeasts, insects, snails, oysters, and seeds, and in the pancreas and epididymis of the rat.[265-269] The enzyme appears to play a part in the degradation of D-mannans. Mold preparations exhibit β-D-mannosidase activity on mannobiose, mannotriose, and methyl β-D-mannopyranoside and are strongly inhibited by D-mannono-1,5-lactone.[270]

F. β-D-FRUCTOFURANOSIDASES (β-D-FRUCTOFURANOSIDE HYDROLASES AND β-D-FRUCTOFURANOSYL TRANSFERASES)

The enzymic hydrolysis of sucrose is catalyzed by two types of enzymes: α-D-glucosidases (discussed earlier) and β-D-fructofuranosidases. The β-D-fructofuranosidases have been given various names such as invertin, invertase, and saccharase. These enzymes are widely distributed in yeasts, molds, and higher plants. In addition to sucrose, these enzymes are concerned with the hydrolysis of α-D-galactosylsucroses (oligosaccharides of the raffinose series), inulins [β-D-(2 → 1)-linked D-fructans], and phleans [β-D-(2 → 6)-linked D-fructans]. Invertases from yeasts, molds, and higher plants also show transferring activity leading to the formation of various oligosaccharides and alkyl or aryl β-D-fructofuranosides, which ultimately may be hydrolyzed into the sugar components. Furthermore, enzyme systems comprising several β-D-fructofuranosidases, known to be present in plant organs containing D-fructans and related oligosaccharides, are involved in the hydrolysis and synthesis of these saccharides. These enzymes appear to be intracellular and under the control of growth hormones.

1. *Purification and Properties*

Despite many efforts, yeast invertase has not yet been obtained crystalline. A method advocated for purifications of yeast enzyme involved precipitation with picric acid and acetone, followed by absorption and elution with aluminum hydroxide. The purified enzyme contained about 70% of a D-mannan and was dissociable by electrophoresis into an inactive polysaccharide and an active protein–carbohydrate complex. It had a nitrogen content of 4 to 5% with an activity of 4000 units per milligram of nitrogen.[21]

Adams and Hudson[271] reported the isolation of an invertase preparation

* *References start on p. 290.*

having a low D-mannan (7%) and a high nitrogen (14.8%) content. Application of chromatography on an anion-exchange derivative of cellulose yielded an enzyme preparation having an activity of 22,000 units per milligram of nitrogen and an approximate molecular weight[272] of 260,000. By using O-(diethylamino)ethylcellulose, Myrbäck and Schilling[273] obtained an enzyme preparation having 10,000 to 14,000 units of activity per milligram of nitrogen. It contained 11.5% of nitrogen and 30% of carbohydrate; most of the latter could be removed without loss of activity.

By a similar chromatographic method, autolyzates of Bakers' yeast were separated into three active invertase components: I, II, and III. Components I and II differed in optimum pH, thermal stability, and K_m values, but had the same sedimentation constant of 4.25S. Incubation at pH 6.2 and 37° for seven days led to conversion of I into II.[274]

Metzenberg[39] obtained a purified invertase of *Neurospora crassa* which was homogeneous on electrophoresis and ultracentrifugation. It had a specific activity of 28,400 units per milligram of protein, and an activation energy of 10.8, 12.3, and 13.9 kcal mole^{-1} for sucrose, raffinose, and methyl β-D-fructofuranoside, respectively. Inhibition of the enzyme by heavy metal and p-mercuribenzoate ions indicated that the enzyme might be an SH enzyme. A characteristic feature is its reversible dissociation into enzymically active subunits at an acidic pH in the presence of high concentrations of salt. The conversion was evidenced by a change in the sedimentation constant from the 10.2S of the polymerized form to 5.3S for the subunits, and by differences in electrophoretic mobilities and in behavior during gel filtration. This phenomenon may also account for the presence of two invertase components that are electrophoretically different in the extracts of certain strains of *N. crassa*.[275]

Arnold[276] achieved a 55-fold purification of a soluble β-D-fructofuranosidase from grapes (*Vitis vinifera* var. Ohanez); the enzyme contained approximately 25% of carbohydrate. To assay the activity of the insoluble fraction of this enzyme, a method was devised with a column of spherical glass beads on which the enzyme particles were retained, and through which a variety of substrates were pumped at a constant rate.[276]

Growth of several fungi and yeasts on a medium containing sucrose monopalmitate has been shown to enhance the production of extracellular β-D-fructosidases. The patterns of enzyme action obtained upon zone electrophoresis showed that most of the microorganisms studied produced only one β-D-fructosidase, but, in some cases, two or more enzyme components were observed, which differed in their specificity toward β-D-(2 $\rightarrow$ 6)-fructosidic linkages.[277]

A sucrose-splitting enzyme isolated from homogenates of rat intestinal mucosa could be solubilized only after treatment with sodium deoxycholate.[278] The enzyme appeared to be a β-D-fructofuranosidase, since a trisaccharide isolated from sucrose digests was composed of two moles of D-fructose and one mole of D-glucose. Rat intestinal invertase thus differs from the intestinal enzymes of other mammals such as hog and rabbit; the latter enzymes are α-D-glucosidases.

2. Inhibition of β-D-Fructofuranosidases

The pH–activity curve of yeast invertase indicated two ionic groups, one on each of the two active sites. One group may be an imidazole group (histidine) having a pK_a of ~ 6.7; the other is of unknown origin and has a pK_b of ~ 3.

The action of iodine on yeast invertase is peculiar in that iodine forms an iodine–invertase complex which retains about 55% of the original activity and exhibits a pH–activity curve identical to that of the native enzyme.[279] The iodine–invertase complex became insensitive to Ag^+ and other metals which inhibit the native enzyme; possibly an interaction occurs between iodine and the sulfhydryl groups. Experiments with radioactive iodine showed the presence of organic iodine in the complex, but the exact binding site is unknown.

Various heavy metals, particularly Ag^+, inhibit yeast invertase strongly.[280,281] The inhibition by Ag^+ is instantaneous, reversible, and noncompetitive, and occurs in the pH range 4 to 7. Silver ions appear to compete with hydronium ions for the imidazole group of histidine, which possibly interacts in a protonated form with the sulfhydryl groups of the enzyme molecule.[280] The rate of inhibition varies considerably for crude and partially purified enzyme preparations because contaminating yeast nucleic acids bind metals strongly. Degraded products of nucleic acids were less inhibitive.[281] Inactivation of one enzyme molecule or of one active site required two moles of Ag^+ or one mole of $Cu.^{2+}$ Cupric ions seem to compete with hydronium ions for the imidazole group and probably the sulfhydryl group of the enzyme.[282]

Both anionic and cationic detergents such as sodium dodecyl sulfate and cetylpyridinium chloride react with the enzyme causing its rapid and irreversible inactivation.[283,284] This inhibition is assumed to arise from ionic effects in a suitable pH range. Similarly, heparin reversibly inactivates the enzyme[285] at pH below 5.

Aniline and other primary aromatic amines are potent inhibitors for yeast invertase.[283] This reversible inhibition seems to result from the formation of salt-like compounds with the enzyme molecule rather than through the formation of Schiff bases. Several heterocyclic amines exert a similar but weaker inhibitory action. In general, strong bases are less inhibitory than weaker ones. Inhibition by 2-amino-2-hydroxymethylpropan-1,3-diol (Tris) increases with increasing pH and appears to be caused by reaction of the cationic form with an anionic group of the enzyme.[286]

* References start on p. 290.

Inhibition of the hydrolysis of sucrose with *Neurospora* invertase by D-fructose was interpreted as the result of a nonspecific binding of D-fructose at the same enzyme site.[287] L-Sorbose or aniline inhibited the enzyme non-competitively. Aniline seemed to bind strongly with the enzyme–D-fructose complex, but much more weakly with the free enzyme form. Streptomycin has been reported to be an inhibitor of yeast invertase.[288]

3. *Specificity of β-D-Fructofuranosidases*

Only few studies of the substrate specificity of β-D-fructofuranosidases have been made, because the substrates available are mostly naturally occurring oligosaccharides containing residues of sucrose.[289,290] However, a few alkyl and aryl β-D-fructofuranosides have been synthesized by transferring reactions with invertase. The β-D-fructofuranosidases from *Neurospora*[287] and from grape exhibit similar K_m values and velocities for the hydrolysis of sucrose, raffinose, and methyl β-D-fructofuranoside, but they do not hydrolyze turanose, melezitose, or benzyl β-D-fructofuranoside.

p-Hydroxyphenyl β-D-fructofuranoside was found to be a better substrate than sucrose, raffinose, and methyl β-D-fructofuranoside for invertases from yeast, Taka-diastase, and wheat.[291]

β-D-Fructofuranosidases are unable to hydrolyze sucrose in which one of the hydroxyl groups in the D-fructose moiety is substituted at C-6, C-3, or C-1. This was shown by the inability of yeast invertase to hydrolyze planteose (substituted at C-6), oligosaccharides of the lychnose series (substituted at C-1), or the isolychnose series (substituted at C-3).[292] A free hydroxyl group at C-2 of the D-glucose moiety also seems essential for the formation of the enzyme–substrate complex, as shown by the resistance of umbelliferose (2-*O*-α-D-galactosylsucrose) to hydrolysis by invertase.[293]

A trisaccharide [*O*-α-D-galactopyranosyl-(1 → 3)-α-D-glucopyranosyl β-D-fructofuranoside] isolated from the seeds of *Festuca* or *Lolium* is hydrolyzed by yeast invertase, but at a much slower rate than raffinose; thus, substitution at C-3 by a D-galactosyl group did not prevent enzymic activity. Comparing the hydrolysis rates of the enzymes of Bakers' yeast and mullein leaves (*Verbascum thapsiforme* Schrad) on several α-D-galactosylsucroses of the raffinose family, Courtois *et al.*[294] observed that the rate of hydrolysis decreases with increasing chain length of α-D-galactosyl residues in the order of raffinose > stachyose > verbascose. Sucrose analogs containing D-xylose, D-galactose, or lactose instead of D-glucose were all susceptible to the action of yeast invertase.[295,296] Invertase also splits the β-D-fructofuranosyl residues of di- and higher oligosaccharides from sucrose digests.[297]

Edelman and Jefford[298] have separated from the tubers of Jerusalem artichoke (*Helianthus tuberosus* L.) an invertase that hydrolyzes sucrose and raffinose, but not melezitose. In addition, they isolated two β-D-fructofuranosidases, which differ in pH optimum, thermostability, and chromatographic behavior. The enzymes are virtually inactive on sucrose, but hydrolyze β-D-fructofuranosyl residues linked by a β-D-(2 → 1) linkage in oligo- and polysaccharides. Better substrates are inulodextrin, inulin, or oligosaccharides containing five to eight β-D-(2 → 1)-linked D-fructosyl residues terminated with sucrose, raffinose, or a hexitol.[298]

Purified invertase preparations from both brewers' and Bakers' yeast showed inulase activity.[271] However, most of the evidence indicates that inulin is normally hydrolyzed by enzymes other than invertase. The inulinase of *Saccharomyces fragilis* hydrolyzes terminal D-fructose residues linked either β-D-(2 → 1) or β-D-(2 → 6) in inulin, levan, and irisan.[299]

4. β-D-*Fructofuranosyl Transfer*

In 1950 Bacon and Edelman[300] and Blanchard and Albon[301] independently reported that, in concentrated sucrose solution, yeast invertase produces other oligosaccharides. With yeast or mold invertases, at least eight oligosaccharides were formed, all of which ultimately yielded free D-glucose and D-fructose. Of these, four were identified as disaccharides, three as trisaccharides, and one as a tetrasaccharide.[302-304]

Yeast invertase catalyzes D-fructosyl transfer from sucrose to the primary alcohol group of free D-glucose and free D-fructose to give rise to three disaccharides: O-β-D-fructosyl-(2 → 6)-D-glucose, O-β-D-fructosyl-(2 → 1)-D-fructose (inulobiose), and O-β-D-fructosyl-(2 → 6)-D-fructose (levanbiose). Instead of free sugars, various alcohols such as methyl, ethyl, propyl, butyl, and benzyl alcohols also can serve as acceptors to form the corresponding alkyl β-D-fructofuranoside; secondary alcohols, such as isopropyl alcohol, are inert as the acceptor.[303,305] Whelan and Jones [306] used methyl β-D-fructofuranoside as a donor and demonstrated that D-glucose, D-mannose, D-galactose, and alditols such as D-mannitol and D-glucitol may function as acceptors for D-fructosyl transfer. However, sugars devoid of a primary alcohol group failed to react. With D-glucose as acceptor, 6-O-β-D-fructosyl-D-glucose was isolated. Nakamura[291] obtained crystalline p-hydroxyphenyl β-D-fructofuranoside from sucrose digests containing hydroquinone as the acceptor by the action of yeast invertase. Of twenty phenols examined, only hydroquinone and resorcinol could serve as acceptors. Methyl β-D-fructofuranoside could not act as a donor in place of sucrose.

Trisaccharides are produced by transfer of a D-fructosyl residue to any of the three primary alcohol groups of sucrose by the enzymes from yeast,[302] molds[303,304] artichoke, and other high plants.[307-309] They are called 6-kestose, 1-kestose, and neokestose, although other names also have been used.

* *References start on p. 290.*

6-Kestose is produced most efficiently when yeast invertase is incubated with a 50% solution of sucrose; its structure is O-β-D-fructofuranosyl-(2 → 6)-O-β-D-fructofuranosyl-(2 → 1)-α-D-glucopyranose.[310] 1-Kestose, often called inulobiosylsucrose or 1-fructosylsucrose, is O-β-D-fructofuranosyl-(2 → 1)-O-β-D-fructofuranosyl-(2 → 1)-α-D-glucopyranose.[311] Neokestose has the structure O-β-D-fructofuranosyl-(2 → 6)-α-D-glucopyranosyl β-D-fructofuranoside.[312] A tetrasaccharide, designated nystose, was isolated upon incubation of sucrose with a fungal D-fructosyl transferase. The structure of this tetrasaccharide is O-β-D-fructofuranosyl-(2 → 1)-O-β-D-fructofuranosyl-(2 → 1)-O-β-D-fructofuranosyl-(2 → 1)-α-D-glucopyranose.[312a]

These trisaccharides are widely distributed in plants, especially in the bulbs and tubers of monocotyledons.[313] Of these, 1-kestose appears to play an essential role in the metabolism of D-fructose polymers in the tubers of Jerusalem artichoke.[314] Transfructosylases of the artichoke catalyze the transfer of D-fructosyl residues of 1-kestose to suitable primers such as inulodextrins, to produce highly polymerized D-fructans; an exchange reaction of D-fructosyl residues with sucrose and other fructosyloligosaccharides also occurs.[298,314]

Pazur[315] reported the formation of some higher oligosaccharides from sucrose by an enzyme of *Aspergillus oryzae*, including a β-D-(2 → 1)-linked trisaccharide, and tetra- and pentasaccharides.

Levansucrase from *Aerobacter levanicum* was found to transfer the β-D-fructosyl group to the hydroxyl groups at C-2 and C-3 of D-glucose to yield the corresponding disaccharides. In this reaction, L-arabinose, D-galactose, D-xylose, melibiose, and lactose could also serve as acceptors.[295,296]

Taka-diastase, when acting on mixtures of raffinose and D-glucose, yields sucrose by transfructosylation.[316]

G. 2-Acetamido-2-deoxy-α,β-D-hexosidases and Related Enzymes*

1. *2-Acetamido-2-deoxy-D-hexosidases*

The enzymes that hydrolyze the glycosides of 2-acetamido-2-deoxy-D-glucose or 2-acetamido-2-deoxy-D-galactose are the 2-acetamido-2-deoxy-α- or β-D-hexosidases, specifically the 2-acetamido-2-deoxy-D-glucosidases and 2-acetamido-2-deoxy-D-galactosidases. No simple, naturally occurring substrates for these enzymes are known, and synthetic aryl glycosides such as

---

* The enzymes discussed in this section hydrolyze glycosides of 2-acetamido-2-deoxy-D-hexosides and polymers. They are called 2-acetamido-2-deoxy-D-hexosidases or 2-acetamido-2-deoxy-D-hexoside hydrolases. The obsolete trivial names *N*-acetylhexosaminidases are also encountered in the literature.

p-nitrophenyl 2-acetamido-2-deoxy-D-hexopyranosides are used as test substrates.[317] It seems probable that 2-acetamido-2-deoxy-D-hexosidases participate, in conjunction with other glycosidases, in some stages of the enzymic breakdown of diverse materials containing 2-amino-2-deoxy-D-hexose such as the milk sugars, glycolipids, chitin, acid mucopolysaccharides, and glycoproteins, by splitting the 2-acetamido-2-deoxy-D-hexosyl residues.

2-Acetamido-2-deoxy-D-glucosidase activity on α-D-glycosides was first demonstrated in extracts of snail hepatopancreas[318] and subsequently in various mammalian tissues.[319] Experiments with rat liver homogenates characterized the enzyme as a lysosomal acid hydrolase.[320]

With *p*-nitrophenyl 2-acetamido-2-deoxy-α-D-glucopyranoside as substrate, the activity of the 2-acetamido-2-deoxy-α-D-glucosidases in guinea pig, beef, lamb, pig, and rat livers was shown to be weaker than that of 2-acetamido-2-deoxy-β-D-glucosidases by a factor of 1000. Centrifugation of the liver homogenates sedimented the α-D and β-D enzymes in the lysosomal fraction.[321]

Weissmann and Friederici[322] showed that homogenates of pig and rat livers contain a specific 2-acetamido-2-deoxy-α-D-galactosidase that appears to be bound to subcellular particles.

Sensitive histochemical methods were developed with 1-naphthyl 2-acetamido-2-deoxy-β-D-glucopyranoside or its derivatives as substrate for the localization of 2-acetamido-2-deoxy-β-D-glucosidase activity in tissues.[323,324]

A fluorometric assay with 4-umbelliferyl 2-acetamido-2-deoxy-β-D-glucopyranoside showed that the enzyme activity in human serum increases progressively during pregnancy.[325]

2-Acetamido-2-deoxy-β-D-hexosidases from porcine epididymis,[326] ovine testis,[327] bovine liver,[328] aortas and spleen[329] have been obtained in purified form. These purified enzyme preparations split both 2-acetamido-2-deoxy-β-D-glucopyranosides and 2-acetamido-2-deoxy-β-D-galactopyranosides in variable ratios; the enzyme activities are increased in the presence of serum albumin, but are strongly inhibited by heavy metals,[330] the lactones of amino aldonic acids,[331] and acid mucopolysaccharides.[332] The apparent lack of influence of the configuration at C-4 upon these enzymes is of particular interest, since these enzymes are presumably related to the liberation of terminal 2-acetamido-2-deoxy-D-glucosyl or 2-acetamido-2-deoxy-D-galactosyl groups from oligosaccharides derived from mucopolysaccharides and glycoproteins. For example, a 2-acetamido-2-deoxy-β-D-hexosidase from bovine spleen was able to hydrolyze not only phenyl 2-acetamido-2-deoxy-β-D-glucopyranoside and the 2-acetamido-2-deoxy-D-hexosyl residues of trisaccharides from hyaluronate and chondroitin sulfates, but also the glycosidic linkage between 2-acetamido-2-deoxy-D-galactose and hydroxyamino acids of glycopeptides derived from ovine submaxillary mucin.[329]

Frohwein and Gatt[333] observed that calf brain 2-acetamido-2-deoxy-β-D-hexosidase is responsible for the enzymic degradation of cerebrosides, as evidenced by the cleavage of the terminal 2-acetamido-2-deoxy-D-galactosyl residue of ceramides.

2. *Lysozymes and Enzymes that Hydrolyze Polymeric 2-Amino-2-deoxy-D-Hexose-Containing Compounds*

Lysozymes of various origins are endo-2-acetamido-2-deoxy-D-glucosidases capable of splitting the N-acetylmuramyl-2-acetamido-2-deoxy-D-glucosyl linkage in cell walls of gram-positive bacteria.[334] (See also Vol. IIB, Chap. 41.) Crystalline egg-white lysozyme was shown[335] to catalyze a disproportionation reaction of the tetrasaccharide O-(2-acetamido-2-deoxy-β-D-glucosyl)-(1 → 4)-O-(N-acetylmuramyl)-(1 → 4)-O-(2-acetamido-2-deoxy-β-D-glucosyl)-(1 → 4)-N-acetylmuramic acid from the cell wall of *Micrococcus lysodeikticus* by successive transfer of disaccharide residues with the formation of hexa-, octa-, and dodecaoligosaccharides:

$$2 \text{ Tetrasaccharides} \longrightarrow \text{Hexasaccharide} + \text{disaccharide}$$

Egg-white lysozyme also hydrolyzes p-nitrophenyl N,N'-diacetylchitobioside but not its β-D-(1 → 6)-linked isomer, nor p-nitrophenyl 2-acetamido-2-deoxy-β-D-glucopyranoside.[336]

2-Acetamido-2-deoxy-D-hexosidases also exist in bacteria, molds, higher plants, and invertebrates. They are valuable not only in the investigation of chitinolytic enzyme systems but also in structural studies regarding the carbohydrate moiety of 2-amino-2-deoxy-D-hexose-containing materials, such as glycoproteins and blood-group substances.

2-Acetamido-2-deoxy-D-hexosidases of *Clostridium tertium*,[337,338] *Trichomonas foetus*,[339] and *Helix pomatia*[340] have been utilized, together with other specific glycosidases, to modify the serological activity of blood group A substances. They were identified as 2-acetamido-2-deoxy-D-galactosidases, since the serological activity was ascribed to the terminal 2-acetamido-2-deoxy-D-galactosyl residue linked through an α-D-glycosidic bond to D-galactose. A 2-acetamido-2-deoxy-α-D-galactosidase isolated from the hepatopancreas of *Helix pomatia*, purified 300-fold, caused almost complete loss of serological specificity.[340]

A strain of *E. coli* contains an enzyme that hydrolyzes specifically the terminal 2-acetamido-2-deoxy-D-glucosyl residue of mucopeptides isolated from the bacteria.[341] A highly purified preparation of 2-acetamido-2-deoxy-β-D-glucosidase obtained from the germinating seeds of *Phaseolus vulgaris* was found to split the nonreducing, terminal 2-acetamido-2-deoxy-D-glucosyl residues from glycopeptides derived from desialyzed fetuin and orosomucoid pretreated with β-D-galactosidase.[341a]

Crude extracts of rat epididymis containing both 2-acetamido-2-deoxy-α- and -β-D-glucosidase, and a partially purified 2-acetamido-2-deoxy-β-D-glucosidase from pig epididymis, were valuable in structural studies of ribitol teichoic acid from various strains of *Staphylococcus aureus*.[342]

A porcine enzyme was useful in elucidating the nature of the linkage in the disaccharides: O-(2-acetamido-2-deoxy-β-D-glucopyranosyl)-(1 $\rightarrow$ 4)-N-acetylmuramic acid and O-(2-acetamido-2-deoxy-β-D-glucopyranosyl)-(1 $\rightarrow$4)-di-N,6-O-acetylmuramic acid isolated from the enzymic digests of the mucopeptides of the cell wall from *Staphylococcus aureus*.[343]

3. Chitobiase

It is generally accepted that the enzymic breakdown of chitin by molds is caused by two enzymes: chitinase, an endo-2-acetamido-2-deoxy-glucosidase hydrolyzing chitin mainly to N,N'-diacetylchitobiose, and chitobiose, which subsequently yields 2-acetamido-2-deoxy-D-glucose. Purified chitobiases from *Streptomyces griseus* and *Streptomyces antibioticus* were found to be highly active toward chitobiose and to lesser extent on aryl 2-acetamido-2-deoxy-D-glucopyranosides; the reverse was true for mammalian enzymes.[344,345]

An enzyme fraction from *Helix pomatia* hydrolyzed the N-acetylmuramyl linkage of a polysaccharide of *Micrococcus lysodeikticus* as well as O-(hydroxyethyl)chitin. Another fraction of the same enzyme showed only muramidase activity.[346] The 2-acetamido-2-deoxy-β-D-glucosidase from *Aspergillus niger* attacks chitodextrins as well as chitobiose.[347] The occurrence of similar enzyme systems has been reported in the digestive juice of hepatopancreas of snails,[346] crustacean hypodermis,[348] insects, and seeds of some higher plants.[349]

Barker and co-workers[350] demonstrated that the enzymic synthesis of 6-O-(2-acetamido-2-deoxy-β-D-glucosyl)-N,N'-diacetylchitobiose from N,N'-diacetylchitobiose takes place by a transferring reaction catalyzed by enzymes from *Myrothecium verrucaria* and other molds:

$$2 \text{ N,N'-Diacetylchitobiose} \longrightarrow \text{Trisaccharide} + \text{2-acetamido-2-deoxy-D-glucose}$$

These enzymes were also able to transfer 2-acetamido-2-deoxy-D-glucosyl residues to D-glucuronic acid, yielding O-(2-acetamido-2-deoxy-D-glucosyl)-D-glucuronic acid.

* References start on p. 290.

H. β-D-GLUCOSIDURONASES*

Most of the information concerning the purification, properties, and physiological significance of β-D-glucosiduronases from mammalian tissues and of plant origin is due to studies by Fishman and by Levvy and their associates.[351,352] Their reviews are important sources of information.

A variety of alcohols, phenols, and steroids, after administration to mammals, are converted into the corresponding D-glucosiduronic acids, which are excreted in the urine or bile. In target organs, steroid hormones may be released from their glycosiduronic acids by β-D-glucosiduronases.[353] These enzymes are often used in the analysis of steroid components that are present as glucosiduronic acids in the urine. β-D-Glucosiduronases are also involved in the catabolism of acid mucopolysaccharides of connective tissues and in the microbial degradation of a number of plant gums and hemicelluloses which contain D-glucosyluronic acid residues.

β-D-Glucosiduronase activity is generally measured by the colorimetric method of Fishman, which uses phenolphthalein β-D-glucopyranosiduronic acid as substrate. The Fishman unit has been generally accepted as a standard for activity. The use of phenyl,[352] p-nitrophenyl,[354] 1-naphthyl,[355] and 4-methylumbelliferone[356] β-D-glucopyranosiduronic acid is also applicable for the measurement of β-D-glucosiduronase activity in serum and other body fluids.

Rich sources of β-D-glucosiduronases are the preputial gland of the female rat, the digestive tract of the snail (*Helix pomatia*), the visceral hump of the limpet (*Patella vulgata*), and the crop fluid of the locust (*Locusta migratoria*).[351] Only in few instances has its presence been reported in microorganisms.[357–360] Some strains of *E. coli*[357] and *Streptococci*[358] contain large amounts of β-D-glucosiduronases. Marsh[360] has reported that the cell-free extracts of microorganisms of the rumen are rich in β-D-glucosiduronases; their presence seems to be related to the digestion of plant hemicelluloses. β-D-Glucosiduronases are distributed universally in mammalian tissues, serum, leucocytes, and body fluids. Like other similar glycosidases, these enzymes are localized in subcellular particles. It was shown[361,362] that the β-D-glucosiduronase of rat and mouse livers is found mainly in the cytoplasmic granules; it can be solubilized by treatment with acetate buffer or Triton X-100, by mechanical disintegration, or by repeated freezing and thawing. In the adult mouse, the β-D-glucosiduronase of liver and kidney is located in the lysosomal fraction, whereas in spleen and cancer tissues, β-D-glucosiduronase occurs in a free state in the cytoplasm.

---

* These enzymes hydrolyze the glycosides of D-glucuronic acids and are called D-glucosiduronases or D-glucosiduronic acid hydrolases. The obsolete trivial name glucuronidase has also been used in the literature.

de Duve *et al.*[362,363] found that a considerable portion of β-D-glucosiduronase activity of rat liver is localized in the lysosomal fraction as well as in the microsomal fraction. The pH–activity curves of these particulate enzymes differ significantly from each other, presumably because of formation of a complex between the enzyme and different tissue constituents.

Histochemical studies with chromogenic substrates such as 6-bromo-3-[(*o*-methoxyphenyl)carbamoyl]-2-naphthyl β-D-glucopyranosiduronic acid have demonstrated that the β-D-glucosiduronase activity is localized in specific cytoplasmic granules of diverse tissues, in agreement with biochemical findings.[364,365]

Kubler and Frieden[366] observed a more than 30-fold increase in the total activity of β-D-glucosiduronase in both lysosomal and soluble fractions of the regressing tadpole tail of the frog (*Rana grylio*) during spontaneous metamorphosis.

In earlier attempts to purify β-D-glucosiduronase, preparations having specific activities of 60,000 to 107,000 and 7900 Fishman units per milligram of protein, respectively, have been obtained from spleen[367] and liver.[368] Levvy and co-workers[369] have achieved a 25-fold purification of β-D-glucosiduronase from female rat preputial gland having a specific activity of 455,000 phenolphthalein units per milligram of protein. Other mammalian β-D-glucosiduronases have been obtained in highly purified states from mouse liver, kidney, and urine by zone electrophoresis.[370] Plapp *et al.*[371] prepared by a similar method a bovine β-D-glucosiduronase which exhibited 240,000 Fishman units per milligram of protein. Musa *et al.*[372] succeeded in purifying from human liver extracts a β-D-glucosiduronase having a specific activity of 570,000 Fishman units; the molecular weight was 218,000. A method was developed for the isolation of a sulfatase-free enzyme from snail digestive juice by heat treatment.[373] Alfsen-Blanc[374] reported the preparation of an apparently homogeneous β-D-glucosiduronase having 106,000 units of activity per milligram of protein.

Studies of the aglycon specificity revealed that β-D-glucosiduronases from different origins show similar rates of hydrolysis for alkyl, aryl, heterocyclic, and steroid β-D-glucopyranosiduronic acids. The enzymes from bovine liver, *E. coli*, *Helix pomatia*, and the common limpet (*Patella vulgata*) showed similar hydrolytic activities toward various steroid β-D-glucopyranosiduronic acids.[375] By using enzyme preparations from ox spleen, a limpet, and a bacterium, Becker[376] found that the β-D-glucopyranosiduronic acids of 5-β-androstane and 3-β-hydroxy-5-androstanes are hydrolyzed faster than the corresponding 5-α isomers and 3-α epimer, respectively. The enzyme baicalinase, extracted from the root of *Scutellaria baicalensis*, hydrolyzes not only the flavone β-D-glucosiduronic acids such as baicalin, scutellarin, and chrysin, but also phenolphthalein β-D-glucopyranosiduronic acids and other D-glucosiduronic acids.[377]

These flavone β-glucosiduronic acids also serve as substrates for mammalian enzymes, which indicates a similar specificity of β-D-glucosiduronases from different origins. Limpet and mammalian β-D-glucosiduronases have a broad aglycon specificity, since they also hydrolyze β-D-galactosiduronic acids, although less efficiently.[378]

Fishman and Green[379] showed that purified β-D-glucosiduronase preparations from various sources catalyze the transfer reactions of β-D-glucosyluronic groups from eight natural D-glucosiduronic acids to various alcohols with the formation of new β-D-glucosiduronic acids. Of nineteen alcohols used as acceptors, methanol, ethylene glycol, and propane-1,2-diol were highly efficient. The influence of pH, temperature, and inhibitors upon both hydrolytic and transferring activities was similar, supporting the view that the β-D-glucosiduronases themselves were responsible for the two reactions.

The hydrolysis of β-D-glucosiduronic acids by enzymes from mammalian and other sources are strongly and competitively inhibited by D-glucaric-1,4-lactone, and to lesser extent by D-glucuronic acid and boiled galactaric acid. Several hydroxylated carboxylic acids and heparin are also known to inhibit β-D-glucosiduronase activity. Inactivation of highly purified β-D-glucosiduronases upon dilution can be prevented by the use of DNA or albumin.[380]

I. β-D-XYLOSIDASES (β-D-XYLOSIDE HYDROLASES)

Some D-xylan-fermenting bacteria produce β-D-xylosidases, which catalyze the hydrolysis of a series of xylo-oligosaccharides leading to complete degradation of D-xylan by synergistic action of several D-xylanases. Howard et al.[381] found in rumen bacteria a β-D-xylosidase that can split xylobiose and phenyl β-D-xylopyranoside; the enzyme also catalyzed the synthesis of xylotriose and 3-O-xylosylxylobiose upon incubation with xylobiose. Morita[382] classified the activity of β-D-xylosidases for p-nitrophenyl β-D-xylopyranoside into two types, based on differences in the mode of inhibition; one is represented by the enzyme of Taka-diastase which is inhibited by D-xylose, phenyl β-D-xylopyranoside, and D-xylono-1,5-lactone; the other is of the almond-emulsin type and is inhibited by phenyl β-D-xylopyranoside and β-D-glucopyranoside. β-D-Xylosidases of limpet and rumen are inhibited by D-xylono-1,4-lactone.[383]

β-D-Xylosidase was found in the subcellular particles of rat liver, in the chick embryo, and in cartilage. In addition to aryl β-D-xylopyranosides, both almond emulsin and mammalian β-D-xylosidases split the serine D-xyloside linkage which connects the carbohydrate and protein moieties in the condroitin sulfate–protein complex.[384]

J. SIALIDASES

Sialidases (neuraminidases) comprise a group of enzymes that cleave the ketosidic linkages through which sialic acids are joined to other sugars in oligosaccharides, and in the carbohydrate portions of glycoproteins, mucins, and glycolipids.[385] (See Vol. IIB, Chap. 43 for discussion of sialic acids.) These enzymes have been isolated from viruses, bacteria, a protozon, *Trichomonas foetus*,[386] and certain animal organs.

Highly purified and crystalline sialidases have been obtained from several strains of influenza viruses,[387,388] *Vibrio cholerae*,[389] *Diplococcus pneumoniae*,[214] *Corynebacterium diphteriae*,[390] and *Clostridium perfringens*.[391,392]

A standard method for the preparation of viral enzymes has been established.[387] The molecular weights of crystallized enzymes range between 40,000 and 60,000. Two different molecular weights of about 20,000 and 90,000 have been reported for the purified neuraminidase of *Vibrio cholerae*.[389]

The neuraminidase activity is generally low but is widely distributed in animal organs.[393,394] It is present in serum glycoprotein fractions. Neuraminidases are generally inhibited to various degrees by (ethylenedinitrilo)-tetraacetic acid (EDTA), and this inhibition can be reversed by the addition of Ca^{2+} and other divalent metal ions.[395] However, neither inhibition by EDTA nor any effect of Ca^{2+} was observed in the case of *Clostridium* neuraminidase.[391]

Substitution of acetyl groups by glycolyl groups does not alter neuraminidase activity. O-Acetylation appears to reduce the enzyme activity significantly.[396] Neuraminidase attacks the O-(N-acetylneuraminyl)-(2 → 3)-O-D-galactosyl bond in N-acetylneuraminyllactose, which is often used as a standard substrate for enzyme assay. It also hydrolyzes the O-(N-acetylneuraminyl)-(2 → 6)-2-acetamido-2-deoxy-D-galactose linkage of ovine and bovine submaxillary mucins.[385,396] However, (2 → 6)-linked N-acetylneuraminyllactose is not attacked by neuraminidase. Colominic acid, a (2 → 8)-linked polymer of N-acetylneuraminic acid, is resistant to enzymic hydrolysis by most neuraminidases, but the *Clostridium* enzyme hydrolyzes it completely, although at a slower rate.[391] Terminal neuraminic acids of the carbohydrate moieties in a number of glycoproteins, mucins, and glycolipids are split by various neuraminidases.[385,392,397] All known neuraminidases appear to be purely hydrolytic enzymes, although no attempt has been made to examine them for transferring activity.

* *References start on p. 290.*

K. α,β-D- AND L-FUCOSIDASES AND α,β-L-RHAMNOSIDASES (α,β-D- AND L-FUCOSIDE HYDROLASES AND α,β-L-RHAMNOSIDE HYDROLASES)

The occurrence of an α-L-fucosidase has been demonstrated in mammalian tissues, where it exists mainly in a latent state, apparently in association with cytoplasmic granules.[398,399] Highest activity is displayed by homogenates of rat epididymis, where the enzyme production is apparently controlled by endocrine glands. Mammalian α-L-fucosidase hydrolyzes α-L-fucosides (6-deoxy-L-galactosides) and L-fucose-containing oligosaccharides, but does not liberate L-fucose from blood group substances. In contrast, the α-L-fucosidase of *Trichomonas foetus*[400] degrades macromolecular mucous substances and L-fucosyloligosaccharides of human milk, but shows no activity toward methyl α-L-fucopyranoside. The visceral hump of the limpet and the hepatopancreas of some molluscs are the best-known sources of α-L-fucosidases.

Two types of α-L-fucosidase have been purified from abalone liver, having optimum pH values of 5 and 2, respectively. The former is active on *p*-nitrophenyl α-L-fucoside only, whereas the latter acts on both the synthetic substrate and porcine submaxillary mucin.[400a]

Usually, mammalian β-D-fucosidases are in many respects indistinguishable from β-D-galactosidases. Thus, highly purified preparations of β-D-galactosidase from *E. coli* and calf intestine hydrolyze β-D-fucosides. Both enzymes exhibit similar behavior in the course of fractionation, partial inactivation, and inhibition by the corresponding aldonolactones. However, differences were observed in the mode of inhibition between the enzymes from rat epididymis and ox liver; the rat enzyme was strongly inhibited by D-galactono-1,4-lactone but only weakly by D-fucono-1,4-lactone, whereas the reverse behavior was observed for the bovine enzyme.[401] Limpet β-D-fucosidase appears to be distinct from β-D-galactosidase, since the two enzymes differ in their rates of inactivation by heat and inhibition by lactones.

Rosenfeld and Wiederschein[402] found appreciable α-L-rhamnosidase activity toward phenyl α-L-rhamnopyranoside in rat, ox, and hog tissues. Highest activity was observed in hog adrenals and hypophysis. Barker *et al.*[403] found in cell-free extracts of *Klebsiella aerogenes* both α- and β-L-rhamnosidases after induction with methyl α- and β-L-rhamnopyranoside.

L. CYCLODEXTRIN TRANSGLUCOSYLASES

French *et al.*[404] isolated a cyclodextrin transglucosylase from *Bacillus macerans* that catalyzes the transfer of α-D-glucosyl groups from cycloamyloses (Schardinger dextrins) (see this volume, Chap. 34 and Vol. IIB, Chap. 38) to a number of carbohydrates. By means of the *B. macerans* enzyme and cyclohexaamylose, Pazur *et al.*[405] synthesized malto-oligosaccharides ending in both methyl α- and β-D-glucosides. Monosaccharides and their derivatives, including *O*-methyl-D-glucoses, methyl α- or β-D-glucopyranosides, D-xylopyranosides, and 6-deoxy-D-glucose, were also acceptors.[406] 6-*O*-Maltosyl-D-glucose was obtained by a coupling reaction between cyclomaltohexaose and isomaltose.[407]

M. D-Enzyme

Peat and associates[408] found a transglycosylase in potato juice which they named D-enzyme. It catalyzes the transfer of two or more D-glucose residues from maltodextrins to suitable acceptors, resulting in a rapid disproportionation between maltodextrins and higher α-D-(1 → 4)-linked polymers of D-glucose. A maltosaccharide disproportionating enzyme similar to potato D-enzyme has been isolated in purified form from extracts of tomatoes and carrots.[408a] The smallest donor is maltotriose, which is converted into maltopentaose and D-glucose:

$$\text{2 Maltotriose} \rightleftharpoons \text{Maltopentaose} + \text{D-glucose}$$

On further reaction with constant removal of the D-glucose, a series of higher maltodextrins is obtained. With increasing size of substrate, the enzyme yields an equilibrated mixture having iodine-staining power. Addition of D-glucose to the reaction mixture of the maltodextrins causes a rapid decrease in iodine color, presumably because of shortening of the chains by transfer of maltodextrinyl residues to D-glucose.[409] Similarly, D-mannose, D-xylose, L-sorbose, methyl α-D-glucopyranoside, and maltodextrins can function as acceptors. By means of this transfer reaction, a series of maltodextrin derivatives terminating in methyl α-D-glucopyranoside was prepared. The use of L-sorbose or methyl L-sorbopyranoside as acceptors leads to the formation of oligosaccharides linked to C-3 of the L-sorbose or of the sorboside.[410]

A cell-bound transglucosylase of *Streptococcus bovis* produced, in a similar fashion to the D-enzyme, iodine-staining polymers from lower maltodextrins when the D-glucose was removed as it was formed.[411] Unlike the D-enzyme, the *Streptococcal* transglucosylase catalyzes the transfer of D-glucosyl residues forming maltose during the disproportionation reaction of higher dextrins.

Brown and Illingworth[412] discovered a trans-α-D-glucosylase in liver capable of transferring maltosyl, or preferentially maltotriosyl, residues from maltodextrins or glycogen to linear or branched oligosaccharides; this enzyme caused elongation of α-D-(1 → 4)-linked D-glucosidic chains in the acceptor

* *References start on p. 290.*

molecules. The enzyme was present in highly purified dextrin-$(1 \rightarrow 6)$-glucosidase preparations from rabbit muscle.[412]

N. Amylomaltase

A strain of *E. coli* which produces an amylose type of polysaccharide was shown by Monod and Torriani[413] to contain an intracellular trans-α-D-glucosylase called amylomaltase that catalyzes the reversible reaction:

$$n(\text{Maltose}) \rightleftharpoons n(\text{D-Glucose}) + (\text{D-glucose})_n$$

Oligosaccharides containing four or six D-glucose residues were the usual transfer products at equilibrium. Removal of D-glucose by D-glucose oxidase shifted the reaction to the right, yielding an iodine-staining amylose type of polymer. Barker and Bourne[414] characterized the reaction products as a homologous series of maltodextrins having as many as five residues of D-glucose. Wiesmeyer and Cohn[415] using *E. coli.* ML 308, obtained a highly homogeneous preparation of amylomaltase having a molecular weight of 124,000 and a turnover number of 1160 moles of maltose per mole of enzyme per minute at 28° and pH 6.9. It is interesting that this enzyme reversibly catalyzes interconversion between the monomeric subunits and the polymerized form, depending upon the ionic strength. With labeled D-glucose as an acceptor, it was shown that the D-glucosyl residues are transferred from the nonreducing end of an amylose chain. D-Mannose and methyl β-maltoside are the only other known acceptors for D-glucose. The enzyme is specific for maltodextrins, having 2 to 5 D-glucose residues.[416]

Burger and Pavlasova[417] were able to show that a mutant isolated from certain strains of *Escherichia coli* is incapable of utilizing maltose because the amylomaltase of this mutant is associated with insoluble cell structures; in contrast, the wild types have the enzyme free in the cytoplasm and can, therefore, utilize maltose.

REFERENCES

1. C. A. Bunton, T. A. Lewis, D. R. Llewellyn, H. Tristram, and C. A. Vernon, *Nature*, **174**, 560 (1954).
2. D. E. Koshland, Jr., and S. S. Stein, *J. Biol. Chem.*, **208**, 138 (1954).
3. M. Halpern and J. Leibowitz, *Bull. Res. Council Israel, Sect.* **A8**, 41 (1959).
4. F. C. Mayer and J. Larner, *J. Amer. Chem. Soc.*, **81**, 188 (1959).
5. F. Eisenberg, *Federation Proc.*, **18**, 221 (1959).
6. D. R. Whitaker, *Arch. Biochem. Biophys.*, **53**, 436 (1954).
6a. D. E. Eveleigh and A. S. Perlin, *Carbohyd. Res.*, **10**, 87 (1969).
7. W. Z. Hassid, E. F. Neufeld, and D. S. Feingold, *Proc. Nat. Acad. Sci. U.S.*, **45**, 905 (1959).

8. D. O. Brummond and A. P. Gibbons, *Biochem. Biophys. Res. Commun.*, **17**, 156 (1964).
9. C. Fitting and M. Doudoroff, *J. Biol. Chem.*, **199**, 153 (1952)
10. W. A. Ayers, *J. Biol. Chem.*, **234**, 2819 (1959).
11. E. J. Hehre, *Science*, **93**, 237 (1941); *Advan. Enzymol.*, **11**, 297 (1950).
12. K. Takano and T. Miwa, *J. Biochem. (Tokyo)*, **37**, 435 (1950).
13. E. Bourquelot and M. Bridel, *Ann. Chim. Phys.*, **29**, 145 (1913).
14. S. Peat, W. J. Whelan, and K. A. Hinson, *Nature*, **170**, 1056 (1952).
15. D. E. Koshland, *Biol. Rev., Cambridge Phil. Soc.*, **28**, 416 (1953); in "Symposium on the Mechanism of Enzyme Action," W. D. McElroy and B. Glass, Eds., Johns Hopkins Press, Baltimore, Maryland, 1954, p. 608.
16. W. W. Pigman, *J. Res. Nat. Bur. Stand.*, **27**, 1 (1941); *Advan. Enzymol.*, **4**, 41 (1944).
16a. K. Wallenfels and O. P. Malhotra, *Advan. Carbohyd. Chem.*, **16**, 239 (1961).
16b. D. E. Koshland, Jr., J. A. Yankeelov, Jr., and J. A. Thoma, *Federation Proc.*, **21**, 1031 (1962).
17. M. A. Jermyn, *Science*, **125**, 12 (1957); *Rev. Pure Appl. Chem.*, **11**, 92 (1961).
18. H. Suzuki, *Sci. Rept. Tokyo Kyoiku Daigaku*, **8B**, 80 (1957).
19. K. Wallenfels, M. L. Zarnitz, G. Laule, H. Bender, and M. Keser, *Biochem. Z.*, **331**, 459 (1959).
20. K. Wallenfels and J. Fischer, *Biochem. Z.*, **321**, 223 (1960).
20a. S. Gascón, N. P. Neumann, and J. O. Lampen, *J. Biol. Chem.*, **243**, 1573 (1968).
21. E. H. Fischer and L. Kohtès, *Helv. Chim. Acta*, **34**, 1123 (1951).
22. M. A. Jermyn, *Aust. J. Biol. Sci.*, **15**, 769 (1962).
23. M. Anai, T. Ikenaka, and Y. Matsushima, *J. Biochem. (Tokyo)*, **59**, 57 (1966).
23a. D. R. Lineback, *Carbohyd. Res.*, **7**, 106 (1968).
24. G. Okada, T. Niwa, H. Suzuki, and K. Nisizawa, *J. Ferment. Technol.*, 44, 682 (1965); G. Okada, K. Nisizawa, and H. Suzuki, *J. Biochem. (Tokyo)*, **63**, 591 (1968).
25. M. R. Pollock and M. H. Richmond, *Nature*, **194**, 446 (1962).
26. E. H. Eyler, *J. Theoret. Biol.*, **10**, 891 (1965).
27. K. Wallenfels, C. Streiffer, and C. Gölker, *Biochem. Z.*, **342**, 495 (1965).
28. G. R. Craven, E. Steers, Jr., and C. B. Anfinsen, *J. Biol. Chem.*, **240**, 2468 (1965).
29. L. H. Li, R. M. Flora, and K. W. King, *Arch. Biochem. Biophys.*, **111**, 439 (1965).
30. T. Iwasaki, R. Ikeda, K. Hayashi, and M. Funatsu, *J. Biochem. (Tokyo)*, 57, 478 (1965).
31. M. A. Jermyn, *Aust. J. Biol. Sci.*, **18**, 417 (1965).
32. B. Helferich and T. Kleinschmidt, *Hoppe-Seyler's Z. Physiol. Chem.*, **324**, 211 (1961).
33. H. Sund and K. Weber, *Biochem. Z.*, **337**, 24 (1963).
34. H. Sund and K. Weber, *Angew. Chem.*, **78**, 217 (1966).
35. D. Zipser, *J. Mol. Biol.*, **7**, 113 (1963).
36. S. H. Appel, D. H. Alpers, and G. M. Tomkins, *J. Mol. Biol.*, **11**, 12 (1965).
37. U. Karlsson, S. Koorajian, I. Zabin, F. Sjøstrand, and A. Miller, *J. Ultrastruct. Res.*, **10**, 451 (1964).
38. E. Steers, Jr., G. R. Craven, and C. B. Anfinsen, *J. Biol. Chem.*, **240**, 2478 (1965).
39. R. L. Metzenberg, *Biochim. Biophys. Acta*, **89**, 291 (1964).
40. J. R. Trevithick and R. L. Metzenberg, *Biochem. Biophys. Res. Commun.*, **16**, 319 (1964).
41. I. D. Fleming and B. A. Stone, *Biochem. J.*, **99**, 13P (1965).
42. A. Gottschalk, in "The Enzymes," J. S. Sumner and K. Myrbäck, Eds., Vol. 1, Part 1, Academic Press, New York, 1950, p. 551.
43. B. Helferich and J. Johannis, *Hoppe-Seyler's Z. Physiol. Chem.*, **320**, 75 (1960).

44. A. N. Hall, S. Hollingshead, and H. N. Rydon, *Biochem. J.*, **84**, 390 (1962).
45. T. Ikenaka, S. Akabori, and S. Matsubara, *J. Biochem. (Tokyo)*, **46**, 425 (1959).
46. S. Matsubara, *J. Biochem. (Tokyo)*, **49**, 232 (1961).
47. H. Halvorson and L. Ellias, *Biochim. Biophys. Acta*, **30**, 28 (1958).
48. A. W. Phillips, *Arch. Biochem. Biophys.*, **80**, 346 (1959).
49. T. Matsubara, *J. Biochem. (Tokyo)*, **48**, 138 (1960).
50. D. H. Hutson and D. J. Manners, *Biochem. J.*, **94**, 783 (1965).
51. O. B. Jørgensen, *Acta Chem. Scand.*, **17**, 1765, 2471 (1963).
52. O. B. Jørgensen, *Acta Chem. Scand.*, **18**, 1975 (1964).
53. R. W. Bailey and A. M. Robertson, *Biochem. J.*, **82**, 272 (1962).
54. S. C. Pan, L. W. Nicholson, and P. Kolachov, *Arch. Biochem. Biophys.*, **42**, 406 (1953).
55. S. C. Pan, L. W. Nicholson, and P. Kolachov, *J. Amer. Chem. Soc.* **73**, 2547 (1951).
56. J. H. Pazur and D. French, *J. Biol. Chem.*, **196**, 265 (1952).
57. K. Saroja, R. Venkataraman, and K. V. Giri, *Biochem. J.*, **60**, 399 (1955).
58. J. H. Pazur, *J. Biol. Chem.*, **216**, 531 (1955).
59. J. H. Pazur and T. Ando, *Arch. Biochem. Biophys.*, **93**, 43 (1961).
60. M. Abdullah and W. J. Whelan, *Biochem. J.*, **75**, 12P (1960).
61. J. H. Pazur, T. Budovich, and C. L. Tipton, *J. Amer. Chem. Soc.*, **79**, 625 (1957).
62. S. A. Barker, A. Gomez-Sánchez, and M. Stacey, *J. Chem. Soc.*, 3264 (1959).
63. S. A. Barker, M. Stacey, and D. B. E. Stroud, *Nature*, **189**, 138 (1961).
64. A. R. Archibald and D. J. Manners, *Biochem. J.*, **73**, 292 (1959).
65. R. W. Bailey and B. H. Howard, *Biochem. J.*, **86**, 446 (1963).
66. W. A. M. Duncan and D. J. Manners, *Biochem. J.*, **69**, 343 (1958).
67. V. N. Nigam and K. V. Giri, *J. Biol. Chem.*, **235**, 947 (1960).
68. J. Edelman and A. J. Keys, *Biochem. J.*, **79**, 12P (1961).
69. L. G. Whitby, *Biochem. J.*, **57**, 390 (1954).
70. H. Katagiri, H. Yamada, and K. Imai, *J. Vitaminol.* (Kyoto), **3**, 264 (1957); **4**, 126 (1958); **5**, 1, 298 (1959).
71. A. Yasumura, *Sci. Rept. Tokyo Kyoiku Daigaku*, **8**, 21 (1957).
72. I. Lieberman and W. H. Eto, *J. Biol. Chem.*, **224**, 899 (1957).
73. K. D. Miller and W. H. Copeland, *J. Biol. Chem.*, **231**, 997 (1958).
74. B. Borgstrøm and A. Dahlqvist, *Acta Chem. Scand.*, **12**, 1997 (1958).
75. A. Dahlqvist, *Acta Chem. Scand.*, **13**, 1817 (1959); **14**, 1 (1960).
76. A. Dahlqvist, *Acta Chem. Scand.*, **14**, 9 (1960).
77. A. Dahlqvist, *Acta Chem. Scand.*, **14**, 1797 (1960).
78. A. Dahlqvist, *Acta Chem. Scand.*, **15**, 808 (1961).
79. J. A. Carnie and J. W. Porteous, *Biochem. J.*, **85**, 620 (1962).
80. J. A. Carnie and J. W. Porteous, *Biochem. J.*, **85**, 450 (1962).
81. A. Dahlqvist, B. Bull, and B. E. Gustafsson, *Arch. Biochem. Biophys.*, **109**, 150 (1965).
82. A. Dahlqvist, B. Bull, and D. L. Thomson, *Arch. Biochem. Biophys.*, **109**, 159 (1965).
83. S. Auricchio, G. Semenza, and A. Rubino, *Biochim. Biophys. Acta*, **96**, 498 (1965).
84. G. Semenza, S. Auricchio, A. Rubino, A. Prader, and J. D. Welsh, *Biochim. Biophys. Acta*, **105**, 386 (1965).
85. M. Davidson, *Pediatric Clin. North America*, **14**, 93 (1967).
85a. A. Dahlqvist, *J. Clin. Invest.*, **41**, 463 (1962).
85b. S. S. Huang and T. M. Bayliss, *Science*, **160**, 83 (1968).

86. N. Lejeune, D. Thinès-Sempoux, and H. G. Hers, *Biochem. J.*, **86**, 16 (1963).
87. H. N. Torrés and J. M. Olavarría, *J. Biol. Chem.*, **239**, 2427 (1964).
88. H. G. Hers, in "Control of Glycogen Metabolism," Ciba Foundation Symposium, W. J. Whelan and M. P. Cameron, Eds. Little, Brown, Boston, Massachusetts, 1964, p. 354.
89. S. Shibko and A. L. Tappel, *Biochem. J.*, **95**, 731 (1965).
90. J. W. White, Jr. and J. Maher, *Arch. Biochem. Biophys.*, **42**, 360 (1953).
91. M. Zimmermann, *Experientia*, **10**, 145 (1954).
92. H. E. Gray and G. Fraenkel, *Science*, **118**, 304 (1953).
93. J. P. Wolf III and W. H. Ewart, *Arch. Biochem. Biophys.*, **58**, 365 (1955).
94. J. S. D. Bacon and B. Dickinson, *Biochem. J.*, **66**, 289 (1957).
95. G. Avigad, *Biochem. J.*, **73**, 587 (1959).
96. R. Weidenhagen and S. Lorenz, *Angew. Chem.*, **69**, 641 (1957); *Z. Zuckerind.*, **82**, 533 (1957).
97. H. M. Tsuchiya, *Bull. Soc. Chim. Biol.*, **42**, 1777 (1960).
98. R. W. Bailey, S. A. Barker, E. J. Bourne, and M. Stacey, *J. Chem. Soc.*, 3536 (1957).
99. R. W. Bailey, *Biochem. J.*, **72**, 42 (1959).
100. E. J. Bourne, D. H. Hutson, and H. Weigel, *Biochem. J.*, **79**, 549 (1961).
101. E. J. Hehre, *J. Biol. Chem.*, **177**, 267 (1949).
102. H. J. Koepsell, H. M. Tsuchiya, N. N. Hellman, A. Kazenko, C. A. Hoffman, E. S. Sharpe, and R. W. Jackson, *J. Biol. Chem.*, **200**, 793 (1953).
103. E. S. Sharpe, F. H. Stodola, and H. J. Koepsell, *J. Org. Chem.*, **25**, 1062 (1960).
104. S. A. Barker, E. J. Bourne, P. M. Grant, and M. Stacey, *J. Chem. Soc.*, 601 (1958).
105. E. J. Bourne, J. Hartigan, and H. Weigel, *J. Chem. Soc.*, 2332 (1959).
106. E. J. Bourne, J. Hartigan, and H. Weigel, *J. Chem. Soc.*, 1088 (1961).
107. S. A. Barker, E. J. Bourne, and O. Theander, *J. Chem. Soc.*, 2064 (1957).
108. W. B. Neely, *Arch. Biochem. Biophys.*, **79**, 154 (1959).
109. Y. Suzuki, *J. Vitaminol. (Kyoto)*, **11**, 95, 320 (1965).
110. Y. Suzuki and K. Uchida, *J. Vitaminol. (Kyoto)*, **11**, 313 (1965).
111. J. H. Pazur and K. Kleppe, *J. Biol. Chem.*, **237**, 1002 (1962); *Biochem. J.*, **89**, 35P (1963).
112. M. Abdullah, I. D. Fleming, P. M. Taylor, and W. J. Whelan, *Biochem. J.*, **89**, 35P (1963).
113. J. H. Pazur, K. Kleppe, and E. M. Ball, *Arch. Biochem. Biophys.*, **103**, 515 (1963).
114. K. Myrbäck, *Ergeb. Enzymforsch.*, **10**, 168 (1949).
115. J. E. Courtois, F. Petek, and M. A. Kolahi-Zanouzi, *Bull. Soc. Chim. Biol.*, **44**, 735 (1962).
116. J. E. Courtois, M. M. Debris, and J. C. Georget, *Bull. Soc. Chim. Biol.*, **44**, 568 (1962).
117. E. P. Hill and A. S. Sussman, *J. Bacteriol.*, **88**, 1556 (1964).
118. T. M. Lukes and H. J. Phaff, *Antonie van Leuwenhoek, J. Microbiol. Serol.*, **18**, 323 (1952).
119. S. Saito, *J. Biochem. (Tokyo)*, **48**, 101 (1960).
120. S. Friedman, *Arch. Biochem. Biophys.*, **87**, 252 (1960).
121. G. F. Kalf and S. V. Rieder, *J. Biol. Chem.*, **230**, 691 (1958).
122. A. E. S. Gussin and G. R. Wyatt, *Arch. Biochem. Biophys.*, **112**, 626 (1965).
123. A. Panek and N. O. Souza, *J. Biol. Chem.*, **239**, 1671 (1964).
124. G. Avigad, O. Ziv, and E. Neufeld, *Biochem. J.*, **97**, 715 (1965).
125. C. Ceccarini, *Science*, **151**, 454 (1966).
126. K. Hansen, *Biochem. Z.*, **344**, 15 (1966).

127. J. E. Courtois and J. F. Demelier, *Bull. Soc. Chim. Biol.*, **48**, 277 (1966).
128. M. Ghione, A. Minghetti, and A. Sanfilippo, *Giorn. Microbiol.*, **7**, 94 (1959).
129. S. Veibel, in "The Enzymes," J. B. Sumner and K. Myrbäck, Eds., Vol. I, Academic Press, New York, 1950, p. 583.
130. B. Helferich and T. Kleinschmidt, *Hoppe-Seyler's Z. Physiol. Chem.*, **340**, 31 (1965).
131. B. Helferich and T. Kleinschmidt, *Naturwissenschaften*, **53**, 132 (1966).
132. T. Miwa and A. Miwa, *Medicine (Tokyo)*, **1**, 229 (1942).
133. K. Nisizawa, *Fac. Textiles Sericult.*, *Shinshu Univ.*, *Ser.* **C1**, 1 (1951).
134. S. Veibel, J. Wangel, and G. Østrup, *Biochim. Biophys. Acta*, **1**, 126 (1947).
135. R. Heyworth and P. G. Walker, *Biochem. J.*, **83**, 331 (1962).
136. B. Helferich and K. H. Jung, *Hoppe-Seyler's Z. Physiol. Chem.*, **311**, 54 (1958).
137. Y. Shibata and K. Nisizawa, *Arch. Biochem. Biophys.*, **109**, 516 (1965).
138. T. Miwa and K. Tanaka, *Symp. Enzyme Chem. (Tokyo)*, **2**, 19 (1949).
139. S. Veibel and H. Lillelund, *Hoppe-Seyler's Z. Physiol. Chem.*, **253**, 55 (1938).
140. B. Helferich and J. Gördeler, *Ber. Verhandl. Saechs Akad. Wiss. Leipzig Math. Naturw. Kl.*, **92**, 75 (1940).
141. W. W. Pigman and N. K. Richtmyer, *J. Amer. Chem. Soc.*, **64**, 374 (1942).
142. G. Wagner and R. Metzner, *Naturwissenchaften*, **52**, 83 (1965).
143. R. Weidenhagen, *Handbuch Enzymol.*, **1**, 512 (1940).
144. K. Yoshida, T. Kamada, N. Harada, and K. Kato, *Chem. Pharm. Bull. (Tokyo)*, **14**, 583 (1966).
145. T. Miwa, C. Cheng, M. Fujisaki, and A. Toishi, *Acta Phytochim. (Japan)*, **10**, 155 (1937).
146. J. A. Barnett, M. Ingram, and T. Swain, *J. Gen. Microbiol.*, **15**, 529 (1956).
147. H. K. Kihara, A. S. L. Hu, and H. O. Halvorson, *Proc. Nat. Acad. Sci. U.S.*, **47**, 489 (1961).
148. A. Herman and H. Halvorson, *J. Bacteriol.*, **85**, 901 (1963).
149. J. D. Duerksen and H. Halvorson, *J. Biol. Chim.*, **233**, 1113 (1958).
150. J. G. Kaplan, *J. Gen. Physiol.*, **48**, 873 (1965).
151. P. R. Mahadevan and B. M. Eberhart, *J. Cell. Comp. Physiol.*, **60**, 281 (1962).
152. P. R. Mahadevan and B. Eberhart, *Arch. Biochem. Biophys.*, **108**, 22, 30 (1964).
153. M. A. Jermyn, *Aust. J. Biol. Sci.*, **11**, 114 (1958).
154. M. A. Jermyn, *Aust. J. Biol. Sci.*, **16**, 926 (1963).
155. G. Youatt, *Aust. J. Biol. Sci.*, **11**, 209 (1958).
156. M. A. Jermyn, *Aust. J. Biol. Sci.*, **18**, 387, 425 (1965).
157. C. R. Krishna Murti and B. A. Stone, *Biochem. J.*, **78**, 715 (1961).
158. F. B. Anderson, W. L. Cunningham, and D. J. Manners, *Biochem. J.*, **90**, 30 (1964).
159. D. Robinson, *Biochem. J.*, **63**, 39 (1956).
160. R. Heyworth and A. Dahlqvist, *Biochim. Biophys. Acta*, **64**, 182 (1962).
161. S. Gatt, *Biochem. J.*, **101**, 687 (1966).
162. J. Rabaté, *Bull. Soc. Chim. Biol.*, **17**, 572 (1935); **20**, 449 (1938).
163. K. Takano, *J. Biochem. (Tokyo)*, **43**, 205 (1956).
164. J. E. Courtois and M. Leclerc, *Bull. Soc. Chim. Biol.*, **38**, 365 (1956).
165. J. B. Pridham and M. J. Saltmarsh, *Biochem. J.*, **74**, 42P (1962).
166. J. D. Anderson, L. Hough, and J. B. Pridham, *Biochem. J.*, **77**, 564 (1960).
167. S. A. Barker, E. J. Bourne, G. C. Hewitt, and M. Stacey, *J. Chem. Soc.*, 3734 (1955); 3541 (1957).
168. E. M. Crook and B. A. Stone, *Biochem. J.*, **65**, 1 (1957).
169. W. A. M. Duncan, D. J. Manners, and J. L. Thompson, *Biochem. J.*, **73**, 295 (1959).

170. D. H. Huston, *Biochem. J.*, **92**, 142 (1964).
171. K. W. Knox, *Biochem. J.*, **94**, 534 (1965).
172. L. S. Berger and B. M. Eberhart, *Biochem. Biophys. Res. Commun.*, **6**, 62 (1961).
173. H. W. Buston and A. Jabbar, *Biochim. Biophys. Acta*, **15**, 543 (1954).
174. K. Nisizawa, Y. Hashimoto, and Y. Shibata, in "Advances in Enzymic Hydrolysis of Cellulose and Related Materials," E. T. Reese, Ed., Macmillan (Pergamon), New York, 1963, p. 171.
175. M. Charpentier, *Ann. Inst. Pasteur*, **109**, 771 (1965).
176. P. A. J. Gorin, K. Horitsu, and J. F. T. Spencer, *Can. J. Chem.*, **43**, 2259 (1965).
177. F. Petek and To Dong, *Enzymologia*, **23**, 133 (1961).
178. J. E. Courtois, F. Petek, and To Dong, *Bull. Soc. Chim. Biol.*, **45**, 95 (1963).
179. J. J. Wohnlich, *Bull. Soc. Chim. Biol.*, **45**, 1171 (1963).
180. P. M. Dey and J. B. Pridham, *Phytochem.*, **7**, 1737 (1968).
181. M. U. Ahmed and F. S. Cook, *Can. J. Biochem. Physiol.*, **42**, 605 (1964).
182. K. Wakabayashi, *J. Fac. Eng., Shinshu Univ. Japan*, **10**, 14, 31 (1960–61); **11**, 17 (1960–61).
183. Y. T. Li and M. R. Shetlar, *Arch. Biochem. Biophys.*, **108**, 523 (1964).
184. Y. T. Li, S. C. C. Li, and M. R. Shetlar, *Arch. Biochem. Biophys.*, **103**, 436 (1963).
185. D. S. Hongess and E. H. Battley, *Federation Proc.*, **16**, 197 (1957).
186. H. Shenin and B. F. Crocker, *Can. J. Biochem. Physiol.*, **39**, 55, 63 (1961).
187. R. Schmidt and B. Rotman, *Biochem. Biophys. Res. Commun.*, **22**, 473 (1966).
188. S. Veibel in "Handbuch der Pflanzenphysiologie," W. Ruhland, Ed., Vol. 6, Springer Verlag, Berlin, 1958, p. 780.
189. A Gottschalk, in "Handbuch der Pflanzenphysiologie," W. Ruhland, Ed., Vol. 6, Springer Verlag, Berlin, 1958, p. 87.
190. R. W. Bailey and B. H. Howard, *Biochem. J.*, **87**, 146 (1963).
191. R. W. Bailey, *Biochem. J.*, **86**, 509 (1963).
192. P. S. Sastry and M. Kates, *Biochemistry*, **3**, 1280 (1964).
193. C. Anagnostopoulos, J. E. Courtois, and F. Petek, *Bull. Soc. Chim. Biol.*, **36**, 1115 (1954).
194. J. E. Courtois, F. Petek, and To Dong, *Bull. Soc. Chim. Biol.*, **43**, 1189 (1961).
195. J. J. Wohnlich, *Bull. Soc. Chim. Biol.*, **44**, 567 (1962).
196. F. Petek and J. E. Courtois, *Bull. Soc. Chim. Biol.*, **46**, 1093 (1964).
197. F. Percheron and E. Guilloux, *Bull. Soc. Chim. Biol.*, **46**, 543 (1964).
198. Y. T. Li and M. R. Shetlar, *Arch. Biochem. Biophys.*, **108**, 301 (1964).
199. W. M. Watkins, *Nature*, **181**, 117 (1958).
200. M. M. Debris, J. E. Courtois, and F. Petek, *Bull. Soc. Chim. Biol.*, **44**, 7 (1962).
201. S. M. Gillman, K. C. Tsou, and A. M. Seligman, *Arch. Biochem. Biophys.*, **98**, 229 (1962).
202. K. Wallenfels and O. P. Malhotra, *Advan. Carbohyd. Chem.*, **16**, 239 (1961).
203. K. Wallenfels and O. P. Malhotra in "The Enzymes," P. D. Boyer, H. Lardy, and K. Myrbäck, Eds., Vol. 4, Academic Press, New York, 1962, p. 409.
204. M. Cohn and J. Monod, *Biochim. Biophys. Acta*, **75**, 890 (1953).
205. S. A. Kuby and H. A. Lardy, *J. Amer. Chem. Soc.*, **71**, 890 (1953).
206. A. S. L. Hu, R. G. Wolfe, and F. J. Reithel, *Arch. Biochem. Biophys.*, **81**, 500 (1959).
207. E. Steers, Jr., G. R. Craven, and C. B. Anfinsen, *Proc. Nat. Acad. Sci. U.S.*, **54**, 1174 (1965).
208. D. Perrin, *Ann. N.Y. Acad. Sci.*, **103**, 1058 (1963).
209. K. Wallenfels and C. Gölker, *Biochem. Z.*, **346**, 1 (1966).

210. J. I. Brown, S. Koorajian, J. Katza, and I. Zabin, *J. Biol. Chem.*, **241**, 2826 (1966).
211. S. G. Korenman, G. R. Craven, and C. B. Anfinsen, *Biochim. Biophys. Acta*, **124**, 160 (1966).
212. D. B. Cowie, S. Spiegelman, R. B. Roberts, and J. D. Duerksen, *Proc. Nat. Acad. Sci. U.S.*, **47**, 114 (1961).
213. Z. Fogel and D. Elson, *Biochim. Biophys. Acta*, **80**, 601 (1964).
214. R. C. Hughes and R. W. Jeanloz, *Biochemistry*, **3**, 1536 (1964).
215. O. E. Landman, *Biochim. Biophys. Acta*, **23**, 558 (1957).
216. P. J. Anema, *Biochim. Biophys. Acta*, **89**, 495 (1964).
217. O. E. Landman, *Arch. Biochem. Biophys.*, **52**, 93 (1954).
218. W. K. Betes and D. O. Woodward, *Science*, **146**, 777 (1964).
219. G. Lester and A. Byers, *Biochem. Biophys. Res. Commun.*, **18**, 725 (1965).
220. K. M. L. Agrawal and O. P. Bahl, *J. Biol. Chem.*, **243**, 103 (1968).
221. O. Z. Sellinger, H. Beaufay, P. Jacques, A. Doyens, and C. de Duve, *Biochem. J.*, **74**, 450 (1960).
222. J. Conchie and A. J. Hay, *Biochem. J.*, **87**, 354 (1963).
223. G. Vaes and P. Jacques, *Biochem. J.*, **97**, 380 (1965).
224. A. K. Hajra, D. H. Bowen, Y. Kishimoto, and N. S. Radin, *J. Lipid Res.*, **7**, 379 (1966); J. Conchie, J. Findlay, and G. A. Levvy, *Biochem. J.*, **71**, 318 (1959).
225. S. Gatt and M. M. Rapport, *Biochim. Biophys. Acta*, **113**, 567 (1966).
226. J. C. Caygill, C. P. J. Roston, and F. R. Jevons, *Biochem. J.*, **98**, 405 (1966).
227. L. Hsu and A. L. Tappel, *Biochim. Biophys. Acta*, **101**, 83 (1965).
228. R. G. Doell and N. Kretchmer, *Biochim. Biophys. Acta*, **62**, 353 (1963); **67**, 516 (1963).
229. O. Koldovský, R. Noack, G. Schenk, V. Jirsová, A. Heringová, H. Braná, F. Chytil, and M. Fridrich, *Biochem. J.*, **96**, 492 (1965). F. Chytil, *Biochem. Biophys. Res. Commun.*, **19**, 630 (1965).
230. G. Semenza, S. Auricchio, and A. Rubino, *Biochim. Biophys. Acta*, **96**, 487 (1965).
231. J. W. Woollen and P. G. Walker, *Clin. Chim. Acta*, **12**, 647 (1965).
232. S. M. Gillman, K. C. Tsou, and A. M. Seligman, *Arch. Biochem. Biophys.*, **98**, 229 (1962).
233. A. Dahlqvist, J. B. Hammond, R. K. Crane, J. V. Dunnophy, and L. A. Littman, *Gastroenterology*, **45**, 488 (1964).
234. D. H. Alpers, *J. Biol. Chem.*, **244**, 1238 (1969).
235. D. Y. Y. Hsia, M. Makler, G. Semenza, and A. Prader, *Biochim. Biophys. Acta*, **113**, 390 (1966).
236. A. Dahlqvist and J. D. Gryboski, *Biochim. Biophys. Acta*, **110**, 635 (1965).
237. A. J. Furth and D. Robinson, *Biochem. J.*, **97**, 59 (1965).
238. R. O. Brady, A. E. Gal, J. N. Kaufer, and R. M. Bradley, *J. Biol. Chem.*, **240**, 3766 (1965).
239. S. Gatt, *J. Biol. Chem.*, **238**, PC 3131 (1963).
240. G. A. Levvy, A. McAllan, and A. J. Hay, *Biochem. J.*, **82**, 225 (1962).
241. J. E. G. Barnett, *Biochem. J.*, **96**, 72P (1965).
242. R. Got, A. Marnay, and P. Jarrige, *Experientia*, **21**, 653 (1965).
243. H. J. Horstmann, *Hoppe-Seyler's Z. Physiol. Chem.*, **337**, 57 (1964).
244. K. Wallenfels, J. Lehmann, and O. P. Malhotra, *Biochem. Z.*, **333**, 209 (1960).
245. K. Wallenfels and J. Lehmann, *Ann. Chem.*, **635**, 166 (1960).
246. B. v. Hofsten, *Biochim. Biophys. Acta*, **48**, 159 (1961).
247. B. Rotman, *Proc. Nat. Acad. Sci. U.S.*, **47**, 1981 (1961).
248. R. J. W. Byrde and A. H. Fielding, *Nature*, **205**, 390 (1965).

249. K. Wallenfels, *Angew. Chem.*, **65**, 137 (1953).
250. K. Wallenfels, E. Bernt, and G. Limberg, *Ann. Chem.*, **584**, 63 (1953).
251. M. Aronson, *Arch. Biochem. Biophys.*, **39**, 370 (1952).
252. J. H. Pazur, J. M. Marsh, and C. L. Tipton, *J. Biol. Chem.*, **233**, 277 (1958).
253. J. H. Pazur, M. Shadaksharaswamy, and A. Cepure, *Arch. Biochem. Biophys.*, **94**, 142 (1961).
254. K. Wallenfels, *Bull. Soc. Chim. Biol.*, **42**, 1715 (1960).
255. J. E. Courtois, F. Percheron, and H. N. Quang, *Bull. Soc. Chim. Biol.*, **46**, 984 (1964).
256. K. Takano, *J. Biochem. (Tokyo)*, **40**, 471 (1953).
257. W. Boos, J. Lehmann, and K. Wallenfels, *Carbohyd. Res.*, **1**, 419 (1966).
258. J. B. Pridham and K. Wallenfels, *Nature*, **202**, 488 (1964).
259. P. A. J. Gorin, J. F. T. Spencer, and H. J. Phaff, *Can. J. Chem.*, **42**, 2307 (1964).
260. F. Zilliken, P. N. Smith, C. S. Rose, and P. György, *J. Biol. Chem.*, **217**, 79 (1955).
261. R. Kuhn, H. H. Baer, and A. Gauhe, *Ber.*, **88**, 1713 (1955).
262. A. Alessandrini, E. Schmidt, F. Zilliken, and P. György, *J. Biol. Chem.*, **220**, 71 (1955).
263. G. A. Levvy and A. McAllan, *Nature*, **195**, 387 (1962).
264. D. H. Hutson and J. D. Manners, *Biochem. J.*, **63**, 545 (1964).
265. G. A. Levvy, A. J. Hay, and J. Conchie, *Biochem. J.*, **91**, 378 (1964).
266. D. J. D. Hockenhull, G. C. Ashton, K. H. Fantes, and B. K. Whitehead, *Biochem. J.*, **57**, 93 (1954).
267. J. E. Courtois, C. Chararas, and M. M. Debris, *Bull. Soc. Chim. Biol.*, **43**, 1173 (1961).
268. T. Nagaoka, *Tohoku J. Exp. Med.*, **51**, 131 (1949).
269. J. E. Courtois, F. Petek, and To Dong, *Bull. Soc. Chim. Biol.*, **44**, 11 (1962).
270. E. T. Reese and Y. Shibata, *Can. J. Microbiol.*, **11**, 167 (1965).
271. M. Adams, N. K. Richtmyer, and C. S. Hudson, *J. Amer. Chem. Soc.*, **65**, 1369 (1943).
272. B. Andersen, *Acta Chem. Scand.*, **14**, 1849 (1960).
273. K. Myrbäck and W. Schilling, *Enzymologia*, **29**, 306 (1965).
274. J. Hoshino and A. Momose, *J. Biochem. (Tokyo)*, **59**, 192 (1966).
275. F. I. Eilers, J. Allen, E. P. Hill, and A. S. Sussman, *J. Histochem. Cytochem.*, **12**, 448 (1964).
276. W. N. Arnold, *Arch. Biochem. Biophys.*, **113**, 451 (1966).
277. E. T. Reese, R. Birzgalis, and M. Mandels, *Can. J. Biochem. Physiol.*, **40**, 273 (1962).
278. J. A. Carnie and J. W. Porteous, *Biochem. J.*, **89**, 100P (1963).
279. K. Myrbäck and E. Willstaedt, *Arkiv Kemi*, **12**, 203 (1958).
280. K. Myrbäck, *Arkiv Kemi*, **12**, 839 (1958); **21**, 221 (1963).
281. K. Myrbäck, *Arch. Biochem. Biophys.*, **83**, 17 (1959).
282. K. Myrbäck, *Arkiv Kemi*, **24**, 471 (1965).
283. E. D. Wills, *Biochem. J.*, **57**, 109 (1954).
284. K. Myrbäck and E. Willstaedt, *Arkiv Kemi*, **15**, 379 (1960); **16**, 221 (1960).
285. K. Myrbäck, *Arkiv Kemi*, **15**, 519 (1960).
286. K. Myrbäck, *Arkiv Kemi*, **25**, 315 (1966).
287. J. R. Trevithick and R. L. Metzenberg, *Arch. Biochem. Biophys.*, **107**, 260 (1964).
288. S. A. Barker, E. J. Bourne, M. Stacey, and R. B. Ward, *Nature*, **175**, 203 (1955).
289. C. Neuberg and I. Mandl, in "The Enzymes," J. B. Sumner and K. Myrbäck, Eds., Vol. 1, Academic Press, New York, 1950, p. 527.

290. K. Myrbäck, in "The Enzymes," P. D. Boyer, H. Lardy, and K. Myrbäck, Eds., Vol. 4, Academic Press, New York, 1961, p. 379.
291. S. Nakamura, *Sci. Rept. Tokyo Kyoiku Daigaku*, **12**, 554 (1965).
292. J. E. Courtois, *Bull. Soc. Chim. Biol.*, **42**, 1451 (1960).
293. R. Somme and A. Wickstrøm, *Acta Chem. Scand.*, **19**, 537 (1965).
294. J. E. Courtois, A. W. Wickstrøm, and P. Le Dizet, *Bull. Soc. Chim. Biol.*, **38**, 863 (1956).
295. G. Avigad, D. S. Feingold, and S. Hestrin, *Biochim. Biophys. Acta*, **20**, 129 (1956).
296. G. Avigad, *J. Biol. Chem.*, **229**, 121 (1957).
297. J. Edelman, *Advan. Enzymol.*, **17**, 195 (1956).
298. J. Edelman and T. G. Jefford, *Biochem. J.*, **93**, 148 (1964).
299. H. E. Synder and H. J. Phaff, *J. Biol. Chem.*, **237**, 2438 (1962).
300. J. S. D. Bacon and J. Edelman, *Arch. Biochem. Biophys.*, **28**, 467 (1950).
301. P. H. Blanchard and N. Albon, *Arch. Biochem. Biophys.*, **29**, 220 (1950).
302. J. S. D. Bacon, *Biochem. J.*, **57**, 320 (1954).
303. F. J. Bealing, *Biochem. J.*, **55**, 93 (1953).
304. J. Edelman, *Biochem. J.*, **57**, 22 (1954).
305. K. Ishizawa, Y. Iriki, and T. Miwa, *Sci. Rept. Tokyo Kyoiku Daigaku*, **8**, 103 (1957).
306. W. J. Whelan and D. M. Jones, *Biochem. J.*, **54**, xxxiv (1953).
307. J. Edelman, *Bull. Soc. Chim. Biol.*, **42**, 1737 (1960).
308. I. S. Bhatia, M. N. Satyanarayana, and M. S. Srinivasan, *Biochem. J.*, **61**, 171 (1955).
309. R. W. Henderson, R. K. Morton, and W. A. Rawlinson, *Biochem. J.*, **72**, 340 (1959).
310. N. Albon, D. J. Bell, P. H. Blanchard, D. Gross, and J. T. Rundell, *J. Chem. Soc.*, 24 (1954).
311. S. A. Barker, E. J. Bourne, and T. R. Carrington, *J. Chem. Soc.*, 2125 (1954).
312. D. Gross, P. H. Blanchard, and D. J. Bell, *J. Chem. Soc.*, 1727 (1954).
312a. W. W. Binkley and F. W. Altenburg, *Int. Sugar J.*, 67, 110 (1965).
313. J. S. D. Bacon, *Biochem. J.*, **73**, 507 (1959).
314. J. Edelman, and A. G. Dickerson, *Biochem. J.*, **98**, 787 (1966).
315. J. H. Pazur, *J. Biol. Chem.*, **199**, 217 (1957).
316. H. J. Breuer and J. S. D. Bacon, *Biochem. J.*, **66**, 462 (1957).
317. P. G. Walker, J. W. Woollen, and R. Heyworth, *Biochem. J.*, **79**, 288, 294 (1961).
318. L. Zechmeister, G. Tóth, and É. Vajda, *Enzymologia*, 7, 170 (1939).
319. S. Roseman and A. Dorfman, *J. Biol. Chem.*, **191**, 607 (1951).
320. B. Weissmann, G. Rowin, J. Marshall, and D. Friederici, *Biochemistry*, **6**, 207 (1967).
321. G. Rowin and B. Weissmann, *Federation Proc.*, **24**, 220 (1965).
322. B. Weissmann and D. Friederici, *Biochim. Biophys. Acta*, **117**, 498 (1966).
323. D. Pugh and P. G. Walker, *J. Histochem. Cytochem.*, **9**, 242 (1961).
324. M. Hayashi, *J. Histochem. Cytochem.*, **13**, 355 (1965).
325. J. W. Woollen and P. Turner, *Clin. Chim. Acta*, **12**, 671 (1965).
326. J. Findlay and G. A. Levvy, *Biochem. J.*, **77**, 170 (1960).
327. J. C. Caygill, C. P. J. Roston, and F. R. Jevons, *Biochem. J.*, **98**, 405 (1966).
328. B. Weissmann, S. Hadjiioannou. and J. Tornheim, *J. Biol. Chem.*, **239**, 59 (1964).
329. B. Buddecke and E. Werries, *Hoppe-Seyler's Z. Physiol. Chem.*, **340**, 257 (1965).
330. J. C. Caygill and F. R. Jevons, *Clin. Chim. Acta*, **11**, 233 (1965).
331. J. Findlay, G. A. Levvy, and C. A. Marsh, *Biochem. J.*, **69**, 467 (1958).
332. J. C. Caygill, *Biochem. J.*, **98**, 9P (1966).
333. Y. Z. Frohwein and S. Gatt, *Israel J. Biochem.*, **3**, 106 (1965); *Biochemistry*, **6**, 2783 (1967).

334. P. Jollès, *Angew. Chem.*, **76**, 20 (1964).
335. N. Sharon and S. Seifter, *J. Biol. Chem.*, **239**, PC 2398 (1964).
336. T. Osawa, *Carbohyd. Res.*, **1**, 435 (1966).
337. G. Schiffman, C. Howe, and E. A. Kabat, *J. Amer. Chem. Soc.*, **80**, 6662 (1958).
338. S. Iseki and K. Furukawa, *Proc. Jap. Acad.*, **35**, 620 (1959).
339. G. J. Harrap and W. M. Watkins, *Biochem. J.*, **93**, 9P (1964).
340. H. Tuppy and W. L. Staudenbauer, *Biochemistry*, **5**, 1742 (1966).
341. D. Maass, H. Pelzer, and W. Weidel, *Z. Naturforsch.*, **19B**, 413 (1964).
341a. O. P. Bahl and K. M. L. Agrawal, *J. Biol. Chem.*, **243**, 98 (1968).
342. S. G. Nathenson, N. Ishimoto, J. S. Anderson, and J. L. Strominger, *J. Biol. Chem.*, **241**, 651 (1966).
343. D. J. Tipper, J. M. Ghuysen, and J. L. Strominger, *Biochemistry*, **4**, 468 (1965).
344. L. R. Berger and D. M. Reynolds, *Biochim. Biophys. Acta*, **29**, 552 (1958).
345. C. Jeuniaux, "Chitine et Chitinolyse," Masson et Cie, Paris, 1963, p. 15.
346. H. Takeda, G. A. Strasdine, D. R. Whitaker, and C. Roy, *Can. J. Biochem.*, **44**, 509 (1966).
347. A. Otakara, *Agr. Biol. Chem. (Tokyo)*, **27**, 454 (1963).
348. R. Kuhn and H. Tiedemann, *Ber.*, **87**, 1142 (1954).
349. R. F. Powning and H. Irzykiewicz, *Comp. Biochem. Physiol.*, **12**, 405 (1964); **14**, 127 (1965).
350. S. A. Barker, A. B. Foster, L. I. Khmelnitshi, and J. M. Webber, *Bull. Soc. Chim. Biol.*, **42**, 1799 (1960).
351. W. H. Fishman, in "Methods in Enzymology," S. P. Colowick and N. O. Kaplan, Eds., Vol. III, Academic Press, New York, 1955, p. 262.
352. G. A. Levvy and C. A. Marsh, *Advan. Enzymol.*, **14**, 381 (1959).
353. R. S. Teague, *Advan. Carbohyd. Chem.*, **9**, 186 (1958).
354. R. H. Nimmo-Smith, *Biochim. Biophys. Acta*, **50**, 166 (1964).
355. M. A. Verity, R. Caper, and W. J. Brown, *Arch. Biochem. Biophys.*, **106**, 386 (1964).
356. J. A. R. Mead, J. N. Smith, and R. T. Williams, *Biochem. J.*, **61**, 569 (1955).
357. M. L. Doyle, P. A. Katzman, and E. A. Doisy, *J. Biol. Chem.*, **217**, 921 (1955).
358. R. F. Jacox, *J. Bacteriol.*, **65**, 700 (1953).
359. M. C. Karunairatnam and G. A. Levvy, *Biochem. J.*, **49**, 210 (1951).
360. C. A. Marsh, *Biochem. J.*, **58**, 609 (1965).
361. P. G. Walker and G. A. Levvy, *Biochem. J.*, **49**, 620 (1951).
362. C. de Duve, B. C. Pressman, R. Gianetto, R. Wattiaux, and F. Appelmans, *Biochem. J.*, **60**, 604 (1955).
363. R. Gianetto, *Can. J. Biochem.*, **42**, 499 (1964).
364. J. P. Manning, L. F. Cavazos, W. M. Feagans, and R. Moss, *J. Histochem. Cytochem.*, **11**, 383 (1963).
365. W. H. Fishman, Y. Nakajima, C. Anstiss, and S. Green, *J. Histochem. Cytochem.*, **12**, 298 (1964).
366. H. Kubler and E. Frieden, *Biochim. Biophys. Acta*, **93**, 635 (1964).
367. P. Bernfeld and W. H. Fishman, *J. Biol. Chem.*, **202**, 757 (1953).
368. P. Bernfeld, J. S. Nisselbaum, and W. H. Fishman, *J. Biol. Chem.*, **202**, 763 (1953).
369. G. A. Levvy, A. McAllan, and C. A. Marsh, *Biochem. J.*, **69**, 22 (1958).
370. O. S. Pettengill and W. H. Fishman, *J. Biol. Chem.*, **237**, 24 (1962).
371. B. V. Plapp, T. R. Hopkins, and R. D. Cole, *J. Biol. Chem.*, **238**, 3315 (1963).
372. B. U. Musa, R. P. Doe, and U. S. Seal, *J. Biol. Chem.*, **240**, 2811 (1965).
373. M. Wakabayashi and W. H. Fishman, *Biochim. Biophys. Acta*, **48**, 195 (1961).
374. A. Alfsen-Blanc, *Bull. Soc. Chim. Biol.*, **41**, 1462 (1959).

375. M. Wakabayashi and W. H. Fishman, *J. Biol. Chem.*, **236**, 996 (1961).
376. J. F. Becker, *Biochim. Biophys. Acta*, **100**, 574, 582 (1965).
377. G. A. Levvy, *Biochem. J.*, **58**, 462 (1954).
378. C. A. Marsh and G. A. Levvy, *Biochem. J.*, **68**, 610 (1958).
379. W. H. Fishman and S. Green, *J. Biol. Chem.*, **225**, 435 (1956).
380. K. Takano and T. Miwa, *J. Biochem.* (*Tokyo*), **40**, 471 (1953).
381. B. H. Howard, G. Jones, and M. R. Purdom, *Biochem. J.*, **74**, 173 (1960).
382. Y. Morita, *J. Biochem.* (*Tokyo*), **43**, 7 (1956).
383. J. Conchie and G. A. Levvy, *Biochem. J.*, **65**, 389 (1957).
384. D. Fisher, M. Higam, P. W. Kent, and P. Pritchard, *Biochem. J.*, **98**, 46P (1966).
385. M. E. Rafelson, Jr., *Exposés Ann. Biochim. Méd.*, **24**, 121 (1963).
386. E. Romanovska and W. M. Watkins, *Biochem. J.*, **87**, 37P (1963).
387. V. W. Wilson, Jr. and M. E. Rafelson, Jr., *Biochem. Prep.*, **10**, 113 (1963).
388. J. T. Seto, R. Drzeniek, and R. Rott, *Biochim. Biophys. Acta*, **113**, 402 (1966).
389. E. Mohr and G. Schramm, *Z. Naturforsch.*, **15B**, 568, 575 (1960).
390. L. Warren and C. W. Spearing, *J. Bacteriol.*, **86**, 950 (1963).
391. J. T. Cassidy, G. W. Jourdian, and S. Roseman, *J. Biol. Chem.*, **240**, 3501 (1965).
392. R. M. Burton, *J. Neurochem.*, **10**, 503 (1953).
393. R. Carubelli, R. E. Trucco, and R. Caputto, *Biochim. Biophys. Acta*, **60**, 196 (1962).
394. G. L. Ada, *Biochim. Biophys. Acta*, **73**, 276 (1963).
395. Th. A. C. Boschman and J. Jacobs, *Biochem. Z.*, **342**, 532 (1965).
396. R. A. Gibbons, *Biochem. J.*, **89**, 380 (1963); M. Bertolini and W. Pigman, *J. Biol. Chem.*, **242**, 3776 (1967).
397. A. Rosenberg, B. Binnie, and E. Chargaff, *J. Amer. Chem. Soc.*, **82**, 4113 (1960).
398. G. A. Levvy and A. McAllan, *Biochem. J.*, **80**, 435 (1961).
399. J. Conchie and A. J. Hay, *Biochem. J.*, **87**, 354 (1963).
400. W. M. Watkins, *Biochem. J.*, **71**, 261 (1959).
400a. K. Tanaka, T. Nakano, S. Noguchi, and W. Pigman, *Arch. Biochem. Biophys.*, **126**, 624 (1968).
401. G. A. Levvy and A. McAllan, *Biochem. J.*, **87**, 206, 361 (1963).
402. E. L. Rosenfeld and G. Y. Wiederschein, *Bull. Soc. Chim. Biol.*, **47**, 1433 (1965).
403. S. A. Barker, P. J. Somers, and M. Stacey, *Carbohyd. Res.*, **1**, 106 (1965).
404. D. French, M. Levine, E. Norberg, P. Nordin, J. H. Pazur, and G. M. Wild, *J. Amer. Chem. Soc.*, **77**, 2387 (1954).
405. J. H. Pazur, J. M. Marsh, and T. Ando, *J. Amer. Chem. Soc.*, **81**, 2170 (1959).
406. J. A. Thoma, S. Dygert, and K. Hsue, *Anal. Biochem.*, **13**, 91 (1965).
407. D. French, P. M. Taylor, and W. J. Whelan, *Biochem. J.*, **90**, 616 (1964).
408. S. Peat, W. J. Whelan, and G. W. F. Kroll, *J. Chem. Soc.*, 53 (1956); G. Jones and W. J. Whelan, *Carbohyd. Res.*, **9**, 483 (1969).
408a. D. J. Manners and K. L. Rowe, *Carbohyd. Res.*, **9**, 441 (1969).
409. S. Peat, W. J. Whelan, and G. Jones, *J. Chem. Soc.*, 2490 (1957).
410. M. Abdullah and W. J. Whelan, *Arch. Biochem. Biophys.*, **112**, 592 (1965).
411. G. J. Walker, *Biochem. J.*, **94**, 299 (1965).
412. D. H. Brown and B. Illingworth, in "Control of Glycogen Metabolism," Ciba Foundation Symposium, W. J. Whelan and M. P. Cameron, Eds., Little, Brown, Boston, Massachusetts, 1964, p. 139.
413. J. Monod and A. M. Torriani, *Ann. Inst. Pasteur*, **78**, 65 (1950).
414. S. A. Barker and E. J. Bourne, *J. Chem. Soc.*, 209 (1952).
415. H. Wiesmeyer and M. Cohn, *Biochim. Biophys. Acta*, **39**, 417, 427 (1960).
416. S. A. Barker and M. A. Farisi, *Carbohyd. Res.*, **1**, 97 (1965).
417. M. Burger and E. Pavlasová, *Biochem. J.*, **93**, 601 (1964).

34. BIOSYNTHESIS OF SUGARS AND POLYSACCHARIDES*

W. Z. HASSID

I. Introduction 302
II. Monosaccharides, Phosphorylated Sugars, and Nucleoside Ortho- and Pyrophosphate Derivatives . 304
III. Mechanisms Involved in Enzymic Formation of Complex Saccharides 312
IV. Reactions Catalyzed by Disaccharide Phosphorylases (D-Glucosyltransferases) 314
   A. Sucrose and Sucrose Analogs 314
   B. Maltose. 315
   C. Cellobiose 316
   D. Laminarabiose 316
V. Synthesis of Disaccharides by Transglucosylation . 316
VI. Synthesis of Oligosaccharides from Sugar Nucleotides 319
   A. Sucrose 319
   B. Raffinose 320
   C. Stachyose 321
   D. Lactose. 322
   E. Trehalose 324
VII. Synthesis of Glycosides from Sugar Nucleotides . 325
VIII. Synthesis of Polysaccharides by Phosphorolysis and Transglycosylation 327
   A. Glycogen and Starch 327
   B. The Amylopectin Type of Polysaccharides from Sucrose 332

* Among the abbreviations used in this chapter are the following: AMP, GMP, IMP, UMP, CMP, TMP = the 5′-phosphates of β-D-ribofuranosyl-nucleosides of adenine, guanine, hypoxanthine, uracil, cytosine, thymine; ADP, etc. = the 5′-pyrophosphates of adenosine, etc.; ATP, etc. = the 5′-triphosphates of adenosine, etc.; ADP-D-glucose = adenosine 5′-(D-glucopyranosyl pyrophosphate); CDP-D-ribitol = cytidine 5′-(1-deoxy-D-ribit-1-yl pyrophosphate); CDP-D-glycerol = cytidine 5′-(1-deoxy-D-glycer-1-yl pyrophosphate); CoA and acetyl CoA = coenzyme A and its acyl derivatives; NAD⁺, NADH = nicotinamide adenine dinucleotide and its reduced form; NADP⁺, NADPH = nicotinamide adenine dinucleotide phosphate and its reduced form; P_i, PP_i = orthophosphate and pyrophosphate; TDP-L-rhamnose = thymidine 5′-(L-rhamnopyranosyl pyrophosphate); UDP-D-glucose = uridine 5′-(α-D-glucopyranosyl pyrophosphate); UDP-D-glucuronic acid = uridine 5′-(D-glucopyranosyluronic acid pyrophosphate). The formulas for many of the nucleoside moieties are given in this volume, Chap. 29.

C. Amylose from Maltose 333
D. Cyclic Amylosaccharides 334
E. Dextran from Sucrose 334
F. Dextran from Amylodextrin 335
G. Levan from Sucrose 335
IX. Synthesis of Polysaccharides from Sugar Nucleotides 336
A. Glycogen 336
B. Chitin 339
C. Starch 340
D. Cellulose 342
E. D-Gluco-D-mannan 345
F. D-Xylan 346
G. Pectin 347
H. (1→3)-β-D-Glucans 348
I. Alginic Acid 349
J. DL-Galactan 350
K. Hyaluronic Acid 351
L. Capsular Pneumococcal Polysaccharides . . 352
M. Bacterial Cell-Wall Lipopolysaccharides . . 353
N. Teichoic Acids 356
O. Glycoproteins 358
P. Sulfated Mucopolysaccharides 361
References 362

I. INTRODUCTION

Plants are considered to be the chief producers of carbohydrates in Nature by the process of photosynthesis. The overall process is represented by the empirical equation

$$CO_2 + H_2O \xrightarrow{\text{light}} (CH_2O) + O_2; \quad \Delta G^0 = +115 \text{ kcal mole}^{-1}$$

The reaction is strongly endergonic and is driven by the absorption of light energy.

Almost all forms of life that are unable to conduct photosynthesis depend either directly or indirectly on the assimilation of carbon dioxide by plants. Exceptions are those few types of bacteria that can fill all their energy requirements from chemical work alone, by oxidizing inorganic substances, such as sulfur or hydrogen sulfide. All the organic substances arising from the photosynthetic process serve the other forms of life as starting materials for their diverse metabolic interconversions.

In the photosynthetic process of plants a carboxylation reaction occurs in which D-erythro-pentulose 1,5-diphosphate serves as the acceptor of carbon dioxide for the formation of glyceric acid 3-phosphate.[1,2] This compound subsequently undergoes a series of enzymic reactions resulting in the formation of a number of phosphorylated monosaccharide derivatives in the photosynthetic carbon-reduction cycle.

In this photosynthetic cycle some of the following phosphorylated monosaccharides are formed: D-erythrose 4-phosphate, D-fructose 6-phosphate, D-sedoheptulose 7-phosphate, D-*threo*-pentulose ("D-xylulose") 5-phosphate, D-ribose 5-phosphate, and D-*erythro*-pentulose ("D-ribulose") 5-phosphate. Plants contain various specific phosphatases capable of hydrolyzing these phosphorylated sugars to free sugars and in some cases causing accumulation of large concentrations of these sugars, such as D-glucose and D-fructose.

A D-mannose phosphate isomerase that interconverts D-mannose 6-phosphate and D-fructose 6-phosphate has been found in muscle, red blood cells, and other mammalian tissues.[3] Also a D-mannose isomerase that interconverts D-fructose and D-mannose was discovered in *Pseudomonas saccharophila*.[3]

Although D-mannose is present in plants in combined form as polysaccharide, there is no known direct mechanism for its formation. But since a D-gluco-D-mannan is formed from GDP-D-mannose-^{14}C by a particulate plant enzyme,[4] it can be assumed that an isomerase is present in the plant that interconverts D-mannose and D-glucose.

D-Glucose 6-phosphate, which is formed in the photosynthetic cycle, may also be diverted into another pathway, forming a variety of sugar phosphate and sugar nucleotide derivatives, some of which are known to be precursors of certain oligosaccharides and polysaccharides. The D-glucose 6-phosphate is converted by phosphate glucomutase into α-D-glucopyranosyl phosphate,[5,6] and the latter reacts with uridine 5′-triphosphate in the presence of a pyrophosphorylase found in plants,[7] yeast, and animal tissues, resulting in the formation of UDP-D-glucose. Enzymes (4-epimerases) are present in living cells from various sources capable of interconverting UDP-D-glucose and UDP-D-galactose.[7,8] Plant as well as animal tissues and microorganisms[9-11] contain a UDP-D-glucose dehydrogenase that oxidizes UDP-D-glucose to UDP-D-glucuronic acid; the latter can be epimerized to UDP-D-galacturonic acid and also decarboxylated to a UDP-pentose in plants. The resulting pentose nucleotide is obtained as a mixture of UDP-D-xylose and UDP-L-arabinose, which indicates the presence of a 4-epimerase capable of interconverting the two pentose nucleotides.

Plant and animal tissues have been shown to contain transglycosylases capable of catalyzing the transfer of D-glucose, D-galactose, 2-acetamido-2-deoxy-D-glucose, or other monosaccharides from sugar nucleotides containing various bases (uracil, guanine, thymine, adenine, cytosine) to sugar acceptors,

* *References start on p. 362.*

forming complex saccharides such as sucrose,[12] sucrose phosphate,[13] trehalose,[14] lactose,[15,16] raffinose,[17] cellulose,[18,19] (1 → 3)-β-D-glucan,[20] starch,[21,22] glycogen,[23] and pectin.[24] There is good reason to believe that most of the other complex saccharides numerous in Nature are formed from sugar nucleotide precursors.

Animals obtain monosaccharide sugars either from ingestion of carbohydrate material or by resynthesis, through reversal of the glycolytic cycle, whereby pyruvate is produced and is converted into glycogen. Some animal polysaccharides may also be formed from lactic acid, amino acids, or glycerol. Microorganisms utilize monosaccharides from the medium on which they grow and synthesize complex saccharides from them.

This chapter will be concerned with the mechanisms involved in the synthesis of monosaccharides and complex saccharides in living cells.

II. MONOSACCHARIDES, PHOSPHORYLATED SUGARS, AND NUCLEOSIDE ORTHO- AND PYROPHOSPHATE DERIVATIVES

As previously mentioned, D-fructose 6-phosphate is produced in the photosynthetic carbon-reduction cycle. Isomerases that catalyze the interconversion of aldose and ketose sugars are present in plants and animal tissues, yeast, and microorganisms.[3]

For example, a hexose phosphate isomerase (D-glucose 6-phosphate aldo-keto-isomerase) catalyzes the following interconversion:

$$\text{D-Fructose 6-phosphate} \rightleftharpoons \text{D-Glucose 6-phosphate}$$

An epimerase, such as UDP-D-galactose 4-epimerase, catalyzes an inversion in which the configuration about C-4 of the D-glucosyl residue is transformed into that of a D-galactosyl residue[25] by the reaction UDP-D-glucose $\rightleftharpoons$ UDP-D-galactose. This epimerase occurs in animal tissues, plants, yeast, and other microorganisms. D-Glucose and D-fructose, which are found as free sugars in higher plants, are probably derived from their phosphorylated derivatives through hydrolysis by phosphatases. D-Glucose occurs in closely controlled concentration in the blood of most animals. Free D-fructose is found in secretion of the seminal vesicle in many mammals.[26] D-Galactose is not found in free form in either plant or animal tissues, but it is widely distributed in nature in combined form. L-Galactose also exists as a polysaccharide constituent in plants[27-29] and in the L-galactan of the snail *Helix pomatia*.[30]

A nonspecific D-mannose isomerase that interconverts D-fructose and D-mannose has been discovered by Palleroni and Doudoroff[31] in mutant strains of *Pseudomonas saccharophila*. Another isomerase, which interconverts D-mannose 6-phosphate and D-fructose 6-phosphate, has been found in muscle, red blood cells, and other mammalian tissues, and in yeast.[32]

Although D-mannose is present in plants in combined form as polysaccharide, there is no known mechanism for formation of D-mannose in plants. However, since it is known that a particulate preparation from mung beans forms a D-gluco-D-mannan when GDP-D-mannose is used as substrate,[4] there is an indication that an epimerase is present in plants capable of interconverting GDP-D-mannose and GDP-D-glucose.

UDP-2-acetamido-2-deoxy-D-glucose was first isolated from yeast[33] and has been detected in mammalian liver,[34-36] in hen oviduct,[37] and in mungbean seedlings.[38] 2-Amino-2-deoxy-D-glucose is also a component of glycolipids in higher plants.[39]

Extracts from *Neurospora crassa* convert hexose phosphate and glutamine into hexosamine.[40] Experiments with purified enzymes[41] from *Escherichia coli*, rat liver, and *Neurospora* indicate that these enzymes catalyze the conversion of D-fructose 6-phosphate and L-glutamine into 2-amino-2-deoxy-D-glucose 6-phosphate and glutamic acid, and that the hexosamine nitrogen is derived from L-glutamine. 2-Amino-2-deoxy-D-glucose 6-phosphate can also be formed from D-fructose 6-phosphate and ammonium ion in the presence of appropriate enzyme and 2-amino-2-deoxy-D-glucose 6-phosphate deaminase.[42] An enzyme has also been found in rat liver extracts[43] that produces 2-amino-2-deoxy-D-glucose 6-phosphate directly from D-glucose 6-phosphate and L-glutamine. An enzyme that acetylates 2-amino-2-deoxy-D-glucose 6-phosphate in the presence of acetyl CoA to form 2-acetamido-2-deoxy-D-glucose 6-phosphate is present in pigeon liver,[44] and in a strain of group A hemolytic streptococcus, rabbit and human liver, and other animal organs.[45]

The general process of nucleoside (glycosyl pyrophosphate) formation usually involves the transfer of a nucleotidyl group from a nucleoside triphosphate to a glycosyl phosphate, with the simultaneous release of pyrophosphate according to the following general reaction:[46,47]

Nucleoside triphosphate + glycosyl phosphate

pyrophosphorylase

Nucleoside (glycosyl pyrophosphate) + PP$_i$

Most of the nucleoside (glycosyl pyrophosphates) containing different bases and different sugar moieties were found to be synthesized by this enzymic process.

A similar reaction[48,49] leads to the synthesis of a nucleoside (glycosyl orthophosphate) which occurs as follows:

Cytidine triphosphate + neuraminic acid $\longrightarrow$ CMP-neuraminic acid + PP$_i$

The enzymic synthesis of the important cell wall component of bacteria, muramic acid (Vol. IIB, Chap. 41), from UDP-*N*-acetylmuramic acid (1) is

* References start on p. 362.

believed to occur in two steps. First, there is a reaction between UDP-2-acetamido-2-deoxy-D-glucose and "phosphopyruvate," with the liberation of P_i. The product formed is then reduced with NADH to give UDP-N-acetylmuramic acid.[50,51]

Synthesis of CMP-N-acetylneuraminic acid (2) has been shown[52] to occur in *E. coli* K-235. Its synthesis in animal tissues was shown[48,53] to take place

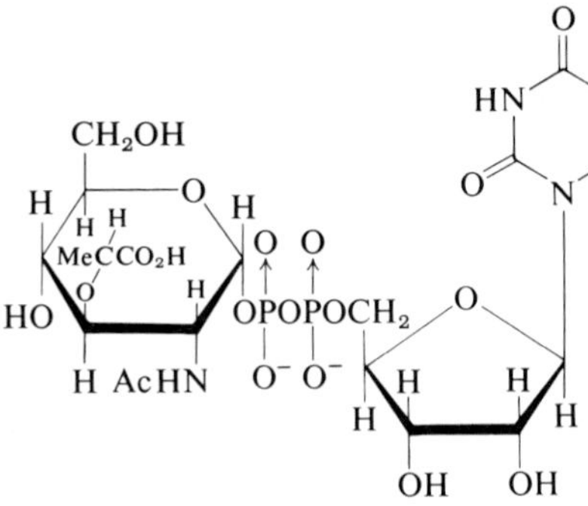

UDP-N-acetylmuramic acid (1)

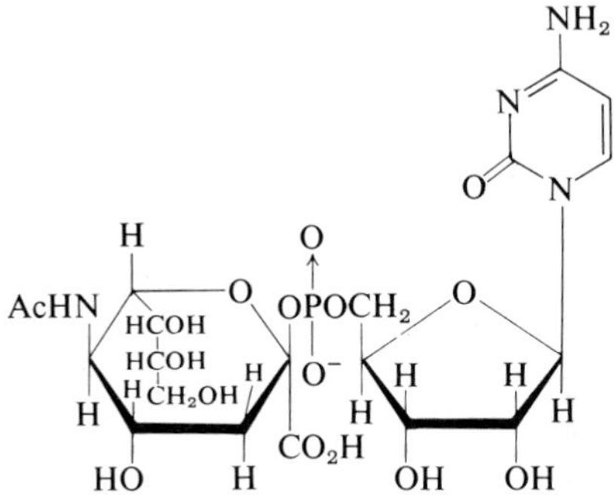

CMP-N-acetylneuraminic acid (2)

by the following pathway:

2-Acetamido-2-deoxy-D-mannose 6-phosphate + enolpyruvate phosphate $\longrightarrow$
N-Acetylneuraminic acid 9-phosphate + P_i
N-Acetylneuraminic acid 9-phosphate $\longrightarrow$ N-Acetylneuraminic acid + P_i
N-Acetylneuraminic acid + CTP $\longrightarrow$ CMP-N-acetylneuraminic acid + PP_i

The acyl-substituted neuraminic acids are referred to as sialic acids. There are several known sialic acids (see Glycoproteins, Vol. IIB, Chap. 40). They occur in complex oligosaccharides and mucolipids in vertebrate and invertebrate tissues.[54-56] They are constituents of a variety of milk oligosaccharides

(this volume, Chap. 30), glycoproteins, and certain glycolipids called gangliosides (Vol. IIB, Chap. 44). The enzymic synthesis of one of these compounds,
N-acetylneuraminic acid, has been shown to take place by the reaction shown
in the following scheme (3)

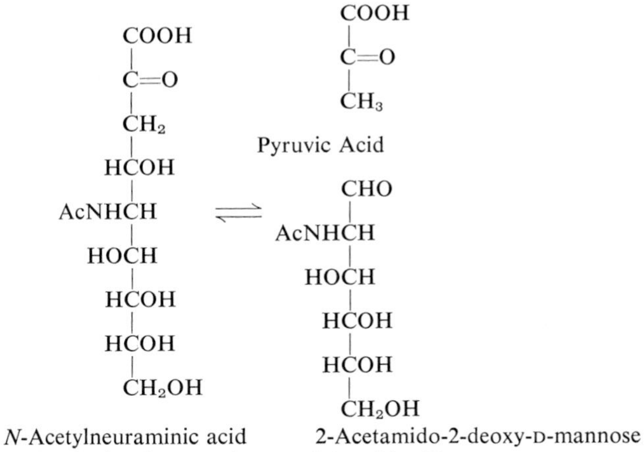

$$
\begin{array}{ccc}
& & \text{COOH} \\
& \text{COOH} & | \\
& | & \text{C}=\text{O} \\
& \text{C}=\text{O} & | \\
& | & \text{CH}_3 \\
& \text{CH}_2 & \\
& | & \text{Pyruvic Acid} \\
& \text{HCOH} & \\
& | & \text{CHO} \\
\text{AcNHCH} & \rightleftharpoons & | \\
& | & \text{AcNHCH} \\
& \text{HOCH} & | \\
& | & \text{HOCH} \\
& \text{HCOH} & | \\
& | & \text{HCOH} \\
& \text{HCOH} & | \\
& | & \text{HCOH} \\
& \text{CH}_2\text{OH} & | \\
& & \text{CH}_2\text{OH}
\end{array}
$$

N-Acetylneuraminic acid   2-Acetamido-2-deoxy-D-mannose
Synthesis of N-acetylneuraminic acid   (3)

This reaction is catalyzed by a specific aldolase isolated from a microorganism[57] and from animal tissues.[58] The enzyme in the reverse reaction splits
sialic acid to yield pyruvate and the corresponding 2-acylamido-2-deoxy-D-
mannose. Besides N-acetylneuramic acid, the following sialic acids (acylneura-
minic acids) have been shown to be synthesized similarly by an aldolase
reaction: N-glycolylneuraminic acid, N,O-diacetylneuraminic acid, and N-
acetyldi-O-acetylneuraminic acid.

The biosynthesis of the carbohydrate constituent of glycoproteins and of
gangliosides results from a sequence of reactions involving the conversion of
D-glucose into other monosaccharides, and polymerization of the glycose
units.[59]

The pathway of reactions leading from D-glucose to the "activated" form of
N-acetylneuraminic acid—that is, CMP-N-acetylneuraminic acid—has
been shown[60] to be as follows: D-fructose 6-phosphate $\rightarrow$ 2-amino-2-deoxy-
D-glucose 6-phosphate $\rightarrow$ various 2-amino-2-deoxy-D-glucose derivatives $\rightarrow$
derivatives of 2-amino-2-deoxy-D-mannose $\rightarrow$ sialic acids and 9-phosphate
esters, and, finally, to CMP-N-acetylneuraminic acid, which serves as the
sialic acid donor for glycoproteins and gangliosides.

The incorporation of sialic acids, N-acetylneuraminic acid, and N-glycolyl-
neuraminic acid into polymers is catalyzed by a family of enzymes known as

* *References start on p. 362.*

sialyltransferases. The latter effect the transfer of sialic acid from the nucleo-tide derivative to an appropriate acceptor.[61] The sialyltransferases can be dis-tinguished from one another on the basis of their specificities toward the acceptor molecules, or on the basis of the chemical structure of the products. Thus, the first sialyltransferase detected in mammalian tissues[62] was obtained from the mammary gland of the lactating rat; it required β-D-galactopyrano-side as acceptor. A similar requirement was found for a transferase isolated from colostrum.[63] However, whereas the enzyme of mammary gland pro-duced one isomer of sialyl-lactose as a product, the sialic acid being linked to C-3 of the D-galactose residue, the colostrum enzyme gave the other known isomer, sialyl-(2 → 6)-lactose.[64]

If CMP-N-acetylneuraminic acid is activated, N-acetylneuraminic acid can serve as a donor in the presence of a sialyltransferase in a polymerization reaction for the formation of colominic acid. This polysaccharide was shown to be formed extracellularly by E. coli.[65] The reaction is as follows:[61]

n(CMP-N-acetylneuraminic acid) + colominic acid ⟶
       (N-Acetylneuraminic acid)$_n$-colominic acid + n(cytidine monophosphate)

The sialyltransferases are known to catalyze several other similar reactions with the production of CMP in each case.
Rat mammary gland:

CMP-N-acetylneuraminic acid + D-galactoside (lactose) ⟶
       Sialyl-D-galactoside   [for example, sialyl-(2 → 3)-lactose]

Goat colostrum:

CMP-N-acetylneuraminic acid + D-galactosyl-glycoprotein ⟶
       Sialyl-D-galactosyl-glycoprotein

Sheep submaxillary gland:

CMP-N-acetylneuraminic acid + sheep submaxillary mucin (pretreated with sialidase)
       ⟶ Sialyl-2-acetamido-2-deoxy-D-galactosyl-mucin

Embryonic chicken brain:

CMP-N-acetylneuraminic acid + ceramide-lactoside ⟶ Sialyl-D-galactosyl-ceramide-glucosyl-ceramide

Levin and Racker[66] synthesized 3-deoxy-D-*manno*-octulosonic acid by an enzymic preparation from E. coli J-5 by the following reaction:

D-Arabinose 5-phosphate + enolpyruvate phosphate ⟶
       3-Deoxy-D-*manno*-octulosonic acid + P_i

CMP-3-deoxy-D-*manno*-octulosonate (**4**) has been obtained[66a] by use of a synthetase from *E. coli* 0111-B$_4$ as follows:

Cytidine triphosphate + 3-deoxy-D-*manno*-octulosonic acid $\longrightarrow$
CMP-3-deoxy-D-*manno*-octulosonic acid + PP$_i$

The glycosyl moiety of the nucleoside sugar 5′-pyrophosphates formed by the pyrophosphorylase reaction is capable of undergoing a number of transformations. The first of these reactions to be studied was the 4-epimerization

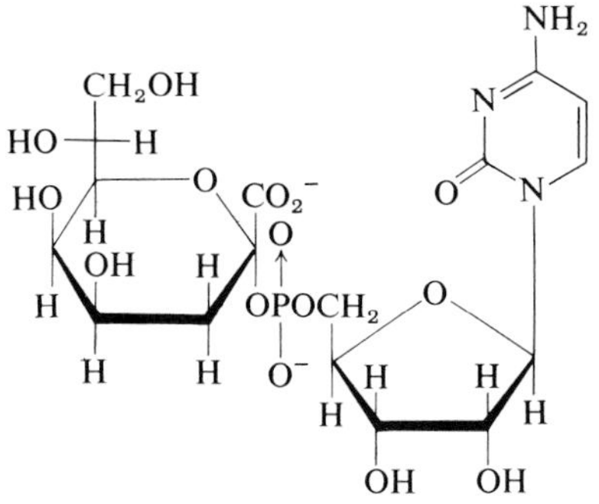

Cytidine 5′-(3-deoxy-D-*manno*-octulosonate phosphate) (**4**)

of UDP-D-glucose to UDP-D-galactose.[67] In spite of extensive study, this reaction[68,69] is not well understood. The process requires catalytic amounts of NAD$^+$ and presumably takes place through an oxidized intermediate in which the asymmetry at C-4 is lost. Attempts to trap such an intermediate have not been successful. It is considered that such an intermediate may exist as a transiently enzyme-bound compound.

Since extensively purified UDP-D-galactose 4-epimerase from *E. coli* contains 1 mole of tightly bound NAD$^+$ per mole of enzyme, and since an absorption peak which is almost identical in shape to that of NADH appears on addition of substrate to the enzyme,[70] the proposed mechanism strongly supports an oxidation–reduction[71] of the UDP hexoses at C-4.

The claim that an intermediate substance (X-UDPG) is formed during the first part of the reaction and that this unidentified intermediate is further oxidized to UDP-D-glucuronic acid[72] could not be substantiated.[72a] Upon reinvestigation of the dehydrogenation of UDP-D-glucose with a homogeneous enzyme of high specific activity, no intermediate product could be demonstrated. The anomalous kinetics of the reaction was explained on the basis of product inhibition. Evidence was also presented which indicated that the uracil moiety of UDP-D-glucose is not directly involved in the action of the UDP-D-glucose dehydrogenase.

* *References start on p. 362.*

Two other types of epimerization have been reported in other positions.[55] One is a 2-epimerization, shown in the following reaction:

UDP-2-acetamido-2-deoxy-D-glucose ⇌ 2-Acetamido-2-deoxy-D-mannose + UDP

It is not known whether UDP-2-acetamido-2-deoxy-D-mannose is an intermediate in this reaction. Another epimerization has been found to take place at C-5 in the conversion of UDP-D-glucuronic acid into UDP-L-iduronic acid.[73]

Conversions have also been discovered in which a change in configuration of several carbon atoms in the hexose unit takes place, leading to the formation of 6-deoxyhexoses. The first conversion of this type, involving the transformation of GDP-D-mannose into GDP-L-fucose by enzymes from *Acetobacter aerogenes*, was studied by Ginsburg.[74,75] The first step is an internal oxidation–reduction reaction, requiring NAD$^+$, in which the OH group at C-4 is converted into a keto group and the CH_2OH group into a CH_3 group. Subsequently, inversions take place at C-3 and C-5, supposedly by an enediol transformation, to produce a second intermediate. Reduction with NADPH at C-4 would result in GDP-L-fucose. These transformations as visualized by Ginsburg are shown in Fig. 1.

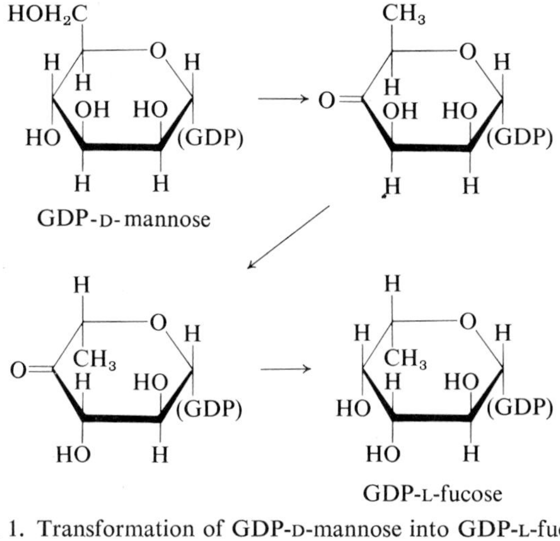

HOH₂C $\qquad$ CH₃

GDP-D-mannose

GDP-L-fucose

FIG. 1. Transformation of GDP-D-mannose into GDP-L-fucose.

Similarly, TDP-D-glucose is transformed into TDP-L-rhamnose by certain microorganisms.[76–78] This transformation involves inversion of C-3, C-4, and C-5 plus reduction at C-6. A corresponding process occurs in plants involving the same steps but with UDP-D-glucose instead of GDP-D-glucose.[79]

A reaction involving an inversion of C-3 and C-5 is postulated to take place in the formation of GDP-L-galactose from GDP-D-mannose;[80] and the synthesis of a 3,6-dideoxy-L-*xylo*-hexose (colitose) from GDP-D-mannose[81] is assumed to involve a reduction at C-3 and C-6 and inversion at C-4.

An enzyme preparation that catalyzes the synthesis of GDP-D-rhamnose and GDP-6-deoxy-D-talose from GDP-α-D-mannose[82] can be obtained from an unclassified bacterium[83] isolated from soil.

The conversion of GDP-α-D-mannose into GDP-D-rhamnose and GDP-D-6-deoxy-D-talose involves the formation of GDP-6-deoxy-D-*lyxo*-4-hexosulose as an intermediate and its subsequent reduction in the presence of NADH.[84]

CDP-3,6-dideoxyhexoses are formed by *Pasteurella pseudotuberculosis* or *Salmonella* strains by the following series of reactions:[85]

$$\text{CDP-D-glucose} \xrightarrow{\text{(NAD}^+)} \text{CDP-6-deoxy-D-}xylo\text{-4-hexosulose} + H_2O \quad (1)$$

$$\text{CDP-6-deoxy-D-}xylo\text{-4-hexosulose} \xrightarrow{\text{2NADH}} \text{CDP-paratose (3,6-dideoxy-D-}ribo\text{-hexose)}$$
$$\text{CDP-abequose (3,6-dideoxy-D-}xylo\text{-hexose)}$$
$$\text{or}$$
$$\text{CDP-ascarylose (3,6-dideoxy-L-}arabino\text{-}$$
$$\text{hexose)} \quad (2)$$

$$\text{CDP-paratose} \xrightarrow{\text{NAD}^+} \text{CDP-tyvelose (3,6-dideoxy-D-}arabino\text{-hexose)} \quad (3)$$

Reaction (1) is similar to the formation of TDP-6-deoxy-D-*xylo*-4-hexosulose from TDP-D-glucose. Reaction (2) involves reduction at C-3 and C-4 of the sugar by NADPH and epimerization at C-4 and C-5. The mechanism of this step is obscure. In reaction (3) CDP-tyvelose is formed from another one of the CDP-3,6-dideoxyhexoses, CDP-paratose, by reversible epimerization at C-2. The 3,6-dideoxyhexoses are characteristic constituents of the lipopolysaccharides of *P. pseudotuberculosis* and *Salmonella* species. Paratose is found in *P. pseudotuberculosis* types I and III and *Salmonella* group A, abequose in *P. pseudotuberculosis* type II and *Salmonella* group B, tyvelose in *P. pseudotuberculosis* type IV and *Salmonella* group D, and ascarylose in *P. pseudotuberculosis* type V. (For additional information and nomenclature of dideoxy sugars, see Vol. IB, Chap. 17, this volume, Chap. 32, and Vol. IIB, Chap. 41.)

TDP-4-Acetamido-4,6-dideoxyhexoses are produced[86] in some strains by the following reactions:

$$\text{TDP-D-glucose} \xrightarrow{\text{(NAD)}^+} \text{TDP-6-deoxy-D-}xylo\text{-4-hexosulose} + H_2O \quad (4)$$

$$\text{TDP-6-deoxy-D-}xylo\text{-4-hexosulose} + \text{L-glutamate} \xrightarrow{\text{pyridoxal phosphate}} \text{TDP-4-}$$
amino-4,6-dideoxy-D-glucose (or corresponding D-galactose) + α-ketoglutarate $\quad (5)$

TDP-4-amino-4,6-dideoxy-D-glucose (or corresponding D-galactose) +
acetyl CoA $\rightleftharpoons$ TDP-4-acetamido-4,6-dideoxy-D-glucose
(or corresponding D-galactose) + CoA $\quad (6)$

Step (4), formation of TDP-6-deoxy-D-*xylo*-4-hexosulose, is derived from TDP-D-glucose. Step (5) involves stereospecific transamination reactions. The transaminase from *E. coli* strain Y-10 and from strains of *P. pseudotuberculosis* and other microorganisms yields TDP-4-amino-4,6-dideoxy-D-galactose. The corresponding enzyme from *E. coli* strain B produces TDP-4-amino-4,6-dideoxy-D-glucose. In the final reaction, acetylation of the 4-amino group by acetyl CoA with catalysis by specific transacetylases involves no change in the configuration of the sugars.

Three of the pentoses, D-xylose, L-arabinose, and D-ribose, are general constituents of higher plants. All are found in complex materials but not usually in the free state. Thus, L-arabinose, and rarely D-arabinose, are found in gums and L-arabinans, D-xylose is found in the xylans of woody material, and D-ribose is a component of nucleotides. As will be described later, the polysaccharides containing the pentose sugars are synthesized in plants from nucleoside glycosyl pyrophosphates. UDP-D-xylose is produced by decarboxylation of UDP-D-glucuronic acid by a specific decarboxylase according the following reaction:[9-11]

$$\text{UDP-D-glucuronic acid} \longrightarrow \text{UDP-D-xylose} + CO_2$$

UDP-L-arabinose is formed by a 4-epimerase as follows:[87]

$$\text{UDP-D-xylose} \longrightarrow \text{UDP-L-arabinose}$$

Although, in general, epimerization takes place with the sugar nucleotide rather than the sugar phosphate as substrate, the interconversion of D-*threo*-pentulose 5-phosphate and D-ribose 5-phosphate does involve the phosphates themselves.[88,89] D-*threo*-Pentulose 5-phosphate may also be formed by the enzyme "transketolase" by transferring C-1 and C-2 of D-fructose 6-phosphate to glyceraldehyde 3-phosphate.[90,91] These C_2 fragments can also be transferred to D-ribose to form sedoheptulose 7-phosphate. D-*erythro*-Pentulose 5-phosphate may also be formed as a result of decarboxylation of D-glucose 6-phosphate, as in the initial reaction of the pentose phosphate shunt pathway of carbohydrate degradation, a pathway which is an alternative one to the Embden–Meyerhof pathway.[91]

III. MECHANISMS INVOLVED IN ENZYMIC FORMATION OF COMPLEX SACCHARIDES

The synthesis of complex saccharides (oligosaccharides, glycosides, and polysaccharides) involves the process of transglycosylation. In this process the glycosyl donor may be sugar phosphate, sugar nucleotide, oligosaccharide, or polysaccharide.

A number of hydrolytic enzymes, chiefly invertases, maltases, and cellobiases, which normally catalyze the hydrolysis of glycosidic linkages, may also transfer glycosyl residues to compounds other than water, resulting in the formation of new glycosyl derivatives.[92–95] (See also this volume, Chap. 33 for further discussion). The nature of the product depends upon the specificity of the particular enzyme. It appears that many hydrolytic enzymes capable of attacking more than one substrate may be expected to act as transglycosylases.[92,95] Certain glycosyl group-transferring enzymes do not produce appreciable hydrolysis; that is, they do not use water as an acceptor. Others catalyze both reactions to various degrees. There is thus a continuous spectrum of transglycosylating enzymes, ranging from those having hydrolytic activity almost exclusively, to others having practically no hydrolytic activity at all.

It has been found[93,94] that, when sucrose is subjected to yeast invertase, several oligosaccharides are produced in addition to the hydrolysis products, D-glucose and D-fructose. Three reducing disaccharides and three non-reducing trisaccharides, composed of D-glucose and D-fructose residues, having different types of glycosidic linkages (α-D or β-D) and different points of attachment of the monosaccharide residues to one another, were isolated from the reaction mixture. Since these oligosaccharides are formed by transglycosylation as a result of competition of the decomposition products of the substrate for the elements of water, they are eventually hydrolyzed in the latter stages of the reaction. Such hydrolytic enzymes do not, however, appear to be of direct importance in the synthesis of complex saccharides in plants, animal tissues, and microorganisms.

In early classification the transferases that formed sugar esters were not considered to be in the same group as the transglycosylases that formed the true glycosidic linkages. However, since the sugar phosphates that participate in the enzymic formation of glycosyl compounds undergo C–O bond cleavage and not P–O bond cleavage, the phosphorylases are classified as transglycosylases. Scission of α-D-glucopyranosyl phosphate between C-1 and O-1 was shown to occur in the enzymic formation of glycogen, and was observed by Cohn[96] through the use of ^{18}O isotope methods. Also, the fact that D-ribosyl phosphate could react with the imidazole nitrogen atom of hypoxanthine in the presence of an enzyme from liver to produce a C–H linkage indicates[97] scission of this pentose phosphate between C-1 and O-1.

From the thermodynamic point of view, nucleoside sugar 5'-pyrophosphates are superior donors for formation of complex saccharides, because they have the highest negative free energy of hydrolysis ($\Delta G°$) of all known compounds containing glycosyl groups that can serve as a monosaccharide donor. Thus, the $\Delta G°$ of UDP-D-glucose (pH 7.4) is -7600 cal mole^{-1} (see Ref. 98) while that of α-D-glucopyranosyl phosphate (pH 8.5) is -4800 cal mole^{-1} (see Ref.

* *References start on p. 362.*

99). The value of $\Delta G°$ of the α-D-glucose-(1 → 4) linkage of glycogen which is produced from these substrates is −4300 cal mole^{-1} (Ref. 99).

Although the relatively high negative $\Delta G°$ value of −7600 cal mole^{-1} applies only to the one sugar nucleotide, UDP-D-glucose, it is assumed that nucleoside sugar pyrophosphate containing glycosyl moieties and bases other than D-glucose and uridine have approximately the same high $\Delta G°$ values.

The most important reaction for complex saccharide formation in animal and plant tissues and in microorganisms appears to be that involving sugar nucleotides as glycosyl donors.

IV. REACTIONS CATALYZED BY DISACCHARIDE
PHOSPHORYLASES (D-GLUCOSYLTRANSFERASES)

A. SUCROSE AND SUCROSE ANALOGS

Certain species of bacteria, namely *Pseudomonas saccharophila*, *Pseudomonas putrifaciens*, and *Leuconostoc mesenteroides*,[100] contain a phosphorylase, which, in the presence of inorganic phosphate, catalyzes the phosphorlytic decomposition of sucrose with the production of α-D-glucopyranosyl phosphate and D-fructose. The reverse reaction, the condensation of α-D-glucopyranosyl phosphate and D-fructose, results in the formation of sucrose with the elimination of phosphoric acid as follows:

$$\text{Sucrose} + P_i \rightleftharpoons \text{α-D-Glucopyranosyl phosphate} + \text{D-fructose}$$

The free energy required for the formation of the glycosidic link in sucrose is available in the α-D-glucopyranosyl phosphate and is transferred with the D-glucose part of the ester to D-fructose. Phosphorylated D-glucose is required because free D-glucose, on account of its low level of free energy, cannot serve as part of the substrate for synthesis of sucrose. The energy level of the D-glucose can be raised through combination with phosphate by utilizing the energy drop from adenosine 5′-triphosphate (ATP) to form D-glucose 6-phosphate through the hexokinase reaction. The latter ester can then be transformed readily by the aid of phosphate glucomutase into α-D-glucopyranosyl phosphate. It therefore appears that the free energy for the synthesis of sucrose is ultimately derived from ATP.

The organism from which sucrose phosphorylase (sucrose glycosyltransferase) is obtained does not accumulate sucrose; rather, it degrades sucrose from the medium to α-D-glucopyranosyl phosphate and D-fructose. The equilibrium constant of the reaction at pH 6.6 and 30° expressed by the mass law equation:

$$K = \frac{\text{(sucrose) (phosphate)}}{\text{(D-fructose) (D-glucopyranosyl phosphate)}}$$

is approximately 0.05 and increases slightly at lower pH values.

Although the presence of sucrose glycosyltransferase in some micro-organisms provides an enzymic mechanism for the formation of sucrose from α-D-glucopyranosyl phosphate and D-fructose, this pathway does not operate in higher plants, which are the main producers of sucrose. As will be seen later, UDP-D-glucose serves as the D-glucose donor to D-fructose or D-fructose 6-phosphate for the formation of sucrose and sucrose 6-phosphate, respectively.[13,101,102-104]

Sucrose phosphorylase is highly specific with regard to the D-glucose moiety of the substrates but less specific toward the D-fructose portion, which may be replaced by a number of other ketoses[100]—for example:

α-D-Glucopyranosyl phosphate + L-sorbose $\rightleftharpoons$
α-D-Glucopyranosyl α-L-sorbofuranoside + PP$_i$

Inasmuch as these disaccharides are nonreducing and their ketose constituents exist in the furanose configuration, they are considered as analogs of sucrose. The sucrose phosphorylase is also capable of combining α-D-glucopyranosyl phosphate with an aldose, L-arabinose, having no obvious structural relation to sucrose.

α-D-Glucopyranosyl phosphate + L-arabinose $\rightleftharpoons$
3-O-α-D-Glucopyranosyl-L-arabinopyranose + P$_i$

B. MALTOSE

Fitting and Doudoroff[105] observed for the first time a new type of enzymic reaction in which the enzyme causes an α,β inversion of the D-glycosidic linkage. They found that the bacterium *Neisseria meningitidis* contains an enzyme, maltose phosphorylase, capable of catalyzing the reversible reaction:[106]

4-O-α-D-Glucosyl-D-glucose + P$_i$ $\underset{\text{maltose phosphorylase}}{\rightleftharpoons}$
β-D-Glucopyranosyl phosphate + D-glucose

Starting with maltose and inorganic phosphate, the enzyme produces β-D-glucopyranosyl phosphate + D-glucose. In the reverse reaction, β-D-glucopyranosyl phosphate and D-glucose form inorganic phosphate and maltose, which contains the α-D linkage. The specificity of the enzyme for β-D-glucopyranosyl phosphate is absolute; however, D-xylose can be substituted for

* References start on p. 362.

D-glucose, resulting in the formation of the reducing disaccharide 4-*O*-α-D-glucosyl-D-xylose, which is analogous in structure to maltose [106] The equilibrium constant, K, for the synthesis of maltose at pH 7.0 and 37° was found to be 4.4; thus energy is released during the formation of maltose from β-D-glucopyranosyl phosphate. (For additional details, see this volume, Chap. 30.)

C. CELLOBIOSE

Three unrelated strains of cellulolytic bacteria—*Clostridium thermocellum*,[107] *Cellovibrio gilvus*,[108] and *Ruminococcus flavefaciens*[109]—which degrade cellulose, contain a phosphorylase capable of catalyzing the reversible phosphorolysis of cellobiose to α-D-glucopyranosyl phosphate:

4-*O*-β-D-Glucosyl-D-glucose + phosphate $\rightleftharpoons$
$$\text{α-D-Glucopyranosyl phosphate + D-glucose}$$

The phosphorolysis of cellobiose, like that of maltose, proceeds with an inversion of the configuration of the glucosidic linkage. The enzyme appears to be specific for D-glucose as glucosyl acceptor.

D. LAMINARABIOSE

Marechal and Goldemberg[110] reported the presence in *Euglena gracilis* of a glucosyltransferase capable of catalyzing the following reversible reaction:

3-*O*-β-D-Glucopyranosyl-D-glucose + P_i $\rightleftharpoons$
$$\text{α-D-Glucopyranosyl phosphate + D-glucose}$$

They tentatively named the enzyme laminarabiose phosphorylase. However, this reaction was not studied in detail.

V. SYNTHESIS OF DISACCHARIDES BY TRANSGLUCOSYLATION

α-D-Glucopyranosyl phosphate does not appear to be an essential product or substrate of sucrose phosphorylase activity for the synthesis of disaccharides. This ester can be regarded as merely one of a number of "D-glucose donors" for the enzyme. The sucrose phosphorylase of *Pseudomonas saccharophila* can act not only as a "phosphorylase" but also as a "transglucosylase" capable of mediating the transfer of the D-glucose portion of substrate to a variety of acceptors.[94] The evidence for the double function of the enzyme is adduced from the observation that, when ^{32}P-labeled inorganic phosphate and nonradioactive α-D-glucopyranosyl phosphate are added to sucrose phosphorylase preparations in the absence of ketose sugars, a rapid redistribution of the isotope occurs between the organic and inorganic fractions without the liberation of D-glucose. This observation led to the assumption that the enzyme combines reversibly with the D-glucose of α-D-glucopyranosyl phosphate, forming a D-glucose–enzyme complex and releasing inorganic phosphate, in accordance with the equation:

$$\text{α-D-Glucopyranosyl phosphate} + \text{enzyme} \rightleftharpoons \text{D-Glucose–enzyme} + P_i$$

The equilibrium reaction would require that the energy of the α-D-glucopyranosyl phosphate linkage be preserved in the D-glucose–enzyme bond. The transfer of phosphate could not involve the formation of free D-glucose, because, if this occurred, approximately 4800 cal mole^{-1} would be released in the decomposition of the ester and would be required for its resynthesis. Since no external source of energy would be available for the resynthesis of the ester, it can be concluded that the original bond energy is conserved in the D-glucose–enzyme complex.

That the enzyme is really a "glucosyl transfer agent" is shown by the catalysis of an exchange of glycosidic bonds in the absence of phosphate:

β-D-Fructofuranosyl α-D-glucopyranoside + L-sorbose $\rightleftharpoons$
(sucrose)   D-Glucosyl L-sorboside + D-fructose

In a similar manner, sucrose can be prepared by a reaction between D-fructose and the corresponding disaccharide containing D-*threo*-pentulose.[111]

Supporting evidence for the formation of a D-glucose–enzyme complex in reactions catalyzed by sucrose phosphorylase from *Pseudomonas saccharophila* was obtained by Voet and Abeles.[112] They showed that, when sucrose phosphorylase is denatured after exposure to uniformly labeled sucrose-^{14}C, the denatured protein contains firmly bound D-glucose or a compound derived from the D-glucose moiety of sucrose. The D-fructose moiety of sucrose is not bound to the protein. The molecular weight of the enzyme as determined by Sephadex chromatography was shown[113] to be 80,000 to 100,000. The kinetics of the reaction were found to be consistent with Koshland's proposed theory[114] of a double displacement mechanism involving formation of a D-glucose–enzyme complex and subsequent reaction of this complex with an acceptor.

A number of other oligosaccharides have been synthesized by the transglycosylation process by the action of various microorganisms. A reducing disaccharide, 5-O-α-D-glucopyranosyl-D-fructopyranose, named leucrose,[115] together with another disaccharide, 6-O-α-D-glucopyranosyl-D-fructofuranose, is formed in the reaction mixture during the synthesis of dextran from sucrose by an enzyme isolated from *Leuconostoc mesenteroides*. This enzyme transfers D-glucose from sucrose to many other monosaccharides including D-galactose, in which case α-D-glucopyranosyl D-galactofuranoside is one of the products.[116]

* References start on p. 362.

The action of β-D-glucosidases from, for example, molds,[117,118] barley,[119] and algae[120] on cellobiose results in the formation of oligosaccharides such as cellotriose, laminarabiose, gentiobiose, gentiotriose, O-β-D-glucosyl-(1 → 6)-O-β-D-glucosyl-(1 → 4)-β-D-glucose and, in the presence of D-xylose,[120] 3-O-β-D-glucosyl-D-xylose.

The transglycosidase action of glycosylases, particularly of microbial ones, is an important tool for the production of oligosaccharides containing different monosaccharide residues and diverse linkages.

The levansucrase enzyme system of *Aerobacter levanicum*, which utilizes the D-fructose moiety of sucrose, forming a polysaccharide (levan) and D-glucose, possesses a complementary property of catalyzing the reversible transfer of the D-fructosyl unit of β-D-fructofuranosyl aldosides of different configurations to the anomeric carbon atom of an aldose.[121] This process, in analogy to the term "transglucosylation," may be termed "transfructosylation."

When a cell-free solution of levansucrase was allowed to act on raffinose in the presence of D-glucose, in addition to the appearance of melibiose and a comparatively small proportion of levan, rapid formation of a nonreducing disaccharide identified as sucrose occurred. This reaction proved to be reversible:

$$\text{Raffinose} + \text{D-glucose} \rightleftharpoons \text{Sucrose} + \text{melibiose}$$

A number of other aldoses were found to react similarly with raffinose in the presence of levansucrase. Thus, by interacting raffinose with D-xylose in the presence of levansucrase enzyme, Avigad *et al.*[122] obtained a nonreducing disaccharide which was totally hydrolyzed with yeast invertase to equimolar amounts of D-xylose and D-fructose. Periodate oxidation together with other data showed that the compound was a sucrose analog, α-D-xylopyranosyl β-D-fructofuranoside, to which the name "xylosucrose" was given.

Similarly, by using the same enzyme with raffinose as substrate and D-galactose as an acceptor, Feingold *et al.*[123] synthesized a crystalline, non-reducing disaccharide, "galsucrose," which could be hydrolyzed with yeast invertase to D-glucose and D-galactose. Data obtained from chemical analysis showed this disaccharide to be another sucrose analog, having the structural configuration α-D-galactopyranosyl β-D-fructofuranoside.

By means of levansucrase-catalyzed transfer of the D-fructose group from sucrose to the anomeric carbon atom of lactose, Avigad[124] synthesized a non-reducing trisaccharide, "lactsucrose." This trisaccharide could be decomposed by yeast invertase to equimolar amounts of D-fructose and lactose, and by β-galactosidase to equimolar amounts of D-galactose and sucrose. Structurally, the trisaccharide was shown to be O-β-D-galactopyranosyl-(1 → 4)-α-D-glucopyranosyl β-D-fructofuranoside.

In addition to D-xylose, D-galactose, and lactose, L-arabinose and melibiose were shown to be converted into the corresponding aldosyl D-fructofurano-sides by reaction with raffinose or sucrose in the presence of levansucrase.

VI. SYNTHESIS OF OLIGOSACCHARIDES FROM SUGAR NUCLEOTIDES

A. SUCROSE

The discovery by Leloir and collaborators[12,13] of the existence of two enzymes in plants capable of synthesizing sucrose and sucrose phosphate was one of the most significant achievements in the area of carbohydrate biochemistry. The synthesis takes place according to the following two reactions:

$$\text{UDP-D-glucose} + \text{D-fructose} \rightleftharpoons \text{Sucrose} + \text{UDP} \qquad (a)$$

$$\text{UDP-D-glucose} + \text{D-fructose 6-phosphate} \rightleftharpoons \text{Sucrose phosphate} + \text{UDP} \, (b)$$

Reaction (a), in which sucrose is directly synthesized, is freely reversible. The equilibrium constant of this reaction at pH 7.4 and 37° was determined[12,13] to be approximately 5, a value corresponding to a $\Delta G°$ of -1000 cal mole^{-1}. By using the known value of -6600 cal mole^{-1} for the $\Delta G°$ of hydrolysis of sucrose,[125] the $\Delta G°$ of hydrolysis of the α-D-glycosyl phosphate bond for UDP-D-glucose was calculated as -7600 cal mole^{-1}.

Reaction (b) results in the formation of sucrose phosphate and is catalyzed by another enzyme which utilizes D-fructose 6-phosphate as an acceptor instead of D-fructose. The equilibrium constant of this reaction was reported to be 3250 at 38° and pH 7.5, indicating a surprisingly low $\Delta G°$ of -2700 cal mole^{-1} for the hydrolysis of the glycosidic bond of sucrose 6′-phosphate.

In reaction (a) the formation of sucrose involves incorporation of D-fructose, which exists mainly (about 80%) as the D-fructopyranose in free form. Its conversion into D-fructofuranose in sucrose involves an additional several hundred calories per mole. Since the 6-hydroxyl group of D-fructofuranose is phosphorylated in reaction (b), no such conversion would be required to occur. This difference may account in part for the considerable difference in the ΔG values of the two reactions.

Although small amounts of sucrose phosphate have been detected among labeled photosynthetic products in plants, this sugar phosphate is not readily obtainable in plant tissue, as it is apparently hydrolyzed by a phosphatase to sucrose as soon as it is formed. However, Bird et al.[126] found that when

* References start on p. 362.

tobacco leaf chloroplasts, isolated from a nonaqueous medium, were used as an enzyme source with UDP-D-glucose and D-fructose 6-phosphate as substrates, sucrose 6-phosphate was formed.[127] Haq and Hassid[128] also showed that acetone-extracted chloroplasts from sugar cane leaves, in addition to forming sucrose from UDP-D-glucose and D-fructose, utilized D-fructose 6-phosphate as an acceptor for the D-glucose, forming small amounts of sucrose phosphate. Furthermore, a phosphatase was present which appeared to hydrolyze sucrose phosphate more readily than D-fructose 6-phosphate. Hatch[129] also demonstrated the presence of enzymes that catalyze the synthesis and breakdown of sucrose phosphate from the stem and leaf tissue of sugar cane. Results of experiments in which ^{14}C-labeled D-glucose was supplied to plants have shown that the D-fructofuranosyl moiety of sucrose becomes highly labeled before any label appears in the free D-fructose pool, suggesting that the monosaccharide is not an intermediate in synthesis of sucrose.[130] It was also found that D-fructose 6-phosphate becomes labeled before sucrose, and small amounts of sucrose 6'-phosphate have been detected among the labeled products.[104,131] These results strongly indicate that sucrose 6'-phosphate is synthesized first and then hydrolyzed to free sucrose. Because the last hydrolytic step of such a pathway would be irreversible, it might account for the large accumulation of sucrose in sugar beets, sugar cane, and some other plants.

Reaction (a) may serve an important function in degradation of sucrose. In spite of the fact that its equilibrium is in the direction of synthesis, complete breakdown of this disaccharide may occur if the resulting UDP-D-glucose is used up in various other metabolic reactions. This assumption is supported by the experiment of Milner and Avigad,[132] who purified the UDP-D-glucose: D-fructose transglucosylase (sucrose synthetase) of sugar beet more than 80-fold, and practically freed it from traces of invertase and phosphatase. The purified sucrose synthetase was effective in the degradation of sucrose. A study of specificity of this enzyme with regard to various nucleoside pyrophosphates showed them to be efficient D-glucosyl acceptors.[133] Chromatographic analysis of reaction systems containing sucrose in the presence of UDP, dTDP, ADP, CDP, or GDP indicated the appearance of new nucleotide components which proved to be UDP-D-glucose, dTDP-D-glucose, ADP-D-glucose, CDP-D-glucose, and GDP-D-glucose, respectively. Based on the value of UDP as 100, the relative effectiveness as D-glucose acceptors of dTDP, ADP, CDP, and GDP was 52, 16, 12, and 6, respectively.

B. RAFFINOSE

Sucrose is found in many higher plants together with the trisaccharide, raffinose, in which D-galactose is attached through an α-D-glycosidic linkage to the C-6 position of the D-glucose moiety of sucrose. Its constitution is therefore: O-α-D-galactopyranosyl-(1 → 6)-α-D-glucopyranosyl β-D-fructofuranoside. Thus, it is logical to assume that this sugar is formed from a nucleoside 5'-(D-galactosyl pyrophosphate) by a transfer of the D-galactosyl moiety to sucrose. However, attempts to synthesize raffinose in this way by using enzymic preparations from germinated mung seedlings and other plants were not successful. It appeared that, if raffinose was formed, the germinated seedlings most probably contained glycosidases which degraded this trisaccharide.

It has been shown[134] that raffinose can be synthesized by an enzyme preparation from seeds of the dormant broad bean (*Vicia faba*), with a mixture of sucrose, α-D-galactopyranosyl phosphate, and UTP as substrates. By using UDP-D-galactose labeled with ^{14}C in the D-galactose moiety, together with sucrose in the presence of an enzyme preparation from mature broad beans, a direct transfer of D-galactose-^{14}C to sucrose was effected, resulting in the production of raffinose.[135] The amount of ^{14}C-labeled D-galactose incorporated into the raffinose was 33% and 39% after 1 and 2.5 hours, respectively. Although the enzyme preparation was crude, the percentage of incorporation of label was relatively high, indicating that this mechanism functions in maturing seeds. Raffinose has also been synthesized from UDP-D-galactose and sucrose with an enzyme preparation from immature soybeans.[136]

Moreno and Cardini[137] found a specific transgalactosylase capable of catalyzing the following reaction:

$$\text{Raffinose} + \text{sucrose-}^{14}C \rightleftharpoons \text{Raffinose-}^{14}C + \text{sucrose.}$$

These results can be interpreted as evidence that this enzyme transfers α-D-galactosyl residues from raffinose to a molecule of sucrose, thus regenerating raffinose.

The possibility that raffinose is also formed by transfer of D-galactose from low-energy donors to sucrose should not be overlooked. The synthesis of raffinose and planteose, by using D-galactose as the donor in the presence of α-D-galactosidase preparations, has been demonstrated.[138] Although the equilibrium for glycosidase-catalyzed reactions normally favors hydrolysis, *in vivo* factors, such as localized high concentrations of substrate or rapid utilization of products, may encourage synthesis of oligosaccharides.

C. Stachyose

A soluble enzyme from ripening seeds of dwarf beans (*Phaseolus vulgaris*) transfers D-galactose with high yield from galactinol (1-O-α-D-galactopyranosyl-*myo*-inositol) to raffinose, giving rise to stachyose and *myo*-inositol.[139]

* *References start on p. 362.*

Galactinol is a major galactoside constituent in the bean during a certain maturation period, and its formation precedes that of stachyose. The enzyme appears to be specific with regard to the acceptor molecule. The labeled D-galactose was not transferred from galactinol-^{14}C to glycerol, D-fructose, sucrose, maltose, cellobiose, gentiobiose, melizitose, or trehalose. Slight acceptor activity was observed with D-glucose, D-galactose, and lactose. However, these compounds were less efficient than raffinose by a factor of 20 to 30. The most efficient acceptor of the sugar tested besides raffinose was melibiose, which was one-fourth as efficient as raffinose.

The crude extract from the ripening bean seeds also contains an enzyme which transfers the labeled D-galactose from UDP-D-galactose-^{14}C, but not from ADP-D-galactose-^{14}C, to *myo*-inositol. This enzyme was discovered previously in extracts of pea seeds.[140] No transfer could be observed from either UDP-D-galactose-^{14}C or ADP-D-galactose-^{14}C, in the case of raffinose biosynthesis.[135]

On the basis of these results, the pathway for synthesis of stachyose is postulated to be as follows:

UDP-D-galactose + *myo*-inositol $\longrightarrow$ Galactinol + UDP

Galactinol + raffinose $\longrightarrow$ Stachyose + *myo*-inositol

D. Lactose

The synthesis of lactose was first accomplished with particulate preparations from lactating guinea pig or bovine mammary glands[141] by a process involving the following reaction:

UDP-D-galactose + D-glucose $\xrightarrow{\substack{\text{D-galactosyl} \\ \text{transferase}}}$ Lactose + UDP

However, attempts to solubilize the preparations containing UDP-D-galactose:D-glucose 1-β-D-galactosyl transferase were not successful. Later, a soluble preparation was obtained from bovine milk capable of synthesizing lactose from UDP-D-galactose and D-glucose by the same reaction.[142,143]

The enzyme appears to be specific for UDP-D-galactose and none of the ^{14}C-labeled D-galactosyl nucleotides containing bases other than uracil (guanine, adenine, cytosine, thymine) can serve as substrate for the formation of lactose. The following compounds, which are listed in decreasing order, inhibit the activity of the enzyme: PP_i, ITP, UTP, UDP, P_i, UMP, TTP, and GTP. No inhibition could be shown by ATP, CTP, or D-galactose. The pattern of enzyme inhibition appears to be in accord with the high enzyme affinity shown for UDP-D-galactose ($K_m = 5.0 \times 10^{-4}M$) and lower affinity shown for D-glucose ($K_m = 2.5 \times 10^{-2}M$).

α-D-Glucopyranosyl phosphate, α-D-galactopyranosyl phosphate, L-glucose, D-xylose, maltose, and methyl α-D-glucopyranoside will not act as acceptors for the D-galactose moiety of UDP-D-galactose to form the corresponding oligosaccharides. However, 2-acetamido-2-deoxy-D-glucose is 25% as effective as D-glucose. The product appears to be O-4-β-D-galactosyl-(2-acetamido-2-deoxy-D-glucose). No reversal of the enzymic reaction can be demonstrated when lactose and UDP are used as substrates for the formation of UDP-D-galactose and D-glucose. The lactose synthetase of bovine milk is activated by several divalent cations, showing maximum activation with Mn^{2+}.

The existence of cellular material appears to be the result of the disintegration of the glandular cells by enzymic activity in the milk during secretion. Apocrine secretion (decapitation of the apical portion of the alveolar cells) has been noted in the mammary glands during the secretory phase.[144,145] The soluble lactose synthetase in milk can thus be presumed to have originated from the autolysis of the particulate enzyme which became part of the secretion of the mammary gland after disintegration of the alveolar cells.

The work of Wood and co-workers[146-149] on lactose synthesis in the cow, or in cow udders, with ^{14}C-labeled sugars shows that blood D-glucose is used directly for the formation of the D-glucose moiety of lactose, but is transformed into the D-galactosyl moiety only by way of hexose phosphate intermediates. All the enzymes required for the formation of UDP-D-galactose from D-glucose have been found in the mammary gland by a number of investigators.[141,150-152]

A study of incorporation of several isomers of D-glucose labeled with ^{14}C or ^{3}H into slices of mammary glands of lactating rats showed that D-glucose (which was incorporated intact) was the only D-galactosyl acceptor for the formation of lactose.[153]

The soluble lactose synthetase from milk can be resolved into two protein components, A and B, which individually do not exhibit any catalytic activity.[154] Recombination of fractions A and B, however, restores full lactose synthetase activity. Microsomal lactose synthetase from mammary gland tissue of lactating cows can be solubilized by sonic disintegration. The B fraction has been crystallized from bovine skim milk and bovine mammary tissue, and was identified as α-lactalbumin.[155] This protein, which has been crystallized from bovine skim milk and bovine mammary tissue, can be substituted for the B protein of lactose synthetase. Lactose synthetases from milk of the sheep, goat, pig, and human were also resolved into the A and B proteins, and the fractions from these species were shown to be qualitatively interchangeable in the rate essay.[155]

Results of determination of the amino acid sequence of α-lactalbumin

* *References start on p. 362.*

(B protein) have shown a remarkable homology in the amino acid sequence of α-lactalbumin and hen's egg-white lysozyme.[155a] This homology suggests that lysozyme and α-lactalbumin have evolved from a common ancestral gene. The A protein may act as a general D-galactosyl transferase, which catalyzes the following reaction:[155b,155c]

UDP-D-galactose + 2-acetamido-2-deoxy-D-glucose ⟶ N-acetyl-lactosamine + UDP

Such a reaction has been previously observed in a study of the D-galactosyl transferase in bovine colostrum and a particulate fraction of rat tissues.[155d]

Under normal assay conditions, α-lactalbumin inhibits the lactosamine reaction from UDP-D-galactose and 2-acetamido-2-deoxy-D-glucose and allows synthesis of lactose in the presence of D-glucose.[155b] Purified A protein of lactose synthetase was shown to be UDP-galactose: 2-acetamido-2-deoxy-D-glucose galactosyltransferase.[155b,155c] In the absence of α-lactalbumin, this catalyzes the formation of N-acetyl-lactosamine, although it possesses a very but significant lactose-synthetase activity. It appears that the α-lactalbumin modifies the D-galactosyl acceptor specificity of the A protein, so that the D-glucose becomes the better acceptor of the D-galactosyl moiety from UDP-D-galactose. It is likely that the A protein is a general D-galactosyl transferase and is important in the biosynthesis of carbohydrates found in glycoproteins and blood group substances.[155d]

E. TREHALOSE

This disaccharide (α-D-glucopyranosyl α-D-glucopyranoside) was synthesized by Cabib and Leloir.[156] Its synthesis takes place in two steps catalyzed by enzyme preparations from yeast and from insects. In the first step, the α-D-glucose residue from UDP-D-glucose is transferred to an acceptor, D-glucose 6-phosphate, producing the 6'-phosphate of α,α-trehalose. The phosphate group is then hydrolyzed by phosphatase to yield α,α-trehalose:

UDP-D-glucose + D-glucose 6-phosphate $\xrightarrow{\text{transferase}}$ α,α-Trehalose phosphate + UDP

α,α-Trehalose phosphate $\xrightarrow{\text{phosphatase}}$ α,α-Trehalose + phosphate

The equilibrium of the reaction is displaced toward the synthesis of trehalose phosphate, and reversibility could not be demonstrated. An upper limit of -4400 cal mole^{-1} was calculated for the $\Delta G°$ of hydrolysis of the trehalose glucosidic bond.

Analysis of the enzymically synthesized, phosphorylated product showed that it was identical with the trehalose phosphate isolated from the fermentation products of dry yeast.[157] The formation of trehalose phosphate from

UDP-D-glucose and D-glucose 6-phosphate and its subsequent dephosphory-lation have also been shown to be catalyzed by enzyme preparations from insects.[158,159]

An enzyme system from *Streptomyces hygroscopicus* catalyzes the transfer of D-glucose-^{14}C from GDP-D-glucose-^{14}C to D-glucose 6-phosphate to form trehalose phosphate.[160] The reaction appears specific for GDP-D-glucose. This disaccharide phosphate is not formed from any nucleoside (D-glucosyl pyrophosphate) containing a base other than guanine. The K_m values for both GDP-D-glucose and α-D-glucose 6-phosphate have been calculated to be about $7 \times 10^{-4} M$.

VII. SYNTHESIS OF GLYCOSIDES FROM SUGAR NUCLEOTIDES

Numerous aliphatic or aromatic alcohols and carboxylic acids are con-verted by some mammalian organs, particularly the liver and kidney, into their respective β-D-glucosiduronic acids.[161] These glycosides can be trans-ported in the blood or excreted through the urine or bile. Dutton and Storey[162] demonstrated the synthesis of *o*-aminophenyl β-D-glucosiduronic acid with a boiled extract which was subsequently shown to contain UDP-D-glucuronic acid[163] and *o*-aminophenol. This synthesis was among the first examples of a transglycosylation reaction from a glycosyl nucleotide:

UDP-D-glucuronic acid + *o*-aminophenol $\longrightarrow$
o-Aminophenyl β-D-glucosiduronic acid + UDP

A large number of acceptors besides *o*-aminophenol may participate in this type of reaction.[164–166] These include aliphatic alcohols and carboxylic acids. The latter produce 1-*O*-acyl D-glucosiduronic acids.[161]

An enzyme preparation from French bean (*Phaseolus vulgaris*) leaves was shown to catalyze the formation of quercetin-3-yl β-D-glucopyranosiduronic acid from UDP-D-glucuronic acid and quercetin.[167] In contrast to the liver enzyme, the transferase from beans appears to be specific for the acceptor.

In plants, phenolic glycosides are generally combined with D-glucose instead of with D-glucuronic acids. Several enzyme systems have been found capable of effecting these conjugations. By using an enzyme preparation from wheat germ, *p*-hydroxyphenyl β-D-glucopyranoside was synthesized by the following reaction:[168]

UDP-D-glucose + hydroquinone $\longrightarrow$ UDP + *p*-hydroxyphenyl β-D-glucopyranoside
(arbutin)

A large variety of di- and triphenols (but no monophenols) could substitute for hydroquinone as acceptors in the purified enzyme system. An enzyme system for the synthesis of gentiobiosides has also been obtained from wheat germ; it could be separated from the enzyme catalyzing the formation of β-D-glucopyranosides.[169] A typical reaction catalyzed by this preparation is the formation of phenyl β-gentiobioside:

UDP-D-glucose + phenyl α-D-glucopyranoside ⟶ UDP + phenyl β-gentiobioside

Acceptors include a variety of aryl β-D-glucopyranosides, but the β-gentiobioside is not acted upon further. However, D-glucosides containing oligosaccharides having a greater number of monosaccharide residues can be formed by the action of a crude wheat-germ preparation.[170]

The formation of aryl D-glucosides is characteristic of insects as well as of plants. An enzyme preparation has been obtained from the fat body of the locust which will catalyze the synthesis of arbutin from hydroquinone and UDP-D-glucose,[171] and homogenates of cockroach hepatic coecum have been shown to catalyze the formation of o-aminophenyl β-D-glucopyranoside from the same glycosyl nucleotide and o-aminophenol.[172]

L-Rhamnose is found most commonly as a glycoside with aromatic alcohols and other aglycons, in higher plants (see this volume, Chap. 32). Sometimes, glycosides composed of both D-glucose and L-rhamnose are encountered, as in a rutin flavonol, in which the disaccharide 6-O-α-rhamnopyranosyl-D-glucose ("rutinose") is attached glycosidically to the 3-hydroxyl group of quercetin.[173] The synthetic process was shown to be catalyzed by enzymes from mung-bean leaves by the following two steps:[174]

TDP (UDP)-D-glucose + quercetin ⟶ 3-Quercetin β-D-glucoside
TDP(UDP)-L-rhamnose + 3-quercetin β-D-glucoside ⟶ Rutin + TDP (UDP)

D-Glucose is first transferred to quercetin from TDP-D-glucose or UDP-D-glucose to form 3-quercetin β-D-glucoside. Both sugar nucleotides appear to be equally effective as D-glucosyl donors. Rutin is then synthesized by transfer of L-rhamnose to the glucoside from TDP-L-rhamnose or from UDP-L-rhamnose. The apparent lack of specificity for the pyrimidine portion of the base might arise from the existence of one nonspecific enzyme or might indicate a mixture of several specific enzymes.

The lack of specificity of the transferases involved in rutin formation *in vitro* brings up the question as to which of the nucleoside (L-rhamnosyl pyrophosphates) is the precursor of rutin *in vivo*. Both UDP-L-rhamnose[175] and TDP-L-rhamnose[176,177] have been isolated from bacteria, but neither has been observed in higher plants. However, since it was found that UDP-L-rhamnose is synthesized from UDP-D-glucose by extracts from mung bean leaves,[178] it can be assumed that UDP-L-rhamnose is produced in the leaf and contributes its L-rhamnosyl moiety to rutin.

Cultures of *Pseudomonas aeruginosa* excrete into the medium an acidic glycolipid which appears to have the following structure:[179,180]

O-L-Rhamnopyranosyl-(1 → 3)-*O*-L-rhamnopyranosyl-3-(3-hydroxydecanoyl)-
(3-hydroxydecanoic acid)

Cell-free extracts of this microorganism are capable of catalyzing the formation of the complete rhamnolipid from TDP-L-rhamnose and (3-hydroxydecanoyl)-coenzyme A.[181]

UDP-D-galactose has been shown to be the immediate precursor of the D-galactose moiety of galactolipids. Microsomal preparations from rat brain were found to transfer the D-galactosyl moiety from this sugar nucleotide to an endogenous lipid acceptor, to yield a cerebroside identified tentatively as cerebronyl-*O*-D-galactopyranosylsphingosine.[182,183]

Glucosyldeoxyribonucleic acids of some bacterial viruses (T2, T4, and T6) are unique in that they contain the base 5-(hydroxymethyl)cytosine instead of cytosine, and the D-glucosyl residues are attached to the hydroxyl group of the pyrimidine.[184] The linkage may be α-D, or β-D-glucose, or two D-glucose residues may be attached to one pyrimidine residue as *O*-β-glucopyranosyl-(1 → 6)-α-D-glucopyranosyl (α-gentiobiosyl) groups.[185]

Infection of *E. coli* by these bacteriophages triggers the synthesis of new enzymes, including D-glucosyl transferases, which had been totally lacking before infection.[186] These enzymes catalyze the synthesis of the type of deoxyribonucleic acid characteristic of that phage.[187,188]

VIII. SYNTHESIS OF POLYSACCHARIDES BY PHOSPHOROLYSIS AND TRANSGLYCOSYLATION

A. Glycogen and Starch (see also Vol. IIB, Chap. 38)

After Cori *et al.*[189] synthesized glycogen *in vitro* in 1939, and Hanes[190] synthesized starch in 1940, it was universally believed that phosphorylase *in vivo* is responsible for the synthesis as well as for the degradation of these polysaccharides. Since the discovery by Leloir and Cardini[191] in 1957 that glycogen can be synthesized *in vitro* from UDP-D-glucose by a synthetase present in animal liver, evidence accumulated which led to the view that glycogen and starch are synthesized *in vivo* by synthetases from nucleoside D-glucopyranosyl pyrophosphates and that the function of phosphorylase is degradative in nature.

* *References start on p. 362.*

Most natural starches consist of two components, amylose and amylopectin, the former component constituting approximately 20 to 30% of the whole.[192,193] The amylose is made up of linear chain molecules of several hundred D-glucose residues joined only through α-D-(1 → 4) linkages; there is little or no branching in these chains. The amylopectin molecules are highly branched, each consisting of several thousand D-glucose residues. The D-glucose residues are combined by α-D-(1 → 4) linkages forming chains joined to similarly linked chains attached by α-D-(1 → 6) linkages at the points of branching. The short amylopectin branches possessing the (1 → 4) type of linkage average approximately twenty D-glucose residues in length. Glycogen has a branched structure[194] similar to that of amylopectin, with the difference that the outer branches are shorter (from twelve to eighteen D-glucose residues) (see Vol. IIB, Chap. 38).

Although the phosphorylases from different sources vary in certain respects, they all catalyze the phosphorolytic cleavage of the α-D-(1 → 4)-glucosidic linkage of the glycogen or starch linear chains. The reaction is reversible and can be formulated as follows:

α-D-glucopyranosyl phosphate

The phosphorolysis reaction results in the formation of α-D-glucopyranosyl phosphate and the loss of one D-glucose residue from the nonreducing end of the polysaccharide chain. In the reverse reaction, inorganic phosphate is liberated from α-D-glucopyranosyl phosphate with a lengthening of the polysaccharide chain at the nonreducing end.

Synthesis of polysaccharide from α-D-glucopyranosyl phosphate does not occur unless a small amount of starch, glycogen, or dextrin is present as a priming agent. In the presence of the primer, the enzyme rapidly adds D-glucose residues to the preexisting polysaccharide chain. The nature of the primer required varies with the source of the phosphorylase. All phosphorylases are specific in that they act only with α-D-glucopyranosyl phosphate and form only the α-D-(1 → 4) maltosidic bond.

Phosphorylase will catalyze the complete degradation of the unbranched amylose chain to α-D-glucopyranosyl phosphate. Branched polysaccharides such as amylopectin are degraded only about 55%; the residue is called limit dextrin. In amylopectin the branching of the polysaccharide chains is accomplished by an α-D-(1 → 6)-glucosidic linkage; this linkage constitutes a barrier upon which the enzyme is inactive.

The equilibrium of the phosphorylase reaction, which is readily reversible, is independent of polysaccharide concentration, provided a certain minimum concentration is exceeded. Thus, in the following expression for the equilibrium constant of the polysaccharide concentration, the number of the non-reducing chain termini does not change in the branched glycogen or starch molecule. It therefore follows that at any pH the value of K is determined by the relative concentrations of α-D-glucopyranosyl phosphate and inorganic phosphate:

$$\frac{[C_6H_{10}O_5]_{n-1}\,[\text{α-D-glucosyl phosphate}]}{[C_6H_{10}O_5]_n\,[P_i]} = \frac{[\text{α-D-glucosyl phosphate}]}{[P_i]}$$

The K value at pH 7.0 is 0.3.

Since naturally occurring glycogen and starches contain (1 → 6) in addition to (1 → 4) linkages, the action of the phosphorylases, whose function *in vivo* is the establishment of (1 → 4) linkages, must obviously be supplemented by another enzyme-catalyzed reaction through which branching is induced. It is assumed that in the process of preparation of the muscle or potato phosphorylase the enzyme responsible for the formation of the (1 → 6) linkages is eliminated, which accounts for the *in vitro* production of the amylose type of polysaccharides. Evidence for the existence of such a supplementary enzyme was first provided by Cori and Cori.[195] They showed that several animal organs, such as heart, brain, and liver, contain a "branching factor" capable of synthesizing (1 → 6) linkages. The combined action of the branching enzyme and crystalline muscle phosphorylase resulted in the formation of a polysaccharide which closely resembled glycogen.

Larner[196] obtained a transglycosylase from liver which cleaves fragments of the glycogen chain at α-D-(1 → 4) linkages and transfers them to the same or another glycogen molecule but in an α-D-(1 → 6) linkage. Apparently the specificity of this "branching enzyme" determines the interbranch distance along the polysaccharide chain.

Haworth et al.[197] reported the isolation from potato juice of an enzyme fraction, termed Q-enzyme [(1 → 4)-α-D-glucan:(1→4)-α-D-glucan 6-glycosyltransferase], which, in association with potato phosphorylase, produced a polysaccharide having the properties of amylopectin. Peat and his collaborators[198] also presented evidence that this enzyme was capable of converting linear amylose into branched amylopectin without the participation of inorganic phosphate in the reaction. They therefore concluded that the Q-enzyme is a nonphosphorolytic enzyme. Besides being present in the potato, Q-enzyme has been found in other plants[199-201] and in a microorganism, *Polytomella coeca*.[202]

The Q-enzyme appears to be capable of converting about one in every twenty (1 → 4) linkages of the amylose into (1 → 6) linkages, forming a branched structure. Like the enzyme present in *Pseudomonas saccharophila* and several other microorganisms,[100] Q-enzyme can be regarded as belonging to the class of transglycosylases.

Although the Q-enzyme (α-D-glucan-branching glycosyltransferase) differs from Cori's "branching enzyme" in that the latter is devoid of action toward amylose, the two enzymes possess a common property: they both act as transglucosylases capable of establishing (1 → 6) linkages.

Peat et al.[203] showed that potatoes contain an enzyme, named D-enzyme, (dextrin transglycosylase) capable of catalyzing the reversible disproportionation of maltodextrins, the D-glucose oligosaccharides containing the α-D-(1 → 4) linkage. For example, when a preparation containing D-enzyme acts on maltotriose, D-glucose and maltopentaose are produced as the first products of the reaction, and at equilibrium D-glucose and a whole series of maltodextrins are present. However, with maltotriose as the initial substrate, none of the synthetic oligosaccharides is of sufficient length to give a color with iodine (the minimum chain length required for color formation is about twelve D-glucose residues.)

Enzymes have been discovered that are capable of degrading α-D-(1—6)-glucosidic linkages in glycogen and amylopectin. Cori and Larner[204] showed that two enzymes, muscle phosphorylase and (1 → 6)-α-D-glucan glucanohydrolase, are required for the complete degradation of the branched polysaccharides, glycogen and amylopectin. Phosphorylase attacks the glycogen molecule from the terminal D-glucose of each chain, releasing successive molecules of D-glucose residues, marked ⊙ (Fig. 2), as α-D-glucopyranosyl phosphate until the branch points are reached, when activity ceases. This results in a limit dextrin similar to that which remains after treatment of glycogen with β-amylase.[194] Hydrolytic removal of the D-glucose present in the (1 → 6) linkages at the branch points by (1 → 6)-α-D-glucan glucanohydrolase then permits phosphorolysis to continue until the next branch is reached and so on. It is to be noticed that, in this process of degradation of glycogen by muscle phosphorylase, only the side chains are reduced to a single (1 → 6)-bonded α-D-glucose residue, which is subsequently released by dextran (1 → 6)-glucosidase, while six to seven D-glucose residues are retained in the branch to which the side chain is attached. Upon reinvestigation of this problem,

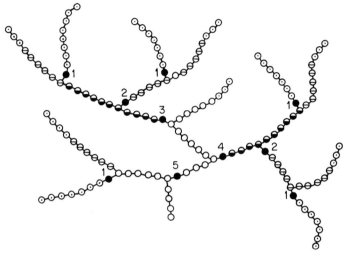

FIG. 2. Fragment of a molecule of glycogen, determined by stepwise enzymic degradation. ○, ◒, ◓, D-glucose residues removed by first, second, and third digestions with phosphorylase, respectively. ●, D-glucose residue split off as free glucose from (1 → 6) linkage by dextran 6-glucanohydrolase.

Walker and Whelan[205] found no evidence for the presence of single (1 → 6)-bonded α-D-glucose residues in glycogen or amylopectin limit dextrins. The length of the shortened chains appeared, on the contrary, to be four D-glucose residues. This was shown by the formation of maltotetraose when amylopectin was treated with the plant-starch debranching R-enzyme (amylopectin 6-glucanohydrolase). They pointed out that a transglycosylase could convert the molecule formulated by Walker and Whelan into the molecule proposed by Larner et al.[206] (Fig. 2) by a transfer of three of the D-glucose residues from the side chain to the main chain or another side chain, with the resynthesis of an α-D-(1 → 4) bond between the transferred units and the acceptor site. This would leave one D-glucose residue on the main chain for hydrolysis with dextran (1 → 6)-glucosidase. Such a transfer is known by the action of the dextrin transglycosylase in plants. Subsequently Brown

* *References start on p. 362.*

and Illingworth[207] found that dextran-(1 → 6)-glucosidase preparations contain transglycosylase activity that catalyzes transfer reactions between glycogen as donor, and maltodextrins as acceptors. This transferase usually transfers three D-glucose residues (maltotriose) at a time. Verhue and Hers[208] also examined glycogen limit-dextrin and reported the coexistence of chains containing one and four units.

Other debranching enzymes, R-enzymes (amylopectin 6-glucanohydrolases), bearing a resemblance to the dextran-(1 → 6)-glucosidase, were shown to be present in several plants[209,210] and in muscle.[211] These enzymes likewise hydrolyze (1 → 6) linkages in amylopectin, but have no action on the (1 → 4) linkages of either amylopectin or amylose. The R-enzymes do not synthesize (1 → 6) or (1 → 4) linkages; their action is purely hydrolytic.

An enzyme obtained by Bender and Wallenfels[212] from a mucoid *Aerobacter* bacterium, called pullulanase, is specific for α-D-(1 → 6) linkages of starch and glycogen. This enzyme is similar to the R-enzyme of plants (see Chapter 38).

B. The Amylopectin Type of Polysaccharide from Sucrose

Hehre and co-workers[213-215] found that cultures, washed cells, and enzyme preparations of *Neisseria perflava*, isolated from the human throat or nasopharynx, produce amylopectin- or glycogen-like polysaccharides from sucrose. Determination of the structure of this polysaccharide shows that it consists of chains averaging eleven to twelve α-D-(1 → 4)-glucopyranose residues in length, and that the branches are (1 → 6)-linked.[216] The enzyme system responsible for the synthesis of this polysaccharide from sucrose is known as "amylosucrase."[217]

Inasmuch as the polysaccharide is branched, it must be assumed that another enzyme is present, which in collaboration with the amylosucrase synthesizes a branched structure. The cell-free enzyme preparation catalyzes the reaction involving the substitution of (1 → 4) linkages in the polysaccharide chain for the (1 → 2) linkage in sucrose:

$$(n)C_{12}H_{22}O_{11} \longrightarrow (C_6H_{10}O_5)_n + (n)C_6H_{12}O_6$$

The reaction tends markedly to go to the right. For this reason it is difficult to demonstrate the reverse reaction. However, Hehre and Hamilton[218] were able to show that a small amount of polysaccharide, possessing the serological properties of dextran, is formed when a mixture of starch and D-fructose is subjected to the action of amylosucrase and dextransucrase, which have the ability to convert sucrose into dextran.

C. AMYLOSE FROM MALTOSE

Monod and Torriani[219-221] obtained a cell-free enzyme preparation from a special variant strain of *Escherichia coli* that converts maltose into a starch-like polysaccharide and D-glucose. They showed that this enzyme, named "amylomaltase" [(1 → 4)-α-D-glucan 4-glucosyltransferase], catalyzes the reversible reaction:

$$(n)\mathrm{C}_{12}\mathrm{H}_{22}\mathrm{O}_{11} \rightleftharpoons (\mathrm{C}_6\mathrm{H}_{10}\mathrm{O}_5)_n + (n)\mathrm{C}_6\mathrm{H}_{12}\mathrm{O}_6$$
(maltose)   (polysaccharide)   (D-glucose)

The nature of the polysaccharide formed by the amylomaltase depends upon the concentration of D-glucose in the reaction. Equilibrium is established when approximately 60% of the maltose has been degraded. The product produced in the reaction stains faintly red with iodine, indicating that the polymeric material consists of short-chain dextrins. However, if the D-glucose is continually removed with D-glucose oxidase, equilibrium can never be established and the conversion of maltose proceeds to completion. Under these conditions, the product stains deep blue with iodine, indicating that it is at least partially amylose. The effect of the presence of D-glucose on the molecular size of the polysaccharide can be interpreted in terms of the reversibility of the above reaction. By removing the D-glucose the equilibrium would be expected to shift to the right, whereas in the presence of D-glucose the reverse reaction would lead to the partial depolymerization of the polysaccharide.

Another variant of *E. coli* found by Doudoroff *et al.*[222] catalyzes the same type of nonphosphorolytic reaction. When D-glucose is allowed to accumulate during the decomposition of maltose, the polysaccharide produced by this enzyme consists of reducing dextrins composed on the average of four to six D-glucose residues. Barker and Bourne[223] obtained similar results with Monod's strain of *E. coli*.

Besides the enzyme, amylomaltase, the *E. coli* cells also contain phosphorylase, phosphate glucomutase, and hexose phosphate isomerase.[222] Owing to the reversible nature of the amylomaltase reaction, amylose and reducing dextrins are produced when α-D-glucopyranosyl phosphate and D-glucose are added to the preparations. In the absence of D-glucose only a starch-like polysaccharide is formed from α-D-glucopyranosyl phosphate:

$$\mathrm{Maltose} \xleftrightarrow{\text{amylomaltase}} \mathrm{Polysaccharide} + \text{D-glucose}$$
$$\Updownarrow$$
α-D-Glucopyranosyl phosphate

* *References start on p. 362.*

It appears that an identical polysaccharide is formed in the same organism by two different enzymic mechanisms—namely, one involving a trans-glucosylase and the other a phosphorylase.

D. CYCLIC AMYLOSACCHARIDES

The Schardinger dextrins, α, β, and γ, which are produced by the action of *Bacillus macerans* on starch solutions, are known to possess cyclic structures consisting of six, seven, and eight $(1 \rightarrow 4)$-α-D-glucosidically linked D-glucose residues, respectively.[224–226] Hudson *et al.*[227,228] showed that the dextrins are produced by an extracellular enzyme of the microorganism and probably arise from the linear amylose starch fraction and from the outer branches of amylopectin.

In forming cyclic dextrins from starch, the *B. macerans* enzyme is apparently capable of promoting the transfer of energy from one α-D-$(1 \rightarrow 4)$-glucosidic bond in a linear chain to another new bond of the same chain in a cyclic dextrin. The enzyme, therefore, belongs in the group of transglucosylases and not to the amylases as classified by the early investigators.

French *et al.*[229] showed that the action of *B. macerans* is reversible. The enzyme is capable of rupturing α- or β-cyclic dextrins and lengthening the linear chains by combining with D-glucose, maltose, methyl α-D-glucopyrano-side, sucrose, cellobiose, or maltobionic acid. Reversal of the action of this enzyme has been confirmed[230] by using cyclic α-dextrin and maltose. This action does not, however, result in the formation of large molecules.

Norberg and French[231] have shown that the *B. macerans* enzyme is not limited to reactions involving cyclic dextrin molecules; these processes represent only one aspect of a more general reaction. The enzyme can cause linear dextrins to enter into glucosidic exchange with one another, resulting in a redistribution of the D-glucose residues among dextrins of longer and shorter chain lengths. These distribution reactions among linear reducing oligosaccharides and dextrins are referred to as "homologizing" reactions.

E. DEXTRAN FROM SUCROSE

Growing cultures of *Leuconostoc mesenteroides, Leuconostoc dextranicum,* and *Betabacterium vermiforme* are capable of synthesizing dextrans from sucrose.[217,218–247] These dextrans are polymers of D-glucose residues, which are combined chiefly by α-D-$(1 \rightarrow 6)$ linkages. Hehre and collaborators[217] showed that dextran can also be synthesized from sucrose by cell-free extracts of *Leuconostoc*. The enzyme of these microorganisms responsible for the synthesis of dextran was named "dextransucrase" [$(1 \rightarrow 6)$-α-D-glucan:D-fructose 2-glucosyltransferase].

Sucrose is a highly specific substrate for the enzyme, which is produced only by bacteria grown on this disaccharide. When a crude dextransucrase preparation is inoculated with sucrose, the synthesis of dextran is indicated by serological testing,[248] development of opalescence, formation of an alcohol-precipitable polysaccharide, and accumulation of D-fructose. The dextrans produced by different species of bacteria vary considerably in their molecular constitution. Thus, the one synthesized by *Leuconostoc dextranicum* appears to be essentially unbranched, whereas those obtained by the action of other microorganisms are highly branched.[223–247] The branches are generally connected by $(1 \rightarrow 4)$-D-glucosidic linkages, but in some cases the $(1 \rightarrow 3)$ type is found.[249–251] In the polysaccharides where branching occurs, it is necessary to assume that in addition to dextransucrase, which forms the main $(1 \rightarrow 6)$ linkage, another enzyme responsible for the establishment of the $(1 \rightarrow 4)$ or $(1 \rightarrow 3)$ linkage is present.

The evidence indicates that dextransucrase acts by a direct transfer of D-glucose residues. The reaction appears to involve the substitution of a $(1 \rightarrow 6)$-glucosidic linkage for a glucose–fructose bond and can be represented by the following equation:

$$(n)C_{12}H_{22}O_{11} \longrightarrow (C_6H_{10}O_5)_n + (n)C_6H_{12}O_6$$
$$\text{(sucrose)} \qquad\qquad \text{(dextran)} \qquad \text{(D-fructose)}$$

F. DEXTRAN FROM AMYLODEXTRIN

Hehre and Hamilton[252] demonstrated that certain acetic acid bacteria, *Acetobacter viscosum* and *A. capsulatum*, are also capable of forming dextran. Unlike *Leuconostoc mesenteroides*, these bacteria do not produce a polysaccharide from sucrose, but require dextrin as a specific substrate. The enzyme responsible for this reaction (dextran 6-glucosyltransferase) has been obtained free from bacterial cells. The conversion of dextrin into dextran presumably involves the conversion of $(1 \rightarrow 4)$-glucosidic bonds to $(1 \rightarrow 6)$ bonds.

G. LEVAN FROM SUCROSE

In 1910 Beijerinck,[253] working with *Bacillus megaterium* and certain other bacteria, was first to observe polysaccharide formation by the agency of an extracellular enzyme. When the microorganisms were grown on agar plates containing sucrose, microscopically visible particles of levan [a polyfructofuranose in which the main glycosidic linkages are of the $(2 \rightarrow 6)$ type] appeared in the agar.

* References start on p. 362.

Hestrin and his co-workers[254,255] succeeded in liberating the enzyme (levansucrase) [(2 → 6)-α-D-fructan:D-glucose 6-fructosyltransferase] from washed cells of *Acetobacter levanicum* by autolysis in the presence of thymol and chloroform, and showed that it converts sucrose into levan and D-glucose. In this reaction a (2 → 6) linkage is substituted for the glycosidic bond in sucrose, forming the levan:

$$(n)C_{12}H_{22}O_{11} \longrightarrow (C_6H_{10}O_5)_n + (n)C_6H_{12}O_6$$
$$\text{(sucrose)} \qquad \text{(levan)} \qquad \text{(D-glucose)}$$

Similarly, raffinose is converted by this enzyme into levan and melibiose.

The molecular weights of levans are extremely high. Feingold and Gehatia[256] determined the molecular weight of native levan of *Acetobacter levanicum* cultures to be 17 million, and of enzymically synthesized levan 67 million. Dedonder and Slizewicz[257] reported molecular weight values for levans produced by *Bacillus subtilis* and for levans that had been subjected to mild acid hydrolysis as being from a 1000 to over 100 million.

IX. SYNTHESIS OF POLYSACCHARIDES FROM SUGAR NUCLEOTIDES

A. GLYCOGEN

In the course of time, considerable experimental evidence accumulated indicating that phosphorylase is not responsible *in vivo* for the formation of the α-D-(1 → 4)-linked glycogen chain from α-D-glucopyranosyl phosphate.[258] It was observed that, although *in vitro* the phosphorylase reaction attains equilibrium at pH 7.5 when the ratio of inorganic phosphate to α-D-glucopyranosyl phosphate is approximately 3, glycogen formation can take place *in vivo* when the ratio is many times as high without depressing the synthesis of glycogen.[259] Further doubts about the synthetic role of phosphorylase came from the studies of the action of glycogenolytic agents, epinephrine and glucagon, or high concentration of sodium ions, which are known to lower the concentration of glycogen.[260,261] An increased breakdown of glycogen was always observed. Conversely, under conditions known to lower the phosphorylase activity, such as high concentration of potassium ions, the glycogen content was increased.[262]

Studies on McArdle's and Hers' diseases demonstrated that there definitely is a pathway for glycogen synthesis that does not involve phosphorylase. In the muscles of patients having these diseases, phosphorylase is not detectable, but the amount of glycogen is normal or slightly above normal.[263-266] These results are inconsistent with the hypothesis that phosphorylase catalyzes the synthesis of glycogen *in vivo*, and led to the view that its role is confined to degradation of glycogen.

An enzyme capable of synthesizing glycogen from UDP-D-glucose was first observed by Leloir and Cardini.[191] Similar glycogen synthetases (UDP-D-glucose: glycogen 4-α-D-glucosyltransferases) have been found in other mammalian tissues, in some invertebrates, in yeast and in bacterial species.[267] The reaction catalyzing the transfer of D-glucose from UDP-D-glucose to an acceptor is expressed by the following equation:

$$[(1 \rightarrow 4)\text{-}\alpha\text{-D-glucosyl}]_n + x(\text{UDP-D-glucose}) \longrightarrow$$
$$\text{(acceptor)} \qquad\qquad [(1 - 4)\text{-}\alpha\text{-D-Glucosyl}]_{n+x} + x \text{ UDP}$$
$$\text{(glycogen chain)}$$

This reaction[268] is similar to the phosphorylase reaction in several respects. The D-glucose residues are united by the same α-D-$(1 \rightarrow 4)$ linkages, and a high-molecular-weight unit is required for the formation of the polysaccharide. However, the equilibrium of the reaction when UDP-D-glucose is used as substrate is more favorable for synthesis than when α-D-glucopyranosyl phosphate is used. Since the $\Delta G°$ of hydrolysis of UDP-D-glucose[98] is -7600 cal mole^{-1}, and that of the α-D-$(1 \rightarrow 4)$ linkage of glycogen[99] is -4300 cal mole^{-1}, the free-energy change during the formation of glycogen from UDP-D-glucose can be calculated as -3300 cal mole^{-1}. This value corresponds to an equilibrium constant of approximately 250, corresponding to a practically quantitative conversion of the nucleotide-bound D-glucose into glycogen.[269]

It is now generally accepted[270] that the synthesis of this polysaccharide begins with a transfer of D-glucose residues from UDP-D-glucose by UDP-D-glucose: glycogen 4-α-D-glucosyltransferase to a $[(1 \rightarrow 4)\text{-}\alpha\text{-D-glucosyl}]_n$ acceptor forming α-D-$(1 \rightarrow 4)$-linked D-glucose chains. When the chains become about ten residues long, these portions are transferred, forming α-$(1 \rightarrow 6)$-D-glucose linkages by a branching enzyme, $(1 \rightarrow 4)$-α-D-glucan:$(1 \rightarrow 4)$-α-D-glucan 6-glucosyltransferase. Degradation takes place by formation of α-D-glucopyranosyl phosphate from the α-D-$(1 \rightarrow 4)$-linked D-glucose residues by phosphorylase, but on reaching the vicinity of an α-D-$(1 \rightarrow 6)$-linked residue the reaction stops. The oligotransferase enzyme at this stage distributes the chains so as to leave a single α-D-$(1 \rightarrow 6)$-linked D-glucose unit, which is then hydrolyzed by dextrin 6-glucanohydrolase.

In considering the question of availability of the primer, Leloir[270] has suggested that under normal physiological conditions some glycogen molecules or molecular fragments are probably always present. Presumably each animal cell retains some glycogen during cell division. But if glycogen were to disappear completely, maltose, the smallest product of degradation, might remain to serve as an acceptor but at a very low rate.[271]

Another possibility is that, starting with α-D-glucopyranosyl phosphate, under conditions where there is apparently no preformed acceptor, an $(1 \rightarrow 4)$-α-D-glucose chain can be formed *de novo*, and this could serve as primer[272] for the formation of glycogen.

The formation of glycogen by the enzymes of yeast or of animal origin is stimulated to a variable extent by the addition of D-glucose 6-phosphate.[273,274] This ester has been shown to protect the glycogen synthetase of liver against denaturation,[275] and to activate the synthetase of yeast at unfavorable pH.[274] It has also been shown that liver synthetase, denatured by heating at 37°, and rabbit muscle synthetase which has lost activity through prolonged storage, are reactivated by the addition of D-glucose 6-phosphate.[276] These results indicate that D-glucose 6-phosphate acts on the enzyme protein in such a manner that it can assume and retain the conformation in which it is catalytically active.

Larner and co-workers[187,277] and Traut and Lipmann[188] found that the glycogen synthetase (dextrin 6-glucosyltransferase) exists in two forms, one requiring D-glucose 6-phosphate for activation, and the other stimulated only slightly by the ester. The form "independent" of D-glucose 6-phosphate, which requires ATP and Mg^{2+}, does not appear to be phosphorylated.[258] An effect of Ca^{2+} similar to that of phosphorylase *b*-kinase has also been observed.

Belocopitow *et al.*[278,279] found that injection of epinephrine in rats led to a decrease in total activity of glycogen synthetase, and that addition of calcium ions to some glycogen synthetase preparations produced a conversion of the "independent" into the dependent form. This conversion does not involve ATP and is not affected by adenosine 3′,5′-cyclic phosphate. It requires a protein factor[261] similar to that involved in the activation of inactive phosphorylase *b*-kinase by Ca^{2+}. It thus appears that calcium exerts an effect on the regulation of glycogen synthetase.

While glycogen is considered to be synthesized by glycogen synthetase from UDP-D-glucose, this polysaccharide can also be synthesized from ADP-D-glucose by what appears to be the same enzyme, but at a rate 50% of that for UDP-D-glucose.[47,271]

Shen and Preiss[280] showed that extracts of *Arthrobacter* sp. NRRLB1973 contain a pyrophosphorylase capable of catalyzing the formation of ADP-D-glucose in the presence of PP_i from ATP and α-D-glucopyranosyl phosphate. This pyrophosphorylase also synthesizes to a small extent CDP-D-glucose, IDP-D-glucose, GDP-D-glucose, and dADP-D-glucose from α-D-glucopyranosyl phosphate and the respective nucleoside triphosphates.

A purified ADP-D-glucose:glycogen transglucosylase was subsequently obtained from the same microorganism;[281] it formed glycogen from ADP-D-glucose and dADP-D-glucose only and from no other nucleoside sugar pyrophosphates. A number of $(1 \rightarrow 4)$-α-D-glucans, as well as some oligosaccharides of the maltodextrin series, were active as acceptors.

B. CHITIN

Chitin is a structural polysaccharide found primarily in the exoskeletons of arthropods and the cell walls of fungi.[282] It consists of 2-acetamido-2-deoxy-D-glucose residues that are combined by β-D-$(1 \rightarrow 4)$ linkages to form the polymer. The structure of chitin is similar to that of cellulose, except that the D-glucose residues have acetamido groups at C-2. (See this volume, Chap. 36 for more details of structure and biosynthesis.)

A particulate preparation from *Neurospora crassa* catalyzes the synthesis of an insoluble polysaccharide from UDP-2-acetamido-2-deoxy-D-glucose (UDP-GlcNAc) according to the following reaction:[234]

$$\text{UDP-GlcNAc} + (\text{GlcNAc})_n \rightleftharpoons \text{UDP} + (\text{GlcNAc})_{n+1}$$

This polysaccharide is formed by glycosyl transfer from the nucleotide-linked sugar, UDP-GlcNAc, to a preformed chitodextrin chain composed of 2-acetamido-2-deoxy-D-glucose $(\text{GlcNAc})_n$. The reaction has been found to be reversible; it is in favor of polysaccharide synthesis and is markedly stimulated by the addition of 2-acetamido-2-deoxy-D-glucose. The requirement for a monosaccharide "activator" which does not serve as a glycosyl acceptor has been encountered in the synthesis of callose.

Inasmuch as the UDP-2-acetamido-2-deoxy-D-glucose is known to contain α-D-glycosidically bonded 2-acetamido-2-deoxy-D-glucose, the chitin synthesis, as in a number of other polysaccharides, must proceed with inversion of the α-D-linkage to produce a β-D-linked polymer.

A particulate enzyme from the mycelium of *Allomyces maerogynus* is capable of catalyzing the synthesis of chitin from UDP-2-acetamido-2-deoxy-D-glucose.[283] The enzyme is associated with both mitochondrial and microsomal fractions but exhibits higher specific activity and greater stability in the latter. The enzyme is activated with chitodextrin and 2-acetamido-2-deoxy-D-glucose. The pH optimum for the reaction is 7.8, and the temperature optimum is 30°. The K_m for the reaction with UDP-2-acetamido-2-deoxy-D-glucose is 1.2×10^{-3} M.

* *References start on p. 362.*

C. Starch

Similar to the problem of glycogen synthesis, starch had long been thought to be formed by the action of a phosphorolytic enzyme on α-D-glucopyranosyl phosphate. It was observed later[284] that the phosphate content in plant cells is such that the equilibrium would be expected to be in the direction of starch breakdown.[285] Also, while phosphorylase is found in the soluble portion of the cytoplasm, starch actually has been found to be synthesized in the plastids.[286] Some other mechanism for starch synthesis, which did not involve phosphorylase, was therefore suspected. The discovery of an enzyme that synthesizes glycogen from UDP-D-glucose initiated a search for a similar enzyme in plant tissues. Such an enzyme was found by Leloir and co-workers;[21] it was associated with starch grains from beans potatoes, and corn seedlings. Freshly isolated starch grains from these sources were found to catalyze the incorporation of radioactivity from UDP-D-glucose labeled with ^{14}C in the D-glucosyl moiety into a polysaccharide containing linkages of the α-D-(1 → 4) type. This was shown by degradation of the polysaccharide with β-amylase to radioactive maltose. The enzyme responsible for the synthesis is closely bound to the starch granule, and attempts to dissociate the enzymic activity from the grain have not been successful. Many other starch granule systems transferring D-glucose from either ADP-D-glucose or UDP-D-glucose to starch have been reported.[287,288]

However, Frydman and Cardini[289] later found a soluble enzyme from sweet corn which catalyzes the transfer of D-glucose from UDP-D-glucose and ADP-D-glucose to phytoglycogen, which has a chemical structure similar to that of amylopectin. Other soluble preparations were obtained by these investigators from tobacco leaves and potato tubers.[290]

The specificity of the soluble (1 → 4)-α-D-glucan transferases differs with regard to primer requirements. Thus, the soluble preparation from sweet corn can use only amylopectin and glycogen as primers. The soluble enzyme from tobacco leaves is capable of using either amylopectin, glycogen, or heated starch granules, and that from potato tubers can use only starch granules as the acceptor. The enzyme that leads to the synthesis of starch is very similar to that involved in glycogen formation, but it has a dissimilar specificity with regard to primers.[21]

With the starch granules as a source of enzyme, transfer of the D-glucose residues is ten times as fast from ADP-D-glucose as from UDP-D-glucose.[291] Since a specific enzyme[292] that catalyzes the reaction ATP + D-glucosyl P → ADP-D-glucose + PP$_i$, and ADP-D-glucose, has been detected in *Chlorella*,[293] corn,[294] and in rice grains,[295] it seems very likely that ADP-D-glucose is a precursor of starch *in vivo*. However, Leloir[47] has indicated that probably both UDP-D-glucose and ADP-D-glucose are involved and probably in about equal proportions, because, although UDP-D-glucose reacts more slowly, its concentration is five to ten times as high in plant tissues. This would result in about the same rate of transfer.

A rapid interconversion of sucrose and starch has been known for a long time to occur *in vivo*. De Fekete and Cardini[296] demonstrated a transfer of ^{14}C from sucrose to starch by an enzyme preparation from corn endosperm when the appropriate nucleotides were added to the reaction mixture. To account for these results obtained *in vitro*, the authors postulated the following sequence of reactions for which all the enzymes were present in the endosperm extract:

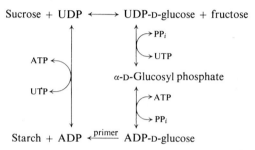

Evidence for the probable operation of the above scheme is supported by the fact that all the participating enzymes could be detected in the corn endosperm.

In a coupled system of sucrose synthetase (a glucosyltransferase) and starch synthetase (a glucosyltransferase), both isolated[297] from ripening rice grains, the more efficient transfer of D-glucose-^{14}C from sucrose-^{14}C to starch occurs in the presence of ADP as compared to UDP, indicating that ADP-D-glucose plays a predominant role in the process. However, UDP-D-glucose was observed to inhibit the ADP-D-glucose–sucrose transglycosylation reaction.[298] Also, the K_m value of the sucrose synthetase for ADP was found to be higher than that for UDP and UDP-D-glucose.[296] Thus, reversal of the enzymic reaction to form ADP-D-glucose is less likely to occur *in vivo*. Similar results were obtained with ripening rice grains.[287,299,300] The results of these authors support the view that the sucrose breakdown proceeds through the reversal of UDP-sucrose transglycosylation rather than directly via the ADP-sucrose transglucosylation.

Other results[301] indicate that the starch granule-bound ADP-D-glucose–starch glycosyltransferase isolated from the embryos of non-waxy corn (maize) seeds is different from that of the similar preparation isolated from the endosperms. Their investigation suggests that the preparations contain two different enzyme systems. This is compatible with the observation that in

* *References start on p. 362.*

the waxy mutants of corn (maize) the activity of the endosperm preparation is reduced to a very low level, while the activity of the embryo preparation is unimpaired.

A soluble enzyme preparation from spinach chloroplasts[302,303] is capable of transferring D-glucose from ADP-D-glucose to a primer to form $(1 \rightarrow 4)$-α-D-glucosyl linkages. When this partially purified enzyme preparation was tested with a number of sugar nucleotides containing different bases, only ADP-D-glucose and dADP-D-glucose could serve as glucosyl donors for polysaccharide formation. The acceptor could be an $(1 \rightarrow 4)$-α-D-glucan or an oligosaccharide of the maltodextrin series. It was also found that, as in the case of mammalian UDP-D-glucose:glycogen transglucosylase[304] and bacterial ADP-D-glucose–transglucosylase,[281] sulfhydryl groups are essential for the spinach chloroplast transglucosylase activity.

Unlike the mammalian or yeast transglycosylases,[304,305] the chloroplast transglucosylase is not activated by D-glucose 6-phosphate. However, it has been previously shown that ADP-D-glucose pyrophosphorylase activity in spinach chloroplasts is stimulated about 50-fold by glyceric acid 3-phosphate.[306] This observation may be significant in the control of starch formation during photosynthesis. It is possible that during CO_2 fixation the accumulation of glyceric acid 3-phosphate causes an increase of ADP-D-glucose synthesis by stimulating the ADP-D-glucose pyrophosphorylase, thus causing an increase in ADP-D-glucose concentration which would then enhance the rate of starch synthesis. It therefore appears that activation of starch synthesis occurs at the pyrophosphorylase level rather than, as in glycogen, at the transglucosylase level.[307]

D. CELLULOSE

The first biosynthetic formation of cellulose was accomplished[308] with an insoluble enzyme preparation from the microorganism *Acetobacter xylinum*, with UDP-D-glucose as substrate. However, work by Hassid and co-workers[18,19,309] with particulate enzyme preparations from mung beans (*Phaseolus aureus*) and other plants indicated that GDP-D-glucose and not UDP-D-glucose was the precursor of cellulose. (See this volume, Chap. 36 for more information on synthesis with enzymes from *A. xylinum*.) GDP-D-glucose was first encountered by Carlson and Hanson[309a] in the mammary gland tissue of the lactating cow. They also synthesized this enzyme chemically and by use of an enzyme preparation from bovine mammary glands.

Ripening peas and various tissues of other species were shown to contain an enzyme capable of forming ^{14}C-labeled GDP-D-glucose from guanosine triphosphate (GTP) and α-D-glucopyranosyl phosphate.[309] Another enzyme,[18,19] which could utilize this radioactive sugar nucleotide as substrate for the formation of a radioactive polysaccharide having chemical properties indistinguishable from those of natural cellulose, was found to exist in root tissues of mung-bean seedlings, peas, corn, squash, and string beans[19] and in the immature seed hairs of cotton.[310] The cellulose-synthesizing enzyme was capable of transferring the activated D-glucose moiety from GDP-D-glucose-^{14}C to an unknown acceptor, forming the polysaccharide chain.

Under the particular experimental conditions, the enzyme system that polymerized the D-glucose to cellulose showed a high degree of specificity for GDP-D-glucose. None of the ^{14}C-labeled D-glucosyl nucleotides containing bases other than guanine (uracil, adenine, cytosine, thymine) served as substrate for the formation of cellulose.

It has not been possible to separate the transferase activity from the endogenous acceptor. No stimulation of incorporation of D-glucose was observed upon addition of D-glucose, 2-amino-2-deoxy-D-glucose, D-mannose, D-galactose, sucrose, cellobiose, soluble or insoluble cellodextrins, or swollen cellulose.

The incorporation of activity into the cellulose (20 to 25% of substrate) was enhanced when certain cations, especially Co^{2+}, Mn^{2+}, or Mg^{2+} were added to the reaction mixture when GDP-D-glucose was used as substrate, but only after the particulate preparation had been treated with ammonium sulfate.

On the basis of these results, the formation of cellulose in plants is postulated to take place by a reaction in which GDP-D-glucose is first formed as follows:

$$\text{GTP} + \alpha\text{-D-glucopyranosyl phosphate} \xrightleftharpoons{\text{pyrophosphorylase}} \text{GDP-D-glucose} + \text{PP}_i$$

The polysaccharide is then formed by another enzyme (GDP-D-glucose: $(1 \rightarrow 4)$-β-D-glucan β-D-4-glucosyltransferase), which catalyzes repetitive transfers of D-glucose residues to an acceptor, as shown by the following equation:

$$n(\text{GDP-D-glucose}) + \text{acceptor} \longrightarrow$$
$$\text{Acceptor-}[(1 \rightarrow 4)\text{-}\beta\text{-D-glucosyl}]_n + n(\text{GDP})$$
$$(\text{cellulose})$$

However, other workers reported[311,312] that an enzyme preparation from *Lupinus albus* is capable of incorporating glucose-^{14}C from UDP-D-glucose-^{14}C as well as from GDP-D-glucose-^{14}C into a mixture of polysaccharides ranging from water-soluble to insoluble in alkali. They found that the alkali-insoluble fraction (about 7%) was synthesized more readily from UDP-D-

* *References start on p. 362.*

glucose than from GDP-D-glucose. However, repetition of this work with *Lupinus albus*[312a] yielded a polysaccharide similar to the one obtained from mung beans, containing only (1 → 3)-β-D-glucosyl linkages.[313,314]

It was reported[314a] that particulate preparations from oat coleoptiles could utilize both GDP-D-glucose or UDP-D-glucose as substrates for polysaccharide formation. Upon degradation of the polysaccharide derived from UDP-D-glucose with cellulase [(1 → 4)-β-D-glucan glucanohydrolase], cellobiose, and a substance identified as a trisaccharide containing mixed β-(1 → 4) and β-(1 → 3)-D-glucosyl linkages were obtained. However, when GDP-D-glucose was used as a substrate, only cellobiose and cellotriose were obtained, indicating the production of a polysaccharide containing only β-D-(1 → 4) linkages.

A similar alkali-insoluble polysaccharide, containing both β-D-(1 → 4) and β-D-(1 → 3) linkages, was obtained[314b] with a particulate mung bean preparation and UDP-D-glucose-¹⁴C as substrate when an incubation mixture containing sucrose and albumin was used; hydrolysis of this polysaccharide with acid produced, in addition to what seemed to be cellobiose, from 10 to 20 % of laminarabiose, in which the D-glucose residues are joined by β-D-(1 → 3) glucosyl linkages. These results appeared to be similar to those obtained with oat coleoptiles,[314a] in which specific hydrolases were used for determination of the linkages in the D-glucan.

Whereas the results[314a] with enzyme preparations from oat coleoptiles and UDP-D-glucose as substrate could be substantiated by chemical methods, a subsequent investigation[312a] showed that particulate enzyme preparations from mung bean seedlings differed in their action from those previously reported,[314b] in that the glucan contained only β-D-(1 → 3) linkages.

Treatment of a D-glucan that was synthesized from UDP-D-glucose by the particulate enzyme from *Phaseolus aureus* with a highly purified exo-(1 → 3)-β-D-glucanase[314c] hydrolyzed the polysaccharide to the extent of 91 % in 24 hours. The alkali-insoluble D-glucan produced from GDP-D-glucose formed by the particulate enzyme system from the same plant, known to be (from chemical data) a β-(1 → 4)-linked D-glucan (cellulose), was not acted upon by this glucanase.[314d]

Digitonin extracts of particulate enzyme preparations proved to be active in the synthesis of cellulose from GDP-D-glucose. A soluble preparation, obtained from this solution and from which the protein was subsequently precipitated by the addition of ammonium sulfate and then dissolved, did not show any synthetic activity. However, when boiled, active mung bean particles were added to the soluble preparation, the cellulose synthetic activity of the soluble preparation became active.[315] This strongly indicates that the primer for the cellulose synthesis resides in the particles. Careful examination of the polymer enzymically produced from GDP-D-glucose also revealed that it contained a small proportion of mannose, indicating the presence of an epimerase that interconverts GDP-D-glucose and GDP-D-mannose.[316]

In general, the data appear to be consistent with the assumption that GDP-D-glucose is the sugar nucleotide from which the β-D-(1 → 4)-glucosidically linked chains of cellulose are synthesized.

Although the experiments *in vitro* with mung beans[19] indicate that GDP-D-glucose is the direct donor of D-glucose for the formation of cellulose in plants, Colvin's work[316a] suggests the presence of a glucose–lipid precursor within the plant cell, which is converted extracellularly into cellulose. The D-glucose is transferred from the nucleoside D-glucose pyrophosphate to the glucolipid at or on the cytoplasmic membrane, and the D-glucose is then translocated outside the cell by the lipid moiety; that is, the glucolipid acts as a carrier from within the cell to the external medium where the D-glucose is polymerized to cellulose.

E. D-GLUCO-D-MANNAN

It has been shown[19] that extracts from mung bean seedlings incorporate the glycosyl portion of GDP-D-glucose-^{14}C into cellulose. It was observed that the addition of GDP-D-mannose to the incubation mixture increased incorporation of D-glucose-^{14}C from the latter sugar nucleotide into an alkali-insoluble polysaccharide, which was not cellulose. However, it was not clear at that time whether this compound was a D-gluco-D-mannan or a mixture of cellulose and a D-mannan. It was later found[4] that when ^{14}C-labeled GDP-D-mannose alone is used as substrate, a radioactive D-gluco-D-mannan is produced, and that Mg^{2+} is required for its synthesis.

When GDP-D-glucose is added to the reaction mixture containing GDP-D-mannose-^{14}C, a marked inhibition of incorporation of D-mannose is observed, but the addition of GDP-D-mannose to the reaction mixture containing GDP-D-glucose-^{14}C results in the stimulation of incorporation of radioactivity into an insoluble product.[316b] The explanation for these results is not clear. However, it is possible that enzyme(s) in the synthesis of D-gluco-D-mannan have a greater affinity for GDP-D-glucose than for GDP-D-mannose, which would explain the inhibition of D-mannose incorporation by GDP-D-glucose.

The D-gluco-D-mannan was characterized by isolation of a number of oligosaccharides after treatment of the product with partially purified β-D-mannanase. Several of these oligosaccharides contained D-glucose and D-mannose in approximate ratios of 1:1, 1:2, 1:3, or 1:4. Since ^{14}C-labeled D-glucose was found in the hydrolysis products of the oligosaccharides, it is possible that the particulate enzyme contains an epimerase that converts GDP-D-mannose into GDP-D-glucose.

* References start on p. 362.

F. D-XYLAN

The work of Neish and his collaborators[317,318] indicated that uronic acids play an important part in synthesis of D-xylan. When D-glucose labeled with ^{14}C in certain positions of the molecule was introduced into wheat seedlings, the D-xylose isolated from the D-xylan of these plants contained most of the activity in the carbon atoms which corresponded to those of the D-glucose. The D-xylose isolated from the hydrolyzed D-xylan when D-glucuronolactone-1-^{14}C was introduced into these plants contained the major ^{14}C activity in C-1 of the pentose. Slater and Beevers[319] showed that D-glucuronolactone labeled with ^{14}C in various positions is a good precursor of D-xylan and appeared to give rise to the D-xylose by a loss of C-6. These *in vivo* experiments indicated that D-xylan originated in plants from D-glucose by a series of reactions leading to the formation of D-glucuronic acid, which was subsequently decarboxylated to pentose and polymerized to the polysaccharide.

The mechanism of synthesis of xylan with enzymic plant preparations *in vitro* proved to be in accord with the mechanism postulated for that polysaccharide *in vivo*. UDP-D-xylose has been isolated from plant seedlings[320] and synthesized enzymically from UDP-D-glucose by the following sequence of reactions:[321-323]

$$\text{UDP-D-glucose} \xrightarrow[\text{NAD}^+]{\text{dehydrogenase}} \text{UDP-D-glucuronic acid} \xrightarrow{\text{decarboxylase}} \text{UDP-D-xylose}$$

A particulate preparation from asparagus shoots was shown to transfer D-xylosyl residues from UDP-D-xylose to water-soluble D-xylose oligosaccharides ranging in size from xylobiose to xylopentaose having the same β-D-(1 → 4) linkages as D-xylan. The product was, however, not polymeric D-xylan, but in each case an oligosaccharide containing one more residue than the acceptor. The failure of the asparagus system to transfer more than one D-xylosyl residue may be attributed to the combined effect of unfavorable experimental conditions and to a low affinity for the oligosaccharide acceptor.

Particulate enzyme preparations from corn shoots[324] and immature corn cobs[325] were found to incorporate D-xylose-^{14}C from UDP-D-xylose-^{14}C into a polysaccharide which appears to be (1 → 4)-linked D-xylan. Hydrolysis of this polysaccharide produced D-xylose and L-arabinose, indicating that the particulate preparation contained an epimerase that interconverted UDP-D-xylose and UDP-L-arabinose. The labeled polymer appears to be a polysaccharide that closely resembles plant D-xylan in practically all its physical and chemical properties. Partial hydrolysis with fuming hydrochloric acid produced a homologous series of oligosaccharides (degree of polymerization 2 to 7) which were shown to be D-xylo-dextrins and which were chromatographically identical with the authentic plant (1 → 4)-β-D-xylo-dextrins.

Total acid hydrolysis yielded labeled D-xylose and L-arabinose in the ratio of about 4:1. Hydrolysis with weak acid (0.01N), which is known to hydrolyze sugars having the furanose configuration, liberated L-arabinose-^{14}C, but scarcely any D-xylose-^{14}C, indicating that some of the arabinose residues have the furanoid ring form.

G. PECTIN

Pectin, which is an important structural component of all higher plants, has as its basic building unit a (1 → 4)-α-D-galactopyranuronan. The carboxyl groups of this compound are methylated to various degrees, and polygalacturonates are associated with other carbohydrates, mainly with D-galactan and L-arabinan.[326]

It has been shown[327] that UDP-α-D-galacturonic acid exists in mung bean shoots and that enzymes are present in this plant, causing the formation of this uronic acid nucleotide by the following pathway:[321-323]

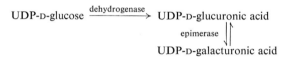

Furthermore, it has also been demonstrated[328] that a kinase is present in plants that catalyzes the formation of D-galactosyluronic acid phosphate from D-galacturonic acid and ATP. UDP-D-galacturonic acid is then formed from the uronic acid 1-phosphate and UTP by a pyrophosphorylase reaction. Thus, the same uronic acid nucleotide can be formed by an alternative mechanism. The sequence of these enzymic reactions led to the hypothesis that UDP-D-galacturonic acid is a precursor of pectin. Experiments performed *in vitro* proved to be in accord with this hypothesis.

It has been found[329] that a particulate preparation from mung bean (*Phaseolus aureus*) seedlings catalyzes the formation of a radioactive D-galacturonan chain from UDP-D-galacturonic acid-[14]C. The biosynthetic D-galacturonan could be completely hydrolyzed with the D-galactosiduronase of *Penicillium chrysogenum* to D-galacturonic acid-[14]C. Partial degradation of this synthetic product with an D-galacturonan transeliminase produced radioactive, unsaturated digalacturonic acid having a double bond at C-4,5.[330,331] The action of this enzyme is known to be specific for the D-galacturonic acid chain. No nucleoside D-galacturonosyl pyrophosphate containing a base other than uracil can be utilized for the formation of D-galacturonan.

The D-galacturonan transferase has an apparent Michaelis constant of 1.7 × 10^{-6} M for UDP-D-galacturonic acid, and at a substrate concentration

* *References start on p. 362.*

of 3.7 $\times$ 10^{-5} will catalyze the polymerization of D-galacturonic acid residues at the rate of 4.7 μmoles per milligram of protein per minute.

The particulate enzyme system does not incorporate D-galacturonic acid residues from UDP-D-galacturonate methyl ester into D-galacturonan. Hence, methylation probably occurs at a later stage in the synthesis of the polymer.[332-334] Kauss *et al.*[334] showed that a particulate enzyme from *Phaseolus aureus* contains an enzyme capable of transferring the methyl group of *S*-adenosyl-L-methionine to the D-galacturonan that is present in the particles.

H. (1 → 3)-β-D-GLUCANS

Particulate preparations from homogenates of mung bean seedlings have been shown[20] to catalyze the formation of insoluble, radioactive material from UDP-D-glucose-[14]C. This polysaccharide, known as "callose,"[335,336] requires the action of fuming hydrochloric acid for solubilization and degradation to oligosaccharides. However, unlike cellulose, it is soluble in hot, dilute alkali.

The polymer could be partially hydrolyzed with fuming hydrochloric acid to a series of oligosaccharides having a degree of polymerization from 2 to 8. These oligosaccharides were chromatographically identified with those present in a partial hydrolyzate of the β-D-(1 → 3)-linked polysaccharide, laminaran. Chemical degradation of the disaccharide with lead tetraacetate produced D-arabinose.[337] This experiment furnished additional proof that the disaccharide is linked between the C-1 and C-3 positions of the D-glucose residues. The oligosaccharides obtained from the synthetic polymer were completely hydrolyzed to D-glucose by almond emulsin, indicating that the D-glycosidic linkages are of the β type.

When the enzyme that catalyzes the synthesis of callose was solubilized with digitonin and partially purified by fractionation with ammonium sulfate, the enzyme was found to be strongly activated by a number of saccharides, such as laminarabiose, cellobiose, laminaratriose, and D-glucose. Inasmuch as related compounds such as D-fructose, D-galactose, D-mannose, and lactose are inert, there appears to be considerable specificity in this activation. Since it has been shown[321] that the activators do not become incorporated into the polymer, their action seems to be similar to that of D-glucose 6-phosphate on glycogen synthetase. It has been suggested by Leloir[338] that the activators presumably combine with some specific site on the enzyme surface, causing a conformational change which increases the activity of the enzyme.

Callose was known to be predominantly localized in the phloem tissue of higher plants;[339] however, it now appears that the polysaccharide is readily produced in other plant tissues as a result of injury to the cells.[340]

The synthesis of the reserve carbohydrate, paramylon, of *Euglena gracilis* has been studied.[341] This polysaccharide is known to be an insoluble polymer consisting of β-D-(1 → 3)-linked D-glucose residues. As in the case of callose, the donor is UDP-D-glucose, and no added acceptor is required. Polysaccharide formation could be carried out with *Euglena* enzyme in the presence of a potent (1 → 3)-β-D-glucan glucanohydrolase from snail gut juice, which caused hydrolysis of the polysaccharide to D-glucose as rapidly as it was formed. However, there was no indication that the initial acceptor is split under these conditions or by preincubation with the (1 → 3)-β-D-glucan glucanohydrolase. These authors assumed that the initial acceptor is not hydrolyzed by this enzyme, either because it has no (1 → 3)-β-D-glucosidic linkages or, if it has, the linkages are not available for hydrolysis by this enzyme.

A further study[342] showed that the particulate enzyme which could be extracted from the particles with sodium deoxycholate results in smaller particles after dialysis and centrifugation at 100,000 × g. The enzyme does not appear to be bound to the polysaccharide fraction. The fact that the synthetic product is hydrolyzed with the D-glucanase of Basidiomycetes is in accord with the finding that the glycosyl residues are β-D-(1 → 3)-linked. This paramylon synthetase shows activity without addition of a primer.

I. ALGINIC ACID

This glycuronan, consisting of a mixture of β-D-(1 → 4)-linked polymers of D-mannuronic acid and its 5-epimer L-guluronic acid, is known to be a major constituent of the structural polysaccharides of brown algae (see Vol. IIB, Chap. 40). The D-mannuronic acid is usually the main component of this polymer. Since it was shown that a number of complex carbohydrates are synthesized from sugar nucleotide precursors, it seemed likely that the uronic acid residues of the alginic acid polymer are derived from nucleotide-linked derivatives.[343,344]

The isolation of GDP-D-mannuronic acid and GDP-L-guluronic acid from the marine brown alga *Fucus gardneri* suggested that these nucleoside uronosyl pyrophosphates are the precursors of the polyuronic acid chain of alginic acid.[344]

Enzyme preparations were obtained which contained the following enzymic activities: hexokinase, phosphate mannomutase, pyrophosphorylase, GDP-D-mannose dehydrogenase, and D-mannuronic acid transferase.[345] These active enzymic preparations could be obtained only when the *Fucus gardneri* homogenate was treated with polyvinylpyrrolidone. Starting with D-mannose,

* *References start on p. 362.*

the enzyme systems carry out several consecutive reactions, resulting in the biosynthesis of the D-mannuronic acid chain:

$$\alpha\text{-D-Mannose} \xrightarrow[\text{ATP}]{\text{kinase}} \alpha\text{-D-Mannose 6-P} \xrightarrow{\substack{\text{phosphate manno-} \\ \text{mutase}}} \alpha\text{-D-Mannosyl phosphate}$$

$$\alpha\text{-D-Mannosyl phosphate} + \text{GTP} \xrightarrow{\text{pyrophosphorylase}} \text{GDP-D-mannose} + \text{PP}_i$$

$$\text{GDP-D-mannose} + 2\text{NAD}^+(\text{P}^+) + \text{H}_2\text{O} \xrightarrow{\text{dehydrogenase}}$$
$$\text{GDP-D-mannuronic acid} + 2\text{NAD(P)H} + 2\text{H}^+$$

$$\text{GDP-D-mannuronic acid} \xrightarrow{\text{transferase}} \text{D-Mannuronan} + \text{GDP}$$

To synthesize the complete alginic acid polymer, L-guluronic acid (the minor component), which is probably synthesized through interconversion of GDP-D-mannuronic acid and GDP-L-guluronic acid by a 5-epimerase, should also be incorporated into the molecule. This, however, has not yet been achieved.

The final stage of alginic acid synthesis probably involves a succession of transfers of uronic acid residues from the two GDP-D-uronic acids to an acceptor molecule, forming β-D-(1 → 4)-linked glycuronan chains:

$$\begin{array}{c}\text{GDP-D-mannuronic acid} \\ \text{GDP-L-guluronic acid}\end{array} \xrightarrow[\text{acceptor}]{\text{glycosyl transferase(s)}} \text{Alginic acid} + \text{GDP}$$

J. DL-GALACTAN

Su et al.[29] investigated the carbohydrates of the marine red alga, *Porphyra perforata*. The ethanol-insoluble fraction of this plant contains a galactan which on hydrolysis gives about equal proportions of D- and L-galactose (see Chapter 40). Analysis of the polysaccharide indicated that it consists of D-galactose, 6-O-methyl-D-galactose, 3,6-anhydro-L-galactose, and sulfate (in ester form) in the molar ratio of approximately 1:1:2:1, respectively.

A search for sugar nucleotides in this alga revealed the presence of UDP-D-galactose, GDP-L-galactose, UDP-D-glucose, GDP-D-mannose, and UDP-D-glucuronic acid. In addition, AMP, UMP, GMP, IMP, ADP, UDP, IDP, NAD$^+$, NADP$^+$, and a new nucleotide, adenosine 3',5'-diphosphate, were isolated.

Inasmuch as *Porphyra perforata* contains complex saccharides composed chiefly of galactose moieties of both D and L configuration, together with sugar nucleotides containing the two enantiomorphs of galactose, it may be assumed that these sugar nucleotides are the precursors of the complex galactose compounds.

Furthermore, since GDP-L-galactose is found together with GDP-D-mannose in this red alga, it is suggested that the formation of the former sugar nucleotide is probably brought about in the plant by a mechanism similar to that for the conversion of GDP-D-mannose into GDP-L-fucose,[74,346] in which the glycosyl moiety has the L-galactose configuration.

Although no attempt was made to isolate enzyme systems from the alga that would suggest a pathway for the formation of the DL-galactan, on the basis of the above considerations this polysaccharide is probably synthesized by the following sequence of reactions:

D-Mannosyl phosphate $\longrightarrow$ GDP-D-mannose $\longrightarrow$ GDP-L-galactose $\longrightarrow$
DL-Galactan

Regarding the origin of the sulfuric acid group in the galactan, it is of interest to compare the structure of the so-called "active" sulfate [(adenylyl 3'-phosphate) sulfate], which is known to be a sulfate donor,[347] with that of adenosine 3',5'-diphosphate discovered in the *Porphyra perforata*. It is conceivable that the adenosine 3',5'-diphosphate in this alga activates inorganic sulfate to form the "active sulfate" which engages in the enzymic sulfation of the polysaccharide.

While the seaweed polysaccharides (alginic acid and L-fucan) appear to be derived by a series of reactions from the common precursor, GDP-D-mannose,[344,345] most of the polysaccharide constituents of higher plants, namely, D-galactose, D-glucuronic acid and D-galacturonic acid, D-xylose, and L-arabinose, are derived from UDP-D-glucose.[348,338]

K. HYALURONIC ACID

UDP-D-glucuronic acid was found to be implicated in the synthesis of mucopolysaccharides. These compounds form an integral part of animal connective tissue. One of this group, hyaluronic acid, has been isolated from umbilical cord, synovial fluid, and skin, and is produced by some bacteria. It is a straight-chain polymer composed of alternating β-D-(1 $\rightarrow$ 4)-linked 2-acetamido-2-deoxy-D-glucose and β-D-(1 $\rightarrow$ 3)-linked D-glucuronic acid residues (see Vol. IIB, Chap. 42).

By using radioactive UDP-2-acetamido-2-deoxy-D-glucose in the presence of homogenates from Rous chicken sarcoma, a transfer of the label into short polysaccharide chains having a hyaluronic acid type of structure has been demonstrated.[235] Markovitz *et al.*[349] showed that UDP-D-glucuronic acid and UDP-2-acetamido-2-deoxy-D-glucose are precursors of hyaluronic acid. Tritiated nucleoside glycosyl pyrophosphates, treated with extracts from a hyaluronic acid-producing strain of *Streptococcus*, incorporated ³H into the product; a net synthesis of hyaluronic acid was also demonstrated.

The mechanism pertaining to the alternating sequence of D-glucuronic acid and 2-acetamido-2-deoxy-D-glucose is not known. It has been suggested that the hyaluronic acid synthetase possesses one site for D-glucuronic acid residues and another for 2-acetamido-2-deoxy-D-glucose residues, and that the two sites can function only in alternation.[349]

* *References start on p. 362.*

L. Capsular Pneumococcal Polysaccharides

Like hyaluronic acid, the capsular polysaccharide of Type III *Pneumococcus* consists of two monosaccharide residues in an alternating sequence. The D-glucose and D-glucuronic acid residues of this polysaccharide are joined by β-D linkages.[350] The D-glucosyl residues are linked to C-3 of the D-glucuronic acid, while the D-glucopyranosyl capsular polysaccharide is one of the eighty known, serologically distinct types of this *Pneumococcus* organism[351] (see Vol. IIB, Chap. 41).

Smith *et al.*[352,353] showed that extracts from this organism are capable of catalyzing the equimolar incorporation of D-glucose-^{14}C and D-glucuronic-^{14}C acid residues from UDP-D-glucose-^{14}C and UDP-D-glucuronic acid-^{14}C into a polysaccharide which was quantitatively precipitated with Type III antiserum. This indicates that the monosaccharide constituents of the polysaccharides are derived from these nucleoside glycosyl pyrophosphates.

The structure of the polysaccharide of Type I *Pneumococcus* has not been completely established. Preliminary data indicate that the polysaccharide is probably composed of a chain of D-galacturonic acid residues to which are linked residues of 2-acetamido-2-deoxy-D-glucose, D-galactose, and L-fucose.[354] The enzyme from Type I *Pneumococcus* catalyzes the synthesis of a polymer consisting of D-galacturonic acid from UDP-D-galacturonic acid. The product is precipitated with Type I antiserum, and the addition of UDP-2-acetamido-2-deoxy-D-glucose to the enzymic reaction mixture doubles the serological reactivity of the product.[355] These results suggest that the chain of D-galacturonic acid residues is synthesized first and that other monosaccharide components are later attached to it. Type VIII *Pneumococcus* polysaccharide, consisting of residues of D-glucose, D-galactose, and D-glucuronic acid, appears to be synthesized from the respective uridine 5'-glycosyl pyrophosphate derivatives.[356]

Enzymic studies in conjunction with genetic investigations have shown that some *Pneumococcus* mutants can form capsular polysaccharides because they lack certain enzymes needed for synthesizing the required glycosyl nucleotides.[357] Type III cells synthesize their capsule by the following enzymic reactions:

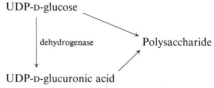

The mutant S_{112}, which produced negligible amounts of polysaccharide *in vivo*, was shown to contain an enzyme which could form the polysaccharide *in vitro* when the proper glycosyl nucleotides were made available. The metabolic lesion of this mutant involves an absence of the dehydrogenase, so that UDP-D-glucuronic acid is not formed.

Fully encapsulated cells of Type I *Pneumococcus* contain enzymes required for catalyzing the following sequence of reactions:

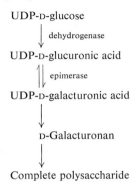

UDP-D-glucose

⬇ dehydrogenase

UDP-D-glucuronic acid

⬍ epimerase

UDP-D-galacturonic acid

⬇

D-Galacturonan

⬇

Complete polysaccharide

Some Type I mutants that do not form capsular polysaccharide lack the dehydrogenase, whereas others lack the epimerase. Such enzyme deficiencies can be repaired by introducing, into cultures of the mutant, deoxyribonucleic acid from pneumococci possessing the required enzyme. For example, if transforming deoxyribonucleic acid from Type I cells that contain epimerase is added to a culture of mutants which contain dehydrogenase but not epimerase, the mutants become capable of forming UDP-D-galacturonic acid. Since such treatment repairs the break in the metabolic sequence described above, it can be assumed that the transformed mutants are capable of synthesizing the capsule.

M. BACTERIAL CELL-WALL LIPOPOLYSACCHARIDES
(see also Vol. IIB, Chap. 41)

The knowledge of the structure and biosynthesis of the lipopolysaccharide macromolecules has been derived primarily from the study of various mutants of microorganisms. These complex macromolecules are responsible for many of the biological properties of the cell surface. They contain the somatic antigen (O-antigen) determinants and bacteriophage receptor sites. They are also responsible for the toxic properties (so-called endotoxin) of heat-killed bacteria or isolated cell-wall preparations.

Elucidating the biosynthesis of lipopolysaccharides began as a result of the observation[358,359] that *Salmonella typhimurium* mutants, which lack epimerase, are unable to form UDP-D-galactose. For this reason, these mutants contain only D-glucose and heptose in their lipopolysaccharide. Presumably D-galactosyl residues are normally transferred to the lipoheptoglycan, and the remaining sugars, D-mannose, L-rhamnose, and abequose, are subsequently attached to D-galactose. When the sequence is interrupted by lack of UDP-D-galactose, the nucleotide derivatives of the last three sugars accumulate in abnormally high concentrations.[358] However, the epimerase-lacking cells are not deficient in D-galactosyl transferase. It has been shown that a particulate enzyme obtained from these cells transfers D-galactosyl residues from UDP-D-galactose to lipoheptoglycan.[360] Mutants unable to synthesize UDP-D-glucose[361] form a lipopolysaccharide containing heptose only. D-Glucosyl residues can be enzymically linked to this heptose "core" when UDP-D-glucose is incubated with a particulate preparation.

Osborn and D'Ari[362] found that when a mutant from *S. typhimurium* lacking UDP-D-galactose-4-epimerase was used, a sequential addition of D-galactose, D-glucose, and 2-acetamido-2-deoxy-D-glucose was obtained, which suggested that the core portion of lipopolysaccharide contains the sequence 2-acetamido-2-deoxy-D-glucosyl-D-glucosyl-D-galactosyl-D-glucose. The enzymes that catalyze these reactions have been designated UDP-D-galactose-LPS (lipopolysaccharide) transferase I, UDP-D-glucose-LPS transferase II, and UDP-2-acetamido-2-deoxy-D-glucose-LPS transferase, respectively. Incorporation of the first D-glucose residue is catalyzed by the enzyme UDP-D-glucose-LPS transferase I. Similar results have been obtained[363] with *E. coli* J5, a mutant of *E. coli* 0111 lacking UDP-D-galactose-4-epimerase.

Two soluble enzymes, UDP-D-glucose-LPS transferase I and UDP-D-galactose-LPS transferase, were obtained[364,365] from *S. typhimurium*. These soluble enzyme preparations made it possible to examine the specificity of the lipopolysaccharide acceptors. The soluble UDP-D-glucose LPS transferase catalyzed the incorporation of D-glucose only into cell-envelope fractions prepared from the glucose-deficient strains. In these strains the core polysaccharide contained only 3-deoxy-D-*manno*-octulosonate, but no D-glucose, D-galactose, or 2-acetamido-2-deoxy-D-glucose. Cell-envelope fractions prepared from the D-galactose-deficient strain, possessing a lipopolysaccharide which already contained D-glucose, or from the wild-type *S. typhimurium* were inactive. Cell-envelope preparations of the glucose-deficient strain, which lacked the glucosyl acceptor sites, were completely inactive. The purified lipopolysaccharide isolated from the active cell-envelope fraction was completely inactive, regardless of the source from which it had been isolated. The lack of activity of purified lipopolysaccharide indicated that the active cell-envelope fractions contain an additional factor which is essential for activity. This factor has been identified[366] as a phospholipid. The active acceptor has been characterized as a lipopolysaccharide–phospholipid complex which can be generated from the separated components. The separated lipopolysaccharide shows full activity.

The results of Rothfield and Takeshita[367] suggest that the mechanism of biosynthesis of the core portion of the lipopolysaccharide may involve selective adsorption of the specific transferases to the unfilled acceptor sites in the lipopolysaccharide–phospholipid complex; the enzyme would be released when the reaction is completed, and the new lipopolysaccharide, containing the additional sugar, would be ready to absorb the next transferase enzyme. The core would thus be formed by the stepwise addition of individual sugars. The biosynthetic studies in conjunction with chemical analyses by a number of investigators[368,369] of rough and mutant strains indicate a tentative structure of the core and backbone portions of *Salmonella* liposaccharide shown in Fig. 3. Considerable information upon which this structure is based has

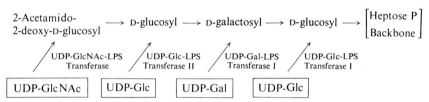

FIG. 3. Scheme for a stepwise synthesis of core (rough) portion of *Salmonella* and *Escherichia coli* lipopolysaccharide.

been obtained from analyses of rough mutants which contain only the core sugars, D-glucose, D-galactose, 2-amino-2-deoxy-D-glucose, and heptose. There appears to be one side chain containing two D-glucose residues, two D-galactose residues, and one 2-acetamido-2-deoxy-D-glucose for every other heptose residue in the backbone.

Mutant strains of *S. typhimurium* which are deficient in UDP-D-galactose-LSP-transferase I have been shown[369] to have a polysaccharide core containing one D-galactose and one D-glucose residue. Partial hydrolysis of the polysaccharide produced melibiose, indicating the presence of a $(1 \rightarrow 6)$-glucosidic linkage. Melibiose has also been isolated from R_I and R_{II} mutants.[368] In addition to melibiose, a number of other oligosaccharides have been isolated, including a pentasaccharide.

The biosynthesis of the O-specific antigenic side chains has been studied with strains of *S. typhimurium*, *S. anatum*, and *E. coli*.[369] In *S. typhimurium* and *S. anatum*, chemical and serological evidence had shown that these antigenic side chains are composed of repeating units having the probable structures shown in Fig. 4. Hexasaccharides corresponding to two repeating units have been isolated from the lipopolysaccharide of *S. anatum*.[370] Biosynthetic studies of the antigenic side chains have proved to be consistent with these structures.[371,372]

* *References start on p. 362.*

Edstrom and Heath[373] found that the mutant strain *E. coli* 0111, which does not contain D-galactose-4-epimerase, produces a lipopolysaccharide lacking in D-galactose, 2-amino-2-deoxy-D-glucose, and 3,6-dideoxy-L-*xylo*-hexose (colitose). Particulate fractions isolated from cell-free extracts catalyze the incorporation of colitose from GDP-colitose, provided that UDP-D-galactose, UDP-D-glucose, and UDP-2-acetamido-2-deoxy-D-glucose are present.

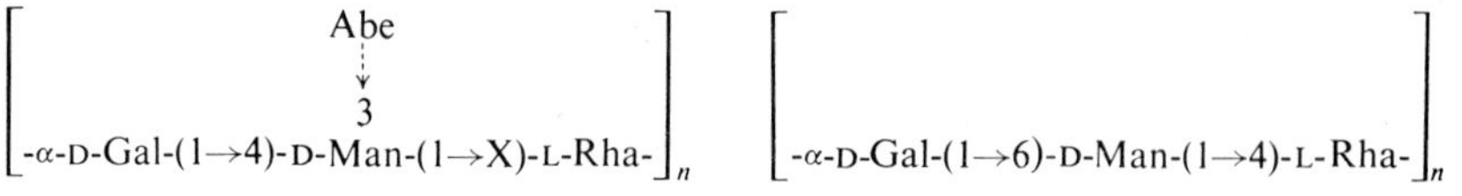

$$
\left[\begin{array}{c} \text{Abe} \\ \vdots \\ 3 \\ \text{-}\alpha\text{-D-Gal-}(1{\to}4)\text{-D-Man-}(1{\to}X)\text{-L-Rha-} \end{array} \right]_n
\qquad
\left[\begin{array}{c} \\ \\ \\ \text{-}\alpha\text{-D-Gal-}(1{\to}6)\text{-D-Man-}(1{\to}4)\text{-L-Rha-} \end{array} \right]_n
$$

Salmonella typhimurium   *Salmonella anatum*

FIG. 4. Repeating units of O-specific antigens of *Salmonella*.

The use of organisms which are unable to form more than a "partial" polysaccharide can greatly simplify studies on the biosynthesis of heteropolysaccharide and can also elucidate the relationship between the genetic constitution of an organism and the structure of the polysaccharide that it produces.

Lee and Ballou[374] studied the structure of the glycolipids of *Mycobacterium phlei* and *M. tuberculosis* cells, consisting of D-mannose derivatives of phosphatidyl *myo*-inositol, and lipid material (see Chapter 44). On the basis of their results, the complete structure of the D-mannopentaoside was formulated as shown in (10), Vol. IIB, Chap. 44.

It was subsequently shown[375] that biosynthesis of the phosphatidyl-*myo*-inositol D-mannosides proceeds from phosphatidyl-*myo*-inositol, presumably with the sequential addition of single D-mannose residues. An enzymic system isolated from this microorganism was capable of incorporating the radioactivity of GDP-D-mannose-^{14}C into a radioactive phospholipid fraction, which proved to be identical with 2-deoxy-1-phosphatidyl-L-*myo*-inosit-2-yl α-D-mannopyranoside. This compound is considered to be the first intermediate in the conversion of 1-phosphatidyl-L-*myo*-inositol into the poly-D-mannosides. The reason for the failure to introduce more than one D-mannose-^{14}C residue into the higher homologs with the enzymic system *in vitro* is not clear.

Neufeld and Hall[376] found that isolated spinach chloroplasts catalyze the transfer of D-galactose from UDP-D-galactose-^{14}C to an endogenous acceptor, yielding alkali-insoluble products similar to the galactolipids isolated from plant material.

N. Teichoic Acids

The teichoic acids were discovered by Baddiley and co-workers[377] in 1958. They are a group of polymers present chiefly in the cell walls of a number of gram-positive bacteria[378] (see Vol. IIB, Chap. 41). Two types of teichoic acids are known, one containing chains of glycerol and the other chains of ribitol, with a variety of sugar and amino acid constituents. The precursors of teichoic acids are considered to be CDP-D-glycerol and CDP-D-ribitol, previously isolated[379,380] from *Lactobacillus arabinosus*. Some of the cell walls of the gram-positive bacteria contain up to 50% (dry weight) of teichoic acid.

In typical cases glycerol or ribitol are joined together through phosphate diester linkages to form a chain, and sugars and D-alanine ester residues are attached to hydroxyl groups on the alditols. The general structures [(6), Vol. IIB, Chap. 41], as formulated by Baddiley and his collaborators,[381–385] are typical of many bacteria. Considerable variations occur in the nature and number of glycosyl residues but only D-glucose, 2-acetamido-2-deoxy-D-glucose, 2-acetamido-2-deoxy-D-galactose, D-galactose, and possibly D-mannose, have been found in these compounds. Both α-D- and β-D-glycosidic linkages and mono-, di-, or trisaccharide units may occur. Variations in both the proportion of sugar residues and the ratio of α to β linkages have been observed.

The degree of regularity in structure varies. All ribitol teichoic acids have sugar substituents in the 4 position of the ribitol 5-phosphate residues, and where the sugar-to-ribitol ratio is unity a regular structure occurs. In cases where the sugar content is higher, it is not known whether mono- and diglycosyl substituents alternate along the chain. In the ribitol teichoic acid, phosphate ester groups join positions 1 and 5 in the ribitol moieties of adjacent units. Most of the glycerol teichoic acids have phosphate diester groups joining positions 1 and 3 of glycerol, and the sugar or alanine usually occupies position 2. An interesting exception is the polymer isolated from the walls of a strain a *Staphylococcus lactis*, where the sugar is joined through phosphate at its 1 position to glycerol[386] as shown in Fig. 5.

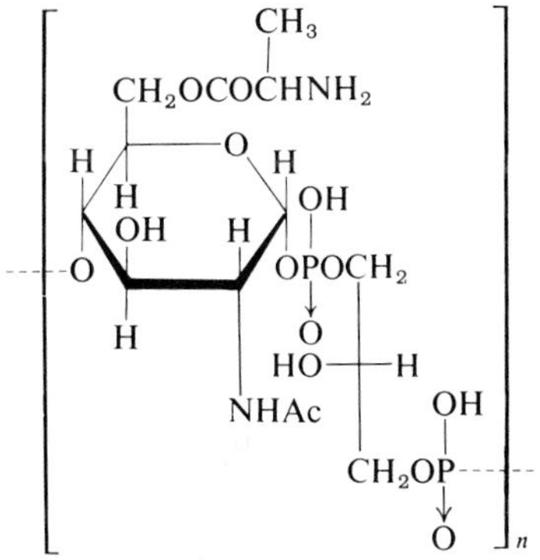

FIG. 5. Sugar joined through phosphate at C-1 of glycerol.

The teichoic acid of *S. aureus* (Copenhagen) similarly contains ribitol residues linked to each other by 1,5-phosphate diester bridges.[387] 2-Acetamido-2-deoxy-D-glucose residues are joined to the ribitol residues, 85% in α-D linkages and 15% in β-D linkages. The polymer contains D-alanine, esterified to the ribitol. Particulate preparations of this strain catalyze the transfer of 2-acetamido-2-deoxy-D-glucose residues from the UDP-derivative to teichoic acid from which D-alanine residues and 2-acetamido-2-deoxy-D-glucose residues have been removed. Intact teichoic acid appears to be inert in this reaction. Teichoic acids from different strains of this microorganism differ in the relative proportion of α- and β-D forms of the 2-acetamido-2-deoxy-D-glucose residues.

Glycerol teichoic acids are much more widely distributed than are the ribitol compounds. They appear to be located mainly in the region between the wall and the underlying protoplast membrane.[388]

The biosynthesis of teichoic acids from cell walls of several microorganisms has been studied,[389] and it was found that particulate enzyme preparations from *Bacillus licheniformis* (ATCC 9945) and *B. subtilis* (NCTC 3610) catalyze the synthesis of polyglycerol phosphate according to the reaction

$$n(\text{CDP-glycerol}) \longrightarrow \text{CMP} + (\text{glycerol phosphate})_n$$

Net synthesis of glycerol phosphate has been demonstrated with this preparation. The synthetic material, like natural teichoic acid, is a $(1 \rightarrow 3)$-linked polymer of glycerol phosphate. A similar preparation from *Lactobacillus plantarum* (ATCC 8014) will catalyze the synthesis of polyribitol phosphate from CDP-ribitol. Both enzymes show an absolute requirement for Ca^{2+} and Mg^{2+}.

In *B. licheniformis* transglycosylation from UDP-D-glucose to the terminal glycerol unit of polyglycerol phosphate has been demonstrated, and the D-glucose shown to be α-D-linked to the primary hydroxyl group of the terminal glycerol phosphate. This D-glucosylation reaction thus prevents the addition of further glycerol phosphate units to the teichoic acid. A particulate enzyme was obtained from *B. subtilis* that catalyzes the transfer of D-glucosyl residues from UDP-D-glucose to every glycerol residue in polyglycerol phosphate, and the D-glucose was shown to be α-D-linked to glycerol.

The gram-positive teichoic acids have been shown to possess immunological properties. Many teichoic acids are serologically active and often constitute the group-specific components of bacteria.[390-392]

O. Glycoproteins

The term glycoprotein is restricted to proteins containing covalently linked carbohydrate residues other than uronic acids. Their occurrence in animals has long been recognized. Recently they have also been found in plants and microorganisms[393-395] (see Vol. IIB, Chap. 43).

Experiments with labeled precursors on tissues and on cellular fractions showed that microsomes are the intercellular site of synthesis of serum glycoproteins in liver[396,397] and of soluble blood-group substances in hog gastric mucosa.[398] Fractionation of liver microsomes, which were preincubated with ^{14}C-labeled sugars, with deoxycholate yielded soluble labeled proteins, but not ribosomal-bound protein.[396] These investigations indicate that the polypeptide chain of glycoproteins is synthesized on ribosomes and then released into the membranous portion of the microsomes, where the glycosyl transfer of sugar takes place.

Nucleotide-linked oligosaccharides, UDP-N-acetyl-lactosamine, and UDP-L-fucosyl-N-acetyl-lactosamine were identified in human milk,[399] and UDP-sialyl-N-acetyl-lactosamine in goat colostrum.[400] Since, as in fetuin[401] and in immunologic glycoproteins,[402] these carbohydrates are identical in structure with those of terminal groups of the carbohydrate chains of glycoproteins, Roseman[55] suggested that oligosaccharides are possibly transferred from nucleotides to incomplete glycoproteins. However, later work by Roseman and collaborators[56,63] indicates that the sialyl-transferases are added singly (or possibly as oligosaccharides) from the sugar nucleotides to the polymer as the last step of the biosynthesis of glycoproteins.

A particulate system from the oviduct of laying hens catalyzes the transfer of D-xylose from UDP-D-xylose to glycoprotein,[403] and it was further found[404] that a preparation from an ascitic mast-cell tumor catalyzes the transfer of D-xylose from UDP-D-xylose-^{14}C to serine residues of an endogenous protein acceptor. The transfer takes place in the absence of protein synthesis, suggesting that glycosylation of proteins begins after synthesis of the polypeptide. The mastocytoma also contains a second D-xylosyl transferase which transfers D-xylose to an unidentified acceptor, to produce a relatively alkali-stable xyloside.

Studies of biosynthesis of intestinal mucins[404a] revealed that ovine colonic mucin contains three types of sialic acids, chiefly N-acetyl- and N-glycolyl-neuraminic acid in different ratios. The third sialic acid molecule appears to be O-acetylated neuraminic acid (5-acetamido-3,5-dideoxy-D-glycero-α-D-galacto-nonulopyranosonic acid). A particle-free enzyme preparation obtained from sheep-colonic mucosa catalyzes a series of reactions involving N-acylamino sugars, leading to the formation of sialic acid in vitro, and incorporation of L-serine-U-^{14}C into the mucin by whole mucosal preparations.

It was also shown[404b] that sheep-colonic, mucosal tissue incorporates L-threonine-U-^{14}C into a well-characterized glycoprotein fraction, and that

* References start on p. 362.

when puromycin is present the incorporation of the radioactive L-threonine is considerably diminished. Furthermore, puromycin also decreases incorporation of radioactivity from D-glucose and $^{35}SO_4{}^{2-}$ into the glycoprotein fraction. It was suggested that puromycin acts at the site of ribosomal protein biosynthesis in the formation of sheep-colonic mucosal mucin and that the biosynthesis of the carbohydrate portion is strongly dependent on the prior formation of peptide. The finding that puromycin affects the sulfate and carbohydrate utilization to a somewhat similar degree suggests that the sulfation step, that is, transferance to the carbohydrate acceptors, takes place during the biochemical synthesis of the glycoprotein rather than later with the macromolecule already formed.

A study of the biosynthesis of fraction I, which is an α-globulin fraction containing acidic glycoproteins isolated from guinea-pig serum,[404c] showed that most of the microsome-bound substances in that fraction lack sialic acid residues, and that these residues appear to be precursors of serum glycoprotein molecules with incomplete prosthetic groups.

Two glycopeptides (Glycopeptides I and II)[404d] have been isolated from a deoxycholate-soluble fraction of rat liver microsomes after injection of 2-amino-2-deoxy-D-glucose-1-^{14}C. Glycopeptide I contained sialic acid, D-galactose, 2-acetamido-2-deoxy-D-glucose, D-mannose, aspartic acid, serine, glutamic acid, proline, and glycine. Its minimum weight appears to be 2950. Glycopeptide II consisted only of aspartic acid, 2-acetamido-2-deoxy-D-glucose, and D-mannose. The minimum molecular weight of this glycopeptide, as determined by centrifugation, was 2360. The data of this author suggest that these glycopeptides isolated from the microsomal membrane may carry mainly the sugar residues that occupy the inner core of the mature plasma glycoprotein.

On the basis of the properties of known systems involved in the biosynthesis of glycoproteins, gangliosides, and related substances, Roseman[404e] suggested a general mechanism for the formation of these polymers that is compatible with their structures. He presumed that the oligosaccharide chains grow at their nonreducing ends or at branch points and that the sugars are added in a stepwise sequence as monosaccharide residues. The enzymes, glycosyltransferases, comprise homologous families, each family transferring only one type of monosaccharide residue. The glycosyl donors appear to be the corresponding sugar nucleotides, where all members of a single family of transferases require the same sugar nucleotide. There is reason to believe that the members of one family of glycosyltransferases can be differentiated by two properties: the transferases may require different acceptor molecules; or when the requirement for the acceptor molecule is the same, they attach the glycose residue at different positions, producing different products. Roseman[404e] has pointed out that the biosynthetic mechanism for the glycoproteins, glycolipids, and so on, differs significantly from that for nucleic acids and proteins in that the acceptor molecules themselves play a key role in the nature of the products.

P. SULFATED MUCOPOLYSACCHARIDES

Sulfated mucopolysaccharides are widely distributed throughout the animal body, especially in connective tissues.[405] They cover a number of related compounds—hyaluronic acid, heparin, chondroitin sulfates, and others (see Chapter 42).

Silbert[406] found that enzymic preparations from mast-cell tumors incorporate equimolar amounts of D-glucuronic acid and 2-acetamido-2-deoxy-D-galactose from UDP-derivatives into a polysaccharide which appears to be a precursor of heparin. He then obtained[407] a cell-free particulate enzyme preparation from chick embryo epiphyses that catalyzes the synthesis of high-molecular-weight polysaccharide from UDP-2-(acetamido-^3H)-2-deoxy-D-galactose and UDP-D-glucuronic acid-^{14}C. In the same system, UDP-2-(acetamido-^3H)-2-deoxy-D-glucose was relatively inactive as a precursor. The radioactive polysaccharide obtained appeared to be similar to chondroitin or chondroitin sulfate of low sulfate content. 2-Acetamido-2-deoxy-D-galactose and the disaccharide, chondrosine, were detected as products of acid hydrolysis of the polysaccharide.

A study of the biosynthesis of the chondroitin sulfate complex by Telser et al.[408] shows that the carbohydrate units are added to preformed protein for the formation of the protein–polysaccharide complex. These results are in accord with those of other investigators.[409-411]

The sulfation of the 2-acetamido-2-deoxy-D-galactose residues of chondroitin sulfate has been studied[347] by using preparations from embryonic chick cartilage. It was found that "active sulfate", [(adenylyl 3'-phosphate) sulfate] (PAPS), was the sulfate donor. A subsequent investigation,[412] with hen oviducts revealed that the sulfate is transferred from PAPS directly to the 2-acetamido-2-deoxy-D-galactose residues of the polysaccharide.

A cell-free enzyme preparation from chick embryo epiphyses will catalyze the synthesis of a high-molecular-weight polysaccharide from UDP-2-acetamido-2-deoxy-D-galactose or UDP-2-acetamido-2-deoxy-D-glucose and UDP-D-glucuronic acid.[413] The crude preparations catalyze the conversion of UDP-2-acetamido-2-deoxy-D-glucose into UDP-2-acetamido-2-deoxy-D-galactose and the incorporation of $^{35}SO_4^{2-}$ from PAPS into the polysaccharide. Inasmuch as chondroitin 4-sulfate exists in covalent association with protein,[414] it is likely that the protein itself is the initial acceptor of the sugar residues.

* References start on p. 362.

The presence of an enzyme in a rat brain which catalyzes the transfer of sulfate from PAPS to mucopolysaccharides was also demonstrated by Balasubramanian and Buchhawat.[415] They found that the sulfate transferase has high activity when heparitin sulfate or dermatan sulfate (chondroitin sulfate B) is used as acceptor, and has less activity when chondroitin 4-sulfate or keratan sulfate is used as an acceptor. Chondroitin 6-sulfate and hyaluronic acid are inactive as acceptors. The products formed appear to be ^{35}S-mucopolysaccharides.

Likewise, Wortman[416] detected activity in beef cornea epithelial extract. Keratan sulfate and chondroitin 4-sulfate served as the sulfate acceptors, and PAPS as the donor. The mucous gland extracts of a marine gastropod, *Charonia lampas*,[417] was found to contain a sulfotransferase of different specificity, which transfers $^{35}SO_4{}^{2-}$ from ^{35}S-PAPS into a D-glucan polysulfate.

REFERENCES

1. J. A. Bassham and M. Calvin, "The Path of Carbon in Photosynthesis," Prentice-Hall, Englewood Cliffs, New Jersey, 1957, pp. 39–66.
2. M. Calvin and J. A. Bassham, pp. 8–75. "The Photosynthesis of Carbon Compounds," Benjamin, New York, 1962, pp. 8–75.
3. Y. J. Topper, in "The Enzymes," Vol. 5, P. D. Boyer, H. Lardy and K. Myrbäck, Eds., Academic Press, New York, 1961, p. 429.
4. A. D. Elbein and W. Z. Hassid, *Biochem. Biophys. Res. Commun.*, **23**, 311 (1966).
5. E. W. Sutherland, S. P. Colowick, and C. F. Cori, *J. Biol. Chem.*, **140**, 309 (1941).
6. C. S. Hanes, *Proc. Roy. Soc.* **B128**, 421 (1940).
7. H. M. Kalckar, W. D. McElroy, and B. Glass, Eds., in "Symposium on the Mechanism of Enzyme Action," Johns Hopkins Press, Baltimore, Maryland, 1954, p. 675; W. Z. Hassid, D. S. Feingold, and E. F. Neufeld, *Proc. Nat. Acad. Sci. U.S.*, **45**, 905 (1959).
8. E. F. Neufeld, V. Ginsburg, E. W. Putman, D. Fanshier, and W. Z. Hassid, *Arch. Biochem. Biophys.*, **69**, 602 (1957).
9. J. L. Strominger, E. S. Maxwell, J. Axelrod, H. M. Kalckar, *J. Biol. Chem.*, **224**, 79 (1957).
10. E. S. Maxwell, H. M. Kalckar, and J. L. Strominger, *Arch. Biochem. Biophys.*, **65**, 2 (1956).
11. J. L. Strominger and L. W. Mapson, *Biochem. J.*, **66**, 567 (1957); E. E. B. Smith, G. T. Mills, H. P. Bernheimer, and R. Austrian, *Biochim. Biophys. Acta*, **28**, 211 (1958).
12. C. E. Cardini, L. F. Leloir, and J. Chiriboga, *J. Biol. Chem.*, **214**, 149 (1955).
13. L. F. Leloir and C. E. Cardini, *J. Biol. Chem.*, **214**, 157 (1955).
14. R. C. Bean, and W. Z. Hassid, *J. Amer. Chem. Soc.*, **77**, 5737 (1955).
15. W. M. Watkins and W. Z. Hassid, *J. Biol. Chem.*, **237**, 1432 (1962).
16. H. Babad and W. Z. Hassid, *J. Biol. Chem.*, **241**, 2672 (1966).
17. J. B. Pridham and W. Z. Hassid, *Plant Physiol.*, **40**, 984 (1965).
18. A. D. Elbein, G. A. Barber, and W. Z. Hassid, *J. Amer. Chem. Soc.*, **86**, 309 (1964).
19. G. A. Barber, A. D. Elbein, and W. Z. Hassid, *J. Biol. Chem.*, **239**, 4056 (1964).

20. D. S. Feingold, E. F. Neufeld, and W. Z. Hassid, *J. Biol. Chem.*, **233**, 783 (1958).
21. L. F. Leloir, M. A. R. de Fekete, and C. E. Cardini, *J. Biol. Chem.*, **236**, 636 (1961).
22. E. Recondo and L. F. Leloir, *Biochem. Biophys. Res. Commun.*, **6**, 85 (1961).
23. L. F. Leloir, *Proc. Int. Biochem., 6th Congr., New York, 1964*, p. 15.
24. C. L. Villemez, Jr., T. S. Lin, and W. Z. Hassid, *Proc. Nat. Acad. Sci. U.S.*, **54**, 1626 (1965).
25. E. S. Maxwell, in "The Enzymes," P. D. Boyer, H. Lardy, and K. Myrbäck, Eds., Vol. 5, Academic Press, New York, 1961, p. 443.
26. T. Mann, "The Biochemistry of Semen," Methuen's Monograph on Biochemical Subjects, Methuen, London, 1954.
27. E. Anderson, *J. Biol. Chem.*, **100**, 249 (1933); E. Anderson, and H. J. Lowe, *ibid.*, **168**, 289 (1947).
28. C. Araki, *J. Chem. Soc. Jap.*, **59**, 424 (1938).
29. J. C. Su and W. Z. Hassid, *Biochemistry*, **1**, 468, 474 (1962).
30. E. Baldwin and D. J. Bell, *J. Chem. Soc.*, 1461 (1938); D. J. Bell and E. Baldwin, *ibid.*, 125 (1941).
31. N. J. Palleroni and M. Doudoroff, *J. Biol. Chem.*, **218**, 535 (1956).
32. Y. J. Topper, see ref. 25, p. 413.
33. E. Cabib, L. F. Leloir, and C. E. Cardini, *J. Biol. Chem.*, **203**, 1055 (1953).
34. R. B. Hulbert, and V. R. Potter, *J. Biol. Chem.*, **209**, 1 (1954).
35. H. G. Pontis, *J. Biol. Chem.*, **216**, 195 (1955).
36. E. E. B. Smith, and G. T. Mills, *Biochim. Biophys. Acta*, **13**, 386 (1954).
37. J. L. Strominger, *Biochim. Biophys. Acta*, **17**, 283 (1955).
38. J. Solms and W. Z. Hassid, *J. Biol. Chem.*, **228**, 357 (1957).
39. H. E. Carter, W. D. Celmer, D. S. Galanos, R. H. Gigg, W. E. M. Land, J. H. Law, K. L. Mueller, T. Nakayama, H. H. Tomizawa, and E. Weber, *J. Amer. Oil Chem. Soc.*, **35**, 338 (1958).
40. L. F. Leloir and C. E. Cardini, *Biochim. Biophys. Acta*, **12**, 15 (1953).
41. S. Ghosh, H. J. Blumenthal, E. A. Davidson, and S. Roseman, *J. Biol. Chem.*, **235**, 1265 (1960).
42. L. F. Leloir and C. E. Cardini, *Biochim. Biophys. Acta*, **20**, 23 (1956).
43. B. M. Pogell and R. M. Gryder, *J. Biol. Chem.*, **228**, 701 (1957).
44. T. C. Chou and M. Soodak, *J. Biol. Chem.*, **196**, 105 (1952).
45. E. A. Davidson, H. J. Blumenthal, S. Roseman, *J. Biol. Chem.*, **226**, 125 (1957).
46. A. Munch-Peterson, H. M. Kalckar, E. Cutolo, and E. E. B. Smith, *Nature*, **172**, 1036 (1953).
47. L. F. Leloir, *Biochem. J.*, **91**, 1 (1964).
48. S. Roseman, *Proc. Nat. Acad. Sci. U.S.*, **48**, 437 (1962).
49. L. Warren and R. S. Blacklow, *Biochem. Biophys. Res. Commun.*, 7, 433 (1962).
50. J. L. Strominger, *Biochim. Biophys. Acta*, **30**, 645 (1958).
51. J. L. Strominger, *Federation Proc.*, **21**, 134 (1962).
52. D. G. Comb, F. Shimizu, and S. Roseman, *J. Amer. Chem. Soc.*, **81**, 5513 (1959).
53. L. Warren and H. Felsenfeld, *J. Biol. Chem.*, **237**, 1421 (1962).
54. A. Gottschalk, "The Chemistry and Biology of the Sialic Acids," Cambridge Univ. Press, Cambridge, Massachusetts, 1960.
55. S. Roseman, *Federation Proc.*, **21**, 1075 (1962).
56. S. Roseman, *Abstr. Sixth Int. Cong. Biochem. I.U.B.*, Section VI, 467 (1964).
57. D. G. Comb and S. Roseman, *J. Biol. Chem.*, **235**, 2529 (1960).
58. P. Brunetti, G. W. Jourdian, and S. Roseman, *J. Biol. Chem.*, **237**, 2447 (1962).
59. S. Roseman, *Ann. Rev. Biochem.*, **28**, 545 (1959).

60. L. Warren, R. S. Blacklow, and C. W. Spearing, *Ann. N. Y. Acad. Sci.*, **106**, 191 (1963).
61. S. Roseman, in "Birth Defects Original Series," Vol. II, No. 1, p. 25 (1966). (Publ. by The Nat. Found. March of Dimes.)
62. G. W. Jourdian, D. M. Carlson, and S. Roseman, *Biochem. Biophys. Res. Commun.*, **10**, 352 (1963).
63. B. Bartholomew, D. M. Jourdian, and S. Roseman, *Abstr. Sixth Int. Cong. Biochem.* VI, 503 (1964).
64. R. Kuhn, *Symp. 4th Meeting Int. Cong. Biochem.* pp. 67–79.
65. D. Aminoff, F. Dodyk, and S. Roseman, *J. Biol. Chem.*, **238**, PC1177 (1963).
66. D. H. Levin and E. Racker, *J. Biol. Chem.*, **234**, 2532 (1959); *cf.* M. A. Ghalambor, E. M. Levine, and E. C. Heath, *J. Biol. Chem.*, **241**, 3207 (1966).
66a. M. A. Ghalambor and E. C. Heath, *J. Biol. Chem.*, **241**, 3216 (1966).
67. L. F. Leloir, *Arch. Biochem. Biophys.*, **33**, 186 (1951).
68. E. S. Maxwell, H. de Robichon-Szulmajster, and H. M. Kalckar, *Arch. Biochem. Biophys.*, **78**, 407 (1958).
69. E. S. Maxwell and H. de Robichon-Szulmajster, *J. Biol. Chem.*, **235**, 308 (1960).
70. D. B. Wilson and D. S. Hogness, *J. Biol. Chem.*, **239**, 2469 (1964).
71. E. S. Maxwell, *J. Biol. Chem.*, **229**, 139 (1957).
72. P. C. Simonart, W. L. Salo, and S. Kirkwood, *Biochem. Biophys. Res. Commun.*, **24**, 120 (1966).
72a. J. Zelitis and D. S. Feingold, *Biochem. Biophys. Res. Commun.*, **31**, 693 (1968).
73. B. Jacobson and E. A. Davidson, *Biochim. Biophys. Acta*, **73**, 145 (1963).
74. V. Ginsburg, *J. Biol. Chem.*, **235**, 2196 (1960).
75. V. Ginsburg, *J. Biol. Chem.*, **236**, 2389 (1961).
76. L. Glaser and S. Kornfeld, *J. Biol. Chem.*, **236**, 1795 (1961).
77. R. Okazaki, T. Okazaki, J. L. Strominger, and M. A. Michelson, *J. Biol. Chem.*, **237**, 3014 (1962).
78. J. H. Pazur and E. W. Shuey, *J. Biol. Chem.*, **236**, 1780 (1961).
79. G. A. Barber, *Arch. Biochem. Biophys.*, **103**, 276 (1963).
80. J. C. Su and W. Z. Hassid, *Biochemistry*, **1**, 468 (1962).
81. E. C. Heath and A. D. Elbein, *Proc. Nat. Acad. Sci. U.S.*, **48**, 1209 (1962).
82. A. Markovitz, *J. Biol. Chem.*, **239**, 2091 (1964).
83. A. Markovitz and S. Sylvan, *J. Bacteriol.*, **83**, 483 (1962).
84. A. Markovitz, *J. Biol. Chem.*, **237**, 1767 (1962).
85. S. Matsuhashi and J. L. Strominger, in "Methods in Enzymology: Complex Carbohydrates," S. P. Colowick and N. O. Kaplan, Eds., Vol. VIII, Academic Press, New York, 1966, p. 310.
86. M. Matsuhashi and J. L. Strominger, *J. Biol. Chem.*, **239**, 2454 (1964); see also ref. 85, p. 317.
87. D. S. Feingold, E. F. Neufeld, and W. Z. Hassid, *J. Biol. Chem.*, **235**, 910 (1960).
88. P. A. Srere, J. R. Cooper, V. Klybas, and E. Racker, *Arch. Biochem. Biophys.*, **59**, 535 (1955).
89. J. Hurwitz, B. L. Horecker, *J. Biol. Chem.*, **223**, 993 (1956).
90. B. L. Horecker, P. Z. Smyrniotis, and H. Klenow, *J. Biol. Chem.*, **205**, 661 (1953).
91. B. Axelrod, R. S. Bandurski, C. M. Greiner, and R. Jang, *J. Biol. Chem.*, **202**, 619 (1953).
92. J. Edelman, *Advan. Enzymol.*, **17**, 189 (1956).
93. J. S. D. Bacon and J. Edelman, *Arch. Biochem. Biophys.*, **28**, 467 (1950).
94. P. H. Blanchard and N. Albon, *Arch. Biochem. Biophys.*, **29**, 220 (1950).
95. R. K. Morton, *Nature*, **172**, 65 (1953).

96. M. Cohn, *J. Biol. Chem.*, **180**, 771 (1949).
97. H. M. Kalckar, *J. Biol. Chem.*, **167**, 477 (1947).
98. L. F. Leloir, C. E. Cardini and E. Cabib, in "Comparative Biochemistry," M. Florkin and H. S. Mason, Eds., Vol. 2, Academic Press, New York, 1960, p. 97.
99. K. Burton and H. A. Krebs, *Biochem. J.*, **54**, 94 (1953).
100. W. Z. Hassid and M. Doudoroff, *Advan. Enzymol.*, **10**, 123 (1950).
101. C. E. Cardini, L. F. Leloir, and J. Chiriboga, *J. Biol. Chem.*, **214**, 149 (1955).
102. D. H. Turner and J. F. Turner, *Biochem. J.*, **69**, 448 (1958).
103. D. P. Burma and D. C. Mortimer, *Arch. Biochem. Biophys.*, **62**, 16 (1956).
104. J. G. Buchanan, *Arch. Biochem. Biophys.*, **44**, 140 (1953).
105. C. Fitting and M. Doudoroff, *J. Biol. Chem.*, **199**, 153 (1952).
106. E. W. Putman, C. F. Litt, and W. Z. Hassid, *J. Amer. Chem. Soc.*, **77**, 4351 (1955).
107. C. J. Sih, N. M. Nelson and R. H. McBee, *Science*, **126**, 1116 (1957).
108. F. H. Hulcher and K. W. King, *J. Bacteriol.*, **76**, 571 (1958).
109. W. A. Ayers, *J. Biol. Chem.*, **234**, 2819 (1959).
110. L. R. Marechal and S. H. Goldemberg, *Biochem. Biophys. Res. Commun.*, **13**, 106 (1963).
111. H. M. Kalckar, in "The Mechanism of Enzyme Action," W. D. McElroy and B. Glass, Eds. Johns Hopkins Press, Baltimore, Maryland 1953, p. 675.
112. J. Voet and R. H. Abeles, *J. Biol. Chem.*, **241**, 2731 (1966).
113. R. Silverstein, J. Voet, D. Reed, and R. H. Abeles, *J. Biol. Chem.* **242**, 1338 (1967).
114. D. E. Koshland, Jr., in "Phosphorus Metabolism," W. D. McElroy and B. Glass, Eds., Vol. 1, Johns Hopkins Press, Baltimore, Maryland 1951, p. 536.
115. F. H. Stodola, E. S. Sharpe, and H. J. Koepsell, *J. Amer. Chem. Soc.*, **78**, 2514 (1956).
116. E. J. Bourne, J. Hartigan, and H. Weigel, *J. Chem. Soc.*, 1088 (1961).
117. E. M. Crook and B. A. Stone, *Biochem. J.*, **65**, 1 (1957).
118. L. S. Berger and B. M. Eberhart, *Biochem. Biophys. Res. Commun.*, **6**, 62 (1961).
119. F. B. Anderson and D. J. Manners, *Biochem. J.*, **71**, 407 (1959).
120. W. A. M. Duncan, D. J. Manners, and J. L. Thompson, *Biochem. J.*, **73**, 295 (1959).
121. S. Hestrin, D. S. Feingold, and G. Avigad, *J. Amer. Chem. Soc.*, **77**, 6710 (1955).
122. G. Avigad, D. S. Feingold, and S. Hestrin, *Biochim. Biophys. Acta*, **20**, 129 (1956).
123. D. S. Feingold, G. Avigad, and S. Hestrin, *J. Biol. Chem.*, **224**, 295 (1957).
124. G. Avigad, *J. Biol. Chem.*, **229**, 121 (1957).
125. H. A. Barker and W. Z. Hassid, in "Bacterial Physiology" C. H. Werkman and P. W. Wilson, Eds., Academic Press, New York, 1951, p. 548.
126. I. F. Bird, H. K. Porter, and C. R. Stocking, *Biochim. Biophys. Acta*, **100**, 366 (1965).
127. R. B. Frydman and W. Z. Hassid, *Nature*, **199**, 382 (1963).
128. S. Haq and W. Z. Hassid, *Plant Physiol.*, **40**, 591 (1965).
129. M. D. Hatch, *Biochem. J.*, **93**, 521 (1964).
130. E. W. Putman and W. Z. Hassid, *J. Biol. Chem.*, **207**, 885 (1954).
131. R. C. Bean, B. K. Barr, H. V. Welch, and G. G. Porter, *Arch. Biochem. Biophys.*, **96**, 524 (1962).
132. Y. Milner and G. Avigad, *Israel J. Chem.*, **2**, 316 (1964).
133. Y. Milner and G. Avigad, *Nature*, **206**, 825 (1965).
134. E. J. Bourne, J. B. Pridham, and M. W. Walter, *Biochem. J.*, **82**, 44P (1962).
135. J. B. Pridham and W. Z. Hassid, *Plant Physiol.*, **40**, 984 (1965).
136. T. Gomyô and M. Nakamura, *Agr. Biol. Chem. (Tokyo)*, **30**, 425 (1966).
137. A. Moreno and C. E. Cardini, *Plant Physiol.*, **41**, 909 (1966).
138. J. E. Courtois, F. Petek, and T. Dong, *Bull. Soc. Chim. Biol.*, **43**, 1189 (1961).

139. W. Tanner and O. Kandler, *Plant Physiol.*, **41**, 1540 (1966).
140. R. S. Frydman and E. F. Neufeld, *Biochem. Biophys. Res. Commun.*, **12**, 121 (1963).
141. W. M. Watkins and W. Z. Hassid, *J. Biol. Chem.*, **237**, 1432 (1962).
142. H. Babad and W. Z. Hassid, *J. Biol. Chem.*, **239**, PC946 (1964).
143. H. Babad and W. Z. Hassid, *J. Biol. Chem.*, **241**, 2672 (1966).
144. W. Bloom and D. W. Fawcett, "A Textbook of Histology," 8th ed., Saunders, Philadelphia, Pennsylvania, 1962, p. 632.
145. S. K. Kon and A. T. Cowie, "Milk: The Mammary Gland and Its Secretion," Vol. I, Academic Press, New York, 1961, p. 90.
146. H. G. Wood, S. Joffe, R. Gillespie, R. G. Hansen, and H. Hardenbrook, *J. Biol. Chem.*, **233**, 1264 (1938).
147. R. G. Hansen, H. G. Wood, G. J. Peeters, B. Jacobson, and J. Wilken, *J. Biol. Chem.*, **237**, 1034 (1962).
148. H. G. Wood, P. Schambye, and G. J. Peeters, *J. Biol. Chem.*, **226**, 1023 (1957).
149. H. G. Wood, G. J. Peeters, R. Verbeke, M. Lauryssens, and B. Jacobson, *Biochem. J.*, **96**, 607 (1965).
150. E. E. B. Smith and G. T. Mills, *Biochim. Biophys. Acta*, **18**, 152 (1960).
151. E. S. Maxwell, H. M. Kalckar, and R. M. Burton, *Biochim. Biophys. Acta*, **18**, 444 (1955).
152. G. W. Kittinger and F. J. Reithel, *J. Biol. Chem.*, **205**, 527 (1953).
153. J. C. Bartley, S. Abraham, and I. L. Chaikoff, *J. Biol. Chem.*, **241**, 1132 (1966).
154. U. Brodbeck and K. Ebner, *J. Biol. Chem.*, **241**, 762 (1966).
155. U. Brodbeck, W. L. Denton, N. Tanahashi, and K. E. Ebner, *J. Biol. Chem.*, **242**, 1391 (1967).
155a. N. Tanahashi, U. Brodbeck, and K. E. Ebner, *Biochim. Biophys. Acta*, **154**, 247 (1968).
155b. K. Brew. T. C. Vanaman, and R. L. Hill, *J. Biol. Chem.*, **242**, 3747 (1967).
155c. K. Brew, T. C. Vanaman, and R. L. Hill, *Proc. Nat. Acad. Sci. U.S.*, **59**, 491 (1968).
155d. E. J. McGuire, G. W. Jourdian, D. M. Carlson, and S. Roseman, *J. Biol. Chem.*, **240**, PC 4112 (1965).
156. E. Cabib and L. F. Leloir, *J. Biol. Chem.*, **231**, 259 (1958).
157. D. Robison and W. T. J. Morgan, *Biochem. J.*, **22**, 1277 (1928).
158. J. Candy and B. A. Kilby, *Nature*, **183**, 1594 (1959).
159. S. Friedman, *Arch. Biochem. Biophys.*, **88**, 339 (1960).
160. A. D. Elbein, *J. Biol. Chem.*, **242**, 403 (1967).
161. R. S. Teague, *Advan. Carbohyd. Chem.*, **9**, 185 (1954).
162. G. J. Dutton and I. D. E. Storey, *Biochem. J.*, **48**, xxix (1951).
163. G. J. Dutton and I. D. E. Storey, *Biochem. J.*, **57**, 275 (1954); I. D. E. Storey, and G. J. Dutton, *ibid.*, **59**, 279 (1955).
164. K. J. Isselbacher and J. Axelrod, *J. Amer. Chem. Soc.*, **77**, 1070 (1955).
165. G. J. Dutton, *Biochem. J.*, **64**, 693 (1956).
166. G. M. Grodsky and J. V. Carbone, *J. Biol. Chem.*, **226**, 449 (1957).
167. C. A. Marsh, *Biochim. Biophys. Acta*, **44**, 359 (1960).
168. T. Yamaha and C. E. Cardini, *Arch. Biochem. Biophys.*, **86**, 127 (1960).
169. T. Yamaha and C. E. Cardini, *Arch. Biochem. Biophys.*, **86**, 133 (1960).
170. J. Conchie, A. Moreno, and C. E. Cardini, *Arch. Biochem. Biophys.*, **94**, 342 (1961).
171. J. C. Trivelloni, *Arch. Biochem. Biophys.*, **89**, 149 (1960).
172. G. J. Dutton and A. M. Duncan, *Biochem. J.*, **77**, 18P (1960).
173. P. A. J. Gorin and A. S. Perlin, *Can. J. Chem.*, **37**, 1930 (1959).
174. G. A. Barber, *Biochemistry*, **1**, 463 (1962).

175. E. E. B. Smith, B. Galloway, and G. T. Mills, *Biochem. Biophys. Res. Commun.*, **5**, 148 (1961).
176. J. L. Strominger and S. S. Scott, *Biochim. Biophys. Acta*, **35**, 552 (1959).
177. R. Okazaki, T. Okazaki, and Y. Kuriki, *Biochim. Biophys. Acta*, **38**, 384 (1960).
178. G. A. Barber, *Biochem. Biophys. Res. Commun.*, **8**, 204 (1962).
179. G. Hauser and M. Karnovsky, *J. Bacteriol.*, **68**, 645 (1954).
180. F. G. Jarvis and M. J. Johnson, *J. Amer. Chem. Soc.*, **71**, 4124 (1949).
181. M. M. Burger, L. Glaser, and R. M. Burton, *Biochim. Biophys. Acta*, **56**, 172 (1962).
182. R. M. Burton, M. A. Sodd, R. O. Brady, *J. Biol. Chem.*, **233**, 1053 (1958).
183. W. W. Cleland and E. P. Kennedy, *J. Biol. Chem.*, **235**, 45 (1960).
184. G. R. Wyatt and S. S. Cohen, *Biochem. J.*, **55**, 774 (1953); R. L. Sinsheimer, *Proc. Nat. Acad. Sci. U.S.*, **42**, 502 (1956).
185. I. R. Lehman and E. A. Pratt, *J. Biol. Chem.*, **235**, 3254 (1960); S. Kuno and I. R. Lehman, *ibid.*, **237**, 1266 (1962).
186. A. Kornberg, S. B. Zimmerman, S. R. Kornberg, and J. Josse, *Proc. Nat. Acad. Sci. U.S.*, **45**, 772 (1959).
187. M. Rosell-Perez and J. Larner, *Biochemistry*, **3**, 81 (1964).
188. R. R. Traut and F. Lipmann, *J. Biol. Chem.*, **238**, 1213 (1963).
189. C. F. Cori, G. Schmidt and G. T. Cori, *Science*, **89**, 464 (1939).
190. C. S. Hanes, *Proc. Roy. Soc. Ser.* **B128**, 421 (1940); **B129**, 174 (1940).
191. L. F. Leloir and C. E. Cardini, *J. Amer. Chem. Soc.*, **79**, 6340 (1957).
192. W. Z. Hassid, in "Organic Chemistry, An Advanced Treatise," H. Gilman, Ed., Vol. 4, New York, 1953, p. 901.
193. K. H. Meyer and G. C. Gibbons, *Advan. Enzymol.*, **12**, 341 (1951).
194. K. H. Meyer, "Natural and Synthetic High Polymers," 2nd ed., Wiley (Interscience), New York, 1950, pp. 468–469.
195. G. T. Cori and C. F. Cori, *J. Biol. Chem.*, **151**, 57 (1943).
196. J. Larner, *J. Biol. Chem.*, **202**, 491 (1953).
197. W. N. Haworth, S. Peat, and E. J. Bourne, *Nature*, **154**, 236 (1944).
198. S. Peat, *Advan. Enzymol.*, **11**, 339 (1951).
199. G. T. Cori, M. A. Swanson, and C. F. Cori, *Federation Proc.*, **4**, 234 (1945).
200. P. N. Hobson, W. J. Whelan, and S. Peat, *J. Chem. Soc.*, 3566 (1950).
201. E. J. Hehre, A. S. Carlson, and J. M. Neill, *Science*, **106**, 523 (1947).
202. A. Babbington, E. J. Bourne, and I. A. Wilkinson, *J. Chem. Soc.*, 246 (1952).
203. S. Peat, W. J. Whelan, and W. R. Rees, *Nature*, **172**, 158 (1953).
204. G. T. Cori and J. Larner, *J. Biol. Chem.*, **188**, 17 (1951).
205. G. J. Walker and W. J. Shelan, *Biochem. J.*, **76**, 264 (1960).
206. J. Larner, B. Illingworth, G. T. Cori, and C. F. Cori, *J. Biol. Chem.*, **199**, 641 (1952). See also ref. 114, Vol. 1, p. 62.
207. D. H. Brown and B. Illingworth, *Proc. Nat. Acad. Sci. U.S.*, **48**, 1783 (1962).
208. W. Verhue and H. G. Hers, *Arch. Int. Physiol. Biochim.*, **69**, 757 (1961).
209. P. N. Hobson, W. J. Whelan and S. Peat, *J. Chem. Soc.*, 1451, (1950).
210. B. Maruo and T. Kobayashi, *Nature*, **167**, 606 (1951); *J. Agr. Chem. Soc. Jap.*, **23**, 115, 120 (1949).
211. A. N. Petrova, *Biokhimiya*, **13**, 244 (1948); **16**, 482 (1951).
212. H. Bender and K. Wallenfels, *Biochem. Z.*, **334**, 79 (1961).
213. E. J. Hehre and D. M. Hamilton, *J. Biol. Chem.*, **166**, 777 (1946).
214. E. J. Hehre and D. M. Hamilton, *J. Bacteriol.*, **55**, 197 (1948).
215. E. J. Hehre, D. M. Hamilton, and A. S. Carlson, *J. Biol. Chem.*, **177**, 267 (1949).
216. S. A. Barker, E. J. Bourne, and M. Stacey, *J. Chem. Soc.*, 2884, (1950).

217. E. J. Hehre, *Advan. Enzymol.*, **11**, 297 (1951).
218. E. J. Hehre and D. M. Hamilton, *J. Biol. Chem.*, **192**, 161 (1951).
219. J. Monod and A. M. Torriani, *C. R. Acad. Sci.*, **227**, 240 (1948).
220. A. M. Torriani and J. Monod, *C. R. Acad. Sci.*, **228**, 718 (1949).
221. J. Monod and A. M. Torriani, *Ann. Inst. Pasteur*, **78**, 65 (1950).
222. M. Doudoroff, W. Z. Hassid, E. W. Putman, A. L. Potter, and J. Lederberg, *J. Biol. Chem.*, **179**, 921 (1949).
223. S. A. Barker and E. J. Bourne, *J. Chem. Soc.*, 209, (1952).
224. K. Freudenberg, E. Schaaf, G. Dumpert, and T. Ploetz, *Naturwissenschaften*, **27**, 850 (1939).
225. D. French, M. L. Levine, J. H. Pazur, and E. Norberg, *J. Amer. Chem. Soc.*, **71**, 353 (1949).
226. K. Freudenberg and P. Cramer, *Ber.*, **83**, 296 (1950).
227. E. B. Tilden and C. S. Hudson, *J. Amer. Chem. Soc.*, **61**, 2900 (1939).
228. E. J. Wilson, T. J. Schoch, and C. S. Hudson, *J. Amer. Chem. Soc.*, **65**, 1380 (1943).
229. D. French, J. H. Pazur, M. L. Levine, and E. Norberg, *J. Amer. Chem. Soc.*, **70**, 3145 (1948).
230. K. Myrbäck and E. Willstaedt, *Acta Chem. Scand.*, **3**, 91 (1949).
231. E. Norberg and D. French, *J. Amer. Chem. Soc.*, **72**, 1202 (1950).
232. F. Schardinger, *Zentr. Bacteriol. Parasitenk. Abt. II*, **14**, 772 (1905); **22**, 98 (1909); **29**, 188 (1911).
233. S. Schwimmer and J. A. Garibaldi, *Cereal Chem.*, **29**, 108 (1952).
234. L. Glaser and D. H. Brown, *J. Biol. Chem.*, **228**, 729 (1957).
235. L. Glaser and D. H. Brown, *Proc. Nat. Acad. Sci. U.S.*, **41**, 253 (1955).
236. L. Glaser, *J. Biol. Chem.*, **232**, 627 (1958).
237. J. R. Colvin, *Nature*, **183**, 1135 (1959).
238. A. J. Charlson and A. S. Perlin, *Can. J. Chem.*, **34**, 1200 (1956).
239. R. L. Whistler and C. L. Smart, "Polysaccharide Chemistry," Academic Press, New York, 1953, pp. 350–353.
240. G. O. Aspinall and G. Kessler, *Chem. Ind. (London)*, 1296 (**1957**). G. Kessler, *Ber. Schweiz. Botan. Ges.*, **68**, 5 (1958).
241. W. Eschrich, *Protoplasma*, **47**, 487 (1956).
242. H. B. Currier, *Amer. J. Botany*, **44**, 478 (1957).
243. F. Fowler, F. L. Buckland, I. K. Brauns, and H. Hibbert, *Can. J. Res.*, **B15**, 486 (1937).
244. M. Stacey and G. Swift, *J. Chem. Soc.*, 1555 (1948).
245. S. Peat, E. Schluchterer, and M. Stacey, *J. Chem. Soc.*, 581 (1939).
246. W. Z. Hassid and H. A. Barker, *J. Biol. Chem.*, **130**, 163 (1940).
247. A. Jeane and C. A. Wilham, *J. Amer. Chem. Soc.*, **72**, 2655 (1950).
248. E. J. Hehre, J. M. Neill, J. Y. Sugg, and E. Jaffe, *J. Exp. Med.*, **70**, 427 (1939).
249. M. Abdel-Akher, J. K. Hamilton, R. Montgomery and F. Smith, *J. Amer. Chem. Soc.*, **74**, 4970 (1952).
250. R. Lohmar, *J. Amer. Chem. Soc.*, **14**, 4974 (1952).
251. M. Stacey and C. R. Ricketts, *Fortschr. Chem. Org. Naturstoffe*, **8**, 28 (1951).
252. E. J. Hehre and D. M. Hamilton, *Proc. Soc. Exp. Biol. Med.*, **71**, 336 (1949).
253. M. W. Beijerinck, *Proc. Koninkl. Ned. Akad. Wetenschap. Sect. Sci.*, **12**, 635 (1910).
254. S. Hestrin, S. Avineri-Shapiro, and M. Aschner, *Biochem. J.*, **37**, 450 (1943).
255. S. Hestrin and S. Avineri-Shapiro, *Biochem. J.*, **38**, 2 (1944).
256. D. S. Feingold and M. Gehatia, *J. Polym. Sci.*, **23**, 783 (1957).
257. R. Dedonder and P. Slizewicz, *Bull. Soc. Chim. Biol.*, **40**, 873 (1958).

258. D. L. Friedman and J. Larner, *Biochim. Biophys. Acta*, **64**, 185 (1962).
259. J. Larner, C. Villar-Palasi and D. J. Richman, *Arch. Biochem. Biophys.*, **86**, 56 (1960).
260. E. W. Sutherland and C. F. Cori, *J. Biol. Chem.*, **188**, 531 (1951).
261. W. L. Meyer, E. H. Fischer, and E. G. Krebs, *Biochemistry*, **3**, 1033 (1964).
262. G. F. Cahill, J. Ashmore, S. Zottu, and A. B. Hastings, *J. Biol. Chem.*, **224**, 237 (1957).
263. W. F. Mommaerts B. Illingworth, C. M. Pearson, R. J. Guillory, and K. Seraydarian, *Proc. Nat. Acad. Sci. U.S.*, **45**, 791 (1959).
264. J. Larner and C. Villar-Palasi, *Proc. Nat. Acad. Sci. U.S.*, **45**, 1234 (1959).
265. H. G. Hers, *Rev. Int. Hepatol.*, **9**, 35 (1959).
266. R. Schmid, in "Control of Glycogen Metabolism," Ciba Foundation Symposium, Churchill, London, 1964, p. 305.
267. W. Z. Hassid and E. F. Neufeld, in "The Enzymes," P. D. Boyer, H. Lardy, and K. Myrbäck, Eds., Vol. 6, Academic Press, New York. 1962, p. 308.
268. L. F. Leloir and C. E. Cardini, see ref. 267, p. 317.
269. P. W. Robbins, R. R. Traut, and F. Lipmann, *Proc. Nat. Acad. Sci. U.S.*, **45**, 6 (1959).
270. L. F. Leloir, Plenary Sessions, *6th Int. Cong. Biochem. I.U.B.*, New York, **35**, p. 15 (1964).
271. S. H. Goldemberg, *Biochim. Biophys. Acta*, **56**, 357 (1962).
272. Y. P. Lee, *Biochim. Biophys. Acta*, **43**, 18, 25 (1960).
273. L. F. Leloir, J. M. Olavarriá, S. H. Goldemberg, and H. Carminatti, *Arch. Biochem. Biophys.*, **81**, 508 (1959).
274. I. D. Algranati and E. Cabib, *Biochim. Biophys. Acta*, **43**, 141 (1960); *J. Biol. Chem.*, **237**, 1007 (1962).
275. D. R. Steiner, *Biochim. Biophys. Acta*, **54**, 206 (1961).
276. R. Kornfeld and D. H. Brown, *J. Biol. Chem.*, **237**, 1772 (1962).
277. M. Rosell-Perez, C. Villar-Palasi, and J. Larner, *Biochemistry*, **1**, 763 (1962).
278. E. Belocopitow, *Arch. Biochem. Biophys.*, **93**, 457 (1961).
279. E. Belocopitow, M. M. Appleman, and H. N. Torres, *J. Biol. Chem.*, **240**, 3473 (1965).
280. L. Shen and J. Preiss, *J. Biol. Chem.*, **240**, 2334 (1965).
281. E. Greenberg and J. Preiss, *J. Biol. Chem.*, **240**, 2341 (1965).
282. A. B. Foster and J. M. Webber, *Advan. Carbohyd. Chem.*, **15**, 371 (1960).
283. C. A. Porter and E. G. Jaworski, *Biochemistry*, **5**, 1149 (1966).
284. L. F. Leloir, *Harvey Lecture Ser.*, **56**, 23 (1961).
285. M. H. Ewart, D. Siminovitch, and D. R. Briggs, *Plant Physiol.*, **29**, 407 (1954).
286. C. R. Stocking, *Amer. J. Botany*, **39**, 283 (1952).
287. T. Akazawa, T. Minamikawa, and T. Murata, *Plant Physiol.*, **39**, 371 (1964).
288. T. Murata and T. Akazawa, *Biochem. Biophys. Res. Commun.*, **16**, 6 (1964).
289. R. B. Frydman and C. E. Cardini, *Biochem. Biophys. Res. Commun.*, **14**, 353 (1964).
290. R. B. Frydman and C. E. Cardini, *Biochem. Biophys. Res. Commun.*, **17**, 407 (1964).
291. E. Recondo and L. F. Leloir, *Biochem. Biophys. Res. Commun.*, **6**, 85 (1961).
292. J. Espada, *J. Biol. Chem.*, **237**, 3577 (1962).
293. H. Kauss and O. Kandler, *Z. Naturforsch.*, **17b**, 858 (1962).
294. E. Recondo, M. Dankert, and L. F. Leloir, *Biochem. Biophys. Res. Commun.*, **12**, 204 (1963).
295. T. Murata, T. Minamikawa, and T. Akazawa, *Biochem. Biophys. Res. Commun.*, **31**, 439 (1963).

296. M. A. R. de Fekete and C. E. Cardini, *Arch. Biochem. Biophys.*, **104**, 173 (1964).
297. T. Murata, T. Sugiyama, T. Minamikawa, and T. Akazawa, *Arch. Biochem. Biophys.*, **113**, 34 (1966).
298. C. E. Cardini and E. Recondo, *Plant Cell Physiol.*, *Tokyo*, **3**, 313 (1962).
299. T. Murata, T. Sugiyama, and T. Akazawa, *Arch. Biochem. Biophys.*, **107**, 92 (1964).
300. T. Murata, T. Sugiyama, and T. Akazawa, *Biochem. Biophys. Res. Commun.*, **18**, 371 (1965).
301. T. Akatsuka and O. E. Nelson, *J. Biol. Chem.*, **241**, 2280 (1966).
302. H. P. Ghosh and J. Preiss, *Biochemistry*, **4**, 1354 (1965).
303. R. Kornfeld and D. H. Brown, *J. Biol. Chem.*, **237**. 1772 (1962).
304. L. F. Leloir and S. H. Goldemberg, *J. Biol. Chem.*, **235**, 919 (1960).
305. I. D. Algranati and E. Cabib, *J. Biol. Chem.*, **237**, 1007 (1962).
306. H. P. Ghosh and J. Preiss, *J. Biol. Chem.*, **240**, PC960 (1965).
307. M. Rosell-Perez and J. Larner, *Biochemistry*, **3**, 81 (1964).
308. L. Glaser, *J. Biol. Chem.*, **232**, 627 (1957).
309. G. A. Barber and W. Z. Hassid, *Biochim. Biophys. Acta*, **86**, 397 (1964).
309a. D. M. Carlson and R. G. Hansen, *J. Biol. Chem.*, **237**, 1260 (1962).
310. G. A. Barber and W. Z. Hassid, *Nature*, **207**, 295 (1965).
311. D. O. Brummond and A. P. Gibbons, *Biochem. Biophys. Res. Commun.*, **17**, 156 (1964).
312. D. O. Brummond and A. P. Gibbons, *Biochem. Z.*, **342**, 308 (1965).
312a. H. M. Flowers, K. K. Batra, J. Kemp, and W. Z. Hassid, *Plant Physiol.*, **43**, 1703 (1968).
313. S. Peat, W. J. Whelan and H. G. Lawley, *J. Chem. Soc.*, 724, 729 (1958).
314. D. S. Feingold, E. F. Neufeld, and W. Z. Hassid, *J. Biol. Chem.*, **233**, 783 (1958).
314a. L. Ordin and M. A. Hall, *Plant Physiol.*, **42**, 205 (1967); *ibid.* **43**, 473 (1968).
314b. C. L. Villemez, Jr., G. Franz, and W. Z. Hassid, *Plant Physiol.* **42**, 1219 (1967).
314c. F. I. Huotari, T. E. Nelson, F. Smith, and S. Kirkwood, *J. Biol. Chem.*, **243**, 952 (1968).
314d. K. K. Batra and W. Z. Hassid, *Plant Physiol.*, **44**, 755 (1969).
315. T. Y. Liu and W. Z. Hassid, unpublished results.
316. H. M. Flowers, K. K. Batra, J. Kemp, and W. Z. Hassid, *J. Biol. Chem.*, **244**, 4969 (1969).
316a. J. R. Colvin, *Can. J. Biochem. Physiol.*, **39**, 1921 (1961).
316b. A. D. Elbein and W. Z. Hassid, *Biochem. Biophys. Res. Commun.*, **23**, 311 (1966).
317. H. A. Altermatt and A. C. Neish, *Can. J. Biochem. Physiol.*, **34**, 405 (1956).
318. A. C. Neish, *Can. J. Biochem. Physiol.*, **36**, 187 (1958).
319. W. G. Slater and H. Beevers, *Plant Physiol.*, **33**, 146 (1958).
320. V. Ginsburg, P. K. Stumpf, and W. Z. Hassid, *J. Biol. Chem.*, **223**, 977 (1956).
321. R. C. Bean and W. Z. Hassid, *J. Biol. Chem.*, **218**, 425 (1956).
322. J. L. Strominger and L. W. Mapson, *Biochem. J.*, **66**, 567 (1957).
323. E. E. B. Smith, G. T. Mills, H. P. Bernheimer, and R. Austrian, *Biochim. Biophys. Acta*, **28**, 211 (1958).
324. J. B. Pridham and W. Z. Hassid, *Biochem. J.*, **100** 21P (1966).
325. R. W. Bailey, *Proc. Nat. Acad. Sci. U.S.*, **56**, 1586 (1966).
326. G. H. Beavan and J. K. N. Jones, *J. Chem. Soc.*, 1218, **1947**.
327. E. F. Neufeld and D. S. Feingold, *Biochim. Biophys. Acta*, **53**, 589 (1961).
328. E. F. Neufeld, D. S. Feingold, S. M. Ilves, G. Kessler, and W. Z. Hassid, *J. Biol. Chem.*, **236**, 3102 (1961).
329. C. L. Villemez, Jr., T. S. Lin, and W. Z. Hassid, *Proc. Nat. Acad. Sci. U.S.*, **54**, 1626 (1965).

330. J. D. Macmillan and R. H. Vaughn, *Biochemistry*, **3**, 564 (1964).
331. J. D. Macmillan, H. J. Phaff, and R. H. Vaughn, *Biochemistry*, **3**, 572 (1964).
332. P. Albersheim and J. Bonner, *J. Biol. Chem.*, **234**, 3105 (1959).
333. P. Albersheim, *J. Biol. Chem.*, **238**, 1608 (1963).
334. H. Kauss, A. L. Swanson, and W. Z. Hassid, *Biochem. Biophys. Res. Commun.*, **26**, 234 (1967).
335. G. O. Aspinall and G. Kessler, *Chem. Ind. (London)*, 1296 (**1957**).
336. G. Kessler, *Ber. Schweiz. Botan. Ges.*, **68**, 5 (1958).
337. A. J. Charlson and A. S. Perlin, *Can. J. Chem.*, **34**, 1200 (1956).
338. L. F. Leloir, Plenary Sessions, *6th Int. Cong. Biochem. I.U.B.*, Vol. **35**, p. 15 (1964).
339. W. Eschrich, *Protoplasma*, **47**, 487 (1956).
340. H. B. Currier, *Amer. J. Botany*, **44**, 478 (1956).
341. S. H. Goldemberg and L. R. Marechal, *Biochim. Biophys. Acta*, **71**, 743 (1963).
342. L. R. Marechal and S. H. Goldemberg, *J. Biol. Chem.*, **239**, 3163 (1964).
343. T. Y. Lin and W. Z. Hassid, *J. Biol. Chem.*, **239**, PC944 (1964).
344. T. Y. Lin and W. Z. Hassid, *J. Biol. Chem.*, **241**, 3283 (1966).
345. T. Y. Lin and W. Z. Hassid, *J. Biol. Chem.*, **25**, 5284 (1966).
346. V. Ginsburg, *J. Amer. Chem. Soc.*, **80**, 4426 (1958).
347. F. D'Abramo and F. Lipmann, *Biochim. Biophys. Acta*, **25**, 211 (1957).
348. W. Z. Hassid, E. F. Neufeld, and D. S. Feingold, *Proc. Nat. Acad. Sci. U.S.*, **45**, 905 (1959).
349. A. Markovitz, J. A. Cifonelli, and A. Dorfman, *J. Biol. Chem.*, **234**, 2343 (1959).
350. R. E. Reeves and W. F. Goebel, *J. Biol. Chem.*, **139**, 511 (1941).
351. M. Heidelberger, *Fortschr. Chem. Org. Naturstoffe*, **18**, 503 (1960).
352. E. E. B. Smith, G. T. Mills, H. P. Bernheimer, and R. Austrian, *J. Biol. Chem.*, **235**, 1876 (1960).
353. E. E. B. Smith, G. T. Mills, and H. P. Bernheimer, *J. Biol. Chem.*, **236**, 2179 (1961).
354. E. E. B. Smith, B. Galloway, and G. T. Mills, *Biochem. Biophys. Res. Commun.*, **4**, 420 (1961).
355. E. E. B. Smith and G. T. Mills, *Biochem. J.*, **82**, 42P (1962).
356. G. T. Mills and E. E. B. Smith, *Biochem. J.*, **82**, 38P (1962).
357. G. T. Mills, *Federation Proc.*, **19**, 991 (1960); G. T. Mills, and E. E. B. Smith, *Brit. Med. Bull.*, **18**, 27 (1962).
358. A. Nikaido and K. Jokura, *Biochem. Biophys. Res. Commun.*, **6**, 304 (1961).
359. T. Fukasawa and H. Nikaido, *Biochim. Biophys. Acta*, **48**, 470 (1961); H. Nikaido, *Proc. Nat. Acad. Sci. U.S.*, **48**, 1337 (1962).
360. H. Nikaido, *Proc. Nat. Acad. Sci. U.S.*, **48**, 1542 (1962); M. J. Osborn. S. M. Rosen, L. Rothfield, and B. L. Horecker, *ibid.*, **48**, 1831 (1962).
361. L. Rothfield, D. Fraenkel, and M. J. Osborn, *Federation Proc.*, **22**, 465 (1963).
362. M. J. Osborn and L. D'Ari, *Biochem. Biophys. Res. Commun.*, **16**, 568 (1964).
363. A. D. Elbein and E. C. Heath, *J. Biol. Chem.*, **240**, 1919 (1965).
364. M. J. Osborn, S. M. Rosen, L. Rothfield, L. D. Zeleznick, and B. L. Horecker, *Science*, **145**, 783 (1964).
365. L. Rothfield, M. J. Osborn, and B. L. Horecker, *J. Biol. Chem.*, **239**, 2788 (1964).
366. L. Rothfield and B. L. Horecker, *Proc. Nat. Acad. Sci., U.S.*, **52**, 939 (1964); L. Rothfield and M. Pearlman, *J. Biol. Chem.*, **241**, 1386 (1966).
367. L. Rothfield and M. Takeshita, *Biochem. Biophys. Res. Commun.*, **20**, 521 (1965).
368. O. Lüderitz, A. M. Staub, and O. Westphal, *Bacteriol. Rev.*, **30**, 192 (1966).
369. B. L. Horecker, *Ann. Rev. Microbiol.*, **20**, 253 (1966).

370. P. W. Robbins, J. M. Keller, A. Wright, and R. L. Bernstein, *J. Biol. Chem.*, **240**, 384 (1965).
371. S. M. Rosen, L. D. Zeleznick, D. Fraenkel, I. M. Weiner, M. J. Osborn, and B. L. Horecker, *Biochem. Z.*, **342**, 375 (1965).
372. A. Wright, M. Dankert and P. W. Robbins, *Proc. Nat. Acad. Sci. U.S.*, **54**, 235 (1965).
373. R. D. Edstrom and E. C. Heath, *Biochem. Biophys. Res. Commun.*, **16**, 576 (1964).
374. Y. C. Lee and C. E. Ballou, *Biochemistry*, **4**, 1395 (1965).
375. D. L. Hill and C. E. Ballou, *J. Biol. Chem.*, **241**, 895 (1966).
376. E. F. Neufeld and C. W. Hall, *Biochem. Biophys. Res. Commun.*, **14**, 503 (1964).
377. J. J. Armstrong, J. Baddiley, J. G. Buchanan, B. Carss, and G. R. Greenberg, *J. Chem. Soc.*, 4344 (1958).
378. A. R. Archibald and J. Baddiley, *Advan. Carbohyd. Chem.*, **21**, 323 (1966).
379. J. Baddiley, J. G. Buchanan, B. Carss, and A. P. Mathias, *J. Chem. Soc.*, 4583 (1956).
380. J. Baddiley and A. P. Mathias, *J. Chem. Soc.*, 2723 (1957).
381. J. Baddiley, in "Current Biochemical Energetics," N. O. Kaplan and E. P. Kennedy, Eds., Academic Press, New York, 1966, p. 371.
382. J. Baddiley, J. G. Buchanan, R. O. Martin and U. L. RajBhandary, *Biochem. J.*, **85**, 49 (1962).
383. P. Critchley, A. R. Archibald, and J. Baddiley, *Biochem. J.*, **85**, 420 (1962).
384. D. H. Ellwood, M. V. Kelemen, and J. Baddiley, *Biochem. J.*, **86**, 213 (1963).
385. A. J. Wicken and J. Baddiley, *Biochem. J.*, **87**, 54 (1963).
386. A. R. Archibald, J. Baddiley, and D. Dutton, *Biochem. J.*, **95**, 8c (1965).
387. S. G. Nathenson and J. L. Strominger, *J. Biol. Chem.*, **237**, PC 3839 (1962).
388. J. Baddiley, *Federation Proc.*, **21**, 1084 (1962).
389. L. Glaser and M. M. Burger, *Abstr. Sixth Int. Congr. Biochem. I.U.B.*, Section VI, **32**, 509 (1964).
390. M. McCarty, *J. Exp. Med.*, **109**, 361 (1959); W. E. Juergens, A. R. Sanderson, and J. L. Strominger, *Bull. Soc. Chim. Biol.*, **42**, 1669 (1960).
391. J. Baddiley and A. L. Davison, *J. Gen. Microbiol.*, **24**, 295 (1961).
392. A. J. Wicken, S. D. Elliot, and J. Baddiley, *J. Gen. Microbiol.*, **31**, 231 (1963).
393. R. W. Jeanloz and E. A. Balazs, Eds., "The Amino Sugars," Vol. IIA, Academic Press, New York, 1965.
394. P. T. Grant and J. L. Simkin, *Ann. Rept. Progr. Chem.* (London), **61**, 491 (1964).
395. A. Gottschalk, Ed., "Glycoproteins," Elsevier, Amsterdam, 1966.
396. E. J. Sarcione, *J. Biol. Chem.*, **239**, 1686 (1964).
397. G. B. Robinson, J. Molnar, and R. J. Winzler, *J. Biol. Chem.*, **239**, 1134 (1964).
398. S. Kornfeld, R. Kornfeld, and V. Ginsburg, *Federation Proc.*, **23**, 274 (1964).
399. A. Kobata, *Biochem. Biophys. Res. Commun.*, **7**, 346 (1962).
400. G. W. Jourdian, F. Shimuzu, and S. Roseman, *Federation Proc.*, **20**, 161 (1961).
401. R. G. Spiro, *J. Biol. Chem.*, **237**, 646 (1962).
402. W. T. J. Morgan, *Bull. Soc. Chim. Biol.*, **46**, 1627 (1964).
403. E. E. Grebner, C. W. Hall, and E. F. Neufeld, *Biochem. Biophys. Res. Commun.*, **22**, 672 (1966).
404. E. E. Grebner, C. W. Hall, and E. F. Neufeld, *Arch. Biochem. Biophys.*, **116**, 391 (1966).
404a. P. W. Kent and P. Draper, *Biochem. J.*, **106**, 293 (1968).
404b. A. Allen and P. W. Kent, *Biochem. J.*, **106**, 301 (1968).
404c. J. L. Simkin and J. C. Jamison, *Biochem. J.*, **106**, 23 (1968).
404d. Y. T. Li, S. H. Li, and M. R. Shetlar, *J. Biol. Chem.*, **243**, 656 (1968).

404e. S. Roseman, *4th Int. Conference on Cystic Fibrosis of the Pancreas (Mucoviseidosis)*, *Berne/Grindwald 1966*, Part II, Karger, Basel, 1968, p. 244.

405. J. S. Brimacombe and J. M. Webber, "Mucopolysaccharides," B.B.A. Library, Elsevier, Amsterdam, 1965.

406. J. E. Silbert, *J. Biol. Chem.*, **238**, 3542 (1963).

407. J. E. Silbert, *J. Biol. Chem.*, **239**, 1310 (1964).

408. A. Telser, H. C. Robinson, and A. Dorfman, *Proc. Nat. Acad. Sci. U.S.*, **54**, 912 (1965).

409. P. J. O'Brien, M. R. Canady, and E. F. Neufeld, *Federation Proc.*, **24**, 231 (1965).

410. E. J. Sarcione, M. Bohne, and M. Leahy, *Biochemistry*, **3**, 1973 (1964).

411. J. Molnar, G. B. Robinson, and R. J. Winzler, *J. Biol. Chem.*, **240**, 1882 (1965).

412. S. Suzuki and J. L. Strominger, *J. Biol. Chem.*, **235**, 257, 267, 274 (1960).

413. R. L. Perlman, A. Telser, and A. Dorfman, *J. Biol. Chem.*, **239**, 3623 (1964).

414. L. Rodén, *Abstr. Am. Chem. Soc., 148th Meeting, Chicago, 1964*, 12C.

415. A. S. Balasubramanian and B. K. Buchhawat, *J. Neurochem.*, **11**, 877 (1964).

416. B. Wortman, *Biochim. Biophys. Acta*, **83**, 288 (1964).

417. H. Yoshida and F. Egami, *J. Biochem. (Tokyo)*, **57**, 215 (1965).

35. INTRODUCTION TO POLYSACCHARIDE CHEMISTRY

I. DANISHEFSKY, ROY L. WHISTLER AND F. A. BETTELHEIM

I. Occurrence, Types, and Functions BY ROY L. WHISTLER . . 375
   A. Classification and Nomenclature 375
   B. Structural Characteristics 377
   C. Functions of Polysaccharides 382
II. Physical Methods of Characterization of Polysaccharides BY F. A.
   BETTELHEIM 383
   A. Electron Microscopy. 383
   B. X-Ray Diffraction 384
   C. Infrared Spectra and Infrared Dichroism 385
   D. Optical Rotatory Dispersion 386
   E. Osmotic Pressure and other Colligative Properties. . . 387
   F. Light Scattering 388
   G. Ultracentrifugation 390
   H. Viscosity. 391
III. Characterization by Chemical, Enzymic, and Immunological
   Methods BY I. DANISHEFSKY 394
   A. Isolation and Purification 394
   B. Identification of the Constituents of the Polysaccharide. . 396
   C. Linkage and Sequence 397
   References 410

I. OCCURRENCE, TYPES, AND FUNCTIONS*

A. CLASSIFICATION AND NOMENCLATURE

Polysaccharides are high-molecular-weight carbohydrates. They may be viewed as condensation polymers in which monosaccharides (or their derivatives such as the uronic acids or amino sugars) have been glycosidically joined, with the elimination of water, according to the empirical equation

$$nC_6H_{12}O_6 \longrightarrow (C_6H_{10}O_5)_n + (n-1)H_2O$$

From the reverse direction this equation states that polysaccharides, on complete hydrolysis, yield only simple sugars (or their derivatives). Polysaccharides differ from oligosaccharides only in molecular weight and in the

---

* This section was prepared by Roy L. Whistler.

unique physical properties characteristic of substances of high molecular weights. Since there are no easily defined upper or lower limits for molecular weight, polysaccharides are arbitrarily considered as those condensation polymers of monosaccharides that contain 10 or more monosaccharide residues. Carbohydrates containing 5 to 15 sugar residues rarely occur naturally. A few natural polysaccharides contain 25 to 75 sugar residues, but the majority of them contain 80 to 100 sugar residues. Some exceed this number; for example, the average number of D-glucopyranose residues in native cellulose is 3000. Polysaccharides differ from biopolymers such as proteins in that they exist as polymer-homologous series with a distribution of molecular weights about a mean value, rather than as discrete macromolecules all of the same molecular weight.

Polysaccharides are usually given names that reflect their origin. Illustrative examples are found in the names cellulose, the principal component of cell walls in plants, and dermatan, a polysaccharide isolated in its sulfated form, found in skin, derma. Early names often reflected some property of the isolated polymer, such as starch, a name derived from the Anglo-Saxon *stercan*, meaning to stiffen.

Progress toward a systematic nomenclature has produced the significant ending -*an* to designate a substance as a polysaccharide. Thus, another word for polysaccharide is the generic term glycan. This term has evolved from the generic word glycose, meaning a simple sugar, and the ending -*an*, signifying a sugar polymer. Though now recommended for use in the coinage of new names, the ending is well established in the early literatures for such names as arabinan for an arabinose polymer, xylan for xylose polymers, mannan for those of mannose, galactan for those of galactose, and galactomannans for galactose–mannose combinations. Many early polysaccharide names, especially those ending with the unsystematic and undesirable -*in* ending, have been changed to end in -*an* as a step toward uniformity. However, some well-established and frequently used names, such as pectin, amylopectin, inulin, chitin, and heparin, presumably will persist in modern terminology.

The systematic name glucan does not refer to a specific polysaccharide but signifies only that the polysaccharide is composed of glucose residues. The manner of linkages, the arrangement, and the configurational series are not specified. The name is a group name only and applies as well to cellulose as to glycogen, laminaran, or other glucose polymers. The polysaccharide can be defined more specifically if a source designation is also employed as part of the name. Thus, a more definite polysaccharide is specified in each case by the designation beechwood xylan, yeast mannan, or peanut arabinan.

Polysaccharides of the same type differ slightly from one source to another. Sometimes the differences are quite marked, as with starches. There is a well-known and readily apparent difference among starches from different plants. A particular starch is meant, however, by banana starch or corn starch.

Ideally, the polysaccharides should be classified according to their chemical composition and structure.[1] In such a classification, polysaccharides that on hydrolysis give only one monosaccharide type are termed homoglycans, while polysaccharides hydrolyzing to two or three or more monosaccharide types are termed heteroglycans, with prefixes of di-, tri-, and so on to designate the number of different types of sugar residues. At present, there is no proof that more than five or six types of sugar residues occur in a single polysaccharide. The number of types of sugar residues contained in a polysaccharide can easily be determined by chromatographic examination of the hydrolyzate, provided, of course, that the polysaccharide is pure. In this structural classification the first logical subclassification separates polysaccharides as linear or branched. This separation can readily be made by performing several simple tests. The easiest test is that of film formation. An aqueous polysaccharide solution, when spread on a glass plate and dried, will be brittle if a branched polysaccharide is present. Films from linear molecules will be strong, will undergo folding without breaking, and when plasticized can be stretched with the development of birefringence and a detectable "fibrous" X-ray pattern. Linear polysaccharides also show streaming birefringence when their solutions are stirred and viewed between crossed polarizing plates. Methylation studies may further reveal whether a molecule is branched.

Unfortunately, some polysaccharides have not been examined in rigorously pure condition, and often examination has not extended to characterization of films. Therefore, at times, it is useful to classify polysaccharides according to source.

B. STRUCTURAL CHARACTERISTICS

Though polysaccharides may be viewed as condensation polymers formed by the combination of monosaccharides with the elimination of the elements of water, the naturally occurring polysaccharides are far less complicated than would occur if the combination of monomers took place at random. In fact, many simplifying features are apparent on careful examination of all known polysaccharide structures. The basic reasons for such simplified and ordered arrangements stem from the action of those specific synthesizing enzymes that connect the monosaccharides, by various and sometimes complex routes, to polymeric structures. Methods by which enzymes produce polysaccharides are given elsewhere (this volume, Chap. 34).

References start on p. 410.

In the condensation of monosaccharides to form natural polymers, the hydroxyl group on the anomeric carbon atom always participates in the condensation. Since most polysaccharides are composed of aldose sugar units, this discussion will be confined to such units. The hydroxyl on C-1 (anomeric carbon) may condense with any hydroxyl group other than that at C-1 of an adjoining monosaccharide unit. In this way, a linear chain can be formed with a free C-1 hydroxyl group at one end. A complete randomness of linkage with the various hydroxyl groups has never been found in nature. Most frequently, a particular mode of linkage is repeated uniformly through the chain. Even the configuration of C-1 remains constant in most polymers. Thus, in amylose, there is a uniform α-D-(1 → 4) linkage, in cellulose a uniform β-D-(1 → 4) linkage, and in laminaran an essentially uniform β-D-(1 → 3) linkage.

A further and highly simplifying fact in polysaccharide chemistry is that, of the multitude of stereoisomeric monosaccharides, only a very few are found in natural polysaccharides. Of the hexoses there are D-glucose, D-mannose, D-fructose, D-galactose, and infrequently L-galactose, and possibly D-idose or L-altrose. Of the pentoses there are D-xylose, L-arabinose, and infrequently D-arabinose. Of the modified simple sugars there are D-apiose, L-fucose, D-galactosamine, D-galacturonic acid, D-glucosamine, D-glucuronic acid, L-guluronic acid, L-iduronic acid, D-mannuronic acid, and L-rhamnose. Even these monosaccharides do not occur at random in polysaccharides but rather are found in a systematic arrangement.

Frequently a polysaccharide contains but a single type of sugar unit (homoglycan). The most abundant polysaccharides are of this type. Paramount, as an example, is cellulose, which is present in the world in a quantity equal to or greater than the quantity of all other polysaccharides. Yet cellulose consists of a chain of D-glucopyranose residues linked uniformly by β-D-(1 → 4) bonds. The presence of another linkage, once in some 700 links, is not ruled out. Essentially cellulose is a linear chain represented by A in Fig. 1.

Sometimes the hydroxyl groups of C-1 from two sugars apparently condense with two hydroxyl groups other than C-1 on a third sugar residue in a polysaccharide. When this occurs, a branch point is produced in the molecule. The molecule may contain a single branch, as in B of Fig. 1, or it may contain numerous branch points. Sometimes the branch may be but a single sugar residue in length. The molecule then is a substituted linear polysaccharide with sugar residues acting as the substituents; C in Fig. 1 is an example. In other instances a branch-on-branch structure may occur which may be likened to a bush; a small section is depicted in D of Fig. 1.

In no known instance do polysaccharides occur as a cage or three-dimensional net structure. They are either linear, cyclic, or branched. It is apparent that when a branch point is introduced the glycosidic bond connects different positions from those connected in the linear portions between branches. It is common to find the same kind of glycosidic linkage at all branch points in a homoglycan. If more than one type of sugar residue is present (heteroglycan), it is usual for all residues of the same sugar to be linked in the chain by the same type of glycosidic bond.

Even in a linear homoglycan, it is possible for more than one type of glycosidic linkage to be prevalent. In such a molecule, the linkages do not occur randomly but are usually in an ordered arrangement.

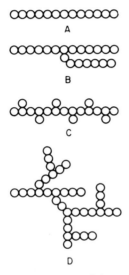

A

B

C

D

Fig. 1. Different types of arrangements of the residues in polysaccharides.

If two types of sugars occur in a polysaccharide, the sugar residues generally seem to be in an ordered arrangement. Thus, in most linear diheteroglycans, polysaccharides composed of two kinds of sugars, the monomer units seemingly are arranged in an alternating and regular fashion. Some diheteroglycans have the structural arrangement illustrated by *C* of Fig. 1. In this structure the principal chain may be composed of one type of sugar linked uniformly throughout, while the branches are composed of a second type of sugar which may be connected to the main chain by identical glycosidic bonds.

When more than two types of sugars are combined to produce a polysaccharide, they usually form a branch-on-branch structure exemplified by *D* in Fig. 1. Even here, some semblance of a simplifying order seems to

* *References start on p. 410.*

exist. Thus, it is common to find hexose sugars and perhaps uronic acids in the main or central branches, while the pentose sugars, D-xylose and L-arabinose, are in the side branches.

There are, then, in polysaccharides certain naturally imposed simplifications which greatly facilitate their structural characterization. However, it is possibly correct to say that no polysaccharide is completely uniform. Even such seemingly regular molecules as cellulose and amylose appear to have irregularities or "anomalous links" in their structures. The irregular linkages are rare and may have a frequency of only 1 in each 700 linkages, as seems to be true in some celluloses. Such rare irregularities are explainable on the basis that the various possible glycosidic bonds do not differ greatly in energy; hence, within the enzyme–substrate system there may at times occur a brief change, influential in causing an irregularity to develop in the chain growth. The irregularity may be due to a brief abnormal action on the part of the principal chain-synthesizing enzyme, or it may be due to the interference or the usurping action of a second enzyme. If enzymes are beyond reproach in the formation of irregular bonds, then quite conceivably such infrequent bonds could be produced by chance chemical synthesis.

Uronic acids are present in about half of the approximately 250 known polysaccharides. In each instance the uronic acid has the pyranoid ring form. The proportion of uronic acid residues to other sugar residues may be low, as in certain plant hemicelluloses and gums, or uronic acid residues may constitute the entire polysaccharide chain, as in alginic and pectic acids. D-Glucuronic acid is the most common natural uronic acid, but D-galacturonic acid constitutes the monomer units of pectic acid and is present in many plant gums and animal polysaccharides. D-Mannuronic acid is infrequently found in polysaccharides, although it is the major sugar component of alginic acid and may constitute all the chain residues of some alginic acid molecules. L-Guluronic acid is also found in alginic acid. L-Iduronic acid, likewise, is infrequently observed in Nature but is present in dermatan sulfate, also called chondroitin sulfate B.

D-Glucuronic acid residues occur in polysaccharides both as interchain units and as nonreducing end units. The D-glucuronic acid residues may also be derivatized. In animal and microbial polysaccharides, D-glucuronic acid usually occurs as a chain unit, sometimes next to a D-glucose residue to which it is joined by a β-D-(1 → 4) linkage to form a cellobiouronic acid unit within the chain. Commonly, it alternates with hexosamine residues. However, in plant gums and hemicelluloses, D-glucuronic acid is found most commonly as nonreducing end units, either as substituents on a chain or as terminal units of branches. Annual plants sometimes contain D-glucuronic acid residues, but in annual plant hemicelluloses and in most wood hemicelluloses the D-glucuronic acid often occurs as the 4-methyl ether. In fact, 4-*O*-methyl-D-glucuronic acid is very widespread in wood hemicelluloses and represents one of the few instances where *O*-methyl groups occur at C-4 instead of at the more common C-2 and C-3 locations in sugars. In heparin, one or two hydroxyl groups of the D-glucuronic acid residue are esterified with sulfuric acid. So far, there is no definite proof that the uronic acid carboxyl group is esterified, either with an aliphatic alcohol or with a hydroxyl group of another polymer molecule.

The presence of sterically accessible carboxyl groups causes the polysaccharide to combine in salt links with cationic polymers, particularly protein. This property of combining with or precipitating proteins can cause difficulty in extraction and isolation procedures, but it has been used advantageously to separate hyaluronic acid from some soluble impurities. In this separation,[1a] a protein may be added to a crude solution of hyaluronic acid to effectively precipitate the hyaluronate–protein salt as a "mucin clot." Most carboxylated polysaccharides will form an insoluble salt on addition of cetyltrimethylammonium halide to their aqueous solution, thus allowing their separation from neutral polysaccharides and other extraneous substances.

Multivalent cations will combine with uronic acid carboxyl groups and can form cross links with carboxyl groups on other polysaccharide molecules to prevent dissolution of the polysaccharide or to cause gelatinization and precipitation of the polysaccharide from solution.

Polysaccharides that are, in effect, copolymers of a glycuronic acid and an amino sugar in approximately equal amounts are somewhat amphoteric, although their basicity is weak because the amino group is usually *N*-acetylated or *N*-sulfated.

The D-glucuronic acid residues in polysaccharides usually are not esterified and exist naturally in salt form with common inorganic cations; a major exception are the pectins, with many methyl esters. The unesterified acids are normally ionized and hence highly hydrated. This extensive hydration, coupled with the formal negative charge of the ionized carboxyl groups, gives the polysaccharides good solubility and solution stability over a wide range of hydrogen-ion concentrations. However, if the acidity of the solution is lowered to such an extent that ionization of carboxylic acid groups is repressed, much hydration is lost, coulombic repulsion between molecules is diminished, and the dissolved molecules can more easily associate to form a gel or precipitate.

When D-glucopyranosyluronic acid residues occur within a polysaccharide chain having a neighboring chain unit joined to C-4 of the uronic acid, as is often observed, the polysaccharide is alkali-labile. The lability is increased if the uronic acid carboxyl is esterified. Lability results from one type of the well-known alkali-catalyzed β-elimination reaction, frequently observed in

* *References start on p. 410.*

polysaccharides.[2] This particular β elimination results from tautomeric shifts in the uronic acid structure, permitting the remainder of the chain to be expelled from position C-4 as an anion. The resultant effect on the polysaccharide is random depolymerization (Fig. 2).

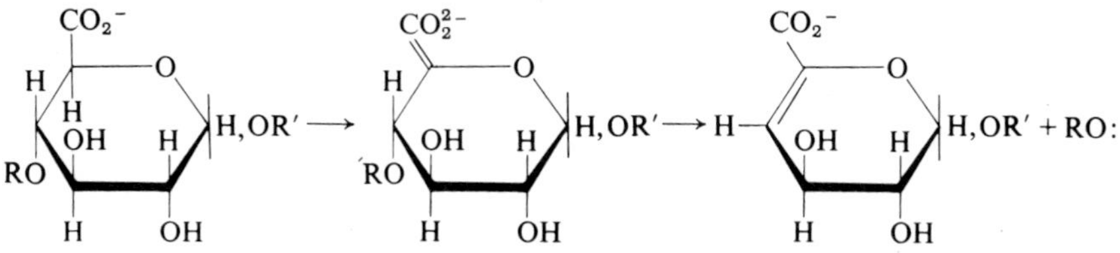

FIG. 2. Alkali-catalyzed cleavage of uronic acid chains.

D-Glucopyranosyluronic acid residues are not readily hydrolyzed from the chain unit to which they are glycosidically bound. The stability of the glycosidic bond to acid-catalyzed hydrolysis is so great that extensive destruction of the uronic acid and of other sugar units, if present, occurs on attempts to force hydrolysis of the polysaccharide to completion. If hydrolysis of a uronic acid-containing polysaccharide is anticipated, the uronic acid residues should be reduced to normal hexose units. Reduction can be done with diborane or by producing the methyl ester with diazomethane and reducing this with borohydride.[3] Resistance of the glycosidic linkage in the uronic acid can be used to advantage, however, when it is joined to a normal hexose unit in the polysaccharide chain. In these structures, acid-catalyzed hydrolysis leads to cleavage of normal glycosidic bonds and results in the accumulation of aldobiouronic disaccharide fragments that can be separated, and the nature of the contained glycosidic linkage can be identified.

C. Functions of Polysaccharides

Polysaccharides serve living organisms in two distinct ways. One is as food reserves, as illustrated by starch, glycogen, and galactomannans. The second service derives from the unique physical properties of macromolecules. Thus, cellulose, hemicelluloses, and chitin serve obviously as structural molecules and as links between other cell wall components and likely as agents that help to control cell wall permeability. Hyaluronic acid and certain mucoidal polysaccharides serve as lubricants and as thickeners, and probably control intercellular permeability.[3a] Many polysaccharides act as shields or protective coatings on the surfaces of microorganisms, and many plant gums are secretions from wounds where they aid in sealing the wound and in offering protection; they are the antigens characteristic of microorganisms and often are highly toxic (endotoxins). Heparin has the unique function of increasing blood clotting time and is hence an important systemic anticoagulant. Polysaccharides are important components in normal connective tissue, in granulation tissue of healing animal wounds, and probably in calcification processes. Those containing large amounts of hexosamine, especially in connective tissues, are known as mucopolysaccharides. Some polysaccharides are antigenic, and some from microbial sources are even highly toxic. These physical and biochemical properties of polysaccharides are for the most part not clearly understood. It is in the elucidation of such functions that much research is needed and is progressing.

II. PHYSICAL METHODS OF CHARACTERIZATION OF POLYSACCHARIDES†

Most of the properties (physical, chemical, and physiological) of polysaccharides depend largely on the size, shape, and dimensions of the individual molecules as well as their interactions. For this reason the characterization by physical methods aims to obtain parameters on molecular size, weight, and distribution in any given preparation.

A. ELECTRON MICROSCOPY

The most direct method of investigating structural conformation would be electron microscopic observations. However, when information of the conformation of a single molecule is sought, there are some difficulties encountered because of problems in resolution. The molecules themselves are too small to scatter electrons, and therefore they do not give enough contrast to the photographic plate. The most frequently used technique of obtaining an image of the particles is to deposit a metal "shadow" on them.[4] Even with shadowing with a heavy metal such as platinum or palladium, only in very rare instances can one obtain single molecular patterns. One such case is a study of O-(hydroxyethyl)cellulose with shadowing by Pt–Pd alloy.[5] In this case the molecular diameters were obtained from the length of the shadow and were calculated to be 10 ± 5 Å. The length of the molecules varied from 2200 to 5000 Å. More frequently, however, the molecular aggregates and their conformational shape can be evaluated from electron microscopic studies. Bettelheim and Philpott[6] have shown with Pt shadowing that hyaluronic acid is a relatively stiff material, and it aggregates into highly anisotropic, needle-like platelets of the three-dimensional network in which the individual strands or microfibrils have an average diameter of about 50 Å.

* References start on p. 410.
† This section was prepared by Frederick A. Bettelheim.

B. X-Ray Diffraction

X-ray diffraction studies on polysaccharides are in a sense different from those on simple sugar moieties (Vol. IB, Chap. 27) because the available information is much more scanty. In the structural studies of simple molecules, the intensities of the scattering from a single crystal are used to obtain a detailed structure. A single crystal is supposed to have perfect or almost perfect three-dimensional lattice structure. However, single crystals of polysaccharides are rarely available for studies, although single crystals of amylose V complexes have been obtained. Usually in polysaccharides the crystallinity of the material is only partial. This means that there are severe dislocations in the crystal lattice or large amounts of amorphous material, which make interpretation of the diffraction pattern difficult. Most of the information on polysaccharides has been obtained on oriented material by using a fiber pattern or an unoriented powder pattern. The information obtainable from a powder pattern is the interplanar spacings and the indexing of this interplanar spacings in a definite unit cell.[7] On the other hand, from a fiber (oriented) pattern, with the assumption that the long axis of the polymer is oriented most of the time in the direction of the stretch or elongation, one can interpret the meridional spacings of layer lines as those characteristic of the repeating units of the polymer backbone and the equatorial lines as those of the interchain distances.

Besides obtaining the unit cell into which the polysaccharide chains are packed, the most important single piece of information obtainable from such studies is the nature of the repeating units, because these are the intrinsic characteristics of the molecular structure. Rundle[8] reported, for instance, that the butyl alcohol–amylose complex is crystallized in an orthorhombic unit cell with the dimensions of $a = 13.0$ Å, $b = 23.0$ Å, and $c = 8.05$ Å for the anhydrous complex. This structure is assumed to be a helical structure with six D-glucose residues per turn, the 8.05 Å dimension being the pitch of the helix and the 13.0 Å spacing being the diameter of the helix. Other types of polysaccharide chain conformations have been found. Certain cellulose derivatives[9] have an apparent threefold screw axis in which the repeating unit is 15.2 Å long—that is, about 5.21 Å per monosaccharide residue. Similarly, Palmer and Hertzog[10] found that sodium pectate also has a threefold screw symmetry, and the fiber axis identity period in this case is 13.1 Å, considerably shorter than the 15.2 Å of cellulose. In the case of sodium hyaluronate and chondroitin 4-sulfate, Bettelheim[11] has found a twofold screw axis. The sodium hyaluronate has a hexagonal unit cell having the dimensions $a = 12.66$ Å and $c = 11.98$ Å and a repeating unit of 11.98 Å consisting of a disaccharide repeating unit. The chondroitin sulfate has an orthorhombic unit cell with dimensions of $a = 14.06$ Å, $b = 8.62$ Å, and

$c = 9.80$ Å, the repeating unit being 9.80 Å, corresponding again to a disaccharide residue.

Additional information derivable from fiber or powder patterns—namely, the lateral interactions between polysaccharide chains—is more prone to controversy in interpretation. However, some X-ray diffractions have been performed on single crystals. Manley[12] and Yamashita[13] obtained amylose V complexes in single crystal lamellae, and they found that the helices were oriented perpendicular to the lamellae surfaces. Since the lengths of the helices are much longer than the thicknesses of the lamellae, this must be interpreted as chain folding or helix folding occurring within lamellae. Possibly, in the future, more work will be done as single crystals of polysaccharides are isolated.

C. INFRARED SPECTRA AND INFRARED DICHROISM

When oriented samples of polysaccharides are exposed to polarized infrared radiation, the absorption spectra of the parallel and perpendicular polarization will differ. Parallel polarization is meant when the polarized infrared radiation is in the same direction as the orientation of the polysaccharide matrix (film or fiber). Molecular groups, the transition vector of which will be in the direction of the polarization, will absorb strongly, whereas those that are perpendicular to the polarization will not absorb. Hence, infrared dichroism, which is the ratio of the parallel to perpendicular absorptivities, is used to a great extent to elucidate the structure of polysaccharides. This can be done from two points of view. On the one hand, it supplements the assignment of the infrared active bands (see Vol. IB, Chap. 27). On the other hand, it facilitates and supplements the interpretation of X-ray diffraction data for polysaccharides by indicating the correct orientation of different atomic groups within crystalline or amorphous structures. The orientation of a film or a fiber is achieved either by extension or by rolling, and, in most cases, it has been proved that linear polysaccharides will orient themselves with their long geometrical axis in the direction of stretching. In this respect, the work of Liang and Marchessault[14] has yielded much structural information on polysaccharides. In all these cases, the authors not only obtained the group orientations in polysaccharides such as cellulose, chitin, and xylans, but also aided the assignment of the absorption bands of the various groups, such as OH in-plane bending, OH out-of-plane bending, antisymmetric bridge, and C—O—C stretching.

The study of molecular orientation of polysaccharides by infrared dichroism is not limited to the solid state only. Polysaccharide solutions can be forced to flow under a variety of velocity gradients, and such flow will produce

* References start on p. 410.

orientations similar to those observed in the solid state. Hence, structural studies can be performed by what are termed streaming dichroism studies.[15] A typical instrument for the studies of such differential flow of dichroism was reported by Wada and Kozawa.[16] The study of dichroism obviously is not limited to the infrared range and can be used in the visible or ultraviolet if suitable chromophores are present in the molecules. However, in the case of polysaccharides, the infrared range is the most widely used one.

D. Optical Rotatory Dispersion

The optical properties of polysaccharides that are above and beyond those arising from the simple sugar moieties (see Vol. IA, Chap. 4 and Vol. IB, Chap. 27) are always associated with the conformation of the macromolecules. The variation of optical rotatory power of a polysaccharide over a wide range of wavelengths, λ, is termed optical rotatory dispersion. Usually, when the polysaccharide molecule is in the random-coil conformation, its behavior can be described by a one-term Drude equation of the form

$$m' = a_0\lambda_0^2/(\lambda^2 - \lambda_0^2)$$

where m' is related to the molecular rotation and is given by

$$m' = \frac{3M_0}{100(\bar{n}^2 + 2)}[\alpha] \quad \text{and} \quad [\alpha] = \frac{\alpha}{dc}$$

where M_0 is the molecular weight of the monomeric unit (if the concentration, c, of the solution is given also in terms of moles of monomeric units per 100ml.), α is the angle of rotation, d is the length of the light path in decimeters and $\bar{n}$ is the average refractive index. This one-term Drude equation yields a straight-line plot when $[\alpha] \lambda^2$ is plotted versus $[\alpha]$ or m'. On the other hand, when a certain number of chromophores in the polysaccharide are in a stable array, and they are interacting in phase-coherent manner, usually the simple one-term Drude equation does not describe the system correctly, and a two-term Drude equation is used. In protein chemistry an enormous amount of work has been done in calculating the so-called "helical content" of proteins from the two-term Drude parameters. However, these interpretations of the parameters in terms of helical contents have now fallen somewhat into disrepute. Rao and Foster[17] studied the conformational properties of amylose and O-(carboxymethyl)amylose by optical rotatory dispersion. They found that the behavior of these polysaccharides can be described by a one-term Drude equation. They interpreted the data to mean that the amylose exists below pH 11 in an imperfect helical coil form and above pH 12 in a more flexible coil, and that a similar phase-transition occurs with the O-(carboxymethyl)amylose at pH 4.

E. OSMOTIC PRESSURE AND OTHER COLLIGATIVE PROPERTIES

The simplest way theoretically to determine molecular weights is to measure the colligative properties of solutions as a function of concentration. Since the colligative properties are strictly due to the number of particles present in a system, from the knowledge of the concentration of the solutions, in terms of grams per liter, and from the knowledge of the number of particles present, the molecular weight can be calculated. The colligative property most frequently used in determining molecular weights of polysaccharides is the osmotic pressure. However, measurements of vapor-pressure lowering are also used, since commercial instruments are available under the name of vapor-phase osmometers. The commercial availability of new dynamic osmometers that reduce the time-consuming equilibration process to 15 to 20 minutes has greatly enhanced the use of osmometers to determine molecular weights. Usually, the osmotic pressure or any colligative property is concentration-dependent and at a constant temperature can be expressed by the Flory[18] equation:

$$\frac{\pi}{c} = \left(\frac{\pi}{c}\right)_0 (1 + \Gamma_2 c + g\Gamma_2^2 c^2 + \cdots)$$

in which π is the colligative property, measured most commonly as osmotic pressure, c is concentration in grams per liter, $(\pi/c)_0$ is the value of π/c extrapolated to zero concentration, and Γ_2 and $g\Gamma_2^2$ are virial coefficients characteristic for the solute–solvent and solute–solute interactions. Usually the plot of π/c against c gives a straight line, and therefore only the second virial coefficient is evaluated from such a plot. Where the line intersects the π/c axis, the $(\pi/c)_0$ value is obtained, and the molecular weight can be calculated by Van't Hoff's equation, namely:

$$\left(\frac{\pi}{c}\right)_0 = \frac{RT}{M}$$

where R is the gas constant, T is the absolute temperature, and M is the number-average molecular weight of the polysaccharide. Colligative properties are best used with nonelectrolytes where the problem of Donnan equilibrium and the number of free counter-ions are of little importance. However, osmotic pressure theories for polyelectrolytes are well developed.[19] In the case of a polyelectrolyte, the usual procedure is to determine colligative properties in the presence of salt in order to minimize the dissociation of the ionizable groups on the polysaccharide. In the case of polydisperse macromolecules, the molecular weights obtained from the extrapolated values are number-average molecular weight (see below). Potter and Hassid[20] determined the molecular weight of acetylated amyloses and amylopectins by osmotic pressure measurements and found that the amylose fractions ranged from 100,000 to 210,000, and amylopectin from 1,000,000 to 6,000,000, in molecular weight. Similarly, Holtzer et al.[21] determined the molecular weights of cellulose nitrate samples in acetone and found molecular weights ranging from 45,000 to 1,300,000. Descriptions of apparatus for measuring colligative properties can be found in Flory's book,[18] and, for the newer techniques, in papers by Burge[22] for vapor-phase osmometers and by McNeill[23] for the dynamic osmometer.

F. Light Scattering

Light scattering by dilute polymer solutions can be used to determine the weight-average molecular weights of polymers. A scattering cell containing a solution of a specific concentration is illuminated with monochromatic polarized or unpolarized light and the scattered light intensity is measured as a function of the scattering angle by a photometer. The primary data are obtained in the form of intensity readings on the galvanometer, as a function of scattering angle for different solutions of various concentrations. The scattering intensity depends mainly on two factors: (1) the concentration of the polymer in a solution, and (2) the shape and the size of the polymer molecules. The heavier particles will scatter light with greater intensity. Hence, the average molecular weight obtained from such measurements will be accentuated by the larger particles, and it is a weight-average molecular weight.

Since the scattering intensity is a function of concentration, to determine molecular weights the data must be extrapolated to infinitely dilute solutions where supposedly only a single molecule will cause scattering and, therefore, the intensity can be interpreted in terms of the parameters of this single molecule. Secondly, the scattering depends on the size and the shape of the molecule, and the scattering envelope may be nonsymmetrical because of intramolecular interference. Models for different shapes can be set up that permit calculation of theoretical scattering intensity as a function of scattering angle for different shapes, once the size of the particle is known. These theoretical functions are called particle-scattering functions $P(\theta)$. These are complex functions,[24] but they reach unity when the scattering angle is 0°. Hence, to evaluate molecular weights from solutions of polymers without complicating factors of size and shape, the experimental data have to be extrapolated also to zero scattering angle. This is done in the so-called Zimm plot (Fig. 3), in which c/R_θ is plotted versus $\sin^2(\theta/2) + kc$. The term c is the concentration of the polymer solutions in grams per milliliter. The

R_θ values are the corrected Rayleigh ratios for unpolarized light obtained from the scattering intensities in the following manner.

The scattering instrument is calibrated with a substance for which the absolute scattering at 90° is well known—for instance, benzene.[25] Knowing the absolute scattering of this substance at a certain wavelength and the instrument response, one can calculate the instrument constant. This instrument constant is then used to obtain the Rayleigh ratios $R(\theta)$ from the

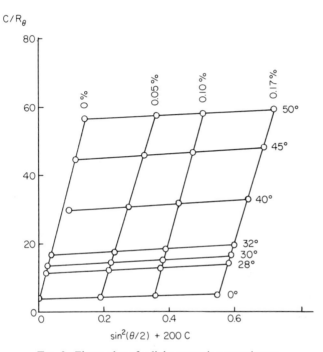

FIG. 3. Zimm plot of a light-scattering experiment.

observed intensities at different angles and at various concentrations. This $R(\theta)$ value has to be corrected, if unpolarized light is used, by multiplying with the factor of $\sin \theta/(1 + \cos^2 \theta)$. Thus, the ordinate of the Zimm plot is the concentration divided by the corrected Rayleigh ratio, and this is plotted against $\sin^2 (\theta/2)$ of the scattering angle and kc concentration, in which k is an arbitrary constant selected so that the two terms $\sin^2 (\theta/2)$ and kc should have about the same magnitude of numerical expression.

The experimental points of such a Zimm plot form a grid-like structure (Fig. 3) which has to be extrapolated to zero concentration and zero angle. This is done by extending the lines of the different concentrations, and,

* References start on p. 410.

where these lines intersect the value for the specific kc values, the points for the zero angle line are obtained. Similarly, extending the lines that connect the points obtained at the same angle but at different concentrations to a point where these lines intercept the value for the specific $\sin^2 (\theta/2)$ gives the points for the zero concentration line. The zero angle and the zero concentration lines are then extrapolated, and this should give an intercept on the ordinate. This intercept, multiplied by

$$K = \frac{2\pi^2 n^2 (dn/dc)^2}{N\lambda^4}$$

gives the reciprocal of the weight-average molecular weight. The parameters included in K are: n, the refractive index of the solvent; (dn/dc), the refractive index increment; N, Avogadro's number; and λ, the wavelength of the light.

The Zimm plot of the light-scattering data yields useful information on the molecular dimension and the interaction parameter as well as the molecular weight. The radius of gyration, R_g, can be obtained from the relationship

$$\bar{R}_g{}^2 = \frac{3\lambda_0{}^2/n^2}{16\pi^2} \times \frac{\text{initial slope of zero concentration line}}{\text{intercept}}$$

When such radii of gyration are obtained, particle-scattering functions for different shapes such as rod, random coil, and compact sphere can be calculated as a function of scattering angle. These functions can be obtained and compared with the experimental points of the zero concentration lines of the Zimm plot. Such a comparison gives the shape of the molecules.

As an example, Erlander and French[26] used light-scattering techniques to study the behavior of starch fractions, and Bettelheim et al.[27] obtained molecular parameters on bovine submaxillary mucin.

G. Ultracentrifugation

Molecular weights and shapes can be obtained also from sedimentation patterns in ultracentrifugal studies. The rate of sedimentation is directly related to the molecular weight by the Svedberg equation:

$$M = \frac{RTS_0}{(1 - \bar{v}\rho)D_0}$$

in which M is the molecular weight, R is the gas constant, T is the absolute temperature, S_0 is the extrapolated sedimentation coefficient, D_0 is the diffusion coefficient, which has to be obtained independently, $\bar{v}$ is the partial specific volume, and ρ is the density. The sedimentation coefficient, S_0, is usually obtained from a plot in which the reciprocal of the sedimentation coefficient at various concentrations is plotted against concentration. This usually gives a straight-line plot, which therefore can be extrapolated to zero concentration to give the S_0 value. However, the rate of sedimentation does not yield any particular type of molecular-weight average. It will depend on the selection of the sedimentation and diffusion boundaries. This was demonstrated very carefully in the work of Laurent et al.,[28] who studied molecular weights of hyaluronic acid by light-scattering and ultracentrifugal techniques. In the case of unfractionated hyaluronic acid samples, there was a discrepancy between the molecular weights obtained, whereas, in the case of the fractionated hyaluronate samples, agreement between the two techniques was good.

The Archibald technique of equilibrium centrifugation yields weight-average molecular weights as well as Z-average molecular weights. In this case, the molecular weight is related to the ratio of concentration c_2/c_1 at two points in the centrifuge cells at distances x_1 and x_2 from the axis of rotation. (For details, see Schachman et al.[29]) The method has been modified by Yphantis.[30]

Beyond obtaining molecular weights, and an indication about the polydispersity, ultracentrifugal techniques are also used as criteria for homogeneity. A single boundary in the sedimentation experiment usually is one of the criteria for homogeneity; others are electrophoretic measurements in which the preparation should also yield a single boundary at different pH values and after chromatography and gel filtration.[31] (See also Vol. IB, Chap. 28.)

H. Viscosity

Among the experiments most frequently performed is the viscosity measurement of dilute polymer solutions. At the same time, this is the least understood phenomenon with regard to interpretation of molecular parameters. In spite of this, however, because of the ease of performance of the experiment, many data are reported in the literature on the viscosity behavior of polysaccharide solutions. In most cases capillary viscometers that involve laminar flow are employed, in which case the efflux time is an indicator of the viscosity coefficient of the solution.

The viscosity of a solution is dependent primarily on concentration, temperature, and velocity gradient. If the temperature and the velocity gradient are kept constant, the viscosity dependence on concentration can be given by Huggins' equation:

$$\frac{\eta_{sp}}{c} = [\eta] + k'[\eta]^2 c$$

References start on p. 410.

or by Martin's equation

$$\log \left(\frac{\eta_{sp}}{c}\right) = \log [\eta] + k'[\eta]c$$

In each of these equations the η_{sp} equals the time of efflux of the solution (t_1) times its density (ρ_1) divided by time of efflux of the solvent (t_0) times its density (ρ_0) minus one:

$$\eta_{sp} = \frac{t_1\rho_1}{t_0\rho_0} - 1$$

$[\eta]$ indicates the intrinsic viscosity, k' is a constant, and c is the concentration. The most frequently used single parameter obtainable from viscosity measurements is the intrinsic viscosity, which, according to the above equations, can be obtained by plotting either η_{sp}/c versus concentration or $\log (\eta_{sp}/c)$ versus concentration; the straight line obtained can be extrapolated to zero concentration. This extrapolated value is the intrinsic viscosity in the case of the first equation and the logarithm of intrinsic viscosity in the case of the second equation.

The intrinsic viscosity is related to a number of molecular parameters—namely, to the molecular weight of the polymer and its shape and size. An empirical relationship (the Staudinger or Mark–Houwink equation) relates the intrinsic viscosity to the molecular weight.

$$[\eta] = KM_v{}^a$$

in which case the $[\eta]$ represents the intrinsic viscosity, K is a constant, characteristic of the polymer and the solvent at the temperature at which the experiment is performed, M_v is the viscosity-average molecular weight, and a is a parameter that relates to the shape of the polymer. If a is approximately 0.5, the shape of the polymer is a compact sphere; if it is near 0.75, it is a random coil; and if it is 1 and above, the shape of the polysaccharide is that of an elongated rod.

Such empirical equations are obtained for a polysaccharide–solvent system by determining the molecular weights of polysaccharide fractions by an independent technique, either osmotic pressure or light scattering, and relating the intrinsic viscosity values obtained for the different fractions to the molecular weights. Once such a relationship is obtained, the empirical Staudinger equation is used to estimate molecular weights of new batches of the same polysaccharide from viscosity measurements. The viscosity-average molecular weight is somewhere between the number-average and the weight-average molecular weights, and it depends on the distribution of the molecular weights in the sample as well as on the shape of the polymer.

The four molecular-weight averages indicated so far—M_n, M_w, M_z, and M_v—are expressed by the following equations:

$$M_n = \frac{w}{\sum\limits_{i=1}^{\infty} N_i} = \frac{\sum\limits_{i=1}^{\infty} N_i M_i}{\sum\limits_{i=1}^{\infty} N_i}$$

$$M_w = \frac{\sum\limits_{i=1}^{\infty} c_i M_i}{c} = \frac{\sum\limits_{i=1}^{\infty} N_i M_i^2}{\sum\limits_{i=1}^{\infty} N_i M_i}$$

$$M_z = \frac{\sum\limits_{i=1}^{\infty} N_i M_i^3}{\sum\limits_{i=1}^{\infty} N_i M_i^2}$$

$$M_v = \left[\frac{\sum\limits_{i=1}^{\infty} N_i M_i^{(1+a)}}{\sum\limits_{i=1}^{\infty} N_i M_i} \right]^{1/a}$$

where M_i is the molecular weight of species i, N_i is the number of moles of species i, w is the total weight of sample, and c_i is the concentration.

There are a number of theories on the viscosity of polymer solutions, which relate in detail the molecular parameters to the flow properties observed. However, the discussion of these is beyond the scope of this review, and the reader is referred to standard textbooks on polymer chemistry such as those by Flory,[18] Tanford,[32] and Morawetz.[33]

An extensive study of the relationship of molecular weights and intrinsic viscosities of cellulosic materials was published by Immergut and Eirich.[34] In these studies, the viscosities were extrapolated to zero concentrations and zero shear gradient, and the indicated molecular weights were compared with molecular weights as obtained by other techniques. It was found that the Staudinger relationship indicated a rod shape for the cellulosic material up to a degree of polymerization of 200 and then a transition followed. For a degree of polymerization of 400 to 600, a coiled model was evident, and for a degree of polymerization of thousands and higher, the data fit the approximation for an impenetrable coil.

In the case of acidic polysaccharides, or indeed any polyelectrolyte, an additional effect, the so-called electroviscous effect, has to be taken into account. This demonstrates itself in that the usual plot of η_{sp}/c versus concentration, which yields a straight-line function down to a certain dilution, suddenly reverses itself and starts to increase with decrease of concentration. This effect can be explained qualitatively by the fact that, at very high dilutions, the interionic interaction decreases between ion pairs; therefore, the polyelectrolyte ionizes to a greater degree, with the result that the similar charges on the polyelectrolyte will repel each other and thereby extend the shape of the molecule. A rod-like shape will contribute more to the viscosity coefficient than a spherical shape.

To avoid electroviscous effects, acid polysaccharides are dissolved in aqueous salt solutions to keep the ionization at a minimum, and in this way a linear plot is obtained. Such studies were performed, for example, by Laurent et al.[28] on the viscosity behavior of hyaluronic acid in $0.1 M$ sodium chloride solutions at various concentrations and shear gradients. The authors compared the viscosity behavior with molecular weights obtained by light-scattering and ultracentrifugal measurements. In the Staudinger equation, they obtained 0.036 for the value of K (hyaluronic acid–solvent interaction parameter) and 0.78 for the a value—that is, the shape factor. This indicates a random-coil shape for the hyaluronic acid under investigation in this solution. The average molecular weight obtained from these measurements indicated two species of hyaluronic acid isolated from the vitreous body, one whose molecular weight was of the order of 1.5×10^6, and the other with a molecular weight below 4.0×10^5.

III. CHARACTERIZATION BY CHEMICAL, ENZYMIC, AND IMMUNOLOGICAL METHODS*

A. ISOLATION AND PURIFICATION

Polysaccharides occur in Nature as components of animal or plant tissue. As a consequence they are usually intimately admixed or chemically bound with various tissue components such as proteins, nucleic acids, other polysaccharides, or materials of lower molecular weight. Before attempting to characterize a polysaccharide it is therefore necessary to separate it from these substances and to isolate it as a homogeneous entity. There is no general procedure for purification of all polysaccharides, and the method employed in each case depends on the possible contaminants and the properties of the substance being studied. The details of a number of purification procedures have been described.[1]

In many instances, a considerable number of impurities may be removed by dissolving the crude material in water and slowly precipitating the polysaccharide with organic solvents—for example, ethanol, acetone, or pyridine.

* This section was prepared by I. Danishefsky.

This generally does not result in complete purification of the desired product because coprecipitation of extraneous material may take place. Certain polysaccharides may also be precipitated from solution by adding different salts or by varying the pH. A number of substances precipitate certain polysaccharides from solution in the form of insoluble salts or complexes. Fehling's solution and various copper salts have been found to be useful for this purpose.[35-37] Barium hydroxide is also a useful precipitant for certain types of polysaccharide.[38]

Acidic polysaccharides, such as the mucopolysaccharides, pectin, or alginic acid, can be precipitated with quaternary ammonium salts.[39] Reagents that have been used extensively for this purpose are cetyltrimethylammonium bromide and cetylpyridinium chloride. The quaternary ammonium complexes that are formed may be fractionated by varying the salt concentration or pH. This allows for the separation of acidic polysaccharides from neutral ones and the fractionation of acidic polysaccharides from each other. Neutral polysaccharides may also be precipitated as quaternary ammonium compounds if they are first converted into the borate complexes.[40]

Selective precipitation of polysaccharides may also be effected with certain proteins, such as antisera[41] or concanavalin;[42] however, this method is applicable to small amounts of material and usually cannot be used for preparation of compounds for structural studies.

In many instances, polysaccharides are found closely associated with protein, and a primary purification step involves the elimination of the latter. Deproteinization can be accomplished by digestion of the crude product with proteolytic enzymes, such as pepsin, trypsin, papain, or pronase. The mixtures of amino acids and peptides that result can be separated from the polysaccharide by dialysis, adsorbants, or by alcohol precipitation. A procedure that is widely used for removal of protein when it is present in comparatively small proportion is to shake an aqueous solution of the crude preparation with chloroform and a small amount of butyl or amyl alcohol.[43] The proteins that are denatured in the process become insoluble and accumulate at the chloroform–water interface. The precipitate is removed, and the process is repeated until no more precipitate is formed. A modification of this method, which has been employed for bacterial polysaccharides, involves the use of trichlorotrifluoroethane instead of chloroform.[44]

Polysaccharides may also be purified by fractionation on ion exchangers such as DEAE-cellulose, ECTEOLA-cellulose, or resins.[45,46] Acidic polysaccharides are adsorbed on these anion exchangers at about pH 6 and can be eluted by increasing the buffer concentration or the pH of the eluent. Neutral polysaccharides may be adsorbed when the column is in the basic

* References start on p. 410.

form. A more efficient method for neutral polysaccharides is to convert the resin into the borate form; polysaccharides that complex with borate will be retained and may subsequently be eluted with borate solutions of increasing concentration.[45,46] Polysaccharides may also be separated from each other, according to their molecular weights, by passing them through columns of various types of Sephadex or poly(acrylamide) gels.

Electrophoretic procedures are also useful for separation and purification of polysaccharides. Several variations of the electrophoretic methods have been found to be applicable.[47-50]

The conclusion that a polysaccharide is completely purified is generally based on the analytical methods available at the time. The results concerning purity, therefore, have to be reevaluated as new techniques appear. The usual criteria of purity are based on results of ultracentrifugation, electrophoresis, and immunochemical procedures whenever applicable.

B. IDENTIFICATION OF THE CONSTITUENTS OF THE POLYSACCHARIDE

1. *Monosaccharide Components*

The first step in the determination of the structure of a purified polysaccharide is to identify the monosaccharide components that serve as the building blocks. This is accomplished by hydrolyzing the polysaccharide with mineral acid and separating and identifying the resulting sugars. The conditions of hydrolysis have to be such as to completely hydrolyze the polysaccharide and not to destroy the products. Since there are variations as to the ease of hydrolysis of different linkages and the stability of the monosaccharides, the optimum conditions for each polysaccharide have to be ascertained by trial experiments. Polysaccharides composed of furanose residues, such as arabinans or fructans, are hydrolyzed under comparatively mild conditions. This is also true for 2-deoxy sugars and glycosylamines. Pentopyranoses are more stable to acid hydrolysis than pentofuranoses, whereas hexopyranoses are more stable than pentopyranoses. Uronic acid and hexosamine residues in a polysaccharide are considerably more difficult to hydrolyze than unsubstituted hexose units. With regard to the effect of configuration on the anomeric carbon atom, α-D-glycosidic linkages are generally more labile than β-D-glycosides. Uronic acids stabilize linkages in neighboring groups (see p. 382).

The mixture of monosaccharides obtained after acid hydrolysis is generally separated by chromatographic procedures, and the components are identified on the basis of their mobilities on paper chromatography and electrophoresis (see Vol. IB, Chap. 28). Since these procedures do not differentiate between D and L enantiomorphs, the components of the hydrolyzate are converted into appropriate derivatives, which can be identified by their melting points and optical rotations.

Certain component sugars in polysaccharides can be identified by specific color reactions without previous direct hydrolysis and separation (see Vol. IIB, Chap. 45). These methods are useful for preliminary studies or for following the purification of a polysaccharide.

2. Specific Groups

Certain types of groups in a polysaccharide may not be detected after total hydrolysis and are analyzed by special methods. Such groups are carboxyl, acetyl, ether, carbonyl or reducing sugar, primary hydroxyl, and amino. Procedures for determination of these functions are described in Vol. IIB, Chap. 45.

C. Linkage and Sequence

Once the individual components of the polysaccharide are known, the next problem is to determine the way they are linked to each other. Thus, the monosaccharides may be joined by $1 \to 3$ or $1 \to 4$ or other types of linkages. The problem becomes even more complicated when several types of linkages are present, when there is branching, and when the polysaccharide contains more than one type of monosaccharide unit. The procedures generally employed to elucidate the linkages are methylation, periodate oxidation, and fragmentation analysis.

If the polysaccharide contains more than one type of monosaccharide unit or more than one type of linkage, studies have to be carried out to determine the sequence of the specific units. This is generally done by partial hydrolysis of the polysaccharide and subsequent characterization of the oligosaccharide units that are obtained, a procedure that can also indicate the configuration of the linkages.

1. Methylation

The most important procedure determining the position of the linkages in polysaccharides involves total methylation of the free hydroxyl groups followed by hydrolysis and identification of the resulting partially O-methylated derivatives (see Vol. IA, Chap. 12). Thus, if one finally obtains a 2,3,6-tri-O-methylhexopyranose, it can be assumed that the linkage to this unit in the original polymer was $1 \to 4$ or $1 \to 5$. On the other hand, if the methylated product in the hydrolyzate is the 2,3,4-trimethyl ether, a $1 \to 5$ or $1 \to 6$ linkage is indicated.

* References start on p. 410.

The critical prerequisite for such a study is that all the free hydroxyl groups in the polysaccharide are fully methylated. The two classical methods for permethylation involve treatment of the polysaccharide with methyl sulfate and sodium hydroxide (Haworth)[51] or with methyl iodide and silver oxide (Purdie and Irvine).[52] Since polysaccharides are generally insoluble in the organic solvents required for the Purdie procedure, they are usually methylated first with methyl sulfate and aqueous alkali. This yields a partially methylated product, which, in many cases, is soluble in methyl iodide or a mixture of methyl iodide and methanol. Further methylation by the Purdie–Irvine method can then be effected.

Another method for methylation, which has not been used extensively, is to dissolve the polysaccharide in liquid ammonia and treat it with methyl iodide and metallic sodium.[53] Unless carefully controlled, this procedure may result in appreciable degradation of the polysaccharide and some demethylation of previously methylated polysaccharide.

Polysaccharides containing uronic acids present considerable difficulty in methylation. One procedure that has been applied with some success is to convert the polysaccharide into the thallium salt and to methylate it with methyl iodide and thallium hydroxide.[54]

A number of variations in the above-mentioned methods have been developed recently that result in more efficient methylation.[55] The polysaccharides are dissolved in N,N-dimethylformamide and then treated with methyl iodide in the presence of barium oxide or with methyl sulfate and barium oxide. Alternatively, methyl sulfoxide may be used as the solvent and the reaction carried out with the same reagents. Another procedure is to treat the polysaccharide in methyl sulfoxide with sodium hydride and methyl iodide.[56,57] The sodium hydride reacts with methyl sulfoxide to generate the methylsulfinyl carbanion. Presumably, when the polysaccharide is added to this solution, the hydroxyl groups are ionized and react with greater facility with the methyl iodide to form the ether.

After the polysaccharide is fully methylated, it is hydrolyzed to the individual methylated monosaccharides. The depolymerization method employed must be such as to minimize any demethylation or destruction of the sugar. Sulfuric acid is the reagent most generally employed, although hydrochloric, formic, oxalic acids, acetyl bromide, and methanolic hydrogen chloride have been found useful under certain conditions. When the polysaccharide is hydrolyzed with methanolic hydrogen chloride, the products are the methyl glycosides of the O-methylated sugars, and they may be identified as such (for example, by gas–liquid chromatography) or subsequently hydrolyzed with aqueous acid. The advantages and disadvantages of various hydrolysis procedures have been reviewed.[58]

The methylated sugars are usually separated from each other by column chromatography on cellulose, Celite, or a mixture of charcoal and Celite, or by thin-layer chromatography. A number of mixtures of methylated monosaccharides can also be separated by paper chromatography and electrophoresis. When the parent sugar of the methyl ether is known, the mobilities on paper can also yield information concerning the identity and degree of methylation of the substance. Methylated sugars can be de-methylated by treatment with boron trichloride for identification of the parent sugar.[58a]

Depending on the amount of material available, the methyl ethers can be identified by their physical properties or those of appropriate derivatives, or by mobilities on paper chromatography and comparison with known derivatives. These substances may also be separated and identified by gas chromatographic methods.[59]

Several examples wherein structural characteristics of polysaccharides were elucidated by methylation procedures will show the usefulness of this technique. Glycogen was converted into the triacetate and treated with methyl sulfate and potassium hydroxide in acetone solution.[60] After the methylation was repeated a number of times, the methoxyl content was almost 44%. The permethylated product was then methanolyzed with metha-nolic hydrogen chloride, and the product obtained after vacuum distillation was found to be methyl 2,3,6-tri-O-methyl-D-glucoside. Hydrolysis with aqueous acid yielded 2,3,6-tri-O-methyl-α-D-glucopyranose. This indicated that the glucose units in glycogen were (1 → 4)-linked. Subsequently, hydrolysis of permethylated glycogen was also found to yield 2,3,4,6-tetra-O-methyl-D-glucopyranose and 2,3-di-O-methyl-D-glucopyranose. These results, and succeeding studies by graded acid and enzymic hydrolysis, showed that glycogen was linked 1 → 4 and also contained branch points having (1 → 6) linkages. This series of reactions is shown schematically in Fig. 4. It should be noted that methylation studies with glycogen still left certain doubts with regard to some linkages because there were indications that the polysaccharide may not have been completely methylated.

Another methylation procedure can be illustrated by the early studies on corn starch.[61] The dried starch was stirred with anhydrous liquid ammonia at 35°, and metallic sodium and methyl iodide were added to the mixture. The reactants were added several times over a designated period. The ammonia was then removed by filtration, and the insoluble methylated poly-saccharide was dried. When this procedure was repeated several times, the methoxyl content of the product was found to be 45.1%. Hydrolysis of methylated starch yielded both 2,3,6-tri-O-methyl-D-glucose and 2,3-di-O-methyl-D-glucose, indicating a (1 → 4)-linked polyglucose chain with branches to some of the glucose units at the 6 position. Actually, starches

* References start on p. 410.

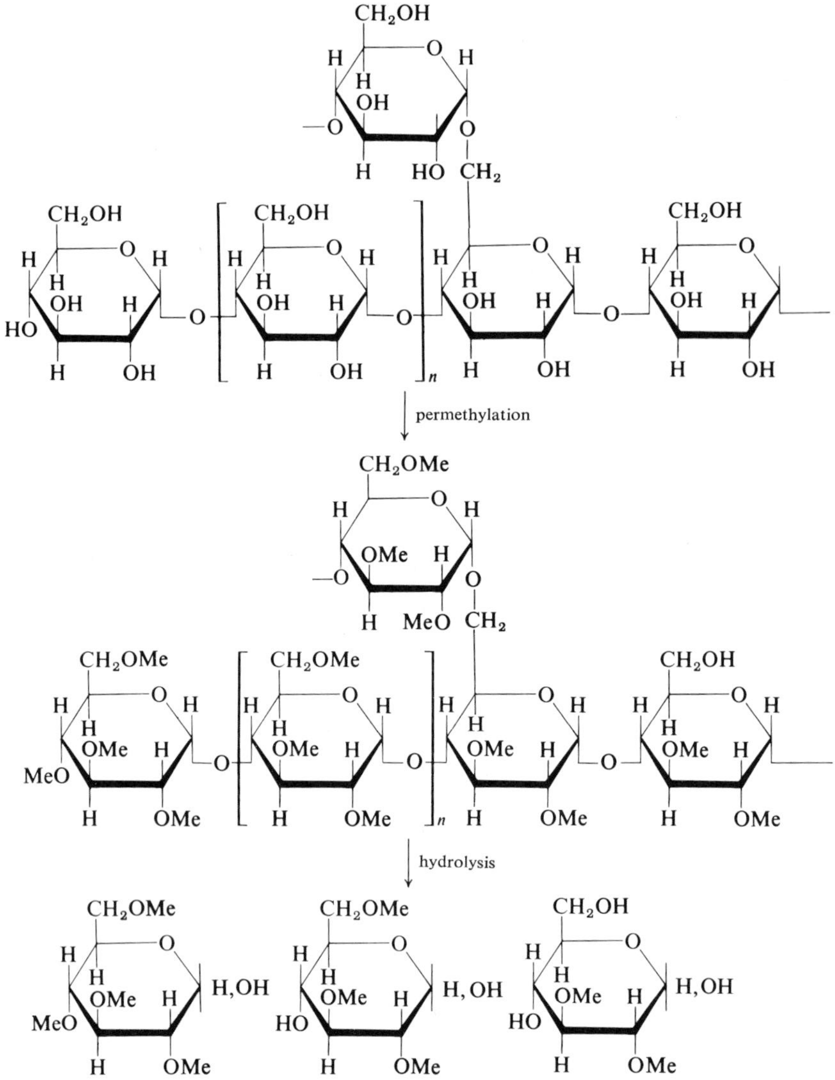

FIG. 4. Glycogen structure by methylation studies.

are mixtures of branched and unbranched polymers (see Vol. IIB, Chap. 38).

Methylation techniques have also been used to determine the structure of cellulose. The polysaccharide is treated with concentrated sodium hydroxide, suspended in toluene, and methylated with methyl sulfate and alkali.[62] Alternatively, the cellulose is first acetylated, dissolved in acetone, and methylated with the same reagents.[63] A third procedure is to dissolve partially methylated cellulose in N,N-dimethylformamide and methylate with methyl iodide and silver oxide.[64] Hydrolysis of the methylated product yields 2,3,6-tri-O-methyl-D-glucose, demonstrating a $(1 \rightarrow 4)$ linkage between the D-glucose residues [other evidence rules out a possible $(1 \rightarrow 5)$ linkage].

Of particular importance for showing the nature of the particular linkage in cellulose were acetolysis experiments. Under acetylating conditions in the presence of an acid catalyst, cellulose is degraded to cellobiose octaacetate.[65] Since cellobiose has a β-D-$(1 \rightarrow 4)$ link between the two D-glucose units, these results were important in establishing the configuration of the linkages in cellulose.

2. Partial Depolymerization

The above procedures provide a fair idea of the gross structure of the polysaccharide. The individual sugars that make up the polysaccharide are now known, and if the methylation analysis was successful the positions of the linkages are determined. If the substance contains only one type of monosaccharide, and if there is no branching, the one primary question that remains to be clarified is the configuration of the linkages—that is, α-D (or L) or β-D (or L). When the polysaccharide contains several different monosaccharides and if branching occurs within the chain, the sequence of the residues also has to be elucidated. One of the methods employed to obtain such informations is partial or graded hydrolysis to di- or oligosaccharides, which can be characterized completely.

The polysaccharide can be depolymerized by various procedures to break the weakest linkages. This is done with different concentrations of mineral acid, acetic acid (acetolysis), or specific enzymes. The products isolated under different conditions of hydrolysis frequently yield fragments of differing composition which can give critical information with regard to the monosaccharide sequence. For example, $(1 \rightarrow 6)$ linkages are more stable to mild hydrolysis with mineral acid than are $(1 \rightarrow 4)$ linkages. On the other hand, when treated with acetic anhydride containing about 5% of sulfuric acid (acetolysis), the $(1 \rightarrow 6)$ linkages are the more labile.

One example wherein the products of graded hydrolysis were important in resolving fundamental aspects of structure was in the study of glycogen. In one series of experiments, hydrolysis of rabbit liver glycogen with $0.05N$ sulfuric acid at 100° for 8 hours yielded D-glucose, maltose, isomaltose, and maltotriose, demonstrating the presence of both $(1 \rightarrow 4)$ and $(1 \rightarrow 6)$ linkages in glycogen.[66] Subsequently a hydrolyzate of glycogen from beef liver was separated by chromatography on a carbon column, to yield oligosaccharide fractions.[67] Acetylation and separation of the disaccharides with Magnesol–Celite yielded the octaacetates of maltose and isomaltose in addition to a minute amount of acetylated 3-O-α-D-glucopyranosyl-D-glucopyranose. Paper chromatographic separation of the trisaccharides yielded 6-O-α-D-glucopyranosyl-4-*O*-α-D-glucopyranosyl-D-glucopyranose (panose), isomaltotriose, and maltotriose. These experiments showed clearly that the principal branch points were composed of α-D-(1 → 6) linkages adjacent to α-D-(1 → 4) linkages. The isolation of isomaltotriose implies that glycogen contains some α-D-(1 → 6) linkages adjacent to each other. The concentration of the material undergoing fragmentation must be kept low (< 0.4%) to prevent formation of artifacts through recombination of sugars (acid reversion).[67a]

Various enzymes may also be employed to degrade polysaccharides to either mono- or oligosaccharide units. Thus, salivary α-amylase [(1 → 4)-α-D-glucan-4-glucanohydrolase] degrades starch to maltose and maltotriose; β-amylase [(1 → 4)-α-D-glucan maltohydrolase] cleaves α-D-(1 → 4)-linked glucose residues in a stepwise manner to yield maltose; and β-D-glucosiduronase (β-D-glucosiduronate glucosiduronate-hydrolase) hydrolyzes β-D-linked glucuronic acid from the nonreducing end of oligosaccharides. Other examples are α- and β-glucosidases (α- and β-D-glucoside glucohydrolase) from animal and plant sources: laminaranase [(1 → 3)-β-D-glucan glucanohydrolase], which hydrolyzes laminaran to 3-*O*-β-D-glucopyranosyl-D-glucose; or D-xylanase [(1 → 4)-β-D-xylan xylanohydrolase] and D-mannosidase (β-D-mannoside mannohydrolase) from plant tissue. These enzymes have a twofold application in that they may be employed for graded hydrolysis to specific units and, when their specificity is known, they may be used to determine the configuration of oligosaccharide linkages. Induced enzymes (Vol. IIA, Chap. 33) may have special application for this purpose.

3. Periodate Oxidation

The oxidation of monosaccharides by periodate has been discussed in Vol. IB, Chap. 25. This reagent reacts with polysaccharides to cleave the linkage between carbons carrying vicinal hydroxyl groups. Theoretically, for each linkage cleaved, one mole of periodate is consumed. Primary hydroxyl groups are oxidized to formaldehyde, and secondary hydroxyl groups give rise to formic acid.

The amount of periodate consumed and the products that are formed will depend on the position of the carbon atoms involved in the linkages with the neighboring residue. In a polysaccharide composed of hexopyranose residues, the reducing unit will consume four moles of periodate and yield four moles of formic acid, if it is linked through C-6 (Fig. 5). If it is linked through any other carbon atom, the reducing unit will consume approximately four moles of periodate (the "active hydrogen atom" of a malondialdehyde fragment is oxidized to a hydroxyl group with consumption of a mole of periodate).[67b] In addition to formic acid, formaldehyde will also be released, unless the linkage is to C-5 (Fig. 5).

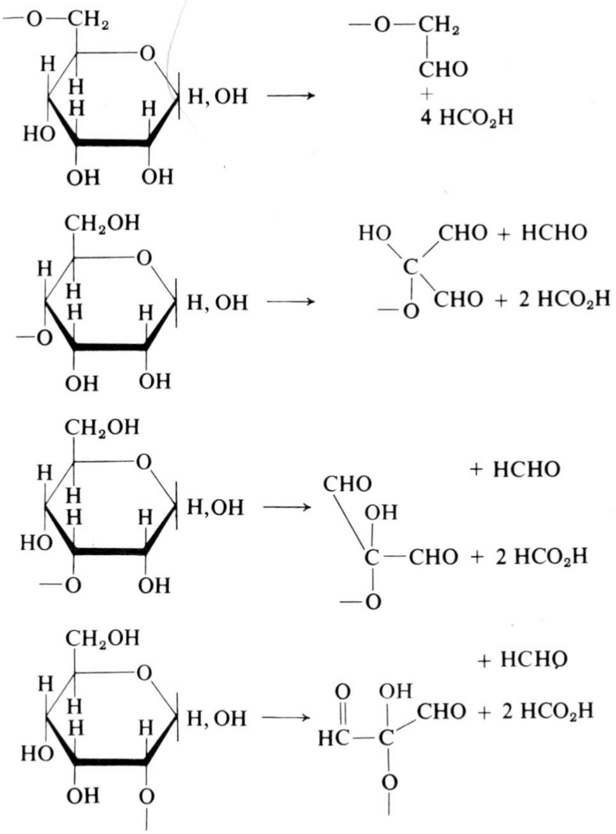

FIG. 5. Reaction of periodate with the reducing unit of a polysaccharide.

The nonreducing pyranose end units of a polysaccharide should each yield one mole of formic acid with the consumption of two moles of periodate (Fig. 6). The reaction of nonterminal hexopyranose residues of linear chains with periodate will also depend on the linkages involved. Residues connected by (1 → 2) or (1 → 4) linkages will react with one mole of periodate, whereas those linked through C-1 and C-3 will not be oxidized at all. Residues connected by (1 → 6) linkages will reduce two moles of periodate and also release a mole of formic acid (Fig. 7).

In a linear polysaccharide of hexopyranose units having (1 → 4) linkages, periodate oxidation should yield three moles of formic acid for each chain, two from the reducing end and one from the nonreducing end. If it can be assumed that oxidation will proceed according to theoretical expectations, the amount of formic acid released per weight of polysaccharide can thus be

* *References start on p. 410.*

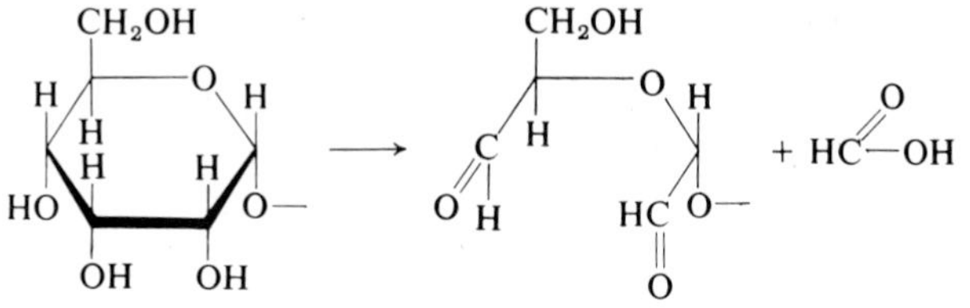

FIG. 6. Reaction of periodate with the terminal (nonreducing) unit of a polysaccharide.

a good indicator of the chain length. If the polysaccharide has a large number of (1 → 6) linkages, this method cannot be used for estimating chain length, since these residues will yield formic acid even when they are nonterminal (Fig. 7).

On the basis of the above-mentioned considerations, it should be possible to obtain structural information through periodate procedures by measuring the amount of periodate consumed and the quantity of formaldehyde and formic acid released. However, several problems have to be considered. The degree and rate of oxidation vary with temperature, pH, concentration of reactants, and a number of other factors. To gain any information by this procedure, the reaction has to be studied under a range of conditions. The most complicating phenomenon in periodate oxidation is overoxidation. This

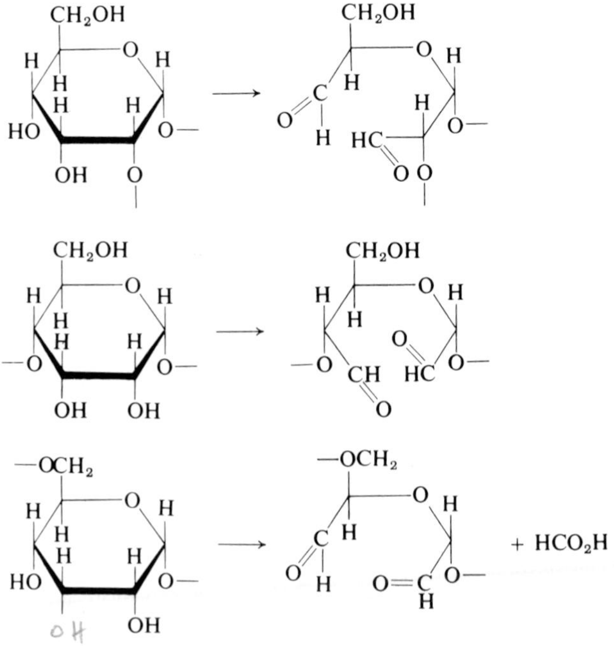

FIG. 7. Reaction of periodate with the nonterminal units of a polysaccharide.

is a slow, nonspecific oxidation of the polysaccharide even after all the vicinal glycol groups are cleaved. When measuring the amount of periodate consumed after different time intervals, it is, therefore, difficult to ascertain when the primary oxidation is complete. The usual procedure is to assume that this occurs when there is a sharp decrease in the rate of periodate reduction. However, this approach is not always reliable, because the ease and rate of oxidation is not the same for all types of vicinal hydroxyl groups, and a decrease in rate does not necessarily indicate that overoxidation is taking place. For example, *cis* vicinal hydroxyl groups in cyclic systems are much more reactive than those that are in a *trans* relationship. Also, the rates of oxidation of certain hydroxyaldehydes are comparatively slow.[68] If the primary oxidation leads to a malondialdehyde group, the active hydrogen atom of that group undergoes oxidation to a hydroxyl group with uptake of a further mole of periodate[67b] (see Vol. IB, Chap. 25).

Although a certain amount of information on polysaccharide structure

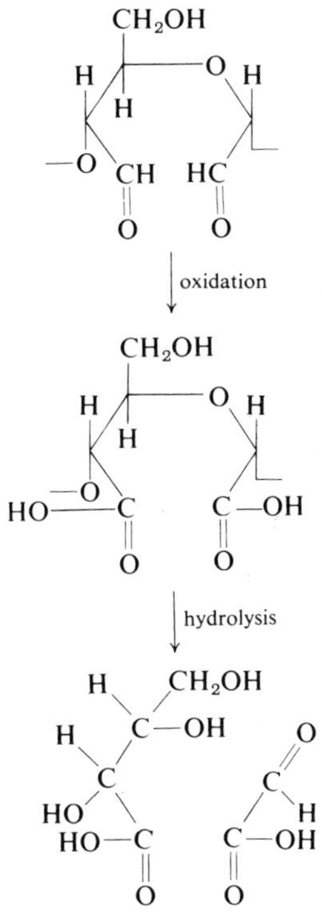

FIG. 8. Oxidation and hydrolysis of a "dialdehyde" obtained after treatment with periodate.

* *References start on p. 410.*

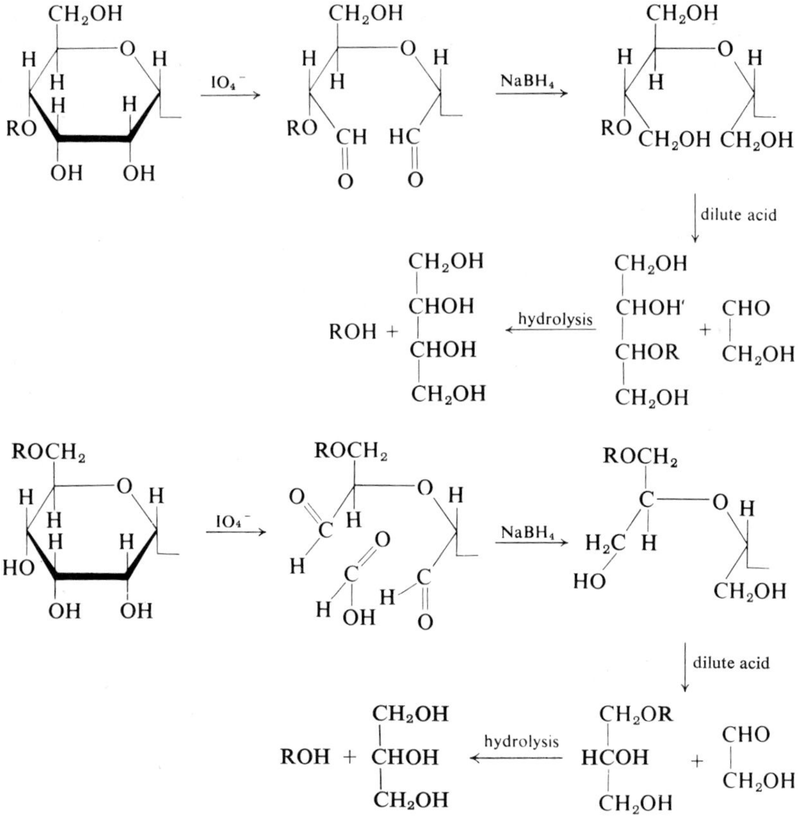

CH$_2$OH $\xrightarrow{\text{IO}_4^-}$ CH$_2$OH $\xrightarrow{\text{NaBH}_4}$ CH$_2$OH

(ring structures with RO, OH, OH; oxidized to dialdehyde with CH=O / HC=O; reduced to CH$_2$OH CH$_2$OH)

$\downarrow$ dilute acid

$$\text{ROH} + \begin{array}{l} \text{CH}_2\text{OH} \\ \text{CHOH} \\ \text{CHOH} \\ \text{CH}_2\text{OH} \end{array} \xleftarrow{\text{hydrolysis}} \begin{array}{l} \text{CH}_2\text{OH} \\ \text{CHOH}' \\ \text{CHOR} \\ \text{CH}_2\text{OH} \end{array} + \begin{array}{l} \text{CHO} \\ \text{CH}_2\text{OH} \end{array}$$

ROCH$_2$ $\xrightarrow{\text{IO}_4^-}$ ROCH$_2$ $\xrightarrow{\text{NaBH}_4}$ ROCH$_2$

(ring structures with HO, OH, OH; oxidized; reduced)

$\downarrow$ dilute acid

$$\text{ROH} + \begin{array}{l} \text{CH}_2\text{OH} \\ \text{CHOH} \\ \text{CH}_2\text{OH} \end{array} \xleftarrow{\text{hydrolysis}} \begin{array}{l} \text{CH}_2\text{OR} \\ \text{HCOH} \\ \text{CH}_2\text{OH} \end{array} + \begin{array}{l} \text{CHO} \\ \text{CH}_2\text{OH} \end{array}$$

FIG. 9. Smith degradation of polysaccharides (R=glycosyl).

may be gained by determining the amount of periodate reduced and formaldehyde and formic acid released, maximal data are obtained only by identification of the oxidation products of the remaining polymer. The "dialdehydes" that are formed are relatively unstable to the treatment required for polysaccharide analyses. It is, therefore, necessary either to oxidize or to reduce these groups prior to further manipulations.

The carbonyl groups resulting from periodate oxidation can be oxidized with bromine in a solution buffered with barium or strontium carbonate to yield the salt of corresponding acid.[69] For example, oxidation of the "dialdehyde" formed by periodate treatment of a (1 → 4)-linked polysaccharide followed by hydrolysis yields the acid shown in Fig. 8. Similarly, various periodate-oxidized residues from within the chain will yield corresponding acids from which the types of linkages in the original polymer may be deduced.

A procedure that has been applied extensively for structural analyses is the reduction of the aldehydic groups of the periodate-oxidized polysaccharide with sodium borohydride to the polyhydroxy compound (Smith degradation [70]). The resulting alcoholic derivative, which is a true acetal, is susceptible to hydrolysis with very dilute acid. On the other hand, the glycosidic linkage of a residue that did not suffer oxidation by periodate is comparatively stable to acid. As a result of this difference in stability of linkages, mild hydrolysis of a reduced, periodate-oxidized polysaccharide yields specific glycosides of oligosaccharides characteristic of the original compound. A $(1 \rightarrow 4)$-linked hexopyranose residue by this procedure yields glycolaldehyde and erythritol, whereas a $(1 \rightarrow 6)$-linked residue gives rise to glycerol and glycolaldehyde. The rationale for this is shown in Fig. 9.

Oligosaccharides obtained by these procedures may again be oxidized, reduced, and hydrolyzed to yield smaller units. This series of reactions can thus be employed for controlled degradation and structure determination of polysaccharides. Examples of such studies are given in this volume, Chap. 37 and Vol. IIB, Chap. 42. A variant of the method applicable on a nanomole scale has been described.[70a]

4. *Oxidation with Lead Tetraacetate*

Lead tetraacetate reacts with vicinal glycols in a manner similar to periodate. The reaction is generally carried out in acetic acid solutions and is apparently catalyzed by sodium or potassium acetate. The oxidation of α-hydroxycarboxylic acids proceeds with ease when lead tetraacetate is the oxidant but only very slowly or not at all with periodate.

By controlled degradations with lead tetraacetate and reduction with sodium borohydride, it is possible to degrade disaccharides to products in which all the asymmetric centers of the reducing unit are destroyed, leaving a fragment wherein the configuration of only the anomeric carbon atom on the nonreducing unit is in question.[71] Reducing units are oxidized most easily, and disaccharides may therefore be oxidized in discrete steps. With maltose (Fig. 10), the reducing end is first oxidized to the formate ester of a four-carbon aldehyde. When the latter is hydrolyzed and reduced with sodium borohydride, a glycoside derivative of erythritol is obtained. Further oxidation with lead tetraacetate followed by reduction yields the 2-*O*-glycosylglycerol. Thus, all the asymmetric centers of the reducing sugar are destroyed, leaving only the asymmetry of the original nonreducing end.

These products are crystalline or can be prepared as crystalline benzoates. Authentic derivatives of the same substances can be prepared either from disaccharides of known configuration or by synthesis.

This procedure may be applied to disaccharides obtained after graded hydrolysis in order to determine the configuration of the glycosidic linkages.

* *References start on p. 410.*

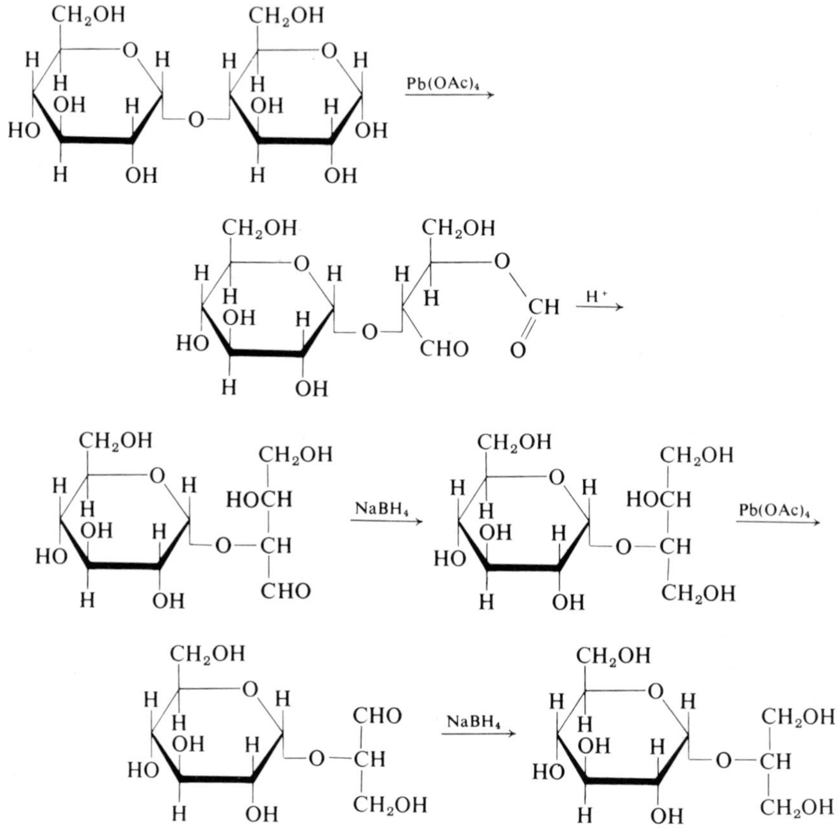

FIG. 10. Controlled degradation of maltose with lead tetraacetate.

An example where this approach was used was with the disaccharide 2-O-D-xylopyranosyl-L-arabinose obtained on partial hydrolysis of xylans. The negative rotation of the 2-O-D-xylopyranosylglycerol finally isolated (Fig. 11) proved that the xylose was connected by a β-D linkage to the arabinose.

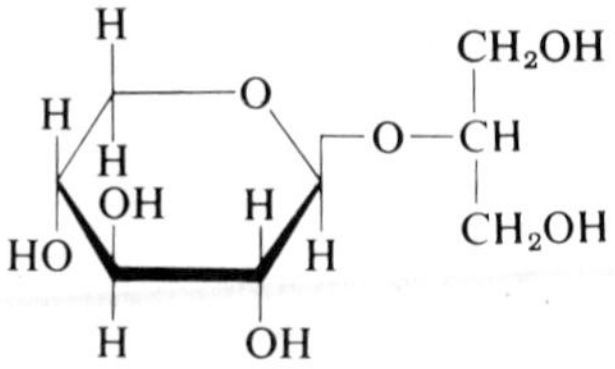

FIG. 11. 2-O-D-xylopyranosylglycerol.

5. *Barry Degradation*

The Barry degradation is a procedure that may be employed for studies of the sequence or arrangement of individual units within the polysaccharide chain.[72] When a "dialdehyde" formed by oxidation of a sugar residue is treated with phenylhydrazine in dilute acetic acid, it forms a complex with an equimolar amount of this reagent. This product, on heating with a mixture of phenylhydrazine and acetic acid, is degraded to osazones and the remaining unoxidized oligosaccharide[73,74] (Fig. 12). Thus, by combining periodate or lead tetraacetate oxidation with subsequent phenylhydrazine treatment, it is possible to remove different sugar residues stepwise. By identifying the osazones and the unoxidized portions of the molecule, many aspects of the fine structure, such as branch points, can be elucidated.

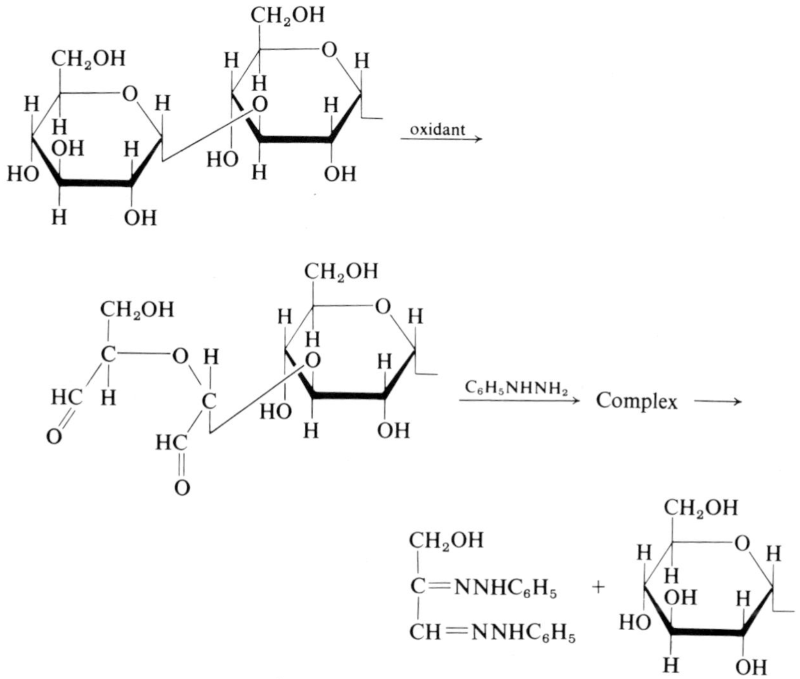

FIG. 12. Barry degradation of a polysaccharide.

6. *Immunochemical Methods*

A number of polysaccharides have been shown to elicit the formation of antibodies when they are injected into animals. Although the primary reaction of the antibody is with the antigen used for immunization, reactions

* References start on p. 410.

also occur with other polysaccharides if they possess structural units or groups in common with the antigen. These are termed cross reactions. Thus, type III pneumococci will cross-react with type VIII presumably because of the structural similarity between their capsular polysaccharides. Since antigenic activity is due only to the antibody-combining site of the polymer, cross reactions may occur between two structurally different polysaccharides as long as they have common antigenic groupings.

Advantage has been taken of this phenomenon to gain evidence as to the sugar units and types of linkages in polysaccharides of undefined structure. If a substance cross-reacts with a polysaccharide that has certain structural units, the assumption is made that it also contains these structures.[75,76]

Immunochemical methods can be employed both in determination of homogeneity of a polysaccharide after purification and in determinations of structure. By immunodiffusion techniques[77] it is possible to determine the purity of a polysaccharide during isolation. This is done by preparing an antiserum to the crude preparation[78] and carrying out immunodiffusion studies against this antiserum with products after different stages of purification. As the purity increases, the number of bands decreases.

Cross-reaction studies of polysaccharides and their degradation products with polysaccharides of known structure are very useful in gaining evidence with regard to structure. This subject is discussed more fully in Vol. IIB, Chapters 41 and 43.

REFERENCES

1. R. L. Whistler and C. L. Smart, "Polysaccharide Chemistry," Academic Press, New York, 1953; W. Pigman and R. M. Goepp, Jr., "Chemistry of the Carbohydrates," Academic Press, New York, 1948; D. Horton and M. L. Wolfrom, in "Comprehensive Biochemistry," M. Florkin and E. H. Stotz, Eds., Vol. 5, Elsevier, Amsterdam, 1963, Chapter VII.

1a. W. Pigman, W. Hawkins, E. Gramling, S. Rizvi, and H. L. Holley, *Arch. Biochem. Biophys.*, **89**, 184 (1960).

2. R. L. Whistler and J. N. BeMiller, *Advan. Carbohyd. Chem.*, **13**, 289 (1958).

3. G. O. Aspinall, *Methods Carbohyd. Chem.*, **5**, 397 (1965).

3a. J. Fabianek, A. Herp, and W. Pigman, *Arch. Int. Physiol. Biochim.*, **71**, 647 (1963).

4. R. W. G. Wyckoff, *Advan. Protein Chem.*, **6**, 1 (1951).

5. N. Gellerstedt, *Arkiv Kemi*, **20**, 147 (1963).

6. F. A. Bettelheim and D. E. Philpott, *Biochim. Biophys. Acta*, **34**, 124 (1959).

7. L. V. Azaroff and M. J. Buerger, "The Powder Method in X-Ray Crystallography," McGraw Hill, New York, 1958.

8. R. E. Rundle, *J. Amer. Chem. Soc.*, **69**, 1769 (1947).

9. J. Gundermann, *J. Phys. Chem.*, **B37**, 387 (1937).

10. K. J. Palmer and M. B. Hertzog, *J. Amer. Chem. Soc.*, **67**, 2122 (1945).

11. F. A. Bettelheim, *J. Phys. Chem.*, **63**, 2069 (1959); F. A. Bettelheim, *Biochim. Biophys. Acta*, **83**, 350 (1964).

12. R. St. J. Manley, *J. Polym. Sci.*, **A2**, 4053 (1964).
13. Y. Yamashita, *J. Polym. Sci.*, **A3**, 3251 (1965).
14. C. Y. Liang and R. H. Marchessault, *J. Polym. Sci.*, **37**, 385 (1959); **39**, 269 (1959); **43**, 71, 85, 101 (1960); **59**, 357 (1962).
15. G. R. Bird, *J. Chem. Phys.*, **28**, 1155 (1958).
16. A. Wada and S. Kozawa, *J. Polym. Sci.*, **A2**, 853 (1964).
17. V. F. Rao and J. F. Foster, *Biopolymers*, **1**, 527 (1963).
18. P. J. Flory, "Principles of Polymer Chemistry," Cornell Univ. Press, Ithaca, New York, 1953.
19. Z. Alexandrowicz, *J. Polym. Sci.*, **56**, 97, 115 (1962); Z. Alexandrowicz and A. Katchalsky, *ibid.*, **A1**, 3231 (1963).
20. A. L. Potter and W. Z. Hassid, *J. Amer. Chem. Soc.*, **70**, 3774 (1948).
21. A. M. Holtzer, H. Benoit, and P. Doty, *J. Phys. Chem.*, **58**, 624 (1954).
22. D. A. Burge, *J. Phys. Chem.*, **67**, 2590 (1963).
23. I. C. McNeill, *Polymer*, **4**, 247 (1963).
24. K. A. Stacey, "Light Scattering in Physical Chemistry," Academic Press, New York, 1956.
25. C. I. Carr and B. H. Zimm, *J. Chem. Phys.*, **18**, 1616, 1624 (1950).
26. S. R. Erlander and D. French, *J. Amer. Chem. Soc.*, **80**, 4413 (1958).
27. F. A. Bettelheim, Y. Hashimoto, and W. Pigman, *Biochim. Biophys. Acta*, **63**, 235 (1962).
28. T. C. Laurent, M. Ryan, and A. Pietruszkiewicz, *Biochim. Biophys. Acta*, **42**, 476 (1960).
29. H. K. Schachman, "Ultracentrifugation in Biochemistry," Academic Press, New York, 1959.
30. D. A. Yphantis, *Biochemistry*, **3**, 297 (1964).
31. W. Banks and C. T. Greenwood, *Advan. Carbohyd. Chem.*, **18**, 357 (1963).
32. C. Tanford, "Physical Chemistry of Macromolecules," Wiley, New York, 1961.
33. H. Morawetz, "Macromolecules in Solution," Wiley (Interscience), New York, 1965.
34. E. H. Immergut and F. R. Eirich, *Ind. Eng. Chem.*, **45**, 2500 (1953).
35. S. K. Chanda, E. L. Hirst, J. K. N. Jones, and E. G. V. Percival, *J. Chem. Soc.*, 1289 (1950).
36. C. T. Bishop, G. A. Adams, and E. O. Hughes, *Can. J. Chem.*, **32**, 999 (1954).
37. A. J. Erskine and J. K. N. Jones, *Can. J. Chem.*, **34**, 821 (1956).
38. H. Meier, *Acta Chem. Scand.*, **12**, 144 (1958); **14**, 749 (1960).
39. J. E. Scott, *Methods Biochem. Anal.*, **8**, 145 (1960).
40. H. O. Bouveng and B. Lindberg, *Acta Chem. Scand.*, **12**, 1973 (1958).
41. M. Heidelberger, *Ann. Rev. Biochem.*, **25**, 641 (1956).
42. J. A. Cifonelli, R. Montgomery, and F. Smith, *J. Amer. Chem. Soc.*, **78**, 2485 (1956).
43. M. G. Sevag, *Biochem. Z.*, **273**, 419 (1934).
44. A. S. Markowitz and J. R. Henderson, *Nature*, **181**, 771 (1958).
45. H. Neukon, H. Deuel, W. J. Henri, and W. Kundig, *Helv. Chim. Acta*, **43**, 64 (1960).
46. N. R. Ringertz and P. Perchard, *Acta Chem. Scand.*, **13**, 1467 (1959).
47. D. H. Northcote, *Biochem. J.*, **58**, 353 (1954).
48. B. A. Lewis and F. Smith, *J. Amer. Chem. Soc.*, **79**, 3929 (1957).
49. I. A. Preece and R. Hobkirk, *Chem. Ind. (London)*, 257 (1955).
50. B. J. Hocevar and D. H. Northcote, *Nature*, **179**, 488 (1957).
51. W. N. Haworth, *J. Chem. Soc.*, **107**, 8 (1915).
52. T. Purdie and J. C. Irvine, *J. Chem. Soc.*, **83**, 1021 (1903).

53. I. E. Muskat, *J. Amer. Chem. Soc.*, **56**, 693 (1934).
54. C. M. Fear and R. C. Menzies, *J. Chem. Soc.*, 937 (1926).
55. K. Wallenfels, G. Bechtler, R. Kuhn, H. Trischmann, and H. Egge, *Angew. Chem. Int. Ed.*, **2**, 515 (1963).
56. S. Hakomori, *J. Biochem.* (*Tokyo*), **55**, 205 (1964).
57. D. M. W. Anderson and G. M. Cree, *Carbohyd. Res.*, **2**, 63 (1966).
58. H. O. Bouveng and B. Lindberg, *Advan. Carbohyd. Chem.*, **15**, 53 (1960).
58a. T. G. Bonner, E. J. Bourne, and S. McNally, *J. Chem. Soc.*, 2929 (1960).
59. C. T. Bishop, *Methods Biochem. Anal.*, **10**, 1 (1962).
60. W. N. Haworth, E. L. Heist, I. J. Webb, *J. Chem. Soc.*, 2479 (1929).
61. J. E. Hodge, S. A. Karjala, and G. E. Hilbert, *J. Amer. Chem. Soc.*, **73**, 3312 (1951).
62. W. S. Denham and H. Woodhouse, *Cellulosechemie*, **1**, 22 (1920).
63. W. N. Haworth, E. L. Hirst, L. N. Owen, S. Peat, and F. J. Averill, *J. Chem. Soc.*, 1885 (1939).
64. R. Kuhn, H. Trischmann, and I. Löw, *Angew. Chem.*, **67**, 32 (1955).
65. H. Hebbert and E. G. V. Percival, *J. Amer. Chem. Soc.*, **52**, 3995 (1930).
66. M. L. Wolfrom, E. N. Lassetre, and A. N. O'Neill, *J. Amer. Chem. Soc.*, **73**, 595 (1951).
67. M. L. Wolfrom and A. Thompson, *J. Amer. Chem. Soc.*, **79**, 4212 (1957).
67a. A. Thompson, M. L. Wolfrom, and E. J. Quinn, *J. Amer. Chem. Soc.*, **75**, 3003 (1953); M. L. Wolfrom, K. Anno, and M. Inatome, *ibid.*, **76**, 1309 (1954).
67b. C. F. Huebner, S. R. Ames, and E. C. Bubl, *J. Amer. Chem. Soc.*, **68**, 1621 (1946).
68. J. W. Platt, N. K. Richtmyer, and C. S. Hudson, *J. Amer. Chem. Soc.*, **74**, 2200 (1952).
69. E. L. Jackson and C. S. Hudson, *J. Amer. Chem. Soc.*, **58**, 378 (1936); **59**, 994 (1937).
70. M. Abdel-Akher *et al.*, *J. Amer. Chem. Soc.*, **74**, 4970 (1952); J. K. Hamilton, G. W. Huffman, and F. Smith, *ibid.*, **81**, 2176 (1959).
70a. P. Nánási and A. Lipták, *Carbohyd. Res.*, **10**, 177 (1969).
71. A. J. Charlson, P. A. J. Gorin, and A. S. Perlin, *Can. J. Chem.*, **34**, 1811 (1956); **35**, 365 (1957).
72. V. C. Barry, *Nature*, **152**, 538 (1943).
73. V. C. Barry, J. E. McCormick, and P. W. D. Mitchell, *J. Chem. Soc.*, 3692 (1954); 4020 (1954).
74. P. S. O'Colla, *Methods Carbohyd. Chem.*, **5**, 382 (1966).
75. M. Heidelberger, "Lectures in Immunochemistry," Academic Press, New York, 1956.
76. M. Heidelberger, *Proc. Chem. Soc.*, 153 (1961).
77. A. J. Crowles, "Immunodiffusion," Academic Press, New York, 1961.
78. G. Schiffman, in "Methods in Enzymology: Complex Carbohydrates," S. P. Colowick and N. O. Kaplan, Eds., Vol. VIII, Academic Press, New York, 1966, p. 79.

36. CELLULOSE, LICHENAN, AND CHITIN

K. WARD, JR. AND P. A. SEIB

I.	Introduction and Chemical Structure	413
II.	Occurrence.	415
III.	Enzymic Synthesis	415
IV.	Isolation	417
V.	Molecular Weight and Fine Structure	418
VI.	Sorption, Swelling, and Solution	420
VII.	Degradation	422
	A. Enzymolysis	422
	B. Hydrolysis	426
	C. Oxidation	426
	D. Miscellaneous Degradations	428
VIII.	Substitution Reactions.	429
	A. Reactivity in General	429
	B. Esters	430
	C. Ethers	432
	D. Cross-linking.	434
IX.	Graft Polymers	434
X.	Lichenan and Isolichenan	435
XI.	Chitin	435
	References	438

I. INTRODUCTION AND CHEMICAL STRUCTURE

Cellulose is a particularly important carbohydrate because of its great abundance and distribution (it is the most abundant of all organic compounds) and because of its very useful properties. These unique properties, arising largely as a result of its chemical structure, are familiar in the characteristics of cellulose products encountered in everyday life.

Cellulose is a linear polymer of D-glucose residues joined by β-D-$(1 \to 4)$-linkages. The molecular chain length is high, and the hydrogen-bonding capacity of the three hydroxyl groups is great, so that cellulose forms fibers of remarkable strength. It is the basis of all the vegetable fibers.

The component residues of the cellulose molecule consist of D-glucose in

413

the pyranose form and in the *C1* (D) chair conformation. The size and arrangement of the molecules in the natural state will be discussed later. The molecular structure itself is as shown in Fig. 1. The average number of D-glucose residues, usually represented by *n*, is several thousand in native cellulose.

Many investigators have interpreted various phenomena as indications of possible irregularities in the simple structure of cellulose as a β-D-$(1 \rightarrow 4)$-linked glucan. Such proposed irregularities have included carboxyl groups,[1] other sugar residues such as D-mannose[2] and D-xylose,[3,4] cross-linking,[5,6] occasional ester linkages[7] and other easily hydrolyzable linkages,[8] and occasional branching of chains.[9] Such slight and random irregularities

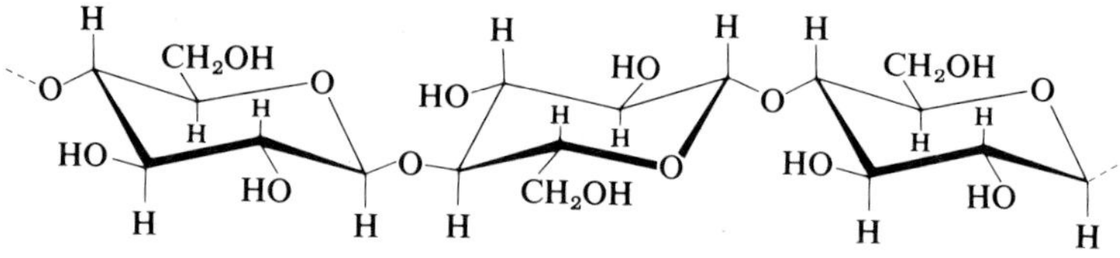

FIG. 1. Cellulose molecule.

apparently do occur in natural polymers; provided that they occur very seldom, and at random, they are not considered to constitute separate molecular species any more than do the different molecular sizes into which most carbohydrate polymers can be fractionated.[10] In the case of cellulose, some of these proposals have been shown to be untenable and others to have been due to some extraneous material in the sample, for cellulose, like many other natural products of high molecular weight, is difficult to purify and is easily modified during purification, especially by oxidation. In the remaining proposals the irregularity constitutes only a very small proportion of the total structure, and its effect on the properties of cellulose is practically negligible.

The preceding remarks apply to isolated cellulose. We cannot be entirely certain of the situation with regard to cellulose in the native state. In wood, the most common source, cellulose occurs in intimate association with lignin and the hemicelluloses. There is a growing body of evidence to indicate that lignin is to some extent chemically bound with the hemicelluloses,[11-13] and hypotheses are not lacking that some similar bonding exists with cellu-lose.[11] There is, however, no direct evidence that anything stronger than hydrogen bonding exists or that anything more than occlusion or sorption is necessary to explain the difficulty of purifying cellulose.

Cellulose in its native state usually exists as strong, tough fibers. Purified, it retains its fibrous nature and is colorless. It can be dissolved, as discussed later, and regenerated to form transparent fibers. An excellent review of cellulose is the three-volume work of Ott, Spurlin, and Grafflin.[14] The X-ray structure[14a] and conformation[14b] of cellulose have been reviewed.

II. OCCURRENCE

Cellulose occurs throughout the vegetable kingdom as a constituent of the cell wall, and in many plants it is the principal cell wall constituent.[15] As the structural component in woody plants and many others, it may be considered the backbone of the vegetable world.

Even the commercially useful sources of cellulose form a long list,[16] headed by the woods, both hardwoods and softwoods. Cellulose occurs in every fibrous part of the plant; the seed hairs of the cotton plant constitute the purest major source in Nature. The grasses have also been important as sources of cellulosic materials—papyrus, bamboo, and esparto, for instance. Many agricultural residues from which cellulosic pulps have been derived also belong in this botanical group. Sugar cane and sorghum bagasse, cornstalks, and straws from rye, wheat, oats, rice, and seed-flax are examples. Bast fibers such as flax, hemp, jute, and ramie are cellulosic, and so are the leaf fibers—for example, abaca, sisal, and henequen. Such miscellaneous raw materials as Spanish moss, extracted roots, marine plants, and coir fiber from coconut husks have also been used. Even bacterial cellulose has been considered.

Cellulose is also found in the mineral and animal kingdoms. It is relatively stable, and its fibers have been isolated from peat,[17] lignite,[18] and other fossilized vegetable matter.[19] Tunicin, a component of the outer mantles of the tunicates, which are marine organisms, has been shown to be identical with cellulose.[20,21] A cellulose–protein complex has also been reported as a normal, although minor, constituent of mammalian connective tissue, including certain human tissues.[21]

III. ENZYMIC SYNTHESIS

Plants are the major source of cellulose, but investigation of cellulose biosynthesis in plant cells is difficult, mainly because of problems associated with the formation of thick cell walls along with a variety of heterogeneous polysaccharides.[22] Consequently, many workers have employed bacteria of genus *Acetobacter*, which produce a relatively pure, extracellular cellulose.[23]

* References start on p. 438.

These studies have given some insight into the enzymic pathway to bacterial cellulose and have resulted in a working hypothesis for the biosynthesis of cellulose in plants.[24]

In vivo investigations with D-glucose specifically labeled with carbon-14 have shown that the carbon skeleton of this hexose is incorporated intact into cotton cellulose.[24a] Whether the same is also true for bacterial cellulose is a matter of controversy.[24a,24b] Cell-free synthesis of cellulose has been successful only when D-glucose is added in activated form to enzyme systems. The discovery that glycosyl esters of nucleotides are active in carbohydrate synthesis[25] led to the isolation[26,27] of a particulate preparation from *Acetobacter xylinum* that catalyzes the transfer of a D-glucopyranosyl moiety from uridine 5'-(α-D-glucopyranosyl-^{14}C pyrophosphate) to a cellodextrin "primer." The radioactive, alkali-insoluble product is not present as microfibrils,[28] but results of its enzymolysis and hydrolysis are consistent with formulation as a β-D-$(1 \rightarrow 4)$-linked glucan.

A different substrate and extracellular enzyme(s) have been isolated from active cultures of *Acetobacter*.[29] When the two water-soluble preparations were mixed, typical cellulose microfibrils were produced. The cellulose precursor, extracted from *Acetobacter xylinum* cells with 80% ethanol, was purified by column chromatography[30] and was reported to contain D-glucose and a lipid identified[24] as a long-chain polyhydroxylic alcohol.

It is probable that the biochemical pathway to cellulose in *Acetobacter xylinum* involves both a sugar nucleotide and D-glucolipid. The hypothesis is summarized as stated by Neufeld and Hassid:[31] "(*a*) inside the bacterial cell, D-glucopyranosyl residues are transferred from uridine 5'-(α-D-glucopyranosyl pyrophosphate) to a lipid; (*b*) the resulting D-glucolipid passes through the cellular membrane into the external medium, where it contributes its D-glucopyranosyl residues to form chains of cellulose; and (*c*) the lipid moiety returns to the cell, and the cycle is repeated."

Cellulose synthesis in higher plants may proceed by a mechanism that is quite different from that in *Acetobacter xylinum*. Cell-free preparations from several plant sources, including cotton bolls[32] and seedlings[33] of mung bean, string bean, squash, pea, and corn were found to catalyze synthesis of cellulose. The reactions were reported initially to be completely specific for guanosine 5'-(D-glucopyranosyl pyrophosphate) as the D-glucosyl donor. These plants are capable of synthesizing the nucleotide sugar from the common intermediary metabolites α-D-glucopyranosyl phosphate and guanosine triphosphate.[34] More recent findings,[35,35a] however, indicate that uridine 5'-(D-glucopyranosyl pyrophosphate) can also serve as a D-glucosyl donor in plants. No D-glucolipid is extractable from bean seedlings,[36] nor does the addition of mung bean lipid extracts stimulate production of cellulose in its particulate enzyme system.[33] There is a possibility, however, that plants contain a certain concentration of D-glucolipid.[36,36a] (See this volume, Chap. 34 for details on biosynthesis by plant enzymes.)

Incorporation of D-glucose from sugar nucleotides into cellulose is usually[36b] less than 20%. This observation, together with kinetic evidence,[24a,36c] supports Marx-Figini's hypothesis that biosynthesis of cellulose in higher plants can be divided into two stages. In the first stage the primary layer of the cell wall is formed slowly by an enzyme that employs nucleoside sugar pyrophosphates, while in the second stage the bulk of the cellulose present is laid down rapidly in the secondary wall. Furthermore, the second stage is thought to proceed by way of a template mechanism, since the degree of polymerization of cellulose formed during this period is independent of the rate of polymerization,[36c] and since the molecular-weight distribution of the cellulose is narrow.[24a] Marx-Figini has proposed that the enzyme system for cellulose synthesis in the secondary layer may not be the same as that in the primary layer, and that Hassid's cell-free biosynthesis[32] of cellulose may correspond to only the first stage.

A complete description of cellulose biosynthesis would include information on the processes controlling the growth and arrangement of the microfibrils that are universal entities in cellulose. A number of postulates have been made[28,37] in this area, but experimental confirmation has been limited.

IV. ISOLATION

The isolation of cellulose from wood is as difficult, or more so, than its isolation from most of its other sources. The pulping processes for wood can be taken as a typical isolative procedure.[38] The methods for other vegetable materials are usually milder variants.

There are two major processes in use—alkaline and acid pulping, if the pulping processes are carried far enough to isolate true cellulose. Other processes remove only a part of the noncellulosic constituents.

The simplest alkaline process consists of the digestion of wood chips with aqueous sodium hydroxide of about 5% concentration at 160° to 180°. The sulfate process, however, is more common; in this process a part of the alkalinity is in the form of sodium sulfide. The name "sulfate" arises from the fact that sodium sulfate is used in preparing the cooking liquor. The process is also called the kraft process from the high strength (German *Kraft*) of the pulp.

The main acid-pulping process is the sulfite process. The cooking liquor in this process is a solution of sulfur dioxide in aqueous calcium bisulfite. Liquors based on sulfites other than that of calcium are also used, and a wide range of pH conditions have been tried out. The temperature used in

* References start on p. 438.

this process is somewhat lower, very rarely above 150°. Dilute nitric acid has also been used for pulping at still lower temperatures, below 100°.

A number of variations and combinations of these methods are used for preparing paper pulps, but they are not pertinent to the isolation of the cellulose component. In paper manufacture it is good economy to keep the yield high, and it is particularly desirable to retain the hemicelluloses, since they improve the properties of the paper sheet. Pulps that are manufactured for chemical conversion must, however, be nearly pure cellulose, and their preparation illustrates the industrial processes used for isolating cellulose free from hemicelluloses.

For the manufacture of dissolving pulps, pulp from the acid processes is usually subjected to various chlorine, hypochlorite, and chlorine dioxide bleaching sequences[39] followed by alkali treatments, either with dilute sodium hydroxide[40] (2% or less in concentration) at 100° to 160°, or with sodium hydroxide of roughly mercerizing strength (5 to 25%) at room temperature or slightly higher.[41] Ordinary kraft pulps do not lend themselves to these treatments because the hemicelluloses are not removed, but kraft pulps may be used if the wood is subjected to a preliminary hydrolysis before the "sulfate" process, in which case the hemicelluloses are removed by the later treatments.

A number of ways can be used in the laboratory to isolate cellulose from vegetable matter. They have been excellently summarized by Browning.[42] The material used should first be freed of organic-soluble extractives by extraction with a suitable solvent. The isolation methods are essentially delignifications and must usually be followed by extraction with cold alkali to obtain the pure cellulose. Many delignification methods are effected with chlorine. Successive treatments with chlorine, alternated by extractions with alkaline alcoholic solutions or with sodium sulfite, have been widely used. Chlorine dioxide or acidified chlorite solutions also yield holocelluloses (cellulose plus hemicellulose). Nitric acid can be used to remove both lignin and hemicelluloses. 2-Aminoethanol and hypochlorite can also be used to remove lignin and much of the hemicellulose.

V. MOLECULAR WEIGHT AND FINE STRUCTURE

Although the average number of D-glucose residues (degree of polymerization, or d.p.) in cellulose as it occurs in Nature is not known, in some cases it may exceed 10,000 or even 15,000, for d.p. values in this range have been found.[43,44] Isolation probably always degrades the cellulose, and the molecular-weight distributions of isolated celluloses are usually wide.[45,46] These distributions may also have been changed during isolation, but all cellulose appears to be polymolecular.

The strong tendency for cellulose chains to form intermolecular and intramolecular hydrogen bonds is responsible for the characteristic molecular arrangement in the vegetable fibers.[47] Cellulose chains are aligned to form microfibrils having the chains parallel to the microfibrillar axis.[14a,14b] The microfibrils are similarly aligned in the fiber, although the orientation here is usually at an angle with the fiber axis.[15] There are difficulties in estimating the diameter of the microfibril, and there are variations with the source of the material, but recent measurements[48,49] indicate the diameter to be 35 Å, although some estimates[50,51] are as low as 16 Å. There is an apparent ellipticity of the cross section. Some investigators prefer to call this unit an "elementary fibril"[52] or "micelle string"[47] and reserve the term "microfibril" for a composite of about 200 Å in diameter.

The arrangement of molecules within the microfibril is regular enough to give an X-ray diffraction pattern, which is essentially the same for cellulose from almost every natural source.[14a] Cellulose is polymorphic, however, and exhibits a different X-ray pattern if it is regenerated from a derivative or from solution. The crystalline arrangement of native cellulose is called cellulose I, and that of regenerated cellulose (which has been found in Nature only in certain marine organisms) is cellulose II. Under certain conditions of regeneration or treatment, variants have been described:[53] cellulose III and cellulose X, which resemble cellulose II, and cellulose IV, which resembles cellulose I.

The X-ray patterns of cellulose are not as distinct as those of crystals of nonpolymers; there is a great deal of background scattering, which has been interpreted as indicating a measure of disorder in the molecular arrangement. Various methods have been suggested for evaluating the degree of order from X-ray data.[53] Other evidence as to variations in molecular order within a cellulose fiber and between different fibers are found on examination of sorption data,[54] density,[54] dielectric properties,[55] and chemical reactivity,[56] to mention a few. The degrees of order calculated from any of these measurements are symbatic with each other, but the exact relationships are not known, and opinions differ about the ultimate details of the submicroscopic structure of cellulose.[14a]

The Meyer–Misch model[57] for the structure of cellulose I, based on ramie cellulose, is given in Fig. 2. Modifications have been proposed to agree better with all the X-ray reflections, and the hydrogen bonding that is indicated by infrared spectroscopic measurements; Frey–Wyssling has summarized some of these.[52]

* References start on p. 438.

Many suggestions for modifications of the unit cell have been offered, but there is even more disagreement about the overall molecular arrangement.[58] No one structure seems to reflect all the evidence. For many years, the fringed micelle, in which regions of good order (crystallites) and poor order (noncrystalline regions) alternate along the microfibril, has been a popular model, explaining as it does the fact that the crystallite length, estimated in various ways, is much shorter than the length of a molecule.[59] Other recent proposals include the fringed fibril,[58] in which the whole microfibril is

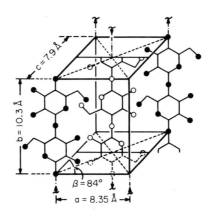

Fig. 2. Unit cell of cellulose. [K. H. Meyer and L. Misch, *Helv. Chim. Acta*, **20**, 232 (1937).]

essentially crystalline and the disorder arises from the sides of the micro-fibril, and possible interfibrillar, molecular connections. Various types of folded-chain systems, in which the folds constitute the disordered regions, also have been suggested.[49,60]

VI. SORPTION, SWELLING, AND SOLUTION

Cellulose has a great affinity for water, although it is insoluble. When dry, it absorbs moisture from the air, eventually reaching an equilibrium with the atmosphere. The proportion adsorbed increases as the relative humidity is progressively increased. If adsorption is carried to saturation and the relative humidity is then progressively decreased, the proportion of moisture sorbed decreases, but the new values at any given relative humidity are a little higher than for the adsorption curve. This "hysteresis loop" is shown in Fig. 3; the explanation of the hysteresis is based on the interconversion of cellulose–water and cellulose–cellulose hydrogen bonds.[61] During desorption, many bonds between cellulose and water are converted into cellulose–cellulose bonds, which, during adsorption, can only be freed for adsorption of water at higher vapor pressures. A similar phenomenon is observed with other polar liquids.

The sorption of water by cellulose is accompanied by swelling,[62] because of the formation of a solid solution. The fact that the X-ray pattern of native cellulose does not change on wetting indicates that the uptake of water is limited to the disordered regions. There are, on the other hand, many liquids, mainly aqueous solutions, that do change the X-ray pattern. Most of these liquids cause much more increase of volume than water alone. The

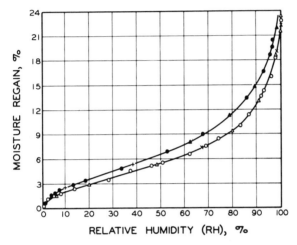

FIG. 3. Water adsorption (lower) and desorption (upper) on purified cotton [A. R. Urquhart, *Text. Res. J.*, **28**, 159 (1958)].

most common example of such a liquid is aqueous 15 to 20% sodium hydroxide. Such swelling agents actually separate chains in the lattice, frequently along a single set of planes, resulting in a new lattice of a more or less stable swelling-complex.

In certain cases, the affinity of cellulose for the liquid is so great that the cellulose molecules are entirely separated from each other, and solution occurs. This sequence of sorption, swelling, and solution is well known with polymers. The solution of cellulose is, in one sense, a misnomer in that the dissolved molecular species is no longer the original cellulose. It is always a new species—either a more or less stable derivative, such as cellulose acetate or cellulose xanthate, or a complex with some component of the dissolving liquid. Some of these complexes are probably oxonium compounds, as, for

* *References start on p. 438.*

instance, with acids[63] or zinc chloride,[64] and some are coordination complexes, as, for instance, with copper–amine solutions[65] or with the iron–sodium–tartrate reagent.[66,67]

VII. DEGRADATION

A. ENZYMOLYSIS

The biological degradation of cellulose is most frequently an enzymic hydrolysis catalyzed by a cellulase. Cellulases occur widely in fungi and bacteria[68,69] and are thereby involved in damage to cellulosic materials such as wood, cotton, and paper. Organisms that catabolize cellulose play an important role in balancing the carbon cycle. Cellulases are probably involved, along with related enzymes, in the germination of seeds and in the ripening as well as the rotting of fruits and vegetables.[70] Cellulase excreted by microflora present in the digestive tracts of herbivores enables these animals to utilize cellulose as a prime source of energy.

Because of their economic impact and potential, the cellulases of microorganisms have been the most widely investigated. When a fungal spore germinates on an intact cotton fiber, there is a general dissolution of the primary wall, and a thread-like hypha bores through the cell wall to the lumen. There the hypha continues to grow by etching away the interior of the fiber wall.[71] The tips of fungal hyphae exude extracellular enzymes that break down cellulose into water-soluble sugars that are absorbed and metabolized intracellularly, to produce mycelium protein, fatty acids, and carbon dioxide.[72] Bacteria also solubilize cotton by excretion of extracellular cellulase, but, in contrast to the boreholes and inner-fiber damage of fungal growth, bacterial attack is characterized at the resolution of the light microscope by surface pitting of the fiber. Bacteria do not generally make a significant contribution[73] to the decay of wood.

The attack of microorganisms on cotton generally gives a product having a much smaller change in degree of polymerization for a given loss in fiber strength than would have been the case after acid hydrolysis.[74] This difference has been known for years and together with other evidence[75] has led to the postulate of localized attack of an organism at the point of contact with the fiber. This difference is, however, consistent[76,77] also with results expected on the basis of enzymic action at a distance from the organism. Selby,[78] for instance, has used cell-free culture filtrates of *Myrothecium verrucaria* to imitate degradation by the intact organism. He found that complete loss of fiber strength, effected by repeated addition of culture filtrate, was accompanied by a 25 to 30% loss of weight and a lowering of the viscosity-average d.p. from 4100 to 3250. In contrast, a 75% loss in strength with acid hydrolysis caused little loss in weight, and the residual cellulose had a d.p. of 260.

Localized attack on cellulose by an enzyme is a result of the limited accessibility of the substrate to the macromolecular catalyst. Highly crystalline cellulose is quite resistant to enzymic degradation, since complexing between hydrolase and substrate is restricted to surfaces of the ordered regions. Several crystalline hydrocelluloses,[79] ranging in number-average d.p. from 225 to 112, were treated with filtrates from *Trichoderma viride*, and it was found that the residues retained their original degree of polymerization as well as particle size, even though 23 to 27% losses in weight occurred. A slightly swollen cellulose[76] was hydrolyzed by cellulase from *Aspergillus niger*, and the initial viscosity-average d.p. of cellulose was observed to fall rapidly at first from 1290 to 830. However, the shorter polymer chains that are mainly responsible for the lowering of the average molecular weight are in accessible regions, and during the next stage of reaction these chains were removed preferentially, and the average d.p. rose to over 1000. In highly amorphous cellulose the size of the enzyme molecule is less of a limitation, and the results of enzymolysis are comparable to those of acid hydrolysis. A cellulose[80] prepared by swelling and degrading cotton in phosphoric acid underwent an initial rapid depolymerization without measurable loss in weight when treated with a cellulase from *Myrothecium verrucaria*. The random hydrolysis was followed by a very slow degradation of cellulose, with a leveling-off d.p. value of approximately 125.

Treatment of fibrous cellulose with most culture filtrates containing cellulases gives a very low yield of water-soluble products.[72,81] The small extent of reaction of highly crystalline cellulose is possibly due to inactivation, through strong adsorption on cellulose, of an enzyme in the culture filtrate that is vital for the process of disaggregation of molecular chains in crystalline regions.[78,82] Alteration of the physical nature of cellulose changes its susceptibility to enzymic hydrolysis, and the magnitude of change varies with filtrates from different organisms.[81] Partially degraded cellulose was treated with preparations from *Aspergillus niger*[83] and *Myrothecium verrucaria*[84] and showed 80% hydrolysis in 3 days and 90% hydrolysis in 10 hours, respectively. On the other hand, culture filtrates that extensively solubilize cotton fibers have been reported recently. Mandels and Reese[85] described an enzyme preparation from *Trichoderma viride* that produces 70 to 80% hydrolysis, while Halliwell[86] reported quantitative conversion of cotton into D-glucose by using a single addition of culture filtrate from *Trichoderma koningii*.

Although the extent of solubilization by isolated enzymes depends largely

* *References start on p. 438.*

on the molecular order of the cellulose, intact organisms do not always preferentially attack accessible linkages in amorphous regions. Cowling[75] reported that white-rot fungus, *Polyporus versicolor*, degraded crystalline and amorphous cellulose of sweet gum holocellulose simultaneously at rates proportional to the amounts originally present, while a brown-rot fungus, *Poria monticola*, preferentially attacked amorphous regions. *Cellulomonas biazotea*[87] is better able to solubilize less-crystalline cellulose from a variety of sources, whereas *Sporocytophaga myxococcoides* is apparently indifferent to the physical properties of the same celluloses and solubilizes them at the same rate.

The problem of limited accessibility of fibrous cellulose has frequently been avoided by use of cellulose ethers and esters, especially as assay substrates.[88–89] As the degree of substitution increases, the susceptibility of the derivative to enzymolysis increases up to the point of complete water-solubility. Additional substitution decreases susceptibility, since substrate specificity becomes the limiting factor. Although heterogeneity in the enzymic reaction is eliminated, the use of water-soluble cellulose derivatives introduces other complications. *O*-(Carboxymethyl)cellulose, which is widely employed as a cellulosic substrate, has been shown to inhibit cellulase activity, probably because of complexing[90] of the protein with the polymeric poly-carboxylic acid. Furthermore, susceptibility of the cellulose derivative varies with the type of substituent as well as with its distribution along the polymer chain. The linkage between two adjacent D-glucose residues is stable to a cellulase from *Helix aspersa* if only one of the two residues is substituted with an *O*-carboxymethyl group, but both residues must be substituted with *O*-ethyl or *O*-methyl groups for the derivative to resist hydrolysis.[91] *O*-(Hydroxyethyl) cellulose treated with a cellulase from *Myrothecium verrucaria* was exclusively hydrolyzed[92] at glycosidic bonds of unsubstituted D-glucosyl residues that had aglycon residues that were either unsubstituted or singly substituted at C-6.

The biodegradation of cotton fibers can be accomplished by only a small number of the hundreds of microorganisms that contain cellulases. This fact resulted in the proposal that, although many organisms contain "C_x-enzyme" [$(1 \rightarrow 4)$-β-D-glucan 4-glucanohydrolase], only a few produce "C_1-enzyme," which is responsible for the production of linear chains from highly crystalline cellulose.[93] A "C_1-enzyme" devoid of hydrolase activity has not been isolated, but enzymes rich in "C_1-activity" have been purified from several cell-free systems. The "C_1-component" has been purified 11-fold over the crude enzyme obtained from culture filtrates of *Trichoderma viride*.[94] Selby and Maitland[95] separated three fractions from culture filtrates of *Myrothecium verrucaria*. One component contained 90% of the total *O*-(carboxymethyl)-cellulase activity, while the other two (termed "A-enzymes") were mainly responsible for the activity of the filtrates toward cotton. The latter two components are possibly related to the swelling factor that was earlier[82,96] demonstrated to be present in filtrates of *M. verrucaria*.

The requirement of "C_1-enzyme" for the enzymic degradation of fibrous cellulose, together with a lack of understanding concerning the changes it causes, has led to confusion as to the meaning of the term cellulase. The Commission on Enzymes of the International Union of Biochemistry defined[97] cellulase as the trivial name for enzymes that catalyze the *hydrolysis* of β-D-$(1 \rightarrow 4)$ D-glucosidic linkages not confined to specific linkages at the end of polysaccharide chains. Several investigators[78,89,98] have used the term cellulase to denote the action of all enzymes involved in the degradation of dried cotton fibers. Whitaker[97] has argued that the decomposition of intact fibers "tends to be a complex, topochemical process which in many fibers is a collaborative effort of several types of enzymes." He further pointed out that the mold *Aspergillus flavus* degrades never-dried cotton fibers, whereas it fails to attack dried cotton fibers wherein physical changes such as redistribution of structural elements may have occurred during drying.

Excluding the disaggregating components that are present in some cellulolytic systems, there has been an increasing number of reports of cellulases that contain a family of enzymes. Multicomponent cellulases have been reported in the culture filtrate including, among others, *Aspergillus niger*,[98] *Trichoderma viride*,[99] *Streptomyces antibioticus*,[100] *Irpex lacteus*,[101] *Cellvibrio gilvus*,[102] and *Polyporus versicolor*.[103] Nevertheless, the reported complexity of a cellulase may be due to anomalous indications of heterogeneity that have been introduced into the system during either isolation or analysis. For example, there is evidence[104] for a single molecular species in the cellulase of *Myrothecium verrucaria*, despite accounts of its resolution into two to sixteen cellulolytic components. A unienzyme structure has also been claimed[105] for the cellulase in filtrates of *Aspergillus terreus*.

The majority of cellulolytic enzymes catalyze random cleavage of glycosidic bonds in accessible cellulosic substrates of high molecular weight. Evidence for random cleavage during enzymolysis includes[97] a very rapid decrease of weight-average d.p. accompanied by a small change in reducing power, and the presence of cellodextrins and D-glucose[81] in reaction mixtures devoid of cellobiase activity. The frequent appearance of cellobiose as the product of the enzymic hydrolysis of cellulose is perhaps due, as illustrated by the cellulase of *Myrothecium*, to the sharp decrease in the rate of hydrolysis of cellodextrins as the degree of polymerization decreases from four to two,[106] since cellobiose is relatively inert to the enzyme.[107] The broad substrate specificity of cellulases, as implied by their activity on celluloses having various chain lengths, is further exemplified by several reports[108,109] of cellulases that catalyze the hydrolysis of xylan. The mechanism of enzymic hydrolysis of cellulose is not known in detail, but the upward direction of mutarotation during enzymolysis of cellopentaose by the cellulase of *Myrothecium* is consistent with a double displacement reaction,[97] most probably at C-1.

B. Hydrolysis

Aqueous acids break the chain of the cellulose molecule at the glycosidic linkage between C-1 and oxygen. The attack is practically random along the chain, with very slightly increased rate near the ends. There has been much discussion of weak chemical links, but there seems to be no valid evidence that these exist in native cellulose,[110] although the inductive effect of carbonyl or carboxyl groups introduced during isolation and purification unquestionably does affect the rate of hydrolysis.[111]

If the hydrolysis is effected in a homogeneous solution, it will, if carried to completion, produce D-glucose. Several processes have been developed for industrial saccharification, and these are the basis of an important industry in the Soviet Union.[112] In the laboratory, the hydrolysis process used in the analysis of wood pulp is typical.[113] The procedure consists in an initial brief treatment with strong sulfuric acid followed by treatment with hot dilute acid, to hydrolyze the reversion products that result from the primary hydrolysis.

The heterogeneous hydrolytic process, hydrolysis of solid cellulose, is complicated. The kinetics indicate an initial rapid reaction followed by a slower one, and are usually interpreted on the basis of accessibility.[114] The easily accessible regions, disordered regions or fibrillar surfaces, react first, and then the less accessible regions react, but more slowly. The molecular chain length decreases, rapidly at first and then more slowly, asymptotically approaching a "leveling off" degree of polymerization that is characteristic of the nature of the cellulose—that is, its origin, past treatment, and crystal structure. Physically, the cellulose loses its strength during the early stages of such hydrolysis and is converted into a friable mass of microcrystalline, molecular aggregates.

C. Oxidation

Cellulose is easily oxidized, but, as might be predicted from its polymeric nature, intermediate oxidation stages do not result in high yields of well-defined products. The hydroxyl groups are the sites most susceptible to attack. Most oxidations are random processes and lead to introduction of carbonyl and carboxyl groups at various positions in the D-glucose residues and along the chain. There are also secondary reactions, particularly chain scission. However, attack that is completely random is by no means general.

The accessibility of some D-glucose residues causes them to be attacked before others, and there is also limited specificity of certain oxidants for individual types of hydroxyl groups.[115]

Hypoiodite and acidified chlorite solutions have little or no action on the hydroxyl groups but oxidize the terminal reducing group (or any extraneous aldehyde groups) to carboxyl groups. These oxidations are sufficiently specific that they have been used for quantitative measurement of such reducing groups, although neither is completely specific.[116,117] The oxidation by alkaline copper(II) reagents, commonly used for the determination of reducing end-groups, is much less stoichiometric.[118]

Nitrogen tetraoxide has some specificity also, its main effect being to convert the primary hydroxyl groups into carboxyl groups. However, the product usually contains reducing groups and also combined nitrogen, present mainly as nitrites or nitrates.[119,120]

Periodic acid and sodium periodate act primarily on the glycol group, breaking the ring to generate aldehyde groups at C-2 and C-3. These groups may react further to form hydrates, acetals, hemiacetals, and hemialdals.[121]

Most oxidants are less specific than the aforementioned and produce a random agglomeration of different kinds of carbonyl groups and carboxyl groups. There is, in many cases, a marked pH effect; in acidic solutions, oxidations produce "reducing oxycelluloses" having a high content of carbonyl groups, and in alkaline solutions, they produce "acidic oxycelluloses" having a high content of carboxyl groups.[115] However, most oxidized celluloses contain aldehyde, ketone, and carboxyl groups. Most attention has been given to the chlorine and hypochlorite systems used in bleaching wood pulp and cellulosic textiles, but oxidations by permanganate are very similar.

Chromium trioxide,[122] acidified dichromate,[123] and *tert*-butyl chromate[124] introduce both carbonyl and carboxyl groups. Peroxides[125] and ozone[126] also react in a complex fashion.

The action of oxygen is very slight at room temperature, but can become important at high temperatures.[127,128] It is also marked in the presence of strong bases. This finds industrial utilization in the aging of "alkali cellulose,"[129] which is shredded and allowed to stand for a day or two as a preliminary step in the manufacture of viscose rayon. This type of autooxidation results in chain scission and the introduction of carboxyl groups.

A free-radical mechanism has been proposed for such autooxidation,[115,129] although ionic mechanisms have also been suggested.[130,131] Other free-radical oxidations are those by ceric ion[132] or by Fenton's reagent.[133]

D. MISCELLANEOUS DEGRADATIONS

The instability of "reducing oxycelluloses" to alkali is well known.[134] This is a special case of the alkaline degradation of β-carbonyl ethers arising from inductive effects of the electronegative group. Since purification processes usually lead to introduction of some carbonyl groups, most cellulose samples are subject to a certain degradation by alkali.

In fact, even if there has been no oxidation, the reducing end-group of the cellulose chain can trigger a somewhat similar reaction with hot alkali.[134] The mechanism corresponds to that involved in the formation of "glucoisosaccharinic" acids from D-glucose, except that a cellulosate ion is split off instead of a hydroxyl ion. This produces another chain having a reducing end-group, and the process repeats itself with progressive shortening of the chain, the so-called "peeling" reaction. A competing reaction, which corresponds to conversion into a "glucometasaccharinic" acid derivative, and which does not involve chain scission, prevents complete degradation of the cellulose.

Alcoholysis of cellulose or of cellulose derivatives is the breakdown by alcohols in the presence of an acid catalyst such as hydrogen chloride, producing the corresponding alkyl D-glucosides as final products.[135] Acetolysis (the action of acetic anhydride and relatively large proportions of sulfuric acid) results in a mixture of acetates ranging from the pentaacetates of D-glucose up to the acetates of cellulose that has been only mildly degraded. Conditions may be so chosen as to obtain good yields of cellobiose octaacetate[136] or of a polymer-homologous mixture of acetates of the β-D-(1 $\rightarrow$ 4)-linked oligosaccharides of D-glucose.[95]

Thermal deterioration of cellulose usually occurs in air and results, in fact, from mild oxidation.[137] At high temperatures and in the absence of oxygen, pyrolysis will occur. 1,6-Anhydro-β-D-glucopyranose (levoglucosan)[138] is a major product in such degradation by heat.[139] Other reactions that occur produce water, carbon monoxide, carbon dioxide, and a number of other compounds.[140] This process is of great importance in the field of flameproofing and fire prevention.[141,142]

The cellulose molecule can also be ruptured by other forms of energy such as visible and ultraviolet light[143,144] or high-energy irradiation.[145] Even mechanical action, such as ball-milling, can break the chain to form free radicals.[146-148] The final products of these homolytic processes are caused by secondary reactions of the radicals, especially with oxygen (see Vol. IB, Chap. 26).

VIII. SUBSTITUTION REACTIONS

A. REACTIVITY IN GENERAL

The nature of cellulose reactivity has been admirably treated by Spurlin,[149] and his discussion should be read by anyone interested in this subject. Only a brief outline of some of the factors to be considered will be given here, with primary reference to substitution reactions of the hydroxyl groups that lead to esters or ethers. However, the same considerations frequently apply to substitution reactions that are less common.

An important point to be made at the outset is that the reaction products from a polysaccharide of high molecular weight (and high polymolecularity) such as cellulose will, of necessity, be much more complex than those of a simple sugar such as D-glucose. One cellulose molecule may contain 600 hydroxyl groups, and another molecule in the same sample may contain 20,000. As a general rule, cellulose derivatives are the products of only partial reaction; that is, they contain a certain proportion of unchanged hydroxyl groups. The properties of the derivatives vary, not only with the proportion of these groups (or with the proportion substituted) but also with the distribution of substituents. Therefore, the expressions "O-methyl-cellulose" or "cellulose acetate" by no means characterize a product; each simply denotes a series of products, although the most common commercial member of that series is often understood by the general term.

The degree of substitution is, of course, important. The abbreviation d.s. refers to the average number of substituents per D-glucose residue. The d.s. ranges from 0 in untreated cellulose to a maximum of 3, which is seldom reached. Most important industrial cellulose derivatives have intermediate d.s. values, and the distribution of substituents becomes important. This distribution may vary in several ways and for several reasons, two of which are very important.

In the first place, distribution may be governed by accessibility. In solution, all D-glucose residues are supposed to be equally accessible, and cellulose derivatives prepared by reaction in solution are more likely to be uniform than those prepared by reaction of the solid fiber. In the case of the fiber, it is obvious that only those hydroxyl groups reached by the reagent can react, and the inaccessible hydroxyl groups will remain unsubstituted. Reaction will be limited to the outer surface of the fiber in a nonswelling medium and

* References start on p. 438.

will extend to an increasingly larger number of hydroxyl groups if the internal surface is opened up by swelling.

In the second place, not all accessible hydroxyl groups are equally reactive. The three hydroxyl groups of the D-glucose residues will differ in their reactivity to a given reagent, and the relative rates are not necessarily the same for other reagents. The inductive effect of carbonyl or carboxyl groups introduced during processing and, in some cases, the effect of substitution of one hydroxyl group on the reactivity of a neighboring hydroxyl group may also be important.

In acetylation with acetic anhydride, the primary hydroxyl group reacts the most rapidly,[149] but in etherification, the C-2 hydroxyl group frequently reacts. Table I gives the relative rate constants of the C-2, C-3, and C-6 hydroxyl groups for the reactions of cellulose with various etherifying agents.[150-159] For most entries in Table I, k_2 is greater than k_6. The base-catalyzed addition of acrylamide[160] or methyl vinyl sulfone[161] to cellulose

TABLE I

RELATIVE RATE CONSTANTS FOR ETHERIFICATION OF CELLULOSE

| | Relative rate constants | | | |
Reagent	k_2	k_3	k_6	References
Methyl sulfate	3.5	1.0	2.0	152, 153
Methyl chloride	5.0	1.0	2.0	154
Methyl iodide	5.0	1.0	2.5	155
Diazomethane	1.2	1.0	1.5	156
Ethyl chloride	4.5	1.0	2.0	157
Ethylene oxide	3.0	1.0	10.0	158
Chloroacetic acid	2.0	1.0	2.5	159

produces a cellulose ether having the greatest extent of substitution at the C-6 hydroxyl group. These two Michael reactions, in contrast to the reactions included in Table I, are reversible, and the arrangement of O-(methylsulfonyl)ethyl or O-(carbamoyl)ethyl groups may often be equilibrium-controlled. It has been pointed out[161] that the distributions of O-(methylsulfonyl)ethyl and O-(carbamoyl)ethyl groups are similar. From the distribution of O-(carbamoyl)ethyl groups the relative equilibrium constants k_2, k_3, and k_6 have been calculated to be 9, 1, and 19.

B. ESTERS

Esters of cellulose with over a hundred acids have been described, but because of the numerous variations arising from degree and distribution of substitution and from other factors (degree of polymerization, preparation of mixed esters or ester-ethers, oxidation, etc.) it is impossible to give much detail in this chapter. The reader is referred to specialized reviews[14] for specific esters; here only the three most important cellulose esters will be mentioned. The molecular mechanisms of cellulose esterifications are well discussed by Rånby and Rydholm.[162]

The cellulose nitrates[163,164] are a versatile material. The very highly nitrated products are used in explosives (propellants), but those of somewhat lower nitrogen content have been used in plastics, lacquers, and adhesives. Industrially, preparation of nitric esters is effected with a mixture of nitric and sulfuric acids. The active agent, which penetrates very rapidly even into the crystalline regions, may be the nitronium ion[162,164,165] as in aromatic nitrations,[166] but this proposal has been questioned.[163,167] The degree of substitution of the nitric ester is determined by an equilibrium with the nitrating liquor and is a function of the composition of the liquor. There is a certain amount of degradation by the sulfuric acid used as catalyst, which may be desirable for uses where solutions of low viscosity are needed. If undegraded nitrates and high nitrogen contents are desired, as for determinations of molecular weight, nitric acid and phosphorus pentaoxide[168] or nitric acid and acetic anhydride[169] may be used.

Cellulose acetate is widely used for plastics, sheeting, films, and rayon. There are several distinct types of industrial cellulose acetate, and the mixed esters, acetate–propionate and acetate–butyrate, are also used.

The commonest type of cellulose acetate is produced in what is essentially a two-stage process. Cellulose is first esterified, most simply with acetic anhydride and sulfuric acid,[149] although there are several modifications. The esterification is allowed to proceed until a more or less completely esterified, mixed ester of acetic and sulfuric acids is formed, which dissolves in the acetylation mixture to form a viscous homogeneous solution. This "primary acetate" is highly acetylated. In the second stage of the process, water and more acid are added, and hydrolysis is allowed to take place to remove the sulfate groups and some of the acetate groups, and to cause some chain scission. There are several types of the resulting "secondary acetate" which usually have d.s. values of 2 to 2.5; these are commonly known as "cellulose acetate." Continued hydrolysis will produce a water-soluble cellulose acetate[170] of d.s. 0.5 to 1.0.

The product known as "cellulose triacetate"[171] has a d.s. that approaches 3 and may be produced by working up the solution of the primary acetate or by the acetylation of fibrous cellulose as described in the next paragraph. At the other extreme of substitution are the high-tenacity textile fibers

---

* References start on p. 438.

produced by complete deacetylation of a highly swollen cellulose acetate fiber under tension.[172]

Still other types of acetylated cellulose are prepared for use in textiles or paper by acetylating cotton,[173] rayon,[174] or wood pulp[175] under conditions that do not destroy the fibrous form. These include low degrees of substitution where only the surface or readily accessible portion is esterified[176] and high degrees of substitution where the fiber structure has reacted throughout.[177] These acetylations are commonly performed in the presence of a nonsolvent to prevent solution of the acetylated fiber, but vapor-phase acetylation has also been used.[174]

The effective acetylating agent is probably the acetylium ion.[162,178] Where sulfuric acid is present, sulfation is a competing and much faster reaction, but the sulfuric esters first formed are later replaced by acetic ester groups.[149]

One of the most important esters of cellulose, cellulose xanthate, is not used at all as such, but is an intermediate in the manufacture of viscose rayon. Cellulose is converted into "alkali cellulose" and aged, as was mentioned briefly in Section VII,C. It is then allowed to react with carbon disulfide to form sodium cellulose xanthate. Relatively few hydroxyl groups need to be converted into xanthate groups in order to induce solubility in aqueous alkali. The viscose rayon process and the manufacture of cellophane involve the regeneration of cellulose from such a solution by extrusion through an orifice or a slit, respectively, into an acid bath where the xanthate ester groups are split off.[179]

C. ETHERS

Most etherification reactions of cellulose are nucleophilic substitutions by the cellulosate ion, which is formed by ionization of the hydroxyl groups. For industrial purposes the etherification reactions are carried out on "alkali cellulose," in which the hydroxyl groups are most effectively ionized.

Properties of the cellulose ethers cover a wide range, depending on the nature of the substituents and their distribution. Many types of cellulose ethers are manufactured and sold today for various purposes.

Methylation of cellulose is performed industrially by the action of methyl chloride on alkali cellulose, but numerous other methylating agents have been used in the laboratory.[151,180–182] The commercial product is only partly etherified (d.s. 1.5 to 2.0) and is soluble in cold water. At substitution levels below a d.s. of about 1.0, alkali-soluble, water-soluble O-methylcelluloses are formed, and at levels from 2.5 to 3.0, the solubility in water decreases and eventually disappears, and O-methylcellulose becomes soluble in organic solvents such as chlorinated hydrocarbons. The distribution of substituents along the chain also affects the solubility.[183-185] The temperature coefficient of solubility in water is negative, so that solutions form gels or precipitates on heating.[14b]

This solubility behavior can be explained on the basis of molecular structure and arrangement. As already discussed in Sections V and VI, the regular arrangement of hydroxyl groups leads to the formation of strongly hydrogen-bonded, ordered regions which cannot be broken apart by water. Even alkali does not separate these regions completely. If the regularity is interrupted occasionally because of random substitution by O-methyl groups, however, dilute alkali, and at higher levels water alone, can penetrate the interchain regions, break the hydrogen bonds, and solvate the hydroxyl groups. The insolubility in hot water is due to thermal breakdown of the hydration complex and consequent molecular aggregation. However, as methylation proceeds, hydroxyl groups become fewer, and the product becomes insoluble in water.

The water-soluble O-methylcellulose of commerce serves a number of purposes, most of which depend on the colloidal properties of its aqueous solutions.[180] It is variously used as a defoamer, an adhesive, a thickener, a dispersant, a stabilizer, a suspending agent, and a base for water-soluble films. It is nontoxic and is used in foods and in pharmaceuticals, where its action is that of a bulk laxative.

O-Ethylcellulose exhibits somewhat similar behavior in solution,[184] but the water-soluble range of d.s. is very narrow and the commercial product has a d.s. of 2 to 2.6. It is insoluble in water but soluble in organic solvents, and is used in protective coatings, films, foils, and plastics.[186]

O-(Hydroxyethyl)cellulose is formed by treating "alkali cellulose" with ethylene oxide or 2-chloroethanol.[186-188] It has many of the properties and uses of O-methylcellulose but is soluble in both cold and hot water.

O-(Carboxymethyl)cellulose[181,189] is produced when "alkali cellulose" is allowed to react with chloroacetic acid. The sodium salt is also a water-soluble, protective colloid and is manufactured for this purpose. It is also prepared in edible grades. The aqueous solutions are not greatly affected by heat but can be modified by pH, since the free acid is less soluble than the salt. The reactive carboxyl group gives this material interesting properties, and other reactive groups have also been so introduced into cellulose. Many such derivatives,[190] including cellulose phosphate, O-(carboxymethyl)cellulose, and various aminoalkyl ethers are manufactured for use as ion-exchange materials.

Mixed derivatives having more than one type of substituent group can be prepared to achieve various solubility effects and various responses to

* References start on p. 438.

temperature and pH. The most widely used of the mixed ethers contains both ethyl and 2-(hydroxyethyl) substitution.[191]

D. Cross-linking

Polyfunctional reagents may react with two different molecules of cellulose to form a single, bridged molecule.[192-194] In actual practice the degree of cross-linking may be very small, but it produces marked effects on some of the properties of the cellulosic material, such as its dimensional stability[195] and, specifically for textiles, its wrinkle resistance[196,197] and related properties of the fabric.[198,199]

Formaldehyde,[200] and also other aldehydes,[201,202] form methylenedioxy bridges. Many of the commercial cross-linking agents are based on formaldehyde or derived reagents, but other types of reagents have also been used, such as dihalides,[203] diisocyanates,[204] diepoxy compounds,[205] and divinyl sulfones.[206,207]

IX. GRAFT POLYMERS

Graft polymers of cellulose are copolymers in which the second polymer forms a long branch to the cellulose molecule. While both condensation and addition polymers are possible, it is the latter, or vinyl-type, of polymer that will be discussed here. This type of reaction is not a substitution reaction and has, therefore, not been covered in Section VIII.

The vinyl type of polymerization is usually initiated by a free radical. Most graft polymers of cellulose have been prepared, therefore, by converting the cellulose molecule into a radical.[208] This can be done by homolytic scission of the chain by application of energy, such as high-energy irradiation—for example, γ-rays[209-213] (see Vol. IB, Chap. 26). Ultraviolet light will initiate grafting in the presence of certain dyestuffs.[214,215] Even mechanical action[216] can produce some initiation sites. Many oxidants that produce free radicals have also been effective—for example, ceric ions,[217,218-221] Fenton's reagent[219,222] (ferrous ions and hydrogen peroxide), and potassium persulfate[217,223] (see also Vol. IB, Chap. 26). Other types of radical initiation,[208,210,224,225] as well as ionizing polymerization,[226,227] have been proposed as methods for forming polymer grafts on cellulose.

Most of the common vinyl monomers have been grafted onto cellulose with various degrees of success. It has been found fairly easy to graft acrylonitrile,[214,223,227] acrylamide[215,218] or the acrylates,[217,221] and methacrylates.[216,222] Vinyl acetate,[210] vinyl chloride,[217,228] vinylidene chloride,[210,217] and the vinylpyridines[210,217] can also be grafted. Styrene, likewise, forms grafts, but with more difficulty.[209,211,212,224]

Graft polymers of cellulose are usually accompanied by large amounts of homopolymers, and the actual yield of the graft should probably not be equated with the total add-on, even after extensive purification. The actual extent of true chemical reaction with the cellulose is extremely low, less than one chain per cellulose molecule in many cases that have been measured.[209,211,212,215,224] This is even lower than with cross-linking or the low-substituted ethers.

X. LICHENAN AND ISOLICHENAN

Lichenan, formerly called lichenin, is a D-glucan from Iceland moss that is very similar to cellulose. The D-glucosidic linkages are all β-D but are not exclusively (1 → 4). The molecule appears to consist of a series of (1 → 3)-linked cellotriose molecules, with an occasional cellobiose or cellotetraose residue[229–232] It also contains about 0.4% of methoxyl groups.[229,233] Lichenan appears to be a linear molecule and has a degree of polymerization variously estimated as from 60 to 365.[129,231,233–235] It is a white amorphous powder that is soluble in water.

There are also other polysaccharides in Iceland moss. One of these, isolichenan, has been shown to be essentially the α-D analog of lichenan.[231] It contains only α-D linkages, both (1 → 4) and (1 → 3). It contains a higher proportion of (1 → 3) linkages than does lichenan and has a somewhat lower d.p., about 42 to 44.

The structure of lichenan has been investigated both by acid[230,232] and by enzymic hydrolysis.[236,237] It was largely as a result of the studies on lichenan that the Parrish–Perlin theory of the action of the glycanases was developed— namely, that specificity is a function of the substitution of the glycosyl unit released rather than of the nature of the bond broken.[238]

XI. CHITIN

Chitin[239] is the most abundant of those polysaccharides that contain amino sugars; it is a linear polymer of β-D-(1 → 4)-linked 2-acetamido-2-deoxy-D-glucose residues (see also Vol. IB, Chap. 16). Chitin is often discussed along with cellulose, since it frequently replaces cellulose as the structural entity in the cell walls of many species of lower plants. Chitin is found[240] in most fungi, mycelial yeasts, green algae, and several species of brown[241]

References start on p. 438.

and red algae;[242] it is not present in bacteria, true yeasts, and *Actinomycetes.* Chitin is also widespread in the animal kingdom, occurring in the form of sheets as in the cuticles of arthropods, annelids, and molluscs, or in the form of well-oriented fibers as in the mandibular tendon of lobster.[243]

Chitin seldom occurs alone in Nature, one exception being the fibrous chitin produced by the diatom *Thalssiosira fluviatilis.*[243a] This extracellular chitin is readily dislodged from the algal cells and can be isolated in 31 to 38% yield.[243b] The "chitinous" exoskeletons of arthropods actually contain only 25 to 30% of chitin, which in many, if not all, species is complexed[244] with or possibly covalently bonded[244,245] to an equal or smaller amount of protein. Furthermore, the proportion of chitin in different parts of the cuticle varies with the hardness of that part; hardened portions contain smaller proportions of chitin and larger proportions of either inorganic salts (principally calcium carbonate) or cross-linked protein, or both. Chitin in fungi is likewise associated intimately with other materials, and yields of purified chitin from fungi are frequently less than 5%. The identity of fungal chitin with animal chitin has been demonstrated by a number of chemical and physical methods.[239]

Prior to the investigations of McLachlan, McInnes, and Falk[243b] the isolation of native, undegraded chitin had not been reported because of the drastic conditions used to remove contaminants from the polysaccharide. Chitin for laboratory[246,247] and commercial[248] use has been commonly obtained from crustacean waste materials by decalcification of the shells with dilute mineral acid and by removal of protein and other organic matter with hot alkali. Fungal chitin isolated by similar methods frequently[249] contains much less nitrogen than the theoretical 6.9%, and further purification can be effected by treatment of crude chitin with permanganate[250] or by dissolution in cold concentrated hydrochloric acid followed by precipitation in cold water.[251,252]

The molecular structure of chitin is well established and can be visualized by replacing the OH groups on C-2 of the cellulose molecule (Fig. 1, Section I) with NHAc groups. Complete acid hydrolysis of purified chitin gives equimolar amounts[253] of acetic acid and an amino sugar that has been proved[254,255] to be 2-amino-2-deoxy-D-glucose. Controlled acid hydrolysis[256] or enzymic hydrolysis[246,257] gives the true repeating residue of chitin, 2-acetamido-2-deoxy-D-glucose. The homologous series of chitin oligosaccharides up to and including the heptasaccharide has been isolated[258] from a partially de-esterified hydrolyzate of chitin that had been selectively re-*N*-acetylated prior to separation of oligomers by column chromatography. Chitin contains β-D-glycosidic linkages, as indicated by the change in its optical rotation during hydrolysis,[259] the value of its energy of activation for hydrolysis,[260] its absorption in the infrared spectrum[258] at 884 to 890

cm^{-1}, and its susceptibility to enzymolysis with emulsin chitinase.[261] The molecular weight of chitin is thought to be of the same order of magnitude as that of cellulose.[240,260] However, as expected for a polymer that is difficult to isolate and purify, a wide range of molecular weights has been reported.[248]

X-Ray diffraction investigations have shown that chitin contains highly ordered regions giving one of several possible crystallographic patterns.[14a] These have been called α-chitin, β-chitin, and native chitin.[243a,262,263] The crystal structure of α-chitin was determined by Meyer and Pankow,[264] who proposed an orthorhombic unit-cell containing eight 2-acetamido-2-deoxy-D-glucose residues. In chitin the repeating period along the fiber axis was found to be the same as that in cellulose. This analysis was verified by other workers.[243,262] In more recent studies, Carlström[265] arrived at a new structure that has four hexose residues in an orthorhombic unit-cell of dimensions $a = 4.76$ Å, b (fiber axis) $= 10.28$ Å, and $c = 18.85$ Å. This latest model adequately explains[266] the polarized infrared spectrum of α-chitin, including the following features: (a) polymer chains in the "bent" conformation, (b) intramolecular hydrogen bonding between a C-3 hydroxyl group and the ring oxygen atom of the adjacent pyranose ring, (c) full intermolecular hydrogen bonding involving the C=O—HN groups along the direction of the a-axis, and (d) no free OH or NH groups and no C=O—HO bonds in the α-chitin crystal. The arrangement of molecules in β-chitin has also been investigated.[267,268]

Chitin is insoluble in water, organic solvents, and cuprammonium reagent[260] but dissolves with some depolymerization[243,269] in concentrated mineral acids. Chitin is also soluble in anhydrous formic acid[263,270] and is dispersed in concentrated aqueous solutions containing certain lithium or calcium salts.[271] Chitin fibers or films have been prepared by regeneration of chitin from dispersions in salt solutions and solutions in mineral acids, and by regeneration from chitin xanthate.[239]

Chitin is less reactive than cellulose because of its general insolubility. Chitin is slowly N-deacetylated and degraded by strong alkali to give a complex mixture of partially deacetylated products collectively termed chitosan. Chitan, completely N-deacetylated chitin, is obtained by retreatment of chitosan with alkali[272] or by fractional precipitation[247] of the amino polymer as its hydrochloride salt. Chitin is difficult to acetylate or methylate by methods commonly used to form derivatives of cellulose, since it does not swell in the usual reaction media. A completely acetylated chitin is obtained[273] only by treatment with acetic anhydride and dry hydrogen chloride, and a product 92% methylated was obtained only after repeated methylation by the Haworth and Kuhn procedures.[274] Other derivatives of chitin that have been reported include sulfate, nitrate, and xanthate esters.[239]

* *References start on p. 438.*

Biosynthetic reactions leading to chitin are known. Candy and Kirby[275] demonstrated in desert locusts the presence of enzymes that catalyze a sequence of reactions leading to the direct precursor[276-279] of chitin, uridine 5′-(2-acetamido-2-deoxy-D-glucopyranosyl pyrophosphate). When this sugar nucleotide,[276] labeled with ^{14}C in the monosaccharide moiety, was treated with a particulate enzyme from *Neurospora crassa* in the presence of water-soluble chitodextrins, there was formed an insoluble β-D-(1 → 4)-linked polymer containing radioactive 2-acetamido-2-deoxy-D-glucopyranosyl residues—that is, a polysaccharide structurally identical with native chitin (see this volume, Chap. 34).

Enzymes that degrade chitin occur widely in microorganisms, and they have also been found in, among other sources, protozoa,[280] earthworms,[281] snails,[282,283] and plant seeds.[284] These enzymes are active in a number of natural phenomena. Chitinase originating from symbiotic organisms[285,286] or from a gland[287] enables insectivorous vertebrates to digest the chitin of their prey. Chitinous debris in the soil and sea is disposed of by chitinolytic microorganisms.[239,240] Chitinase in crustaceans[288,289] and insects[290] is associated with the moulting process of these animals. Chitinase has also been shown[291] to be involved in enzyme-induced lysis, a common mechanism of interaction between microorganisms wherein one organism eliminates another by digesting the cells or hyphae of its competitor.

The biological degradation of chitin, like that of cellulose, is an enzymic hydrolysis generally involving at least two enzymes,[292,293] one of which catalyzes the random[284,294-296] depolymerization of chains with the production of oligomers, and the other the cleavage of chitin oligosaccharides, especially chitobiose, to yield 2-acetamido-2-deoxy-D-glucose. Each type of enzyme has been purified: polysaccharidase from, for example, bean seeds[284] and cockroaches,[295] and the chitobiase from *Aspergillus niger*.[297] Chitinolytic enzymes do not act on *N*-deacetylated substrates[257,294,296] or on chitin analogs[298] that contain a high percentage of *N*-acyl groups other than *N*-acetyl groups.

REFERENCES

1. E. Husemann and O. H. Weber, *J. Prakt. Chem.*, **159**, 334 (1942).
2. L. E. Wise, *Tappi*, **41** (No. 9), 14A (1958).
3. G. V. Schulz and E. Husemann, *Z. Phys. Chem.* (*Leipzig*), **52(B)**, 23 (1942).
4. D. B. Das, M. K. Mitra, and J. F. Wareham, *Nature*, **174**, 1058 (1954).
5. W. N. Haworth, *J. Soc. Chem. Ind.* (*London*), **58**, 917 (1939).
6. K. Hess and E. Steurer, *Ber.*, **73**, 669 (1940).
7. H. Staudinger and A. W. Sohn, *J. Prakt. Chem.*, **155**, 177 (1940).
8. E. Pacsu and L. A. Hiller, Jr., *Text. Res. J.*, **16**, 243 (1946).
9. A. J. A. VanderWyk and J. Schmorak, *Helv. Chim. Acta*, **36**, 385 (1953).
10. International Union of Pure and Applied Chemistry, *J. Polym. Sci.*, **8**, 257 (1952).
11. J. W. T. Merewether, *Holzforschung*, **11**, 65 (1957).

12. H. Meier, *Acta Chem. Scand.*, **12**, 1911 (1958).
13. K. Kringstad and Ø. Ellefsen, *Papier*, **18**, 583 (1964).
14. E. Ott, H. M. Spurlin and M. W. Grafflin, Eds., "Cellulose and Cellulose Derivatives," Wiley (Interscience), New York, 1954.
14a. R. H. Marchessault and A. Sarko, *Advan. Carbohyd. Chem.*, **22**, 421 (1967).
14b. D. A. Rees, *Advan. Carbohyd. Chem.*, **24**, 267 (1969).
15. P. A. Roelofsen, "The Plant Cell Wall," Vol. 3, Part 4 of Encyclopedia of Plant Anatomy, Borntraeger, Berlin, 1959.
16. K. Ward, Jr., see ref. 14, Chapter II.
17. W. A. P. Black, W. J. Cornhill, and F. N. Woodward, *J. Appl. Chem. (London)*, **5**, 484 (1955).
18. K. Siwek, *Papier*, **18**, 10 (1964).
19. C. Alexanian, *C. R. Acad. Sci.*, **246**, 1192 (1958): *Chem. Abstr.*, **52**, 11674 (1958).
20. H. Krässig, *Makromol. Chem.*, **13**, 21 (1954).
21. D. A. Hall, F. Happey, P. F. Lloyd, and N. Saxl, *Proc. Roy. Soc. Ser.*, **B155**, 202 (1961).
22. D. T. Dennis and J. R. Colvin, *Paper Pulp Mag. Canada*, **65**, T-395 (1964).
23. S. Hestrin and M. Schramm, *Biochem. J.*, **58**, 345 (1954).
24. J. R. Colvin, in "The Formation of Wood in Forest Trees," M. H. Zimmermann, Ed., Academic Press, New York, 1964, pp. 169–188.
24a. M. Marx-Figini, in "Encyclopedia of Polymer Science and Technology," H. Mark, N. Gaylord, and N. Bikales, Eds., Vol. 3, Wiley, New York, 1965, pp. 230–242.
24b. R. G. Everson and J. R. Colvin, *Can. J. Biochem.*, **44**, 1567 (1966).
25. E. J. Hehre, *Advan. Enzymol.*, **11**, 297 (1951).
26. L. Glaser, *J. Biol. Chem.*, **232**, 627 (1958).
27. S. Klungsöyr, *Nature*, **185**, 104 (1960).
28. G. Ben-Hayyim and I. Ohad, *J. Cell Biol.*, **25** (Part 2, No. 2), 191 (1965).
29. J. R. Colvin, *Nature*, **183**, 1135 (1959); A. M. Brown and J. A. Gascoigne, *ibid.*, **187**, 1010 (1960).
30. A. W. Kahn and J. R. Colvin, *Science*, **133**, 2014 (1961).
31. E. F. Neufeld and W. Z. Hassid, *Advan. Carbohyd. Chem.*, **18**, 309 (1963).
32. G. A. Barber and W. Z. Hassid, *Nature*, **207**, 295 (1965).
33. G. A. Barber, A. D. Elbein, and W. Z. Hassid, *J. Biol. Chem.*, **239**, 4056 (1964).
34. G. A. Barber and W. Z. Hassid, *Biochim. Biophys. Acta*, **86**, 397 (1964).
35. D. O. Brummond and A. P. Gibbons, *Biochem. Z.*, **342**, 308 (1965).
35a. L. Ordin and M. A. Hall, *Plant Physiol.*, **42**, 205 (1967).
36. J. R. Colvin, *Can. J. Biochem. Physiol.*, **39**, 1921 (1961).
36a. R. W. Bailey, S. Haq, and W. Z. Hassid, *Phytochemistry*, **6**, 293 (1967).
36b. D. J. Manners, *Ann. Rept. Progr. Chem. (Chem. Soc. London)*, **64**, 611 (1967).
36c. M. Marx-Figini, *Nature*, **210**, 754, 755 (1966).
37. J. R. Colvin, *Can. J. Botany*, **43**, 1478 (1965).
38. S. A. Rydholm, "Pulping Processes," Wiley (Interscience), New York, 1965.
39. F. Kraft, in "Bleaching of Pulp," TAPPI Monograph No. 27, W. H. Rapson, Ed., TAPPI, New York, 1963, Chapter II, especially p. 241.
40. A. Meller, in "Bleaching of Pulp," TAPPI Monograph 10, TAPPI, New York, 1953, Chapter III.
41. R. S. Hatch, see ref. 40, Chapter III, Appendix, p. 75.
42. B. L. Browning, in "Wood Chemistry" L. E. Wise and E. C. Jahn, Eds., Reinhold, New York, 1952, Chapter 29.
43. D. A. I. Goring and T. E. Timell, *Tappi*, **45**, 454 (1962).
44. M. Marx-Figini and E. Penzel, *Makromol. Chem.*, **87**, 307 (1965).

45. H. Vink and R. Wikström, *Svensk Papperstidn.*, **66**, 55 (1963).
46. H. Vink and H. Johansson, *Svensk Papperstidn.*, **66**, 711 (1963).
47. B. G. Rånby, in "Fundamentals of Papermaking Fibres," Techn. Sect. Brit. Paper & Board Makers Assoc. Kenley, Surrey, 1958.
48. K. Mühlethaler, *Papier*, **17**, 546 (1963).
49. R. St. J. Manley, *Nature*, **204**, 1155 (1964).
50. I. Ohad, D. Danon, and S. Hestrin, *J. Cell Biol.*, **12**, 31 (1962).
51. I. Ohad and D. Mejzler, *J. Polym. Sci.*, **3A**, 399 (1965).
52. A. Frey-Wyssling and K. Mühlethaler, "Ultrastructural Plant Cytology," Elsevier, Amsterdam, 1949.
53. Ø. Ellefsen, K. Kringstad, and B. A. Tønnesen, *Norsk Skogind.*, **18**, 419 (1964).
54. P. H. Hermans, "Physics and Chemistry of Cellulose Fibres," Elsevier, Amsterdam, 1949.
55. A. Venkateswaran and J. A. Van den Akker, *J. Appl. Polym. Sci.*, **9**, 1149, 1167 (1965).
56. B. Philipp and J. Baudisch, *Faserforsch. Textiltech.*, **16**, 173 (1965).
57. K. H. Meyer and L. Misch, *Helv. Chim. Acta*, **20**, 232 (1937).
58. J. W. S. Hearle, in "Fibre Structure," J. W. S. Hearle and R. H. Peters, Eds., Butterworths, Washington, D.C. and London, 1963, Chapter VI.
59. J. A. Howsmon and W. A. Sisson, see ref. 14, Chapter IV, B.
60. B. A. Tønnesen and Ø. Ellefsen, *Norsk Skogind.*, **14**, 266 (1960).
61. A. R. Urquhart, *Text. Res. J.*, **28**, 159 (1958).
62. J. A. Howsmon and W. A. Sisson, see ref. 14, especially pp. 317–334.
63. A. af Ekenstam, *Ber.*, **69**, 549 (1936).
64. G. S. Kasbekar, *Curr. Sci.*, **9**, 411 (1940).
65. H. Vink, *Makromol. Chem.*, **76**, 66 (1964).
66. G. Jayme, *Tappi*, **44**, 299 (1961).
67. G. F. Bayer, J. W. Green, and D. C. Johnson, *Tappi*, **48**, 557 (1965).
68. R. G. H. Siu, "Microbial Decomposition of Cellulose," Reinhold, New York, 1951, p. 99.
69. J. A. Gascoigne and M. M. Gascoigne, "Biological Degradation of Cellulose," Butterworths, Washington, D.C. and London, 1960, p. 52.
70. W. Pigman, in "The Enzymes," J. B. Summer and K. Myrbäck, Eds., Vol. I, Part 2, Academic Press, New York, 1952, p. 725.
71. R. G. H. Siu, see ref. 14, Chapter III, C, 5.
72. J. A. Gascoigne, *J. Soc. Dyers Colour.*, **77**, 53 (1961).
73. R. G. H. Siu and E. T. Reese, *Botan. Rev.*, **19**, 377 (1953).
74. K. Selby, in "Enzymatic Hydrolysis of Cellulose and Related Materials," E. T. Reese, Ed., Macmillan, New York, 1963, p. 33.
75. E. B. Cowling, see ref. 74, pp 1–32.
76. C. S. Walseth, *Tappi*, **35**, 233 (1952).
77. R. Blum and W. H. Stahl, *Text. Res. J.*, **22**, 178 (1952).
78. K. Selby, C. C. Maitland, and K. V. A. Thompson, *Biochem. J.*, **88**, 288 (1963).
79. E. T. Reese, L. Segal, and V. W. Tripp, *Text. Res. J.*, **27**, 626 (1957).
80. D. R. Whitaker, *Can. J. Biochem. Physiol.*, **35**, 733 (1957).
81. E. T. Reese, *Appl. Microbiol.*, **4**, 39 (1956).
82. E. T. Reese and W. Gilligan, *Text. Res. J.*, **24**, 663 (1954).
83. C. S. Walseth, *Tappi*, **35**, 228 (1952).
84. G. Halliwell, *Biochem. J.*, **68**, 605 (1958).
85. M. Mandels and E. T. Reese, *Develop. Ind. Microbiol.*, **5**, 5 (1964).

86. G. Halliwell, *Biochem. J.*, **95**, 270 (1965).
87. G. Youatt, *Aust. J. Biol. Sci.*, **13**, 188 (1960).
88. H. S. Levinson and E. T. Reese, *J. Gen. Physiol.*, **33**, 601 (1950).
89. G. Halliwell, see ref. 74, pp. 71–92.
90. P. K. Datta, K. R. Hanson, and D. R. Whitaker, *Biochim. Biophys. Acta*, **50**, 113 (1961).
91. M. Holden and M. V. Tracey, *Biochem. J.*, **47**, 407 (1950).
92. W. Klop and P. Kooiman, *Biochim. Biophys. Acta*, **99**, 102 (1965).
93. E. T. Reese, R. G. H. Siu, and H. S. Levinson, *J. Bacteriol.*, **59**, 485 (1950).
94. L. H. Li, R. M. Flora, and K. W. King, *Arch. Biochem. Biophys.*, **111**, 439 (1965).
95. K. Selby and C. C. Maitland, *Biochem. J.*, **94**, 578 (1965).
96. P. B. Marsh, G. V. Merola, and M. E. Simpson, *Text. Res. J.*, **23**, 831 (1953).
97. D. R. Whitaker, see ref. 74, pp. 51–70.
98. L. H. Li and K. W. King, *Appl. Microbiol.*, **11**, 320 (1963).
99. W. Gilligan and E. T. Reese, *Can. J. Microbiol.*, **1**, 90 (1954).
100. M. D. Enger and B. P. Sleeper, *J. Bacteriol.*, **89**, 23 (1965).
101. K. Nisizawa, Y. Hashimoto, and Y. Shibata, see ref. 74, p. 171.
102. W. O. Storvick and K. W. King, *J. Biol. Chem.*, **235**, 303 (1960).
103. G. Pettersson and J. Porath, *Biochim. Biophys. Acta*, **67**, 9 (1963).
104. D. R. Whitaker, K. R. Hanson, and P. K. Datta, *Can. J. Biochem. Physiol.*, **41**, 671 (1963).
105. P. N. Pal and B. L. Ghosh, *Can. J. Biochem.*, **43**, 81 (1965).
106. E. G. Hanstein and D. R. Whitaker, *Can. J. Biochem. Physiol.*, **41**, 707 (1963); D. R. Whitaker, *Arch. Biochem. Biophys.*, **53**, 439 (1954).
107. R. A. Aitken, B. P. Eddy, M. Ingram, and C. Weurman, *Biochem. J.*, **64**, 63 (1956).
108. C. T. Bishop and D. R. Whitaker, *Chem. Ind. (London)*, 119 (1955).
109. P. Kooiman, *Enzymologia*, **18**, 371 (1957).
110. A. Sharples, *J. Polym. Sci.*, **14**, 95 (1954).
111. R. H. Marchessault and B. G. Rånby, *Svensk Papperstidn.*, **62**, 230 (1959).
112. W. Sandermann, "Chemische Holzverwertung," BLV Verlags., München, 1963.
113. J. F. Saeman, W. E. Moore, R. L. Mitchell, and M. A. Millett, *Tappi*, **37**, 336 (1954).
114. L. F. McBurney, see ref. 14, Chapter III, C,1 and C,2.
115. L. F. McBurney, see ref. 14, Chapter III, C,3.
116. R. L. Colbran and T. P. Nevell, *J. Text. Inst. Trans.*, **49**, T333 (1958).
117. B. Alfredsson, W. Czerwinsky, and O. Samuelson, *Svensk Papperstidn.*, **64**, 812 (1961).
118. P. O. Bethge and T. P. Nevell, *Svensk Papperstidn.*, **67**, 37 (1964).
119. P. A. McGee, W. F. Fowler, Jr., E. W. Taylor, C. C. Unruh, and W. O. Kenyon, *J. Amer. Chem. Soc.*, **69**, 355 (1947).
120. T. P. Nevell, *J. Text. Inst. Trans.*, **42**, T91 (1951).
121. H. Spedding, *J. Chem. Soc.*, 3147 (1960).
122. E. K. Gladding and C. B. Purves, *Paper Trade J.*, **116** (No. 14), 26 (1943).
123. G. F. Davidson, *J. Text. Inst. Trans.*, **32**, T132 (1941).
124. C. B. Roth, U.S. Patent 2,758,111 (Aug. 7, 1956).
125. J. F. Haskins and M. S. Hogsed, *J. Org. Chem.*, **15**, 1264 (1950).
126. C. Dorée and A. C. Healey, *J. Text. Inst. Trans.*, **29**, T27 (1938).
127. W. D. Major, *Tappi*, **41**, 530 (1958).
128. H. G. Higgins, *J. Polym. Sci.*, **28**, 645 (1958).
129. D. Entwistle, E. H. Cole, and N. S. Wooding, *Text. Res. J.*, **19**, 527, 609 (1949).
130. J. A. Mattor, *Tappi*, **46**, 586 (1963).

131. D. M. MacDonald, *Tappi*, **48**, 708 (1965).
132. G. Mino and S. Kaizerman, *J. Polym. Sci.*, **31**, 242 (1958).
133. G. N. Richards, *J. Appl. Polym. Sci.*, **5**, 539 (1961).
134. A. Meller, *Holzforschung*, **14**, 78, 130 (1960).
135. J. C. Irvine and E. L. Hirst, *J. Chem. Soc.*, **121**, 1585 (1922).
136. P. E. Robbins and D. M. Hall, *Ind. Eng. Chem., Prod. Res. Develop.*, **1**, 285 (1962).
137. M. L. Wolfrom and S. Haq, *Tappi*, **47**, 692 (1964).
138. A. Pictet and J. Sarasin, *Helv. Chim. Acta*, **1**, 87 (1918).
139. S. L. Madorsky, V. E. Hart, and S. Straus, *J. Res. Nat. Bur. Stand.*, **56**, 343 (1956); **60**, 343 (1958).
140. R. F. Schwenker, Jr. and L. R. Beck, Jr., *J. Polym. Sci.*, **2C**, 331 (1963).
141. R. C. Laible, *Amer. Dyestuff Reptr.*, **47**, 173 (1958).
142. F. H. Holmes and C. J. G. Shaw, *J. Appl. Chem. (London)*, **11**, 210 (1961).
143. G. O. Phillips and J. C. Arthur, Jr., *Text. Res. J.* **34**, 497, 572 (1964).
144. J. H. Flynn and W. L. Morrow, *J. Polym. Sci.*, **2A**, 81 (1964).
145. R. E. Florin and L. A. Wall, *J. Polym. Sci.*, **1A**, 1163 (1963).
146. K. Hess and E. Steurer, *Z. Phys. Chem. (Leipzig)*, **193**, 234 (1944).
147. E. Steurer and K. Hess, *Z. Phys. Chem. (Leipzig)*, **193**, 248 (1944).
148. R. L. Ott, *J. Polym. Sci.*, **2A**, 973 (1964).
149. H. M. Spurlin, see ref. 14, Chapter IX, A.
150. C. J. Malm and L. J. Tanghe, *Ind. Eng. Chem.*, **47**, 995 (1955).
151. I. Croon, *Svensk Papperstidn.*, **63**, 247 (1960).
152. I. Croon and B. Lindberg, *Svensk Papperstidn.*, **60**, 843 (1957).
153. I. Croon, B. Lindberg, and A. Ros, *Svensk Papperstidn.*, **61**, 35 (1958).
154. I. Croon, *Svensk Papperstidn.*, **61**, 919 (1958).
155. R. W. Lenz, *J. Amer. Chem. Soc.*, **82**, 182 (1960).
156. I. Croon, *Svensk Papperstidn.*, **62**, 700 (1959).
157. I. Croon and E. Flamm, *Svensk Papperstidn.*, **61**, 963 (1958).
158. I. Croon and B. Lindberg, *Svensk Papperstidn.*, **59**, 794 (1956).
159. I. Croon and C. B. Purves, *Svensk Papperstidn.*, **62**, 876 (1959).
160. G. Touzinsky, *J. Org. Chem.*, **30**, 426 (1965).
161. S. P. Rowland, V. O. Cirino, and A. L. Bullock, *Can. J. Chem.*, **44**, 1051 (1966).
162. B. G. Rånby and S. A. Rydholm, in "Polymer Processes," C. E. Schildknecht, Ed., Wiley (Interscience), New York, 1956, pp. 379–386.
163. F. D. Miles, "Cellulose Nitrate," Wiley (Interscience), New York, 1955.
164. J. Barsha, see ref. 14, Chapter IX, B.
165. R. Klein and M. Mentser, *J. Amer. Chem. Soc.*, **73**, 5888 (1951).
166. J. Hine, "Physical Organic Chemistry," 2nd. ed., McGraw-Hill, New York, 1962, pp. 344–347.
167. T. Urbanski and J. Hackel, *Tetrahedron*, **2**, 300 (1958).
168. R. L. Mitchell, *Ind. Eng. Chem.*, **45**, 2526 (1953).
169. C. F. Bennett and T. E. Timell, *Svensk Papperstidn.*, **58**, 281 (1955).
170. C. J. Malm, K. T. Barkey, M. Salo, and D. C. May, *Ind. Eng. Chem.*, **49**, 79 (1957).
171. D. Finlayson and E. B. Thomas, *Chem. Ind. (London)*, 928 (1957).
172. R. W. Work, *Text. Res. J.*, **19**, 381 (1949).
173. E. V. Anderson and A. S. Cooper, Jr., *Ind. Eng. Chem.*, **51**, 608 (1959).
174. T. Takagi and J. B. Goldberg, *Mod. Textiles Mag.*, **41** (No. 4), 49 (1960).
175. L. E. Herdle and W. H. Griggs, *Tappi*, **48** (No. 7), 103A (1965).
176. A. V. Bailey, E. Honold, and E. L. Skau, *Text. Res. J.*, **28**, 861 (1958).
177. E. M. Buras, Jr., S. R. Hobart, C. Hamalainen, and A. S. Cooper, Jr., *Text. Res. J.*, **27**, 214 (1957).

178. P. Howard and R. S. Parikh, *Indian J. Chem.*, **3**, 191 (1965).
179. E. Kline, see ref. 14, Chapter IX, F.
180. G. K. Greminger, Jr., and A. B. Savage, in "Industrial Gums" R. L. Whistler and J. N. BeMiller, Eds., Academic Press, New York, 1959, Chapter XXIV.
181. T. E. Timell, "Studies on Cellulose Reactions," Esselte, Stockholm, 1950.
182. J. W. Weaver, C. A. MacKenzie, and D. A. Shirley, *Ind. Eng. Chem.*, **46**, 1490 (1954).
183. L. H. Bock, *Ind. Eng. Chem.*, **29**, 985 (1937).
184. E. J. Lorand, *Ind. Eng. Chem.*, **30**, 527 (1938); **31**, 891 (1939).
185. J. F. Mahoney and C. B. Purves, *J. Amer. Chem. Soc.*, **64**, 15 (1942).
186. A. B. Savage, A. E. Young, and A. T. Maasberg, see ref. 14, Chapter IX, E.
187. A. W. Schorger and M. J. Shoemaker, *Ind. Eng. Chem.*, **29**, 114 (1937).
188. R. T. K. Cornwell, see ref. 180, Chapter XXV.
189. J. R. Batdorf, see ref. 180, Chapter XXVII.
190. E. A. Peterson and H. A. Sober, *J. Amer. Chem. Soc.*, **78**, 751 (1956).
191. I. Jullander, see ref. 180, Chapter XXVI.
192. J. L. Gardon and R. Steele, *Text. Res. J.*, **31**, 160 (1961).
193. D. D. Gagliardi and F. B. Shippee, *Text. Res. J.*, **31**, 316 (1961).
194. T. C. Allen, *Text. Res. J.*, **34**, 331 (1964).
195. A. J. Stamm, *Tappi*, **42**, 39, 44 (1959).
196. H. Tovey, *Text. Res. J.*, **31**, 185 (1961).
197. S. J. O'Brien and W. J. VanLoo, Jr., *Text. Res. J.*, **31**, 276, 340 (1961); **32**, 292 (1962).
198. W. A. Reeves, *J. Text. Inst. Proc.*, **53**, P22 (1962).
199. W. A. Reeves, R. M. H. Kullman, J. G. Frick, Jr., and R. M. Reinhardt, *Text. Res. J.*, **33**, 169 (1963).
200. W. J. Roff, *J. Text. Inst. Trans.*, **49**, T646 (1958).
201. F. S. H. Head, *J. Text. Inst. Trans.*, **49**, T345 (1958).
202. M. D. Hurwitz and L. E. Conlon, *Text. Res. J.*, **28**, 257 (1958).
203. V. B. Chipalkatti, R. M. Desai, N. B. Sattur, J. Varghese, and J. C. Patel, *Text. Res. J.*, **33**, 282 (1963).
204. N. Tokita and K. Kanamaru, *J. Polym. Sci.*, **27**, 255 (1958).
205. E. W. Jones and J. A. Rayburn, *J. Appl. Polym. Sci.*, **5**, 714 (1961).
206. G. C. Tesoro, *J. Appl. Polym. Sci.*, **5**, 721 (1961).
207. G. C. Tesoro, P. Linden, and S. B. Sello, *Text. Res. J.*, **31**, 283 (1961).
208. J. J. Hermans, *Pure Appl. Chem.*, **5**, 147 (1962).
209. Y. Kobayashi, *J. Polym. Sci.*, **51**, 359, 368 (1961).
210. Kh. U. Usmanov, B. I. Aïkhodzhaev, and U. Azizov, *J. Polym. Sci.*, **53**, 87 (1961).
211. R. Y.-M. Huang, B. Immergut, E. H. Immergut, and W. H. Rapson, *J. Polym. Sci.*, **1A**, 1257 (1963).
212. J. C. Arthur, Jr. and D. J. Daigle, *Text. Res. J.*, **34**, 653 (1964).
213. V. Stannett, *Tappi*, **47** (No. 3), 58A (1964).
214. N. Geacintov, V. Stannett, E. W. Abrahamson, and J. J. Hermans, *J. Appl. Polym. Sci.*, **3**, 54 (1960).
215. H. Yasuda, J. A. Wray, and V. Stannett, *J. Polym. Sci.*, **2C**, 387 (1963).
216. R. J. Ceresa, *Polymer*, **2**, 213 (1961).
217. D. J. Bridgeford, *Ind. Eng. Chem.*, *Prod. Res. Develop.*, **1**, 45 (1962).
218. H. Kamogawa and T. Sekiya, *Text. Res. J.*, **31**, 585 (1961).
219. G. N. Richards, *J. Appl. Polym. Sci.*, **5**, 539 (1961).
220. J. H. Daniel, Jr., S. T. Moore, and N. R. Segro, *Tappi*. **45**, 53 (1962).
221. S. Kaizerman, G. Mino, and L. F. Meinhold, *Text. Res. J.*, **32**, 136 (1962).

222. G. N. Richards, *J. Soc. Dyers Colour.*, **80**, 640 (1964).
223. A. Y. Kulkarni, A. G. Chitale, B. K. Vaidya, and P. C. Mehta, *J. Appl. Polym. Sci.*, **7**, 1581 (1963).
224. D. K. R. Chaudhuri and J. J. Hermans, *J. Polym. Sci.*, **48**, 159 (1960); **51**, 373, 381 (1961).
225. J. Schurz, *Papier*, **16**, 525 (1962).
226. R. F. Schwenker, Jr., and E. Pacsu, *Tappi*, **46**, 665 (1963).
227. B.-A. Feit, A. Bar-Nun, M. Lahav, and A. Zilkha, *J. Appl. Polym. Sci.*, **8**, 1869 (1964).
228. F. K. Guthrie, *Tappi*, **46**, 656 (1963).
229. K. H. Meyer and P. Gürtler, *Helv. Chim. Acta*, **30**, 751 (1947).
230. R. A. Boissonnas, *Helv. Chim. Acta*, **30**, 1703 (1947).
231. N. B. Chanda, E. L. Hirst, and D. J. Manners, *J. Chem. Soc.*, 1951 (1957).
232. S. Peat, W. J. Whelan, and J. G. Roberts, *J. Chem. Soc.*, 3916 (1957).
233. H. Staudinger and B. Lantzsch, *J. Prakt. Chem.*, **156(2)**, 65 (1940).
234. S. R. Carter and B. R. Record, *J. Chem. Soc.*, 664 (1939).
235. K. H. Meyer, G. Noelting, and P. Bernfeld, *Helv. Chim. Acta*, **31**, 103 (1948).
236. A. S. Perlin and S. Suzuki, *Can. J. Chem.*, **40**, 50 (1962).
237. W. L. Cunningham and D. J. Manners, *Biochem. J.*, **90**, 596 (1964).
238. A. S. Perlin, see ref. 74, p. 185.
239. For reviews, see A. B. Foster and J. M. Webber *Advan. Carbohyd. Chem.*, **15**, 371 (1960); J. S. Brimacombe and J. M. Webber, "Mucopolysaccharides," Elsevier, Amsterdam, 1964, Chapter 2.
240. R. L. Whistler and C. L. Smart, "Polysaccharide Chemistry," Academic Press, New York, 1953, Chapter XXVIII.
241. M. Quillet, *C. R. Acad. Sci.*, **258**, 3349 (1964); *Chem. Abstr.*, **61**, 2182b (1964).
242. M. Quillet and M. L. Priou, *C. R. Acad. Sci.*, **256**, 2903 (1963); *Chem. Abstr.*, **59**, 902h (1963).
243. G. L. Clark and A. F. Smith, *J. Phys. Chem.*, **40,** 863 (1936).
243a. M. Falk, D. G. Smith, J. McLachlan, and A. G. McInnes, *Can. J. Chem.*, **44**, 2269 (1966).
243b. J. McLachlan, A. G. McInnes, and M. Falk. *Can. J. Botany*, **43**, 707 (1965).
244. K. M. Rudall, *Advan. Insect Physiol.*, **1**, 257 (1963).
245. R. H. Hackman, *Aust. J. Biol. Sci.*, **13**, 568 (1960).
246. R. H. Hackman, *Aust. J. Biol. Sci.*, **7**, 168 (1954).
247. S. T. Horowitz, S. Roseman, and H. J. Blumenthal, *J. Amer. Chem. Soc.*, **79**, 5046 (1957).
248. W. H. McNeely, see ref. 180, Chapter IX.
249. A. G. Norman and W. H. Peterson, *Biochem. J.*, **26**, 1946 (1932).
250. M. Schmidt, *Arch. Mikrobiol.*, **7**, 241 (1936).
251. P. Karrer and A. Hoffman, *Helv. Chim. Acta*, **12**, 616 (1929).
252. L. Zechmeister and G. Tóth, *Hoppe-Seyler's Z. Physiol. Chem.*, **223**, 53 (1934).
253. H. Brach and O. Fuerth, *Biochem. Z.*, **38**, 468 (1912); *Chem. Abstr.*, **6**, 1433 (1912).
254. W. N. Haworth, W. H. G. Lake, and S. Peat, *J. Chem. Soc.*, 271 (1939).
255. E. G. Cox and G. A. Jeffrey, *Nature*, **143**, 894 (1939).
256. S. Fränkel and A. Kelly, *Monatsh. Chem.*, **23**, 123 (1902).
257. P. Karrer and G. François, *Helv. Chim. Acta*, **12**, 986 (1929).
258. S. A. Barker, A. B. Foster, M. Stacey, and J. M. Webber, *J. Chem. Soc.*, 2218 (1958).
259. J. C. Irvine, *J. Chem. Soc.*, **95**, 564 (1909).
260. K. H. Meyer and H. Wehrli, *Helv. Chim. Acta*, **20**, 353 (1937).

261. L. Zechmeister and G. Tóth, *Naturwissenschaften*, **27**, 367 (1939).
262. W. Lotmar and L. E. R. Picken, *Experientia*, **6**, 58 (1950).
263. K. M. Rudall, *Symp. Soc. Exp. Biol.*, **9**, 49 (1955).
264. K. H. Meyer and G. W. Pankow, *Helv. Chim. Acta*, **18**, 589 (1935).
265. D. Carlström, *J. Biophys. Biochem. Cytol.*, **3**, 669 (1957).
266. F. G. Pearson, R. H. Marchessault, and C. Y. Liang, *J. Polym. Sci.*, **43**, 101 (1960).
267. N. E. Dweltz, *Biochim. Biophys. Acta*, **51**, 283 (1961).
268. D. Carlström, *Biochim. Biophys. Acta*, **59**, 361 (1962).
269. R. H. Hackman, *Aust. J. Biol. Sci.*, **15**, 526 (1962).
270. P. Schulze and G. Kunike, *Biol. Zentr.*, **43**, 556 (1923).
271. P. P. von Weimarn, *Ind. Eng. Chem.*, **19**, 109 (1927).
272. M. L. Wolfrom, G. G. Maher, and A. Chaney, *J. Org. Chem.*, **23**, 1990 (1958).
273. P. Schorigin and E. Hait, *Ber.*, **68**, 971 (1935).
274. M. L. Wolfrom, J. R. Vercellotti, and D. Horton, *J. Org. Chem.*, **29**, 547 (1964).
275. D. J. Candy and B. A. Kilby, *J. Exp. Biol.*, **39**, 129 (1962).
276. L. Glaser and D. H. Brown, *J. Biol. Chem.*, **228**, 729 (1957).
277. M. R. Lunt and P. W. Kent, *Biochem. J.*, **78**, 128 (1961).
278. E. G. Jaworski, L. Wang, and G. Marco, *Nature*, **198**, 790 (1963).
279. C. A. Porter and E. G. Jaworski, *Biochemistry*, **5**, 1149 (1966).
280. M. V. Tracey, *Nature*, **175**, 815 (1955).
281. M. V. Tracey, *Nature*, **167**, 776 (1951).
282. G. A. Strasdine and D. R. Whitaker, *Can. J. Biochem. Physiol.*, **41**, 1621 (1963).
283. M. P. Thirlwell, G. A. Strasdine, and D. R. Whitaker, *Can. J. Biochem. Physiol.*, **41**, 1603 (1963).
284. R. F. Powning and H. Irzykiewicz, *Comp. Biochem. Physiol.*, **14**, 127 (1965).
285. C. E. ZoBell and S. C. Rittenberg, *J. Bacteriol.*, **35**, 275 (1938).
286. R. Spencer, *Nature*, **190**, 938 (1961).
287. C. Jeuniaux, *Nature*, **192**, 135 (1961).
288. M. R. Lunt and P. W. Kent, *Biochim. Biophys. Acta*, **44**, 371 (1960).
289. C. Jeuniaux, *Arch. Int. Physiol. Biochim.*, **68**, 837 (1960).
290. M. L. Bade and G. R. Wyatt, *Biochem. J.*, **83**, 470 (1962).
291. J. J. Skujins, H. J. Potgieter, and M. Alexander, *Arch. Biochem. Biophys.*, **111**, 358 (1965).
292. L. Zechmeister and G. Tóth, *Enzymologia*, **7**, 165 (1939).
293. C. Jeuniaux, *Arch. Intern. Physiol. Biochim.*, **67**, 115 (1959).
294. A. Otakara, *Agr. Biol. Chem. (Tokyo)*, **26**, 30 (1962).
295. R. F. Powning and H. Irzykiewicz, *Nature*, **200**, 1128 (1963).
296. L. R. Berger and D. M. Reynolds, *Biochim. Biophys. Acta*, **29**, 522 (1958).
297. A. Otakara, *Agr. Biol. Chem. (Tokyo)*, **28**, 745 (1964).
298. P. Karrer and S. M. White, *Helv. Chim. Acta*, **13**, 1105 (1930).

37. HEMICELLULOSES

Roy L. Whistler and E. L. Richards

I. Introduction. 447
II. Isolation 449
   A. Extraction 449
   B. Chemical Modification during Isolation 450
   C. Purification 450
III. Properties 451
   A. Molecular Weight 451
   B. Crystal Structure 452
IV. Hemicelluloses 452
   A. D-Xylans. 452
   B. D-Mannans 458
   C. D-Gluco-D-mannans 459
   D. D-Galacto-D-gluco-D-mannans 461
   E. L-Arabino-D-galactans 462
V. Enzymic Hydrolysis of Hemicelluloses 464
   A. D-Xylans 464
   B. D-Mannans, D-Gluco-D-mannans, and D-Galacto-D-gluco-D-
     mannans 466
   References 467

I. INTRODUCTION

Cell walls of land plants consist of large amounts of cellulose microfibrils embedded in a continuous phase of lignin, pectin, and hemicellulose, of which the hemicellulose predominates. Pectin, present in 1 to 4% concentration, is localized largely in the primary wall. Some lignin and most of the hemicellulose are found mixed with cellulose in both the primary and secondary cell wall. The physical properties of the plant cell wall depend largely on interactions between these three main components. These substances are in close association, physically entangled and held by secondary forces such as Van der Waals forces and hydrogen bonding, and there are probably also some interconnecting primary bonds. At a very early stage of cell differentiation, cellulose, hemicellulose, and pectic substances are found as the main components. During growth of the primary wall, quantitative differences in these components occur, since they are not formed at equal rates. For example, analysis of cambial tissue during differentiation into phloem and xylem in sycamore (*Acer pseudoplatanus*) and in pine (*Pinus sylvestris*) shows that the ratio of D-xylose to 4-*O*-methyl-D-glucuronic acid decreases in the D-xylans.[1] Summative analysis[2] of developing cell walls of silver birch, Norway spruce, and Scots pine show that the middle lamella (intercellular substance) and primary wall are rich in pectic material (see Fig. 1). Lignin is the predominant substance of the middle lamella. The

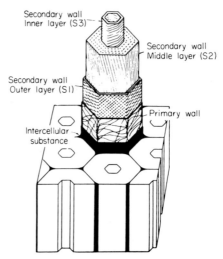

FIG. 1. Schematic representation of structure in a typical cell wall of wood.

cellulose content is highest in the S_1 and S_2 layers and in the S_3 layer of birch, while most of the uronic acid-containing D-xylan is found in the S_1 and S_2 layers. The S_3 layer is rich in L-arabino-D-glucurono-D-xylan, while the content of D-mannan increases gradually from the outer to the inner cell wall sections.

The name hemicellulose was proposed by Schulze[3] in 1891 to designate those polysaccharides extractable from plants by aqueous alkali. The name seemed appropriate, since these polysaccharides were found in close association with cellulose in the cell wall and were thought to be intermediates in cellulose biosynthesis. It is now known that the hemicelluloses are not precursors of cellulose and have no part in cellulose biosynthesis but rather represent a distinct and separate group of plant polysaccharides. Most workers limit the term hemicellulose to designate cell wall polysaccharides of land plants, except cellulose and pectin, and classify them according to the type of sugar residues present. Thus, D-xylan is a polymer of D-xylose residues, D-mannan of D-mannose residues, and D-galactan of D-galactose residues. Such homoglycans do not occur in large proportions. Most hemicelluloses are heteroglycans containing two to four, and rarely five or six, different types of sugar residues. Commonly occurring heteroglycans are L-arabino-D-xylans, L-arabino-D-glucurono-D-xylans, 4-*O*-methyl-D-glucurono-D-xylans, L-arabino-(4-*O*-methyl-D-glucurono)-D-xylans, D-gluco-D-mannans, D-galacto-D-gluco-D-mannans, and L-arabino-D-galactans. The hemicelluloses usually have branched structures, and the molecules are often partially acetylated. While most of the hemicelluloses examined have been from woods, cereals, and grasses, hemicelluloses also occur in ferns,[4] in the primitive gymnosperm *Ginkgo biloba* L.,[5] and in other primitive woods.[6]

II. ISOLATION

A. EXTRACTION

Hemicelluloses are usually extracted from plant tissue after removal of lipid and lignin. Removal of lignin exposes the hemicellulose for easy dissolution in alkali and permits isolation of pure hemicellulose.[7] Lipids are usually removed by extraction with a hot, azeotropic mixture of benzene and ethanol.[8]

Although some hemicelluloses are water-soluble after isolation, they are generally not water-extractable from plants prior to delignification, and sometimes not until after the use of strong swelling agents such as alkali or methyl sulfoxide. The water-soluble L-arabino-D-glucurono-D-xylan and D-galacto-D-gluco-D-mannan of softwoods can be extracted only with alkaline solution after delignification. *O*-Acetyl-(4-*O*-methyl-D-glucurono)-D-xylan is extractable from woods with methyl sulfoxide. One of the few major polysaccharides of wood that can be isolated in high yield by extracting the fully lignified wood with water is the L-arabino-D-galactan of larch.

The commonest method[9] for removing lignin is by treatment of the plant material with chlorous acid at 70 to 75°. Such delignified plant material is known as holocellulose. The hemicelluloses are then extracted from the holocellulose[10] with alkaline solution. Various concentrations of alkali, from 2% to 18% may be used, but 10% is usual. When the extracts are neutralized, the fraction that precipitates is called hemicellulose A; it is composed mainly of L-arabino-D-xylans, with some D-glucuronic acid residues usually present. A group of lower-molecular-weight acidic polysaccharides, termed hemicellulose B, remain dissolved and can be precipitated from the supernatant with ethanol. By taking advantage of differences in solubility it is possible to effect broad separations by alkaline extraction. For example, 24% potassium hydroxide solution extracts the L-arabino-(4-*O*-

* *References start on p. 467.*

methylglucurono)-D-xylans and D-galacto-D-gluco-D-mannans from the holo-cellulose of both gymnosperm[11] and angiosperm[12] woods but does not dissolve the D-gluco-D-mannans. The latter can be extracted from the residue with 17.5% sodium hydroxide solution containing 4% of borate.

B. Chemical Modification during Isolation

Delignification of plant tissue with acidified sodium chlorite causes slight degradation of the hemicelluloses. Reducing end groups are oxidized to aldonic acids,[13] and slight depolymerization and oxidation of 2,3-glycol groups may occur.[14] Chlorine, which is also used for delignification, also appears to cause some degradation of the hemicelluloses. Tissue of annual plants is converted into holocellulose more readily than is wood, possibly because these tissues contain less lignin. Delignification with either chlorine or chlorous acid yields hemicellulose of high quality. Chlorous acid is the preferred delignifying agent because of its convenience in handling.

Alkaline extraction of plant material or of holocellulose can result in many changes in the polysaccharide[15] even under oxygen-free conditions. Alkaline degradation occurs at the reducing groups of the polysaccharide and gives rise to saccharinic acid units by way of a β-elimination reaction. This leads to the exposure of new reducing groups, and the "peeling" reaction proceeds in a stepwise manner unless interrupted by a "stopping" reaction. In a $(1 \to 4)$-linked polysaccharide the action of alkali on a reducing-sugar end bearing a branch at C-3 has been thought to cause elimination of the side chain as well as the reducing sugar, which simultaneously rearranges to an alkali-stable metasaccharinic acid residue. However, it now appears[16] that alkaline degradation of $(1 \to 4)$-linked polysaccharides is not necessarily arrested at such branching points. D-Xylans are found to be degraded slowly from the reducing end of the chain when treated with cold, dilute alkali under the usual extraction conditions. At higher temperatures the 4-O-methyl-D-glucuronic acid residues undergo loss of O-methyl groups by β-elimination, and there is simultaneous disappearance of the uronic acid residue.[17] Alkaline degradation is greatly reduced, and the β-elimination reaction is prevented, if the polysaccharide is first reduced with borohydride.

C. Purification

It is often difficult to isolate a homogeneous polysaccharide from the complex mixture of polysaccharides obtained by the extraction of plant tissues.

The homogeneity or number of components in polysaccharide mixtures may be determined by ultracentrifugation[18] and by free-boundary electro-phoresis,[19] which may also be used for preparative purposes.[20]

Most methods for fractionation of hemicellulose on a macro scale make use of precipitation. Alkaline solutions of copper salts are often added to hemicellulose solutions to precipitate those containing significant amounts of D-mannose or D-xylose. The insoluble polysaccharide–copper complex is isolated, and the copper is removed by use of acidified alcohol solution or by a chelating agent. Pure polysaccharides are frequently obtained from dilute aqueous solutions by the gradual addition of ethanol.[21] An example is the fractional precipitation of hemicellulose B of corn cob, as shown in Fig. 2. A plot of weight of precipitated material against concentration of ethanol gives a characteristic two-component precipitation curve.

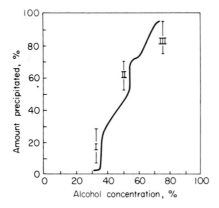

FIG. 2. Fractional precipitation curve for the hemicellulose B fraction of corn cob at 2% concentration and pH 2.0. (Whistler and Lauterbach.[22])

Acidic hemicelluloses are easily separated from neutral hemicelluloses by precipitation with quaternary ammonium salts[23] such as cetyltrimethyl-ammonium bromide, generally in neutral or mildly alkaline solution. Neutral polysaccharides can be fractionated after conversion into the slightly acidic complex, which forms an insoluble salt with certain quaternary ammonium ions.[24] Under strongly alkaline conditions, quaternary ammonium ions can precipitate neutral polysaccharides directly.[25]

III. PROPERTIES

A. MOLECULAR WEIGHT

Hemicelluloses from hardwoods have average degrees of polymerization (d.p.) of 150 to 200. Many hemicelluloses have a d.p. in the range of 50 to 100, low enough to make molecular-weight measurements difficult.[26] Their molecular weights can, however, be measured with osmometers that have membranes with very small pores. They can also be measured in a vapor-phase osmometer, particularly if the polymer is neutral and can be solubilized in organic solvents after conversion into the acetate or benzoate. The degree of polymerization may also be measured by light-scattering, sedimentation equilibrium and sedimentation–diffusion.

B. Crystal Structure

Some hemicelluloses are crystalline, and their crystal spacings have been determined.[27] In examination of (4-O-methyl-D-glucurono)-D-xylan[28] it is apparent[28] that the D-xylan backbone possesses a threefold screw axis, with a rotation of 120° for each D-xylose residue and a repeating length of 15 Å (see Fig. 3).

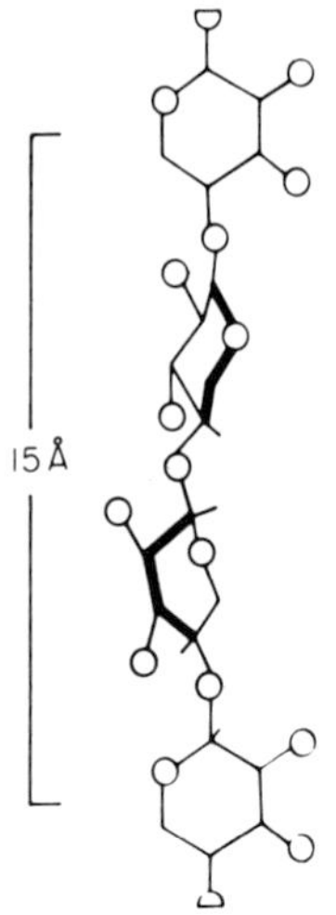

Fig. 3. Repeating unit and conformation of a (4-O-methyl-D-glucurono)-D-xylan. [R. H. Marchessault and C. Y. Liang, *J. Polym. Sci.*, **59**, 357 (1962).]

These conclusions are confirmed by spectroscopic examination of crystalline films of (4-O-methyl-D-glucurono)-D-xylans[28] with polarized infrared light. This work also suggests a definite orientation and conformation of the 4-O-methyl-D-glucuronic acid side chains and indicates that the intramolecular hydrogen bond 3–OH · · ·O′–5, previously shown to occur in cellulose,[29] is also present in the D-xylan.

IV. HEMICELLULOSES

A. D-Xylans

A group of polysaccharides, known as D-xylans, having backbone chains of (1 → 4)-linked β-D-xylopyranosyl residues occurs in all land plants[30] and in almost all plant parts.[31] Annual plants contain about 15 to 30% of these polymers, hardwoods 20 to 25%, and softwoods 7 to 12%. The nature of the glycosidic linkage between the D-xylopyranose residues has been determined by partial hydrolysis of the polymer to oligosaccharide fragments whose structures can be determined by classical means. Thus, from D-xylan a polymer-homologous series is obtained consisting of xylobiose, xylotriose, and so on, up to xyloheptaose, in which only (1 → 4)-linked β-D-xylopranosyl residues occur.[32] True D-xylans, composed exclusively of D-xylose residues, are neither common nor abundant. Esparto grass D-xylan is an example of such a simple homoglycan. D-Xylan from esparto grass can be fractionated as the copper complex to produce a polysaccharide free from L-arabinose,[33] which on further hydrolysis gives D-xylose in 95 to 98% yield. Hydrolysis of the methylated D-xylan produces 2,3-di-O-methyl-D-xylose (92%), 2,3,4-tri-O-methyl-D-xylose (2.6%), and 2-O-methyl-D-xylose (5%). Since acetolysis of the methylated D-xylan gives the disaccharide (1), the glycosidic link is (1 → 4) and the D-xylose residues are in the pyranose form.[34]

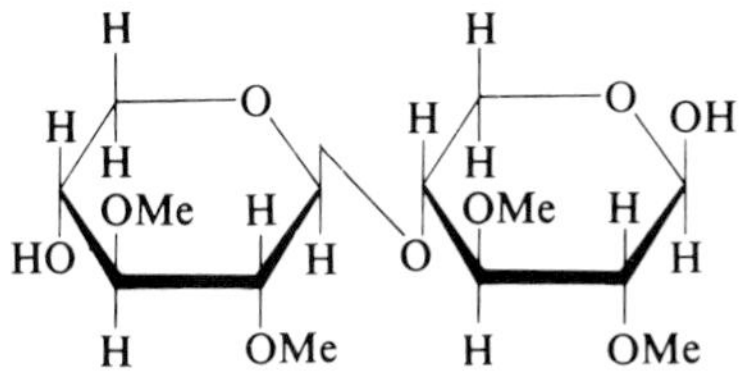

1

The yield of 2-O-methyl-D-xylose and 2,3,4-tri-O-methyl-D-xylose, and the molecular-weight measurements, suggest that the molecule has a single branch located at C-3 of one D-xylose residue. The partial structure shown in **2** summarizes the main features. Thus, the basic chain of the D-xylans is like

$$\beta\text{-}D\text{-}Xylp\text{-}(1 \longrightarrow 4)\text{-}(\beta\text{-}D\text{-}Xylp)_x\text{-}(1 \longrightarrow 4)\text{-}(\beta\text{-}D\text{-}Xylp)\text{-}(1 \longrightarrow 4)\text{-}(\beta\text{-}D\text{-}Xylp)_y\text{-}$$

$$3$$
$$\uparrow$$
$$1$$

$$\beta\text{-}D\text{-}Xylp\text{-}(1 \longrightarrow 4)\text{-}(\beta\text{-}D\text{-}Xylp)_z\text{-}(1 \longrightarrow 4)\text{-}\beta\text{-}D\text{-}Xylp$$

Xylp = Xylopyranose $x + y + z = 75 \pm 5$

2

that of cellulose except that each sugar ring lacks a projecting primary alcohol group, the molecule has points of branching, and the molecular weight is lower.

Few D-xylans are neutral molecules containing D-xylose residues only, and most bear side chains of other sugars. The most common side chain is

* References start on p. 467.

4-O-methyl-α-D-glucopyranosyluronic acid. Polysaccharides having this side chain are abundant in hardwoods, which are richer in hemicelluloses that contain D-xylose residues than are softwoods. Many hardwoods also have substantial amounts of O-acetyl-L-arabino-(4-O-methyl-D-glucurono)-D-xylan. The major softwood hemicellulose that contains D-xylose residues is an L-arabino-(4-O-methyl-D-glucurono)-D-xylan. Sometimes, and especially in annual plants, unmethylated D-glucuronic acid occurs. Each of these residues is usually joined to the D-xylopyranose chain residues by α-D-(1 → 2) linkages. Wood D-xylans, for example, give high yields of 2-O-(4-O-methyl-α-D-glucopyranosyluronic acid)-D-xylose on extended acid hydrolysis.[35a] This aldobiouronic acid, which is quite stable to acid hydrolysis, has a high positive rotation of +110°, which indicates the presence of an α-D linkage. Its structure (3) has been proved by classical methylation analysis.[35b,36] On esterification followed by acetylation, it produces a fully acetylated, crystalline methyl ester methyl glycoside.[37] An aldotriouronic acid, consisting of 4-O-methyl-β-D-glucopyranosyluronic acid residues joined (1 → 2) to a xylobiose residue, has also been obtained.[38] More recently an aldotetraouronic acid, which

2-O-(4-O-methyl-α-D-glucopyranosyluronic acid)-D-xylose (3)

appears to be O-β-D-xylopyranosyl-(1 → 4)-O-[4-O-methyl-α-D-glucuronopyranosyl-(1 → 2)]-O-β-D-xylopyranosyl-(1 → 4)-D-xylopyranose, has been obtained from the partial hydrolysis of paper-birch 4-O-methylglucuronoxylan.[39] Confirmation of sugar linkages was established by enzymic hydrolysis of birch D-xylan and isolation of five acidic oligosaccharides, three of which have been identified[40] as 4-O-methyl-α-D-glucuronopyranosyl-(1 → 2)-β-D-xylopyranosyl-(1 → 4)-β-D-xylopyranose (4, n = 0) and its two higher D-xylopolymer homologs (4, n = 1 or 2).

$$\beta\text{-D-Xyl}p\text{-(1} \longrightarrow 4)\text{-(}\beta\text{-D-Xyl}p)_n\text{-(1} \longrightarrow 4)\text{-}\beta\text{-D-Xyl}p$$
$$2$$
$$\uparrow$$
$$1$$
$$4\text{-}O\text{-Me-}\alpha\text{-D-Glc}p\text{A}$$

4

The number of 4-*O*-methyl-D-glucuronic acid groups along the chain varies considerably, but most hardwood D-xylans have approximately one acidic side chain per ten D-xylose residues. The distribution of the groups along the D-xylan chain is not known fully, but the isolation of xylan oligomers and the observation that none of the oligosaccharides contain two carboxyl groups suggest that the side chains are fairly widely separated and do not occur on adjacent D-xylose residues.

Although D-glucuronic acid and 4-*O*-methyl-D-glucuronic acid are most often linked at C-2 of the D-xylose residues, C-3 linkages have also been observed. Thus, acid hydrolysis of the hemicellulose from sunflower heads yields 3-*O*-(α-D-glucopyranosyluronic acid)-D-xylose,[41] and the hemicellulose from Monterey pine (*Pinus radiata*) and possibly maritime pine (*Pinus pinaster*)[42] and wheat straw[43] yields 3-*O*-(4-*O*-methyl-α-D-glucopyranosyluronic acid)-D-xylose. A hemicellulose from plum leaf gives 2-*O*-methyl-D-xylose on hydrolysis.[44] This is the first recognition of a methylated pentose in Nature.

Acetyl groups occur to about 3 to 5% in wood. Extraction of angiosperm wood with methyl sulfoxide yields hemicelluloses having 16.9% of acetate groups, corresponding to 7.1 ester groups per ten D-xylose residues.[45] Most of the *O*-acetyl groups are linked at C-3 and the remainder at C-2 of the D-xylose residues. These acetylated hemicelluloses are soluble in water and in organic solvents such as methyl sulfoxide, formamide, and *N,N*-dimethylformamide, but the acetyl groups migrate under acid or alkaline conditions and are easily saponified.

Another common side chain on xylan is L-arabinose. Perlin[46] showed that all the L-arabinose in a wheat-flour xylan is present as nonreducing furanose end groups linked to D-xylose residues. Oligosaccharides obtained from partial enzymic hydrolysis of L-arabino-D-xylans still contain a terminal

5

L-arabinofuranose residue. Thus, enzymic degradation of wheat-straw xylan by an enzyme from the mold *Myothecium verrucaria*[47] gives a series of oligosaccharides, one of which is the trisaccharide O-α-L-arabinofuranosyl-$(1 \rightarrow 3)$-O-β-D-xylopyranosyl-$(1 \rightarrow 4)$-D-xylopyranose,[48] while degradation of wheat L-arabino-D-xylan with an enzyme from *Streptomyces xylanose* gives a mixture of oligosaccharides that include a tetrasaccharide[49] O-α-L-arabino-furanosyl-$(1 \rightarrow 3)$-O-$[\alpha$-D-xylopyranosyl-$(1 \rightarrow 4)]$-O-β-D-xylopyranosyl-$(1 \rightarrow 4)$-D-xylose (**5**).

Characterization of these compounds and methylation analysis show that L-arabinose is linked to the main chain of the D-xylan, as depicted in formula **6**.

-$(1 \longrightarrow 4)$-β-D-Xyl$p$-$(1 \longrightarrow 4)$-β-D-Xyl$p$-$(1 \longrightarrow 4)$-$(\beta$-D-Xyl$p)_n$-$(1 \longrightarrow 4)$-

$$3$$
$$\uparrow$$
$$1$$

L-Ara$f$

6

Although the L-arabinose residues are generally in the furanose ring form, the more stable pyranose ring is sometimes present. Often the L-arabinosyl groups occur as single-unit side chains, but they may constitute side chains several sugar residues in length. Such branches may be terminated with a 4-O-methyl-D-glucopyranosyluronic acid group or even with a D-xylopyrano-syl group.

Determination of the fine structure of L-arabino-D-xylans requires evalua-tion of the side-chain distribution. D-Xylose residues are not oxidized by periodate at branch points. Since other sugar residues are readily attacked, the isolation of the periodate-resistant portions of the polysaccharide provides a method for assessing the arrangement of the side chains. A modification of Barry's[50] method for the degradation of periodate-oxidized polysaccharides with phenylhydrazine permits the isolation of the resistant branch points from wheat-flour L-arabino-D-xylan.[51] D-Xylose and xylobiose are major products, and xylotriose a minor product. Hydrogenation of the periodate-oxidized polysaccharide gives isolable fragments of glycerol glycosides.[52] Further evidence of side-chain distribution has been obtained by selective enzymolysis of L-arabino-D-xylans.[49]

The results of these and other experiments provide detailed information on the structure of highly branched L-arabino-D-xylans, which can be represented in general by structure **7**.

Ara$f$		Ara$f$	Ara$f$		Ara$f$		Ara$f$	Ara$f$

-Xyl$p$-Xyl$p$-Xyl$p$-Xyl$p$-Xyl$p$-Xyl$p$-Xyl$p$-Xyl$p$-Xyl$p$-Xyl$p$-

(a)   (b)   (a)   (b)

7

Sometimes widely separated D-xylopyranosyl residues (*a*) serve as branch points, but often two or, infrequently, more than two consecutive branched points (*b*) are separated by a single unbranched unit. At irregular intervals, averaging 20 to 25 residues, these sequences are interrupted by "open regions" of two to five unbranched D-xylopyranosyl residues. Occasionally an L-arabinosyl branch is separated by at least two D-xylopyranosyl units from branches at either side. Since the ratio of L-arabinose to D-xylose in the L-arabino-D-xylans varies considerably[46] for different plants, some molecules probably contain large proportions of highly branched sequences, while others contain frequent "open" regions.

L-Arabinose is most commonly found as a nonreducing end group, but some D-xylans contain nonterminal L-arabinofuranose residues. Since such L-arabinofuranosyl linkages are easily hydrolyzed, oligosaccharides produced from them by mild hydrolysis have L-arabinose residues at the reducing end. Such D-xylans are found in corn cob,[53] barley husk,[54] and perennial rye-grass,[55] and they release 2-O-β-D-xylopyranosyl-L-arabinose (**8**) on mild hydrolysis.

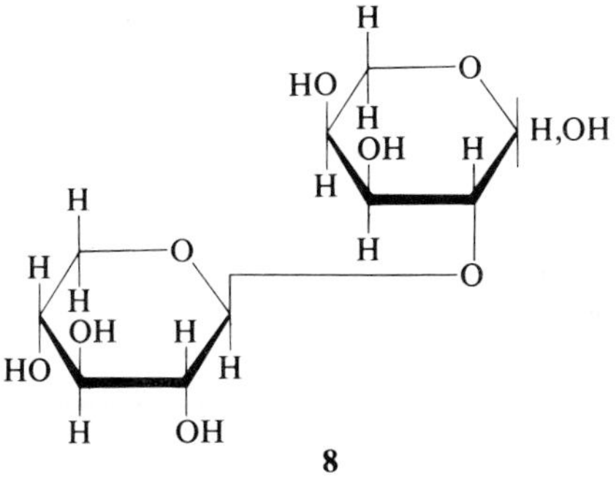

8

This nonterminal position of L-arabinose is confirmed by the isolation of 3,5-di-O-methyl-L-arabinose after methylation and hydrolysis. Application of the Smith degradation to barley husk L-arabino-D-xylan[52] has provided evidence that the 2-O-β-D-xylopyranosyl-L-arabinofuranose side chains are directly linked to the main chain, as shown in **9**. The backbone chains of

-(1 ⟶ 4)-β-D-Xyl*p*-(1 ⟶ 4)-
3
↑
1
-β-D-Xyl*p*-(1 ⟶ 2)-L-Ara*f*
9

D-xylans from the hemicellulose A fraction of Gramineae are similar to those from Leguminosae, but their substituent sugars differ.[56] The cereal D-xylans have L-arabinose and D-glucuronic acid residues attached to the D-xylan chain, whereas those from the legumes show a slightly higher content of uronic acid and contain no L-arabinose. Similar differences exist in the heteroxylans from the hemicellulose B fraction.[57] In both plant groups the main constituent is D-xylose. In hemicellulose B fractions, there are more L-arabinose and fewer uronic acid side chains than in the corresponding hemicellulose A fractions. Furthermore, the D-xylans of hemicellulose B fractions from Leguminosae contain less L-arabinose and more uronic acid than do those from the Gramineae. While there are no marked structural differences between the D-xylans of the cereals and grasses and those of woods, the D-xylans from softwoods and hardwoods and from other dicotyledonous plants are characterized by side chains of 4-O-methyl-D-glucuronic acid with some L-arabinofuranose side chains.

B. D-MANNANS

Polysaccharides composed almost entirely of D-mannose residues are the chief constituents of palm seed endosperm, occurring as food reserves that disappear on germination. The prime source of such D-mannans is the endosperm of the tagua palm seed, which is called vegetable ivory. Two morphologically distinct D-mannans are isolated from vegetable ivory. D-Mannan A is extracted with alkaline solution and gives crystalline X-ray patterns in the native state and when precipitated from alkaline solution.[58] D-Mannan B occurs as microfibrils analogous to those of cellulose. It has a higher molecular weight than D-mannan A and is not readily soluble in alkaline solutions, but it is soluble in cuprammonium hydroxide solution. On precipitation it gives an amorphous X-ray pattern.[58] Methylation and hydrolysis of these polysaccharides yields 2,3,6-tri-O-methyl-D-mannose as the main product,[59] while partial acetolysis followed by deacetylation yields a β-D-$(1 \rightarrow 4)$-linked disaccharide (mannobiose), mannotriose, and higher homologs.[60] Thus the D-mannans must be linear chains of D-mannopyranose residues linked by β-$(1 \rightarrow 4)$-glycosidic bonds (10). D-Mannans have been defined[61]

10

as polysaccharides containing 95% or more of D-mannose residues. In addition to their occurrence in ivory nuts, they are found in green coffee beans,[62] in tubers of various species of orchids (salep mannan),[63] and in the seaweed alga *Porphyra umbilicalis*.[64] All have the same general chemical structure,[62] but they appear to differ in chain length.

C. D-GLUCO-D-MANNANS

Wood of angiosperms contains 3 to 5% of D-gluco-D-mannans, which can be extracted from the holocellulose with aqueous sodium hydroxide containing borate after prior removal of 4-*O*-methylglucurono-D-xylan with aqueous potassium hydroxide.[12] The ratio of D-glucose to D-mannose is usually 1:2 in hardwood D-gluco-D-mannans, and their copolymer structure is indicated by the isolation of 4-*O*-β-D-glucopyranosyl-D-mannopyranose and 4-*O*-β-D-mannopyranosyl-D-glucopyranose along with β-D-(1 $\rightarrow$ 4)-linked mannobiose and mannotriose.[65a] The isolation of these fragments indicates that the hexose residues occur in the β-D form, which configuration is confirmed by their low specific optical rotation. Methylation and hydrolysis of these hardwood polymers yields mainly 2,3,6-tri-*O*-methyl-D-glucose and 2,3,6-tri-*O*-methyl-D-mannose,[65b, 66] a further indication that the polymers are composed of β-D-(1 $\rightarrow$ 4)-linked D-glucopyranose and D-mannopyranose residues. These hardwood D-gluco-D-mannans contain no D-galactose, unlike softwood D-gluco-D-mannans and D-galacto-D-mannans. Glycans of the same general structural type are also present in the seeds of various land plants. For example, Iles D-mannan, isolated from tubers of *Amorphophallus hearts*, has been shown to be a mixture of a D-gluco-D-mannan and a starch-like α-D-glucan. The D-gluco-D-mannan, which can be separated as its insoluble copper complex, has been shown by methylation analysis[67] to contain linear chains of (1 $\rightarrow$ 4)-linked β-D-mannopyranose (70%) and β-D-glucopyranose (30%) residues. Acetolysis followed by deacetylation gives[68] cellobiose, 4-*O*-β-D-glucopyranosyl-α-D-mannopyranose, and 4-*O*-β-D-mannopyranosyl-α-D-glucopyranose. Although neither mannobiose nor mannotriose can be detected after partial hydrolysis, evidence favors the presence of adjacent D-mannose residues in a partial structure such as **11**. Similar D-gluco-D-mannans are present in iris seeds,[69] lily bulbs,[70] and blue-bell seeds.[71]

-(1 $\longrightarrow$ 4)-(β-D-Man$p$)$_4$-(1 $\longrightarrow$ 4)-(β-D-Glc$p$)$_2$-(1 $\longrightarrow$ 4)-(β-D-Man$p$)$_4$-(1 $\longrightarrow$ 4)-
-(β-D-Glc$p$)$_2$-(1 $\longrightarrow$ 4)-
|
Glc$p$

11

* References start on p. 467.

Wood of the gymnosperms contains a larger proportion of D-gluco-D-mannans than that of angiosperms. These polysaccharides are structurally similar to cellulose and occur in association with cellulose molecules. The most widely applied technique for isolation involves the pre-extraction of the holocellulose with 24% potassium hydroxide to remove the L-arabino-(4-O-methyl-D-glucurono)-D-xylans and the D-galacto-D-gluco-D-mannans. The residue is then extracted exhaustively with 17.5% sodium hydroxide solution containing 4% of borate.[11] The majority of D-gluco-D-mannans extracted in this way from softwoods contain a few D-galactose residues, mostly as nonreducing end groups. D-Galactose residues are integral parts of most softwood D-gluco-D-mannans. For example, almost all D-gluco-D-mannans so far isolated have been found to contain D-galactose, D-glucose, and D-mannose in a fairly constant ratio of 1 to 2:10:30. Hence the so-called D-gluco-D-mannans of softwoods should be considered as triheteropolymers (D-galacto-D-gluco-D-mannans),[11] which, however, occur in a partially acetylated form. They contain approximately 6% of O-acetyl groups, which are linked singly to D-mannose residues. Polymers containing acetyl groups can be extracted from wood holocellulose with methyl sulfoxide. Osmotic pressure measurements on these polymers indicate that they contain about 100 hexose residues. However, light-scattering measurements made on a D-gluco-D-mannan from black spruce[72] indicate a d.p. of 400.

Hydrolysis of the methylated polysaccharides yields 2,3,6-tri-O-methyl-D-mannose and 2,3,6-tri-O-methyl-D-glucose as major products.[73] Small proportions of 2,3-di-O-methyl-D-glucose, 2,3-di-O-methyl-D-mannose, 2,3,4,6-tetra-O-methyl-D-galactose, 2,3,4,6-tetra-O-methyl-D-glucose, and 2,3,4,6-tetra-O-methyl-D-mannose are also obtained. These results suggest that the original glycan is a linear or slightly branched chain of (1 → 4)-linked D-mannopyranose and D-glucopyranose residues with D-galacto-pyranose residues as single-unit substituents joined to the main chain by (1 → 6) linkages, as shown in 12. Substantiation of this structure is obtained

⟶ 4)-β-D-Manp-(1 ⟶ 4)-β-D-Glcp-(1 ⟶ 4)-(β-D-Manp)₃-(1 ⟶ 4)-β-D-Glcp-
(1 ⟶ 4)-(β-D-Manp)-(1 ⟶ 4)-β-D-Manp
6
↑
1
α-D-Galp

12

by fragmentation analysis wherein the glycan is partially depolymerized by acetolysis or hydrolysis to oligosaccharides whose structures can be established by classical methods. Oligosaccharide fragments so obtained are shown in Table I. The large number of chain fragments obtained indicates a random arrangement of the sugar residues in the original polysaccharide. Isolation of cellobiose from the D-gluco-D-mannans of Western hemlock,[74] Amabilis fir,[75] and European larch[73] shows the contiguity of D-glucopyranose residues in the glycans, but no cellobiose was isolated by fragmentation of the D-gluco-D-mannan from loblolly pine, which must contain only isolated D-glucose residues.

TABLE I

OLIGOSACCHARIDE FRAGMENTS FROM D-GLUCO-D-MANNANS

Disaccharides	Trisaccharides	Others
M → M	M → M → M	M → M → M → M
M → G	M → M → G	M → M → G → M
G → M	M → G → M	M → M₃ → M
G → G	M → G → G	M → M₃ → G
	G → M → M	M → M₄ → M

Key: M →: β-D-mannopyranosyl-(1 → 4)-; G →: β-D-glucopyrano-syl-(1 → 4)-.

D. D-GALACTO-D-GLUCO-D-MANNANS

The alkali-soluble D-gluco-D-mannans, as well as the wood of all gymnosperms, contain a low-molecular-weight water-soluble polysaccharide composed of D-galactose, D-glucose, and D-mannose residues in the ratio of 1:1:3. The softwood D-galacto-D-gluco-D-mannans can be extracted from chlorite holocellulose with aqueous potassium hydroxide solution and are precipitated from solution by addition of barium hydroxide.[11] On methylation and hydrolysis, the major products are 2,3,6-tri-O-methyl-D-glucose and 2,3,6-tri-O-methyl-D-mannose,[75,76] indicating that the polysaccharides are composed of (1 → 4)-linked β-D-glucopyranose and β-D-mannopyranose residues. The number of 2,3-di-O-methyl-D-glucose and 2,3-di-O-methyl-D-mannose residues corresponds approximately to the number of 2,3,4,6-tetra-O-methyl-D-galactose residues, indicating that the D-galactose residues may be single-unit side chains joined by (1 → 6) links. D-Glucose and D-mannose occur also as nonreducing end groups.

The structures of the oligosaccharides obtained on partial hydrolysis (Table II)[75,76c,77] show that the linkages in the backbone are β-D-(1 → 4) and that D-galactose is linked to D-mannose, and probably also to D-glucose residues, by way of α-D-(1 → 6)-glycosidic bonds. For reasons still not

* References start on p. 467.

adequately explained, the α-D-(1 → 6)-galactosidic linkages are very acid-labile, so that D-galactose is the first sugar to appear in the hydrolyzate when D-galacto-D-gluco-D-mannans are subjected to mild hydrolysis.[76a,b] The D-galacto-D-gluco-D-mannans are partially O-acetylated.[78] The general structure of these polysaccharides is similar to that shown in structure 12 for the softwood D-gluco-D-mannans. They differ, however, in being of lower molecular weight and in having a relatively large number of D-galactose side chains. The side chains are undoubtedly responsible for conferring water solubility; they probably prevent the macromolecules from forming strong,

TABLE II

OLIGOSACCHARIDES OBTAINED FROM D-GALACTO-D-GLUCO-D-MANNAN

Disaccharides	Trisaccharides	Higher oligosaccharides
M → M	M → M → M	M → M → M → M
M → G	M → M → G	G → M → M → M
G → M	M → G → M	Gal → M → M → M
G → G	G → M → M	Gal → M → M → M → M
Gal → M	G → G → M	
	Gal → M → M	
	Gal → G → M (?)	

Key: G →: β-D-glucopyranosyl-(1 → 4)-; M →: β-D-mannopyranosyl-(1 → 4)-; Gal →: α-D-galactopyranosyl-(1 → 6).

intermolecular hydrogen bonds. In this respect the D-galactose residues serve the same function as the O-acetyl groups, the L-arabinose side chain, and the 4-O-methyl-D-glucuronic acid side chains in other hemicelluloses.

E. L-ARABINO-D-GALACTANS

Water-soluble, highly branched L-arabino-D-galactans are present in the wood of conifers and in the sap of at least one angiosperm, the maple tree. While most gymnosperm woods contain only small proportions of this polysaccharide, wood of the genus *Larix* may contain 25%. Although larches are deciduous, other deciduous gymnosperms such as bald cypress and *Ginkgo biloba* do not contain more than small amounts of L-arabino-D-galactan. Much work has been done on larch glycan by Wise *et al.*,[79] White,[80] and Bouveng *et al.*[81]

L-Arabino-D-galactans are the only major wood glycans that can be isolated in good yield by extraction of wood with water before delignification.

Their ease of extraction and their useful qualities as gums have brought them into commercial production marketed as the commercial gum, Stractan.

Larch L-arabino-D-galactan is a mixture of two components of similar structures,[82] having molecular weights of 100,000 and 160,000. The mixture can be separated by electrophoresis on glass-fiber sheets in a borate buffer. Among the larch species, the ratio of L-arabinose to D-galactose residues varies from 1:4 to 1:8, with 1:6 the most prevalent.

Mild hydrolysis of L-arabino-D-galactan from *Larix occidentalis* (Western larch) removes all the L-arabinose residues.[83] Although most of the sugar is obtained as a monosaccharide, some disaccharide, 3-*O*-β-L-arabino-pyranosyl-L-arabinose, is present.[84] Surprisingly, a small proportion of 6-*O*-β-D-galactopyranosyl-D-galactopyranosyl-D-galactose is also obtained with the D-galactose. Isolation of the disaccharide shows that not all the L-arabinose residues occur in the furanose form but suggests that the groups linked to D-galactose residues may be in this acid-labile form. All the D-galactose occurs in the pyranoid ring form.

Structural analysis of the glycans has followed the conventional procedures of methylation, hydrolysis, and identification of methylated sugars, and characterization of oligosaccharides obtained on partial hydrolysis. Such results, considered with hydrolytic fragmentation of the natural glycan and with Barry and Smith degradations, lead to generalized structures for these highly branched polymers.

The L-arabino-D-galactans of Western larch and those of other species of larch are structurally similar, with a basic framework of (1 → 3)-linked β-D-galactopyranose residues, each of which carries various side chains linked to C-6 positions of the principal chain. The structure of all the side chains is not known, but the majority seem to contain an average of two (1 → 6)-linked β-D-galactopyranose residues, whereas others consist of 3-*O*-β-D-arabinopyranosyl-L-arabinofuranose residues. In addition, a few terminal residues of L-arabinofuranose and D-glucuronic acid occur, linked

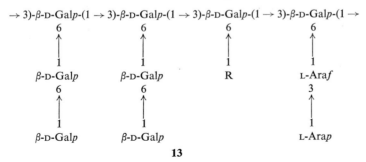

13

to C-6 positions of the main chain. There is also evidence for some doubly branched and some $(1 \rightarrow 4)$-linked D-galactose residues. A generalized structure may be written as in **13**, where R = β-D-galactopyranose or, less frequently, L-arabinofuranose, or a D-glucopyranosyluronic acid residue.

L-Arabino-D-galactans from other sources have similar properties and structures, but the detailed structures appear to differ from species to species. As one example the generalized structure of the glycan from maritime pine is shown (**14**):

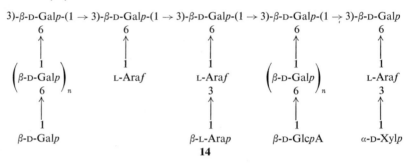

An interesting L-arabino-L-rhamno-D-galactan has been isolated from maple sap.[85] After removal of sucrose by dialysis, the polymer was electrophoretically uniform, containing D-galactose, L-arabinose, and L-rhamnose in the ratio of 51:45:5. Like the softwood L-arabino-D-galactan, this polymer has a backbone consisting mainly of $(1 \rightarrow 3)$-linked β-D-galactopyranose residues.

V. ENZYMIC HYDROLYSIS OF HEMICELLULOSES[86]

Enzymes from a wide variety of sources hydrolyze hemicelluloses, and their action provides important structural information. (See Vol. IIB, Chap. 47 for systematic names of enzymes.) The low yields of oligomers produced, a former experimental problem, has been largely overcome by the introduction of new techniques such as conducting the hydrolysis in a dialysis bag so that oligomers may diffuse from the enzyme system, escaping further hydrolysis.[87]

A. D-XYLANS

Enzymes that catalyze the hydrolysis of D-xylans are found in animals, plants, and a large number of microorganisms.[86] The products of hydrolysis depend on the source and purity of the enzyme preparation. Although all xylanases yield xylo-oligosaccharides from D-xylans, some do not produce significant amounts of D-xylose. Certain bacteria[88] contain at least two enzymes active in the hydrolysis of D-xylans—an exoenzyme able to act on D-xylopyranose chains having three or more D-xylose units, and an endo-enzyme that is able to split xylobiose. Other enzymes act on D-xylan in a way similar to the action of α- and β-amylase on starch. For example, two enzymes have been isolated from a culture of *Aspergillus batatae*.[89] Xylanase A, which has been obtained in a crystalline form, hydrolyzes the $(1 \rightarrow 4)$-β-D-xylosidic linkages of poly-D-xylosaccharides at random, whereas xylanase B attacks poly-D-xylosaccharides and D-xylo-oligosaccharides from the end of the chain, liberating D-xylose units.[90] A xylanase, similar in action to xylanase A, has been isolated from cell-free extracts of the rumen ciliate *Epidinium ecaudatum*.[91] This enzyme hydrolyzes D-xylodextrins randomly to xylobiose and D-xylose. The enzyme does not hydrolyze D-xylan or xylobiose.

The presence of side-chain linkages affects enzymic degradation. For example, commercial "pectinase" degrades 4-*O*-methyl-D-glucurono-D-xylan, yielding both D-xylose and its oligomers, and also a series of acidic oligosacch-arides extending from an aldotetra- to an aldo-octauronic acid. The latter series has at the nonreducing end a 4-*O*-methyl-D-glucuronic acid residue linked $(1 \rightarrow 2)$ to a D-xylose residue that is joined in the linear series of $(1 \rightarrow 4)$-linked β-D-xylose residues.[40] The isolation of oligosaccharides having a 4-*O*-methyl-D-glucuronic acid residue on the nonreducing D-xylose end unit shows that a D-xylosidic linkage attached to this D-xylose residue is cleaved by an enzyme in the "pectinase," as shown in the schematic formula **15**. Likewise, the absence of 4-*O*-methyl-D-glucuronic acid and any aldobio-uronic or aldotriouronic acids in the hydrolyzate shows that the α-D-glucu-ronosyl bond and the two subsequent β-D-xylopyranosyl links are stable.[40] These bonds are also somewhat stable to acid hydrolysis.[92]

$$-X-X-X-X-X-X-X-X-$$
$$\quad\quad | \quad\quad\quad\quad |$$
$$\quad\quad A \quad\quad\quad\quad A$$

15

Possible action of "pectinase" on 4-*O*-methyl-D-glucurono-D-xylan

X = D-xylose; A = 4-*O*-methyl-D-glucuronic acid; — stable linkages; — labile linkages

Enzymes from many sources degrade L-arabino-D-xylans randomly to D-xylose, L-arabinose, and D-xylo-oligosaccharides.[90, 93] Arabinofuranosidase removes L-arabinose from highly branched L-arabino-D-xylans before the xylan chain is hydrolyzed.[91] That this is not the only mode of action is shown by the isolation of a series of oligosaccharides, containing both L-arabinose and D-xylose, from the partial enzymic hydrolysis of wheat-straw L-arabino-D-xylan[47] with an enzyme preparation from the mold *Myrothecium verrucaria*. One of these oligosaccharides is *O*-L-arabinofuranosyl-$(1 \rightarrow 3)$-O-β-D-xylopyranosyl-$(1 \rightarrow 4)$-D-xylose,[48] and its isolation constitutes proof that L-arabinofuranose residues are an integral part of the L-arabino-D-xylans. Partial degradation to oligosaccharides is also induced by a xylanase from *Streptomyces*. This enzyme readily hydrolyzes bonds joining D-xylopyranosyl residues at branch points. The minimal spatial requirement for this enzyme appears to be two adjacent D-xylopyranosyl residues between branches or, for the formation of xylobiose, at least four, consecutive, unbranched residues.[94] As described earlier, the use of this enzyme has provided much information on the interbranch sequences in wheat-straw L-arabino-D-xylan. The enzyme of *Myrothecium verrucaria*, which acts on wheat-straw L-arabino-D-xylan,[47] and the enzyme "pectinase," which acts on 4-O-methyl-D-glucurono-D-xylan,[40] appear able to catalyze hydrolysis nearer to branch points than does the *Streptomyces* enzyme. Both former enzymes yield oligosaccharides in which L-arabinose or 4-O-methyl-D-glucuronic acid is linked to the nonreducing end of a xylobiose or xylotriose residue.

B. D-Mannans, D-Gluco-D-mannans, and D-Galacto-D-gluco-D-mannans

Known enzymes that catalyze the degradation of D-mannans are distributed erratically in Nature; they occur in malt, a culture filtrate of a bacterium that splits agar–agar, the mold *Neurospora sitophila*, and the marine alga *Cladsphora oupestois*.[86a,b] The algal mannanase appears to catalyze the random hydrolytic cleavage of ivory-nut D-mannan A, since the products include D-mannose, mannobiose, and other saccharides.

Information on the linkages in D-gluco-D-mannans can be obtained by enzymic hydrolysis. Hydrolysis of a D-gluco-D-mannan from jack pine (*Pinus banksiana*) by a crude hemicellulase gives products[87b] that, with a few exceptions, are similar to those obtained by partial acid hydrolysis. The major products, mannobiose and mannotriose, are obtained in 25% yield, indicating the presence of long chains of $(1 \rightarrow 4)$-linked β-D-mannopyranose residues. Other hydrolysis products are β-D-$(1 \rightarrow 4)$-linked disaccharides of D-glucose and D-mannose. These disaccharide residues appear to be positioned randomly in a linear polymer chain. The concept is confirmed by periodate oxidation, which shows that the D-gluco-D-mannan has no branches originating at C-2 or C-3. No cellobiose is found in the hydrolysis products, but two trisaccharides are obtained wherein two D-glucose residues are joined by a presently unestablished linkage. In the hydrolysis of D-gluco-D-mannan from Norwegian spruce with a commercial cellulase from *Penicillium chrysogenum notatum*[95] a D-mannopentaose was produced identical with that previously isolated from ivory-nut D-mannan.[60]

The use of purified enzymes should lead to specific hydrolysis of D-gluco-D-mannans and the production of fragments from which the structure of the natural polymer may be presented more clearly.

REFERENCES

1. J. P. Thornber and D. H. Northcote, *Biochem. J.*, **82**, 340 (1962); D. H. Northcote, *Symp. Soc. Exp. Biol.*, **17**, 157 (1963).
2. H. Meier and K. C. B. Wilkie, *Holzforschung*, **13**, 177 (1959); H. Meier, *J. Polym. Sci.* **51**, 11 (1961); *Pure Appl. Chem.*, **5**, 37 (1962).
3. E. Schulze, *Ber.*, **24**, 2277 (1891).
4. T. E. Timell, *Chem. Ind. (London)*, 474 (1961).
5. T. E. Timell, *Svensk Papperstidn.*, **63**, 652 (1960).
6. D. J. Brasch and J. K. N. Jones, *Tappi*, **42**, 913 (1959).
7. S. C. Rogers, R. L. Mitchell, and G. J. Ritter, *Anal. Chem.*, **19**, 1029 (1947).
8. E. F. Kurth, *Ind. Eng. Chem. Anal. Ed.*, **11**, 203 (1939).
9. R. L. Whistler, J. Bachrach, and D. R. Bowman, *Arch. Biochem. Biophys.*, **19**, 25 (1948).
10. R. L. Whistler and M. S. Feather, *Methods Carbohyd. Chem.*, **5**, 144 (1965).
11. T. E. Timell, *Tappi*, **44**, 88 (1961).
12. T. E. Timell, *Svensk Papperstidn.*, **63**, 472 (1960).
13. A. Jeanes and H. S. Isbell, *J. Res. Nat. Bur. Stand.*, **A27**, 125 (1941).
14. E. S. Becker, J. K. Hamilton, and W. E. Lucke, *Tappi*, **48**, 60 (1965).
15. R. L. Whistler and J. N. BeMiller, *Advan. Carbohyd. Chem.*, **13**, 289 (1958).
16. G. O. Aspinall, C. T. Greenwood, and R. J. Sturgeon, *J. Chem. Soc.*, 3667 (1961).
17. R. J. Ross and N. S. Thompson, *Tappi*, **48**, 376 (1965).
18. G. A. Adams, *Can. J. Chem.*, **38**, 280 (1960).
19. D. H. Northcote, *Biochem. J.*, **58**, 353 (1954).
20. B. J. Hocevar and D. H. Northcote, *Nature*, **179**, 488 (1957).
21. R. L. Whistler and J. L. Sannella, *Methods Carbohyd. Chem.*, **5**, 34 (1965).
22. R. L. Whistler and G. E. Lauterbach, *Arch. Biochem. Biophys.*, **77**, 62 (1958).
23. J. E. Scott, *Chem. Ind. (London)*, 168 (1955).
24. H. O. Bouveng and B. Lindberg, *Acta Chem. Scand.*, **12**, 1977 (1958).
25. E. L. Hirst, D. A. Rees, and N. G. Richardson, *Biochem. J.*, **95**, 453 (1965).
26. C. T. Greenwood, *Advan. Carbohyd. Chem.*, **7**, 289 (1952); **11**, 335 (1956).
27. R. H. Marchessault and A. Sarko, *Advan. Carbohyd. Chem.*, **22**, 421 (1967).
28. R. H. Marchessault and C. Y. Liang, *J. Polym. Sci.*, **59**, 357 (1962); R. H. Marchessault, *Pure Appl. Chem.*, **5**, 107 (1962).
29. C. Y. Liang and R. H. Marchessault, *J. Polym. Sci.*, **37**, 385 (1959); R. H. Marchessault and C. Y. Liang, *ibid.*, **43**, 71 (1960).
30. T. E. Timell, *Svensk Papperstidn.*, **65**, 266 (1962); K. Matsuzaki, M. Moriza, A. Ishida, and H. Sobue, *J. Chem. Soc. Jap. Ind. Chem. Sect.*, **65**, 987 (1962).
31. D. H. Northcote, *Biochem. Soc. Symp.*, **22**, 105 (1963).
32. R. L. Whistler and C. C. Tu, *J. Amer. Chem. Soc.*, **75**, 3609 (1952).
33. S. K. Chanda, E. L. Hirst, J. K. N. Jones, and E. G. V. Percival, *J. Chem. Soc.*, 1289 (1950).
34. W. N. Haworth and E. G. V. Percival, *J. Chem. Soc.*, 2850 (1931).

35a. J. K. N. Jones and L. E. Wise, *J. Chem. Soc.*, 3389 (1952).
35b. T. E. Timell, *Methods Carbohyd. Chem.*, **1**, 301 (1965).
36. C. P. J. Glaudemans and T. E. Timell, *J. Amer. Chem. Soc.*, **80**, 941, 1209 (1958).
37. T. E. Timell, *Can. J. Chem.*, **37**, 827 (1959).
38. J. K. Hamilton and N. S. Thompson, *J. Amer. Chem. Soc.*, **79**, 6464 (1957).
39. W. H. Bearce, *J. Org. Chem.*, **30**, 1613 (1965).
40. T. E. Timell, *Can. J. Chem.*, **40**, 22 (1962); *Acta Chem. Scand.*, **16**, 1027 (1962); *J. Org. Chem.*, **27**, 1804 (1962) and *Svensk Papperstidn.*, **65**, 435 (1962).
41. C. T. Bishop, *Can. J. Chem.*, **33**, 1521 (1955).
42. A. Roudier and L. Eberhard, *Tappi*, **38**, 156A (1955).
43. A. Roudier, *Assoc. Tech. Ind. Papetière Bull.*, **2**, 53 (1954).
44. P. Andrews and L. Hough, *Chem. Ind. (London)*, 1278 (1956).
45. H. O. Bouveng, P. J. Garegg, and B. Lindberg, *Chem. Ind. (London)*, 1727 (1958); *Acta Chem. Scand.*, **14**, 742 (1960).
46. A. S. Perlin, *Cereal Chem.*, **28**, 382 (1951).
47. C. T. Bishop and D. R. Whitaker, *Chem. Ind. (London)*, 119 (1955).
48. C. T. Bishop, *J. Amer. Chem. Soc.*, **78**, 2840 (1956).
49. H. R. Goldschmid and A. S. Perlin, *Can. J. Chem.*, **41**, 2272 (1963).
50. V. C. Barry and P. W. D. Mitchell, *J. Chem. Soc.*, 4020 (1954).
51. C. M. Ewald and A. S. Perlin, *Can. J. Chem.*, **37**, 1254 (1959).
52. G. O. Aspinall and K. M. Ross, *J. Chem. Soc.*, 1681 (1963).
53. R. L. Whistler and D. I. McGilvray, *J. Amer. Chem. Soc.*, **77**, 1884 (1955).
54. G. O. Aspinall and R. J. Ferrier, *J. Chem. Soc.*, 4188 (1957).
55. G. O. Aspinall, I. M. Cairncross, and K. M. Ross, *J. Chem. Soc.*, 1721 (1963).
56. R. L. Whistler and B. D. E. Gaillard, *Arch. Biochem. Biophys.*, **93**, 332 (1961).
57. B. D. E. Gaillard, *Phytochemistry*, **4**, 631 (1965).
58. H. Meier, *Biochim. Biophys. Acta*, **28**, 229 (1958).
59. F. Klages, *Ann.*, **509**, 159 (1934); **512**, 185 (1934).
60. G. O. Aspinall, R. B. Rashbrook, and G. Kessler, *J. Chem. Soc.*, 215 (1958).
61. G. O. Aspinall, *Advan. Carbohyd. Chem.*, **14**, 429 (1959).
62. M. L. Wolfrom, M. L. Laver, and D. L. Patin, *J. Org. Chem.*, **26**, 4533 (1961).
63. H. Pringsheim and A. Genin, *Hoppe-Seyler's Z. Physiol. Chem.*, **140**, 229 (1964); F. Klages and R. Maurenbrecher, *Ann.* **535**, 175 (1938).
64. J. K. N. Jones, *J. Chem. Soc.*, 3292 (1950).
65a. T. E. Timell, *Tappi*, **43**, 844 (1960).
65b. A. Jabbar Mian and T. E. Timell, *Can. J.Chem.*, **38**, 1511 (1960).
66. J. K. N. Jones, E. Merler, and L. E. Wise, *Can. J. Chem.*, **35**, 634 (1957); G. A. Adams, *ibid.*, **39**, 2423 (1961).
67. P. A. Rebers and F. Smith, *J. Amer. Chem. Soc.*, **76**, 6097 (1954).
68. F. Smith and H. C. Srivastava, *J. Amer. Chem. Soc.*, **78**, 1404 (1956).
69. P. Andrews, L. Hough, and J. K. N. Jones, *J. Chem. Soc.*, 1186 (1953).
70. P. Andrews, L. Hough, and J. K. N. Jones, *J. Chem. Soc.*, 181 (1956).
71. J. L. Thompson and J. K. N. Jones, *Can. J. Chem.*, **42**, 1088 (1964).
72. H. A. Swenson, A. J. Morak, and S. Kurath, *J. Polym. Sci.*, **51**, 231 (1961).
73. G. O. Aspinall, R. Begbie, and J. E. McKay, *J. Chem. Soc.*, 214 (1962).
74. J. K. Hamilton and H. W. Kircher, *J. Amer. Chem. Soc.*, **80**, 4703 (1958).
75. E. C. A. Schwarz and T. E. Timell, *Can. J. Chem.*, **41**, 1381 (1963).
76a. J. K. Hamilton, E. V. Partlow, and N. S. Thompson, *J. Amer. Chem. Soc.*, **82**, 451 (1960).
76b. T. E. Timell, *Tappi*, **45**, 734 (1962).

76c. A. R. Mills and T. E. Timell, *Can. J. Chem.*, **41**, 1389 (1963).

77. H. Meier, *Acta Chem. Scand.*, **14**, 749 (1960).

78. G. E. Annergoen, I. Croon, B. F. Enström, and S. A. Rydholm, *Svensk Papperstidn.*, **64**, 386 (1961).

79a. L. E. Wise and F. C. Peterson, *Ind. Eng. Chem.*, **22**, 362 (1930).

79b. L. E. Wise and H. C. Unkauf, *Cellulosechemie*, **14**, 20 (1933).

79c. L. E. Wise, P. L. Hammer, and F. C. Peterson, *Ind. Eng. Chem.*, **25**, 184 (1933).

79d. F. C. Peterson, M. Maughan, and L. E. Wise, *Cellulosechemie*, **15**, 109 (1934).

80. E. V. White, *J. Amer. Chem. Soc.*, **63**, 2871 (1941); **64**, 302, 1507, 2838 (1942).

81. H. O. Bouveng and B. Lindberg, *Acta Chem. Scand.*, **12**, 1977 (1958); *Advan. Carbohyd. Chem.*, **15**, 53 (1960).

82. H. Mosimann and T. Svedberg, *Kolloid.-Z.*, **100**, 99 (1942).

83. H. O. Bouveng, *Acta Chem. Scand.*, **13**, 1869 (1959).

84. H. O. Bouveng and B. Lindberg, *Acta Chem. Scand.*, **10**, 1515 (1956).

85. G. A. Adams and C. T. Bishop, *Can. J. Chem.*, **38**, 2380 (1960).

86a. W. Pigman in "The Enzymes," J. B. Sumner and K. Myrbäck, Eds., Vol. I, Part 2, Academic Press, 1951, p. 725.

86b. D. J. Manners, *Quart. Rev.*, (*London*), **9**, 73 (1955).

86c. T. E. Timell, *Svensk Papperstidn.*, **65**, 435 (1962).

87a. T. J. Painter, *Can. J. Chem.*, **37**, 497 (1959).

87b. O. Perila and C. T. Bishop, *Can. J. Chem.*, **39**, 815 (1961).

88. H. Sorenson, *Nature*, **172**, 305 (1953).

89. S. Fukui and M. Sato, *Bull. Agr. Chem. Soc. Jap.*, **21**, 392 (1957).

90. R. L. Whistler and E. Masak, *J. Amer. Chem. Soc.*, **77**, 1241 (1955); B. H. Howard, *Biochem. J.*, **67**, 643 (1957).

91. R. W. Bailey and B. D. E. Gaillard, *Biochem. J.*, **95**, 758 (1965).

92. S. C. McKee and E. E. Dickey, *J. Org. Chem.*, **28**, 1561 (1963).

93. J. A. Gascoigne and M. M. Gascoigne, *J. Gen. Microbiol.*, **22**, 242 (1960).

94. A. S. Perlin and E. T. Reese, *Can. J. Biochem. Physiol.*, **41**, 1842 (1963).

95. H. O. Bouveng, T. Iwasaki, B. Lindberg, and H. Meier, *Acta Chem. Scand.*, **17**, 1796 (1963).

AUTHOR INDEX

Volume IIA

Numbers in parentheses are reference numbers and indicate that an author's work is referred to although his name is not cited in the text. Numbers in italics show the page on which the complete reference is listed.

A

Abdel-Akher, M., 83(86), *131*, 335(249), *368*, 407(70), *412*
Abdullah, M., 128(338), *137*, 253, 257(112), 289(410), *292, 293, 300*
Abe, J., 142(300), *211*
Abe, T., 165(116c), *204*
Abeles, R. H., 317(113), *365*
Abraham, S., 323(153), *366*
Abrahamson, E. W., 434(214), *443*
Acree, T. E., 94(156a), *133*
Acton, E. M., 13(56), 18(91, 92), 53(56), *61, 62*, 188(256a), *209*
Ada, G. L., 287(394), *300*
Adams, G. A., 78(41), 111(237), *130, 134*, 395(36), *411*, 450(18), 459(66), 464(85), *467, 468*
Adams, M., 90(144), *132*, 275, 279(271), *297*
af Ekenstam, A., 422(63), *440*
Agahigian, H., 143(186), 174(186), *206*
Agarival, K. L., 26(139), 35(194), *63, 65*
Agrawal, K. M. L., 270(320), 283(341a), *296, 299*
Ahmed, M. U., 266(181), *295*
Aïkhodzhaev, B. I., 434(210), 435(210), *443*
Aitken, R. A., 113(246), *135*, 425(107), *441*
Akabori, S., 252(45), *292*
Akatsuka, T., 341(301), *370*
Akazawa, T., 340(287, 288, 295), 341(287, 297, 299, 300), *369, 370*
Akita, E., 143(161), 145(264), 171(160, 161, 162, 163), 193(264, 265), 195(265), *205, 206, 210*
Albano, E. L., 144(195), 154(55), 175(195), 182(195), 188(195), *202, 207*
Albersheim, P., 348(332, 333), *371*
Albon, N., 279, 280(310), *298*, 313(94), 316(94), *364*
Albrecht, H. P., 39(214), *65*
Alessandrini, A., 274(262), *297*

Alexander, M., 438(291), *445*
Alexandrowicz, Z., 387(19), *411*
Alexanian, C., 415(19), *439*
Alford, E. F., 145(268), 194(268), *210*
Alfredsson, B., 427(117), *441*
Alfsen-Blanc, A., *248*, 285, *299*
Algranati, I. D., 338(274), 342(305), *369, 370*
Alivisatos, S. G. A., 45(246), *66*
Allen, A., 359(404b), *372*
Allen, J., 276(275), *297*
Allen, T. C., 434(194), *443*
Alpers, D. H., 250(36), 269(36), 270(234), *291, 296*
Altenburg, F. W., 280(312a), *298*
Altermatt, H. A., 114(256), *135*, 346(317), *370*
Ames, S. R., 402(67b), 405(67b), *412*
Aminoff, D., 308(65), *364*
Anagnostopoulos, C., 267(193), *295*
Anai, M., *248*, 250(23), *291*
Anand, N., 198(283), *210*
Andersen, B., 276(272), *297*
Anderson, C. D., 53(273), *67*, 172(172), *206*
Anderson, D. M. W., 398(57), *412*
Anderson, E., 304(27), *363*
Anderson, E. V., 432(173), *442*
Anderson, F. B., 264(158), *294*, 318(119), *365*
Anderson, J. D., 265(166), *294*
Anderson, J. S., 283(342), *299*
Anderson, W., 4(5), 28(154), *59, 64*
Ando, T., 99(168), *133*, 251, 253, 289(405), *292, 300*
Andrews, P., 76(32), 114(258), 115(258, 262, 263), *130, 135*, 455(44), 459(69, 70), *468*
Anema, P. J., 269(216), *296*
Anfinsen, C. B., *249*, 250(28, 38), *251*, 268(207), 269(38, 211), *291, 295, 296*
Anliker, R., 142(25), 148(25), 149(30), *201*

1

Annergoen, G. E., 462(78), *469*
Anno, K., 82(78), 111(78), 112(78), *131*, 402(67a), *412*
Anstiss, C., 285(365), *299*
Anzai, K., 179(210), *207*
Appel, S. H., 250(36), 269, *291*
Appelmans, F., 284(362), 285(362), *299*
Appleman, M. M., 338(279), *369*
Araki, C., 304(28), *363*
Arcamone, F., 145(255, 259, 260), 188(254, 254a, 254b, 255), 190(259, 260), *209*
Archibald, A. R., 253(64), *292*, 357(378, 383, 386), *372*
Argoudelis, A. D., 143(114, 124, 127), 145(263a), 162(96, 98), 163(96, 98), 164 (114), 165(114), 167(122, 124, 127), 168 (122), 192(263, 263a), *203, 204, 209*
Arison, B., 143(174), 172(174), *206*
Armentrout, S., 198(280), *210*
Armstrong, E. F., 105(189), *133*
Armstrong, H. E., 218(24), *238*
Armstrong, J. J., 356(377), *372*
Armstrong, R. F., 218(24), *238*
Arnold, W. N., *248*, 276, *297*
Aronson, M., 273, *297*
Arthur, J. C., Jr., 428(143), 434(212), 435 (212), *442, 443*
Asai, M., 145(258c), 189(258c), *209*
Asbun, W., 20(106, 107), 55(106), *63*
Aschner, M., 336(254), *368*
Ashmore, J., 336(262), *369*
Ashton, G. C., 275(266), *297*
Aso, K., 14(59), *61*
Aspinall, G. O., 75(19), 76(34), *77*, 78(34, 45), 85(114), 114(257, 259), 115(259), 116 (271), *129, 130, 131, 135*, 334(240), 335 (240), 348(335), *368, 371*, 382(3), *410*, 450(16), 456(52), 457(52, 54, 55), 458 (60, 61), 460(73), 461(73), 467(60), *467, 468*
Assarsson, A., 112(242), *135*
Augier, J., 216(8), 217(13), *237, 238*
Auricchio, S., 255(83, 84), 270(230), *292, 296*
Austrian, R., 303(11), 312(11), 346(323), 347(323), 352(352), *362, 370, 371*
Autrey, R. L., 164(115a), *204*
Averill, F. J., 400(63), *412*
Avigad, G., 83(84), 116(268), 124(316), *131, 135, 136*, 256, 258(124), 278(295, 296), 280(295, 296), *293, 298*, 318(121, 123), 320, *365*
Avineri-Shapiro, S., 336(254, 255), *368*
Axelrod, B., 312(91), *364*
Axelrod, J., 303(9), 312(9), 325(164), *362, 366*
Axen, U., 142(297, 298), *211*
Ayers, W. A., 244(10), *291*, 316(109), *365*
Azaroff, L. V., 384(7), *410*
Azizov, U., 434(210), 435(210), *443*

B

Babad, H., 304(16), 322(142, 143), *362, 366*
Babbington, A., 330(202), *367*
Bachli, P., 112(241), 113(241), *135*
Bachrach, J., 449(9), *467*
Bacon, E. E., 110(223), *134*
Bacon, J. S. D., 85(111), 103(178), 110(223), 123(314), *131, 133, 134, 136*, 232(79), *240*, 256(94), 279(302), 280(313, 316), *293, 298*, 313(93), *364*
Baddiley, J., 6(23), 8(23), 52(268), *60, 67*, 356, 357(378, 379, 380, 383, 386), 358(388, 391, 392), *372*
Bade, M. L., 438(290), *445*
Baer, H. H., 80(57, 58), *130*, 274(261), *297*
Bär, H. P., 23(123), 26(123), *63*
Bagchi, P., 116(270), *135*
Bahl, O. P., 270(220), 283(341a), *296, 299*
Bailey, A. V., 432(176), *442*
Bailey, J. M., 76(36), *77*, 125(36), 128(36), 129(36), *130*
Bailey, R. W., 69(1), 71(1), 73(1), 74(12), 75(17, 22, 29), 85(103), 90(142), *129, 131, 132*, 252(53), 253(65), 256(98, 99), 267, *292, 293, 295*, 346(325), *370*, 416(36a), *439*, 465(91), *469*
Baker, B. R., 8(30a), 28(156), 32(180), 35(30a), 36(200), 37, 41(156), 53(273), *60, 64, 65, 67*, 143(183), 172(172), 174(183), *206*
Balasubramanian, A. S., 362, *373*
Balazs, E. A., 359(393), *372*
Baldwin, E., 304(30), *363*
Ball, D. H., 115(263), 117(278), *135*
Ball, E. M., 257(113), *293*
Ball, S., 147(17), *201*
Ballou, C. E., 51(266a), *67*, 217(14, 15), *238*, 356(375), *372*

Bandurski, R. S., 312(91), *364*
Banks, W., 391(31), *411*
Bannister, B., 33(185), 57(185), 65, 145(261), 162(96, 98), 163(96, 98), 190(261, 261b), *203, 209*
Barber, G. A., 235(94), *240*, 304(18, 19), 310(79), 326(174, 178), 342(18, 19, 309), 343(19, 310), 345(19), *362, 364, 366, 367, 370*, 416(32, 33, 34), 417(32), *439*
Barber, G. L., 110(220), 114(220), *134*
Barber, M., 146(14), 197(14), *200*
Barbieri, W., 188(254), *209*
Bardos, T. J., 13(58), *61*
Barker, A., 265(167), *294*
Barker, G. R., 5(18), 6(18), *60*
Barker, H. A., 103(176), *133*, 319(125), 334(246), 335(246), *365, 368*
Barker, S. A., *77*, 78(46), 83(87), 85(112), 91(150), *130, 131, 132*, 253(62, 63), 256 (98), 257(104, 107), 278(288), 280(311), 283, 288, 290(416), *292, 293, 297, 298, 299, 300*, 332(216), 333, 334(223), 335(223), *367, 368*, 436(258), *444*
Barkey, K. T., 431(170), *442*
Barnes, C. C., 51(266), *67*
Barnett, J. A., 262, *294*
Barnett, J. E. G., 272(241), *296*
Bar-Nun, A., 434(227), *444*
Barr, B. K., 320(131), *365*
Barry, V. C., 77, 112(240), *135*, 409(73), *412*, 456, *468*
Barsha, J., 431(164), *442*
Bartholomew, B., 308(63), 359(63), *364*
Bartley, J. C., 323(153), *366*
Bartuska, V. J., 27(147), 38(204), *64, 65*
Bartz, Q. R., 142(71), 143(136, 137, 138), 155(69), 156(71), 157(72), 168(135, 136, 137, 138), 175(191), 176(191), 197(71, 135, 138), *202, 204, 205, 206*
Bass, L. E., 5(16), *60*
Bassham, J. A., 303(1, 2), *362*
Batdorf, J. R., 433(189), *443*
Bates, F. J., 85(102), 101(170), 107(102), 109(102), 111(102), *131, 133*
Batra, K. K., 344(312a, 314d), 345(316), *370*
Battley, E. H., 266(185), *295*
Baudisch, J., 419(56), *440*
Bauer, H. F., 107(202), *134*
Bax, P. C., 52(269a), *67*
Bayer, G. F., 422(67), *440*

Bayliss, T. M., 255(85b), *292*
Bealing, F. J., 279(303), *298*
Bean, R. C., 304(14), 320(131), 346(321), 347(321), 348(321), *362, 365, 370*
Bear, H. H., 169(148a), *205*
Bearce, W. H., 454(39), *468*
Beaufay, H., 270(221), *296*
Beauquesne, L., 227(72), *239*
Beavan, G. H., 347(326), *370*
Bechmann, F., 15(71), *62*
Bechtler, G., 398(55), *412*
Beck, J. R., 182(227, 228), 188(227), *208*
Beck, L. R., Jr., 428(140), *442*
Becker, B., 221(47), *239*
Becker, E. S., 450(14), *467*
Becker, J. F., 285, *300*
Beevers, H., 346, *370*
Begbie, R., 460(73), 461(73), *468*
Beijerinck, M. W., 335, *368*
Belenky, B. G., 142(299), *211*
Bell, D. J., 85(111), 116(267), *131, 135*, 280 (310, 312), *298*, 304(30), *363*
Belocopitow, E., 338, *369*
BeMiller, J. N., 98(164a), *133*, 382(2), *410*, 450(15), *467*
Bender, H., 75(28), *129*, 248(19), 268(19), 273(19), *291*, 332, *367*
Bendich, A., 5(15), 6(24a), *60*
Benedict, R. G., 164(113), *204*
Ben-Hayyim, G., 416(28), 417(28), *439*
Benitez, A., 33(186), 36(200), 53(273), 57(186), 65, 67, 172(172), *206*
Benn, M. H., 231(77a), *240*
Bennett, C. F., 431(169), *442*
Benoit, H., 388(21), *411*
Benson, B. W., 142(52), 153(52), 154(52), *202*
Bentley, H. R., 143(171), 172(171), *206*
Benz, E., 42(226), *66*
Beránek, J., 27(140), 32(179), 39(213), 42 (224b), *63, 64, 65, 66*
Berger, L. R., 283(344), *299*, 438(296), *445*
Berger, L. S., 266(172), *295*, 318(118), *365*
Bergmann, M., 113(250), *135*
Berlin, Y. A., 144(242), 185(242, 243), 186 (247b), *208, 209*
Bernfeld, P., 285(367, 368), *299*, 435(235), *444*
Bernheimer, H. P., 303(11), 312(11), 346 (323), 347(323), 352(352, 353), *362, 370, 371*

Bernstein, R. L., 355(370), *372*
Bernt, E., 273(250), *297*
Berry, M., 141(6), *200*
Bertolini, M., 287(396), *300*
Bertrand, G., 218(22), *238*
Bessell, C. J., 147(17), *201*
Betes, W. K., 269(218), *296*
Bethge, P. O., 427(118), *441*
Bettelheim, F. A., 383, 384, 390, *410*, *411*
Bhacca, N. S., 91(152), *132*, 185(244), *208*, 330(202), *367*
Bhatia, I. S., *298*
Biedron, S. I., 198(278), *210*
Biemann, K., 197(277), *210*
Bilow, A., 151(41), *201*
Bines, B. J., 75(16), *129*
Binkley, S. B., 20(106, 107), 55(106), *63*
Binkley, W. W., 85(106), 91(152), *131*, *132*, 280(312a), *298*
Binnie, B., 287(397), *300*
Birch, A. J., 141(7a), 148(7a), *200*
Bird, G. R., 386(15), *411*
Bird, I. F., 319, *365*
Birkenmeyer, R. D., 145(261), 150(261, 261d), *209*
Birkhofer, L., 12, *61*
Birzgalis, R., 276(277) *297*
Bishop, C. T., 75(18, 21), 76(33), 86(121), 114(33), *129*, *130*, *132*, 395(36), 399(59), *411*, *412*, 426(108), *441*, 455(41), 456(47, 48), 464(85), 465(47), 466(47, 48, 87b), *468*, *469*
Bishop, M. N., 143(123), 167(123), *204*
Bizioli, F., 145(260), 190(260), *209*
Black, W. A. P., 415(17), *439*
Blackford, R. W., 18(85), *62*
Blacklow, R. S., 305(49), 307(60), *363*, *364*
Blanchard, P. H., 279, 280(310, 312), *298*, 313(94), 316(94), *364*
Blank, H. U., 22(116), 23(116), *63*
Blindenbacher, F., 152(48), *201*
Bloch, A., 172(163a), *206*
Block, R. J., 85(104), 88(104), *131*
Blomqvist, G., 98(165), *133*
Bloom, W., 323(144), *366*
Blum, R., 78(47), *130*, 422(77), *440*
Blumbergs, P., 43(232), *66*, 144(195, 196a), 175(194, 195), 176(194, 196a, 198), 182(195), 188(195), *207*

Blumenthal, H. J., 85(115), 113(115), *132*, 305(41, 45), *363*, 436(247), 437(247), *444*
Bobbitt, J. M., 86(123), 91(147), *132*
Bobek, M., 21(109, 110), 63(109a), *63*
Bock, L. H., 433(183), *443*
Bock, R. M., *251*
Bodley, J. W., 198(282), *210*
Böeseken, J., 8(27a), *60*
Boesenberg, H., 219(35), *238*
Bognár, R., 80(62), *130*
Bohne, M., 361(410), *373*
Bohonos, N., 142(85), 159(85), 174(182), 195(85), 197(85), *203*, *206*
Boissonnas, R. H., 435(230), *444*
Bolton, C. H., 150(35), *201*
Bonner, J., 348(332), *371*
Bonner, T. G., 399(58a), *412*
Boos, W., 274(257), *297*
Borders, D. B., 182(227), 183(229), 188(227), *208*
Borgstrom, B., 254, *292*
Borud, A. M., 164(113), *204*
Boschman, Th. A. C., 287(395), *300*
Bottomley, W., 227(71), *239*
Botvinick, I. J., 14(62), *61*
Bouquet, A., 232(84), *240*
Bourdier, L., 232(81), *240*
Bourne, E. J., 77, 85(112), 91(150), *131*, *132*, 256(98, 100), 257(104, 105, 106, 107), 265(167), 278(288), 280(311), 290, *293*, *294*, *297*, *298*, *300*, 317(116), 321(134), 330(197, 202), 332(216), 333, 334(223), 335(223), *365*, *367*, *368*, 399(58a), *412*
Bourquelot, E., 106(195), 111(230), 128(333), *134*, *137*, 218(25, 28), 219(36), 232(82), *238*, *240*, 244(13), *291*
Boutron-Charlard, A. F., 217, *238*
Bouveng, H. O., 395(40), 398(58), *411*, *412*, 451(24), 455(45), 462, 463(83, 84), 467(95), *467*, *468*, *469*
Bowen, D. H., 270(224), 271(224), *296*
Bowles, W. A., 11(48), 12(48, 50), 54(48), *61*
Bowman, D. R., 449(9), *467*
Boxer, G. E., 8(28), 19(100), 53(100), 55(100), *60*, *62*, 199(290), *210*
Brach, H., 436(253), *444*
Bradley, R. M., 271(238), *296*
Brady, R. O., 271(238), *296*, 327(182), *367*
Braná, H., 270(229), *296*
Brasch, D. J., 449(6), *467*

Braun, G., 109(218), *134*
Braun, G. A., 113(248), *135*
Brauns, I. K., 334(243), 335(243), *368*
Bravo, P., 142(16), 146(16), *201*
Brazhnikova, M. G., 145(270), 194(270), *210*
Bredereck, H., 3(1), *59*, 80(61), *130*
Breuer, H. J., 103(178), *133*, 280(316), *298*
Brew, K., 324(155b, 155c), *366*
Bricker, H. M., 74(9), *129*
Bridel, M., 124(315), 128(333), *136, 137*, 219 (34, 37), 221(45, 46), *238, 239*, 244(13), *291*
Bridgeford, D. J., 434(217), 435(217), *443*
Briggs, D. R., 340(285), *369*
Briggs, L. H., 225(64), *239*
Brigl, P., 81(74), *131*
Brimacombe, J. S., 155(70), 157(70), 186 (246, 247), *202, 208, 209*, 361(405), *373*, 435(239), 436(239), 437(239), 438(239), *444*
Brockmann, H., 142(22, 23a), 144(248, 250, 251), 145(248, 301), 148(22, 23a), 149 (23a, 29), 150(33), 186(248, 249), 187(248, 250, 251), 188(248, 252), 195(251), *201, 209, 211*
Brodbeck, U., 323(154, 155), 324(155a), *366*
Bromer, W. W., 171(158), *205*
Brook, F. X., 145(304a), *211*
Broom, A. D., 27(148), 55(280), *64, 67*
Broschard, R. W., 145(266), 193(266), 196 (266), *210*
Brown, A. M., 416(29), *439*
Brown, D. H., 289, 290(412), *300*, 332, 334 (234, 235), 335(234, 235), 338(276), 339 (234), 342(303), 351(235), *367, 368, 369, 370*, 438(276), *445*
Brown, D. M., 20, 45(245), *62, 66*
Brown, G. B., 50(263), *67*, 172(166), *206*
Brown, J. I., 269(210), *296*
Brown, R., 4(10), 58(10), *59*
Brown, R. J., 217(14), *238*
Brown, W. J., 284(355), *299*
Browning, B. L., 418, *439*
Brownstein, A. M., 111(226), 112(226), *134*
Brufani, M., 142(70c), 155(70c), *202*
Brummond, D. O., 244, *291*, 343(311, 312), *370*, 416(35), *439*
Brunetti, P., 307(58), *363*
Bubl, E. C., 402(67b), 405(67b), *412*
Buchanan, J. G., 315(104), 320(104), 356 (377), 357(379, 382), *365, 372*

Buchhawat, B. K., 362, *373*
Buckland, F. L., 334(243), 335(243), *368*
Buddecke, B., 281(329), 282(329), *298*
Budovich, T., 76(37), *77*, 93(156), *130, 133*, 253(61), *292*
Budovskii, E. I., 35(193), *65*
Buehler, E., 17(77), *62*
Buerger, M. J., 384(7), *410*
Bull, B., 255(81, 82), *292*
Bullock, A. L., 430(161), *442*
Bungay, H. R., 145(269), 194(269), *210*
Bunker, J. E., 14(62), *61*
Bunones, E. P., 142(297), *211*
Bunton, C. A., 96(160), *133*, 243, *290*
Buras, E. M., Jr., 432(177), *442*
Burdon, M. G., 20(105), *62*
Burge, D. A., 388, *411*
Burger, M. M., 290, *300*, 327(181), 358(389), *367, 372*
Burma, D. P., 315(103), *365*
Burrows, E. P., 142(297, 298), *211*
Burstein, C., 236(96), *240*
Burton, K., 314(99), 337(99), *365*
Burton, R. M., 287(392), *300*, 323(151), 327 (181, 182), *366, 367*
Buston, H. W., 266(173), *295*
Butler, K., 47(251), *66*
Button, D., 357(386), *372*
Byers, A., 269(219), *296*
Byrde, R. J. W., 273(248), *296*

C

Cabib, E., 107(201), *134*, 305(33), 313(98), 324, 337(98), 338(274), 342(305), *363, 365, 366, 369 370*
Cahill, G. F., 336(262), *369*
Cairncross, I. M., 75(19), 78(45), *129, 130*, 457(55), *468*
Calendi, E., 199(294), *210*
Calkins, D. F., 23(122), 29(162), 31(169), 36(122), 38(162, 204), *63, 64, 65*
Calvin, M., 303(1, 2), *362*
Campbell, N., 226(68), 227(72), *239*
Campbell, R., 217(21), *238*
Canady, M. R., 361(409), *373*
Candy, D. J., 325(158), *366*, 438, *445*
Canevazzi, G., 145(259), 190(259), *209*
Caper, R., 284(355), *299*

Caputto, R., 104(185), 124(185), 125(185), *133*, 287(393), *300*
Carbon, J. A., 15(68, 72), *61, 62*
Carbone, J. V., 325(166), *366*
Cardini, C. E., 103(175), *133*, 235(93), *240*, 304(12, 13, 21), 305(33, 40, 42), 313(98), 315(13, 101), 319(12, 13), 321, 324(168), 326(169, 170), 327, 337(98, 268), 340(21), 341(298), *362, 363, 365, 366, 367, 369, 370*
Carlson, A. S., 330(201), 332(215), *367*
Carlson, D. M., 308(62), 324(155d), 342, *364, 366, 370*
Carlström, D., 437(268), *445*
Carminatti, S. H., 338(273), *369*
Carnie, J. A., 254(79, 80), 276(278), *292, 297*
Caron, E. L., 175(189, 190), 183(231), 184(231), *206, 208*
Carr, C. I., 389(25), *411*
Carrington, T. R., 280(311), *298*
Carroll, P. M., 24(126), *63*
Carss, B., 356(377), 357(379), *372*
Carter, H. E., 144(222), 167(122), 168(122, 130), 181(220, 221, 222, 223), *204, 208*, 217(10), *238*, 305(39), *363*
Carter, S. R., 435(234), *444*
Carter, W., 198(278d), (279a), *210*
Carubelli, R., 287(393), *300*
Casparis, P., 221(45), *239*
Cassidy, F., 20(102), *62*, 180(217), *207*
Cassidy, J. T., 287(391), *300*
Cassinelli, G., 145(255), 188(254, 254b, 255), *209*
Cavalieri, L. F., 58(291a), *68*
Cavazos, L. F., 285(364), *299*
Caygill, J. C., 270(226), 281(327, 330, 332) *296, 298*
Ceccarini, C., 258(125), *293*
Ceder, O., 142(13), 146(13), 147(13), 195(13), *200*
Celmer, W. D., 142(47), 146(15), 150(15), 152(15, 46, 47), 153(15), 154(15), 155(15), 156(15), 158(15), *200, 201*, 305(39), *363*
Cepure, A., 86(120), 117(120), *132*, 273(253), 274(253), *297*
Ceresa, R. J., 434(216), *443*
Černý, M., 69(2), *129*, 215(4), *237*
Chaikoff, I. L., 323(153), *366*
Chambers, R. W., 20(104), 28(155), *62, 64*
Chanda, N. B., 435(231), *444*

Chanda, S. K., 395(35), *411*, 453(33), *467*
Chaney, A., 437(272), *445*
Chang, C., 164(108), *203*
Chang, F. N., 200(296), *210*
Chang, N., 58(291a), *68*
Chararas, C., 237(98), *240, 297*
Charaux, C., 220(40), 221(45, 46), *238, 239*
Chargaff, E., 46(247), *66*, 287(397), *300*
Charlson, A. J., 76(31), 83(91), 91(148), 115(261), 117(261), *130, 131, 132, 135*, 334(238), 335(238), 348(337), *368, 371*, 407(71), *412*
Charney, W., 143(157), 145(303), 170(157), *205, 211*
Charpentier, M., 266(175), *295*
Chaudhuri, D. K. R., 434(224), 435(224), *444*
Chauvette, R. R., 142(42), 151(42), 152(42), *201*
Cheng, C., 262(145), *294*
Cheymol, J., 219(36), 232(81, 85), *238, 240*
Chilton, W. S., 167(122), 168(122, 133), 169(133), *204*
Chipalkatti, V. B., 434(203), *443*
Chiriboga, J., 304(12), 315(101), 319(12), *362, 365*
Chitale, A. G., 434(223), *444*
Chizov, O. S., 91(152a), *132*, 197(273a), *210*
Chládek, S., 23(125), 24(129), 25(135, 136), *63*
Chou, T. C., 305(44), *363*
Choudhury, D., 116(270), *135*
Christensen, B. W., 231(77b), *240*
Christensen, J. J., 58(296, 297), *68*
Christensen, T. B., *77*
Chu, P., 169(145), 170(145), *205*
Chua, J., 41(224), 42(227, 228, 230), 43(231), *66*
Chuprunova, O. A., 185(243), *208*
Chytil, F., 270(229), *296*
Ciaudelli, J. P., 186(245), *208*
Cifonelli, J. A., 351(349), *371*, 395(42), *411*
Cirino, V. O., 430(161), *442*
Clark, E. P., 122(310), *136*
Clark, G. L., 436(243), 437(243), *444*
Clark, J. M., 178(207), *207*
Clark, R. K., Jr., 150(34), *201*
Clark, V. M., 28(153), *64*
Clarke, R. T. J., 74(12), *129*
Claussen, U., 154(65), *202*

Clegg, J. S., 106(194), *133*
Cleland, W. W., 327(183), *367*
Clements, G. C., 6(24a), *60*
Codington, J. F., 22(117), 34(190), 35(197, 198, 199), 38(198), 54(277b), *63, 65, 67*
Cohen, A. I., 143(186), 174(186), *206*
Cohen, S. S., 15(70), *61*, 327(184), *367*
Cohn, M., 108(211), 125(211), *134*, 232(81), *240, 251*, 268(204), 290, *295, 300*, 313, *365*
Cohn, W. E., *4*(6), 6(25), 7(27), 20(101), 31 (172), 45(244), *59, 60, 62, 64, 66*, 233, *240*
Colbran, R. L., 427(116), *441*
Cole, E. H., 427(129), 428(129), 435(129), *441*
Cole, R. D., 285(371), *299*
Colin, H., 216(8), 217(13), *237, 238*
Colowick, S. P., 51(266), *67*, 303(5), *362*
Colvin, J. R., 334(237), 335(237), 345, *368, 370*, 415(22), 416(24, 24b, 29, 30, 36), 417 (37), *439*
Comb, D. G., 306(52), 307(57), *363*
Conchie, J., 107(206), *134*, 270(222, 224), 271(224), 274, 275(265), 286(383), 288 (399), *296, 297, 300*, 326(170), *366*
Coniglio, C. T., 143(157), 170(157), *205*
Conlon, L. E., 434(202), *443*
Connell, J. J., 84(99), *131*
Cook, A. F., 39(210), *65*
Cook, F. S., 266(181), *295*
Cooper, A. S., Jr., 342(173, 177), *442*
Cooper, D. J., 143(157a), 170(157a), *205*
Cooper, J. R., 312(88), *364*
Cope, A. C., 142(297, 298), *211*
Copeland, W. H., 254(73), *292*
Corbaz, R., 142(31, 70b), 150(31), 153(50), 155(70b), *201, 202*
Corbett, W. M., 83(83), *131*
Cori, C. F., 303(5), 327, 329, 336(260), *362, 367, 369*
Cori, G. T., 327(189), 329, 330(199), 331 (206), *367*
Cornhill, W. J., 415(17), *439*
Cornwell, R. T. K., 433(188), *443*
Cory, J. G., 172(175), *206*
Cosulich, D. B., 145(266), 193(266), 196 (266), *210*
Côté, R. H., *77*
Courtois, J. E., 90(140), *132*, 218(23), 237 (98), *238, 240*, 258(115, 116), 264(164), 265 (164), 266(178), 267(193, 194, 196), 268

(200), 274(255), 275(269), 278(292), *293, 294, 295, 297, 298*, 321(138), *365*
Covill, M. J., *61*
Cowie, A. T., 323(145), *366*
Cowie, D. B., 269(212), *296*
Cowling, E. B., 422(75), 424, *440*
Cox, E. C., *210*
Cox, E. G., 436(255), *444*
Cramer, F., 23(123, 124), 26(123), *63*
Cramer, P., 334(226), 335(226), *368*
Crane, R. K., 270(233), 271(233), *296*
Craven, G. R., *249*, 250(28, 38), *251*, 268(28, 38, 207), 269(211), *291, 295, 296*
Cree, G. M., 398(57), *412*
Critchley, P., 357(383), *372*
Crocker, B. F., 266(186), *295*
Cron, M. J., 118(287), 119(287), *136*, 163 (101), 169(145, 147), 170(145, 153), *203, 205*
Crook, E. M., 111(235), 112(235), 126(235), *134*, 265(168), *294*, 318(117), *365*
Croon, I., 430(151, 152, 153, 154, 156, 157, 158, 159), 432(151), *442*, 462(78), *469*
Crowles, A. J., 410(77), *412*
Culbertson, T. P., 167(128), *204*
Cunningham, K. G., 143(171), 172(171), *206*
Cunningham, W. L., 264(158), *294*, 435 (237), *444*
Currier, H. B., 334(242), 335(242), 348(340), *368, 371*
Cushley, R. J., 33(183), 57(183), 53(275), 54(277b), *65, 67*, 168(131), *204*
Cutolo, E., 305(46), *363*
Czerwinsky, W., 427(117), *441*

D

D'Abramo, F., 351(347), 361(347), *371*
Dacons, J. C., 78(49), 88(133), 91(133), 98 (48), 126(49, 133), *130, 132*
Dahlqvist, A., 96(159), *133*, 254(75, 76, 77, 78), 255(81, 82, 85a), 258(76), 264(160), 270(160, 233), 271(233, 236), *292, 294, 296*
Daigle, D. J., 434(212), 435(212), *443*
Daniel, J. H., Jr., 434(220), *443*
Daniels, E. E., 144(222), 181(220, 222), *208*
Daniher, F. A., 144(196a), 175(194), 176 (194, 196a), *207*
Danjou, E., 218(25), *238*
Dankert, M., 340(294), 355(372), *369, 372*

Danon, D., 419(50), *440*
D'Ari, L., 354, *371*
Darnall, K. R., 14(66), 20(103), 53(66), 55(66), 58(66), *61, 62*, 180(217b), *207*
Da Rooge, M. A., 42(227), 43(231), *66*
Das, D. B., 414(4), *438*
Dasdia, T., 188(253), *209*
Datta, P. K., 424(90), 425(104), *441*
Dauben, W. G., 101(174), *133*
Davidson, E. A., 118(285), *136*, 305(41, 45), 310(73), *363, 364*
Davidson, G. F., 427(123), *441*
Davidson, M., 255(85), *292*
Davie, E. W., 198(282), *210*
Davies, J. E., 198(285, 287), 199(287, 288), *210*
Davis, B., 198(287), 199(287), *210*
Davis, B. D., 198(283), 200(283, 295), *210*
Davison, A. L., 358(391), *372*
Davoll, J., 8(29b), *60*
Dean, J., 78(47), *130*
DeBoer, C., 175(189), *206*
Debris, M. M., 237(98), *240*, 258(116), 268 (200), *293, 295, 297*
Dedonder, R., 103(177), *133*, 336, *368*
de Duve, C., 270(221), 284(362), 285, *296, 299*
de Fekete, M. A. R., 304(21), 340(21), 341, *363, 370*
de Garilhe, M. P., 15(69), *61*
DeJongh, D. C., 172(176), 197(273, 274, 275), 198(176, 278a), *206, 210*
Dekker, C. A., 4(5, 9), *5*, 8(29a), 27(150), 32 (176), 34(176), 44(240), 51(265a), *59, 60, 64, 66, 67*
de Koninck, L., 219(37), *238*
de la Haba, G. L., 199(291), *210*
Delbrück, K., 231(77), *240*
Dellweg, H., 4(4), *59*
DeLorenzo, S., 145(303), *211*
Demain, A. L., 74(11), *129*
Demelier, J. F., 258, *294*
Demmer, E., 220(43), *239*
Denham, W. S., 400(62), *412*
Dennis, D. T., 415(22), *439*
Denton, W. L., 323(155), *366*
Derieg, M. E., 142(297), *211*
de Robichon-Szulmajster, H., 309(68, 69), *364*
de Rudder, J., 15(69), *61*

Desai, R. M., 434(203), *443*
Desmarest, M., 124(315), *136*
de Souza, R., 80(66), *130*
Deuel, H., 395(45), 396(45), *411*
Deulofeu, V., 145(302), *211*
DeVoe, S. E., 142(23), 148(23), *201*
DeWalt, C. W., 163(101, 102), *203*
Dey, P. M., 266(180), *295*
Deyrup, J. A., 58(295), *68*
Dhar, M. M., 26(139), 31(171), 35(194), *63, 64, 65*
Dickerson, A. G., 280(314), *298*
Dickerson, J. P., 154(55), *202*
Dickey, E. E., 88(133), 91(133), *92*, 126(133), *132*, 465(92), *469*
Dickinson, B., 123(314), *136*, 256(94), *293*
Dickinson, M. J., 54(277a), *67*
Dierickx, L., 113(252), *135*
Dillemann, G., 217(16), *238*
Di Marco, A., 188(253), 199(294), *209, 210*
Dimler, R. J., 98(166), *133*
Dimroth, K., 3(2), *59*
Dion, H. W., 142(71), 150(38), 155(69), 156 (71), 157(72, 73), 158(73), 195(38, 73), 197(71), *201, 202*
Djerassi, C., 141(7), 142(7), 149(26, 27, 28), 197(7), *200, 201*
Dodyk, F., 308(65), *364*
Doe, R. P., 285(372), *299*
Doebel, K., 184(238), *208*
Doell, R. G., 270(228), 271(228), *296*
Doerr, I. L., 34(189, 190), 35(198, 199), 38 (198, 207), *65*, 172(167), *206*
Doherty, D. G., 6(25), *60*
Doisy, E. A., 284(357), *299*
Dolliven, M. A., 164(111), *203*
Donatti, J. T., 42(228), *66*
Dong, To, 266(177, 178), 267(194), 275 (269), *295, 297*, 321(138), *365*
Donin, M. N., 142(24), 143(119), 148(24), 166(119), *201, 204*
Dorée, C., 427(126), *441*
Dorfman, A., 281(319), *298*, 351(349), 361 (408, 413), *371, 373*
Doty, P., 388(21), *411*
Doudoroff, M., 103(176), 108(209), *133, 134*, 244(9), *291*, 304, 314(100), 315(100), 330 (100), 333, 334(222), *363, 365, 368*
Dow, J., 49(259), *67*
Doyens, A., 270(221), *296*

Doyle, M. L., 284(357), *299*
Draper, P., 359(404a), *372*
Drummond, G. I., 48(257), *49, 67*
Drzeniek, R., 287(388), *300*
Dube, S. K., 74(13), *129*
Duerksen, J. D., *251*, 263(149), 269(212), *294, 296*
Dufay, P., 9(38), *61*
Duff, R. B., 232(79), *240*
du Merac, L., 216(8), *237*
Dumpert, G., 334(224), 335(224), *368*
Duncan, A. M., 326(172), *366*
Duncan, W. A. M., 253(66), 265(169), *292, 294*, 318(120), *365*
Dunn, D. B., 6(19), *60*
Dunnophy, J. V., 270(233), 271(233), *296*
Duphorn, J., 145(304, 304a), *211*
Durham, L., 150(38), 195(38), *201*
Durr, G. J., 54(277), *67*
Durrum, E. L., 85(104), 88(104), *131*
Durso, D. F., 85(107), *131*
Duschinsky, R., 28(152), *64*
Dutcher, J. D., 140(2), 141(8, 9, 10, 11), 142 (24, 298), 143(119), 148(24), 163(2), 166 (2, 119), 174(185, 187), 175(2), 183(2), *200, 201, 204, 206, 211*
Dutriev, J., 106(200), *134*
Dutta, P. C., 142(59), 154(59), *202*
Dutton, D., 357(386), *372*
Dutton, G. J., 237(99), *240*, 325(165), 326 (172), *366*
Dvornik, D., 142(25), 148(25), *201*
Dweltz, N. E., 437(267), *445*
Dyer, J. R., 144(222), 164(103, 110), 168 (130), 181(220, 221, 222, 223), *203, 204, 208*
Dygert, S., 289(406), *300*

E

Eberhard, L., 455(42), *468*
Eberhart, B., 263(152), 266(172), *294, 295*, 318(118), *365*
Eble, T. E., 192(263), *209*
Ebner, K. E., 323(154, 155), 324(155a), *366*
Eckstein, E., 27(144), *64*
Eddy, B. P., 113(246), *135*, 425(107), *441*
Edelman, J., 116(267), *135*, 253, 278(297), 279(298, 304, 307), 280(298, 314), *292, 298*, 313(92, 93), *364*

Edstrom, R. D., 356, *372*
Edwards, T. E., 78(51), 82(79), 111(233), 127(327), *130, 131, 134, 137*
Effenberger, J. A., 128(338), *137*
Egami, F., 362(417), *373*
Egge, H., 398(55), *412*
Eggers, S. H., 198(278), *210*
Eich, H., 81(73), *131*
Eilers, F. I., 276(275), *297*
Eirich, F. R., 393, *411*
Eisenberg, F., 243(5), *290*
Eisenhuth, W., 145(258), 188(258), *209*
Elbein, A. D., 110(220), 114(220), *134*, 303 (4), 304(18, 19), 305(4), 325(160), 343(19), 345(19, 316b), 354(363), *362, 366, 370, 371*, 416(33), *439*
El Khadem, H., 117(279), *135*
Ellefsen, Ø. 414(13), 419(53), 420(60), *439, 440*
Ellestad, G. A., 142(85a), 159(85a), 195 (85a), *203*
Ellias, L., *251*, 252(47), *292*
Elliot, S. D., 358(392), *372*
Ellwood, D. H., 357(384), *372*
Elmore, D. T., 4(7), *59*
Elmore, N. F., 42(226), *66*
Els, H., 142(47), 152(46, 47), *201*
Elson, D., 269(213), *296*
Emerson, T. R., 8(29), 55(29, 282), 56(285, 287), 57(287), *60, 67*
Endo, T., 144(202), 176(200, 201, 202), 177 (205), *207*
Enger, M. D., 425(100), *441*
Enström, B. F., 462(78), *469*
Entwistle, D., 427(129), 428(129), 435(129), *441*
Eppenberger, U., 28(152), *64*
Epstein, R., *251*
Epstein, W. W., 182(228), *208*
Erlander, S. R., 390, *411*
Erskine, A. J., 395(37), *411*
Eschrich, W., 334(241), 335(241), 348(339), *368, 371*
Esipov, S. E., 144(242), 185(242), *208*
Espada, J., 340(292), *369*
Eto, W. H., 254, *292*
Ettlinger, L., 142(31, 70b), 150(31), 153(50), 155(70b), 159(83), *201, 202, 203*
Ettlinger, M. G., 230(76a), *240*
Etzold, G., 32(182), 38(182), 55(284), *65, 67*

Evans, D. L., 169(145), 170(145), *205*
Evans, W. L., 79(54), 80(56, 65), 101(174), 112(54), *130, 133*
Eveleigh, D. E., 243(6a), *290*
Everest, A. E., 226(69), *239*
Everson, R. G., 416(24b), *439*
Ewald, C. M., 456(51), *468*
Ewart, M. H., 340(285), *369*
Ewart, W. H., 256(93), *293*
Eylar, E. H., *77*
Eyler, E. H., 250(26), *291*

F

Fabianek, J., 382(3a), *410*
Falk, M., 436(243a), 437(243a), *444*
Fanshier, D., 303(8), *362*
Fantes, K., H., 275(266), *297*
Fardig, O. B., 169(147), 170(153), *205*
Farisi, M. A., 290(416), *300*
Farkaš, J., 17(78), 18(87), 21(109, 110, 111), 32(179), 40(218, 219), 51(267), 56(286), 63(109a), *62, 63, 64, 66, 67*, 173(179), *206*
Farley, F. F., 88(128), *132*
Farmer, V. C., 232(79), *240*
Fawcett, D. W., 323(144), *366*
Feagans, W. M., 285(364), *299*
Fear, C. M., 398(54), *412*
Feather, M. S., 96(163), *133*, 449(10), *467*
Fecher, R., 35(197), *65*
Feingold, D. S., 113(247), 116(268), *135*, 243(7), 278(295), 280(295), *290, 298*, 303 (7), 304(20), 309(72a), 312(87), 318(121, 122), 336, 344(314), 347(327, 328), 348(20), 351(348), *362, 363, 364, 365, 368, 370, 371*
Feit, B.-A., 434(227), *444*
Fellig, T., *248*
Felsenfeld, H., 306(53), *363*
Ferrier, R. J., 457(54), *468*
Fichtenholz, A., 218(28), *238*
Fielding, A. H., 273(248), *296*
Fields, D. L., 88(133), 91(133), 126(133), *132*
Fieser, L. F., 223(55), *239*
Fieser, M., 223(55), *239*
Filosa, M. F., 106(194), *133*
Findlay, J., 270(224), 271(224), 281(326, 331), *296, 298*
Finlayson, D., 431(171), *442*
Fischer, E., 105(189), 111(227), *133, 134*, 231(77), *240*

Fischer, E. H., *248*, 250(21), 275(21), *291*, 336(261), 338(261), *369*
Fischer, H. O. L., 197(276), *210*, 217(14), *238*
Fischer, J., 250(20), *248*
Fisher, D., 287(384), *300*
Fisher, L. V., 31(169), *64*
Fishman, W. H., 125(322), *137*, 284, 285 (365, 367, 368, 370, 373, 375), 286, *299, 300*
Fitting, C., 83(85), 108(85, 209), *131, 134*, 244(9), *291*, 315, *365*
Flaks, J. G., *210*
Flamm, E., 430(157), *442*
Fleming, I. D., 251(41), 257(41, 112), *291, 293*
Fletcher, H. G., Jr., 14(61), 18(80b), *61, 62*
Fleury, P., 90(140), *132*
Flodin, P., 85(117), *132*
Flora, R. M., 250(29), *291*, 424(94), *441*
Florin, R. E., 428(145), *442*
Flory, P. J., 387, 388, 393, *411*
Flowers, H. M., 118(282), 121(305), *136*, 344(312a), 345(316), *370*
Flynn, E. H., 142(42), 151(42, 43), 152(42), 175(190), *201, 206*
Flynn, J. H., 428(144), *442*
Fogel, Z., 269(213), *296*
Folkers, K., 143(123, 174), 144(232), 164 (104), 167(123), 172(174), 184(232, 233, 237), *203, 204, 206, 208*
Fontaine, J., 27(143), *64*
Forrest, H. S., 6(24a), *60*
Forrester, A. R., 232(79), *240*
Foster, A. B., *77*, 78(46), *130*, 150(35), 154 (58, 63, 64), 155(58, 67), 157(58, 74), *201, 202*, 283(350), *299*, 339(282), *369*, 435 (239), 436(239, 258), 437(239), 438(239), *444*
Foster, J. F., 88(31), 93(131), *94, 132*, 386, *411*
Fowler, F., 334(243), 335(243), *368*
Fowler, W. F., Jr., 427(119), *441*
Fox, J. J., 9(32), 34(32), 57, 22(115, 117), 32(177, 181), 33(183, 186a), 34(177, 181, 186a, 189, 190), 35(192, 197, 198, 199), 38(198, 207), 39(211, 213), 53(211, 275), 54(277b), 57(183), 58(291a), *60, 63, 64, 65, 67, 68*, 144(209), 172(163a, 167), 175(188), 177(203), 178(203, 209), 192(263), *206, 207, 209*
Fraenkel, D., 354(361), 355(371), *371, 372*

Fraenkel, G., 256(92), *293*
Fränkel, S., 436(256), *444*
Franceschi, G., 145(255), 188(254, 254a, 254b, 255), *209*
Francois, G., 436(257), 438(257), *444*
Frank, W., 52(268), *67*
Franz, G., 344(314b), *370*
Fratantoni, J. C., 24(126), *63*
French, D., 75(15, 24), 86(118), 89(24), 90 (137), 91(155), *93*, 101(24), 108(210), 122(24), 125(210), 127(328), 128(331, 336, 338, 339, 341), 129(341), *129*, *132*, *133*, *134*, *137*, 253(56), 288, 289(407), *292*, *300*, 334(225), 335(225, 229, 231), *368*, 390, *411*
French, J. C., 143(136, 137, 138), 168(135, 136, 137, 138), 197(135, 138), *204*, *205*
Freudenberg, K., 81(73), 98(165), 110(221), 126(325), 128(235), *131*, *133*, *134*, *137*, 219 (35), *238*, 334(224, 226), 335(224, 226), *368*
Frey-Wyssling, A., 419(52), *440*
Frič, I., 56(286), *67*
Frick, J. G., Jr., 434(199), *443*
Fridrich, M., 270(229), *296*
Friebolin, H., 154(53), 196(53), *202*
Frieden, E., 285, *299*
Friederici, D., 281(320), *298*
Friedkin, M., 6(22), *60*
Friedman, D. L., 336(258), 338(258), *369*
Friedman, H. A., 39(211, 213), 53(211)' *65*
Friedman, S., 258(120), *293*, 325(159), *366*
Friis, P., 231(77b), *240*
Frohardt, R. D., 175(191), 176(191), *206*
Frohwein, Y. Z., 282, *298*
Fromageot, H. P. M., 26(137a), 32(175), 54 (278), *63*, *64*, *67*
Frush, H. L., 43(237), *66*
Frydman, R. B., 320(127), 322(140), 340, *365*, *366*, *369*
Fuerth, O., 436(253), *444*
Fujii, K., 9(34), 10(42, 43), *60*, *61*
Fujii, T., 142(300), *211*
Fujisaki, M., 262(145), *294*
Fukasawa, T., 353(359), *371*
Fukatsu, S., 170(154), *205*
Fukui, S., 465(89), *469*
Fukumi, H., 10(40), 11(45), *61*
Fulmor, W., 143(170), 172(170), *206*
Funakoshi, R., 41(220), *66*
Funatsu, M., 250(30), *291*
Furberg, S., 49(259), 50, *67*

Furth, A. J., 271(237), *296*
Furukawa, K., 282(338), *299*
Furukawa, S., 233(53), *239*
Furukawa, Y., 33(184), *65*
Fusari, S. A., 175(191), 176(191), *206*

G

Gadamer, J., 230(76), *240*
Gaetani, M., 188(253), *209*
Gagliardi, D. D., 434(193), *443*
Gaillard, B. D. E., 75(22), *129*, 458(56, 57), 465(91), *468*, *469*
Gakhokidze, A. M., 83(90), *131*
Gal, A. E., 271(238), *296*
Galanos, D. S., 305(39), *363*
Galkowski, T. T., 90(138), 127(138), *132*
Galloway, B., 326(175), 352(354), *367*, *371*
Galmarini, O. L., 145(302), *211*
Ganguly, A. K., 145(258a), 189(258a,b), *209*
Gardon, J. L., 434(192), *443*
Garegg, P. J., 455(45), *468*
Garg, H. G., *61*
Garibaldi, J. A., 334(233), 335(233), *368*
Garrett, E. R., 45(243), 47, 48(248), 51(248), *66*
Gascoigne, J. A., 416(29), 422(69, 72), 423 (72), *439*, *440*, 465(93), *469*
Gascoigne, M. M., 422(69), *440*, 465(93), *469*
Gascón, S., 249(20a), 273(20a), *291*
Gasser, M. M., 144(232), 184(232), *208*
Gasser, R. J., 144(192), 175(192), *206*
Gatt, S., 264(161), 270(161, 225), 271(239), 282, *294*, *296*, *298*
Gäumann, E., 142(31, 70b), 150(31), 153 (50), 155(70b), 158(78), 159(83), *201*, *202*, *203*
Gauhe, A., 80(58), *130*, 274(261), *297*
Gauthier, H., 143(186), 174(186), *206*
Gauze, G. F., 185(240a), *208*
Geacintov, N., 434(214), *443*
Gee, M., 121(299), *136*
Gehatia, M., 336, *368*
Gellerstedt, N., 383(5), *410*
Genin, A., 459(63), *468*
Georges, L. W., 110(222), 111(222), 125 (321), *134*, *136*
Georget, J. C., 258(116), *293*
Georgiadis, M. P., 167(122), 168(122), *204*

Gerber, N. N., 174(184), *206*
Gerecs, A., 81(71), 103(180), *130*, *133*
Gerster, J. F., 20(102), *62*, 180(217), *207*
Gerzon, K., 142(42), 151(42, 43), 152(42), *201*
Ghalambor, M. A., 308(66), 309(66a), *364*
Ghione, M., 145(259), 190(259), *209*, 258 (128), *294*
Ghosh, B. L., 425(105), *441*
Ghosh, H. P., 342(302, 306), *370*
Ghosh, S., 305(41), *363*
Ghuysen, J. M., 113(252), 121(303), *135*, *136*, 283(343), *299*
Giaja, J., 217(19), *238*
Gianetto, R., 284(362), 285(362, 363), *299*
Gibbons, A. P., 244, *291*, 343(312), *370*, 416 (35), *439*
Gibbons, A. P., 343(311), *370*
Gibbons, G. C., 328(193), *367*
Gibbons, R. A., 287(396), *300*
Gigg, R. H., 305(39), *363*
Gilbert, V. E., 81(69), *130*
Gilbert, W., 199(288), *210*
Gillam, I. C., 80(63), 111(63), *130*
Gillespie, R., 323(146), *366*
Gilligan, W., 423(82), 425(82, 99), *440*, *441*
Gillman, S. M., 268(201), 270(232), *295*, *296*
Gilmour, D., 219(38), 237(97), *238*, *240*
Gin, J., 5, 8(29a), 27(150), 48(252), *60*, *64*, *66*
Ginsburg, V., 303(8), 310, 346(320), 350 (74, 346), 359(398), *362*, *364*, *370*, *371*, *372*
Giri, K. V., 111(236), 128(334), *134*, *137*, 253 (57), *292*
Gladding, E. K., 427(122), *441*
Glaser, L., 310(76), 327(181), 334(234, 235, 236), 335(234, 235, 236), 339(234), 342 (308), 351(235), 358(389), *364*, *367*, *368*, *370*, *372*, 416(26), 438(276), *439*, *445*
Glass, B., 303(7), *362*
Glass, C. A., 91(151), *132*
Glaudemans, C. P. J., 14(61), *61*, 454(36), *468*
Glinski, R. P., 154(55), *202*
Goebel, W. F., 80(60), 119(60), 120(292, 294), *130*, *136*, 352(350), *317*
Göhring, K., 154(65), *202*
Gölker, C., *249*, 250(27), 268(209), *291*, *295*
Goepp, R. M., Jr., 377(1), 394(1), *410*
Gördeler, J., 260(140), 262(140), *294*
Gold, A. M., 142(32), 150(32), *201*

Goldberg, I. H., 22(112), 23(112), 24(112), *63*, 198(279), *210*
Goldberg, J. B., 432(174), *442*
Goldberg, M. W., 188(257), *209*
Goldemberg, S. H., 316, 338(271, 273), 342 (304), 349(341, 342), *365*, *369*, *370*, *371*
Golding, B. T., 142(16), 146(14, 16), 184 (236), 196(236), 197(14), *200*, *201*, *208*
Goldman, L., 42(226), *66*
Goldschmid, H. R., 80(64), *130*, 456(49), *468*
Goldstein, I. J., 80(66), *130*
Gómez-Sánchez, A., 83(87), *131*, 253(62), *292*
Gomyô, T., 321(136), *365*
Goodman, L., 10(39), 13(56), 18(80c, 84, 85, 91, 92, 93), 23(121, 122), 27(147), 28 (156), 29(161, 162), 31(169), 32(180), 33 (186), 36(121, 122, 200, 202), 38(162, 204), 41(156), 48(252), 53(56, 273), 54(161), 57 (161, 186), *61*, *62*, *63*, *64*, *65*, *66*, *67*, 172 (172), 188(256a), *206*, *209*
Gordon, A. L., 77, 78(43), 115(43), 116(43), *130*
Gordon, M. P., 50(263), *67*
Gorin, P. A. J., 77, 78(50), 81(72), 83(91), 91 (149), 115(261), 117(261), *130*, *131*, *132*, *135*, 266(176), 274(176, 259), *295*, *297*, 326(173), *366*, 407(71), *412*
Goring, D. A. I., 418(43), *439*
Gorini, L., 198(287), 199(287), *210*
Gorman, M., 142(68), 155(68), *202*
Got, R., 272(242), *296*
Gottlieb, D., 140(4a), 198(287c), *200*, *210*
Gottschalk, A., 125(319), *136*, 252, 267(189), *291*, *295*, 306(54), 359(395), *363*, *372*
Gourlay, R. H., 164(115a), *204*
Grafflin, M. W., 415, 431(14), *439*
Gramling, E., 381(1a), *410*
Grant, P. M., 257(104), *293*
Grant, P. T., 359(394), *372*
Gray, H. E., 256(92), *293*
Grebner, E. E., 359(403, 404), *372*
Green, J. W., 422(67), *440*
Green, S., 285(365), 286, *299*, *300*
Greenberg, G. R., 356(377), *372*
Greenberg, E., 339(281), 342(281), *369*
Greenwood, C. T., 391(31), *411*, 450(16), 451(26), *467*
Gregory, J. D., 85(117), *132*

Greiner, C. M., 312(91), *364*
Greminger, G. K., Jr., 432(180), 433(180), *443*
Griffin, B. E., 24(128), 27(141, 146), 54(278), *63, 64, 67*
Griggs, W. H., 432(175), *442*
Grimmett, M. R., 90(142), *132*
Grisebach, H., 150(37), 152(44), 154(53), 155(44), 196(53), *201, 202*
Grodsky, G. M., 325(166), *366*
Groetsch, H., 15(67), *61*
Gross, D., 280(310, 312), *298*
Grundy, W. E., 145(268), 194(268), *210*
Gryboski, J. D., 271(236), *296*
Gryder, R. M., 305(43), *363*
Guadiano, G., 142(16), 146(16), *201*
Gubler, K., 142(25), 148(25), 149(30), *201*
Gueffroy, D. E., 18(84, 85), *62*
Gürtler, P., 435(229), *444*
Guillory, R. J., 337(263), *369*
Guilloux, E., 267(197), *295*
Gunar, V. I., 19(95), *62*
Gundermann, J., 384(9), *410*
Gunter, J. K., 178(207), *207*
Gussin, A. E. S., 258(122), *293*
Gustafsson, B. E., 255(81), *292*
Guthrie, F. K., 434(228), *444*
Gutowski, G. E., 154(55), *202*
György, P., 113(248), 124(318), 125(318), *135, 136*, 274(260, 262), *297*

H

Hackel, J., 431(167), *442*
Hackman, ·R. H., 436(245, 246), 437 (269), *444, 445*
Hadjiioannou, S., 281(328), *298*
Hahnn, F. E., 198(281), *210*
Hairs, E., 218(27), *238*
Hait, E., 437(273), *445*
Hajra, A. K., 270(224), 271(224), *296*
Hakomori, S., 398(56), *412*
Hall, A. N., 252(44), *292*
Hall, C. W., 356, 359(403, 404), *372*
Hall, D. A., 415(21), *439*
Hall, D. M., 428(136), *442*
Hall, M. A., 344(314a), *370*, 416(35a), *439*
Hall, R. H., 3(3), 4(8), 6(8, 20), *59, 60*
Halliwell, G., 423(84), 424(89), 425(89), *440, 441*

Halpern, M., 243(3), *290*
Halpern, O., 149(27, 28), *201*
Halvorson, H. O., *251*, 252(47), 263(148, 149), *292, 294*
Hamalainen, C., 432(177), *442*
Hamilton, D. M., 332(213, 214, 215), 334 (218), 335, *367, 368*
Hamilton, J. K., 83(86), *131*, 335(249), *368*, 407(70), *412*, 450(14), 454(38), 461(74), 462(76a), *467, 468*
Hammer, P. L., 462(79c), *469*
Hammond, J. B., 270(233), 271(233), *296*
Hampton, A., 24(126), 25, 56, 57(288), *63, 68*, 172(167), *206*
Hanes, C. S., 303(6), 327, *362, 367*
Hanessian, S., 140(4), 144(195), 169(139, 140), 172(176), 175(4, 195), 176(197), 185 (4), 197(4, 139, 274, 275, 276a), 188(195), 197(276a), 198(176, 276a, 278a), *200, 205, 206, 210*
Hann, R. M., 80(67), 81(68), 105(67), 110 (68), *130*
Hans, M., 122(311), *136*
Hansen, K., 258, *293*
Hansen, L. D., 58(296), *68*
Hansen, P., 142(23a), 148(23a), 149(23a), *201*
Hansen, R. G., 323(146, 147), 342, *366, 370*
Hanson, K. R., 424(90), *441*
Hanson, K. R., 424(90), 425(104), *441*
Hanstein, E. G., 425(106), *441*
Happey, F., 415(21), *439*
Haq, S., 78(41), 81(70, 77), 111(237), *130, 131, 134*, 320, *365*, 416(36a), 428(137), *439, 442*
Hara, T., 171(160), *205*
Harada, N., 262(144), *294*
Harada, S., 145(258c), 189(258c), *209*
Harborne, J. B., 227(72), 228(74), *239*
Hardenbrook, H., 323(146), *366*
Harding, T. S., 75(23), 106(198), *129, 134*
Hargie, M. P., 145(267), 194(267), *210*
Harrap, G. J., 282(339), *299*
Harris, D. R., 151(42a), *201*
Harris, J. F., 96(163), *133*
Harrison, W. H., 145(269), 194(269), *210*
Hart, P. A., 28(156), 41(156), *64*
Hart, V. E., 428(139), *442*
Hartigan, J., 257(105, 106), *293*, 317(116), *365*

Haschemeyer, A. E. V., 59(301), *68*
Hasegawa, A., 165(116a), 170(156a), 171 (159a), *204, 205*
Hashimoto, Y., 69(1), 71(1), 73(1), *129*, 266 (174), *295*, 390(27), *411*, 425(101), *441*
Hashizume, T., 9(35), 57(35), 58(298), *60, 68*, 172(169), *206*
Haskell, T. H., 143(136, 137, 138), 144(192, 193, 196), 168(135, 136, 137, 138), 169 (139, 140), 175(191, 192, 193, 196), 176 (191, 193, 196, 197), 197(135, 138, 139), *204, 205, 206, 207*
Haskins, J. F., 427(125), *441*
Haskins, W. T., 80(67), 81(68), 105(67), 110(68), *130*
Hassid, W. Z., 103(176), 110(220), 113(247), 114(220), *133, 134, 135*, 216(8), 235, *237, 240*, 243, *290*, 303(4, 7, 8), 304(14, 15, 16, 17, 18, 19, 20, 24, 29), 305(4, 38), 310(80), 312(87), 314(100), 315(100, 106), 316 (106), 319(125), 320(127, 130), 321(135), 322(135, 141, 142, 143), 323(141), 328 (192), 330(100), 333(222), 334(222, 246), 335(246), 337(267), 342, 343(19, 310), 344 (312a, 314, 314b, 314d, 315), 345(19, 316, 316b), 346(320, 321, 324), 347(321, 328, 329), 348(20, 321, 334), 349(343, 344, 345), 350(29), 351(344, 345, 348), *362, 363, 364, 365, 366, 367, 368, 369, 370, 371*, 387, *411*, 416(31, 32, 33, 34, 36a), 417, *439*
Hastings, A. B., 336(262), *369*
Hata, T., 142(70a, 300a), 155(70a), *202, 211*
Hatch, M. D., 320, *365*
Hatch, R. S., 418(41), *439*
Hatfield, D. L., 6(24a), *60*
Hattori, K., 142(18), 147(18, 20), *201*
Hauser, G., 327(179), *367*
Hausmann, W. K., 142(23), 148(23), *201*
Hawkins, W., 381(1a), *410*
Haworth, G. B., 152(45a), 154(45a), 158 (81a), *201, 202*
Haworth, W. N., 89(136), 107(203), 108 (214, 215), 109(216), 111(231), *132, 134*, 217(21), *238*, 330, *367*, 398, 399(60), 400 (63), *411, 412*, 414(5), 436(254), *438, 444*, 453 (34), *467*
Hawtrey, A. O., 198(278), *210*
Hay, A. J., 270(222), 271(240), 274, 275 (265), 288(399), *296, 297, 300*

Hay, G. W., 75(26), *129*
Hayashi, K., 250(30), *291*
Hayashi, M., 281(324), *298*
Hayes, D. H., 28(154), *64*
Haynes, L. J., 232(86), *240*
Head, F. S. H., 434(201), *443*
Healey, A. C., 427(126), *441*
Hearle, J. W. S., 420(58), *440*
Heath, E. C., 308(66), 309(66a), *311*(81) 354(363), 356, *364*, 371, *372*
Hebbert, H., 401(65), *412*
Hebting, J., 119(288), *136*
Hedström, H., 143(165), 172(165), *206*
Hehre, E. J., 101(173), *133*, 244(11), 256 (101), *291, 293*, 330(201), 332(217), 334, 335(248), *367, 368*, 416(25), *439*
Heidelberger, M., 119(290), 120(294), *136*, 352(351), *371*, 395(41), 410(75, 76), *411, 412*
Heinz, R., 151(41), *201*
Heinzel, D., 3(2), *59*
Heist, E. L., 399(60), *412*
Helfenstein, A., 221(47), *239*
Helferich, B., 80(61), 90(143), *130, 132, 248*, 250(32), *251*, 252, 259, 260, 261(136), 262 (140), *291, 294*
Hellman, N. N., 256(102), *293*
Henderson, J. R., 395(44), *411*
Henderson, M. E., 116(272), *135*
Henderson, R. W., 279(309), *298*
Henkel, W., 142(22), 148(22), *201*
Henri, W. J., 395(45), 396(45), *411*
Herdle, L. E., 432(175), *442*
Heringová, A., 270(229), *296*
Herissey, H., 111(230), *134*, 218(26), 219 (36), 232(82, 85), *238, 240*
Herman, A., 263(148), *294*
Hermans, J. J., 434(208, 214, 224), 435(224), *443, 444*
Hermans, P. H., 419(54), *440*
Herp, A., 382(3a), *410*
Herr, R. R., 145(261, 263a), 190(261, 261a, 261d), 192(263a), *209*
Hers, H. G., 255(86, 88), *293*, 332, 337(265), *367, 369*
Hertzog, M. B., 384, *410*
Herzog, H. L., 143(157), 145(303), 170(157), *205, 211*
Hess, K., 414(6), 429(146, 147), *438, 442*
Hesseltine, C. W., 174(182), *206*

Hestrin, S., 78(44), 116(268), 124(316), *130, 135, 136,* 278(295), 280(295), *298,* 318 (121, 122, 123), 336, *365, 368,* 415(23), 419(50), *439, 440*
Heusen, L. J., 164(111), *203*
Heusser, H., 142(25), 148(25), *201*
Heutter, R., 158(78), 159(83), *202, 203*
Hewitt, G. C., 265(167), *294*
Hewitt, R. J., 174(182), *206*
Hewson, K., 13(57), *61*
Heyl, F. W., 223(54), *239*
Heyworth, R., 261, 264(160), 270(160), 281 (317), *294, 298*
Hibbert, H., 334(243), 335(243), *368*
Hibbits, W., 142(298), *211*
Hichens, M., 167(122, 128), 168(122, 130, 132, 133), 169(132, 133), 170(132), *204*
Hickinbottom, W. J., 107(203), *134*
Hickson, J. L., 129(342), *137*
Higam, M., 287(384), *300*
Higashide, E., 178(206), *207*
Higgins, H. G., 427(128), *441*
Higgins, H. M., 145(269), 194(269), *210*
Hilbert, G. E., 111(225), *134,* 399(61), *412*
Hill, D. L., 356(375), *372*
Hill, E. P., 258(117), 276(275), *293, 297*
Hill, R. L., 324(155b, 155c), *366*
Hiller, L. A., Jr., 414(8), *438*
Hine, J., 431(166), *442*
Hineno, H., 170(156c), *205*
Hinman, J. W., 175(189, 190), 183(231), 184 (231), *206, 208*
Hinson, K. A., 108(213), 110(213), 111(232), *134,* 244(14), *291*
Hintsche, R., 32(182), 38(182), *65*
Hirano, S., 10(40), *61*
Hirasaka, Y., 118(281), 119(281), *135*
Hirata, M., 35(191), *65*
Hirata, Y., 142(18), 147(18), *201*
Hirayama, K., 176(199), *207*
Hirose, Y., 41(220), *66*
Hirst, E. L., 84(99), 85(114), 90(141), 114 (259), 115(259), 117(273), *131, 132, 135,* 395(35), 400(63), *411, 412,* 428(135), 435 (231), *442, 444,* 451(25), 453(33), *467*
Hitomi, H., 167(128), *204*
Hixon, R. M., 44, *66,* 88(128, 131), 93(131), *94, 132*
Hoban, N., 85(109), *131*

Hobart, S. R., 432(177), *442*
Hobkirk, R., 396(49), *411*
Hobson, P. N., 330(200), 332(209), *367*
Hocevar, B. J., 396(50), *411,* 450(20), *467*
Hochstein, F. A., 142(47), 152(47), 154(60, 62), 155(66), *201, 202*
Hockenhull, D. J. D., 275(266), *297*
Hodes, M. E., 46(247), *66*
Hodge, J. E., 399(61), *412*
Hoeksema, H., 43(236), *66,* 143(114, 177, 181, 145(261, 262), 164(114), 165(114), 173(177, 181), 174(181), 183(230, 231), 184(231), 190(261, 261b), 191(262, 262a), *204, 206, 208, 209*
Hörhammer, L., 228(74), *239*
Hoffer, M., 55(281), *67*
Hoffhine, C. E., Jr., 164(104), *203*
Hoffman, A., 436(251), *444*
Hoffman, C. A., 256(102), *293*
Hoffman, P., 118(284), *136*
Hofheintz, W., 150(37), 152(44), 154(53), 155(44), 196(53), *201, 202*
Hofmann, A., 222(51), *239*
Hogness, D. S., 309(70), *364*
Hogsed, M. S., 427(125), *441*
Holden, M., 424(91), *441*
Holley, H. L., 381(1a), *410*
Hollingshead, S., 252(44), *292*
Holly, F. W., 8(28), 18(83, 94), 19(100), 53(100), 54(277a), 55(100), *60, 62, 67,* 172(173), 184(233, 237), *206, 208*
Holmes, F. H., 428(142), *442*
Holmes, R. E., 28(157), 30(157), *64*
Holper, J. C., 145(268), 194(268), *210*
Holtzem, H., 244(58), *239*
Holtzer, A. M., 388, *411*
Hongess, D. S., 266(185), *295*
Honjo, M., 10(41), 33(184), 38(206), 53(41), *61, 65*
Honold, E., 432(176), *442*
Hooper, I. R., 169(145, 147), 170(145, 153), *205*
Hopkins, T. R., 285(371), *299*
Horecker, B. L., 312(89, 90), 354(360, 364, 365, 366), 355(369, 371), *364, 371, 372*
Horii, S., 164(106, 107), 167(128), 168(134), 169(106, 142), *203, 204, 205*
Horitsu, K., 266(176), 274(176), *295*
Horowitz, S. T., 85(115), 113(115), *132,* 436 (247), 437(247), *444*

Horstmann, H. J., 272(243), *296*
Horton, D., 91(152), 117(280a), *132, 135,*
  144(195), 154(55), 174(3), 175(195), 182
  (195), 188(195), *202, 207,* 230(76), *240,*
  377(1), 394(1), *410,* 437(274), *445*
Horton, E., 218(24), *238*
Horwitz, J. P., 41(224), 42(227, 228, 230),
  43(231), *66*
Hosansky, N., 143(119), 166(119), *204*
Hoshino, J., 276(274), *297*
Hosono, A., 9(34), 10(42, 43, 44), *60, 61*
Hotchkiss, R. D., 80(60), 119(60), *130*
Hough, L., 86(122), 114(258), 115(258),
  116(272), *132, 135,* 265(166), *294,* 455(44),
  459(69, 70), *468*
Howard, B. H., 114(255), *135,* 253(65), 267,
  286, *292, 295, 300,* 465(90), *469*
Howard, P., 432(178), *443*
Howarth, G. W., 191(261e), *209*
Howe, C., 282(337), *299*
Howsmon, J. A., 420(59), 421(62), *440*
Hribar, J. D., 198(278a), *210*
Hsia, D. Y. Y., 270(235), *296*
Hsu, L., 270(227), *296*
Hsue, K., 289(406), *300*
Hu, A. S. L., *251,* 263(147), 268(206), *294,*
  295
Huang, R. Y.-M., 434(211), 435(211), *443*
Huang, S. S., 255(85b), *292*
Huber, G., 81(75), 103(75), *131,* 145(304),
  211
Huebner, C. F., 402(67b), 405(67b), *412*
Hudson, C. S., 75(23), 80(67), 81(68), 83
  (81), 84(97), 90(144, 146), 99(167), 104
  (183, 184), 105(67), 106(192, 196), 107
  (196, 205), 109(205), 110(68), 111(228),
  112(228), 123(313), *129, 130, 131, 132,*
  133, 134, 136, 215(3), *237,* 275, 279(271),
  297, 334, 335(227, 228), *368,* 405(68),
  406(69), *412*
Huetter, R., 141(6), *200*
Huffman, G. W., 407(70), *412*
Hughes, E. O., 395(36), *411*
Hughes, N. A., 52(268), *67*
Hughes, R. C., 269(214), 287(214), *296*
Hulbert, R. B., 305(34), *363*
Hulcher, F. H., 316(108), *365*
Hungerford, E. H., 122(308), *136*
Huotari, F. I., 344(314c), *370*
Hursthouse, M. B., 164(112a), *204*

Hurwitz, J., 312(89), *364*
Hurwitz, M. D., 434(202), *443*
Husband, R. M., 163(101), *203*
Husemann, E., 414(1, 3), *438*
Hussey, E. W., 224(61), *239*
Hutchings, B. L., 161(95), *203*
Hutson, D. H., 75(17), *129,* 230(76), *240,*
  252, 256(100), 265(170), 274(264), *292,*
  293, 295, 297
Huttunen, J. K., 124(318), 125(318), *136*

I

Ierardi, P. A., III, 54(277), *67*
Iitaka, Y., 161(92), 169(143a), *203, 205*
Ikeda, M., 141(7), 142(7), 197(7), *200*
Ikeda, R., 250(30), *291*
Ikehara, M., 29(158, 160), 31(165, 166, 167,
  168, 170), 57(290), 58(292), *64, 68,* 179
  (212), *207*
Ikekawa, T., 161(92), *203*
Ikenaka, T., 248, 250(23), 252(45), *291, 292*
Illingworth, B., 289, 290(412), *300,* 331(206),
  332, 337(263), *367, 369*
Ilves, S. M., 347(328), *370*
Imai, K., 10(41), 38(206), 53(41), *61, 65,*
  253(70), *292*
Immergut, B., 434(211), 435(211), *443*
Immegut, E. H., 393, *411,* 434(211), 435
  (211), *443*
Inatome, M., 82(78), 111(78), 112(78), *131,*
  402(67a), *412*
Inch, T. D., 154(58, 63, 64), 155(58), 157(58),
  202
Ingram, M., 113(246), *135,* 262(146), *294,*
  425(107), *441*
Inoue, S., 169(146, 148, 149, 151), 170(146,
  148, 149, 152), *205*
Inouye, S., 145(263c), 192(263c), *210*
Inouye, Y., 83(96), *131*
Iriki, Y., 279(305), *298*
Irvine, J. C., 103(181), *133,* 219(30), *238,*
  398, *411,* 428(135), 436(259), *442, 444*
Irzykiewicz, H., 283(349), *299,* 438(284, 295),
  445
Isbell, H. S., 43(237), *66,* 83(93), 107(204),
  131, 134, 450(13), *467*
Iseki, S., 282(338), *299*
Ishida, A., 452(30), *467*
Ishida, N., 145(263b), 192(263b), *210*

Ishido, Y., 9(33, 34), 10(42, 43, 44), 14(60), 60, 61
Ishikawa, T., 117(280), 135
Ishikawa, Y., 41(222), 66
Ishimoto, N., 283(342), 299
Ishizawa, K., 279(305), 298
Isome, S., 10(42, 43, 44), 61
Isono, K., 142(94), 144(209a), 161(94), 179 (209a), 203, 207
Isselbacher, K. J., 325(164), 366
Ito, T., 143(156), 145(263c), 169(146, 148, 149, 151), 170(146, 148, 149, 152, 156), 192 (263c), 205, 210
Ito, Y., 170(154), 205
Itoh, T., 26(138), 27(138), 29(158, 160), 63, 64
Iwabuchi, K., 10(42), 14(60), 61
Iwaguchi, T., 59(303), 68
Iwai, I., 12(52), 13(53), 53(53), 55(283a), 61, 67
Iwamoto, R. H., 18(91), 62
Iwamura, H., 9(35), 57(35), 60, 172(169), 206
Iwasaki, E., 178(208), 207
Iwasaki, T., 250(30), 291, 467(95), 469
Izatt, R. M., 58(296, 297), 68

J

Jabbar, A., 266(173), 295
Jabbar Mian, A., 468
Jackson, E. L., 406(69), 412
Jackson, R. F., 101(170), 133
Jackson, R. W., 256(102), 293
Jacobi, R., 128(335), 137
Jacobs, J., 287(395), 300
Jacobson, B., 310(73), 323(147, 149), 364, 366
Jacox, R. F., 284(358), 299
Jacques, P., 270(221, 223), 296
Jaenicke, L., 3(2), 4(12), 59, 60
Jaffe, E., 335(248), 368
Jahn, W., 30(164), 41(164), 64
Jahnke, H. K., 145(263a), 192(263a), 209
Jakubowski, Z. L., 175(191), 176(191), 206
James, W. J., 75(24), 89(24), 101(24), 122 (24), 129
Jamieson, J. C., 360(404c), 372
Jang, R., 312(91), 364
Janot, M. M., 232(84), 240

Jansen, E. F., 121(300), 136
Jardetzky, C. D., 49, 52(270, 271), 67
Jarman, M., 25(133), 63
Jarrige, P., 272(242), 296
Jarvis, F. G., 327(180), 367
Jaworski, E. G., 339(283), 369, 438(278, 279), 445
Jayle, M. F., 248
Jayme, G., 422(66), 440
Jeanes, A., 74(9, 10), 99(169), 111(10), 129, 133, 334(247), 335(247), 368, 450(13), 467
Jeanloz, R. W., 77, 80(59), 83(92), 118(282), 121(305), 130, 131, 136, 160(90), 203, 269 (214), 287(214), 296, 359(393), 372
Jefford, T. G., 278, 279(298), 280(298), 298
Jeffrey, G. A., 436(255), 444
Jenkins, S. R., 18(83, 94), 62, 172(173), 206
Jennings, J. P., 55(283), 67
Jensen, L. H., 49(259), 50(260), 67
Jermyn, M. A., 85(108), 131, 247, 250(22, 31), 263(153, 154), 264(156), 270(22), 291, 294
Jeuniaux, C., 283(345), 299, 438(287, 289, 293), 445
Jevons, F. R., 270(226), 281(327, 330), 296, 298
Jirsová, V., 270(229), 296
Jørgensen, O. B., 252(51, 52), 292
Joffe, S., 323(146), 366
Johannis, J., 252, 291
Johansson, H., 418(46), 440
Johnson, A. L., 164(115a), 204
Johnson, A. W., 52(269), 67, 181(219), 208
Johnson, D. C., 422(67), 440
Johnson, D. L., 169(147), 170(153), 205
Johnson, J. M., 104(184), 106(192), 107 (205), 109(205), 111(228), 112(228), 133, 134
Johnson, L. F., 157(73), 158(73), 195(73), 202
Johnson, M. J., 327(180), 367
Johnston, J. A. R., 23(120), 26(120), 63
Jokura, K., 353(358), 354(358), 371
Jollès, P., 282(334), 299
Jones, A. S., 38(208), 51(265), 65, 67
Jones, D. M., 279, 298
Jones, E. W., 434(205), 443
Jones, G., 114(255), 135, 286(381), 289(408, 409), 300

Jones, J. K. N., 76(32), 77, 85(116), 86(122), 90(141), 114(254), 115(260, 262, 263, 264), 117(277, 278), 120(291), 121(298, 299), 130, 132, 135, 136, 152(45a), 154(45a), 158 (81a), 201, 202, 347(326), 370, 395(35, 37), 411, 449(6), 453(33), 454(35a), 459(64, 66, 69, 70, 71), 468
Jones, R. W., 74(10), 98(166), 99(169), 111 (10), 129, 133
Jorissen, A., 218(27), 238
Joseph, J. P., 143(183), 174(183), 206
Josse, J., 327(186), 367
Joubert, F. J., 248
Jourdian, D. M., 308(63), 359(63), 364
Jourdian, G. W., 287(391), 300, 307(58), 308(62), 324(155d), 359(400), 363, 364, 366, 372
Jucker, E., 221(48), 223(55), 239
Juergens, W. E., 358(390), 372
Jullander, I., 434(191), 443
Jung, K. H., 261(136), 294

K

Kabat, E. A., 78(42), 130, 217(14), 238, 282 (337), 299
Kaczka, E. A., 143(174), 144(232), 172(174), 184(232), 206, 208
Kagan, F., 33(185), 57(185), 65, 145(261), 190(261, 261d), 209
Kahn, A. W., 416(30), 439
Kaiser, S., 188(257), 209
Kaizerman, S., 428(132), 434(221), 442, 443
Kalachov, P., 253(54, 55), 292
Kalckar, H. M., 6(22), 60, 303(7, 9, 10), 305 (46), 309(68), 312(9, 10), 313(97), 317 (111), 323(151), 362, 363, 364, 365, 366
Kalf, G. F., 106(199), 134, 258(121), 293
Kaltenbach, U., 86(124), 132
Kamada, T., 262(144), 294
Kameswaramma, A., 215(1), 237
Kamiya, K., 33(184), 58(300), 65, 68
Kamogawa, H., 434(218), 443
Kanamaru, K., 434(204), 443
Kanazawa, T., 41(222), 66
Kandler, O., 321(139), 340(293), 366, 369
Kaneko, M., 31(168, 170), 57(290), 64, 68
Kaneo, M., 35(191), 65
Kano, H., 58(299), 68, 180(217c), 207

Kanzaki, T., 178(206), 207
Kaplan, J. G., 263(150), 294
Kaplan, L., 38(207), 65
Kaplan, N. O., 51(266), 67
Karabinos, J. V., 117(280), 135
Karasawa, I., 83(96), 131
Karjala, S. A., 399(61), 412
Karlsson, U., 250(37), 291
Karnovsky, M., 327(179), 367
Karrer, P., 44(239), 66, 232(78, 81), 240, 436 (251, 257), 438(257, 298), 444, 445
Karunairatnam, M. C., 284(359), 299
Kasbekar, G. S., 422(64), 440
Katabira, Y., 117(280), 135
Katagiri, H., 253(70), 292
Katagiri, M., 142(300a), 211
Katchalsky, A., 387(19), 411
Kates, M., 267(192), 295
Kato, H., 144(217a), 180(217a), 207
Kato, K., 262(144), 294
Katze, J., 269(210), 296
Katzman, P. A., 284(357), 299
Kaufer, J. N., 271(238), 296
Kauss, H., 340(293), 348, 369, 371
Kawada, M., 118(281), 119(281), 135
Kawaguchi, H., 144(239), 162(97), 184(239, 240), 203, 208
Kawalier, A., 218(28), 238
Kawamatsu, Y., 144(241), 185(241, 244), 186(247a), 208, 209
Kazenko, A., 256(102), 293
Keiser, J. F., 54(277), 67
Kelemen, M. V., 357(384), 372
Keller, F., 14(62), 61
Keller, J. M., 355(370), 372
Keller-Schierlein, W., 141(6), 142(31, 70b, 70c, 80, 81), 150(31), 153(50), 155(70b, 70c), 158(78, 79, 80, 81), 159(81, 83), 200, 201, 202, 203
Kelly, A., 436(256), 444
Kemp, J., 344(312a), 345(316), 370
Kendall, F. E., 119(290), 136
Kennedy, E. P., 327(183), 367
Kenner, G. W., 44(238), 46(238), 66
Kenner, J., 83(83), 131
Kent, P. W., 287(384), 300, 359(404a, 404b), 372, 438(277, 288), 445
Kenyon, W. O., 427(119), 441
Kersten, H., 199(293), 200(293), 210
Kersten, W., 199(293), 200(293), 210

Keser, M., 75(28), *129*, 248(19), 268(19), 273 (19), *291*

Kessler, G., 116(271), *135*, 334(240), 335 (240), 347(328), 348(335, 336), *368*, *370*, *371*, 458(60), 467(60), *468*

Keys, A. J., 253, *292*

Khmelnitshi, L. I., 283(350), *299*

Khorana, H. G., 22(112, 113), 23(112), 24 (112), 28(155), 48(257), *49*, *63*, *64*, *67*

Khwaja, T. A., 27(149), *64*

Khym, J. X., 45(244), *66*

Kihara, H. K., 263(147), *294*

Kikuchi, Y., 9(34), 10(42, 43), *60*, *61*

Kilby, B. A., 325(158), *366*, 438, *445*

King, F. E., 227(71), *239*

King, K. W., 250(29), *291*, 316(108), *365*, 424(94), 425(98, 102), *441*

Kinstle, T. H., 182(227), 188(227), *208*

Kircher, H. W., 461(74), *468*

Kirkwood, S., 117(274), *135*, 309(72), 344 (314c), *364*, *370*

Kishi, T., 58(300), *68*, 145(258c), 189(258c), *209*

Kishimoto, Y., 270(224), 271(224), *296*

Kiss, J., 184(235, 238), *208*

Kitagawa, S., 224(59), *239*

Kitahara, K., 161(92a), *203*

Kitazawa, E., 166(121a), *204*

Kittinger, G. W., 323(152), *366*

Kjær, A., 231, *240*

Klages, F., 458(59), 459(63), *468*

Klee, W. A., 56(289), 57(289), *68*

Klein, R., 431(165), *442*

Klein, W., 4(5), *59*

Kleinschmidt, T., *248*, 250(32), *251*, 259, 261 (130), *291*, *294*

Kleinschmidt, W. J., 145(271), 194(271), *210*

Klenow, H., 199(289), *210*, 312(90), *364*

Kleppe, K., 83(82), *131*, *248*, *251*, 257(111, 113), *293*

Kline, E., 432(179), *443*

Klingensmith, C. W., 80(56), *130*

Klop, W., 424(92), *441*

Klundt, I. L., 43(231), *66*

Klungsöyr, S., 416(27), *439*

Klyashchitskii, B. A., 185(243), *208*

Klybas, V., 312(88), *364*

Klyne, W., 55(283), *67*

Knoevenagel, C., 81(73), *131*

Knox, K. W., 265(171), 267(171), *295*

Kobata, A., 359(399), *372*

Kobayashi, A., 232(86a, 86b), *240*

Kobayashi, T., 35(191), *65*, 332(210), *367*

Kobayashi, Y., 434(209), 435(209), *443*

Koch, M., 232(84), *240*

Kochetkov, N. K., 35(193), *65*, 91(152a), *132*, 157(75), 197(273a), *202*, *210*

König, H. B., 150(33), *201*

Koeppen, B. H., 234(88), *240*

Koepsell, H. J., 256(102), 257(103), *293*, 317 (115), *365*

Koh, C., *4*

Kohn, P., 18(86), *62*

Kohtès, L., *248*, 250(21), 275(21), *291*

Koike, M., 143(161), 171(160, 161, 163), *205*, *206*

Koizumi, K., 80(66a), 127(66a), *130*

Kolachov, P., 127(326), *137*

Kolahi-Zanouzi, M. A., 258(115), *293*

Kolb, A., 112(238), *134*

Koldovský, O., 270(229), *296*

Kolodkina, I. I., 24(127), *63*

Kolosov, M. N., 144(242), 185(242, 243), *208*

Komatsu, Y., 144(217a), 180(217a), *207*

Kon, S. K., 323(145), *366*

Kondo, S., 143(143, 161), 159(84), 169(143, 146, 148, 149, 151), 170(146, 148, 149, 152), 171(160, 161, 162, 163), *203*, *205*, *206*

Kooiman, P., 424(92), 426(109), *441*

Koorajian, S., 250(37), 269(210), *291*, *296*

Korenman, S. G., 269(211), *296*

Kornberg, A., 327(186), *367*

Kornberg, S. R., 327(186), *367*

Kornfeld, R., 338(276), 342(303), 359(398), *369*, *370*, *372*

Kornfeld, S., 310(76), 359(398), *364*, *372*

Korte, F., 151(41), 154(65), *201*, *202*, 232 (80), *240*

Koshland, D. E., Jr., 243, 244, 246, 247, *290*, *291*, 317, *365*

Kostic, R. B., 223(56), *239*

Kotera, K., 165(116c), *204*

Kotick, M. P., 13(58), *61*

Koto, S., 170(156b, 156c, 156d, 156e), *205*

Kowashima, K., 144(241), 185(241), *208*

Kowollik, G., 42(224a), *66*

Koyama, G., 144(215), 169(143a), 180(215, 216), *205*, *207*

Koyama, H., 58(299), *68*, 180(217c), *207*

Kozawa, S., 386, *411*

Kradolfer, F., 142(31), 150(31), 153(50), 159 (83), *201, 202, 203*
Krässig, H., 415(20), *439*
Kraft, F., 418(39), *439*
Kramer, A., 219(37), *238*
Krauss, D., 182(227), 188(227), *208*
Kraut, J., 50(260), *67*
Krebs, E. G., 336(261), 338(261), *369*
Krebs, H. A., 314(99), 337(99), *365*
Kreis, W., 222(49, 50, 51), *239*
Kretchmer, N., 270(228), 271(228), *296*
Krevoy, M. M., 48(255), *66*
Kringstad, K., 414(13), 419(53), *439, 440*
Krishna Murti, C. R., 264(157), *294*
Kroll, G. W. F., 289(408), *300*
Krupta, G., 174(182), *206*
Kubler, H., 285, *299*
Kubota, T., 232(83), *240*
Kuby, S. A., 268(205), *295*
Kuehl, F. A., Jr,. 143(123), 164(104), 167 (123), *203, 204*
Kuehne, M. E., 142(52), 153(52), 154(52), *202*
Kuehthau, H. P., 12(51), *61*
Kuhn, R., 80(57, 58), 104(186), 105(186a), 112(238), 124(318), 125(318), *130, 133, 134, 136,* 217(20, 21), 225(65, 66), 229, *238, 239,* 274(261), 283(348), *297, 299,* 308(64), *364,* 398(55), 401(64), *412*
Kulkarni, A. Y., 434(223), *444*
Kullman, R. M. H., 434(199), *443*
Kumagai, K., 145(263b), 192(263b), *210*
Kumashiro, I., 12(49), 16(74, 75), *61, 62*
Kundig, W., 395(45), 396(45), *411*
Kunike, G., 437(270), *445*
Kuno, S., 327(185), *367*
Kunstmann, M. P., 142(85, 85a, 86), 159 (82, 85, 85a, 86), 195(85, 85a), 197(85), *203*
Kunz, A., 217(21), *238*
Kurath, S., 460(72), *468*
Kurihara, N., 165(116a), 170(156a), 171 (159a), *204, 205*
Kuriki, Y., 326(177), *367*
Kurokawa, T., 165(116a), *204*
Kurth, E. F., 449(8), *467*
Kusaka, T., 58(300), *68*
Kussmaul, W., 221(47), *239*
Kuwada, Y., 144(209), 178(209), *207*
Kyburz, E., 142(31), 150(31), 153(50), *201, 202*

L

Lagowski, J. M., 6(24a), *60*
Lahav, M., 434(227), *444*
Laible, R. C., 428(141), *442*
Lake, W. H. G., 436(254), *444*
Laland, S., 47(251), *66*
LaMantia, L., 45(246), *66*
Lamorre, Y., 9(38), *61*
Lampen, J. O., 249(20a), 273(20a), *291*
Lancaster, J. E., 142(85a), 145(266), 159 (85a), 193(266), 195(85a), 196(266), *203, 210*
Land, W. E. M., 305(39), *363*
Landman, O. E., 269(215), *296*
Langen, P., 32(182), 38(182), 42(224a), 55 (284), *65, 66, 67*
Lantzsch, B., 435(233), *444*
Lardy, H. A., 268(205), *295*
Larner, J., 96(158), *133,* 243(4), 247, *290,* 327(187), 329, 330, 331, 336(258, 259), 337(264), 338, 342(307), *367, 369, 370*
Larsen, D., 223(54), *239*
Lassettre, E. N., 78(39), *130,* 401(66), *412*
Laule, G., 75(28), *129,* 248(19), 268(19), 273(19), *291*
Laurant-Hubé, H., 237(98), *240*
Laurent, T. C., 391, 394, *411*
Laursen, R. A., 14(63), 53(63), 58(63), *61*
Lauryssens, M., 323(149), *366*
Lauterbach, G. E., *451, 467*
Laver, M. L., 459(62), *468*
Lavrova, M. F., 145(270), 194(270), *210*
Law, J. H., 305(39), *363*
Lawley, H. G., 77, 344(313), *370*
Lawley, P. D., 51(264a), *67*
Leach, B. E., 166(120), *204*
Leahy, M., 361(410), *373*
Lechevalier, H. A., 165(117), 174(184), *204, 206*
Leclerc, M., 264(164), 265(164), *294*
Lederberg, J., 333(222), 334(222), *368*
Lederer, E., 85(105), 88(105), *131,* 217(15), 224(62), *238, 239*
Lederer, M., 85(105), 88(105), *131*
Le Dizet, P., 278(294), *298*
Lee, C. H., 142(20a), 147(20a), *201*
Lee, H. J., 19(95a), *62*
Lee, W. W., 10(39), 18(80c, 92, 93), 29(161), 36(200, 202), 53(273), 54(161), 57(161), *61, 62, 64, 65, 67,* 172(172), *206*

Lee, Y. C., 217(15), *238*, 356, *372*
Lee, Y. P., 338(272), *369*
Lehman, I. R., 327(185), *367*
Lehmann, J., 154(58, 63, 64), 155(58, 67), 157(58), *202*, 272(244, 245), 273(244), 273 (244), 274(257), *296, 297*
Leibowitz, J., 243(3), *290*
Leinert, H., 165(116b), *204*
Lejeune, N., 255(86), *293*
Leloir, L. F., 103(175), 107(201), *133, 134*, 235, *240*, 304(12, 13, 21, 22, 23), 305(33, 40, 42, 47), 309(67), 313(98), 315(13, 101), 319, 324, 327, 337(98, 268, 270), 338(47, 273), 340(284, 291, 294), 342(304), 348, 351(338), *362, 363, 364, 365, 366, 367, 369, 370, 371*
Lemal, D. M., 152(45), 154(45), *201*
Le Men, J., 232(84), *240*
Lemieux, R. U., 8, 52(272), 55(281), *60, 67*, 80(55), 81(75, 76), 86(121), 91(153), 103 (75), 107(202), *130, 131, 132, 134*, 143 (100), 163(100, 102), 168(131), 169(145, 147), 170(145, 153), *203, 204, 205*
Lenz, R. W., 430(155), *442*
Leonard, N. J., 14(63, 64), 53(63), 55(279), 57(279), 58(63, 295), *61, 67, 68*
Leone, R. E., 27(143), *64*
Lerner, L. M., 18(86), *62*
Leskowitz, S., 78(42), *130*
Lester, G., 269(219), *296*
Letsinger, R. L., 23(119), 27(142, 143), *63, 64*
Letters, R., 31(173), *64*
Leudemann, G. M., 143(157), 170(157), *205*
Leuk, W., 186(249), *209*
Leutzinger, E. E., 11(48), 12(48), 54(48), *61*
Levene, P. A., 5, *60*, 118(286), 119(286), *136*
Levi, I., 104(182), *133*
Levin, D. H., 308, *364*
Levine, E. M., 308(66), *364*
Levine, M. L., 88(131), 93(131), *94*, 108 (210), 125(210), 128(336, 341), 129(341), *132, 134, 137*, 288(404), *300*, 334(225, 229), 335(225, 229), *368*
Levinson, H. S., 424(88, 93), *441*
Levvy, G. A., 107(206), *134*, 270(224), 271 (224, 240), 274(263), 275(265), 281(326, 331), 284(359, 361), 285, 286(377, 378, 383), 288(398, 401), *296, 297, 298, 299, 300*
Lewak, S., 141(6), *200*

Lewis, A. F., 20(102), *62*, 180(217), *207*
Lewis, B. A., 396(48), *411*
Lewis, T. A., 96(160), *133*, 243(1), *290*
Li, L. H., 250(29), *291*, 424(94), 425(98), *441*
Li, S. C. C., 266(184), *295*
Li, S. H., 360(404d), *372*
Li, Y. T., 75(27), *129*, 266(183), 267(198), *295*, 360(404d), *372*
Liang, C. Y., 385, *411*, 437(266), *445, 452* (29), *467*
Lichtenthaler, F. W., 39(212, 214), *65*, 165 (116b), *204*
Lidner, F., 142(61), 154(61), *202*
Lieberman, I., 254, *292*
Liebig, J., 217, *238*
Lillelund, H., 261(139), *294*
Limberg, G., 273(250), *297*
Lin, T. S., 304(24), 347(329), *363, 370*
Lin, T. Y., 349(343, 344, 345), 351(344, 345), *371*
Lin, W., 164(108), *203*
Lindberg, B., *77*, 83(95), 106(197), 111(95, 234), 120(293), *131, 134, 136*, 216(9), 217 (11), *238*, 395(40), 398(58), *411, 412*, 430 (152, 153, 158), *442*, 451(24), 455(45), 462 (81), 463(84), 467(95), *467, 468, 469*
Linden, P., 443(207), *443*
Lindgren, B. O., 117(276), *135*
Lineback, D. R., 8(28a), *60*, 80(63), 111(63), *130*, 250(23a), 257(23a), *291*
Lingappa, B. T., 106(193), *133*
Linker, A., 85(113), 118(284), *131, 136*
Linstead, R. P., 101(174), *133*
Lipmann, F., 327(188), 337(269), 338, 351 (347), 361(347), *367, 369, 371*
Lipták, A., 407(70a), *412*
Litt, C. F., 315(106), 316(106), *365*
Littman, L. A., 270(233), 271(233), *296*
Liu, T. Y., 344(315), *370*
Llewellyn, D. R., 96(160), *133*, 243(1), *290*
Lloyd, P. F., 415(21), *439*
Loach, J. V., 108(214), *134*
Löfgren, N., 143(165), 172(165), *206*
Loeppky, R. N., 55(279), 57(279), *67*
Loercher, R., 121(306), *136*
Löw, I., 104(186), 105(186a), *133*, 225(65, 66), *239*, 401(64), *412*
Loh, L., 164(108), *203*
Lohmar, R., 335(250), *368*
Lohrmann, R., 22(113), *63*

Lomakina, N. N., 145(270), 194(270), *210*
Long, C. W., 89(136), 108(214), 109(216), *132, 134*
Long, R. A., 53(274), *67*
Lorand, E. J., 433(184), *443*
Lorenz, S., 256, *293*
Lotmar, W., 437(262), *445*
Lowe, H. J., 304(27), *363*
Lowery, J. A., 174(182), *206*
Lucke, W. E., 450(14), *467*
Lüderitz, O., 355(368), *371*
Lüning, B., 143(165), 172(165), *206*
Lünzmann, G., 15(71), *62*
Luh, B. S., 121(297), *136*
Lukes, T. M., 258(118), *293*
Lundeen, J. A., 230(76a), *240*
Lunt, M. R., 438(277, 288), *445*
Lythgoe, B., 8(29b), *60*

M

Maasberg, A. T., 433(186), *443*
Maass, D., 283(341), *299*
McAllan, A., 271(240), 274(263), 285(369), 288(398, 401), *296, 297, 299, 300*
McBee, R. H., 316(107), *365*
McBurney, L. F., 426(114), 427(115), 428(115), *441*
McCarthy, J. R., Jr., 41(223), 42(229), 43(234, 235), *66*
McCarty, K. S., 198(278d), (279a), *210*
McCarty, M., 358(390), *372*
McChia, R. H., 217(10), *238*
McCloskey, J. A., 172(176), 197(277), 198(176), *206, 210*
McComb, E. A., 121(301), *136*
McCormick, J. E., 77, 409(73), *412*
McCormick, M. H., 145(269), 194(269), *210*
McCready, R. M., 121(299, 301), *136*
MacDonald, D. L., 217(14), *238*, 197(276), *210*
MacDonald, D. M., 428(131), *442*
MacDonnell, L. R., 121(300), *136*
McDowell, H. D., 80(65), *130*
McElroy, W. D., 303(7), *362*
McEvoy, F. J., 37(203), *65*
McGeachin, S. G., 151(42a), *201*
McGee, P. A., 427(119), *441*
McGilveray, I. J., 163(99), 164(109), *203*

McGilvray, D. I., 457(53), *468*
McGonigal, W. E., 164(110), *203*
McGuire, E. J., 324(155d), *366*
McInnes, A. G., 436(243a), 437(243a), *444*
McKay, J. E., 460(73), 461(73), *468*
McKee, C. M., 142(24), 148(24), *201*
McKee, S. C., 465(92), *469*
MacKellar, F. A., 145(261), 165(116), 190(261, 261c), 196(116), *204, 209*
MacKenzie, C. A., 432(182), *443*
McLachlan, J., 436(243a), 437(243a), *444*
Macmillan, J. D., 347(330, 331), *371*
McNally, S., 157(76), *202*, 399(58a), *412*
McNary, J. E., 181(220, 223), *208*
McNeely, W. H., 85(106), *131*, 436(248), 437(248), *444*
McNeill, I. C., 388, *411*
McNickle, C. M., 96(158), *133*
McPherson, J., 77, 111(234), *134*
Madison, R. K., 118(287), 119(287), *136*
Madorsky, S. L., 428(139), *442*
Maeda, K., 142(87, 92b), 143(143, 150), 145(264), 160(87, 87a, 88, 89, 91), 161(87, 91, 92b), 169(143, 143a, 144, 150), 170(150), 180(216), 193(264, 265), 195(265), *203, 205, 207, 210*
Maeder, R., 221(45), *239*
Maehr, H., 143(157b), 170(157b), *205*
Magerlein, B. J., 145(261), 190(261, 261d), *209*
Mahadevan, P. R., 263(151, 152), *294*
Mahadevan, V., 27(143), *64*
Maher, G. G., 437(272), *445*
Maher, J., 256(90), *293*
Mahoney, J. F., 433(185), *443*
Maitland, C. C., 422(78), 423(78), 424, 425(78), 428(95), *440, 441*
Major, W. D., 427(127), *441*
Makler, M., 270(235), *296*
Malhotra, O. P., 105(188), *133*, 246, 268, 272(244), 273(244), *291, 295, 296*
Malm, C. J., 430(150), 431(170), *442*
Malyshkina, M. A., 142(299), *211*
Mandels, M., 276(277), *297*, 423, *440*
Mandelstam, P., 121(306), *136*
Mandl, I., 278(289), *297*
Manley, R. St. J., 385, *411*, 419(49), 420(49), *440*
Mann, R. L., 142(93), 161(93), 171(158), *203, 205*

Mann, T., 304(26), *363*
Manners, D. J., 252, 253(64, 66), 264(158), 265(169), 274(264), 289(408a), *292, 294, 297, 300,* 318(119, 120), *365,* 417(36b), 435(231, 237), *439, 444,* 466(86b), *469*
Manning, J. P., 285(364), *299*
Mapson, L. W., 303(11), 312(11), 346(322), 347(322), *362, 370*
Marchessault, R. H., 385, *411,* 415(14a), 419 (14a), 426(111), 437(14a, 266), *441, 445,* 452(27, 28, 29), *467*
Marco, G., 438(278), *445*
Marechal, L. R., 316, 349(341, 342), *365, 371*
Markovitz, A., 311(82, 83, 84), 351, *364, 371*
Markowitz, A. S., 395(44), *411*
Marnay, A., 272(242), *296*
Marquez, J. A., 143(157,) 170(157), *205*
Marsh, C. A., 107(206), *134,* 281(331), 284 (352, 360), 285(369), 286(378), *298, 299, 300,* 325(167), *366*
Marsh, J. M., 99(168), *133,* 273(252), 289 (405), *297, 300*
Marsh, J. P., 188(256a), *209*
Marsh, P. B., 425(96), *441*
Marshall, J., 281(320), *298*
Marshall, R., 198(284), 199(284), *210*
Martin, A., 3(1), *59*
Martin, D. M. G., 27(151), *64*
Martin, R. O., 357(382), *372*
Martinez, A. P., 10(39), 18(80c), 29(161), 54 (161), 57(161), *61, 62, 64*
Marumo, S., 144(211), 179(211), *207*
Maruo, B., 332(210), *367*
Maruyama, A., 10(42, 43, 44), *61*
Marx-Figini, M., 416(24a), 417(24a, 36c), 418(44), *439*
Masak, E., Jr., 74(14), *129,* 465(90), *469*
Masamune, H., 117(280), *135*
Mason, D. J., 192(263), *209*
Masuda, F., 10(40), *61*
Matheson, N. K., 85(114), *131*
Mathias, A. P., 357(379, 380), *372*
Matijevitch, B. L., 45(246), *66*
Matsuba, T., 10(42, 43), *61*
Matsubara, S., 252(45, 46, 49), *292*
Matsuda, K., 14(59), *61*
Matsuhashi, S., 311(85, 86), *364*
Matsushima, Y., *248,* 250(23), *291*
Matsuzaki, K., 452(30), *467*
Mattor, J. A., 428(130), *441*

Maughan, M., 462(79d), *469*
Maurenbrecher, R., 459(63), *468*
Mawatari, H., 143(150), 169(144, 150), 170 (150), *205*
Maxwell, E. S., 303(9, 10), 304(25), 309(68, 69, 71), 312(9, 10), 323(151), *362, 363, 364, 366*
May, D. C., 431(170), *442*
Mayama, M., 144(217a), 180(217a), *207*
Mayer, F., 216(7), *237*
Mayer, F. C., 243(4), 247, *290*
Mayron, L. W., 124(317), 125(317), *136*
Mead, J. A. R., 284(356), *299*
Mebane, A. D., 141(6), *200*
Mehta, P. C., 434(223), *444*
Meier, H., 395(38), *411,* 414(12), *439,* 448 (2), 458(58), 461(77), 467(95), *467, 468, 469*
Meinhold, L. F., 434(221), *443*
Meir, H., 448(2), *467*
Mejzler, D., 419(51), *440*
Meller, A., 418(40), 428(134), *439, 442*
Menster, M., 431(165), *442*
Menzies, R. C., 398(54), *412*
Merewether, J. W. T., 414(11), *439*
Merler, E., 459(66), *468*
Merola, G. V., 425(96), *441*
Mertes, M. P., 20(108), *63*
Meseck, E., 145(303), *211*
Metzenberg, R. L., *248,* 251(39, 40), 276, 278(287), *291, 297*
Metzner, R., 262(142), *294*
Meyer, K., 85(113), 118(283, 284, 285), *131, 136*
Meyer, K. H., 328(193, 194), 330(194), *367,* 419, *420,* 435(229, 235), 436(260), 437 (260), *440, 444, 445*
Meyer, W. E., 143(170), 146(13a, 14), 172 (170), 197(14), *200, 206*
Meyer, W. L., 336(261), 338(261), *369*
Meyer zu Reckendorf, W., 167(125, 129), 168(125), *204*
Mian, A. M., 51(265), *67*
Micheel, F., 157(77), *202*
Michelson, A. M., 6(24), 8(24), 28(154), 31 (172, 173), *60, 64,* 310(77), *364*
Miescher, F., 5, *60*
Mikhailopulo, I. A., 19(95), *62*
Miles, F. D., 431(163), *442*
Militzer, W. E., 71(3), *129*

Miller, A., 250(37), *291*
Miller, G. L., 78(47), 126(323), *130, 137*
Miller, I. L., 110(222), 111(222), 125(321), *134, 136*
Miller, K. D., 254(73), *292*
Miller, N. C., 33(183), 35(192), 57(183), *65*
Miller, R. L., 20(102), *62*, 180(217), *207*
Millett, M. A., 426(113), *441*
Mills, A. R., 461(76c), *468*
Mills, G. T., 303(11), 305(36), 312(11), 323 (150), 326(175), 346(323), 347(323), 352 (352, 353, 354, 355, 356, 357), *362, 363, 366, 367, 370, 371*
Mills, H. H., 151(42a), *201*
Milner, Y., 320, *365*
Minamikawa, T., 340(287, 295), 341(287, 297), *369, 370*
Mineyama, K., 57(290), *68*
Minghetti, A., 258(128), *294*
Mino, G., 428(132), 434(221), *442, 443*
Minor, 220(41), *238*
Misch, L., 419, *420, 440*
Mitchell, P. W. D., 409(73), *412*, 456(50), *468*
Mitchell, R. L., 426(113), 431(168), *441, 442*, 449(7), *467*
Mitra, M. K., 414(4), *438*
Mitscher, L. A., 142(85, 85a, 86), 159(82, 85, 85a, 86), 195(85, 85a), 197(85), *203*
Mitts, E., 44, *66*
Miwa, A., *259*, 260(132), *294*
Miwa, T., 244(12), *259*, 260(132), 261, 262, 279(305), 286(380), *291, 294, 298, 300*
Miyake, A., 167(128), 178(206), *204, 207*
Miyaki, M., 14(65), 40(215, 215a), *61, 65*
Miyaki, T., 162(97), 184(240), *203, 208*
Miyamoto, M., 144(241), 185(241, 244), 186 (247a), *208, 209*
Miyoshi, F., 144(225, 225a), 181(225, 225a), *208*
Mizuno, K., 58(300), *68*, 145(258c), 189 (258c), *209*
Mizuno, S., 59(303), *68*
Mizuno, Y., 26(138), 27(138), 29(158, 160), 30(163), 32(174), 35(174), *63, 64*, 179 (212), *207*
Moelwyn-Hughes, E. A., 75(30), *129*
Moffatt, J. G., 28(155), 38(209), 39(210), 42 (225), 43(233), 47(250), *64, 65, 66*
Mohr, E., 287(389), *300*

Molnar, J., 359(397), 361(411), *372, 373*
Molodtsov, N. V., 91(152a), *132*
Mommaerts, W. F., 337(263), *369*
Momose, A., 276(274), *297*
Monahan, R., 142(42), 151(42), 152(42), *201*
Mondelli, R., 145(255), 188(254, 254a, 254b, 255), *209*
Monod, J., 268(204), 290, *295, 300*, 333, 334 (219, 220, 221), *368*
Montgomery, E. M., 83(81), 111(225), *131, 134*
Montgomery, J. A., 9(31), 10(31), 13(57), 18 (81, 82), *60, 61, 62*, 172(164), *206*
Montgomery, R., 117(280), *135*, 335(249), *368*, 395(42), *411*
Moon, S., 142(297), *211*
Moore, S. T., 434(220), *443*
Moore, W. E., 426(113), *441*
Morak, A. J., 450(72), *468*
Morawetz, H., 393, *411*
Moreno, A., 321, 326(170), *365, 366*
Morgan, W. T. J., 77, 78(53), 124(318), 125 (318, *130, 136*, 324(157), 359(402), 366, *372*
Morin, R. B., 142(68), 155(68), *202*
Morita, Y., 286, *300*
Moriza, M., 452(30), *467*
Morris, R. J., 224(61), *239*
Morrow, W. L., 428(144), *442*
Mortimer, A., 147(17), *201*
Mortimer, D. C., 315(103), *365*
Morton, G., 142(85a), 159(85a), 195(85a), *203*
Morton, G. O., 172(170a), *206*
Morton, R. K., 279(309), *298*, 313(95), *364*
Mosettig, E., 223(56), *239*
Mosher, C. W., 188(256a), *209*
Mosher, H. S., 150(38), 195(38), *201*
Mosimann, H., 463(82), *469*
Moss, R., 285(364), *299*
Mowat, J. H., 145(266), 193(266), 196(266), *210*
Moyse, H., 220(44), *239*
Mudd, S. H., 56(289), 57(289), *68*
Mühlethaler, K., 419(48, 52), *440*
Mueller, K. L., 305(39), *363*
Mukherjee, A. K., 116(270), *135*
Mukherjee, T. K., 144(192), 175(192), *206*
Munch-Peterson, A., 305(46), *363*
Mundie, C. M., 232(79), *240*

Muneyama, K., 31(166, 168), *64*
Murai, K., 152(46), 154(60), *201, 202*
Murase, M., 143(150, 155), 169(144, 150), 170(150, 155), *205*
Murata, T., 340(287, 288, 295), 341(287, 297, 299, 300), *369, 370*
Murawaski, A., 145(303), *211*
Muroi, M., 58(300), *68*, 145(258c), 189(258c), *209*
Murray, D. H., 18, *62*
Musa, B. U., 285, *299*
Muskat, I. E., 398(53), *412*
Muto, R., 190(260a), *209*
Muxfeldt, H., 142(23a), 148(23a), 149(23a), *201*
Myrbäck, K., 258(114), 276, 277(279, 280, 281, 282, 284, 285, 286), 278(290), *293, 297, 298*, 334(230), 335(230), *368*

N

Nagahama, T., 112(243), *135*, 232(86a, 86b), *240*
Nagai, W., 110(221), 126(325), *134, 137*
Nagaoka, T., *297*
Nagarajan, K., 144(193), 175(193), 176(193), *206*
Nagarajan, R., 91(153), *132*
Nagasawa, N., 12(49), *61*
Nagpal, K. L., 31(171), *64*
Nagyvary, J., 35(196), *65*
Naito, T., 35(191), *65*, 144(239), 184(239), *208*
Nakabayashi, T., 164(107), *203*
Nakadaira, Y., 185(244), 186(247a), *208, 209*
Nakagawa, T., 39(212), *65*
Nakagawa, Y., 58(299), *68*, 180(217c), *207*
Nakai, Y., 35(191), *65*
Nakajima, M., 162(95a), 165(116a), 170(156a), 171(159a), *203, 204, 205*
Nakajima, Y., 285(365), *299*
Nakamura, G., 179(210), *207*
Nakamura, M., 321(136), *365*
Nakamura, S., 278(291), 279, *298*
Nakamura, Y., 232(78), *240*
Nakanishi, K., 144(241), 181(218), 185(241, 244), 186(247a), *208, 209*
Nakano, H., 142(18, 19), 147(18, 19), *201*
Nakano, T., 288(400a), *300*

Nakayama, T., 305(39), *363*
Nakazawa, K., 178(206), *207*
Namiki, M., 142(94), 161(94), *203*
Nánási, P., 407(70a), *412*
Nathans, D., 199(292), *210*
Nathenson, S. G., 283(342), *299*, 358(387), *372*
Nawata, Y., 178(205a), *207*
Neakin, D., 231(77a), *240*
Neely, W. B., 257(108), *293*
Nees, A. R., 122(308), *136*
Neidl, S., 164(112a), *204*
Neill, J. M., 330(201), 335(248), *367, 368*
Neipp, L., 142(31, 70b), 150(31), 153(50), 155(70b), 158(78), 159(83), *201, 202, 203*
Neish, A. C., 114(256), *135*, 346, *370*
Nelson, N., 88(129), *132*
Nelson, N. M., 316(107), *365*
Nelson, O. E., 341(301), *370*
Nelson, T. E., 344(314c), *370*
Ness, R. K., 18(80b), *62*
Neuberg, C., 278(289), *297*
Neufeld, E., 258(124), *293*
Neufeld, E. F., 113(247), *135*, 235(91), *240*, 243(7), *290*, 303(7, 8), 304(20), 312(87), 322(140), 337(267), 344(314), 347(327, 328), 348(20), 351(348), 356, 359(403, 404), 361(409), *362, 363, 364, 366, 369, 370, 371, 372, 373*, 416(31), *439*
Neukon, H., 395(45), 396(45), *411*
Neumann, N. P., 249(20a), 273(20a), *291*
Nevell, T. P., 427(116, 118, 120), *441*
Newman, H., 142(36), 150(36, 39), 151(39), *201*
Nichol, A. W., 56(288), 57(288), *68*
Nichols, S. H., Jr., 80(65), *130*
Nicholson, L. W., 127(326), *137*, 253(54, 55), *292*
Nicholson, W. H., 115(264), *135*
Nicolson, A., 114(259), 115(259), *135*
Nielsen, N. A., 43(232), *66*, 176(198), *207*
Nigam, V. N., 111(236), 128(334), *134, 137*, 253, *292*
Niida, T., 145(263b, 263c), 192(263b, 263c), *210*
Nikaido, A., 353(358, 359), 354(358, 360), *371*
Nimmo-Smith, R. H., 284(354), *299*
Nirenberg, M., 198(284), 199(284), *210*

Nishano, H., 19(100a), *62*
Nishida, K., 112(243), *135*, 232(86a, 86b), *240*
Nishikawa, M., 33(184), 58(300), *65*, *68*
Nishimura, D., 170(156a), *205*
Nishimura, H., 144(217a), 180(217a), *207*
Nishimura, T., 12, 13(53, 54), 53(53, 54), 55(283a), *61*, *67*
Nishimura, Y., 170(156c), *205*
Nishio, M., 143(156), 170(156), *205*
Nishisawa, Y., 83(96), *131*
Nisizawa, K., *248*, 250(24), *259*, 260(133), 261(137), 266(174), 272(133), *291*, *294*, *295*, 425(101), *441*
Nisselbaum, J. S., 285(368), *299*
Nitta, K., 143(143), 169(143), *205*
Niwa, T., 250(24), *291*
Noack, R., 270(229), *296*
Noel, M., 41(224), 42(227, 228, 230), 43(231), *66*
Noelting, G., 435(235), *444*
Noguchi, S., 288(400a), *300*
Nohara, A., 10(41), 53(41), *61*
Norberg, E., 108(210), 125(210), 128(336), *134*, *137*, 288(404), *300*, 334(225, 229), 335(225, 231), *368*
Nordenson, E., 223(52), *239*
Nordin, P., 74(13), 108(210), 125(210), *129*, *134*, 288(404), *300*
Norman, A. G., 436(249), *444*
Northcote, D. G., 396(47), *411*
Northcote, D. H., 396(50), *411*, 448(1), 450(19, 20), 453(31), *467*
Nottbohm, E., 216(7), *237*
Numata, T., 112(243), *135*, 232(86b), *240*
Nunn, J. R., 76(31), *130*
Nutt, R. F., 8(28), 18(83), 19(100), 53(100), 54(277a), 55(100), *60*, *62*, *67*, 172(173), *206*

O

O'Brien, P. J., 361(409), *373*
O'Brien, S. J., 434(197), *443*
Ochiai, E., 224(59), *239*
O'Colla, P. S., 409(74), *412*
Oden, E. M., 143(157), 170(157), *205*
Østrup, G., 260(134), *294*
Ogawa, H., 143(156), 169(146, 148, 149, 151), 170(146, 148, 149, 152, 156), *205*
Ogawa, S., 165(116b, 116c), *204*

Ogilvie, K. K., 23(119), 27(142), *63*, *64*
Ogura, H., 142(70a, 300a), 155(70a), *202*, *211*
Ohad, I., 416(28), 417(28), 419(50, 51), *439*, *440*
Ohaski, M., 181(218), *208*
Ohkuma, K., 144(214), 180(214), *207*
Ohno, M., 142(87), 160(87, 87a, 89), 161(87), *203*
Okada, G., *248*, 250(24), *291*
Okami, Y., 143(143), 169(143), *205*
Okano, T., 59(303), *68*
Okazaki, R., 310(77), 326(177), *364*, *367*
Okazaki, T., 310(77), 326(177), *364*, *367*
Olavarría, J. M., 255(87), *293*, 338(273), *369*
Oldham, J. W. H., 103(181), *133*
Ollendorff, G., 83(88), *131*
Ōmura, S., 142(70a, 300a), 153(52a), 154(52a), 155(70a), *202*
O'Neill, A. N., 78(39), *130*, 401(66), *412*
Onodera, K., 10(40), 11(45), *61*, 83(96), *131*
Onuma, S., 178(205a), *207*
Ordin, L., 344(314a), *370*, 416(35a), *439*
Orezzi, P., 145(255), 188(253, 254, 254a, 254b, 255), *209*
Oroshnik, W., 141(6), *200*
Osato, Y., 143(143), 169(143), *205*
Osawa, T., 80(59), *130*, 282(336), *299*
Osborn, M. J., 354(360, 361, 364, 365), 355(371), *371*, *372*
Oser, B. L., 95(157), *133*
Osiecki, J. H., 32(180), *65*
Oster, R., 149(29), 150(33), *201*
Otakara, A., 283(347), *299*, 438(294, 297), *445*
Otake, N., 144(202), 176(200, 201, 202), 177(204, 205), 178(204), *207*
Ott, E., 415, 431(14), *439*
Ott, R. L., 429(148), *442*
Otter, B. A., 33(186a), 34(186a), *65*
Otterbach, D. H., 144(196a), 176(196a), *207*
Overend, W. G., 6(21), 47(251), *60*, *66*, 157(76), *202*
Owen, L. N., 400(63), *412*
Owen, O., 82(79), *131*

P

Pacák, J., 69(2), *129*, 215(4), *237*
Pacht, P. D., 152(45), 154(45), *201*

Pacsu, E., 83(94), 84(97), *131*, 414(8), 434 (226), *438*, *444*
Pagano, J., 142(24), 148(24), *201*
Painter, T. J., 78(52, 53), 116(272), *130*, *135*, *469*
Pal, P. N., 425(105), *441*
Palermiti, F. M., 169(145), 170(145), *205*
Palleroni, N. J., 304, *363*
Palmer, K. J., 384, *410*
Pan, S. C., 127(326), *137*, 253(54, 55), *292*
Pan, Y. H., 54(276), *67*
Panek, A., 258(123), *293*
Pankow, G. W., 437, *445*
Pansy, F. E., 174(185), *206*
Parikh, R. S., 432(178), *443*
Paris, R., 220(44), 227(72), 234(89), *239*, *240*
Partlow, E. V., 462(76a), *468*
Patchett, A. A., 32(178), *64*
Patel, J. C., 434(203), *443*
Patin, D. L., 459(62), *468*
Patrick, J. B., 143(170), 146(12, 14), 172 (170), 161(95), 197(14), *200*, *203*, *206*
Patterson, E. L., 159(82), *203*
Paul, A. V., 52(269a), *67*
Paul, R., 142(49, 54, 56), 153(49, 51), 154 (49, 54, 56), *202*
Paulsen, H., 169(141), *205*
Pavlasová, E., 290, *300*
Pazur, J. H., 76(37), *77*, 78(43), 83(82), 84 (98), 86(120), 93(156), 99(168), 108(210, 212), 115(43), 116(43), 117(120), 125 (98, 210), 126(98), 127(328), 128(336, 341), 129(341), *130*, *131*, *132*, *133*, *134*, *137*, *248*, *251*, 253(56, 58, 61), 257(111, 113), 273, 274(253), 280, 288(404), 289, *292*, *293*, *297*, *298*, *300*, 310(78), 334(225, 229), 335(225, 229), *364*, *368*
Pearlman, M., 354(366), *371*
Pearson, C. M., 337(263), *369*
Pearson, F. G., 437(266), *445*
Peat, S., *77*, 78(51), 82(79), 108(213, 215), 110(213), 111(232, 233), 112(245), 127 (327), *130*, *131*, *134*, *135*, *137*, 237(110), *240*, 244(14), 289(409), *291*, *300*, 330 (197, 200), 332(209), 334(245), 335(245), 344(313), *367*, *368*, *370*, 400(63), *412*, 435 (232), 436(254), *444*
Péaud-Lenoël, C., 103(177), *133*
Peck, G. Yu., 185(243), *208*

Peck, R. L., 164(104), *203*
Peeters, G. J., 323(147, 148, 149), *366*
Pelletier, G. E., 86(121), *132*
Pelzer, H., 283(341), *299*
Penco, S., 188(254a), *209*
Penzel, E., 418(44), *439*
Perchard, P., 395(46), 396(46), *411*
Percheron, F., 218(23), 228(73), *238*, *239*, 267(197), 274(255), *295*, *297*
Percival, E. G. V., 84(99), 112(241), 113 (241), *131*, *135*, 395(35), 401(65), *411*, *412*, 453(33, 34), *467*
Perila, O., 75(21), *129*, 466(87b), *469*
Perkin, A. G., 88(130), *132*
Perkins, H. R., 121(304), *136*
Perlin, A. S., 78(50), 80(64), 81(72), 83(91), 90(139), 91(148, 149), 115(261), 117(261, 273), 126(324), *130*, *131*, *132*, *135*, *137*, 243(6a), *290*, 326(173), 334(238), 335 (238), 348(337), *336*, *368*, *371*, 407(71), *412*, 435(236, 238), *444*, 455, 456(49, 51), 457(46), 466(94), *468*, *469*
Perlman, R. L., 361(413), *373*
Perlog, V., 158(78), *202*
Perreira, A., 220(39), *238*
Perrin, D., 268, *295*
Perry, M. B., 120(291), *136*
Pestka, S., 198(284), 199(284), *210*
Petek, F., 258(115), 266(177, 178), 267(193, 194, 196), 268(200), 275(269), *293*, *295*, *297*, 321(138), *365*
Petersen, C. S., 50(261), *67*
Peterson, E. A., 433(190), *443*
Peterson, F. C., 109(217), *134*, 462(79a, 79c, 79d), *469*
Peterson, W. H., 436(249), *444*
Petersson, G., 425(103), *441*
Petrova, A. N., 332(211), *367*
Pettengill, O. S., 285(370), *299*
Pfitzner, K. E., 38(209), 47(250), *65*, *66*
Pfleiderer, W., 11(46), 17(77), 22(116), 23 (116), *61*, *62*, *63*
Phaff, H. J., 74(11), 121(297), *129*, *136*, 258 (118), 274(259), 279(299), *293*, *297*, *298*, 347(331), *371*
Philip, J. E., 145(267), 194(267), *210*
Philipp, B., 419(56), *440*
Phillips, A. W., 252(48), *292*
Phillips, G. O., 428(143), *442*
Philpott, D. E., 383, *410*

Pichat, L., 9(38), *61*
Picken, L. E. R., 437(262), *445*
Pictet, A., 82(80), 101(171), 103(179), *131, 133*, 428(138), *442*
Pierce, J. V., 144(222), 181(222, 223), *208*
Pietruszkiewicz, A., 391(28), 394(28), *411*
Pigman, W., 69(1), 71(1), 73(1), 90(143, 145), 107(204), *129, 132, 134*, 219(32), *238*, 245, 252, 260, 262(141), 263(16), 274, 287 (396), 288(400a), *291, 294, 300*, 377(1), 381 (1a), 382(3a), 390(27), 394(1), *410, 411*, 422(70), *440*, 466(86a), *469*
Pike, J. E., 179(213), *207*
Pillar, C., 20(108), *63*
Piotrovich, L. A., 185(243), *208*
Pískala, A., 17(79), 58(294), *62, 68*
Pitha, J., 58(294), *68*
Pithkova, P., 58(294), *68*
Pittet, A. O., 80(63), 85(116), 111(63), *130, 132*
Plant, J. H. G., 109(216), *134*
Plapp, B. V., 285, *299*
Plat, M., 232(84), *240*
Platt, J. W., 405(68), *412*
Plessas, N. R., 144(195), 175(195), 182(195), 188(195),
Pliml, J., 17(80), *62*
Ploetz, T., 334(224), 335(224), *368*
Plouvier, V., 216(6), *237*
Pogell, B. M., 305(43), *363*
Pollman, W., 15(67), *61*
Pollock, M. R., 250(25), *291*
Polonsky, J., 224(62), *239*
Pontis, H. G., 305(35), *363*
Porath, J., 425(103), *441*
Porteous, J. W., 254(79, 80), 276(278), *292, 297*
Porter, C. A., 339(283), *369*, 438(279), *445*
Porter, G. G., 320(131), *365*
Porter, H. K., 319(126), *365*
Porter, J. W., 174(182), *206*
Portsmouth, D., 186(246, 247), *208, 209*
Potgieter, H. J., 438(291), *445*
Potter, A. L., 333(222), 334(222), *368*, 387, *411*
Potter, V. R., 305(34), *363*
Powell, D. B., 114(258), 115(258), *135*
Powning, R. F., 283(349), *299*, 438(284, 295), *445*
Prader, A., 255(84), 270(235), *292, 296*

Pratt, E. A., 327(185), *367*
Preece, I. A., 396(49), *411*
Preiss, J., 338, 339(281), 342(281, 302, 306), *369, 370*
Prelog, V., 142(25, 31, 32, 70b), 148(25), 150(31, 32), 153(50), 155(70b), 159(83), *201, 202, 203*
Preobrazhenskii, N. A., 24(127), *63*
Pressman, B. C., 284(362), 285(362), *299*
Pridham, J. B., 85(103), *131*, 235(92, 95), *240*, 265(165, 166), 266(180), 274(258), *294, 295, 297*, 304(17), 321(134, 135), 322(135), 346(324), *362, 365, 370*
Pringsheim, H., 459(63), *468*
Prins, W., 88(135), *132*
Priou, M. L., 436(242), *444*
Pritchard, P., 287(384), *300*
Probst, G. W., 145(271), 194(271), *210*
Procter, W., 219(34), *238*
Prokop, J., 18(88, 89, 90), *62*
Prystaš, M., 19, *62*
Puejo, G., 217(12), *238*
Pugh, D., 281(323), *298*
Pulley, A. O., 128(338), *137*
Purdie, T., 398, *411*
Purdom, M. R., 114(255), *135*, 286(381), *300*
Purves, C. B., 104(182), *133*, 231(77), *240*, 427(122), 430(159), 433(185), *441, 442, 443*
Putman, E. W., 83(85), 108(85), *131*, 216(8), *237*, 303(8), 315(106), 316(106), 320(130), 333(222), 334(222), *362, 365, 368*

Q

Quang, H. N., 218(23), *238*, 274(255), *297*
Quarck, J. C., 142(42), 151(42), 152(42), *201*
Quilico, A., 142(16), 146(16), *201*
Quillet, M., 435(241), 436(242), *444*
Quinn, E. J., 402(67a), *412*

R

Rabaté, J., 219, *238*, 264, *294*
Rachlin, A. I., 184(238), *208*
Racker, E., 308, 312(88), *364*
Radin, N. S., 270(224), 271(224), *296*
Rafelson, M. E., Jr., 287(385, 387), *300*
RajBhandary, U. L., 357(382), *372*
Rammler, D. H., 22(112), 23(112), 24(112), 52(269a), *63, 67*

Rånby, B. G., 419(47), 426(111), 431, 432 (162), *440, 441, 442*
Randerath, K., 4(13), *60*
Rao, K. V., 144(214b), *207*
Rao, V. F., 386, *411*
Rapport, M. M., 270(225), 271, *296*
Rapson, W. H., 434(211), 435(211), *443*
Rasenack, D., 219(35), *238*
Rashbrook, R. B., 116(271), *135*, 458(60), 467(60), *468*
Rasmussen, M., 14(64), *61*
Rathgeb, P., 117(280), *135*
Rauchenberger, W., 105(190), *133*
Rauen, H. M., 86(119), *132*
Rawlinson, W. A., 279(309), *298*
Raybin, H. W., 101(172), *133*
Rayburn, J. A., 434(205), *443*
Raymond, A. L., 230(76), *240*
Rebers, P. A., 459(67), *468*
Recondo, E., 304(22), 340(291, 294), 341 (298), *363, 369, 370*
Record, B. R., 435(234), *444*
Reed, D., 317(113), *365*
Rees, D. A., 237(100), *240*, 415(14b), 419 (14b), 433(14b), *439*, 451(25), *467*
Rees, W. R., 330(203), *367*
Reese, C. B., 24(128), 25(130, 133, 134), 26(137, 137a), 27(141, 145, 146, 151), 32 (175), 38(205), 54(278), *63, 64, 65, 67*
Reese, E. T., 126(324), *137*, 275(270), 276 (277), *297*, 422(73), 423(79, 81, 82), 424 (88, 93), 425(81, 82, 99), *440, 441*, 466(94), *469*
Reeves, R. E., 91(153), *132*, 164(105, 168 (105), *203*, 352(350), *371*
Reeves, W. A., 434(198, 199), *443*
Reggiani, M., 199(294), *210*
Regna, P. P., 154(62), 155(66), *202*
Rehorst, K., 224(60), *239*
Reich, H., 225(63), *239*
Reichstein, T., 152(48), *201*, 215(2), 221 (48), 225(63), *237, 239*
Reid, J., 141(9, 10), 143(186), 174(186, 187), *200, 206*
Reid, W. W., 121(298), *136*
Reimann, H., 145(258a), 189(258a), *209*
Reinhardt, R. M., 434(199), *443*
Reist, E. J., 18(84, 85), 23(121, 122), 27 (147), 28(156), 29(159, 162), 31(169), 32

(180), 33(186), 36(121, 122), 38(162, 204), 57(159, 186), *62, 63, 64, 65*
Reithel, F. J., 104(187), *133*, 268(206), *295*, 323(152), *366*
Renfroe, H. B., 142(23), 148(23), *201*
Renz, J., 220(39), 222(50), *238*, 239
Reusser, R., 142(31,70b), 150(31), 155(70b), 159(83), *201, 202, 203*
Reynolds, D. D., 79(54), 112(54), *130*
Reynolds, D. M., 283(344), *299*, 438(296), *445*
Reznik, H., 219(35), *238*
Rhaese, H. J., 23(123), 26(123), *63*
Rice, K. C., 164(110), *203*
Rich, A., 59(301), *68*
Richards, E. L., 90(142), *132*
Richards, G. N., 48, *66*, 428(133), 434(219, 222), *442, 443, 444*
Richardson, A. C., 151(40), 154(57), 155 (57), 188(256), *201, 202, 209*
Richardson, N. G., 451(25), *467*
Richman, D. J., 336(259), *369*
Richmond, M. H., 250(25), *291*
Richter, F., 3(1), *59*
Richter, J. W., 144(232), 184(232), *208*
Richtmyer, N. K., 106(196), 107(196), *134*, 219(32), *238*, 260, 262(141), 275(271), 279 (271), *294, 297*, 405(68), *412*
Rickards, R. W., 142(16), 146(14, 16), 184 (236), 196(236), 197(14), *200, 201, 208*
Ricketts, C. R., 335(251), *368*
Rickher, C. J., 145(268), 194(268), *210*
Rieder, S. V., 106(199), *134*, 258(121), *293*
Riley, A. C., Jr., 164(113), *204*
Rinehart, K. L., Jr., 143(118, 124, 127), 163 (99), 164(107), 166(118, 121), 167(122, 124, 127, 128), 168(118, 122, 130, 132, 133), 169(118, 132, 133), 170(132), 182 (227, 228), 183(229), 188(227), *203, 204, 208*
Ringertz, N. R., 395(46), 396(46), *411*
Rist, C. E., 74(10), 98(166), 111(10), *129, 133*
Rittenberg, S. C., 438(285), *445*
Ritter, A., 12(51), *61*
Ritter, G. J., 449(7), *467*
Rivkind, L., 218(22), *238*
Rizvi, S., 381(1a), *410*
Robbins, P. E., 428(136), *442*

Robbins, P. W., 337(269), 355(370, 372), 369, *372*
Roberts, J. G., 112(245), *135*, 435(232), *444*
Roberts, P. J. P., 76(36), *77*, 125(36), 128 (36), 129(36), *130*
Roberts, R. B., 269(212), *296*
Roberts, W. K., 32(176), 34(176), *64*
Robertson, A., 218(29), 226, *238, 239*
Robertson, A. M., 252(53), *292*
Robins, M. J., 11(47), 41(223), 42(229), 43 (234, 235), 48(256), 53(47), *61, 66*
Robins, R. K., 9(36), 11(46, 47, 48), 12(48, 50), 16(76), 20(102, 103), 27(148, 149), 28(157), 30(157), 41(223), 42(229), 43 (234, 235), 48(256), 53(47, 274), 54(48, 276), 55(279), 57(279, 291), *60, 61, 62, 64, 66, 67, 68*, 144(214a), 180(214a, 217, 217b), *207, 210*
Robinson, D., 264(159), 271(237), *294, 296*
Robinson, G. B., 359(397), 361(411), 372, *373*
Robinson, G. M., 227(71), *239*
Robinson, H. C., 361(408), *373*
Robinson, R., 226(68), 227(71), *239*
Robiquet, P. J., 217, *238*
Robison, R., 324(157), *366*
Robyt, J. F., 88(132), *132*
Rodén, L., 85(117), *132*, 361(414), *373*
Rodin, J. O., 184(233, 237), *208*
Roelofsen, P. A., 415(15), 419(15), *439*
Roff, W. J., 434(200), *443*
Rogers, D., 164(112a), *204*
Rogers, G. T., 18(80a), 40(216, 217), *62, 65*
Rogers, S. C., 449(7), *467*
Romanovska, E., 287(386), *300*
Romming, C. H. R., 50(261), *67*
Roncarni, G., 142(80, 81), 158(79, 80, 81), 159(81), *202*
Ros, A., 430(153), *442*
Rose, C. S., 113(248), *135*, 274(260), *297*
Rose, R. E., 219(30), *238*
Rosell-Perez, M., 327(187), 338(187, 277), 342(307), *367, 369, 370*
Roseman, S., 85(115), 113(115), 125(320), *132, 136*, 281(319), 287(391), *298, 300*, 305(41, 45, 48), 306(48, 52, 55, 56), 307 (57, 58, 59), 308(61, 62, 63, 65), 310(55), 324(155d), 359(400), 360, *363, 364, 366, 372, 373*, 436(247), 437(247), *444*

Rosen, S. M., 354(360, 364), 355(371), *371, 372*
Rosenberg, A., 287(397), *300*
Rosenfeld, E. L., 288, *300*
Ross, K. M., 114(257), *135*, 456(52), 457 (52, 55), *468*
Ross, R. J., 450(17), *467*
Rosselet, J. P., 143(157), 145(303), 170(157) *205, 211*
Roston, C. P. J., 270(226), 281(327), *296, 298*
Roth, C. B., 427(124), *441*
Roth, J. S., 35(196), *65*
Rothfield, L., 354(360, 361, 364, 365, 366), 355, *371*
Rotman, B., 267(187), 272(247), *295, 296*
Rott, R., 287(388), *300*
Roudier, A., 455(42, 43), *468*
Rougvie, M. A., 128(338), *137*
Rousseau, R. J., 16(76), *62*
Roux, D. G., 234(88), *240*
Rover, K. R., 167(128), *204*
Rowe, K. L., 289(408a), *300*
Rowell, R. M., 96(162), *133*
Rowin, G., 281(320, 321), *298*
Rowland, S. P., 430(161), *442*
Roy, C., 283(346), *299*
Rubins, A., 255(83, 84), 270(230), *292, 296*
Rudall, K. M., 436(244), 437(263), *444, 445*
Ruff, O., 83(88), *131*
Rundell, J. T., 280(310), *298*
Rundle, R. E., 384, *410*
Russell, J. D., 232(79), *240*
Rutenberg, A., 101(174), *133*
Ruyle, W. V., 32(178), 34(187), *64, 65*
Ryan, K. J., 13(56), 18(92), 53(56), *61, 62*
Ryan, M., 391(28), 394(28), *411*
Ryder, A., 175(191), 176(191), *206*
Rydholm, S. A., 417(38), 431, 432(162), *439, 442*, 462(78), *469*
Rydon, H. N., 252(44), *292*
Rytting, J. H., 58(296, 297), *68*

S

Sach, E., 224(62), *239*
Saeman, J. F., 426(113), *441*
Saenger, W., 23(123, 124), 26(123), *63*
Saffhill, R., 25(130), *63*
Saito, K., 29(160), *64*

Saito, S., 258(119), *293*
Saito, Y., 178(205a), *207*
Sakai, H., 176(199), *207*
Sakamoto, J. M. J., 159(84), *203*
Sakurai, H., 142(300), *211*
Salo, M., 431(170), *442*
Salo, W. L., 309(72), *364*
Saltmarsh, M. J., 235(95), *240*, 265(165), *294*
Salton, M. R. J., 121(303), *136*, 139(1a), *200*
Sampson, P., 85(113), *131*
Samuelson, O., 84(101), *131*, 427(117), *441*
Sandermann, W., 426(112), *441*
Sanderson, A. R., 358(390), *372*
Sanfilippo, A., 258(128), *294*
Sannella, J. L., 451(21), *467*
Sano, H., 165(116c), *204*
Sarasin, J., 428(138), *442*
Sarcione, E. J., 359(396), 361(410), *372, 373*
Sarko, A., 415(14a), 419(14a), 437(14a), *439*
Saroja, K., 253(57), *292*
Sarre, O. Z., 145(258a), 189(258a, b), *209*
Sasaki, F., 142(92b), 161(92b), *203*
Sasaki, T., 30(163), 32(174), 35(174), *64*
Sasaki, Y., 58(298), *68*
Sastry, P. S., 267(192), *295*
Satake, K., 142(300), *211*
Sato, M., 465(89), *469*
Sato, T., 9(33, 34), 10(42, 43, 44), 14(60), 41(222), *60, 61, 66*
Satoh, K., 171(160), *205*
Sattur, N. B., 434(203), *443*
Satyanarayana, M. N., 308, *298*
Savage, A. B., 432(180), 433(180, 186), *443*
Saxl, N., 415(21), *439*
Scarpinato, B. M., 188(253), 199(294), *209, 210*
Schaaf, E., 334(224), 335(224), *368*
Schachman, H. K., 391, *411*
Schacht, U., 145(304), *211*
Schaeffer, H. J., 172(168), *206*
Schaffer, R., 83(93), *131*
Schaffner, C. P., 142(20a), 143(157b), 147(20a), 167(122), 168(122), 170(157b), 181(220), *201, 204, 205, 208*
Schambye, P., 323(148), *366*
Schardinger, F., 128(337), *137*, 334(232), 335(232), *368*
Schaub, R. E., 143(183), 174(183), *206*
Scheer, I., 223(56), *239*

Scheffler, A., 115(265), *135*
Scheibler, C., 122(307), *136*
Scheit, K. H., 23(123, 124), 26(123), *63*
Schenck, J. R., 145(267), 194(267), *210*
Schenk, G., 270(229), *296*
Schexnayder, D. A., 27(143), *64*
Schiffman, G., 78(42), *130*, 282(337), *299*, 410(78), *412*
Schilling, W., 276, *297*
Schillings, R. T., 167(122), 168(122), *204*
Schloz, V., 38(164a), 41(164a), *64*
Schlubach, H. H., 105(190), 115(265), *133, 135*
Schluchterer, E., 334(245), 335(245), *368*
Schmid, H., 145(258), 188(258), *209*, 232(78), *240*
Schmid, R., 337(266), *369*
Schmid, W., 220(44), *239*
Schmidt, E., 274(262), *297*
Schmidt, G., 40(218, 219), *66*, 327(189), *367*
Schmidt, M., 436(250), *444*
Schmidt, R., 267(187), *295*
Schmidt, R. R., 38(164a), 41(164a), *64*
Schmidt-Thomé, J., 145(304), *211*
Schmitz, H., 169(147), *205*
Schmorak, J., 414(9), *438*
Schneider, G., 23(123), 26(123), *63*
Schneider, W., 231(77), *240*
Schoch, T. J., 99(167), *133*, 334(228), 335(228), *368*
Schofield, K., 38(205), *65*
Schorger, A. W., 433(187), *443*
Schorigin, P., 437(273), *445*
Schramm, G., 15, *61, 62*, 287(389), *300*
Schramm, M., 415(23), *439*
Schroeder, W., 143(177), 145(261), 173(177) 190(261, 261b), *206, 209*
Schulz, G. V., 414(3), *438*
Schulze, E., 448, *467*
Schulze, P., 437(270), *445*
Schurz, J., 434(225), *444*
Schwarz, E. C. A., 461(75), *468*
Schweizer, M. P., 55(280), *67*
Schwenker, R. F., Jr., 428(140), 434(226), *442, 444*
Schwille, D., 38(164a), 41(164a), *64*
Schwimmer, S., 334(233), 335(233), *368*
Scopea, P. M., 55(283), *67*
Scott, J. E., 395(39), *411*, 451(23), *467*
Scott, S. S., 326(176), *367*

Seal, V. S., 285(372), *299*
Segal, L., 423(79), *440*
Segro, N. R., 434(220), *443*
Seifter, S., 282(335), *299*
Seki, M., 142(18), 147(18), *201*
Sekiya, M., 14(60), *61*
Sekiya, T., 36(201), *65*, 434(218), *443*
Selby, K., 422(74), 423(78), 424, 425(78), 428(95), *440, 441*
Seligman, A. M., 268(201), 270(232), *295, 296*
Selleby, L., 120(293), *136*
Sellinger, O. Z., 270(221), *296*
Sello, S. B., 434(207), *443*
Semenza, G., 255(83, 84), 270(230), *292, 296*
Sepp, J., 231(77), *240*
Seraydarian, K., 337(263), *369*
Serro, R. F., 217(14), *238*
Seto, J. T., 287(388), *300*
Sevag, M. G., 395(43), *411*
Seydel, J. K., 45(243), *66*
Sezaki, M., 171(160, 162, 163), *205, 206*
Shadaksharaswamy, M., 86(120), 117(120), *132*, 273(253), 274(253), *297*
Shafizadeh, F., 48(253), *66*, 96(161), 109 (182a), *133*
Shallenberger, R. S., 94(156a), *134*
Shapiro, B. L., 142(47), 152(47), *201*
Shapiro, R., 20(104), 38(205), *62, 65*
Sharon, N., 121(305), *136*, 160(90), *203*, 282 (335), *299*
Sharpe, E. S., 256(102), 257(103), *293*, 317 (115), *365*
Sharpen, A. J. 45(243), *66*
Sharples, A., 426(110), *441*
Shaw, C. J. G., 428(142), *442*
Shaw, E., 6(26), 51(264), *60, 67*
Shaw, N., 52(269), *67*
Shaw, P. D., 140(4a), 168(130), 198(278c), *200, 204, 210*
Shelan, W. J., 331, *367*
Shemyakin, M. M., 144(242), 185(242, 243), *208*
Shen, L., 338, *369*
Shen, T. Y., 32(178), 34(187), *64, 65*
Shenin, H., 266(186), *295*
Sherman, J. H., 142(298), *211*
Sherwood, S. F., 123(313), *136*
Sheshadri, T. R., 215(1), *237*

Shetlar, M. R., 75(27), *129*, 266(183), 267 (198), *295*, 360(404d), *372*
Shibaev, V. N., 35(193), *65*
Shibata, H., 170(156a), *205*
Shibata, M., 178(206), *207*
Shibata, Y., 261(137), 266(174), 275(270), *294, 295, 297*, 425(101), *441*
Shibko, S., 256, *293*
Shier, W. T., 169(142a), *205*
Shigeura, H. T., 199(290), *210*
Shimadate, T., 9(33, 37), *60, 61*
Shimaoka, N., 144(217a), 180(217a), *207*
Shimizu, B., 12(52), 13(53, 54), 14(65), 40 (215, 215a), 53(53, 54), 55(283a), *61, 65, 67*
Shimizu, F., 306(52), 359(400), *363, 372*
Shimura, M., 171(160), *205*
Shinohara, M., 144(241), 185(241, 244), 186 (247a) *208, 209*
Shippee, F. B., 434(193), *443*
Shirley, D. A., 432(182), *443*
Shoemaker, M. J., 433(187), *443*
Shotwell, D. L., 164(113), *204*
Shrader, S., 142(23a), 148(23a), 149(23a), *201*
Shuey, E. W., 310(78), *364*
Shugar, D., 50(262), *67*
Shunk, C. H., 144(232), 184(232), *208*
Siddiqui, I. R., 77
Sie, H. G., 125(322), *137*
Sigal, M. V., Jr., 142(42), 151(42, 43), 152 (42), 171(159), *201, 205*
Sih, C. J., 200(296), *210*, 316(107), *365*
Silberkweit, E., 113(250), *135*
Silbert, J. E., 361, *373*
Silverstein, R., 317(113), *365*
Silvestrini, R., 188(253), *209*
Siminovitch, D., 340(285), *369*
Simkin, J. J., 359(394), 360(404c), *372*
Simon, W., 88(134), *132*
Simonart, P. C., 309(72), *364*
Simonitsch, E., 145(258), 188(258), *209*
Simpson, F. J., 75(26), *129*
Simpson, M. E., 424(96), *441*
Sinsheimer, R. L., 327(184), *367*
Sisson, W. A., 420(59), 421(62), *440*
Siu, R. G. H., 422(68, 71, 73), 424(93), *440, 441*
Siwek, K., 415(18), *439*
Sjøstrand, F., 250(37), *291*
Skau, E. L., 432(176), *442*

Skujins, J. J., 438(291), *445*
Slatcher, R. P., 20(105), *62*
Slater, W. G., 346, *370*
Slechta, L., 179(213), *207*
Sleeper, B. P., 425(100), *441*
Slifer, E. D., 217(10), *238*
Slizewicz, P., 336, *368*
Slomp, G., 43(236), *66*, 143(181), 145(261), 165(116), 173(181), 174(181), 190(261, 261a, 261c), 196(116), *204, 206, 209*
Smakula, E., 126(324), *137*
Smart, C. L., 109(219), *134*, 334(239), 335(239), *368*, 377(1), 394(1), *410*, 435(240), 437(240), 438(240), *444*
Šmejkal, J., 17(78), 56(286), *62, 67*
Smith, A. F., 436(243), 437(243), *444*
Smith, C. G., 183(230), *208*
Smith, D. G., 436(243a), 437(243a), *444*
Smith, D. H., 200(295), *210*
Smith, D. R., 98(164a), *133*
Smith, E. E. B., 303(11), 305(36, 46), 312(11), 323(150), 326(175), 346(323), 347(323), 352(354, 355, 356, 357), *362, 363, 366, 367, 370, 371*
Smith, F., 78(40), 81(69), 83(86), 117(274), *130, 131, 135*, 335(249), 344(314c), *368, 370*, 395(42), 396(48), 407(70), *411, 412*, 459(67, 68), *468*
Smith, J. D., 4(11), 6(19), *59, 60*
Smith, J. N., 284(356), *299*
Smith, M., 22(112), 23, 24(112), 48, *49, 63, 67*
Smith, P. N., 274(260), *297*
Smrt, J., 22(114), 23(125), 24(129), 27(140), 35(195), *63, 65*
Smyrniotis, P. Z., 312(90), *364*
Snyder, H. E., 279(299), *298*
So, A. G., 198(282), *210*
Sober, H. A., 433(190), *443*
Sobotka, H., 217(21), 224(57), *238, 239*
Sobue, H., 452(30), *467*
Sodd, M. A., 327(182), *367*
Sohn, A. W., 414(7), *438*
Soldati, M., 188(253), *209*
Solms, J., 305(38), *363*
Soloviev, S. N., 142(299), *211*
Somers, P. J., 288(403), *300*
Somme, R., 278(293), *298*
Soodak, M., 305(44), *363*
Sorenson, H., 465(88), *469*

Šorm, F., 17(78, 79, 80), 18(87), 19, 21(109, 110), 22(114), 32(179), 34(188), 42(224b), 51(267), 58(294), 63(109a), *62, 63, 64, 65, 66, 67, 68*, 173(179), *206*
Sosa, F., 228(73), *239*
Souza, N. O., 258(123), *293*
Sowden, J. C., 106(191), *133*
Spearing, C. W., 287(390), *300*, 307(60), *364*
Spedding, H., 427(121), *441*
Spencer, C. C., 109(217), *134*
Spencer, C. F., 184(237), *208*
Spencer, J. F. T., *77*, 266(176), 274(176, 259), *295, 297*
Spencer, R., 438(286), *445*
Spicer, L. D., 182(228), *208*
Spiegelberg, H., 184(234, 235, 238), *208*
Spiegelman, S., 269(212), *296*
Spiro, R. G., 359(401), *372*
Spohler, E., 144(250, 251), 187(250, 251), 195(251), *209*
Spriggs, A. S., 106(191), *133*
Spring, F. G., 143(171), 172(171), *206*
Sprinzl, M., 21(111), *63*
Spurlin, H. M., 415, 430(149), 431(14, 149), 432(149), *439, 442*
Srere, P. A., 312(88), *364*
Srinivasan, K. S., 111(236), *134*
Srinivasan, M. S., *298*
Srivastava, H. C., 459(68), *468*
Stacey, K. A., 388(24), *411*
Stacey, M., 6(21), 47(251), *60, 66, 77, 78* (46), 81(69), 83(87), 91(150), *130, 131, 132*, 150(35), 154(58), 155(58, 67, 70), 157(58, 70, 74), 186(246), *201, 202, 208*, 253(62, 63), 256(98), 257(104), 265(167), 278(288), 288(403), *292, 293, 294, 297, 300*, 332(216), 334(244, 245), 335(244, 245, 251), *367, 368*, 436(258), *444*
Stahl, E., 86(124), *132*
Stahl, W. H., 422(77), *440*
Stambouli, A., 234(89), *240*
Stamm, A. J., 434(195), *443*
Stamm, O. A., 145(258), 188(258), *209*
Stammer, C. H., 184(233), *208*
Staněk, J., 69(2), *129*, 215, *237*
Stannett, V., 434(213, 214, 215), 435(215), *443*
Staub, A. M., 355(368), *371*
Staudenbauer, W. L., 282(340), *299*
Staudinger, H., 414(7), 435(233), *438, 444*

Steele, R., 434(192), *443*
Steers, E., Jr., *249*, 250(28, 38), *251*, 268 (28, 38, 207), 269(38), *291*, *295*
Stein, J. Z., 116(269), *135*
Stein, S. S., 243(2), *290*
Steiner, D. R., 338(275), *369*
Steiner, M., 224(58), *239*
Stenlake, J. B., 164(109), *203*
Stephen, A. M., 76(31), 78(40), 117(274), *130*, 135
Stephenson, G. F., 27(146, 151), *64*
Sternbach, L. H., 188(257), *209*
Steurer, E., 414(6), 429(146, 147), *438*, *442*
Stevens, C. L., 43(232), *66*, 144(192, 193, 195, 196a), 154(55), 175(192, 193, 194, 195), 176(193, 194, 196a, 198), 182(195), 188(195), 189(256d), *202*, *206*, *207*
Stevens, J. D., 18(80b), *62*
Stewart, L., 128(340), *137*
Stewart, L. C., 106(196), 107(196), *134*
Stichand, L., 228(74), *239*
Stiehler, O., 231(77), *240*
Stiller, E. T., 164(111), *203*
Stocking, C. R., 319(126), 340(286), *365*, *369*
Stodola, F. H., 164(113), *204*, 257(103), *293*, 317(115), *365*
Stoffyn, P. J., 83(92), *131*
Stoll, A., 220(39), 221(47, 48), 222(49, 50, 51), 223(55), *238*, *239*
Stone, B. A., 111(235), 112(235), 126(235), *134*, 251(41), 257(41), 264(157), 265(168), *291*, *294*, 318(117), *365*
Storey, I. D. E., 325, *366*
Storvick, W. O., 425(102), *441*
Strasdine, G. A., 283(346), *299*, 438(282, 283), *445*
Straus, S., 428(139), *442*
Streiffer, C., *249*, 250(27), *291*
Strepkov, S. M., 116(266), *135*
Striegler, K., 167(128), *204*
Stringer, C. S., 99(169), *133*
Strominger, J. L., 119(289), 121(306), *136*, 283(342, 343), *299*, 303(9, 10, 11), 305 (37), 306(50, 51), 310(77), 311(85, 86), 312(9, 10, 11), 326(176), 346(322), 347 (322), 358(387, 390), 361(412), *362*, *363*, *364*, *367*, *370*, *372*, *373*
Stroud, D. B. E., 253(63), *292*

Stumpf, P. K., 346(320), *370*
Sturgeon, R. J., 75(19), *129*, 450(16), *467*
Sturm, K., 220(42), *238*
Su, J. C., 304(29), 310(80), 350, *363*, *364*
Suami, T., 165(116b), 116c), *204*
Sugg, J. Y., 101(173), *133*, 335(248), *368*
Sugiyama, T., 341(297, 299, 300), *370*
Suhadolnik, R. J., 172(175), *206*
Suhara, Y., 142(87, 92b), 160(87, 87a, 88, 89, 91), 161(87, 91, 92b), *203*
Sulston, J. E., 25(130, 134), 26(137a), 54 (278), *63*, *67*
Sund, H., 250(33, 34), *251*, 268(33), 269(33), *291*
Sundaralingham, M., 50(260), 59(302), *67*, *68*
Sussman, A. S., 106(193), *133*, 258(117), 276(275), *293*, *297*
Sutherland, E. W., 303(5), 336(260), *362*, *369*
Suzaki, S., 29(158), *64*, 179(212), *207*
Suzuki, H., 75(25), *129*, 247, *248*, 250(24), *291*
Suzuki, M., 141(7), 142(7), 197(7), *200*
Suzuki, S., 119(289), *136*, 142(94), 144 (209a, 211), 161(94), 179(209a, 210, 211), *203*, *207*, 361(412), *373*, 435(236), *444*
Suzuki, T., 47(249), *66*
Suzuki, Y., 257, *293*
Svedberg, T., 463(82), *469*
Swain, T., 262(146), *294*
Swan, R. J., 8(29), 55(29), 56(285, 287), 57 (287), *60*, *67*
Swanson, A. L., 348(334), *371*
Swanson, M. A., 330(199), *367*
Sweeley, C. C., 76(35), *130*, 181(220), *208*
Swenson, H. A., 460(72), *468*
Swift, G., 334(244), 335(244), *368*
Sylvan, S., 311(83), *364*
Sylvester, J., 145(268), 194(268), *210*
Szantay, C., 13(58), *61*
Szarek, W. A., 158(81a), *202*
Szer, W., 50(262), *67*
Szybalski, W., 199(293), 200(293), *210*

T

Tada, H., 31(165, 166, 167, 168), 58(292), *64*, *68*
Taft, R. W., Jr., 48(255), *66*
Tagawa, H., 26(138), 27(138), *63*

Takagi, T., 432(174), *442*
Takahashi, S., 162(95a), *203*
Takano, K., 244(12), 264(163), 274(256), 286(380), *291, 294, 297, 300*
Takeda, H., 283(346), *299*
Takemura, S., 144(226), 181(224), 182(226), *208*
Takenishi, T., 12(49), 16(74, 75), *61, 62*
Takeshita, M., 355, *371*
Takeuchi, S., 144(202), 176(199, 200, 201, 202), 177(205), *207*
Takeuchi, T., 143(143), 169(143), *205*
Talbot, G., 142(32), 150(32), *201*
Tamm, C., 46(247), *66*
Tanahashi, N., 323(155), 324(155a), *366*
Tanaka, H., 10(42), 14(60), 57(290), *61, 68*
Tanaka, K., 261(138), 288(400a), *294, 300*
Tanaka, Y., 144(217a), 180(217a), 207
Tanford, C., 393, *411*
Tanghe, L. J., 430(150), *442*
Taniyama, H., 144(225, 225a), 181(224, 225, 225a), *208*
Tanner, W., 321(139), *366*
Tanret, G., 122(309), 123(312), 127(330), *136, 137*
Tappel, A. L., 256, 270(227), *293, 296*
Tarbell, D. S., 164(115a), *204*
Tatsuoka, S., 164(106, 107), 167(128), 168(134), 169(106), *203, 204*
Tatsuta, K., 166(121a), 170(156b, 156c, 156d, 156e), 190(260a), *204, 205, 209*
Taylor, E. W., 427(119), *441*
Taylor, K. G., 144(196a), 176(196a, 198), *207*
Taylor, P. M., 85(110), *131*, 257(112), 289(407), *293, 300*
Tchelitcheff, S., 142(49, 54, 56), 153(49, 51), 154(49, 54, 56), *202*
Teague, R. S., 284(353), *299*, 325(161), *366*
Teeters, C. M., 166(120), *204*
Telfer, R. G. J., 76(34), 78(34), *130*
Telser, A., 361(413), *373*
Tennigkeit, J., 23(123, 124), 26(123), *63*
ter Kuile, J., 44(239), *66*
Tesoro, G. C., 434(206, 207), *443*
Thacker, D., 41(221), *66*
Thannhauser, S. J., 4(5), *59*
Theander, O., 85(112), 112(242), *131, 135*, 257(107), *293*
Thinès-Sempoux, D., 255(86), *293*

Thirlwell, M. P., 438(283), *445*
Thoma, J. A., 86(118), 128(340), *132, 137*, 246(16b), 289(406), *291, 300*
Thomas, E. B., 431(171), *442*
Thomas, G. H. S., 117(277), *135*
Thomas, H. J., 9(31), 10(31), 18(81, 82), *60, 62*, 172(164, 168), *206*
Thomas, L. F., 154(64), *202*
Thompson, A., 82(78), 87(126), 90(138), 98(164), 111(78, 126, 226), 112(78, 226), 125(321), 127(138, 329), *131, 132, 133, 134, 136, 137*, 401(67), 402(67a), *412*
Thompson, J. L., 265(169), *294*, 318(120), *365*, 459(71), *468*
Thompson, K. V. A., 422(78), 423(78), 425(78), *440*
Thompson, N. S., 462(76a), *468*
Thompson, N. S., 450(17), 454(38), *467, 468*
Thomsen, H., 231(77b), *240*
Thomson, D. L., 255(82), *292*
Thornber, J. P., 448(1), *467*
Tiedemann, H., 283(348), *299*
Tilden, E. B., 334(227), 335(227), *368*
Timberlake, C. E., 98(164), *133*
Timell, T. E., 75(20), *77*, 120(295), *129, 136*, 418(43), 431(169), 432(151), 433(181), *439, 442, 443*, 449(4, 5), 450(11, 12), 452(30), 454(36, 37, 40), 459(12, 65a), 460(11), 461(11, 75, 76c), 462(76b), 465(40), 466(40), *467, 468, 469*
Tipper, D. J., 283(343), *299*
Tipson, R. S., 5(17), 8(30), 35(30), *60*, 86(125), *132*
Tipton, C. L., 93(156), *133*, 253(61), 273(252), *292, 297*
Todd, A. R., 4(5), 8, 28(153, 154), 38(205), *59, 60, 64, 65*, 227(70), *239*
Todd, A. W., 167(103), *203*
Tønnesen, B. A., 419(53), 400(60), *440*
Toishi, A., 262(145), *294*
Tokes, Z. A., 124(317), 125(317), *136*
Tokita, N., 434(204), *443*
Tollin, P., *68*
Tolman, R. L., 9(36), 57(291), *60, 68*, 144(214a), 180(214a), *207*
Tomita, Y., 232(83), *240*
Tomkins, G. H., 250(36), 269(36), *291*
Tomlinson, C., 88(134), *132*
Tong, G. L., 10(39), 18(92, 93), 36(202), *61, 62, 65*

Tonizawa, H. H., 305(39), *363*
Topper, Y. J., 303(3), 304(3, 32), *362, 363*
Tornheim, J., 281(328), *298*
Torrés, H. N., 255(87), *293*, 338(279), *369*
Torriani, A. M., 290, *300*, 333, 334(219, 220, 221), *368*
Tóth, G., 113(251), *135*, 281(318), *298*, 436 (252), 437(261), 438(292), *444, 445*
Touzinsky, G., 430(160), *442*
Tovey, H., 434(196), *443*
Townsend, L. B., 9(36), 11(48), 12(48), 14 (66), 16(76), 20(102, 103), 42(229), 53 (66, 274), 54(48, 276), 55(66, 279), 57 (279, 291), 58(66), *60, 61, 62, 66, 67, 68*, 144(214a), 180(214a, 217, 217b), *207, 210*
Tracey, M. V., 424(91), 438(280, 281), *441, 445*
Träger, L., 44, 45(242), 46(242), *66*
Traut, R. R., 327(188), 337(269), 338, *367, 369*
Trenner, N. R., 143(174), 172(174), *206*
Trentham, D. R., 26(137), 27(145, 146), 54 (278), *63, 64, 67*
Trevithick, J. R., 251(40), 278(287), *291, 297*
Tripp, V. W., 423(79), *440*
Trischmann, H., 225(65), *239*, 398(55), 401 (64), *412*
Tristram, H., 243(1), *290*
Trivelloni, J. C., 326(171), *366*
Trucco, R. E., 104(185), 124(185), 125(185), *133*, 287(393), *300*
Tschesche, R., 145(304, 304a), *211*
Ts'o, P. O., 55(280), *67*
Tsou, K. C., 268(201), 270(232), *295, 296*
Tsuchiya, H. M., 74(9, 10), 99(169), 111(10, *129, 133*, 256(97, 102), *293*
Tsuda, K., 224(59), *239*
Tsuiki, S., 69(1), 71(1), 73(1), *129*
Tsukiura, H., 144(239), 162(97), 184(239, 240), *203, 208*
Tsukuda, Y., 58(299), *68*, 180(217c), *207*
Tsumara, T., 170(156c), *205*
Tsumura, T., 170(156d, 156e), *205*
Tsuruoka, T., 145(263b, 263c), 192(263b, 263c), *210*
Tu, C. C., 78(48), 91(154), *92*, 114(48, 253), *130, 133, 135*, 453(32), *467*
Tucker, L. C. N., 155(70), 157(70), *202*
Tuppy, H., 282(340), *299*
Turner, D. H., 315(102), *365*

Turner, J. F., 315(102), *365*
Turner, P., 281(325), *298*
Turvey, J. R., 77(38), 110(38), *130*
Tyminski, A., *77*

U

Uchida, K., 257(110), *293*
Ueda, K., 176(199), *207*
Ueda, M., 143(143), 169(143), *205*
Ueda, T., 19(100a), *62*
Ueno, T., 170(156a), *205*
Ukita, T., 36(201), 41(220), *65, 66*
Ulbricht, T. L. V., 8(29), 40(216, 217), 41 (221), 55(281, 283), 56(285, 287), 57(287), 58(293), *60, 61, 65, 66, 67, 68*
Umezawa, H., 139(1), 140, 142(87, 92b), 143 (5, 143, 150), 144(215), 145(5, 264), 160 (87, 87a, 88, 89, 91), 161(87, 91, 92, 92b), 169(144, 150), 170(150), 180(215, 216), 193(264, 265), 195(265), *200, 203, 205, 207, 210*
Umezawa, S., 166(121a), 170(154, 156b, 156c, 156d, 156e), 190(260a), *204, 205, 209*
Unkauf, H. C., 462(79b), *469*
Unruh, C. C., 427(119), *441*
Urbanski, T., 431(167), *442*
Urquhart, A. R., 420(61), *421, 440*
Usmanov, Kh. U., 434(210), 435(210), *443*
Usov, A. I., 157(75), *202*
Utahara, R., 143(143), 169(143), *205*
Uziel, M., *4*

V

Vaes, G., 270(223), *296*
Vaidya, B. K., 434(223), *444*
Vajda, É., 281(318), *298*
Valentini, L., 188(253), 199(294), *209, 210*
Vanaman, T. C., 324(155b, 155c), *366*
van Dam, J., 88(135), *132*
Van den Akker, J. A., 419(55), *440*
van der Veen, J. M., 91(151), *132*
VanderWyk, A. J. A., 414(9), *438*
VanLoo, W. J., Jr., 434(197), *443*
Van Tamelen, E. E., 43(236), *66*, 143(181), 144(222), 173(181), 174(181), 181(220, 221, 222, 223), *206, 208*
Varghese, J., 434(203), *443*
Vasina, I. V., 185(243), *208*

Vaterlaus, B. P., 184(234, 238), *208*
Vaughn, R. H., 347(330, 331), *371*
Veibel, S., 258(129), 260, 261(139), 267(188), 272, 274(129), *294, 295*
Veksler, V. I., 140(2), 163(2), 166(2), *200*
Venkataraman, R., 104(187), *133*, 253(57), *292*
Venkateswaran, A., 419(55), *440*
Verbeke, R., 323(149), *366*
Vercellotti, J. R., 117(279, 280a), *135*, 437 (274), *445*
Verheyden, J. P. H., 42(225), 43(233), *66*
Verhue, W., 332, *367*
Verity, M. A., 284(355), *299*
Vernon, C. A., 96(160), *133*, 243(1), *290*
Vickers, G. D., 143(186), 174(186), *206*
Vilkas, E., 217(15), *238*
Villar-Palasi, C., 336(259), 337(264), 338 (277), *369*
Villemez, C. L., Jr., 304(24), 344(314b), 347 (329), *363, 370*
Vining, L. C., 225(64), *239*
Vink, H., 418(45, 46), 422(65), *440*
Vir, E., 232(81), *240*
Voet, J., 317(113), *365*
Vogel, H., 82(80), 101(171), 103(179), *131, 133*
Vojtko, C. M., 145(268), 194(268), *210*
Vollbrechtshausen, I., 4(12), *60*
von Euler, H., 223(52), *239*
von Hofsten, B., 272(246), *296*
von Phillipsborn, W., 168(133), 169(133), *204*
von Saltza, M. H., 141(9, 10), 143(186), 174 (185, 186, 187), *200, 206*
von Ulbricht, T. L. 18(80a), *62*
von Weimarn, P. P., 437(271), *445*

W

Wacker, A., 4(4), 44, 45(242), 46(242), *59, 66*
Wada, A., 386, *411*
Wadman, W. H., 86(122), *132*
Waehneldt, T., 144(251), 145(301), 187(251), 188(252), 195(251), *209, 211*
Wagman, G. H., 143(157), 170(157), *205*
Wagner, G., 262(142), *294*
Wagner, H., 228(74), *239*
Wagner, R. L., Jr., 154(62), *202*

Wagnières, W., 231(77b), *240*
Wakabayashi, M., 266(182), 285(373, 375), *295, 299, 300*
Wakae, M., 162(97), *203*
Waksman, S. A., 165(117), *204*
Walker, B., 76(35), *130*
Walker, C. W., 161(95), *203*
Walker, G. J., 289(411), *300*, 331, *367*
Walker, P. G., 261, 270(231), 281(317, 323), 284(361), *294, 296, 298, 299*
Walker, R. T., 51(265), *67*
Walker, R. W., 143(174), 172(174), *206*
Wall, L. A., 428(145), *442*
Wall, R. A., 85(116), *132*
Wallace, W. S., 174(182), *206*
Wallenfels, K., 75(28), 105(188), *129, 133*, 235(95), *240*, 246, *248, 249*, 250(20, 27), 268(209), 272, 273(19, 244, 254), 274(257, 258), *291, 295, 296, 297*, 332, *367*, 398(55), *412*
Waller, C. W., 143(170), 172(170), *206*
Walseth, C. S., 422(76), 423(76, 83), *440*
Walter, M. W., 321(134), *365*
Walters, D. R., 141(8), 142(298), *200, 211*
Walton, E., 8(28), 18(83, 94), 19(100), 53 (100), 54(277a), 55(100), *60, 62, 67*, 172 (173), 184(233, 237), *206, 208*
Wang, L., 438(278), *445*
Wang, S., 24(126), *63*
Wang, Y., 164(108), *203*
Wangel, J., 260(134), *294*
Ward, K., Jr., 415(16), *439*
Ward, R. B., 278(288), *297*
Wareham, J. F., 414(4), *438*
Warren, L., 287(390), *300*, 305(49), 306(53), 307(60), *363, 364*
Warsi, S. A., 112(244), *135*
Watanabe, K., 142(300), 179(212), *207, 211*
Watanabe, K. A., 29(158), 32(181), 34(181), 39(211, 213), 53(211, 275), *64, 65, 67*, 144 (209), 172(163a), 177(203), 178(203, 209), *206, 207*
Watanabe, T., 142(300), *211*
Watenpaugh, K., 49(259), *67*
Waters, R. B., 218(29), *238*
Watkins, W. M., 78(53), 124(318), 125(318), *130, 136*, 267(199), 282(339), 287(386), 288(400), *295, 299, 300*, 304(15), 322 (141), 323(141), *362, 366*
Wattiaux, R., 284(362), 285(362), *299*

Wattiez, N., 122(311), *136*, 216(5), *237*
Weakley, F. B., 111(225), *134*
Weaver, J. W., 432(182), *443*
Weaver, O., 142(42), 151(42), 152(42), 171 (159), *201, 205*
Webb, I. J., 399(60), *412*
Webb, J. S., 145(266), 146(12), 193(266), 196(266), *200, 210*
Webber, J. M., 78(46), *130*, 150(35), 154 (58, 63, 64), 155(58), 157(58, 74), *201, 202*, 283(350), *299*, 339(282), 361(405), *369, 373*, 435(239), 436(239, 258), 437(239), 438(239), *444*
Weber, D. J., 47(249), *66*
Weber, E., 305(39), *363*
Weber, K., 250(33, 34), *251*, 268(33), 269 (33), *291*
Weber, O. H., 414(1), *438*
Wehrli, H., 436(260), 437(260), *444*
Weidel, W., 283(341), *299*
Weidenhagen, R., 219, *238*, 256, 262(143), *293, 294*
Weidenmüller, H. L., 145(304), *211*
Weidmann, H., 167(126), *204*
Weigel, H., 75(17), *129*, 256(100), 257(105, 106), *293*, 317(116), *365*
Weiler, L. S., 142(59), 154(59), *202*
Weiner, I. M., 355(371), *372*
Weinland, H., 117(275), *135*
Weinlich, J., 142(298), *211*
Weinstein, M. J., 143(157), 170(157), *205*
Weisberger, A. S., 198(280), *210*
Weisblum, B., 200(296), *210*
Weiss, M. J., 37(203), *65*
Weissmann, B., 85(113), 118(283), *131, 136*, 281(320, 321, 328), *298*
Weisweiler, G., 218(22), *238*
Welch, H. V., 320(131), *365*
Weliky, V. S., 50(263), *67*, 172(166), *206*
Welsh, J. D., 255(84), *292*
Wempen, I., 9(32), 32(177), 34(32, 177), 38 (207), 57, *60, 64, 65*, 172(167), 175(188), *206*
Werries, E., 281(329), 282(329), *298*
Wessely, F., 220(42, 43), *238, 239*
West, C. A., 181(220), *208*
Westlake, D. W. S., 75(26), *77, 129*
Westley, J. W., 181(219), *208*
Westphal, O., 355(368), *371*
Westphal, W., 81(73), *131*

Westwood, J. H., 157(74), *202*
Wettstein, A., 153(50), *202*
Weurman, C., 113(246), *135*, 425(107), *441*
Weygand, F., 4(4), *59*
Whaley, H. A., 181(220, 221, 223), *208*
Whelan, W. J., 73(7), 75(16), 76(36), 77(38), 78(51), 81(70, 77), 82(79), 85(110), 88 (132), 108(213), 110(38, 213), 111(232, 233), 112(244, 245), 125(36), 127(327), 128(36), 129(36), *129, 130, 131, 132, 134, 135, 137*, 244(14), 253, 257(112), 279, 289 (407, 408, 409, 410), *291, 292, 293, 298, 300*, 330(200, 203), 332(209), 344(313), *367, 370*, 435(232), *444*
Whiffen, D. H., 91(150), *132*
Whistler, R. L., 74(8, 14), 78(48), 85(8, 107), 91(154), *92*, 96(162), 109(219), 114(48, 253), 116(269), 127(8), 129(342), *129, 130, 131, 133, 134, 135, 137*, 334(239), 335(239), *368*, 377(1), 382(2), 394(1), *410*, 435(240), 437(240), 438(240), *444*, 449(9, 10), 450 (15), *451*(21), 453(32), 457(53), 458(56), 465(90), 466(47), *467, 468, 469*
Whitaker, D. R., 75(18), *129*, 243(6), 283 (346), *290, 299*, 423(80), 424(90), 425 (97, 104, 106), 426(97, 108), 438(282, 283), *440, 441, 445*, 456(47), 465(47), 466(47), *468*
Whitby, L. G., 253(69), *292*
White, E. V., 462, *469*
White, J. R., *210*
White, J. W., Jr., 85(109), *131*, 256(90), *293*
White, S. M., 438(298), *445*
Whitefield, G. B., 181(221, 223), *208*
Whitehead, B. K., 275(266), *297*
Whitehead, D. F., 169(145, 147), 170(145, 153), *205*
Wickberg, B., 217(10, 11), *238*
Wicken, A. J., 357(385), 358(392), *372*
Wickstrøm, A. W., 278(293, 294), *298*
Wieczorkowski, J., 52(268), *67*
Wiederschein, G. Y., 288, *300*
Wiesmeyer, H., 108(211), 125(211), *134, 251*, 290, *300*
Wigler, P. W., 19(95a), *62*
Wikström, R., 418(45), *440*
Wild, G. M., 75(24), 89(24), 91(155), *93*, 101(24), 108(210), 122(24), 125(210), *129, 133, 134*, 145(269), 194(269), *210*, 288 (404), *300*

Wiley, P. F., 142(42), 143(114), 151(42, 43), 152(42), 164(114, 115), 165(114), 171 (159), 179(213), *201, 204, 205, 207*

Wilham, C. A., 74(9, 10), 111(10), *129*, 334 (247), 335(247), *368*

Wilken, J., 323(147), *366*

Wilkie, K. C. B., 75(19), *129*, 448(2), *467*

Wilkinson, I. A., 330(202), *367*

Williams, J. H., 143(183), 174(182, 183), *206*

Williams, R. P., 146(12), 161(95), *200, 203*

Williams, R. T., 284(356), *299*

Williamson, A. R., 38(208), *65*

Wills, E. D., 277(283), *297*

Willstaedt, E., 277(279, 284), *297*, 334(230), 335(230), *368*

Willstätter, R., 226, *239*

Wilson, D. B., 309(70), *364*

Wilson, E. J., Jr., 99(167), *133*, 334(228), 335 (228), *368*

Wilson, H. R., *68*

Wilson, V. W., Jr., 287(387), *300*

Winkley, M., 38(208), *65*

Winterstein, A., 229(75), *239*

Wintersteiner, O., 141(8, 9, 10, 11), 143(119), 166(119), *200, 204*

Winzler, R. J., 359(397), 361(411), *372, 373*

Wirth, W. D., 142(297), *211*

Wise, L. E., *77*, 114(254), *135*, 414(2), *438*, 454(35a), 459(66), 462, *468, 469*

Wittenburg, E., 13(55), *61*

Wöhler, F., 217, *238*

Wohnlich, J. J., 266(179), 267(195), *295*

Wolf, F. J., 144(232), 184(232), *208*

Wolf, J. P., III, 256(93), *293*

Wolfe, A. D., 198(281), *210*

Wolfe, R. G., 268(206), *295*

Wolfe, S., 198(280), *210*

Wolfrom, M. L., 74(8), 78(39, 49), 80(63, 66a), 82(78), 85(8, 106), 87(126), 88(133), 90(138), 91(133), *92*, 98(49, 164), 104 (182a), 110(222), 111(63, 78, 126, 222, 226), 112(78, 226), 117(279, 280, 280a), 118 (287), 119(287), 125(321), 126(49, 133), 127(8, 66a, 138, 329), *129, 130, 131, 132, 133, 134, 135, 136, 137*, 143(100), 163(100, 101, 102), *203*, 377(1), 394(1), 401(66, 67), 402(67a), *410, 412*, 428(137), 437(272, 274), *442, 445*, 459(62), *468*

Woo, P. W. K., 142(71), 143(124, 127), 150 (38), 155(69), 156(71), 157(72, 73), 158

(73), 166(121), 167(124, 127), 195(38, 73), 197(71), 198(278a), *201, 202, 204, 210*

Wood, D. L., 144(195), 175(195), 182(195), 188(195), *207*

Wood, H. G., 323, *366*

Woodhouse, H., 400(62), *412*

Wooding, N. S., 427(129), 428(129), 435 (129), *441*

Woodward, D. O., 269(218), *296*

Woodward, F. N., 415(17), *439*

Woodward, R. B., 142(47, 59), 148(21), 152 (45, 47), 154(21, 45, 59, 62), *201, 202*

Woolf, D. O., 142(93), 161(93), 175(190), *203, 206*

Woollen, J. W., 270(231), 281(317, 325), *296, 298*

Work, R. W., 432(172), *442*

Wortman, B., 362, *373*

Wray, J. A., 434(215), 435(215), *443*

Wright, A., 355(370, 372), *372*

Wright, D. E., 74(12), *129*

Wyatt, G. R., 4(10), 58(10), *59*, 258(122), *293*, 327(184), *367*, 438(290), *445*

Wyckoff, R. W. G., 383(4), *410*

Wyer, J. A., 154(64), *202*

Wylam, B., 111(231), *134*

Y

Yagishita, K., 143(143), 169(143), *205*

Yamada, H., 253(70), *292*

Yamaguchi, T., 167(128), *204*

Yamaha, T., 235(93), *240*, 326(169), *325, 366*

Yamamoto, H., 178(206), *207*

Yamaoka, N., 14(59), *61*

Yamashita, Y., 385, *411*

Yamazaki, A., 16(74, 75), *62*

Yankeelov, J. A., Jr., 246(16b), *291*

Yarmolinsky, M. B., 199(291), *210*

Yasuda, H., 434(215), 435(215), *443*

Yasumura, A., 253(71), *292*

Yelland, L., 231(77a), *240*

Yengoyan, L., 52(269a), *67*

Yonehara, H., 144(202), 176(199, 200, 201, 202), 177(204, 205), 178(204), *207*

Yoshida, H., 362(417), *373*

Yoshida, K., 262, *294*

Yoshino, T., 14(60), *61*

Yoshioka, Y., 33(184), *65*

Youatt, G., 264(155), *294*, 424(87), *441*

Young, A. E., 433(186), *443*
Young, B., 75(24), 89(24), 101(24), 122(24), *129*
Young, D. P., 164(112), *204*
Young, D. W., *68*
Young, M. B., 142(298), *211*
Yphantis, D. A., 391, *411*
Yudis, M. D., 143(157a), 170(157a), *205*
Yüntsen, H., 143(178), 173(178, 180), *206*
Yumoto, H., 145(263b), 159(84), 192(263b), *203, 210*
Yung, N. C., 22(115), *63*
Yurina, M. S., 145(270), 194(270), *210*
Yurkevich, A. M., 24(127), *63*

Z

Zabin, I., 250(37), 269(210), *291, 296*
Zähner, H., 141(6), 142(31, 70b), 150(31), 153(50), 155(70b), 158(78), 159(83), *200, 201, 202, 203*
Zamojski, A., 142(32), 150(32), *201*
Zarnitz, M. L., 75(28), *129*, 248(19), 268(19), 273(19), *291*
Zav'yalov, S. I., 19(95), *62*
Zechmeister, L., 113(251), *135*, 281(318), *298*, 436(252), 437(261), 438(292), *444, 445*

Zedric, J. A., 149(26), *201*
Zeleznick, L. D., 354(364), 355(371), *371, 372*
Zelikson, R., 78(44), *130*
Zelitis, J., 309(72a), *364*
Žemlička, J., 22(118), 25(132, 135, 136), 27(140), 34(188), 35(195), *63, 65*
Zemplén, G., 80(62), 81(71), 83(89), 103(180), 111(229), *130, 131, 133, 134*, 217(21), *238*
Zervas, L., 113(250), *135*
Zielinski, J., 20(108), *63*
Zilkha, A., 434(227), *444*
Zilliken, F., 113(248), *135*, 274(260, 262), *297*
Zimm, B. H., 389(25), *411*
Zimmerman, H. K., Jr., 167(126), *204*
Zimmerman, M., 18(83), *62*
Zimmerman, S. B., 327(186), *367*
Zimmermann, M., 256(91), *293*
Zipser, D., 250(35), *291*
Ziv, O., 258(124), *293*
ZoBell, C. E., 438(285), *445*
Zorbach, W. W., 186(245), *208*
Zottu, S., 336(262), *369*
Zussman, J., 28(153), *64*
Zweig, G., 85(104), 88(104), *131*

SUBJECT INDEX

Volume IIA

A

Abequose, 311
Acacipetalin, 218
Acetals, as blocking groups for nucleosides, 23
Acetolysis, of cellulose, 401
Actinamine, 164
  N,*N*-dimethyl-, conformation of, 196
Actinoidin, 194
Actinospectacin, 164
Actinospectoic acid, 165
Actinospectose, 165
Adenine, 5
  9-(2-deoxy-β-D-*erythro*-pentofuranosyl)-, 3
  9-(3-deoxy-β-D-*erythro*-pentofuranosyl)-, *see* Cordycepin
  9-(D-psicofuranosyl)-, *see* Psicofuranine
  9-β-D-ribofuranosyl- (adenosine), 2, 5
Adenosine, 2, 5
  3'-amino-3'-deoxy-, 174
  2'-deoxy-, 3
Aesculin, 220
Ajugose, 267
Aldgamycin E, 159
Aldgarose, 159
Aldgaroside A, 159
  conformation of, 195
Aldgaroside B, 159
Aldolase, 307
Alginic acid, 349, 351
Allose, 6-deoxy-2,3-di-*O*-methyl-D-, *see* Mycinose
Aloinoside B, 234
Amicetamine, 175
Amicetin, 175
Amicetose, 175
Amino acids, of β-D-galactosidases of *Escherichia coli*, 249, 250
Aminosidin, 168
Amosamine, 175
Amygdalin, 217
Amylodextrin, dextran from, 335
Amyloglucosidase, 257
Amylomaltase, 290, 333

Amylopectin, 328
  polysaccharide from sucrose, 332
Amylosaccharides, cyclic, 334
Amylose, 328, 380
  from maltose, 333
  optical rotatory dispersion of, 386
  O-(carboxymethyl)-, 386
Amylose V, 385
Amylosucrase, 256, 332
Angolamycin, 155
Angolosamine, 155
Angustmycin A, 173
Angustmycin C, *see* Psicofuranine
Anhydronucleosides, *see* Cyclonucleosides
Anthocyanidin, glycosides, 226
Anthocyanins, 226
Anthracene, glycosides of derivatives of, 220
Anthracyclines, 199
Anthraquinone, *C*-glycosides, 233
Antibiotics
  aristeromycin, 58
  aromatic-group containing, 183–189
  biological concepts, 198
  characterization of, by physical means, 194–198
  cyclitol, 159–172
  formycin, 20
  "γ-activity X," 193
  glycosylamine type, 180–183
  macrolide, 140–159
  nucleoside, 172–180
  nucleotide, 199
  showdomycin, 20, 58
  sugar-containing, 139–211
Apiin, 215
Apiose, D-, 215
L-Arabino-D-galactans, 462
Arabinopyranose, 3-*O*-β-L-arabinopyranosyl-L-, 115
Arabinose, L-, in xylans, 455
  3-*O*-β-L-arabinofuranosyl-L-, 114
L-Arabino-D-xylans
  enzymic hydrolysis of, 465
  structure of, 456

Arbutin, 218, 326
  methyl-, 218
Arcanose, 158
Aristeromycin, 58
Ascarylose, 311
Asiaticoside, 224
Aucubin, 232

B

Bakankoside, 232
Bamicetamine, 176
Bamicetin, 176
Bandamycin A, 159
Barbaloin, 233
Barry degradation, of polysaccharides, 409
Biosynthesis
  of cellulose, 342, 415
  of chitin, 438
  of glycosides, 235
  of nucleosides, 305
  of starch, 340
  of sugars and polysaccharides, 301–373
Blasticidin S, 176
Blastidic acid, 177
Bluensidine, 162
Bluensomycin, 162
Borate complexing,
  of sugars, 7
Bufadienolides, 222

C

Callose, 348
Carbohydrates, photosynthesis of, 302
Carbomycin, 154
Carbomycin B, 154
Cardenolides, 221
Cardiotonics, 221
Carimbose, 154
Carminic acid, 234
Carotenoid glycosides, 229
Cascarosides, 234
Catalpol, 232
  methyl-, 232
Catenulin, 168
Celesticetin, 191
Cellobiases, 264, 265, 313
Cellobiose, 109
  phosphorolysis of, 316
  synthesis of, and octaacetate, 81
Cellobiouronic acid, 119
Cellobiulose, synthesis of, 83

Cello-oligosaccharides, hydrolysis rate constants for, 98
Cellophane, 432
Cellotriose, 126, 318
Cellulases, 422, 425
Cellulose, 378, 413–435
  acetolysis of, 401
  biosynthesis of, 342, 415
  copolymers (graft), 434
  cross-linking of, 434
  degradation of, 422–426, 428
  etherification of, 430, 432
  hydrolysis of, 426
  hydrolysis rate constants of, 98
  isolation and purification of, 417
  methylation of, 400
  molecular weight and fine structure of, 418
  occurrence of, 415
  oxidation of, 426–428
  reactivity of, 429
  sorption, swelling and solution of, 420
  structure of, 413
Cellulose, O-(carboxymethyl)-, 433
  O-ethyl-, 433
  O-(hydroxyethyl)-, 433
    electron microscopy of, 383
  O-methyl-, 429, 432
Cellulose I, 419
Cellulose II, 419
Cellulose III, 419
Cellulose IV, 419
Cellulose X, 419
Cellulose acetate, 429, 431
Cellulose esters, 430
Cellulose ethers, 432
Cellulose nitrates, 431
Cellulose xanthate, 432
Cerebrosides, enzymic degradation of, 282
Chalcomycin, 156, 159
Chalcose, 157, 158
  methyl glycoside, conformation of, 195
Chartreusin, 188
Chitan, 437
Chitin, 339, 435–438
Chitinase, 283, 438
Chitobiase, 283, 438
Chitobiose, 113
  6-O-(2-acetamido-2-deoxy-β-D-glucosyl)-
    N,N'-diacetyl-, 283
Chondroitin 4-sulfate, 384

Chondrosine, 118
Chromatography, of oligosaccharides, 85
Chromomycin A₃, 185
Chromomycins, 185, 199
Chromose A, 186
Chromose B, 186
Chromose C, 185
Chromose D, 186
Cinerubins, 186
Cladinose, 151, 154
Cobalamins, 232
Colominic acid, 287
Conformation
    of antibiotic sugars, 194
    of oligosaccharides, 91
Coniferin, 219
Cordycepin, 172, 199
Cotton effect, of nucleosides, 55
Coumarin, glycosides, 220
Coumermic acid, 184
Coumermycin A-1, 184
Coumerose, 185
Crocin, 229
Crystallization, of oligosaccharides, 86
Cucurbitacins, 225
Curacose, 145
Curamycin, 145
Cyanin, 226
Cycasin, 232
Cyclitols, antibiotics containing, 159–172
Cyclodextrin transglucosylases, 288
Cyclohexaamylose, 128
Cyclonucleosides, 8, 28–35
Cytidine, 2, 5
    5-aza-, 17
    2′-deoxy-, 3
    5′-(3-deoxy-D-*manno*-octulosonate phosphate, 309
Cytimidine, 175
Cytomycin, 177
Cytosamine, 175
Cytosine, 5
    1-(2-deoxy-β-D-*erythro*-pentofuranosyl)-(2′-deoxycytidine), 3
    1-β-D-ribofuranosyl- (cytidine), 2, 5
Cytosinine, 177

D

Daphnin, 220
Daucosterin, 223
Daunomycin, 188, 199

Daunosamine, 188
Decoyinine, 173
Degradation, of cellulose, 422–429
Degree of polymerization, of oligosaccharides, 92, 93
Delphinin, 227
Demissin, 225
Depolymerization, of polysaccharides, 401
Desosamine, 148, 150, 152
    conformation of, 195
Destomic acid, 171
Destomycin A, 171
Dextran
    from amylodextrin, 335
    from sucrose, 334
Dextran 6-glucosyltransferase, 335
Dextransucrases, 256, 334
Dextrins, Schardinger, 334
Dhurrin, 218
Dialysis, of oligosaccharide solution, 84
Digitonin, 223
Digitoxin, 222
Digitoxose, 188
Diosgenin, 224
Disaccharide phosphorylases, 314
Disaccharides, 101–121
    hydrolysis rate for, 97
    synthesis by transglucosylation, 316
    synthesis of, 82
Drude equation, 386

E

Electron microscopy, in polysaccharide characterization, 383
Endo-hydrolases, hydrolysis of polysaccharides by, 74
Enzymes
    A-, 424
    cellulose synthesis by, 415
    C₁-, for cellulose degradation, 424
    D-, 289, 330
    degradation of polysaccharides by, 402
    hydrolysis of 2-amino-2-deoxy-D-hexose-containing compounds by, 282
    of hemicelluloses by, 464
    of polysaccharides, 74, 95
    in oligosaccharide glycoside synthesis, 99
    Q-, 330
    R-, 332
    T-, 253
    viral, 287

Enzymolysis, of cellulose, 422–426
Epimelibiose, 267
Epimerases, 304
Erythromycin, 151, 198
Erythromycin B, 152
Esters, acyl, as blocking groups for nucleosides, 26
Etherification, of cellulose, 430, 432
Everninomycins, 189
Evernitrose, 189
Exo-hydrolases, hydrolysis of polysaccharides by, 74

F

Fishman unit, 284
Flavones
  C-glycosides, 234
  glycosides of hydroxy derivatives of, 227
Floridoside, 216
Formycin, 20, 180
Formycin B, 180
Foromacidin, 153
Forosamine, 154
Frangularoside, 221
Frangulin, 221
Fraxin, 220
Fructofuranosidase, β-D-, 250
  inhibition of, 277
  purification and properties of, 275
Fructofuranoside, β-D-, specificity of, 278
  α-D-galactopyranosyl β-D-, *see* Galsucrose
  O - α - D - galactopyranosyl - (1→6) - α - D - glucopyranosyl β-D-, *see* Raffinose
  O-β-D-galactopyranosyl-(1→4)-O-α-D-glucopyranosyl-(1→2) β-D-, *see* Lactsucrose
  α-D-xylopyranosyl β-D-, *see* Xylosucrose
β-D-Fructofuranosyl transfer, 279
Fructopyranose, 5-O-α-D-glucopyranosyl-D-, *see* Leucrose
Fructose, D-, 6-phosphate, 304
  1-O-β-D-fructofuranosyl-D-, *see* Inulobiose
Fucosidases, α, β-D-, and α-L-, 288

G

Galactan, DL-, 350
D-Galacto-D-gluco-D-mannans, 461
Galactopyranose, 2-amino-2-deoxy-3-O-(β-D-glucopyranuronosyl)-D-, *see* Chondrosine

Galactopyranose—*continued.*
  3-O-β-D-galactopyranosyl-D-, 116
  6-O-β-D-glucopyranuronosyl-D-, 119
Galactopyranuronic acid, 4-O-α-D-galactopyranuronosyl-D-, 120
Galactose, 6-deoxy-2,4-di-O-methyl-D-, *see* Labilose
Galactosidase, 2-acetamido-2-deoxy-α-D-, 281, 282
Galactosidases
  α-D-, 266
  β-D-, 246, 259, 260
    amino acid composition of, 249
    mammalian, 270
    molecular weights of, 250
    purification and properties of, 268
    specificity of, 272
Galacturonic acid, D-, in polysaccharides, 380
Galsucrose, 318
Garosamine, 170
Gaultherin, 219
Gein, 219
Gel filtration, of oligosaccharides, 84, 85
Gentamycin, 170, 199
Gentianose, 124
Gentiobiose, 111, 217, 318
  octaacetate, synthesis of, 81
Gentiobioside, phenyl β-, 326
Gentiopicrin, 232
Gentiotriose, 318
Gitonin, 223
Glebidine, 162
Glebomycin, 162
α-D-Glucan glucanohydrolase, 330
Glucans, (1→3)-β-D-, biosynthesis of, 348
Glucoamylase, 257
Glucocheirolin, 230
Glucocleomin, 231
Glucofrangulin, 221
D-Gluco-D-mannans, 459
  biosynthesis of, 345
  enzymic hydrolysis of, 466
Glucopyranose, 2-acetamido-2-deoxy-3-O-(D-1-carboxyethyl)-6-O-(2-acetamido-2-deoxy-β-D-glucopyranosyl)-D-, 121
  O,N-acetylneuraminyl-(2→3)-O-β-D-galactopyranosyl-(1→4)-D-, *see* Neuraminlactose
  2 - amino - 2 - deoxy - 4 - O - (2 - amino - 2 -

Glucopyranose—*continued.*

deoxy-β-ᴅ-glucopyranosyl-ᴅ-, *see* Chitobiose

2-amino-2-deoxy-3-*O*-(β-ᴅ-glucopyranuronosyl)-ᴅ-, *see* Hyalobiouronic acid

4-*O*-β-ᴅ-galactopyranosyl-ᴅ-, *see* Lactose

3-*O*-β-ᴅ-glucopyranosyl-ᴅ-, *see* Laminarabiose

4-*O*-α-ᴅ-glucopyranosyl-ᴅ-, *see* Maltose

4-*O*-β-ᴅ-glucopyranosyl-ᴅ-, *see* Cellobiose

6-*O*-α-ᴅ-glucopyranosyl-ᴅ-, *see* Isomaltose

6-*O*-β-ᴅ-glucopyranosyl-ᴅ-, *see* Gentiobiose

O-α-ᴅ-glucopyranosyl-(1→4)-α-ᴅ-glucopyranosyl-(1→4)-ᴅ-, *see* Maltotriose

O-α-ᴅ-glucopyranosyl-(1→4)-[*O*-α-ᴅ-glucopyranosyl-(1→4)]₅-ᴅ-, *see* Maltoheptaose

O-α-ᴅ-glucopyranosyl-(1→6)-*O*-α-ᴅ-glucopyranosyl-(1→4)-ᴅ-, *see* Panose

O-β-ᴅ-glucopyranosyl-(1→4)-*O*-β-ᴅ-glucopyranosyl-(1→4)-ᴅ-, *see* Cellotriose

4-*O*-β-ᴅ-glucopyranuronosyl-ᴅ-, *see* Cellobiouronic acid

Glucopyranoside, ethyl α-ᴅ-, 216

β-ᴅ-fructofuranosyl *O*-α-ᴅ-galactopyranosyl-(1→6)-*O*-α-ᴅ-, *see* Raffinose

β-ᴅ-fructofuranosyl *O*-α-ᴅ-galactopyranosyl-(1→6)-*O*-α-ᴅ-galactopyranosyl-(1→6)-*O*-α-ᴅ-, *see* Stachyose

β-ᴅ-fructofuranosyl *O*-α-ᴅ-galactopyranosyl-(1→6)-[*O*-α-ᴅ-galactopyranosyl-(1→6)]₂-*O*-α-ᴅ-, *see* Verbascose

β-ᴅ-fructofuranosyl *O*-β-ᴅ-glucopyranosyl-(1→6)-*O*-α-ᴅ-, *see* Gentianose

O-α-ᴅ-galactopyranosyl-(1→6)-*O*-β-ᴅ-fructofuranosyl α-ᴅ-, *see* Planteose

α-ᴅ-glucopyranosyl α-ᴅ-, *see* α,α-Trehalose

O-α-ᴅ-glucopyranosyl-(1→3)-*O*-β-ᴅ-fructofuranosyl α-ᴅ-, *see* Melezitose

p-hydroxyphenyl β-ᴅ-, 325

methyl β-ᴅ-, 216

Glucopyranosiduronic acid, quercetin-3-yl β-ᴅ-, 325

Glucose, 2-amino-2-deoxy-ᴅ-, and 6-phosphate, 305

Glucose—*continued.*

3-amino-3-deoxy-ᴅ-, *see* Kanosamine

6-amino-6-deoxy-ᴅ-, 169

2-amino-2-deoxy-4-*O*-(α-ᴅ-glucopyranosyl)-ᴅ-, 117

β-*O*-cellobiosyl-ᴅ-, 80

6-deoxy-5-*O*-β-ᴅ-glucosyl-ᴅ-, 81

2,6-diamino-2,6-dideoxy-ᴅ-, *see* Neosamine C

3,6-dideoxy-3-dimethylamino-β-ᴅ-, *see* Mycaminose

4,6-dideoxy-4-dimethylamino-ᴅ-, *see* Amosamine

2-*O*-α-ᴅ-glucopyranosyl-ᴅ-, *see* Kojibiose

3-*O*-α-ᴅ-glucopyranosyl-ᴅ-, 81

O-β-ᴅ-glucosyl-(1→6)-*O*-β-ᴅ-glucosyl-(1→4)-β-ᴅ-, 318

6-*O*-β-maltosyl-ᴅ-, 80

6-*O*-α-rhamnopyranosyl-ᴅ-, *see* Rutinose

6-*O*-ʟ-rhamnosyl-ᴅ-, 81

UDP-2-acetamido-2-deoxy-ᴅ-, 305

Glucosidases,

α-ᴅ-, mammalian, 254

α-ᴅ-, of plant origin, 251

β-ᴅ-, 258–266

2-acetamido-2-deoxy-ᴅ-, 281

endo-2-acetamido-2-deoxy-ᴅ-, 282

Glucosides

enzymic hydrolysis of, 244

1-thio-ᴅ-, 231

Glucosiduronases, β-ᴅ-, 284–286

Glucosiduronic acid, *o*-aminophenyl β-ᴅ-, 325

Glucosisaustricin, 230

Glucosisymbrin, 231

Glucosyltransferases, ᴅ-, 312

Glucuronic acid, ᴅ-, in polysaccharides, 380

Glycals, ozonolysis of, 83

Glycans, 376

structural analysis of, 463

Glycogen

depolymerization of, 401

methylation of, 399

synthesis of, 327, 336

Glycogen synthetases, 337

Glycolipids, 305

Glycoproteins, 358–361

Glycosidases, 241–300

carbohydrate content of, 248

classification of, 242

Glycosidases—*continued.*
  mechanism of action of, 243–248
  molecular weights of, 250
  nomenclature of, 242
Glycosides
  of aliphatic alcohols and of alditols, 216
  anthocyanidin, 226
  of anthracene derivatives, 220
  biogenesis and metabolism, 235
  C-, 232–234
  carbohydrate constituents of naturally
    occurring, 214
  cardiac, 216, 221
  carotenoid, 229
  complex, 213–240
  coumarin, 220
  cyanogenetic, 217
  hydroxyflavone, 227
  indole, 226
  of natural pigments, 225–230
  oligosaccharides from, by hydrolysis, 74,
    75
  phenanthrene, 221–225
  of phenols, 218
  steroid, 222
  steroid saponins, 223
  synthesis of, 312, 325
  1-thio-, 230
  transglycosylation of, 235
  triterpenoid saponin, 224
Gougerotin, 178
Guanine, 5
  9-(2-deoxy-β-D-*erythro*-pentofuranosyl)-
    (2′-deoxyguanosine), 3
  9-β-D-ribofuranosyl- (guanosine), 2, 5
Guanosine, 2, 5
  2′-deoxy-, 3
Guluronic acid, L-, 380

H

Hemicelluloses, 447–469
  chemical modification during isolation,
    450
  crystal structure of, 452
  enzymic hydrolysis of, 464
  extraction of, 449
  D-glucuronic acids in, 380
  molecular weights of, 451
  purification of, 450
Heparin, 381, 383

Hesperidin, 228
Heteroglycans, 377, 379
Heterosides, 71
Hex-2-enopyranuronic acid, 4-amino-4-
  deoxy-D-*erythro*-, 177
Hexos-2,3-diulose, 4,6-dideoxy-D(or L)-
  glycero-, 165
Hexose, 4-amino-2,3,4,6-tetradeoxy-L-
  erythro-, *see* Tolyposamine
  3-amino-2,3,6-trideoxy-L-*lyxo*-, *see*
    Daunosamine
  6-deoxy-5-*C*-methyl-4-*O*-methyl-3-*O*-
    (5-methyl-2-pyrrolyl)-L-*lyxo*-, *see*
    Coumarose
  2,4-diamino-2,3,4,6-tetradeoxy-D-*arab*-
    ino-, *see* Kasugamine
  3,6-dideoxy-L-*arabino*-, *see* Ascarylose
  3,6-dideoxy-D-*ribo*-, *see* Paratose
  3,6-dideoxy-D-*xylo*-, *see* Abequose
  2,6-dideoxy-3-*O*-methyl-α-L-*arabino*-, *see*
    Oleandrose
  2,6-dideoxy-3-*C*-methyl-L-*ribo*-, *see* My-
    carose
  4,6-dideoxy-3-*O*-methyl-β-D-*xylo*-, *see*
    Chalcose
  2,6-dideoxy-3-*C*-methyl-3-*O*-methyl-L-
    ribo-, *see* Cladinose
  2,6-dideoxy-3-*C*-methyl-3-*O*-methyl-L-
    xylo-, *see* Arcanose
  4-dimethylamino-2,3,4,6-tetradeoxy-D-
    erythro-, *see* Forosamine
  3-dimethylamino-2,3,6-trideoxy-L-*lyxo*-,
    see Rhodosamine
  3-dimethylamino-2,3,6-trideoxy-D-*xylo*-,
    see Angolosamine
  3-dimethylamino-3,4,6-trideoxy-β-D-
    xylo-, *see* Desosamine
  2,3,6-trideoxy-D-*erythro*-, *see* Amicetose
Hexosidases, 2-acetamido-2-deoxy-D-, 280,
  282
Hilbert–Johnson synthesis, of nucleosides,
  19
Holocellulose, 449
Holosides, 71
Holothurin, 224
Homoglycans, 377, 378
Homomycin, 161
Homonataloin, 233
Hyalobiouronic acid, 118
Hyaluronic acid, 351, 383, 384

Hydrocelluloses, 423
Hydrolases
    glycoside, 241–300
    nucleoside, 6
Hydrolysis
    of cellulose, 426
    of hemicelluloses by enzymes, 464
    of nucleosides, 43–52
    of oligosaccharides, 95
    oligosaccharides prepared by, 74, 75
Hydroxymycin, 168
Hygromycin A, 161
Hygromycin B, 199
Hygromycin B₂, 171
Hyosamine, 171
Hyperin, 215, 228

I

Iceland moss, lichenan and isolichenan from, 435
Idose, 2,6-diamino-2,6-dideoxy-L-, *see* Neosamine B, Paromose
Iduronic acid, L-, 380
Imidazoline-5-carboxylic acid, 4-(2-amino-1-hydroxyethyl)-2-amino-, *see* Streptolidine
Immunochemistry, and polysaccharide structure, 409
Indican, 226
Indole glycosides, 226
Infrared dichroism, of polysaccharides, 385
Infrared spectroscopy
    of glycosides, 233
    in oligosaccharide structure determination, 91
    of polysaccharides, 385
Inositol, antibiotics, 160
Inulobiose, 115
Invertases
    honey, 256
    yeast, 278, 279, 313
Ipuranol, 223
Isolichenan, 435
Isomaltose, 110
    octaacetate, 83
Isomaltulase, 254
Isomerases, 304
Isoorientin, 234

K

Kanamycin, 169, 199

Kanamycin A, 169
Kanamycin B, 170
Kanamycin C, 170
Kanosamine, 169
Kasugamine, 160
Kasugamycin, 160
Kasugamycinic acid, 160
Kasuganobiosamine, 160
Kestose, 279
1-Kestose, 279
Ketoses, synthesis of, 83
Koenigs–Knorr reaction, oligosaccharide synthesis by, 79
Kojibiose, synthesis of, 81

L

Labilomycin, 193
Labilose, 193
Labilose methyl glycoside, conformation of, 195
α-Lactalbumin, 323
Lactose, 104, 322
    structure determination of, 89
    synthesis of, 81
Lactose synthetase, 323
Lactsucrose, 318
Lactulose, synthesis of, 83
Laminarabiase, 264
Laminarabiose, 112, 316, 318
Laminarabiose phosphorylase, 316
Lanatoside A, 222
Lankamycin, 158
Lankavose, *see* Chalcose
Lead tetraacetate, polysaccharide oxidation by, 407
Leucomycin A3, 155
Leucrose, 317
Levan, from sucrose, 335
Levansucrase, 280, 318, 336
Lichenan, 435
Light scattering, by polysaccharides, 388
Linamarin, 218
Lincomycin, 190, 200
Lincosamine, 190
Lipopolysaccharides, biosynthesis of, 353–356
Lotaustralin, 218
Lucenin, 234
Lucensomycin, 146
Lupin, 216

Lycotetraose, 215
Lysozymes, 282, 324

M

Macrolides, 140–159
Magnamycin, 154
Magnamycin B, 154
Maltases, 313
Maltoheptaose, 128
Maltose, 107
  amylose from, 333
  phosphorylase action on, 315
  synthesis of, 81
Maltose phosphorylase, 315
Maltoside, methyl hepta-*O*-acetyl-α-, 83
Maltotriose, 125
Maltotriulose, synthesis of, 83
Maltulose, synthesis of, 83
Malvin, 227
Mangiferin, 234
Mannans, D-, 458
  enzymic hydrolysis of, 466
Mannobiose, 116
Mannopyranose, 4-*O*-β-D-mannopyrano-
  syl-D-, *see* Mannobiose
Mannose, 3-amino-3,6-dideoxy-D-, *see* Myco-
  samine
  4-amino-4,6-dideoxy-D-, *see* Perosamine
Mannosidases, α-D- and β-D-, 274
Mannuronic acid, D-, in alginic acids, 380
Mass spectrometry, of antibiotics and sugar
  components, 196
Megacidin, 159
Melezitose, 123
Metabolism, of glycosides, 236
  of cellulose, 432
  of polysaccharides, 397–401
  of sugars, 7
Methymycin, 148
Moenomycin, 145
Molecular weights
  of cellulose, 418
  of glycosidases, 250
  of hemicelluloses, 451
  of polysaccharides, 387, 388, 390, 393
Monomycin, 168
Monosaccharides, biosynthesis of, 304
Mucins
  biosynthesis of, 359
  light scattering by, 390

Mucopolysaccharides, 351, 383
  sulfated, 361
Muramic acid, 305
  6-*O*-(2-acetamido-2-deoxy-β-D-glucosyl)-
    N-acetyl-, 121
Muscle phosphorylase, 330
Mycaminose, 154
Mycarose, 154
  diacetate, 196
  4-*O*-isovaleroyl-, 154
Mycinose, 155, 157, 159
Mycosamine, 141, 146
α-Mycosamine tetraacetate, conformation
  of, 195

N

Narbomycin, 148, 150
Naringin, 219
Neamine, 166
Nebularine, 172
Neobiosamine C, 169
Neobiosaminide B, methyl, 166
Neobiosaminide C, methyl, 166
Neocycasin A, 232
Neokestose, 279
Neomethymycin, 148
Neomycin, 199
Neomycin B, 165
Neomycin C, 165
Neosamine B, 167
Neosamine C, 167
Neuraminic acid, *N*-acetyl-, 307
  N-acetyl-*O*-diacetyl, 307
  CMP-*N*-acetyl-, 306
  N,*O*-diacetyl-, 307
  N-glycolyl-, 307
Neuraminidases, 287
Neuramin-lactose, 124
Neutramycin, 159
Ninhydrin oxidation, 83
Nojirimycin, 192
Nomenclature
  of glycosidases, 242
  of nucleosides, 2
  of oligosaccharides, 71
  of polysaccharides, 376
Noviose, 184
  3-*O*-carbamoyl-, 184
    methyl glycoside, conformation of, 196
Novobiocin, 183, 200

Nuclear magnetic resonance spectroscopy
  of antibiotics, 194
  of glycosides, 233
  of nucleosides, 8, 52
  in oligosaccharide structure determination, 91
Nucleic acids, 3, 5
Nucleocidin, 172
Nucleosides, 1–68, 232
  anhydro-, *see* Cyclonucleosides
  antibiotic, 172–180
  biosynthesis of, 305
  blocking groups for, 22–28
  C-, 20
  characterization of, 5
  definition, 1
  hydrolysis of, 43–52
  isolation and fractionation of, 3
  neighboring-group reactions, 34
  nomenclature, 2
  optical rotatory dispersion measurements of, 8, 55
  oxidation of, 38
  proton magnetic resonance spectra, 52
  structure, 2
  substitution and elimination reactions, 41
  synthesis of, 9–21
  transglycosylation of, 39
  ultraviolet spectra of, 57
Nucleotides
  antibiotic, 199
  glycoside synthesis from sugar, 325
  in marine red alga, 350
  oligosaccharides from, 319
  polysaccharide synthesis from sugar, 336–362
  sugars from, in Nature, 303
Nystatin, 141

O

Octose, 6-amino-6,8-dideoxy-D-*erythro*-D-*galacto*-, *see* Lincosamine
  anhydro-, 196
Octulosonic acid, CMP-3-deoxy-D-*manno*-, 309
  3-deoxy-D-*manno*-, 308
Oleandromycin, 152
Oleandrose, L-, 152
Oligosaccharides, 69–137
  of biological origin, 101

Oligosaccharides—*continued.*
  cello-, 98
  chemical reactions of, 99
  chromatography of, 85
  classification of, 70
  concentration and crystallization of, 86
  degree of polymerization of, 92, 93
  from D-galacto-D-gluco-D-mannans, 462
  hydrolysis by acids, susceptibility to, 96
    by enzymes, susceptibility to, 95
  impurities, 83
  1-^{14}C-labeled, 83
  nomenclature, 71
  physical properties of, 91
  precipitation and extraction of, 84
  relative rates of acid hydrolysis of, 76
  solubilities of, 93
  structure of, determination of, 88
  sweetness of, 94
  synthesis of, 73, 79, 312
    from sugar nucleotides, 319
    by transglycosylation, 317
Oliose, acetyl, 186
Olivomose, 186
Olivomycin, 185
Olivomycose, acetyl-, 186
Olivose, 186
Optical rotatory disperson
  of nucleosides, 8, 55
  of polysaccharides, 386
Orientin, 234
Ortho esters, as blocking groups for nucleosides, 25
Osmotic pressure
  of polysaccharides, 387
Oxidation
  of cellulose, 426
  by lead tetraacetate, of polysaccharides, 407
  of nucleosides, 38
  periodate, of polysaccharides, 402–406

P

Pajaneelin, 215
Panose, 126
  synthesis of, 80
Paramylon, 349
Paratose, 311
Paromamine, 168
Paromobiosaminide, methyl, 168

Paromomycin, 199
Paromomycin I, 168
Paromomycin II, 169
Paromose, 168
Pectin, biosynthesis of, 347
Pelargonin, 227
Pentose, 2-deoxy-D-*erythro*-, 6
Peonin, 227
Periodate oxidation
  of polysaccharides, 402–406
  of sugars, 7, 8
Perosamine, 147
Phenanthrene, glycosides of derivatives of, 221–225
Phenols, glycosides, 218
Phloretin, 219
Phloridzin, 219, 228
Phosphorolysis,
  of cellobiose, 316
  polysaccharide synthesis by, 327–336
Phosphorylases, 6
  disaccharide, 314
  in polysaccharide synthesis, 327
Photosynthesis, of carbohydrates, 302
Phyllanthin, 218
Phytosterolins, 223
Picrocin, *see* Desosamine
Picrocrocin, 229
Picromycin, 148
Pigments, glycosides of natural, 225–230
Pimaricin, 146
Planteose, 89, 122, 321
Plicacetin, 176
Polyoxin C, 179
Polysaccharidase, 438
Polysaccharides
  amylopectin type, from sucrose, 332
  biosynthesis, 301–373
  capsular pneumococcal, 352
  characterization by chemical, enzymic and immunological methods, 394–410
    by physical means, 383–394
  chemistry of, introduction to, 375–412
  classiflcation of, 375
  depolymerization of, 401
  electron microscopy of, 383
  functions of, 382
  identification of constituents of, 396
  infrared spectra and dichroism of, 385
  isolation and purification of, 394

Polysaccharides—*continued.*
  lead tetraacetate oxidation of, 407
  light scattering by, 388
  linkage and sequence of, 397
  molecular weights of, 387, 388, 390, 393
  nomenclature, 376
  oligosaccharides from, by hydrolysis, 74, 75
  optical rotatory dispersion by, 386
  osmotic pressure and other colligative properties of, 387
  periodate oxidation of, 402–406
  reaction conditions for hydrolysis to oligosaccharides, 77
  structural characteristics, 377–382
  synthesis
    by phosphorolysis and transglycosylation, 327–336
    from sugar nucleotides, 336–362
  transglycosylation, 312
  ultracentrifugation of, 390
  viscosity of, 391
  X-ray diffraction by, 384
Populin, 219
Procrocin, 229
Prulaurasin, 218
Prunasin, 218
Pseudoblastidone, 178
Pseudouridine, 233
Psicofuranine, 173
Pullulanase, 332
Purgatives, glycosides, 221
Purine, 9-(β-D-ribofuranosyl)-, *see* Nebularine
Purine cyclonucleosides, 28
Puromycin, 174, 199
Pyrimidine cyclonucleosides, 31
Pyrromycins, 186

Q

Quercitrin, 228

R

Racemomycin A, 181
Racemomycin O, 181
Raffinose, 121, 320
Rhamnetin, 228
Rhamnosidases, α,β-L-, 288
Rhodinose, 188
Rhodomycins, 186
Rhodosamine, 187

β-Rhodosamine diacetate, conformation of, 195
Ristocetin A, 194
Ristocetin B, 194
Rohferment, 259
Roseonine, *see* Streptolidine
Ruberythric acid, 220
Ruff degradation, 83
Rutin, 228, 326
Rutinose, 326

S

Salicin, 219
Salicinase, 264
Sambunigrin, 218
Sangivamycin, 9, 180
Saponins
  glycosides, 223
  triterpenoid, glycosides, 224
Sarsasaponin, 223
Schardinger dextrins, 334
Scillaren A, 222
Scoparin, 234
Sennoside, 221
Septacidin, 174
Showdomycin, 20, 58, 180
Sialic acids, 306
Sialidases, 287
Sialyltransferases, 308
Sinalbin, 231
Sinigrin, 230
Smith degradation, of polysaccharides, 407
Solanin, 225
Solasonins, 225
Spiromycin, 153
Stachyose, 127, 321
Starch
  biosynthesis of, 340
  corn, methylation of, 399
  light scattering by, 390
  synthesis of, 327
Statolon, 194
Steroid saponins, glycosides, 223
Steroids, glycosides, 222
Stractan, 463
Streptamine, deoxy-, 166, 171
Streptidine, 163
Streptobiosamine, 164
Streptobiosaminide, methyl dihydro-, 162
Streptolic acid, 182

Streptolidine, 180
Streptolin, 181
Streptolydigin, 182, 188
Streptomycin, 163, 198
  dihydro-, 164
Streptose, 163
Streptothricin, 181
Streptozotocin, 192
k-Strophanthoside, 222
Sucrose
  amylopectin type polysaccharides from, 332
  dextran from, 334
  α-D-glucosyl transfer from, 256
  invertase action on, 313
  levan from, 335
  occurrence and properties of, 101
  phosphorylase action on, 314
  preparation of, 102
  synthesis of, 81
    phosphate and, 319
    structure and, 103
Sucrose glycosyltransferase, 315
Sucrose phosphorylase, 314, 316
Sucrose transglucosylase, 316
Sugars
  anhydro, oligosaccharides from, 81
  antibiotics containing, 139–211
  biosynthesis of, 301–373
  deoxy, of cardiac glycosides, 215, 216
Sweetness, of oligosaccharides, 94
Swertiamarin, 232
Syringin, 219

T

Taka-diastase, 280, 286
Teichoic acids, 356–358
Tennecetin, 146
l-Thioglycosides, 230
Thymidine, 3
Thymine, 1-(2-deoxy-β-D-*erythro*-pentofuranosyl)- (thymidine), 3
Tigonin, 223
Tiliroside, 228
Tolypomycin γ, 189
Tolyposamine, 189
Tomatin, 215, 225
Toyocamycin, 9, 180
Transferases, glycosyl, 241–300
Transfructosylases, 280

Transfructosylation, 318
Transgalactosylase, 321
Transglucosylases, cyclodextrin, 288
Transglucosylation, disaccharide synthesis
    by, 316
Transglycosylases, 289, 303
Transglycosylation, 312, 325, 327–336
    of glycosides, 235
    mechanism of, 247
Trehalases, 258
Trehalosamine, 190
α,α-Trehalose, 106, 324
Trisaccharides, 121–127
    synthesis of, 80
Triterpenoid saponins, glycosides, 224
Tritylation, of sugars, 7
Trityl ethers, as blocking group for nucleo-
    sides, 22
Tubercidin, 179, 199
Turanose, 84
Tylosin, 155
Tyvelose, 311

U

Ultracentrifugation, of polysaccharides, 390
Ultraviolet spectroscopy
    of glycosides, 233
    of nucleosides, 57
Umbilicin, 217
Uracil, 5
    1-β-D-ribofuranosyl- (uridine), 2, 5
Uridine, 2, 5
Uronic acids
    in polysaccharides, 380
    in D-xylan biosynthesis, 346

V

Vancomycin, 194
Verbascose, 128, 267

Verbenalin, 232
Vicianin, 218
Vicianose, 218
Vicine, 232
Vitexin, 234
Violanthin, 234
Viscose rayon, 432
Viscosity, of polysaccharides, 391
Vitamin P factor, 228

W

Wohl–Zemplén degradation, 83

X

Xanthorhamnin, 228
X-Ray diffraction of polysaccharides, 384
Xylanase A, 465
Xylanase B, 465
Xylans, D-, 452–458
    biosynthesis of, 346
    enzymic hydrolysis of, 464
Xylobiose, 114
Xylopyranose, 2-O-(4-O-methyl-α-D-gluco-
    pyranuronosyl)-D-, 120
    4-O-β-D-xylopyranosyl-D-, *see* Xylobiose
Xylose, 3-O-β-D-glucosyl-D-, 318
    2-O-methyl-D-, 455
Xylosidases, β-D-, 286
Xylosucrose, 318

Y

Ydiginic acid, 183

Z

Zierin, 218
Zygomycin A₁, 169
Zygomycin A₂, 169

C
D 8
E 9
F 0
G 1
H 2
I 3
J 4